AF595093

Aquatic Organisms Research with DNA Barcodes

Aquatic Organisms Research with DNA Barcodes

Editor

Manuel Elias-Gutierrez

Basel • Beijing • Wuhan • Barcelona • Belgrade • Novi Sad • Cluj • Manchester

Editor
Manuel Elias-Gutierrez
ECOSUR Chetumal
Chetumal, Mexico

Editorial Office
MDPI
St. Alban-Anlage 66
4052 Basel, Switzerland

This is a reprint of articles from the Special Issue published online in the open access journal *Diversity* (ISSN 1424-2818) (available at: https://www.mdpi.com/journal/diversity/special_issues/Aquatic_Barcodes).

For citation purposes, cite each article independently as indicated on the article page online and as indicated below:

Lastname, A.A.; Lastname, B.B. Article Title. *Journal Name* **Year**, *Volume Number*, Page Range.

ISBN 978-3-0365-8530-7 (Hbk)
ISBN 978-3-0365-8531-4 (PDF)
doi.org/10.3390/books978-3-0365-8531-4

Cover image courtesy of Manuel Elias-Gutierrez

Contents

About the Editor

Manuel Elias-Gutierrez

Dr. Manuel Elias-Gutierrez is a biologist at the National Autonomous University of Mexico (UNAM, 1982). He holds a Master of Science in Ecology and Doctor of Science from the National Schoolof Biological Sciences, Instituto Politécnico Nacional (Mexico, 1996). He carried out lake zooplankton training at the University of Ghent, Belgium (1993). Since his first studies, he has devoted himself to the study of freshwater zooplankton. He soon discovered that most of Mexico's species were unique and began to describe them. Among his relevant contributions, along with his working group, are the discovery of the first blind cladoceran in the Americas, a new subfamily of freshwater copepods, and many species endemic to crater lakes and the neotropics. He has collaborated with some of the best-known specialists such as Henri Dumont (Belgium), Nikolay Smirnov and Alexey Kotov (Russia), SSS Sarma (Mexico), Rosa Miracle (Spain) and Paul Hebert (Canada), among many others. He was a Professor at the National Autonomous University of Mexico for 18 years. From April 1998 he is a senior researcher at El Colegio de la Frontera Sur, where is continuing his studies on freshwater zooplankton. He is the author of more than 100 scientific articles, 10 book chapters and 3 books, and he was the first researcher to study the DNA barcode of the zooplankton in Mexico and first researcher to describe a cladoceran after integrative taxonomy.

Preface

Since the original inception of DNA barcoding by Paul Hebert in 2003, the methodologies involving these standardized genes have become common in all studies dealing with our planet's biodiversity. However, the development of studies for aquatic animals, except for fishes, has been slower due to difficulties with amplification. This Special Issue was developed to promote and visualize the advances in DNA barcoding for all aquatic life and demonstrate the relevance of the construction of baselines for future studies, such as species description, biomonitoring, and other essential analyses of natural aquatic environments. This book will be helpful for all people dealing with biodiversity in any region of the world, from algae to invertebrates and vertebrates. I hope it will be interesting and inspire others to continue using these powerful tools that will help taxonomists and other scientists interested in biodiversity. These results will be beneficial now that aquatic biodiversity is under intense pressure from human activities, especially in hotspots such as the Caribbean and the Indian Ocean, and in freshwater habitats such as the Amazon basin and all tropical and subtropical lakes, due to a lack of interest from local governments and society.

Manuel Elias-Gutierrez
Editor

Review

Aquatic Organisms Research with DNA Barcodes

Manuel Elías-Gutiérrez [1,*], Nicolas Hubert [2], Rupert A. Collins [3] and Camilo Andrade-Sossa [4]

1 Departamento de Sistemática y Ecología Acuática, El Colegio de la Frontera Sur, Av. Centenario Km.5.5, Chetumal 77014, Mexico
2 UMR 5554 ISEM, IRD, Universite Montpellier, CNRS, EPHE, Université de Montpellier, Place Eugène Bataillon, CEDEX 05, 34095 Montpellier, France; nicolas.hubert@ird.fr
3 School of Biological Sciences, University of Bristol, Life Sciences Building, Bristol BS8 1T, UK; rupertcollins@gmail.com
4 Grupo de Estudios en Recursos Hidrobiológicos Continentales, Departamento de Biología, Universidad del Cauca, Popayán 190002, Colombia; ceandrade@unicauca.edu.co
* Correspondence: melias@ecosur.mx

Abstract: Since their inception, DNA barcodes have become a powerful tool for understanding the biodiversity and biology of aquatic species, with multiple applications in diverse fields such as food security, fisheries, environmental DNA, conservation, and exotic species detection. Nevertheless, most aquatic ecosystems, from marine to freshwater, are understudied, with many species disappearing due to environmental stress, mostly caused by human activities. Here we highlight the progress that has been made in studying aquatic organisms with DNA barcodes, and encourage its further development in assisting sustainable use of aquatic resources and conservation.

Keywords: aquatic life; biodiversity; freshwater; marine; brackish; invertebrate; fish; crustacea; insecta

Citation: Elías-Gutiérrez, M.; Hubert, N.; Collins, R.A.; Andrade-Sossa, C. Aquatic Organisms Research with DNA Barcodes. *Diversity* **2021**, *13*, 306. https://doi.org/10.3390/d13070306

Academic Editor: Eric Buffetaut

Received: 14 May 2021
Accepted: 5 July 2021
Published: 6 July 2021

1. Introduction

Since its inception as an ambitious global bioidentification system [1], DNA barcoding—the use of a standardized gene fragment as an internal tag for species identification—has established itself as an important method in biodiversity sciences, with more than 12,000 papers published (Web of Science search "DNA" and "barcod*", 10 June 2021). The initial proposal by Hebert and collaborators recommended the mitochondrial cytochrome c oxidase I (COI) marker for animals. However, in the case of plants and fungi, other more effective markers have been proposed, such as the maturaseK (matK) and ribulose biphosphate carboxylase large subunit (rbcL) choloroplast markers for flowering plants [2]. Several markers have been suggested as DNA barcodes for diatoms, for example, from 5.8S + ITS-2 [3] to rbcL [4], but studies on these taxa have been limited. For fungi, the ITS has been broadly accepted [5]; however, its implementation also has several problems, particularly in some aquatic species [6], and despite its importance, we found only six papers for DNA barcoding aquatic fungi.

DNA barcoding has been repeatedly demonstrated as a fit-for-purpose method of biodiversity surveying, showing high rates of congruence with traditional taxonomy in well-known groups such as fishes and birds [7–10], while its power as a predictive tool in biodiversity sciences also quickly became apparent, spearheading new molecular frameworks for de novo species discovery [11–13]. Here, some striking examples of overlooked diversity have been observed [14,15], and similar trends have been depicted in numerous aquatic ecosystems. Currently, DNA barcoding can accelerate biodiversity inventories and assist the work of dwindling numbers of taxonomists in many countries. The importance of data sharing and potential for collaborative research was recognized early on, resulting in the creation of the Barcode of Life Data System (BOLD) [16]. Sequence data could be associated with detailed specimen metadata and photographs, supporting trace files, and most importantly vouchered specimens in museum collections [16]. The online workbench also provides the Barcode Index Number (BIN) system, equivalent to a Molecular Operational Taxonomic Unit (MOTU) for all specimens that cover minimal data standards [12], creating a standardized referencing tool for unidentified organisms.

In this overview, we will cover recent trends in the study of aquatic life with DNA barcodes and highlight examples illustrating its utility.

2. Progress in Aquatic DNA Barcoding Studies

General assessments of the use of DNA barcodes in the marine realm were provided in 2011 [17] and 2016 [18], when the number of DNA barcoding studies on aquatic biotas was somewhat less than around 160 per year (Figure 1A). Since then, that number has increased. As such, there is a clear upward trend of DNA barcoding studies, with more than 2500 hits during the last two years (Figure 1A). The words "barcod*" and "DNA" are becoming increasingly used, from zero to more than 1400 hits per year in 2019 and 2020 in the Web of Science database (consulted 10 June 2021) (Figure 1A). However, when we restrict the search to aquatic environments, this figure lowers to 320, with an increment since 2014 (Figure 1B), in comparison with the other trends (Figure 1A) (search "DNA" and "barcod*" and "marin*" or "aquat*" or "freshwat*" or "estuar*" or "fish*" in the Web of Science). Considering that more than 75% of our planet is represented by aquatic environments, this is a modest increment of DNA barcoding studies on aquatic organisms. This modest increment has had mostly the fishes as focal group, where the use of DNA barcoding has been more widespread than invertebrates or aquatic plants (Figure 1A).

The majority of studies included marine and freshwater environments (Figure 1B). It is evident that estuarine systems are almost lacking in analyses with DNA barcoding (Figure 1B).

A good account of the barcoding progress in crustaceans was made in 2015 by Raupach and Radulovici [19], who reported a total of 164 studies, with most studies focusing on Decapoda. Progress in understanding the biodiversity of crustaceans is more advanced for marine environments than in freshwater [19].

Barcoding of aquatic insects has seen most progress in Europe, with Germany the most advanced country, although the diversity there is not high [20,21] compared with the neotropics or the tropics. Here, new species are regularly discovered [22], for example in Cameroon [23]. However, currently DNA barcoding in these and other regions is limited [24,25]. Between North America and Mexico, where the transition between the Nearctic and Neotropics is found, important studies of aquatic insect diversity have been conducted [24,26]. In China, aquatic insects are now being studied in greater depth [27], where 176 species from four rivers in the northwest were analyzed.

Possibly due to their relevance, charismatic nature, and more advanced taxonomic coverage, 37 DNA barcoding studies of Odonata have been published. Among the most recent, the development of reference libraries, mostly in Europe, have been the focus [28–31]. These studies have been started in other regions of the world as well, such as the Andes [32] and Philippines [33].

Another complex group with aquatic immature stages, the chironomids, are starting to be studied with DNA barcodes, where the focus has been in different speciose genera. such as *Tanytarsus*, found almost everywhere [34,35], and other speciose groups [36–38].

In some invertebrate groups such as Polychaeta, with more than 10,000 species described, a total of 65 barcoding studies have been published prior to 2020. Copepoda, which are some of the most abundant organisms in our planet [39], encompassing 14,300 species (World Association of Copepoda; @copepodology), have been targeted by only 87 studies so far, with eight and ten in 2019 and 2020, respectively (Figure 1C), despite the seminal study from Bucklin et al. comparing DNA barcodes to the Rosetta stone of marine biodiversity [40].

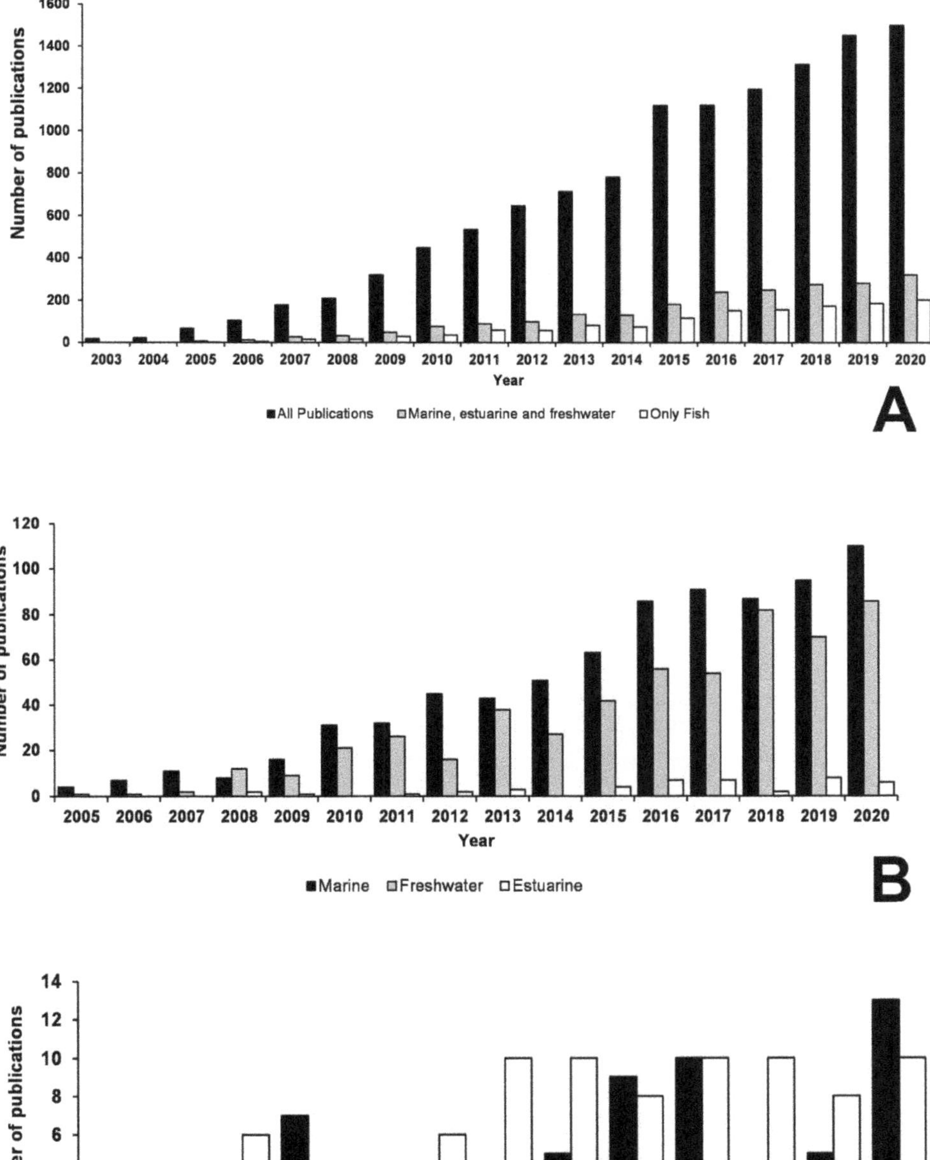

Figure 1. Studies on DNA barcoding on aquatic life. (**A**) Total studies, including all types of aquatic environments and fishes; (**B**) Studies in three different aquatic environments; (**C**) Studies with focus on two of the major groups of aquatic invertebrates.

This trend likely arises from an assortment of shortcomings hindering the development of DNA barcode as a routine survey tool for several groups. In many specialist groups, there are likely issues obtaining funding and taxonomic expertise required to identify the voucher specimens. In others, there are problems with availability of universal primers for amplifying COI. In copepods, for instance, the target marker has been difficult to amplify reliably, and some researchers proposed to adopt the nuclear 28S gene as an alternative marker for COI barcoding [41]. However, 28S fails to distinguish between the species of several groups of crustaceans, such as ostracods, and therefore is of limited use as a species diagnostic [42]. Preservation methods also negatively impacted the acquisition of COI sequences in some cases, such as the shift from liquid nitrogen, which damages voucher specimens, to ethanol, which resulted in a lower yield of mitochondrial DNA, as with the cladoceran *Holopedium* [43]. However, sequencing based on frozen samples with liquid nitrogen proved to increase sequencing success in this case [44]. Due to these problems, new protocols involving cold ethanol were developed with good success in many freshwater zooplankton species, using only a single pair of primers [45].

3. Species Discovery

Discovery of unknown biodiversity has been one of the most important contributions of DNA barcoding studies, including in regions hosting seemingly well-known faunas. In Europe, for instance, DNA barcoding revealed that the gammarid *Gammarus fossarum* encompasses at least 84 MOTUs distributed in 19 European countries [46]. Such levels of cryptic diversity have also been suggested in European odonats [47]. In freshwater fishes, DNA barcodes helped resolve taxonomy of overlooked groups, including some emblematic species such as pike, *Esox lucius*, which has been split into three distinct species [48], minnows from the genus *Phoxinus*, with six species now recognized in France [49], as well as putative new species in *Telestes pleurobipunctatus* [50] and other cypriniformes [51].

The situation is more complicated in the tropics, where biodiversity knowledge gaps are larger and DNA barcoding studies have been conducted in a more piecemeal fashion. Two exceptions are Mexico [52–62] and Brazil [63–72], although both countries are megadiverse hotspots and require much additional effort. Most DNA barcoding studies of tropical aquatic ecosystems have focused on fishes, and frequently report high levels of cryptic and overlooked diversity. These findings include freshwaters of South America [65,73–78], Asia [79–92], and to some extent Africa [93–95]. Similar trends have been observed for tropical marine fishes, particularly in the Indo-Pacific Ocean [96–104]. Overall, efforts to DNA barcode aquatic biotas have been mixed, and the situation in Africa is of most concern. Here, some authors have highlighted that impoverished local scientific capacities curb conservation efforts [105], at the same time as worrying extinction predictions have been observed in some cyprinid fishes [106].

Regarding aquatic invertebrates, studies indicated similar trends. In freshwater mites of Yucatan Peninsula (Mexico), a single DNA barcoding campaign across 24 karstic environments yielded 77 MOTUs, most of them new to science [107]. Similarly, in Panama, a study of invertebrate communities in four streams, with an effort of two hours sampling, yielded ~100–106 MOTUs [108]. Similarly, García-Morales et al. [109] detected a complex of 13 species within the rotifera *Lecane bulla* across 25 localities from south of the United States to Mexico. Elías-Gutiérrez et al. [45] detected a total of 325 BINs among zooplanktonic invertebrates from lakes of Canada and Mexico, with only three BINs (two cladocerans and one copepod) shared between these two countries, suggesting much narrower species distribution ranges in North America freshwater zooplankton than previously thought. Moreover, in an important oligotrophic lake hosting the largest stromatolites in the world [110], the number of possible species increased from about 20 to more than 80, with a projection near to 120 [45]. The closeness (about 100 m) of this lake to a nearby deep sinkhole (64 m) shows almost an entirely different zooplanktonic fauna, explained by a different chemistry of the water [111–113].

In Asia, multiple cases of cryptic diversity have been detected among freshwater shrimps [114–117], other crustaceans and invertebrates [118–123], and parasites [124,125] with implications in conservation and phylogeny [126,127]. Some cases of species under high fishing pressure have been discovered to be complexes of species [128] or unexpected species [129], with important implications for fisheries management. Among African invertebrates, studies are much more limited, mostly focusing on fish parasite communities [124,130,131]. Other parasites studied are helminths, which use aquatic invertebrates as intermediate hosts and are of medical importance [23,132–134], and the use of DNA barcoding to identify larval phases has been explored with success [135]. Due to their medical or invasive importance, some studies focused on molluscs [136,137], highlighting three cryptic species in *Etheria*, for instance [138].

4. Integrative Taxonomy

Numerous taxonomic revisions have been conducted using DNA barcodes [139], guiding the detection of diagnostic characters in new species descriptions for marine [140–144] and freshwater organisms [145–150], or assisting in understanding species range distributions [88,90,118]. Some important complexes of species-groups, which have been used as indicators of toxicants or live food, have also been explored by integrating DNA barcodes with other source of information such as morphology, biogeography and ecology as part of integrative taxonomy [151,152]. Such studies have focused on widely used organisms as biological indicators such as *Moina micrura*, one of the most ubiquitous freshwater cladocera of the world. Widely used in ecotoxicological studies, DNA barcoding indicated that it constitutes a complex of species, with the nominal species being limited to Eurasia [152]. Species descriptions based on integrative taxonomy of freshwater zooplankton provide an enriched framework, allowing not only the delimitation of species, but also access to a wealth of information guiding the acquisition of additional knowledge about their distributions and biology [44,119,145–148,153–155]. Some other groups of marine invertebrates, such as polychaetes, have also been described in an integrative framework [140,142], but a substantial amount of work remains in uncovering the full diversity.

5. Applications

Once DNA barcode libraries are available, several applications have been readily demonstrated. In terms of aquatic ecology, DNA barcoding has been used to identify fish larvae [52,56,156–163] and eggs [54] to the species level, with important implications for fisheries or breeding areas' management. In case of invertebrates, DNA barcoding enabled linking early life stages and adults in aquatic insects [25,164,165]. Along the same line of application, DNA barcoding has been used for food security [166]. Species substitution of food products has been one of the most studied applications, with more than 50 papers devoted to this topic (search: "food mislabel*" and "coi", Web of Science as of 3 March 2021).

The first study of market substitution was published in 2008, focusing on North American seafood, which evidenced 25% of mislabelling [167], this frequency of replacement being lower in Mexico [168]. Papers dealing with this topic have come from Taiwan [169] and Europe [145–148], mostly devoted to fraud in seafood, and also including some other Latin American countries such as Argentina [149].

Detection and impacts of exotic species has been another promising application, such as the invasion of the lion fish (*Pterois volitans*) in the Caribbean [55,150], or the Amazon suckermouth catfish *Pterygoplichthys* in Mexico [151].

6. Future Trends

High throughput sequencing (HTS) methods have had a significant impact in DNA barcoding. DNA barcode reference libraries can now be assembled at larger scale and at lower cost [152] and can even be generated on the lab bench or in the field without the requirement for expensive sequencing equipment [153,154]. However, the main advances have been in DNA metabarcoding; combining HTS with the principals of DNA barcoding has opened a diverse array of applications in terms of species detection and biodiversity monitoring. Of particular relevance in aquatic ecosystems are studies investigating diet ecology and trophic interactions between organisms [155], where in short time passed from Sanger sequencing to study this topic [150] to metabarcoding [170], the structuring and dynamics of plankton communities [156,157], marine benthic biomonitoring [158], and freshwater invertebrate water quality assessment [159]. As well, gut contents can lead to the discovery of unknown biodiversity, as demonstrated in marine and freshwater fishes [150,171]. COI barcode reference libraries for animals have now become standard resources for DNA metabarcoding applications, and have been recommended as the standard metabarcode for metazoans [160]. Environmental DNA (eDNA) metabarcoding techniques have also further transformed biodiversity research by extending metabarcoding to include indirect sequencing of animal communities via their trace DNA [161]. These advances have opened up numerous novel applications in aquatic sciences and ecological monitoring [162,163], but further work is required to optimize the use of the COI barcode for these applications [164].

Automation and big data scientific initiatives also have the potential to provide deep insights into aquatic biodiversity and environmental functioning over extended spatial and temporal scales. Here, DNA metabarcoding methods can be combined with machine learning to predict ecological quality [165], or as an automated plankton recorder [156]. In situations where taxonomic information is not available, such as in many understudied invertebrate groups, taxonomy-free MOTUs can be generated and standardized across studies [166]. New developments in public platforms are under development (e.g., mbrave.net), offering solutions in scalability and standardization for HTS-based approaches to biodiversity, biomonitoring, and biosecurity science. The need to expand publicly available databases applies not only to biodiversity discovery, but is also an essential tool in monitoring traded animals [167,168], exotic species, parasites, pathogens, and almost any species present in our planet [169].

Undoubtedly, most aquatic ecosystems on our planet are tragically understudied, particularly in the tropics, and efforts to understand interactions between anthropogenic pressures and global climate change will be only partial, if not flawed, without accurate biodiversity knowledge. With the use of the new bioinformatic tools and DNA barcoding workflows, important contributions to the conservation of marine, brackish and freshwater organisms will be achieved.

Finally, we must clarify that DNA barcodes should never replace the need for taxonomists. On the contrary, DNA barcodes are an additional suite of characters that can be used in taxonomy, and can also assist in the identification of species by non-specialists that require accurate identification of their specimens.

Author Contributions: All authors contributed equally to this review. All authors have read and agreed to the published version of the manuscript.

Funding: This research received no external funding.

Acknowledgments: The authors appreciate all comments from the four reviewers to improve this review.

Conflicts of Interest: The authors declare no conflict of interest.

References

1. Hebert, P.D.N.; Cywinska, A.; Ball, S.L.; Dewaard, J.R. Biological identifications through DNA barcodes. *Proc. R. Soc. B Boil. Sci.* **2003**, *270*, 313–321. [CrossRef] [PubMed]
2. Hubert, N.; Hanner, R.; Holm, E.; Mandrak, N.E.; Taylor, E.; Burridge, M.; Watkinson, D.; Dumont, P.; Curry, A.; Bentzen, P.; et al. Identifying Canadian Freshwater Fishes through DNA Barcodes. *PLoS ONE* **2008**, *3*, e2490. [CrossRef]
3. Kerr, K.C.R.; Stoeckle, M.Y.; Dove, C.J.; Weigt, L.A.; Francis, C.M.; Hebert, P.D.N. Comprehensive DNA barcode coverage of North American birds. *Mol. Ecol. Notes* **2007**, *7*, 535–543. [CrossRef] [PubMed]

4. Hebert, P.D.N.; Dewaard, J.R.; Zakharov, E.V.; Prosser, S.W.J.; Sones, J.E.; McKeown, J.T.A.; Mantle, B.; La Salle, J. A DNA 'Barcode Blitz': Rapid Digitization and Sequencing of a Natural History Collection. *PLoS ONE* **2013**, *8*, e68535. [CrossRef] [PubMed]
5. Schmidt, S.; Schmid-Egger, C.; Moriniere, J.; Haszprunar, G.; Hebert, P.D.N. DNA Barcoding Largely Supports 250 Years of Classical Taxonomy: Identifications for Central European Bees (Hymenoptera, Apoidea Partim). *Mol. Ecol. Resour.* **2015**, *15*, 985–1000. [CrossRef] [PubMed]
6. Monaghan, M.T.; Wild, R.; Elliot, M.; Fujisawa, T.; Balke, M.; Inward, D.J.; Lees, D.C.; Ranaivosolo, R.; Eggleton, P.; Barraclough, T.; et al. Accelerated Species Inventory on Madagascar Using Coalescent-Based Models of Species Delineation. *Syst. Biol.* **2009**, *58*, 298–311. [CrossRef]
7. Ratnasingham, S.; Hebert, P. A DNA-Based Registry for All Animal Species: The Barcode Index Number (BIN) System. *PLoS ONE* **2013**, *8*, e66213. [CrossRef]
8. Kekkonen, M.; Hebert, P.D.N. DNA Barcode-Based Delineation of Putative Species: Efficient Start for Taxonomic Workflows. *Mol. Ecol. Resour.* **2014**, *14*, 706–715. [CrossRef]
9. Hebert, P.; Penton, E.H.; Burns, J.M.; Janzen, D.H.; Hallwachs, W. Ten species in one: DNA barcoding reveals cryptic species in the neotropical skipper butterfly *Astraptes fulgerator*. *Proc. Natl. Acad. Sci. USA* **2004**, *101*, 14812–14817. [CrossRef]
10. Smith, M.A.; Rodriguez, J.J.; Whitfield, J.B.; Deans, A.; Janzen, D.H.; Hallwachs, W.; Hebert, P. Extreme diversity of tropical parasitoid wasps exposed by iterative integration of natural history, DNA barcoding, morphology, and collections. *Proc. Natl. Acad. Sci. USA* **2008**, *105*, 12359–12364. [CrossRef]
11. Ratnasingham, S.; Hebert, P.D.N. Bold: The Barcode of Life Data System (www.Barcodinglife.Org). *Mol. Ecol. Notes* **2007**, *7*, 355–364. [CrossRef] [PubMed]
12. Bucklin, A.; Steinke, D.; Blanco-Bercial, L. Barcoding of Marine Metazoa. *Annu. Rev. Mar. Sci.* **2011**, *3*, 471–508. [CrossRef]
13. Trivedi, S.; Aloufi, A.A.; Ansari, A.A.; Ghosh, S.K. Role of DNA barcoding in marine biodiversity assessment and conservation: An update. *Saudi J. Biol. Sci.* **2016**, *23*, 161–171. [CrossRef]
14. Mauchline, J.; Blaxter, J.H.S.; Southward, A.J.; Tyler, P.A. The Biology of Calanoid Copepods—Introduction. In *Advances in Marine Biology*; Academic Press: San Diego, CA, USA, 1998; Volume 33, Reprint, Not in File.
15. Bucklin, A.; Ortman, B.D.; Jennings, R.M.; Nigro, L.M.; Sweetman, C.J.; Copley, N.; Sutton, T.; Wiebe, P.A. "Rosetta Stone" for metazoan zooplankton: DNA barcode analysis of species diversity of the Sargasso Sea (Northwest Atlantic Ocean). *Deep Sea Res. Part II Top. Stud. Oceanogr.* **2010**, *57*, 2234–2247. [CrossRef]
16. Hirai, J.; Shimode, S.; Tsuda, A.; Hirai, J.; Shimode, S.; Tsuda, A. Evaluation of ITS2-28S as a molecular marker for identification of calanoid copepods in the subtropical western North Pacific. *J. Plankton Res.* **2013**, *35*, 644–656. [CrossRef]
17. Karanovic, I.; Huyen, P.T.M.; Yoo, H.; Nakao, Y.; Tsukagoshi, A. Shell and Appendages Variability in Two Allopatric Ostracod Species Seen through the Light of Molecular Data. *Contrib. Zool.* **2020**, *89*, 247–269. [CrossRef]
18. Jeffery, N.W.; Elías-Gutiérrez, M.; Adamowicz, S.J. Species Diversity and Phylogeographical Affinities of the Branchiopoda (Crustacea) of Churchill, Manitoba, Canada. *PLoS ONE* **2011**, *6*, e18364. [CrossRef]
19. Rowe, C.L.; Adamowicz, S.J.; Hebert, P. Three new cryptic species of the freshwater zooplankton genus Holopedium (Crustacea: Branchiopoda: Ctenopoda), revealed by genetic methods. *Zootaxa* **2007**, *1656*, 1–50. [CrossRef]
20. Elías-Gutiérrez, M.; Valdez-Moreno, M.; Topan, J.; Young, M.R.; Cohuo-Colli, J.A. Improved protocols to accelerate the assembly of DNA barcode reference libraries for freshwater zooplankton. *Ecol. Evol.* **2018**, *8*, 3002–3018. [CrossRef]
21. Wattier, R.; Mamos, T.; Copilaş-Ciocianu, D.; Jelić, M.; Ollivier, A.; Chaumot, A.; Danger, M.; Felten, V.; Piscart, C.; Žganec, K.; et al. Continental-scale patterns of hyper-cryptic diversity within the freshwater model taxon *Gammarus fossarum* (Crustacea, Amphipoda). *Sci. Rep.* **2020**, *10*, 1–16. [CrossRef]
22. Galimberti, A.; Assandri, G.; Maggioni, D.; Ramazzotti, F.; Baroni, D.; Bazzi, G.; Chiandetti, I.; Corso, A.; Ferri, V.; Galuppi, M.; et al. Italian Odonates in the Pandora's Box: A Comprehensive DNA Barcoding Inventory Shows Taxonomic Warnings at the Holarctic Scale. *Mol. Ecol. Resour.* **2021**, *21*, 183–200. [CrossRef] [PubMed]
23. Denys, G.P.J.; Dettai, A.; Persat, H.; Hautecoeur, M.; Keith, P. Morphological and Molecular Evidence of Three Species of Pikes *Esox* spp. (Actinopterygii, Esocidae) in France, Including the Description of a New Species. *C. R. Biol.* **2014**, *337*, 521–534. [CrossRef]
24. Denys, G.P.J.; Dettai, A.; Persat, H.; Daszkiewicz, P.; Hautecoeur, M.; Keith, P. Revision of Phoxinus in France with the Description of Two New Species (Teleostei, Leuciscidae). *Cybium* **2020**, *44*, 205–237.
25. Buj, I.; Šanda, R.; Zogaris, S.; Freyhof, J.; Geiger, M.F.; Vukic, J. Cryptic diversity in *Telestes pleurobipunctatus* (Actinopterygii; Leuciscidae) as a consequence of historical biogeography in the Ionian Freshwater Ecoregion (Greece, Albania). *Hydrobiology* **2019**, *835*, 147–163. [CrossRef]
26. Epitashvili, G.; Geiger, M.; Astrin, J.J.; Herder, F.; Japoshvili, B.; Mumladze, L. Towards retrieving the Promethean treasure: A first molecular assessment of the freshwater fish diversity of Georgia. *Biodivers. Data J.* **2020**, *8*, e57862. [CrossRef] [PubMed]
27. Valdez-Moreno, M.; Vásquez-Yeomans, L.; Elías-Gutiérrez, M.; Ivanova, N.V.; Hebert, P. Using DNA barcodes to connect adults and early life stages of marine fishes from the Yucatan Peninsula, Mexico: Potential in fisheries management. *Mar. Freshw. Res.* **2010**, *61*, 655–671. [CrossRef]
28. Valdez-Moreno, M.; Ivanova, N.V.; Elías-Gutiérrez, M.; Contreras-Balderas, S.; Hebert, P.D.N. DNA Barcodes in Freshwater Fishes from Mexico and Guatemala. *J. Fish Biol.* **2007**, submitted.

29. Leyva-Cruz, E.; Vásquez-Yeomans, L.; Carrillo, L.; Valdez-Moreno, M. Identifying pelagic fish eggs in the southeast Yucatan Peninsula using DNA barcodes. *Genome* **2016**, *59*, 1117–1129. [CrossRef]
30. Vásquez-Yeomans, L.; Carrillo, L.; Morales, S.; Malca, E.; Morris, J.A.; Schultz, T.; Lamkin, J.T. First larval record of *Pterois volitans* (Pisces: Scorpaenidae) collected from the ichthyoplankton in the Atlantic. *Biol. Invasions* **2011**, *13*, 2635–2640. [CrossRef]
31. Victor, B.C.; Vasquez-Yeomans, L.; Valdez-Moreno, M.; Wilk, L.; Jones, D.L.; Lara, M.R.; Caldow, C.; Shivji, M. The larval, juvenile, and adult stages of the Caribbean goby, *Coryphopterus kuna* (Teleostei: Gobiidae): A reef fish with a pelagic larval duration longer than the post-settlement lifespan. *Zootaxa* **2010**, *2346*, 53–61. [CrossRef]
32. Victor, B.C. *Hypoplectrus floridae* n. sp. and *Hypoplectrus ecosur* n. sp., Two New Barred Hamlets from the Gulf of Mexico (Pisces: Serranidae): More Than 3% Different in COI MtDNA Sequence from the Caribbean Hypoplectrus Species Flock. *J. Ocean Sci. Found.* **2012**, *5*, 2–19.
33. Ahern, A.; Gómez-Gutiérrez, J.; Aburto-Oropeza, O.; Saldierna-Martínez, R.; Johnson, A.F.; Harada, A.; Sánchez-Uvera, A.; Erisman, B.; Arvizú, D.C.; Burton, R. DNA sequencing of fish eggs and larvae reveals high species diversity and seasonal changes in spawning activity in the southeastern Gulf of California. *Mar. Ecol. Prog. Ser.* **2018**, *592*, 159–179. [CrossRef]
34. Bearez, P.; Dettai, A.; Gomon, M.F. *Polylepion russelli* (Labridae), a Trans-Indo-Pacific Species. *Cybium* **2013**, *37*, 305–306.
35. Victor, B.C.; Alfaro, M.E.; Sorenson, L. Rediscovery of *Sagittalarva inornata* n. gen., n. comb. (Gilbert, 1890) (Perciformes: Labridae), a long-lost deepwater fish from the eastern Pacific Ocean: A case study of a forensic approach to taxonomy using DNA barcoding. *Zootaxa* **2013**, *3669*, 551–570. [CrossRef] [PubMed]
36. Mejía, O.; León-Romero, Y.; Soto-Galera, E. DNA barcoding of the ichthyofauna of Pánuco–Tamesí complex: Evidence for taxonomic conflicts in some groups. *Mitochondrial DNA* **2012**, *23*, 471–476. [CrossRef]
37. Venegas, R.D.L.P.; Hueter, R.; Cano, J.G.; Tyminski, J.; Remolina, J.G.; Maslanka, M.; Ormos, A.; Weigt, L.; Carlson, B.; Dove, A. An Unprecedented Aggregation of Whale Sharks, *Rhincodon typus*, in Mexican Coastal Waters of the Caribbean Sea. *PLoS ONE* **2011**, *6*, e18994. [CrossRef]
38. Nogueira, A.F.; Oliveira, C.; Langeani, F.; Netto-Ferreira, A.L. Overlooked biodiversity of mitochondrial lineages in Hemiodus (Ostariophysi, Characiformes). *Zool. Scr.* **2021**, *50*, 337–351. [CrossRef]
39. Adelir-Alves, J.; Spier, D.; Gerum, H.L.N.; Machado, L.F.; Spach, H.L.; Boza, B.R.; Oliveira, C. *Plectorhinchus macrolepis* (Actinopterygii: Haemulidae) in the western Atlantic Ocean. *J. Fish Biol.* **2019**, *95*, 1156–1160. [CrossRef] [PubMed]
40. Arruda, P.S.S.; Ferreira, D.C.; Oliveira, C.; Venere, P.C. DNA Barcoding Reveals High Levels of Divergence among Mitochondrial Lineages of Brycon (Characiformes, Bryconidae). *Genes* **2019**, *10*, 639. [CrossRef] [PubMed]
41. Caires, R.A.; Santos, W.C.R.d.; Machado, L.; Oliveira, C.; Cerqueira, N.; Rotundo, M.M.; Oliveira, C.; Marceniuk, A.P. The Tonkin Weakfish, *Cynoscion similis* (Sciaenidae, Perciformes), an Endemic Species of the Amazonas-Orinoco Plume. *Acta Amaz.* **2019**, *49*, 197–207. [CrossRef]
42. Carvalho, C.O.; Marceniuk, A.P.; Oliveira, C.; Wosiacki, W.B. Integrative Taxonomy of the Species Complex *Haemulon steindachneri* (Jordan and Gilbert, 1882) (Eupercaria; Haemulidae) with a Description of a New Species from the Western Atlantic. *Zoology* **2020**, *141*, 125782. [CrossRef]
43. de Queiroz, L.J.; Cardoso, Y.P.; Jacot-Des-Combes, C.; Bahechar, I.A.; Lucena, C.A.; Py-Daniel, L.R.; Soares, L.M.S.; Nylinder, S.; Oliveira, C.; Parente, T.E.; et al. Evolutionary units delimitation and continental multilocus phylogeny of the hyperdiverse catfish genus Hypostomus. *Mol. Phylogenetics Evol.* **2020**, *145*, 106711. [CrossRef]
44. Ferrette, B.L.D.S.; Domingues, R.R.; Ussami, L.H.F.; Moraes, L.; Magalhães, C.D.O.; De Amorim, A.F.; Hilsdorf, A.W.S.; Oliveira, C.; Foresti, F.; Mendonça, F.F. DNA-based species identification of shark finning seizures in Southwest Atlantic: Implications for wildlife trade surveillance and law enforcement. *Biodivers. Conserv.* **2019**, *28*, 4007–4025. [CrossRef]
45. Mateussi, N.T.B.; Melo, B.F.; Oliveira, C. Molecular Delimitation and Taxonomic Revision of the Wimple Piranhacatoprion (Characiformes: Serrasalmidae) with the Description of a New Species. *J. Fish Biol.* **2020**, *97*, 668–685. [CrossRef]
46. Mattox, G.M.T.; Souza, C.S.; Toledo-Piza, M.; Britz, R.; Oliveira, C. A New Miniature Species of Priocharax (Teleostei: Characiformes: Characidae) from the Rio Madeira Drainage, Brazil, with Comments on the Adipose Fin in Characiforms. *Vertebr. Zool.* **2020**, *70*, 417–433.
47. Hashimoto, S.; Py-Daniel, L.H.R.; Batista, J.S. A molecular assessment of species diversity in *Tympanopleura* and *Ageneiosus catfishes* (Auchenipteridae: Siluriformes). *J. Fish Biol.* **2020**, *96*, 14–22. [CrossRef] [PubMed]
48. Pereira, L.H.G.; Pazian, M.; Hanner, R.; Foresti, F.; Oliveira, C. DNA barcoding reveals hidden diversity in the Neotropical freshwater fish *Piabina argentea* (Characiformes: Characidae) from the Upper Paraná Basin of Brazil. *Mitochondrial DNA* **2011**, *22*, 87–96. [CrossRef] [PubMed]
49. Castro Paz, F.P.; Batista, J.D.; Porto, J.I.R. DNA Barcodes of Rosy Tetras and Allied Species (Characiformes: Characidae: Hyphessobrycon) from the Brazilian Amazon Basin. *PLoS ONE* **2014**, *9*, e98603. [CrossRef]
50. Benzaquem, D.C.; Oliveira, C.; Batista, J.D.S.; Zuanon, J.; Porto, J.I.R. DNA Barcoding in Pencilfishes (Lebiasinidae: *Nannostomus*) Reveals Cryptic Diversity across the Brazilian Amazon. *PLoS ONE* **2015**, *10*, e0112217. [CrossRef]
51. Rossini, B.C.; Oliveira, C.A.M.; de Melo, F.A.G.; Bertaco, V.D.; de Astarloa, J.M.D.; Rosso, J.J.; Foresti, F.; Oliveira, C. Highlighting Astyanax Species Diversity through DNA Barcoding. *PLoS ONE* **2016**, *11*, e0167203. [CrossRef]
52. Machado, V.N.; Collins, R.A.; Ota, R.P.; Andrade, M.C.; Farias, I.; Hrbek, T. One thousand DNA barcodes of piranhas and pacus reveal geographic structure and unrecognised diversity in the Amazon. *Sci. Rep.* **2018**, *8*, 1–12. [CrossRef]

53. Carvalho, A.P.C.; Collins, R.A.; Martinez, J.G.; Farias, I.P.; Hrbek, T. From Shallow to Deep Divergences: Mixed Messages from Amazon Basin Cichlids. *Hydrobiologia* **2019**, *832*, 317–329. [CrossRef]
54. Hubert, N.; Hadiaty, R.K.; Paradis, E.; Pouyaud, L. Cryptic Diversity in Indo-Australian Rainbowfishes Revealed by DNA Barcoding: Implications for Conservation in a Biodiversity Hotspot Candidate. *PLoS ONE* **2012**, *7*, e40627. [CrossRef]
55. Keith, P.; Hadiaty, R.; Hubert, N.; Busson, F.; Lord, C. Three New Species of Lentipes from Indonesia (Gobiidae). *Cybium* **2014**, *38*, 133–146.
56. Lim, H.; Abidin, M.Z.; Pulungan, C.P.; de Bruyn, M.; Nor, S.A.M. DNA Barcoding Reveals High Cryptic Diversity of the Freshwater Halfbeak Genus Hemirhamphodon from Sundaland. *PLoS ONE* **2016**, *11*, e0163596. [CrossRef]
57. Beck, S.V.; Carvalho, G.R.; Barlow, A.; Rüber, L.; Tan, H.H.; Nugroho, E.; Wowor, D.; Nor, S.A.M.; Herder, F.; Muchlisin, Z.A.; et al. Plio-Pleistocene phylogeography of the Southeast Asian Blue *Panchax killifish, Aplocheilus panchax*. *PLoS ONE* **2017**, *12*, e0179557. [CrossRef]
58. Dahruddin, H.; Hutama, A.; Busson, F.; Sauri, S.; Hanner, R.; Keith, P.; Hadiaty, R.; Hubert, N. Data from: Revisiting the ichthyodiversity of Java and Bali through DNA barcodes: Taxonomic coverage, identification accuracy, cryptic diversity and identification of exotic species. *Mol. Ecol. Resour.* **2017**, *17*, 288–299. [CrossRef]
59. Hutama, A.; Dahruddin, H.; Busson, F.; Sauri, S.; Keith, P.; Hadiaty, R.K.; Hanner, R.; Suryobroto, B.; Hubert, N. Identifying spatially concordant evolutionary significant units across multiple species through DNA barcodes: Application to the conservation genetics of the freshwater fishes of Java and Bali. *Glob. Ecol. Conserv.* **2017**, *12*, 170–187. [CrossRef]
60. Conte-Grand, C.; Britz, R.; Dahanukar, N.; Raghavan, R.; Pethiyagoda, R.; Tan, H.H.; Hadiaty, R.K.; Yaakob, N.S.; Rüber, L. *Barcoding snakeheads* (Teleostei, Channidae) revisited: Discovering greater species diversity and resolving perpetuated taxonomic confusions. *PLoS ONE* **2017**, *12*, e0184017. [CrossRef]
61. Farhana, S.N.; Muchlisin, Z.A.; Duong, T.Y.; Tanyaros, S.; Page, L.M.; Zhao, Y.H.; Adamson, E.A.S.; Khaironizam, M.Z.; de Bruyn, M.; Azizah, M.N.S. Exploring Hidden Diversity in Southeast Asia's *Dermogenys* spp. (Beloniformes: Zenarchopteridae) through DNA Barcoding. *Sci. Rep.* **2018**, *8*, 1–11.
62. Shen, Y.; Hubert, N.; Huang, Y.; Wang, X.; Gan, X.; Peng, Z.; He, S. DNA barcoding the ichthyofauna of the Yangtze River: Insights from the molecular inventory of a mega-diverse temperate fauna. *Mol. Ecol. Resour.* **2019**, *19*, 1278–1291. [CrossRef]
63. Hubert, N.; Lumbantobing, D.; Sholihah, A.; Dahruddin, H.; Delrieu-Trottin, E.; Busson, F.; Sauri, S.; Hadiaty, R.; Keith, P. Revisiting Species Boundaries and Distribution Ranges of *Nemacheilus* spp. (Cypriniformes: Nemacheilidae) and *Rasbora* spp. (Cypriniformes: Cyprinidae) in Java, Bali and Lombok through DNA Barcodes: Implications for Conservation in a Biodiversity Hotspot. *Conserv. Genet.* **2019**, *20*, 517–529. [CrossRef]
64. Sholihah, A.; Delrieu-Trottin, E.; Condamine, F.L.; Wowor, D.; Rüber, L.; Pouyaud, L.; Agnèse, J.-F.; Hubert, N. Impact of Pleistocene Eustatic Fluctuations on Evolutionary Dynamics in Southeast Asian Biodiversity Hotspots. *Syst. Biol.* **2021**, syab006. [CrossRef] [PubMed]
65. Sholihah, A.; Delrieu-Trottin, E.; Sukmono, T.; Dahruddin, H.; Risdawati, R.; Elvyra, R.; Wibowo, A.; Kustiati, K.; Busson, F.; Sauri, S.; et al. Disentangling the taxonomy of the subfamily Rasborinae (Cypriniformes, Danionidae) in Sundaland using DNA barcodes. *Sci. Rep.* **2020**, *10*, 1–14. [CrossRef] [PubMed]
66. Delrieu-Trottin, E.; Durand, J.; Limmon, G.; Sukmono, T.; Sugeha, H.Y.; Chen, W.-J.; Busson, F.; Borsa, P.; Dahruddin, H.; Sauri, S.; et al. Biodiversity inventory of the grey mullets (Actinopterygii: Mugilidae) of the Indo-Australian Archipelago through the iterative use of DNA-based species delimitation and specimen assignment methods. *Evol. Appl.* **2020**, *13*, 1451–1467. [CrossRef]
67. Rüber, L.; Tan, H.H.; Britz, R. Snakehead (Teleostei: Channidae) diversity and the Eastern Himalaya biodiversity hotspot. *J. Zool. Syst. Evol. Res.* **2019**, *58*, 356–386. [CrossRef]
68. Chakona, A.; Kadye, W.T.; Bere, T.; Mazungula, D.N.; Vreven, E. Evidence of hidden diversity and taxonomic conflicts in five stream fishes from the Eastern Zimbabwe Highlands freshwater ecoregion. *ZooKeys* **2018**, *768*, 69–95. [CrossRef] [PubMed]
69. Iyiola, O.A.; Nneji, L.M.; Mustapha, M.K.; Nzeh, C.G.; Oladipo, S.; Nneji, I.C.; Okeyoyin, A.O.; Nwani, C.D.; Ugwumba, O.A.; Ugwumba, A.A.A.; et al. DNA barcoding of economically important freshwater fish species from north-central Nigeria uncovers cryptic diversity. *Ecol. Evol.* **2018**, *8*, 6932–6951. [CrossRef]
70. Sonet, G.; Snoeks, J.; Nagy, Z.T.; Vreven, E.; Boden, G.; Breman, F.C.; Decru, E.; Hanssens, M.; Zamba, A.I.; Jordaens, K.; et al. DNA barcoding fishes from the Congo and the Lower Guinean provinces: Assembling a reference library for poorly inventoried fauna. *Mol. Ecol. Resour.* **2018**, *19*, 728–743. [CrossRef]
71. Ward, R.D.; Zemlak, T.S.; Innes, B.H.; Last, P.R.; Hebert, P.D.N. DNA Barcoding Australia's Fish Species. *Philos. Trans. R. Soc. B Biol. Sci.* **2005**, *360*, 1847–1857. [CrossRef]
72. Ward, R.D.; Holmes, B.H.; White, W.; Last, P.R. DNA barcoding *Australasian chondrichthyans*: Results and potential uses in conservation. *Mar. Freshw. Res.* **2008**, *59*, 57–71. [CrossRef]
73. Hubert, N.; Paradis, E.; Bruggemann, H.; Planes, S. Community assembly and diversification in Indo-Pacific coral reef fishes. *Ecol. Evol.* **2011**, *1*, 229–277. [CrossRef]
74. Hubert, N.; Meyer, C.P.; Bruggemann, H.J.; Guérin, F.; Komeno, R.J.L.; Espiau, B.; Causse, R.; Williams, J.T.; Planes, S. Cryptic Diversity in Indo-Pacific Coral-Reef Fishes Revealed by DNA-Barcoding Provides New Support to the Centre-of-Overlap Hypothesis. *PLoS ONE* **2012**, *7*, e28987. [CrossRef]
75. Jaafar, T.N.A.M.; Taylor, M.I.; Nor, S.A.M.; de Bruyn, M.; Carvalho, G.R. DNA Barcoding Reveals Cryptic Diversity within Commercially Exploited Indo-Malay Carangidae (Teleosteii: Perciformes). *PLoS ONE* **2012**, *7*, e49623. [CrossRef] [PubMed]

76. Winterbottom, R.; Hanner, R.H.; Burridge, M.; Zur, M. A cornucopia of cryptic species—A DNA barcode analysis of the gobiid fish genus Trimma (Percomorpha, Gobiiformes). *ZooKeys* **2014**, *381*, 79–111. [CrossRef] [PubMed]
77. Durand, J.-D.; Hubert, N.; Shen, K.-N.; Borsa, P. DNA barcoding grey mullets. *Rev. Fish Biol. Fish.* **2017**, *27*, 233–243. [CrossRef]
78. Delrieu-Trottin, E.; Williams, J.T.; Pitassy, D.; Driskell, A.; Hubert, N.; Viviani, J.; Cribb, T.; Espiau, B.; Galzin, R.; Kulbicki, M.; et al. A DNA barcode reference library of French Polynesian shore fishes. *Sci. Data* **2019**, *6*, 1–8. [CrossRef] [PubMed]
79. Steinke, D.; Zemlak, T.S.; Hebert, P. Barcoding Nemo: DNA-Based Identifications for the Ornamental Fish Trade. *PLoS ONE* **2009**, *4*, e6300. [CrossRef] [PubMed]
80. Skelton, P.H.; Swartz, E.R. Walking the tightrope: Trends in African freshwater systematic ichthyology. *J. Fish Biol.* **2011**, *79*, 1413–1435. [CrossRef]
81. Adeoba, M.I.; Tesfamichael, S.G.; Yessoufou, K. Preserving the tree of life of the fish family Cyprinidae in Africa in the face of the ongoing extinction crisis. *Genome* **2019**, *62*, 170–182. [CrossRef]
82. Montes-Ortiz, L.; Elías-Gutiérrez, M. Water Mite Diversity (Acariformes: Prostigmata: Parasitengonina: Hydrachnidiae) from Karst Ecosystems in Southern of Mexico: A Barcoding Approach. *Diversity* **2020**, *12*, 329. [CrossRef]
83. De León, L.F.; Cornejo, A.; Gavilán, R.G.; Aguilar, C. Hidden biodiversity in Neotropical streams: DNA barcoding uncovers high endemicity of freshwater macroinvertebrates at small spatial scales. *PLoS ONE* **2020**, *15*, e0231683. [CrossRef] [PubMed]
84. García-Morales, A.E.; Domínguez-Domínguez, O. Cryptic species within the rotifer *Lecane bulla* (Rotifera: Monogononta: Lecanidae) from North America based on molecular species delimitation. *Rev. Mex. Biodivers.* **2020**, *91*, 913116. [CrossRef]
85. Gischler, E.; Gibson, M.A.; Oschmann, W. Giant Holocene Freshwater Microbialites, Laguna Bacalar, Quintana Roo, Mexico. *Sedimentology* **2008**, *55*, 1293–1309. [CrossRef]
86. Elías-Gutiérrez, M.; Montes-Ortiz, L. Present Day Kwnoledge on Diversity of Freshwater Zooplancton (Invertebrates) of the Yucatan Peninsula, Using Integrated Taxonomy. *Teor. Y Prax.* **2018**, *14*, 31–48.
87. Montes-Ortiz, L.; Elías-Gutiérrez, M. Faunistic survey of the zooplankton community in an oligotrophic sinkhole, Cenote Azul (Quintana Roo, Mexico), using different sampling methods, and documented with DNA barcodes. *J. Limnol.* **2018**, *77*, 428–440. [CrossRef]
88. Perry, E.; Velazquez-Oliman, G.; Marin, L. The Hydrogeochemistry of the Karst Aquifer System of the Northern Yucatan Peninsula, Mexico. *Int. Geol. Rev.* **2002**, *44*, 191–221. [CrossRef]
89. von Rintelen, K.; von Rintelen, T.; Glaubrecht, M. Molecular Phylogeny and Diversification of Freshwater Shrimps (Decapoda, Atyidae, Caridina) from Ancient Lake Poso (Sulawesi, Indonesia)—The Importance of Being Colourful. *Mol. Phylogenetics Evol.* **2007**, *45*, 1033–1041. [CrossRef]
90. Castelin, M.; De Mazancourt, V.; Marquet, G.; Zimmerman, G.; Keith, P. Genetic and morphological evidence for cryptic species in Macrobrachium australe and resurrection of *M. ustulatum* (Crustacea, Palaemonidae). *Eur. J. Taxon.* **2017**, *289*, 1–27. [CrossRef]
91. de Mazancourt, V.; Klotz, W.; Marquet, G.; Mos, B.; Rogers, D.C.; Keith, P. The complex study of complexes: The first well-supported phylogeny of two species complexes within genus Caridina (Decapoda: Caridea: Atyidae) sheds light on evolution, biogeography, and habitat. *Mol. Phylogenetics Evol.* **2019**, *131*, 164–180. [CrossRef]
92. Hernawati, R.; Nurhaman, U.; Busson, F.; Suryobroto, B.; Hanner, R.; Keith, P.; Wowor, D.; Hubert, N. Exploring community assembly among Javanese and Balinese freshwater shrimps (Atyidae, Palaemonidae) through DNA barcodes. *Hydrobiology* **2019**, *847*, 647–663. [CrossRef]
93. Garibian, P.G.; Neretina, A.N.; Taylor, D.J.; Kotov, A.A. Partial Revision of the Neustonic Genus Scapholeberis Schoedler, 1858 (Crustacea: Cladocera): Decoding of the Barcoding.Results. *PeerJ* **2020**, *8*, e10410. [CrossRef]
94. Yamamoto, A.; Makino, W.; Urabe, J. The taxonomic position of Asian Holopedium (Crustacea: Cladocera) confirmed by morphological and genetic analyses. *Limnology* **2019**, *21*, 97–106. [CrossRef]
95. Liu, Y.; Fend, S.V.; Martinsson, S.; Erséus, C. Extensive cryptic diversity in the cosmopolitan sludge worm *Limnodrilus hoffmeisteri* (Clitellata, Naididae). *Org. Divers. Evol.* **2017**, *17*, 477–495. [CrossRef]
96. Chuluunbat, S.; Morse, J.C.; Boldbaatar, S. Caddisflies of Mongolia: Distribution and diversity. *Zoosymposia* **2016**, *10*, 96–116. [CrossRef]
97. Wu, R.-W.; Liu, Y.-T.; Wang, S.; Liu, X.-J.; Zanatta, D.; Roe, K.J.; Song, X.-L.; An, C.-T.; Wu, X.-P. Testing the utility of DNA barcodes and a preliminary phylogenetic framework for Chinese freshwater mussels (Bivalvia: Unionidae) from the middle and lower Yangtze River. *PLoS ONE* **2018**, *13*, e0200956. [CrossRef] [PubMed]
98. Makino, W.; Machida, R.J.; Okitsu, J.; Usio, N. Underestimated species diversity and hidden habitat preference in Moina (Crustacea, Cladocera) revealed by integrative taxonomy. *Hydrobiology* **2020**, *847*, 857–878. [CrossRef]
99. Locke, S.A.; Al-Nasiri, F.S.; Caffara, M.; Drago, F.; Kalbe, M.; Lapierre, A.R.; McLaughlin, J.D.; Nie, P.; Overstreet, R.M.; Souza, G.T.; et al. Diversity, specificity and speciation in larval Diplostomidae (Platyhelminthes: Digenea) in the eyes of freshwater fish, as revealed by DNA barcodes. *Int. J. Parasitol.* **2015**, *45*, 841–855. [CrossRef] [PubMed]
100. Locke, S.A.; Caffara, M.; Barčák, D.; Sonko, P.; Tedesco, P.; Fioravanti, M.L.; Li, W. A new species of Clinostomum Leidy, 1856 in East Asia based on genomic and morphological data. *Parasitol. Res.* **2019**, *118*, 3253–3265. [CrossRef]
101. Ng, T.H.; Annate, S.; Jeratthitikul, E.; Sutcharit, C.; Limpanont, Y.; Panha, S. Disappearing Apple Snails (Caenogastropoda: Ampullariidae) of Thailand: A Comprehensive Update of Their Taxonomic Status and Distribution. *J. Molluscan Stud.* **2020**, *86*, 290–305. [CrossRef]

102. Vikhrev, I.V.; Konopleva, E.S.; Gofarov, M.Y.; Kondakov, A.V.; Chapurina, Y.E.; Bolotov, I.N. A Tropical Biodiversity Hotspot under the New Threat: Discovery and DNA Barcoding of the Invasive Chinese Pond Mussel Sinanodonta Woodiana in Myanmar. *Trop. Conserv. Sci.* **2017**, *10*, 1–11. [CrossRef]
103. Lima, F.D.; Berbel-Filho, W.M.; Leite, T.S.; Rosas, C.; Lima, S.M.Q. Occurrence of *Octopus insularis* Leite and Haimovici, 2008 in the Tropical Northwestern Atlantic and implications of species misidentification to octopus fisheries management. *Mar. Biodivers.* **2017**, *47*, 723–734. [CrossRef]
104. Rosas-Luis, R.; Badillo, M.D.L.J.; Elena, L.M.; Morillo-Velarde, P.S. Food and feeding habits of *Octopus insularis* in the Veracruz Reef System National Park and confirmation of its presence in the southwest Gulf of Mexico. *Mar. Ecol.* **2019**, *40*, e12535. [CrossRef]
105. Caffara, M.; Locke, S.A.; Echi, P.C.; Halajian, A.; Benini, D.; Luus-Powell, W.J.; Tavakol, S.; Fioravanti, M.L. A morphological and molecular study of *Clinostomid metacercariae* from African fish with a redescription of *Clinostomum tilapiae*. *Parasitology* **2017**, *144*, 1519–1529. [CrossRef]
106. Chibwana, F.D.; Blasco-Costa, I.; Georgieva, S.; Hosea, K.M.; Nkwengulila, G.; Scholz, T.; Kostadinova, A. A first insight into the barcodes for *African diplostomids* (Digenea: Diplostomidae): Brain parasites in *Clarias gariepinus* (Siluriformes: Clariidae). *Infect. Genet. Evol.* **2013**, *17*, 62–70. [CrossRef] [PubMed]
107. Laidemitt, M.R.; Brant, S.V.; Mutuku, M.W.; Mkoji, G.M.; Loker, E.S. The diverse echinostomes from East Africa: With a focus on species that use Biomphalaria and Bulinus as intermediate hosts. *Acta Trop.* **2019**, *193*, 38–49. [CrossRef] [PubMed]
108. Webster, B.; Webster, J.P.; Gouvras, A.; Garba, A.; Lamine, M.S.; Diaw, O.T.; Seye, M.M.; Tchuenté, L.-A.T.; Simoonga, C.; Mubila, L.; et al. DNA 'barcoding' of *Schistosoma mansoni* across sub-Saharan Africa supports substantial within locality diversity and geographical separation of genotypes. *Acta Trop.* **2013**, *128*, 250–260. [CrossRef] [PubMed]
109. Stothard, J.R.; Ameri, H.; Khamis, I.S.; Blair, L.; Nyandindi, U.S.; Kane, R.A.; Johnston, D.A.; Webster, B.; Rollinson, D. Parasitological and malacological surveys reveal urogenital schistosomiasis on Mafia Island, Tanzania to be an imported infection. *Acta Trop.* **2013**, *128*, 326–333. [CrossRef]
110. Alcántar-Escalera, F.A.; García-Varela, M.; Vázquez-Domínguez, E.; de León, G.P. Using DNA Barcoding to Link Cystacanths and Adults of the Acanthocephalan *Polymorphus brevis* in Central Mexico. *Mol. Ecol. Resour.* **2013**, *13*, 1116–1124.
111. Lawton, S.P.; Allan, F.; Hayes, P.M.; Smit, N.J. DNA barcoding of the medically important freshwater snail *Physa acuta* reveals multiple invasion events into Africa. *Acta Trop.* **2018**, *188*, 86–92. [CrossRef]
112. Angyal, D.; Balázs, G.; Krízsik, V.; Herczeg, G.; Fehér, Z. Molecular and morphological divergence in a stygobiont gastropod lineage (Truncatelloidea, Moitessieriidae, Paladilhiopsis) within an isolated karstic area in the Mecsek Mountains (Hungary). *J. Zool. Syst. Evol. Res.* **2018**, *56*, 493–504. [CrossRef]
113. Elderkin, C.L.; Clewing, C.; Ndeo, O.W.; Albrecht, C. Molecular Phylogeny and DNA Barcoding Confirm Cryptic Species in the African Freshwater Oyster *Etheria elliptica* Lamarck, 1807 (Bivalvia: Etheriidae). *Biol. J. Linn. Soc.* **2016**, *118*, 369–381. [CrossRef]
114. Carr, C.M.; Hardy, S.M.; Brown, T.M.; Macdonald, T.A.; Hebert, P. A Tri-Oceanic Perspective: DNA Barcoding Reveals Geographic Structure and Cryptic Diversity in Canadian Polychaetes. *PLoS ONE* **2011**, *6*, e22232. [CrossRef] [PubMed]
115. Villalobos-Guerrero, T.F.; Carrera-Parra, L.F. Redescription of *Alitta succinea* (Leuckart, 1847) and reinstatement of *A. acutifolia* (Ehlers, 1901) n. comb. based upon morphological and molecular data (Polychaeta: Nereididae). *Zootaxa* **2015**, *3919*, 157–178. [CrossRef]
116. Salazar-Silva, P.; Carrera-Parra, L.F. Revision of *Lepidonopsis humilis* (Augener, 1922) and description of *L. barnichae* sp. nov. (Annelida: Polychaeta: Polynoidae) based upon morphological and molecular characters. *Zootaxa* **2014**, *3790*, 555–566. [CrossRef] [PubMed]
117. Carrera-Parra, L.F.; Salazar-Vallejo, S.I. Redescriptions of *Eunice filamentosa* and *E. denticulata* and Description of *E. tovarae* n. sp. (Polychaeta: Eunicidae), Highlighted with Morphological and Molecular Data. *Zootaxa* **2011**, *2880*, 51–64. [CrossRef]
118. Keith, P.; Dahruddin, H.; Limmon, G.; Hubert, N. A New Species of Schismatogobius (Teleostei: Gobiidae) from Halmahera (Indonesia). *Cybium* **2018**, *42*, 195–200.
119. Keith, P.; Lord, C.; Darhuddin, H.; Limmon, G.; Sukmono, T.; Hadiaty, R.; Hubert, N. Schismatogobius (Gobiidae) from Indonesia, with Description of Four New Species. *Cybium* **2017**, *41*, 195–211.
120. Gutiérrez-Aguirre, M.A.; Cervantes-Martínez, A.; Elías-Gutiérrez, M.; Lugo-Vázquez, A. Remarks on Mastigodiaptomus (Calanoida: Diaptomidae) from Mexico using integrative taxonomy, with a key of identification and three new species. *PeerJ* **2020**, *8*, e8416. [CrossRef] [PubMed]
121. Andrade-Sossa, C.; Buitron-Caicedo, L.; Elías-Gutiérrez, M. A new species of Scapholeberis Schoedler, 1858 (Anomopoda: Daphniidae: Scapholeberinae) from the Colombian Amazon basin highlighted by DNA barcodes and morphology. *PeerJ* **2020**, *8*, e9989. [CrossRef]
122. Elías-Gutiérrez, M.; Valdez-Moreno, M. A New Cryptic Species of Leberis Smirnov, 1989 (Crustacea, Cladocera, Chydoridae) from the Mexican Semi-Desert Region, Highlighted by DNA Barcoding. *Hidrobiologica* **2008**, *18*, 63–74.
123. Keith, P.; Mennesson, M.I.; Sauri, S.; Busson, F.; Delrieu-Trottin, E.; Limmon, G.; Dahruddin, H.; Hubert, N. Giuris (Teleostei: Eleotridae) from Indonesia, with Description of a New Species. *Cybium* **2020**, *44*, 317–329.
124. Escobar L, M.D.; Ota, R.P.; Machado-Allison, A.; Andrade-López, J.; Farias, I.P.; Hrbek, T. A new species of Piaractus (Characiformes: Serrasalmidae) from the Orinoco Basin with a redescription of *Piaractus brachypomus*. *J. Fish Biol.* **2019**, *95*, 411–427. [CrossRef]

125. Ota, R.P.; Machado, V.N.; Andrade, M.C.; Collins, R.A.; Farias, I.P.; Hrbek, T. Integrative Taxonomy Reveals a New Species of Pacu (Characiformes: Serrasalmidae: *Myloplus*) from the Brazilian Amazon. *Neotrop. Ichthyol.* **2020**, *18*. [CrossRef]
126. Bekker, E.I.; Karabanov, D.; Galimov, Y.R.; Kotov, A.A. DNA Barcoding Reveals High Cryptic Diversity in the North Eurasian *Moina* Species (Crustacea: Cladocera). *PLoS ONE* **2016**, *11*, e0161737. [CrossRef]
127. Elías-Gutiérrez, M.; Juracka, P.J.; Montoliu-Elena, L.; Miracle, M.R.; Petrusek, A.; Korinek, V. Who Is Moina micrura? Redescription of One of the Most Confusing Cladocerans from Terra Typica, Based on Integrative Taxonomy. *Limnetica* **2019**, *38*, 227–252.
128. Quiroz-Vázquez, P.; Elías-Gutiérrez, M. A New Species of the Freshwater Cladoceran Genus Scapholeberis Schoedler, 1858 (Cladocera: Anomopoda) from the Semidesert Northern Mexico, Highlighted by DNA Barcoding. *Zootaxa* **2009**, *2236*, 50–64. [CrossRef]
129. Mercado-Salas, N.F.; Khodami, S.; Kihara, T.C.; Elías-Gutiérrez, M.; Arbizu, P.M. Genetic Structure and Distributional Patterns of the Genus Mastigodiaptomus (Copepoda) in Mexico, with the Description of a New Species from the Yucatan Peninsula. *Arthropod Syst. Phylogeny* **2018**, *76*, 487–507.
130. Gutiérrez-Aguirre, M.A.; Cervantes-Martínez, A. A new species of Mastigodiaptomus Light, 1939 from Mexico, with notes of species diversity of the genus (Copepoda, Calanoida, *Diaptomidae*). *ZooKeys* **2016**, *637*, 61–79. [CrossRef]
131. Pegg, G.G.; Sinclair, B.; Briskey, L.; Aspden, W.J. MtDNA barcode identification of fish larvae in the southern Great Barrier Reef—Australia. *Sci. Mar.* **2006**, *70*, 7–12. [CrossRef]
132. Hubert, N.; Delrieu-Trottin, E.; Irisson, J.-O.; Meyer, C.; Planes, S. Identifying coral reef fish larvae through DNA barcoding: A test case with the families Acanthuridae and Holocentridae. *Mol. Phylogenetics Evol.* **2010**, *55*, 1195–1203. [CrossRef] [PubMed]
133. Hubert, N.; Espiau, B.; Meyer, C.; Planes, S. Identifying the ichthyoplankton of a coral reef using DNA barcodes. *Mol. Ecol. Resour.* **2015**, *15*, 57–67. [CrossRef]
134. Ko, H.-L.; Wang, Y.-T.; Chiu, T.-S.; Lee, M.-A.; Leu, M.-Y.; Chang, K.-Z.; Chen, W.-Y.; Shao, K.-T. Evaluating the Accuracy of Morphological Identification of Larval Fishes by Applying DNA Barcoding. *PLoS ONE* **2013**, *8*, e53451. [CrossRef]
135. Azmir, I.A.; Esa, Y.; Amin, S.M.N.; Yasin, I.S.M.; Yusof, F.Z.M.; Azmir, I.A.; Esa, Y.; Amin, S.M.N.; Yasin, I.S.M.; Yusof, F.Z.M. Identification of larval fish in mangrove areas of Peninsular Malaysia using morphology and DNA barcoding methods. *J. Appl. Ichthyol.* **2017**, *33*, 998–1006. [CrossRef]
136. Collet, A.; Durand, J.-D.; Desmarais, E.; Cerqueira, F.; Cantinelli, T.; Valade, P.; Ponton, D. DNA barcoding post-larvae can improve the knowledge about fish biodiversity: An example from La Reunion, SW Indian Ocean. *Mitochondrial DNA Part A* **2018**, *29*, 905–918. [CrossRef] [PubMed]
137. Mariac, C.; Vigouroux, Y.; Duponchelle, F.; García-Dávila, C.; Nunez, J.; Desmarais, E.; Renno, J. Metabarcoding by capture using a single COI probe (MCSP) to identify and quantify fish species in ichthyoplankton swarms. *PLoS ONE* **2018**, *13*, e0202976. [CrossRef] [PubMed]
138. Steinke, D.; Connell, A.D.; Hebert, P.D. Linking adults and immatures of South African marine fishes. *Genome* **2016**, *59*, 959–967. [CrossRef] [PubMed]
139. Zhou, X.; Jacobus, L.M.; DeWalt, R.E.; Adamowicz, S.J.; Hebert, P.D.N. Ephemeroptera, Plecoptera, and Trichoptera fauna of Churchill (Manitoba, Canada): Insights into biodiversity patterns from DNA barcoding. *J. N. Am. Benthol. Soc.* **2010**, *29*, 814–837. [CrossRef]
140. Carew, M.E.; Pettigrove, V.; Cox, R.L.; Hoffmann, A.A. DNA identification of urban *Tanytarsini chironomids* (Diptera: Chironomidae). *J. N. Am. Benthol. Soc.* **2007**, *26*, 587–600. [CrossRef]
141. Kusche, H.; Hanel, R. Consumers of mislabeled tropical fish exhibit increased risks of ciguatera intoxication: A report on substitution patterns in fish imported at Frankfurt Airport, Germany. *Food Control* **2021**, *121*, 107647. [CrossRef]
142. Wong, E.H.K.; Hanner, R. DNA Barcoding Detects Market Substitution in North American Seafood. *Food Res. Int.* **2008**, *41*, 828–837. [CrossRef]
143. Sarmiento-Camacho, S.; Valdez-Moreno, M. DNA barcode identification of commercial fish sold in Mexican markets. *Genome* **2018**, *61*, 457–466. [CrossRef]
144. Chang, C.H.; Lin, H.Y.; Ren, Q.; Lin, Y.S.; Shao, K.T. DNA Barcode Identification of Fish Products in Taiwan: Government-Commissioned Authentication Cases. *Food Control* **2016**, *66*, 38–43. [CrossRef]
145. Günther, B.; Raupach, M.J.; Knebelsberger, T. Full-length and mini-length DNA barcoding for the identification of seafood commercially traded in Germany. *Food Control* **2017**, *73*, 922–929. [CrossRef]
146. Christiansen, H.; Fournier, N.; Hellemans, B.; Volckaert, F.A. Seafood substitution and mislabeling in Brussels' restaurants and canteens. *Food Control* **2018**, *85*, 66–75. [CrossRef]
147. Vandamme, S.G.; Griffiths, A.M.; Taylor, S.-A.; Di Muri, C.; Hankard, E.A.; Towne, J.A.; Watson, M.; Mariani, S. Sushi barcoding in the UK: Another kettle of fish. *PeerJ* **2016**, *4*, e1891. [CrossRef] [PubMed]
148. Cawthorn, D.-M.; Baillie, C.; Mariani, S. Generic names and mislabeling conceal high species diversity in global fisheries markets. *Conserv. Lett.* **2018**, *11*, e12573. [CrossRef]
149. Delpiani, G.; Delpiani, S.; Antoni, M.D.; Ale, M.C.; Fischer, L.; Lucifora, L.; de Astarloa, J.D. Are we sure we eat what we buy? Fish mislabelling in Buenos Aires province, the largest sea food market in Argentina. *Fish. Res.* **2020**, *221*, 105373. [CrossRef]
150. Valdez-Moreno, M.; Quintal-Lizama, C.; Gómez-Lozano, R.; García-Rivas, M.D.C. Monitoring an Alien Invasion: DNA Barcoding and the Identification of Lionfish and Their Prey on Coral Reefs of the Mexican Caribbean. *PLoS ONE* **2012**, *7*, e36636. [CrossRef]

151. Valdez-Moreno, M.; Ivanova, N.V.; Elías-Gutiérrez, M.; Pedersen, S.L.; Bessonov, K.; Hebert, P.D.N. Using eDNA to biomonitor the fish community in a tropical oligotrophic lake. *PLoS ONE* **2019**, *14*, e0215505. [CrossRef]
152. Hebert, P.D.N.; Braukmann, T.W.A.; Prosser, S.W.J.; Ratnasingham, S.; Dewaard, J.R.; Ivanova, N.V.; Janzen, D.H.; Hallwachs, W.; Naik, S.; Sones, J.E.; et al. A Sequel to Sanger: Amplicon sequencing that scales. *BMC Genom.* **2018**, *19*, 219. [CrossRef] [PubMed]
153. Pomerantz, A.; Penafiel, N.; Arteaga, A.; Bustamante, L.; Pichardo, F.; Coloma, L.A.; Barrio-Amoros, C.L.; Salazar-Valenzuela, D.; Prost, S. Real-Time DNA Barcoding in a Rainforest Using Nanopore Sequencing: Opportunities for Rapid Biodiversity Assessments and Local Capacity Building. *Gigascience* **2018**, *7*, giy033. [CrossRef]
154. Srivathsan, A.; Baloglu, B.; Wang, W.; Tan, W.X.; Bertrand, D.; Ng, A.H.Q.; Boey, E.J.H.; Koh, J.J.Y.; Nagarajan, N.; Meier, R. A Minion-Based Pipeline for Fast and Cost-Effective DNA Barcoding. *Mol. Ecol. Resour.* **2018**, *18*, 1035–1049. [CrossRef] [PubMed]
155. Berry, T.E.; Osterrieder, S.K.; Murray, D.C.; Coghlan, M.L.; Richardson, A.J.; Grealy, A.K.; Stat, M.; Bejder, L.; Bunce, M. DNA metabarcoding for diet analysis and biodiversity: A case study using the endangered Australian sea lion (*Neophoca cinerea*). *Ecol. Evol.* **2017**, *7*, 5435–5453. [CrossRef]
156. Suter, L.; Polanowski, A.; Clarke, L.J.; Kitchener, J.; Deagle, B.E. Capturing open ocean biodiversity: Comparing environmental DNA metabarcoding to the continuous plankton recorder. *Mol. Ecol.* **2020**. [CrossRef]
157. Zamora-Terol, S.; Novotny, A.; Winder, M. Reconstructing Marine Plankton Food Web Interactions Using DNA Metabarcoding. *Mol. Ecol.* **2020**, *29*, 3380–3395. [CrossRef]
158. Mauffrey, F.; Cordier, T.; Apothéloz-Perret-Gentil, L.; Cermakova, K.; Merzi, T.; Delefosse, M.; Blanc, P.; Pawlowski, J. Benthic monitoring of oil and gas offshore platforms in the North Sea using environmental DNA metabarcoding. *Mol. Ecol.* **2020**. [CrossRef] [PubMed]
159. Elbrecht, V.; Vamos, E.E.; Meissner, K.; Aroviita, J.; Leese, F. Assessing strengths and weaknesses of DNA metabarcoding-based macroinvertebrate identification for routine stream monitoring. *Methods Ecol. Evol.* **2017**, *8*, 1265–1275. [CrossRef]
160. Andújar, C.; Arribas, P.; Yu, D.W.; Vogler, A.P.; Emerson, B.C. Why the COI barcode should be the community DNA metabarcode for the metazoa. *Mol. Ecol.* **2018**, *27*, 3968–3975. [CrossRef]
161. Deiner, K.; Bik, H.M.; Mächler, E.; Seymour, M.; Lacoursière-Roussel, A.; Altermatt, F.; Creer, S.; Bista, I.; Lodge, D.M.; De Vere, N.; et al. Environmental DNA metabarcoding: Transforming how we survey animal and plant communities. *Mol. Ecol.* **2017**, *26*, 5872–5895. [CrossRef]
162. Rees, H.C.; Maddison, B.C.; Middleditch, D.J.; Patmore, J.R.M.; Gough, K.C. Review the Detection of Aquatic Animal Species Using Environmental DNA—A Review of eDNA as a Survey Tool in Ecology. *J. Appl. Ecol.* **2014**, *51*, 1450–1459. [CrossRef]
163. Valentini, A.; Taberlet, P.; Miaud, C.; Civade, R.; Herder, J.; Thomsen, P.F.; Bellemain, E.; Besnard, A.; Coissac, E.; Boyer, F.; et al. Data from: Next-generation monitoring of aquatic biodiversity using environmental DNA metabarcoding. *Mol. Ecol.* **2016**, *25*, 929–942. [CrossRef] [PubMed]
164. Collins, R.A.; Bakker, J.; Wangensteen, O.S.; Soto, A.Z.; Corrigan, L.; Sims, D.W.; Genner, M.J.; Mariani, S. Non-specific amplification compromises environmental DNA metabarcoding with COI. *Methods Ecol. Evol.* **2019**, *10*, 1985–2001. [CrossRef]
165. Cordier, T.; Esling, P.; Lejzerowicz, F.; Visco, J.A.; Ouadahi, A.; Martins, C.I.M.; Cedhagen, T.; Pawlowski, J. Predicting the Ecological Quality Status of Marine Environments from eDNA Metabarcoding Data Using Supervised Machine Learning. *Environ. Sci. Technol.* **2017**, *51*, 9118–9126. [CrossRef]
166. Mächler, E.; Walser, J.; Altermatt, F. Decision-making and best practices for taxonomy-free environmental DNA metabarcoding in biomonitoring using Hill numbers. *Mol. Ecol.* **2020**. [CrossRef]
167. Van Der Walt, K.; Mäkinen, T.; Swartz, E.; Weyl, O. DNA barcoding of South Africa's ornamental freshwater fish—Are the names reliable? *Afr. J. Aquat. Sci.* **2017**, *42*, 155–160. [CrossRef]
168. Collins, R.A.; Armstrong, K.; Meier, R.; Yi, Y.; Brown, S.; Cruickshank, R.H.; Keeling, S.; Johnston, C. Barcoding and Border Biosecurity: Identifying Cyprinid Fishes in the Aquarium Trade. *PLoS ONE* **2012**, *7*, e28381. [CrossRef] [PubMed]
169. Curry, C.J.; Gibson, J.F.; Shokralla, S.; Hajibabaei, M.; Baird, D.J. Identifying North American freshwater invertebrates using DNA barcodes: Are existing COI sequence libraries fit for purpose? *Freshw. Sci.* **2018**, *37*, 178–189. [CrossRef]
170. Jo, H.; Gim, J.A.; Jeong, K.S.; Kim, H.S.; Joo, G.J. Application of DNA Barcoding for Identification of Freshwater Carnivorous Fish Diets: Is Number of Prey Items Dependent on Size Class for Micropterus Salmoides? *Ecol. Evol.* **2014**, *4*, 219–229. [CrossRef]
171. Jo, H.; Ventura, M.; Vidal, N.; Gim, J.S.; Buchaca, T.; Barmuta, L.A.; Jeppesen, E.; Joo, G.J. Discovering Hidden Biodiversity: The Use of Complementary Monitoring of Fish Diet Based on DNA Barcoding in Freshwater Ecosystems. *Ecol. Evol.* **2016**, *6*, 219–232. [CrossRef]

Article

Revealing the Differences in *Ulnaria acus* and *Fragilaria radians* Distribution in Lake Baikal via Analysis of Existing Metabarcoding Data

Alexey Morozov *, Yuri Galachyants, Artem Marchenkov, Yulia Zakharova and Darya Petrova

Limnological Institute, Siberian Branch of the Russian Academy of Sciences, 3 Ulan-Batorskaya St., Irkutsk 664033, Russia
* Correspondence: morozov@lin.irk.ru; Tel.: +7-902-1765206

Abstract: Two diatom species, *Ulnaria acus* and *Fragilaria radians*, are morphologically very similar and often coexist, which makes it difficult to compare their abundances. However, they are easily separated by molecular data; thus, in this work, we attempted to estimate the differences in their spatial and temporal distribution from existing metabarcoding datasets. Reanalyzing published sequences with an ASV-based pipeline and ad hoc classification routine allowed us to estimate the relative abundances of the two species, increasing the precision compared to usual OTU-based analyses. Existing data permit qualitative comparisons between two species that cannot be differentiated by other methods, detecting the distinct seasonal peaks and spatial distributions of *F. radians* and *U. acus*.

Keywords: metabarcoding; hidden diversity; diatoms; phytoplankton

Citation: Morozov, A.; Galachyants, Y.; Marchenkov, A.; Zakharova, Y.; Petrova, D. Revealing the Differences in *Ulnaria acus* and *Fragilaria radians* Distribution in Lake Baikal via Analysis of Existing Metabarcoding Data. *Diversity* **2023**, *15*, 280. https://doi.org/10.3390/d15020280

Academic Editor: Manuel Elias-Gutierrez

Received: 23 November 2022
Revised: 31 January 2023
Accepted: 6 February 2023
Published: 15 February 2023

1. Introduction

In Lake Baikal, like in other freshwater ecosystems, diatoms play a significant role in primary production, sediment deposition and biogeochemical cycles. Among the phytoplanktonic species of the lake, one of the dominant species was originally identified as *Synedra acus* subsp. *radians* Skabitsch. and later renamed *Fragilaria radians* (Kützing) D.M.Williams and Round [1]. This diatom (referred to as *F. radians* in the Introduction, regardless of the name used in other papers) is not only a major player in the lake ecosystem, but also a model object for multiple studies. In particular, it was successfully axenized [2], which allowed it to become the first freshwater diatom to have its nuclear genome sequenced [3].

Since *F. radians* is a key element in the Lake Baikal ecosystem, its dynamics and ecology have been thoroughly studied. This alga blooms under ice, both in littoral and pelagic areas, dominating the eukaryotic phytoplankton community associated with the lower ice surface [4,5]. After the ice-breaking period (April to May), the *F. radians* population decreases; however, it still remains a significant member of the phytoplankton community until mid-summer [6–8]. Many correlations were found between *F. radians* abundance (as estimated by either microscopy or metabarcoding studies) and various biotic and abiotic factors at different times and sampling stations [8–10]. Usually, its abundance and biomass correlate negatively with Si availability and positively with abundances of other common Baikalian diatoms, although there are exceptions [11].

In a recent study, this population, previously thought to consist of a single *Fragilaria* species, was found to include members of two species from different genera: *Fragilaria radians sensu stricto* and *Ulnaria acus* (Kützing) Aboal. The morphology of these two species is nearly identical: *F. radians* has a cell length of 105–239 μm, cell width of 2.5–5.2 and 12–22 rows of areolae per 10 μm; U. acus has a cell length of 60–251 μm, cell width 2.2–5.4 μm and 12–14 rows of areolae per 10 μm. They can also be cultured under identical conditions and have been isolated together from natural samples. This makes separating them in routine microscopy-based phytoplankton monitoring next to impossible, since it requires either sophisticated EM-based analyses or DNA sequencing [12]. All of the ecological studies listed above were also based on the assumption of a single species, using either light microscopy or wider *Synedra* sp./*Fragilaria* sp. OTUs that may have included a mixture of reads from *U. acus, F. radians* or other related species. Further, the taxonomy of these OTUs was typically not identified below the genus.

These similarities explain why the two species have not been separated until recently. However, similar morphology and overlapping ranges of acceptable conditions do not imply exactly identical autecological features of the two species. It is possible that, although overlapping, the optimal temperatures or other factors are somewhat different for the two species. On the other hand, freshwater benthic *Fragilaria* and *Ulnaria* strains identified from molecular data in various streams and lakes in Europe showed considerable overlap in geographical distribution, habitat and ecological preferences [13]. Although multiple OTUs of Fragilaria sp. and Ulnaria sp. were observed in most studies on Lake Baikal, the issues outlined above render them unsuitable for discussing the ecology of these two species.

Thus, a goal of this work was to develop a method to separately estimate the relative abundances of *U. acus* and *F. radians* based on metabarcoding data, and to apply this method to the existing sequences from Lake Baikal.

2. Materials and Methods

2.1. Selecting Amplicons for Distinguishing F. radians and U. acus

In order to study whether 18S rRNA or rbcL metabarcoding analysis can distinguish all groups within *Ulnaria/Fragilaria* species complex, metabarcoding libraries were produced from two samples of phytoplankton taken near the settlement of Bolshiye Koty in March 2020, as well as a mock community consisting of four strains isolated from Lake Baikal. In order to extract DNA, integral water samples of 20 L (equal volumes of samples from different depths) were collected. Samples were first pre-filtered using a 27 μm sieve and then were filtered through 0.2 μm analytical track membranes (Reatrack, Obninsk City, Russia). Biomass was washed off the filters into sterile TE buffer (10 mM Tris–HCl, 1 mM EDTA, pH 8.0) and stored at −80 °C until analysis. DNA was extracted using lysozyme (1 mg mL^{-1}), proteinase K, 10% SDS and phenol:chloroform:isoamyl alcohol mixture (25:24:1) according to the protocol based on Rusch et al. [14].

The mixture included strains L150 (*F. radians*), MM244 (*U. acus*), BZ264 (*U. ulna*) and ChZ411 (*Aulacoseira islandica*) from Lake Baikal. In order to isolate the *Fragilaria* and *Ulnaria* monoclones, phytoplankton samples were collected in different parts of Lake Baikal during 2017 and 2018. The latter two strains were included because *U. ulna* and *A. islandica* commonly coexist with our species of interest; therefore, any practically useful method should be able to distinguish them from *U. acus* and *F. radians*. Cultures of diatoms were obtained via isolation of individual cells. The isolated strains were grown in 96-well plates with Diatom Medium (DM) in a mini-incubator at 8 °C and illuminated with 16 μL Einstein m^{-2} s^{-1} at a photoperiod of 12:12 h light:darkness, and then transferred into Erlenmeyer flasks with a volume of 100 mL for further growth. The strains were grown for three months to obtain sufficient biomass for DNA extraction.

Approximately 300,000 cells were taken for each of the three *Ulnaria* and *Fragilaria* strains. The cell number for the colonial species *A. islandica* was not known precisely; however, a roughly similar number of cells was taken. DNA was isolated as described above and amplified using two primer pairs for each target gene. V3-V4 18S rRNA (418 bp) was amplified with TAReuk454FWD1 5′-CCAGCASCYGCGGTAATTCC and TAReukREV3 5′-ACTTTCGTTCTTGAT [15] primers. For V8-V9 18S rRNA (368 bp), we used V8F 5′-ATAACAGGTCTGTGATGCCCT and 1510R 5′-CCTTCYGCAGGTTCACCTAC [16]. Both 18S rRNA fragments were amplified with Phusion Hot Start II High-Fidelity DNA polymerase (Thermo Fisher Scientific, Waltham, MA, USA). PCR mix consisted of 1× Phusion buffer HF, 1 unit of DNA polymerase, 0.2 mM dNTP mixture, 1.5 mM free Mg^{2+}, 0.2 μM primers and 10 ng of DNA. Temperature profile was as follows: 98 °C for 1 min, 29 cycles of (98 °C for 30 s, 50 °C for 30 s, 72 °C for 30 s) and 72 °C for 3 min. PCR product was purified by AMPure XP magnetic beads (Beckman Coulter, Brea, CA, USA) according to the manufacturer's protocol.

A 312 bp rbcL fragment amplified by pairDiat_rbcL_708F 5′-AGGTGAAGTTAAAGGT TCATACTTDAA [17] and R3 5′-CCTTCTAATTTACCAACAACTG [18] primers was previously proposed for diatom metabarcoding [19]. Since this primer pair has some mismatches with rbcL sequences produced in previous work [12], we designed an additional primer pair (bar_S_rbcL_665F 5′-GCAACAGGTGAAGTTAAAGGTTCT and bar_S_rbcL_867R 5′-GAGTTACCTGCACGGTGTAAGT) to amplify the Baikalian *Fragilaria* and *Ulnaria*. These two primer pairs are referred to as rbcL606 and rbcL708, respectively, in the remaining text. PCR with both rbcL primer pairs was performed using Tersus polymerase (Evrogen, Russia). PCR mix consisted of one Tersus Red buffer, 1 unit of DNA polymerase, 0.2 mM dNTP mixture, 0.2 μM of both primers and 100 ng of DNA. Temperature profile was as follows: 95 °C for 3 min, 30 cycles of (95 °C for 30 s, 60 °C for 30 s, 72 °C for 30 s), 72 °C for 5 min. PCR product was analyzed by electrophoresis in 1.5% agarose gel and purified with Monarch Gel Extraction Kit (NEB, USA) according to the manufacturer's protocol.

Libraries were sequenced on Illumina Miseq with MiSeq® Reagent Kit v3 (2×300 bp) in the Core Centrum "Genomic Technologies, Proteomics and Cell Biology" in ARRIAM (All-Russia Research Institute for Agricultural Microbiology, Russia). Thus, 18S rRNA amplicon libraries were analyzed in mothur 1.44.11 [20] to produce 97% identity OTUs and ASVs, as well as in Usearch 11.0.667 to produce ASVs using the unoise3 algorithm. In both cases, ASVs were generated at a cutoff of 4 substitutions. In mothur-based analysis, the Silva nr v138.1 database was used as a reference for alignment and taxonomic classifications. Since this database does not offer taxonomic resolution below genus, all OTUs/ASVs assigned to genera *Ulnaria* and *Fragilaria* were BLASTed against 18S rRNA sequences sequenced from Baikalian diatoms [12] using blastn 2.2.31+ [21]. Those with sequence identities exceeding 97% with all sequences from one clade, but not others, were classified as belonging to corresponding groups. OTUs and ASVs, which had high-quality hits with both *Ulnaria* clades, were classified as *Ulnaria* sp.; any other combination of hits was considered unclassified. For the purposes of this classification, *U. ulna* and *U. danica* reference sequences were treated as a single group, since these two species are hard to distinguish from either morphological or genetic data and elucidation of their relationship is beyond the scope of this paper.

The rbcL amplicon libraries were analyzed with Usearch only; *Fragilaria*/*Ulnaria* ASVs were classified in a similar way using 98% identity cutoff. All sequencing data are available at NCBI SRA (project ID PRJNA666300).

2.2. Analysis of Published Metabarcoding Data

Raw reads and sample metadata were downloaded from the public databases (ENA Project ID PRJEB24415 for 2013 spatial dataset [22], NCBI SRA project IDs PRJNA657482 and PRJNA662681 for 2013 spatial dataset [23] and 2017 time series [8]). Only V4 amplicons were selected from the latter dataset; otherwise, all available data were used. Usearch and vsearch [24] were used to filter reads (maximum expected error 1.0, minimum assembled length 400 bp), produce ASVs with usearch UNOISE algorithm and remove the chimerae with vsearch UCHIME. To estimate the abundances of these ASVs, filtered reads were mapped to them at 99% identity cutoff using usearch. Preliminary taxonomic annotation was produced by kmer-based naive Bayesian classifier implemented in mothur [20] with SILVA v138.1 reference alignment and taxonomy [25]. These analyses were performed separately for each dataset; the pipeline was identical to that described above for testing datasets.

All ASVs identified as *Ulnaria* or *Fragilaria* were classified as described above. Data processing and plotting were performed in Python3 using matplotlib [26] and Basemap packages.

3. Results

3.1. Amplicon Selection and Testing

Clustering the reads from a mock community into OTUs produced a questionable result, with multiple OTUs per species. These OTUs were absent from natural samples. We assumed that these are clustering artifacts. Further, the total abundance of classified OTUs was very low in the mock sample. Because of this, as well as the conceptual arguments in favor of ASVs/zOTUs over OTUs (28), further analysis was carried out in terms of ASVs. Usearch-produced ASVs were classified with more precision than mothur-produced ones (no Usearch mock community ASVs were assigned to "*Ulnaria* sp." and "unknown"); therefore, only the results of the Usearch ASV pipeline are documented below (all ASVs and abundances are available in Supplementary Table S1).

Complete taxonomies for all 18S rRNA amplicons are available in Supplementary Table S2. In a mock community and in natural samples, V8-V9 variable regions of the 18S rRNA gene were not able to distinguish between the *Ulnaria acus* and *Ulnaria ulna/danica* clades. Proportions of studied groups in the libraries of V3-V4 18S rRNA and both rbcL amplicons are shown in Figure 1. Analysis of a mock community does not recover all three groups at equal abundances; *Aulacoseira islandica* is also strongly overrepresented in V3-V4 and V8-V9 libraries (Supplementary Table S2). The four tested marker/pipeline combinations also do not produce exactly identical results. However, all tested markers—except V8-V9 18S rRNA regions—appear to be applicable for studying the relative abundance of *U. acus* and *F. radians*, and they do not wildly contradict each other.

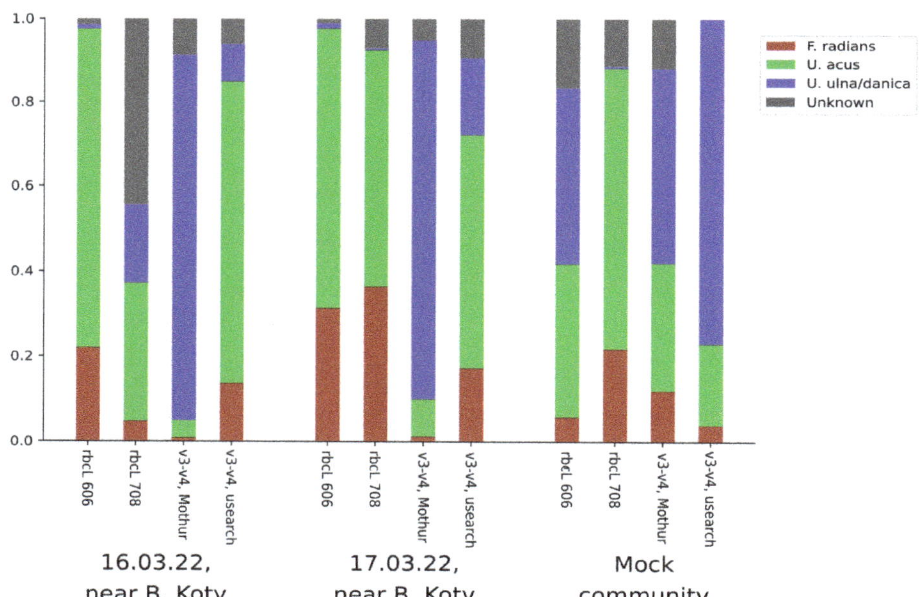

Figure 1. Relative abundances of *F. radians*, *U. acus* and *U. ulna/U. danica* in the sample from Lake Baikal, near Bolshiye Koty settlement, and culture mixture, as revealed by V3-V4 18S rRNA and rbcL 606 amplicons. Although the mock community contains only *U. ulna*, the classification pipeline does not distinguish it from *U. danica*; therefore, the ASVs of this species are marked as *U. ulna/U. danica*.

3.2. Analysis of Existing V4 Datasets

As shown above, V8-V9 variable regions of 18S rRNA are not suitable for studying Baikalian populations of *U. acus* and *F. radians*. rbcL could potentially be useful, but no sequencing data for this amplicon in Lake Baikal samples are publicly available. As for V4 18S rRNA amplicons, three relatively large datasets exist for Lake Baikal: a time series of water column samples from 0–25 m depths during March–September 2017 taken in the pelagic zone of the Southern Basin of Lake Baikal [8], and two datasets from multiple sites and depths within the lake sampled in July 2013 [22] and in the summer of 2017 [23].

Although all three datasets consist of Illumina MiSeq reads, they were amplified with three different primer pairs targeting slightly different 18S rRNA fragments. The amplicon used in [23] failed to produce ASVs that map with 99%+ identity to *Fragilaria radians*. Although there is a number of ASVs that align to *F. radians* better than they do to the two *Ulnaria* species (at roughly 97.5% identity), these sequences could potentially belong to *Fragilaria* species other than *F. radians*, which are known to be present in Lake Baikal [27,28]. Therefore, this dataset was excluded from further analysis, leaving us with one time series from 2017 and one geographical series from 2013.

As Figure 2 shows, the seasonal dynamics of both species follow the general pattern previously documented for *Fragilaria radians sensu lato* (see Introduction), with a spring bloom followed by a decrease in summer, and the near-complete absence of these diatoms in autumn. However, *U. acus* lags behind *F. radians sensu stricto* by roughly a month. Figure 3 shows that the peaks of both populations are positioned very close to the end of the ice season. There are no samples available for the melting period itself (late April to mid-May), but the highest relative abundance of *F. radians sensu stricto* predates this period, while the *U. acus* population peaks in open water.

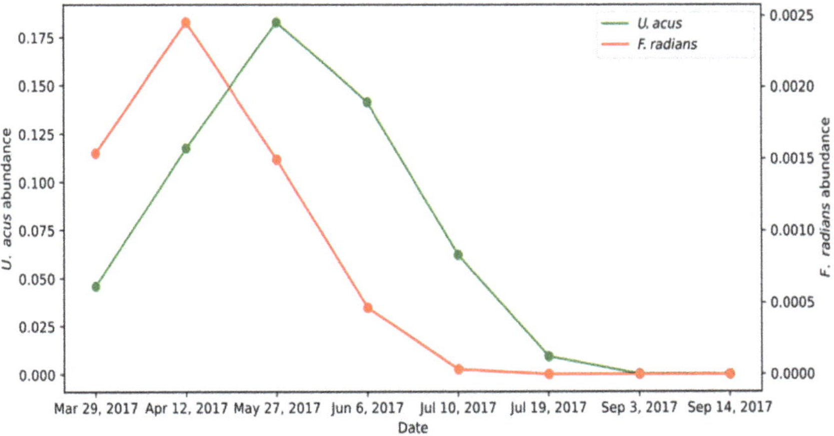

Figure 2. The relative abundance of *U. acus* and *F. radians* in the 2017 time series. Relative abundances of both species (the number of reads identified as one or the other, divided by library size) are plotted on separate Y axes against the time.

Distribution in the lake, as recovered from the 2013 dataset [22] (Figure 3), is also not identical. The highest abundance of both species is identified at the edge of Southern and Central Baikal, near the estuary of Selenga River. Both are also present, although in lower numbers, between Olkhon Island and Svyatoy Nos Peninsula; however, neither has been identified in the Southern Basin. The species distribution differs in the North: while *U. acus* is barely present in this area, *F. radians* populations are similar in size to most of those in Central Baikal. There is no clear pattern for their distribution along depth; however, this distribution is also not identical for most sampling sites.

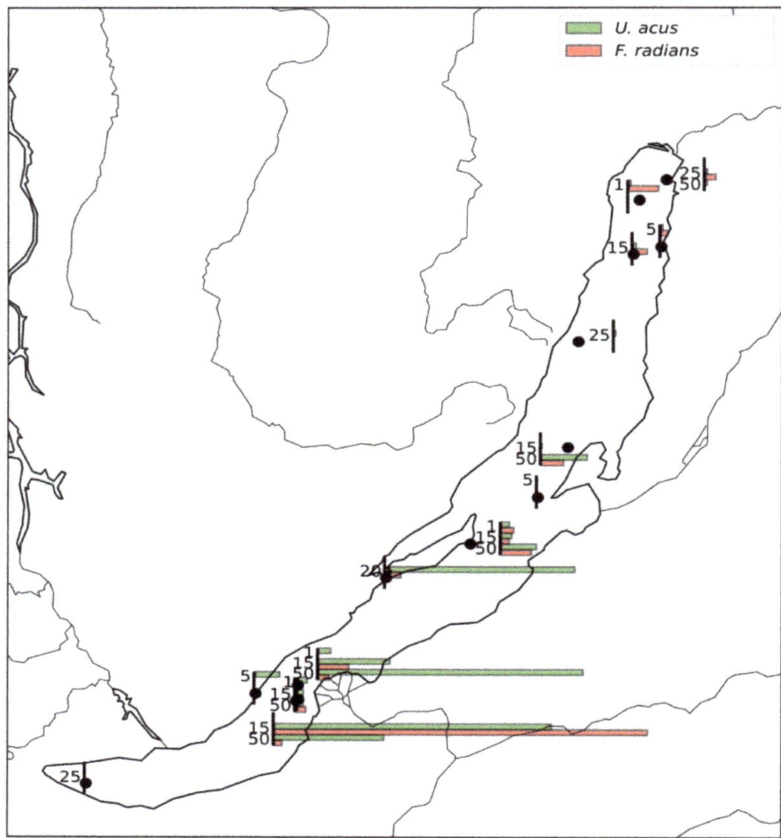

Figure 3. The relative abundance of *U. acus* and *F. radians* in the spatial distribution series; sampling depths are marked. Relative abundances of two species are not to scale with each other.

4. Discussion

4.1. Identifying the Target Species in Existing Data: Methodological Discussion

We estimated the relative abundances of two otherwise nearly indistinguishable species using available metabarcoding data. In addition to the result itself, the study may also be relevant because we encountered several pitfalls that may appear in similar works in the future.

First of all, a commonly used V8-V9 18S rRNA amplicon failed to separate *U. acus* from related *U. ulna* and *U. danica* species usually coexisting with it in Lake Baikal. This is unsurprising, because primer design in metabarcoding studies involves a tradeoff between taxonomic coverage (amplifying the barcode from as many taxa as possible), which requires a conservative sequence region, and precision (ability to separate closely related species), which benefits from having as many substitutions as possible. Commonly used primer pairs are designed to hit the sweet spot of amplifying the majority of eukaryotes while still being able to identify at least some genera [19]. OTU-based bioinformatics pipelines also usually target this kind of resolution, using 97% or 99% identity cutoffs that roughly correspond to species or genus but may also include several taxa (which is, in fact, where the term "Operational Taxonomic Unit" comes from—there is no guarantee that all OTUs generated at a given identity threshold correspond to taxa of a certain taxonomic rank).

This framework is useful for describing the overall composition of communities because ecological differences within genera are usually considered less important, while missing some large distant group altogether will heavily affect the conclusions. However, the requirements of our study are exactly the opposite; we are interested only in a small group of taxa, but with as high a resolution as possible. Practically, this difference in requirements means two things. First, any given amplicon may or may not work, regardless of how useful it was for previous studies of the same ecosystem. Second, the bioinformatics pipeline needs to be optimized for precision.

The obvious first step in this optimization is to use ASVs/zOTUs rather than larger OTUs [29]. Although they do not directly correspond to the taxa as well, ASVs are intended to be as small as possible, which means that any given ASV will correspond to some genotype or strain within species, rather than a group of species. Even with ASVs, the identification procedure has to involve an ad hoc pipeline with a custom reference database, because existing wide-range taxonomic databases do not have resolution below a genus level [25].

Even if relative abundances of the taxa in question have been estimated, it is well known, both from the literature [29] and from the mock community test in this work, that read counts correlate poorly with the actual biomass or cell count of the corresponding species. Further, the different datasets (even with the same marker gene) are produced with somewhat different methods and, therefore, are poorly compatible. On the other hand, it should be noted that although these differences are likely to be smaller in magnitude than those between different taxonomists using microscopy [30], they are not guaranteed to be small enough for quantitative comparison. Although it is tempting to call for the unification of methods that would allow for seamless co-analysis of datasets, this unification may have an unexpected downside if a universally accepted amplicon is unsuitable for some narrow problem (as V8-V9 18S rRNA was in our work). Diversity of methods, on the other hand, provides the chance that at least some part of the existing data would fit the requirements of any study.

4.2. Autecology of *U. acus* and *F. radians*

U. acus and *F. radians* were shown to exhibit both temporal and spatial differences in distribution. In 2017, in Lake Baikal, they follow a similar annual trend (Figure 2), but *F. radians* passes each stage of this trend before *U. acus*. Both under-ice blooms and post-melting populations are probably formed by a mixture of both species, but it appears that *F. radians* numbers start decreasing approximately when the ice starts melting. *U. acus*, on the other hand, continues the bloom and peaks in open water. In summer, both populations continue to decline, with *U. acus* lagging behind *F. radians*.

Spatial distribution is only observed in July, which is a period of decline for both species (as can be seen both in published data [9] and from the 2017 time series). However, both this decline and the lag identified from the time series fail to explain the observed spatial distribution in July 2013. Neither species monotonously decreases along the north–south axis, as would be predicted by a simple model where the Northern Basin lags behind Central and Southern Baikal in seasonal changes. There is also a difference in their distribution along the depth of the water column, but it is not similar across stations. Further, all samples are taken within the photic layer, ignoring the sub-photic zone, which makes it difficult to discuss the possible vertical migration.

In both datasets, the ASV abundance ratio is skewed towards *U. acus* by one or two orders of magnitude (Figure 4). Although analysis of 18S amplicons from the mock community was shown to overestimate *U. acus* abundance (or, equivalently, underestimate *F. radians* abundance), this overestimate was below an order of magnitude. Analysis of natural samples from the Bolshiye Koty settlement has shown a similar bias, although without a better estimate of real abundances, this bias could not be quantified. Further biases could be introduced by the ad hoc classification procedure used in this work. If, for example, some subpopulation of either species is sufficiently divergent for its 18S sequence to be less than 99% identical to reference strains, this would also lead to underestimating the abundance of the species in question.

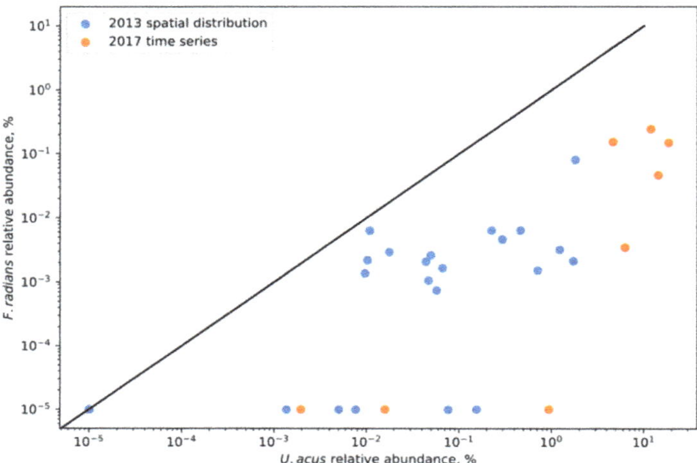

Figure 4. The ratio of *F. radians* and *U. acus* relative abundances in all studied samples. Both axes are in log scale; a pseudocount of 10^{-5}% was added to all values to make zeroes visible in log axes. Black line shows the 1:1 ratio.

Another reason to believe that the difference in abundance is an overestimate of orders of magnitude comes from the fact that the number of strains of both species in the laboratory collection of Limnological Institute is roughly similar [12]. At the time that these strains were isolated, the object of study was considered a single species, so no specific measures were taken to preferentially cultivate *F. radians*. However, it is possible that *U. acus* is significantly less viable in culture, leading to an unintentional enrichment in *F. radians*.

Between the unknown biases introduced by amplicon sequencing, ASV identification and culturing, our amplicon testing data cannot be directly used to calibrate the point estimates of a *Fragilaria*/*Ulnaria* relative abundance ratio. Even if such a calibration was possible, its results would only apply to the tested amplicon (identical to the one in the 2017 dataset [8]), not necessarily extending to the amplicon used in the 2013 data [22]. Thus, we cannot produce quantitative estimates of the two species' abundances, or even their ratio, from metabarcoding data. We can only qualitatively conclude that *U. acus* is likely more abundant than *F. radians*, but we cannot make claims about the magnitude of this difference.

However, we can at least assume that the same species within the same analysis is subject to roughly the same artifacts in all samples. Using this assumption, it is possible to compare the distribution of the species throughout space and time. In other words, one can use the 2017 time series to see whether the seasonal dynamics of the two species were similar in 2017 at the Listvyanka–Tankhoy transect, while the 2013 spatial dataset can be used to see whether they were distributed similarly between various sampling sites and depths across Lake Baikal in July 2013.

Using this assumption, we can observe that the two species exhibit differences in distribution, which, in turn, implies autecological differences. It is tempting to suggest that *F. radians*, which blooms earlier and remains abundant for a longer time in Northern Baikal, is more psychrophilic (or at least psychrotolerant, considering that both are cultured successfully at higher temperatures).

However, any ecological conclusions made from the two relatively small datasets used in this work would be speculative at best, and they may be compromised by the methodological concerns discussed above. In addition, metabarcoding data do not distinguish active and resting cells, and there is a precedent of inactive *F. radians* cells observed in near-surface water in July 2019. These cells have probably finished their bloom and sunk below the photic layer, only to be returned via upwelling [11]. Although such events are thought to be rare, there is no guarantee that something similar did not happen in Northern Baikal in 2017.

A more detailed investigation of these species' distribution requires larger volumes of data than are currently available. Further, the existing data only focus on a small number of abiotic factors, while the difference between species may be something obscure such as: a resistance to an unevenly distributed pathogen or grazer; different light requirements; or sensitivity to minor changes in water chemistry. Similar conclusions were reached in [13], which also used existing metabarcoding datasets to study the abundances of *Fragilaria* and *Ulnaria*, so that they may be generalized at least to all freshwater diatoms, and likely to the majority of non-model unicellular life.

In conclusion, we showed that V3-V4 or V4 18S rRNA amplicons can be reliably used for in situ distinguishing between closely related diatoms, although not quantitatively. Qualitative comparison shows that Baikalian populations of *Ulnaria acus* and *Fragilaria radians* differ in their distribution in both space and time.

Supplementary Materials: The following supporting information can be downloaded at: https://www.mdpi.com/article/10.3390/d15020280/s1: Table S1: Sequences and abundances of *Fragilaria/Ulnaria* ASVs produced from samples taken near the settlement of Bolshiye Koty, March 2020, and in an artificial mock community, by various methods and amplicons. Table S2: Taxonomy and distribution of 18S ASVs in samples taken near the settlement of Bolshiye Koty, March 2020, and in an artificial mock community.

Author Contributions: Conceptualization, A.M. (Alexey Morozov); methodology, A.M. (Alexey Morozov), A.M. (Artem Marchenkov), Y.G.; software, A.M. (Alexey Morozov); formal analysis, A.M. (Alexey Morozov); resources, Y.Z.; writing—original draft preparation, A.M. (Alexey Morozov) and A.M. (Artem Marchenkov); writing—review and editing, A.M. (Alexey Morozov), A.M. (Artem Marchenkov), Y.G., Y.Z., D.P.; visualization, A.M. (Alexey Morozov); supervision, Y.G. and D.P. All authors have read and agreed to the published version of the manuscript.

Funding: The study was funded by the Ministry of Science and Higher Education of the Russian Federation, project number 0279-2021-0009.

Institutional Review Board Statement: Not applicable.

Data Availability Statement: The sequencing data for strain mixture and Bolshiye Koty samples are available at NCBI SRA under project ID PRJNA666300.

Acknowledgments: The authors are grateful to Natalia Annenkova and Ivan Mikhailov for providing their metadata and for help with the technical details of their studies. The authors would like to thank Irkutsk Supercomputer Center of SB RAS for providing access to HPC-cluster "Akademik V.M. Matrosov".

Conflicts of Interest: The authors declare no conflict of interest. The funders had no role in the design of the study; in the collection, analyses or interpretation of data; in the writing of the manuscript; nor in the decision to publish the results.

References

1. Williams, D.M.; Round, F.E. *Fragilariforma* nom. nov., a new generic name for *Neofragilaria* Williams and Round. *Diatom Res.* **1988**, *3*, 265–267. [CrossRef]
2. Shishlyannikov, S.M.; Zakharova, Y.R.; Volokitina, N.A.; Mikhailov, I.S.; Petrova, D.P.; Likhoshway, Y.V. A procedure for establishing an axenic culture of the diatom *Synedra acus* subsp. radians (Kütz.) Skabibitsch. from Lake Baikal. *Limnol. Oceanogr. Meth.* **2011**, *9*, 478–484. [CrossRef]
3. Galachyants, Y.P.; Zakharova Yu, R.; Petrova, D.P.; Morozov, A.A.; Sidorov, I.A.; Marchenkov, A.M.; Logacheva, M.D.; Markelov, M.L.; Khabudaev, K.V.; Likhoshway, Y.V.; et al. Sequencing of the complete genome of an araphid pennate diatom *Synedra acus* subsp. radians from Lake Baikal. *Dokl. Biochem.* **2015**, *461*, 84–88. [CrossRef] [PubMed]
4. Popovskaya, G.I.; Likhoshway, Y.V.; Genkal, S.I.; Firsova, A.D. The role of endemic diatom algae in the phytoplankton of Lake Baikal. *Hydrobiologia* **2006**, *568*, 87–94. [CrossRef]
5. Bashenkhaeva, M.V.; Zakharova, Y.R.; Petrova, D.P.; Khanaev, I.V.; Galachyants, Y.P.; Likhoshway, Y.V. Sub-ice microalgal and bacterial communities in freshwater Lake Baikal, Russia. *Microb. Ecol.* **2020**, *70*, 751–765. [CrossRef]
6. Izmest'eva, L.R.; Moore, M.V.; Hampton, S.E.; Silow, E.A. Seasonal dynamics of common phytoplankton in Lake Baikal. *Proc. Russ. Acad. Sci. Sci. Cent.* **2006**, *8*, 191–196.

7. Bondarenko, N.A.; Logacheva, N.F. Structural changes in phytoplankton of the littoral zone of Lake Baikal. *Hydrobiol. J.* **2017**, *53*, 16–24. [CrossRef]
8. Mikhailov, I.S.; Galachyants, Y.P.; Bukin, Y.S.; Petrova, D.P.; Bashenkhaeva, M.V.; Sakirko, M.V.; Blinov, V.V.; Titova, L.A.; Zhakharova, Y.R.; Likhoshway, Y.V. Seasonal Succession and Coherence Among Bacteria and Microeukaryotes in Lake Baikal. *Microb. Ecol.* **2022**, *84*, 404–422. [CrossRef]
9. Mikhailov, I.S.; Bukin, Y.S.; Zakharova, Y.R.; Usoltseva, M.V.; Galachyants, Y.P.; Sakirko, M.V.; Blinov, V.V.; Likhoshway, Y.V. Co-occurrence patterns between phytoplankton and bacterioplankton across the pelagic zone of Lake Baikal during spring. *J. Microbiol.* **2019**, *57*, 252–262. [CrossRef]
10. Mikhailov, I.S.; Zakharova, Y.R.; Bukin, Y.S.; Galachyants, Y.P.; Petrova, D.P.; Sakirko, M.V.; Likhoshway, Y.V. Co-occurrence networks among bacteria and microbial eukaryotes of Lake Baikal during a spring phytoplankton bloom. *Microb. Ecol.* **2019**, *77*, 96–109. [CrossRef]
11. Grachev, M.; Bukin, Y.; Blinov, V.; Khlystov, O.; Firsova, A.; Bashenkhaeva, M.; Kamshilo, O.; Titova, L.; Bairamova, E.; Bedoshvili, Y.; et al. Is a High Abundance of Spring Diatoms in the Photic Zone of Lake Baikal in July 2019 Due to an Upwelling Event? *Diversity* **2021**, *13*, 504. [CrossRef]
12. Zakharova, Y.; Marchenkov, A.; Petrova, D.; Bukin, Y.; Morozov, A.; Bedoshvili, Y.; Podunay, Y.; Davidovich, O.; Davidovich, N.; Bondar, A.; et al. Delimitation of some taxa of Ulnaria and Fragilaria (Bacillariophyceae) based on genetic, morphological data and mating compatibility. *Diversity*, 2022; *submitted*.
13. Kahlert, M.; Karjalainen, S.M.; Keck, F.; Kelly, M.; Ramon, M.; Rimet, F.; Schneider, S.; Tapolczai, K.; Zimmermann, J. Co-occurrence, ecological profiles and geographical distribution based on unique molecular identifiers of the common freshwater diatoms Fragilaria and Ulnaria. *Ecol. Indic.* **2022**, *141*, 109114. [CrossRef]
14. Rusch, D.B.; Halpern, A.L.; Sutton, G.; Heidelberg, K.B.; Williamson, S.; Yooseph, S.; Wu, D.; Eisen, J.A.; Hoffman, J.M.; Remington, K.; et al. The Sorcerer II Global Ocean Sampling expedition: Northwest Atlantic through eastern tropical Pacifc. *PLoS Biol.* **2007**, *5*, e77. [CrossRef]
15. Stoeck, T.; Bass, D.; Nebel, M.; Christen, R.; Jones, M.D.; Breiner, H.W.; Richards, T.A. Multiple marker parallel tag environmental DNA sequencing reveals a highly complex eukaryotic community in marine anoxic water. *Mol. Ecol.* **2010**, *19*, 21–31. [CrossRef]
16. Bradley, I.M.; Pinto, A.J.; Guest, J.S. Design and evaluation of Illumina MiSeq-compatible, 18S rRNA gene-specific primers for improved characterization of mixed phototrophic communities. *Appl. Environ. Microbiol.* **2016**, *82*, 5878–5891. [CrossRef]
17. Stoof-Leichsenring, K.R.; Epp, L.S.; Trauth, M.H.; Tiedemann, R. Hidden diversity in diatoms of Kenyan Lake Naivasha: A genetic approach detects temporal variation. *Mol. Ecol.* **2012**, *21*, 1918–1930. [CrossRef]
18. Bruder, K.; Medlin, L. Molecular assessment of phylogenetic relationships in selected species/genera in the naviculoid diatoms (Bacillariophyta). I. The genus Placoneis. *Nova Hedwigia* **2007**, *85*, 331. [CrossRef]
19. Kermarrec, L.; Franc, A.; Rimet, F.; Chaumeil, P.; Humbert, J.F.; Bouchez, A. Next-generation sequencing to inventory taxonomic diversity in eukaryotic communities: A test for freshwater diatoms. *Mol. Ecol. Resour.* **2013**, *13*, 607–619. [CrossRef]
20. Schloss, P.D. Reintroducing mothur: 10 years later. *Appl. Environ. Microbiol.* **2020**, *86*, e02343–e02419. [CrossRef]
21. Camacho, C.; Coulouris, G.; Avagyan, V.; Ma, N.; Papadopoulos, J.; Bealer, K.; Madden, T.L. BLAST+: Architecture and applications. *BMC Bioinform.* **2009**, *10*, 421. [CrossRef]
22. Annenkova, N.V.; Giner, C.R.; Logares, R. Tracing the origin of planktonic protists in an ancient lake. *Microorganisms* **2020**, *8*, 543. [CrossRef] [PubMed]
23. David, G.M.; Moreira, D.; Reboul, G.; Annenkova, N.V.; Galindo, L.J.; Bertolino, P.; López-Archilla, A.I.; Jardillier, L.; López-Garcia, P. Environmental drivers of plankton protist communities along latitudinal and vertical gradients in the oldest and deepest freshwater lake. *Environ. Microbiol.* **2021**, *23*, 1436–1451. [CrossRef] [PubMed]
24. Rognes, T.; Flouri, T.; Nichols, B.; Quince, C.; Mahé, F. VSEARCH: A versatile open source tool for metagenomics. *Peer J.* **2016**, *4*, e2584. [CrossRef] [PubMed]
25. Quast, C.; Pruesse, E.; Yilmaz, P.; Gerken, J.; Schweer, T.; Yarza, P.; Peplis, J.; Glöckner, F.O. The SILVA ribosomal RNA gene database project: Improved data processing and web-based tools. *Nucleic Acids Res.* **2012**, *41*, D590–D596. [CrossRef]
26. Hunter, J.D. Matplotlib: A 2D graphics environment. *Comput. Sci. Eng.* **2007**, *9*, 90–95. [CrossRef]
27. Pomazkina, G.V.; Belykh, O.I.; Domysheva, V.M.; Sakirko, M.V.; Gnatovskii, R.Y. Structure and dynamics of the phytoplankton in Southern Baikal (Russia). *Int. J. Algae* **2010**, *12*, 64–79. [CrossRef]
28. Pomazkina, G.V.; Rodionova, Y.V.; Khanaev, I.V.; Scherbakova, T.A. The State of Benthic Diatom Communities in Listvennichnyi Bay of Lake Baikal (Russia). *Int. J. Algae* **2018**, *20*, 387–392. [CrossRef]
29. Callahan, B.J.; McMurdie, P.J.; Holmes, S.P. Exact sequence variants should replace operational taxonomic units in marker-gene data analysis. *ISME J.* **2017**, *11*, 2639–2643. [CrossRef]
30. Kahlert, M.; Albert, R.-L.; Anntila, E.-L.; Bengtsson, R.; Bigler, C.; Eskola, T.; Gälman, V.; Gottschalk, S.; Herlitz, E.; Jarlman, A.; et al. Harmonization is more important than experience—Results of the first Nordic-Baltic diatom intercalibration exercise 2007 (stream monitoring). *J. Appl. Phycol.* **2009**, *21*, 471–482. [CrossRef]

 *diversity* 

Article

A Demonstration of DNA Barcoding-Based Identification of Blade-Form *Ulva* (Ulvophyceae, Chlorophyta) Species from Three Site in the San Juan Islands, Washington, USA

Gabrielle M. Kuba [1,2,3], Brenda Carpio-Aguilar [1,4], Jason Eklund [1,5] and D. Wilson Freshwater [1,6,*]

1 Friday Harbor Marine Laboratory, University of Washington, Friday Harbor, WA 98250, USA
2 Department of Biology, College of Charleston, Charleston, SC 29424, USA
3 Department of Biological and Environmental Sciences, University of Rhode Island, Kingston, RI 02881, USA
4 School of Biology, Universidad de Costa Rica, San Jose 10102, Costa Rica
5 Department of Botany, Connecticut College, New London, CT 06320, USA
6 Center for Marine Science, University of North Carolina at Wilmington, Wilmington, NC 28403, USA
* Correspondence: freshwaterw@uncw.edu; Tel.: +1-910-962-2375

Abstract: Marine macroalgae are foundation species that play a critical ecological role in coastal communities as primary producers. The macroalgal genus *Ulva* is vital in intertidal communities, serving as a food source and shelter for organisms, but these species also form environment-damaging nuisance blooms. This project aimed to demonstrate the utility of DNA barcoding for determining the diversity of *Ulva* species in the San Juan Islands (Washington, DC, USA). Blade-form *Ulva* (Ulvophyceae) specimens were collected from the lower, mid, and upper intertidal zones at three sites experiencing different levels of wave exposure. Sequences of plastid-encoded *tufA* were generated for each specimen and cluster analyses revealed the presence of four species at the collection sites. Two species were positively identified as *Ulva expansa* and *Ulva fenestrata* based on their sharing identical *tufA* sequences with those of the holotype specimens. Sequences of plastid-encoded *rbc*L and the nuclear-encoded ribosomal ITS regions of representative specimens were used to identify the other two species as *Ulva prolifera* and *Ulva californica* based on their similarity to epitype and topotype specimen sequences, respectively. Additional types of specimen sequencing efforts are needed to increase the number of *Ulva* species that can be accurately identified and realize their true biodiversity.

Keywords: ITS; macroalgae; *rbc*L; *tufA*; *Ulva californica*; *Ulva expansa*; *Ulva fenestrata*; *Ulva prolifera*

Citation: Kuba, G.M.; Carpio-Aguilar, B.; Eklund, J.; Freshwater, D.W. A Demonstration of DNA Barcoding-Based Identification of Blade-Form *Ulva* (Ulvophyceae, Chlorophyta) Species from Three Site in the San Juan Islands, Washington, USA. *Diversity* **2022**, *14*, 899. https://doi.org/10.3390/d14110899

Academic Editors: Manuel Elias-Gutierrez and Jun Sun

Received: 25 July 2022
Accepted: 20 October 2022
Published: 24 October 2022

1. Introduction

The northeast Pacific Ocean, from the coasts of Southeast Alaska to Oregon, is characterized by a diverse community of marine algae, including 671 taxa and 284 genera [1]. The San Juan Islands within the Salish Sea are a particularly rich area within this region that experience mixed semidiurnal tides that cause intense tidal flows with vigorous vertical mixing, especially at sills [2,3]. The characteristics of channels through the islands are highly influenced by the Fraser River from the Strait of Georgia [4], and the succession of spring and neap tides modulates the mixing over the sills, regulating the estuarine exchange of water [3]. The mixture of cold ocean waters of high salinity with brackish surface waters, seasonality, and physical factors further supports the diversity of the marine community and affect the interaction among resident organisms [5]. This is especially true for marine macroalgae, which have highly diverse intertidal and subtidal communities in this region. The diversity of these organisms can be masked by the high frequency of cryptic and phenotypically plastic species [6,7].

Marine macroalgae are foundation species that play a critical ecological role in coastal communities as primary producers and habitat-defining organisms [8]. *Ulva* Linnaeus species are important components of biodiversity and bioindicators [9] However, they have also been associated with the majority of blooms of free-floating green algae responsible for 'green tides' because *Ulva* species can rapidly grow in nutrient-rich habitats and have a high tolerance range for abiotic factors such as temperature and salinity [6,10–13]. Eutrophication-driven green tides in shallow waters have a direct economic impact on coastal communities, making it essential to identify the species involved for bloom characterization and control [10,14]. In addition, it is important to understand their potential uses in pharmaceutical applications for drug development [12], as well as in biotechnological and industrial processes as bioremediators, biofuels, and food sources [14]. However, their simple morphology and phenotypic plasticity means that diversity assessments and identifications of *Ulva* species based on morphological characters range from challenging to impossible e.g., [15,16].

The genus *Ulva* is constituted of nearly 100 taxonomically accepted species [17] including those species previously placed in *Enteromorpha* Link [18]. This green algal genus is present in both freshwater and marine environments. In the latter, it is ubiquitous along coasts, rocky shores, and protected bays and estuaries, growing attached to substrata or found as drift. The morphological characterization of *Ulva* species has traditionally included both macro- and microscopic features. Macroscopic features include having distromatic blade-form or monostromatic tubular thalli, thallus shape, size, extent of branching and presence or absence of marginal dentation. Cellular features considered key to identification include cellular shape and dimensions, number of pyrenoids, arrangement of the cells in regular or irregular patterns, and thallus thickness e.g., [19–22]. Although previous studies used these characters for identification, they have been found to vary within species depending on thallus age, reproductive state, wave exposure, tidal factors, temperature, salinity, light, life-history stage, and biological factors such as herbivory and associated microbiome e.g., [9,23,24]. In addition, the morphological plasticity of *Ulva* species results in a variety of forms and ecotypes. Therefore, the taxonomic status of species in this genus remains uncertain and difficult to assess [9,11,16,24,25].

Molecular analysis of *Ulva* spp. is greatly expanding our understanding of their taxonomic and phylogenetic status [25]. Studies utilizing DNA sequence data have defined many molecular-based species e.g., [11,16,26,27], but sequences from type specimens have been generated for relatively few historical species [25,28–30] and only recently have type sequences been included in new species descriptions [16,31,32]. The historical types that have been sequenced demonstrate that very few of the specimen identifications for sequences in public databases are correct [30,33]. Accordingly, while *Ulva* species can be easily delimited with DNA sequence data, the identification of most species remains problematic.

Up to 17 species and varieties of *Ulva* (including taxa formerly classified as *Enteromorpha*) have been reported in the northeast Pacific [34]. Hayden and Waaland [6] reported 12 species based on molecular and morphological analyses in the most recent treatment of the genus from this region. Little is known about *Ulva* species in the San Juan Islands; however, multiple studies have focused on the surrounding Salish Sea ecosystem [6,35–38]. *Ulva* species within this area proliferate into blooms comprised of multiple species in the intertidal zone, similar to many other anthropogenically influenced coastal ecosystems [39–42]. They were found to outcompete other macroalgae within these zones, exhibiting harmful characteristics that alter species interactions [42,43]. A better understanding of the species involved is needed for these reasons.

DNA barcoding was originally envisioned as a utilitarian method that could be simply applied for the identification of species by a non-specialist using a single universal marker, the mitochondria-encoded cytochrome c oxidase subunit 1 gene 5′ region [44,45]. While this vision has been realized for many groups of organisms, others have been found to require different and multiple markers [46,47]. *Ulva* and other green algae are part of the latter group, but studies have shown that plastid-encoded *rbc*L and *tufA*, as well as nuclear-encoded ITS, are useful singly or in combination for barcoding these algae [48–50]. The objectives of this study were to demonstrate the utility of DNA barcoding in its simplest application to determine the number and, if possible, identity of *Ulva* species. This was achieved by exploring the diversity of blade-form *Ulva* species present at three environmentally different study sites in the San Juan Islands, Washington.

2. Materials and Methods

2.1. Sample Collection

Thirty-five blade-form *Ulva* specimens were collected from the intertidal zone at three sites of differing relative wave exposure within the San Juan Islands, Washington (Table 1). Collections were made within the low, mid, and high intertidal zones at each location. Specimens were chosen at each location based on observed macromorphological variation, and two algal specimens of representative morphologies identified in each intertidal zone were collected at each sampled site. Specimens were only collected if attached and not as drift, and transported on ice back to the lab, where they were placed into a running seawater table until processed. Each specimen was morphologically identified using the Gabrielson and Lindstrom [1] key and vouchers were made and deposited in the University of Washington herbarium (WTU). All specimen data, including photographic images, are available from the Barcode of Life Database system website (dx.doi.org/10.5883/DS-MASJI08).

Table 1. *Ulva* specimen collection site information.

Site Name	Wave Exposure Level	Latitude, Longitude	Date
Iceberg Point, Lopez Is.	High	48.42° N, 122.90° W	25 June 2021
Cattle Point, San Juan Is.	Mid	48.45° N, 122.96° W	27 June 2021
Friday Harbor Lab, San Juan Is.	Low	48.55° N, 123.01° W	29 June 2021

2.2. DNA Extraction, Amplification, and Sequencing

Total DNA was extracted from specimens using a Bioline Extract-PCR Kit (Bioline, Taunton, MA, USA) following the protocol of Taylor et al. [50], with small modifications as follows. Approximately 0.5 cm^2 of healthy blade tissue was chopped into small pieces, and incubated at 75 °C in 50 µL of Extract-PCR kit enzymatic solution for 1–20 h before enzyme deactivation by heating at 95 °C for 10 min. Cellular debris was pelleted by centrifugation and samples were diluted 1:10 and stored at −20 °C.

The plastid-encoded *tufA* locus was amplified for each *Ulva* specimen using MyTaq HS Red Mix following the manufacturer's protocol (Bioline) with primers described in Fama et al. [51]. Cycling conditions were as follows: an initial denaturing step of 95 °C for 2:45 min, followed by 35 cycles of 95 °C for 15 s, 45 °C for 15 s, and 72 °C for 1 min, with a final extension at 72 °C for an additional 4 min. PCR products were enzymatically cleaned using Exo-Sap (Thermo Fisher Scientific, Waltham, MA, USA) and sent to Genewiz for DNA sequencing (Azenta Life Sciences, South Plainfield, NJ, USA). Based on initial analyses of *tufA* sequences, nuclear-encoded ITS and plastid-encoded *rbc*L sequences were generated from representative specimens of the detected species. ITS and *rbc*L were amplified and sequenced following the protocols of Freshwater et al. [52] but using a MyTaq HS Red DNA Polymerase Kit (Bioline), and the ITS and *rbc*L primers described by Shimada et al. [53]. Individual sequence reactions were compiled and edited using Sequencher (v. 5.4, Gene Codes Corporation, Ann Arbor, MI, USA).

2.3. DNA Sequence Analyses and Species Identifications

Alignments of DNA sequences were generated using MUSCLE [54] as implemented in MEGA (v. 7.0.26, [55]) or Geneious (v. 9, Biomatters Limited, Aukland, New Zealand). Species were molecularly delineated through barcode sequence clustering. Initially a UPGMA cluster diagram was generated from an alignment of the 35 *tufA* sequences for the newly collected San Juan Islands *Ulva* specimens to establish specimen clusters. Inter- and intra-cluster sequence divergence values were then assessed to determine if there were barcode gaps, as defined by Meier et al. [56] between clusters and whether these barcode gaps fit the *tufA* species divergence threshold ranges of Saunders and Kucera [49] and Kirkendale et al. [57]. GenBank BLAST analyses [58] of the *tufA* and, where needed, ITS and *rbc*L sequences were used to explore the identifications of the resulting molecularly defined species.

3. Results

The 35 *Ulva* specimens were grouped into four species based on UPGMA cluster analysis of *tufA* sequences (Figure 1). Intraspecific variation in *tufA* sequences was only seen in Species-1 (0.0–0.7%; 3 haplotypes) and interspecific variation among the four species ranged from 3.2–3.3% to 7.5% (Table 2). BLAST searches revealed that *tufA* sequences of Species-3 and Species-4 were exact matches to those of the *U. expansa* (Setchell) Setchell & N.L. Gardner (GenBank # MH731007) and *U. fenestrata* Postels & Ruprecht (GenBank # MK456404) type specimens, respectively. BLAST searches of the Species-1 and Species-2 *tufA* sequences returned close matches to specimens within the *U. linza-procera-prolifera* (LPP) complex clade for Species-1 and specimens predominantly identified as *U. californica* Wille for Species-2.

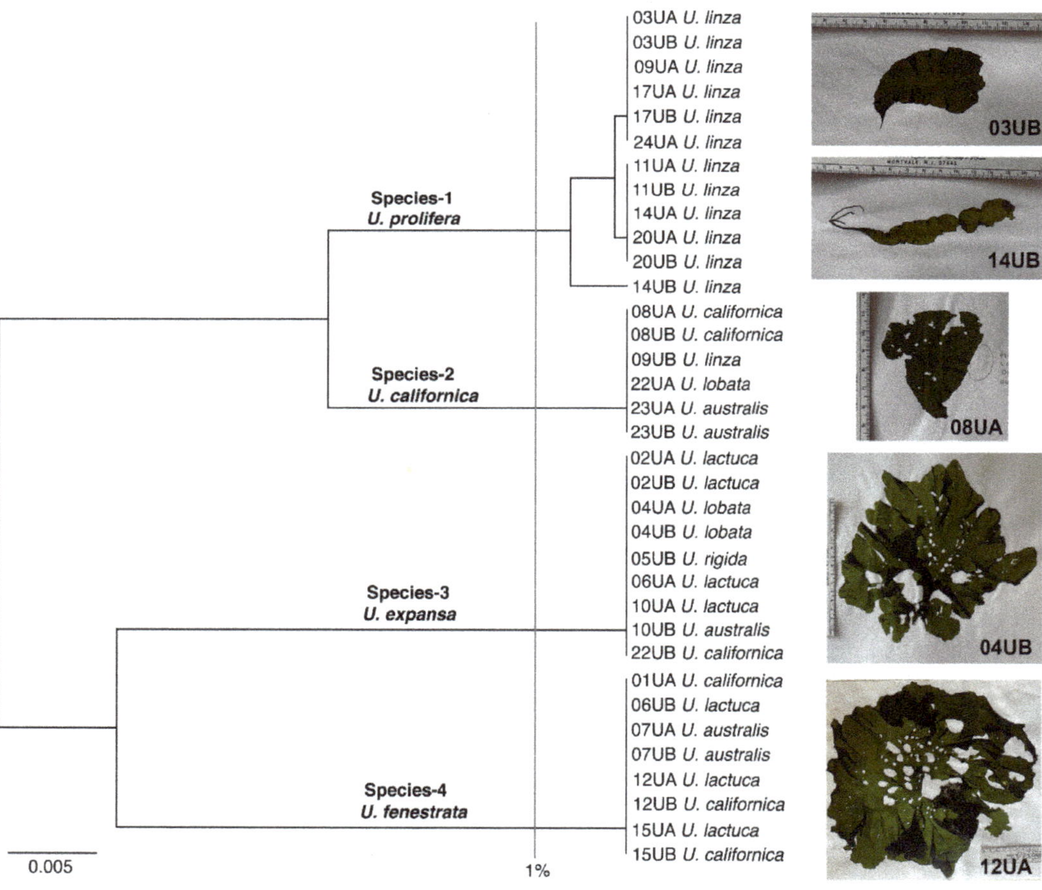

Figure 1. UPGMA *tufA* cluster analysis for 35 blade-form *Ulva* specimens collected at different intertidal zones from three sites in the San Juan Islands. Specimen labels include the collection number (e.g., '03UA') followed by the morphological identification of the specimen based on the Gabrielson and Lindstrom [1] key. The 1% sequence divergence level is indicated by the grey vertical line and example images of specimens are shown on the right.

Table 2. Intra- and interspecific divergences among *tufA* sequences from 35 specimens of four blade-form *Ulva* species collected in the San Juan Islands, WA. Gray background = intraspecific divergences; white background = interspecific divergences.

	Species-1 *U. prolifera* $n = 12$	**Species-2** *U. californica* $n = 6$	**Species-3** *U. expansa* $n = 9$	**Species-4** *U. fenestrata* $n = 8$
Species-1 *U. prolifera* $n = 12$	0.0–0.7%			
Species-2 *U. californica* $n = 6$	3.2–3.3%	0.00%		
Species-3 *U. expansa* $n = 9$	6.2–6.4%	6.2–6.3%	0.00%	
Species-4 *U. fenestrata* $n = 8$	6.9–7.2%	7.50%	5.4–5.5%	0.00%

The ITS sequences of Species-1 specimens representative of *tufA* haplotype 1 (specimen 03UA) and haplotype 2 (specimen 20UA) were identical. There was only a single base-pair difference between the ITS-2 region of this sequence and the ITS-2 sequences of 17 *U. prolifera* O.F. Müller topotype specimens (GenBank# AJ012276, but see discussion), including the epitype designated by Cui et al. [59]. The ITS-2 region sequence of the Species-1 specimen representative of *tufA* haplotype 3 (specimen 14UB) is two base pairs different from that of the *U. prolifera* epitype.

The *rbc*L sequences of two representative specimens of Species-2 (09UB; 23UB) were identical to each other and a topotype specimen identified by Hayden et al. [18] as *U. californica* (GenBank #AY255866). Similarly, ITS sequences of specimens 09UB and 23UB were identical and only 0.7% different from the ITS sequence of the Hayden et al. [18] topotype specimen identified as *U. californica* (GenBank #AY260560).

4. Discussion

Distinguishing *Ulva* species is a well-known problem in phycology. They have a very simple morphology and the few morphological characters that have been used to describe species exhibit intraspecific variation e.g., [15,16,21,60,61]. Analyses of DNA sequences are currently popular for delineating *Ulva* species e.g., [9,62,63]. However, as clearly demonstrated in a series of papers by Hughey et al. [25,29,30], *Ulva* specimens can only confidently be identified to species if DNA sequence data from those specimens can be matched to that of type specimens. These papers, as well as the overall analysis of *Ulva* sequences in GenBank by Fort et al. [33], demonstrated that many of the names assigned to *Ulva* sequences in public databases are incorrect. As an extreme example, all sequences in GenBank assigned to *U. rigida* were incorrectly identified [30]. Fort et al. [33] identified accessions that could be used for species identifications when a query sequence was homologous, and Hughey et al. [30] provide a table with all sequenced historical types and sequence determined synonyms.

Analyses of *tufA* sequences for blade-form *Ulva* specimens from different tidal heights at three different locations in the San Juan Islands revealed the presence of four species (Figure 2). Two of these species could be positively identified through the homology of their *tufA* sequences to that from type specimens. One was identified based on the holotype sequences published by Hughey et al. [28] as *U. expansa* (type locality: Monterey, CA, USA), a species reported in an older floristic survey of nearby Whidbey Island [35] and by Scagel [34] in his flora of British Columbia and northern Washington. Although reported as far north as British Columbia in these and other treatments of Northeast Pacific macroalgae e.g., [64–66], Tanner [36] synonymized *U. expansa* with *U. fenestrata* based on the morphological variation observed in herbarium specimens, field collections and culture studies. This synonymy was followed in the recent keys to the marine algae from southeastern Alaska to Oregon that have functioned as de facto floras for this region in recent years [1,67,68]. However, Hughey et al. [25,28] demonstrated the distinction of these two species. and *U. expansa* was once again recognized in the Northeast Pacific flora e.g., [69].

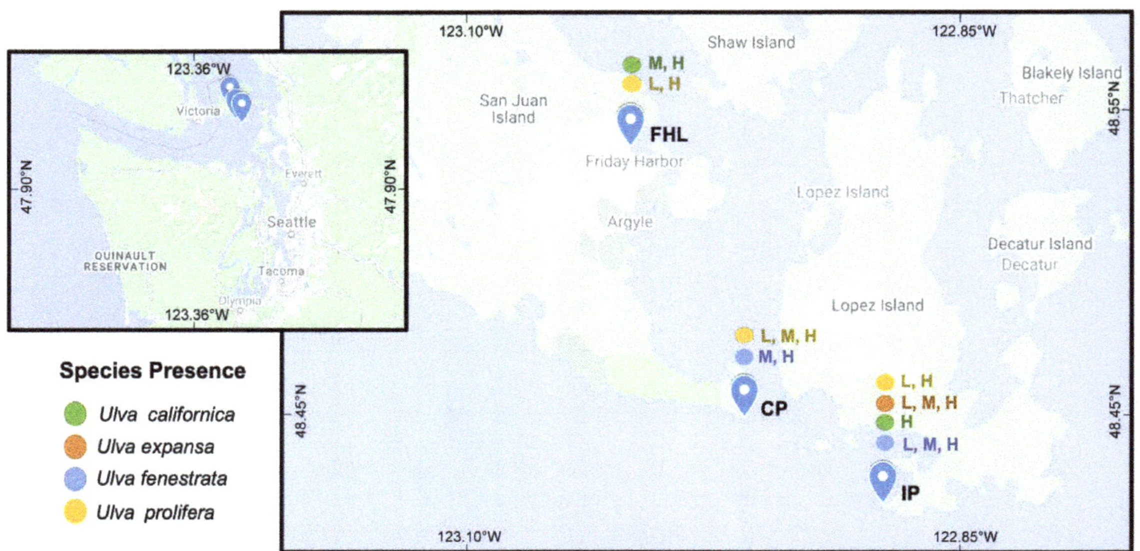

Figure 2. Map showing the distribution of blade-form *Ulva* species collected from different intertidal heights at three San Juan Island sites that experience different wave exposures. Species are indicated by color and intertidal zones by letters: L = low; M = middle; H = high. Locations: FHL = Friday harbor Laboratory beach front; CP = Cattle Point; IP = Iceberg Point.

The second positively identified species was determined to be *U. fenestrata* (type locality: Kamchatka, Russia) based on the *tufA* sequence for its holotype published by Hughey et al. [25]. Similar to *U. expansa*, *U. fenestrata* was included in the Scagel [34] flora and Tanner's [36] treatment of Northeast Pacific *Ulva*, but not in the more recent comprehensive keys [1,67,68]. *Ulva fenestrata* has generally been identified based upon the presence of perforations in the blade e.g., [65,70], but whether the presence or absence of perforations is a true developmental characteristic of species has been questioned, and both perforated and non-perforated specimens have been included in this species [36,70]. Gabrielson et al. [68] included *U. fenestrata*-type perforated blades within an unidentified *Ulva* species given the place-holder name *U. "lactuca"*. Further unpublished observations of San Juan Islands *Ulva* specimens led to this species being considered to represent *U. fenestrata* [71], and the current study verifies this identification.

The remaining two species revealed by the *tufA* analysis did not have close homology to any currently available *tufA* sequences from an *Ulva* type or topotype specimen. The *tufA* sequences from specimens of one of these species included three different haplotypes that had close homology to GenBank sequences from specimens placed in the *Ulva linza-procera-prolifera* (LPP) complex clade, a group composed of specimens variously identified as *U. linza* Linnaeus, *U. procera* (K.Ahlner) H.S.Hayden, Blomster, Maggs, P.C.Silva, Stanhope & Waaland, and *U. prolifera* O.F.Müller. Cui et al. [59] used morphological, molecular and crossing studies to examine the status of LPP complex specimens collected from Lolland Island, Denmark, the type locality of *U. prolifera*. Combining their results with those of previous LPP-complex-related studies e.g., [62,72,73], it was determined that *U. prolifera* was best represented by tubular branched specimens with sexual or asexual life histories [59]. The lectotype of *U. prolifera* is a drawing in Müller [74], thus an epitype was designated and sequences for the ITS-2 and 5S rDNA spacer regions generated. The ITS-2 sequence of this specimen was not made publicly available, but was reported to be identical to a previously published sequence with GenBank accession number AJ012276. The AJ012276 sequence includes not only ITS-2 but also the ITS-1 and 5.8S rRNA regions and, therefore, the epitype ITS-2 sequence is only the 215 bp portion of the AJ012276 sequence between the annealing sites of the two primers used by Cui et al. [59] to amplify and sequence this region in their specimens. ITS-2 sequences for representative specimens of the LPP complex species collected in the San Juan Islands were only 1–2 base pairs different (0.47–0.93%) from that of the *U. prolifera* epitype. This divergence from the epitype ITS-2 sequence is less than or equal to that of any LPP complex specimens included in the Cui et al. [59] study.

The San Juan Islands specimens were identified as *U. prolifera* based on these results. However, an unquestioned identification will require an understanding of whether any taxa within the LPP complex clade represent *U. linza*. Interestingly, the San Juan Islands specimens molecularly identified in this study as *U. prolifera*, were not branched tubes, the characteristic morphology of the species, but distromatic blades that became tubular where they were basally narrow near the point of attachment. This latter morphology has been identified as *U. linza* in the Northeast Pacific e.g., [1,64,68], and all these specimens were morphologically identified as *U. linza* (Figure 1). Similar to *U. prolifera* the lectotype of *U. linza* is an illustration [75] (pl. 9, Figure 6), but there is an epitype in OXF. Unfortunately, requests for the minimal type specimen material needed for current DNA sequence generation techniques have not been fulfilled, and the status of *U. linza* remains unresolved [76]. Regardless, the findings herein indicate that the concept of *U. prolifera* needs to be expanded to include blade-form thalli.

Representative specimens of the fourth species resolved in this study by the *tufA* analysis had *rbc*L sequences that were identical to that generated by Hayden et al. [18] from a La Jolla, California specimen of *U. californica*, the type locality for this species. The ITS sequences of San Juan Island specimens and that of this topotype specimen were also closely homologous and varied by only four base pairs (0.7%). The Hayden et al. [18] topotype specimen (WTU 344798) agrees with the type specimen (US 57108) in being ca. 2 cm or less and having turfy blades, and provides a basis for molecularly identifying specimens as *U. californica* in lieu of sequence data from the type specimen.

Tanner [60] conducted field, herbarium and culture studies of *U. californica*, *U. angusta* Setchell & N.L. Gardner, and *U. scagelii* Chihara, three morphologically similar Northeast Pacific species that differed in size, habit and distribution. The results of these studies led to the synonymy of *U. angusta* and *U. scagelii* with *U. californica*, increasing the size range and geographic distribution of *U. californica*. Specimens of *U. californica* sequenced in this study were also variable in size and distribution (Figure 2; dx.doi.org/10.5883/DS-MASJI08). The close homology of the topotype DNA sequences with those of specimens from this and other studies, e.g., [6,49,77], verifies the wider Northeast Pacific distribution and the environmentally determined morphological variation found in the study of Tanner [60].

As demonstrated in this study, simple analyses of DNA barcode sequences can be a useful tool for quickly distinguishing *Ulva* species, and only with an understanding of the number and extent of these species can applied questions concerning their diversity, ecology and physiology be addressed. For example, determining the species composition of blooms, or establishing monocultures or mixed cultures for industrial applications. However, this in no way diminishes the importance of extensive specimen collection combined with thorough phylogenetic and species delimitation analyses, e.g., [16,26], to establish the species and sequence characteristics upon which utilitarian DNA barcoding methods are based. The application of names to barcode-defined species, however, remains problematic. Two of the four species included in this study could be positively identified because DNA sequences were publicly available for their holotype specimens, and the best current identifications were possible for the other two species based on DNA sequences of epitype and topotype specimens. This fortuitous result is unusual because so few historical *Ulva* types have been sequenced, and only additional type specimen sequencing efforts and cooperation of the herbaria housing *Ulva* types can ensure that the application of additional species names is accurate.

Author Contributions: Conceptualization, G.M.K., B.C.-A. and D.W.F.; methodology, G.M.K., B.C.-A. and D.W.F.; validation, G.M.K., B.C.-A., J.E. and D.W.F.; formal analysis, G.M.K., B.C.-A., J.E. and D.W.F.; investigation, G.M.K., B.C.-A., J.E. and D.W.F.; resources, D.W.F.; data curation, G.M.K., B.C.-A., J.E. and D.W.F.; writing—original draft preparation, G.M.K., B.C.-A. and J.E.; writing—review and editing, G.M.K., B.C.-A. and D.W.F.; visualization, G.M.K. and D.W.F.; supervision, D.W.F.; project administration, G.M.K., B.C.-A., J.E. and D.W.F.; funding acquisition, D.W.F. All authors have read and agreed to the published version of the manuscript.

Funding: Research conducted at the University of Washington's Friday Harbor Marine Laboratory was funded by the Marine Botany class budget and research conducted the University of North Carolina at Wilmington's Center for Marine Science was supported by the CMS DNA-Algal Trust.

Institutional Review Board Statement: Not applicable.

Informed Consent Statement: Not applicable.

Data Availability Statement: The data presented in this study are openly available in the BOLD system database at dx.doi.org/10.5883/DS-MASJI08 and GenBank accessions OP347101-OP347108; OP347119-OP347153; OP347156-OP347160.

Acknowledgments: We would like to thank Tom Mumford for his guidance and mentorship during the Friday Harbor Laboratory Marine Botany class. We would also like to thank the Friday Harbor Laboratory for their facilities, resources, and accessibility during this study. Multiple reviewers provided valuable guidance for this publication.

Conflicts of Interest: The authors declare no conflict of interest.

References

1. Gabrielson, P.W.; Lindstrom, S.C. *Keys to the Seaweeds and Seagrasses of Southeast Alaska, British Columbia, Washington and Oregon*; Phycological Contribution Number 9; Island Blue/Printorium Bookworks: Victoria, BC, Canada, 2018; 180p.
2. Klinger, T.; Fluharty, D.; Evans, K.; Byron, C. *Assessment of Coastal Water Resources and Watershed Conditions at San Juan Island National Historical Park*; Technical Report NPS/NRWRD/NRTR-2006/360; U.S. Department of the Interior: Washington, DC, USA, 2006; p. 144.
3. Masson, D.; Cummins, P.F. Fortnightly modulation of the estuarine circulation in Juan de Fuca Strait. *J. Mar. Res.* **2000**, *58*, 439–463. [CrossRef]
4. Zamon, J.E. Tidal changes in copepod abundance and maintenance of a summer *Coscinodiscus* bloom in the southern San Juan Channel, San Juan Islands, USA. *Mar. Ecol. Prog. Ser.* **2002**, *226*, 193–210. [CrossRef]
5. Burnaford, J.L. Habitat modification and refuge from sublethal stress drive a marine plant-herbivore association. *Ecology* **2004**, *85*, 2837–2849. [CrossRef]
6. Hayden, H.S.; Waaland, J.R. A molecular systematic study of *Ulva* (Ulvaceae, Ulvales) from the northeast Pacific. *Phycologia* **2004**, *43*, 364–382. [CrossRef]
7. Steffensen, D.A. Morphological variation of *Ulva* in the Avon-Heathcote Estuary, Christchurch. *N. Z. J. Mar. Fresh. Res.* **2010**, *10*, 329–341. [CrossRef]
8. Burke, C.; Thomas, T.; Lewis, M.; Steinberg, P.; Kjelleberg, S. Composition, uniqueness and variability of the epiphytic bacterial community of the green alga *Ulva australis*. *ISME J.* **2011**, *5*, 590–600. [CrossRef]
9. Wolf, M.A.; Sciuto, K.; Andreoli, C.; Moro, I. *Ulva* (Chlorophyta, Ulvales) biodiversity in the North Adriatic Sea (Mediterranean, Italy): Cryptic species and new introductions. *J. Phycol.* **2012**, *48*, 1510–1521. [CrossRef]
10. Duan, W.; Guo, L.; Sun, D.; Zhu, S.; Chen, X.; Zhu, W.; Xu, T.; Chen, C. Morphological and molecular characterization of free-floating and attached green macroalgae *Ulva* spp. in the Yellow Sea of China. *J. Appl. Phycol.* **2012**, *24*, 97–108. [CrossRef]
11. Guidon, M.; Thornber, C.; Wysor, B.; O'Kelly, C. Molecular and morphological diversity of Narragansett Bay (RI, USA) *Ulva* (Ulvales: Chlorophyta) populations. *J. Phycol.* **2013**, *49*, 979–995. [CrossRef]
12. Ismail, M.M.; Mohamed, S.E. Differentiation between some *Ulva* spp. by morphological, genetic and biochemical analyses. *Vavilovskii Zh. Genet. Sel.* **2017**, *21*, 360–367. [CrossRef]
13. Rybak, A.S. Species of *Ulva* (Ulvophyceae, Chlorophyta) as indicators of salinity. *Ecol. Indic.* **2018**, *85*, 253–261. [CrossRef]
14. Wichard, T.; Charrier, B.; Mineur, F.; Bothwell, J.H.; De Clerck, O.; Coates, J.C. The green seaweed *Ulva*: A model system to study morphogenesis. *Front. Plant Sci.* **2015**, *6*, 72. [CrossRef] [PubMed]
15. Hofmann, L.C.; Nettleton, J.C.; Neefus, C.D.; Mathieson, A.C. Cryptic diversity of *Ulva* (Ulvales, Chlorophyta) in the Great Bay Estuarine System (Atlantic, USA): Introduced and indigenous distromatic species. *Eur. J. Phycol.* **2010**, *45*, 230–239. [CrossRef]

16. Lagourgue, L.; Gobin, S.; Brisset, M.; Vandenberghe, S.; Bonneville, C.; Jauffrais, T.; Van Wynsberge, S.; Payri, C.E. Ten new species of *Ulva* (Ulvophyceae, Chlorophyta) discovered in New Caledonia: Genetic and morphological diversity, and bloom potential. *Eur. J. Phycol.* **2022**, 1–21. [CrossRef]
17. Guiry, M.D.; Guiry, G.M. AlgaeBase. World-Wide Electronic Publication, National University of Ireland, Galway. Available online: https://www.algaebase.org (accessed on 29 June 2021).
18. Hayden, H.S.; Blomster, J.; Maggs, C.A.; Silva, P.C.; Stanhope, M.J.; Waaland, J.R. Linnaeus was right all along: *Ulva* and *Enteromorpha* are not distinct evolutionary entities. *Eur. J. Phycol.* **2003**, *38*, 277–294. [CrossRef]
19. Bliding, C. A critical survey of European taxa in Ulvales. Part I: *Capsosiphon, Percursaria, Blidingia, Enteromorpha. Opera Bot.* **1963**, *8*, 1–160.
20. Bliding, C. A critical survey of European taxa in Ulvales. Part II: *Ulva, Ulvaria, Monostroma, Kornmannia. Bot. Not.* **1968**, *121*, 535–629.
21. Kapraun, D.F. Field and cultural studies of *Ulva* and *Enteromorpha* in the vicinity of Port Aransas, Texas. *Contrib. Mar. Sci.* **1970**, *15*, 205–285.
22. Kapraun, D.F. An illustrated guide to the benthic marine algae of coastal North Carolina. II. Chlorophyta and Phaeophyta. *Bibloth. Phycol.* **1984**, *58*, 1–173.
23. Matsuo, Y.; Imagawa, H.; Nishizawa, M.; Shizuri, Y. Isolation of an algal morphogenesis inducer from a marine bacterium. *Science* **2005**, *307*, 1598. [CrossRef]
24. Kazi, M.A.; Kavale, M.G.; Singh, V. Morphological and molecular characterization of *Ulva chaugulii* sp. nov. *U. lactuca* and *U. ohnoi* (Ulvophyceae, Chlorophyta) from India. *Phycologia* **2016**, *55*, 45–54.
25. Hughey, J.R.; Maggs, C.A.; Mineur, F.; Jarvis, C.; Miller, K.A.; Shabaka, S.H.; Gabrielson, P.W. Genetic analysis of the Linnaean *Ulva lactuca* (Ulvales, Chlorophyta) holotype and related type specimens reveals name misapplications, unexpected origins, and new synonymies. *J. Phycol.* **2019**, *55*, 503–508. [CrossRef]
26. Fort, A.; McHale, M.; Cascella, K.; Potin, P.; Usadel, B.; Guiry, M.D.; Sulpice, R. Foliose *Ulva* species show considerable inter-specific genetic diversity, low intra-specific genetic variation, and the rare occurrence of inter-specific hybrids in the wild. *J. Phycol.* **2021**, *57*, 219–233. [CrossRef]
27. Melton, J.T.; Lopez-Bautista, J.M. Diversity of the green macroalgal genus *Ulva* (Ulvophyceae, Chlorophyta) from the east and gulf coast of the United States based on molecular data. *J. Phycol.* **2021**, *57*, 551–568. [CrossRef]
28. Hughey, J.R.; Miller, K.A.; Gabrielson, P.W. Mitogenome analysis of a green tide forming *Ulva* from California, USA confirms its identity as *Ulva expansa* (Ulvaceae, Chlorophyta). *Mitochondrial DNA B Resour.* **2018**, *3*, 1302–1303. [CrossRef]
29. Hughey, J.R.; Gabrielson, P.W.; Maggs, C.A.; Mineur, F.; Miller, K.A. Taxonomic revisions based on genetic analysis of type specimens of *Ulva conglobata, U. laetevirens, U. pertusa* and *U. spathulata* (Ulvales, Chlorophyta). *Phycol. Res.* **2021**, *69*, 148–153. [CrossRef]
30. Hughey, J.R.; Gabrielson, P.W.; Maggs, C.A.; Mineur, F. Genomic analysis of the lectotype specimens of. European *Ulva rigida* and *Ulva lacinulata* (Ulvaceae, Chlorophyta) reveals the ongoing misapplication of names. *Eur. J. Phycol.* **2021**, *57*, 143–153. [CrossRef]
31. Hiraoka, M.; Shimada, S.; Uenosono, M.; Masuda, M. A new green-tide-forming alga, *Ulva ohnoi* Hiraoka et *Shimada* sp. nov. (Ulvales, Ulvophyceae) from Japan. *Phycol. Res.* **2003**, *51*, 17–29. [CrossRef]
32. Spalding, H.L.; Conklin, K.Y.; Smith, C.M.; O'Kelly, C.J.; Sherwood, A.R. New Ulvaceae (Ulvophyceae, Chlorophyta) from mesophotic ecosystems across the Hawaiian archipelago. *J. Phycol.* **2016**, *52*, 40–53. [CrossRef]
33. Fort, A.; McHale, M.; Cascella, K.; Potin, P.; Perrineau, M.-M.; Kerrison, P.D.; da Costa, E.; Calado, R.; Domingues, M.R.; Azevedo, I.C.; et al. Exhaustive reanalysis of barcode sequences from public repositories highlights ongoing misidentifications and impacts taxa diversity and distribution. *Mol. Ecol. Resour.* **2021**, *22*, 86–101. [CrossRef]
34. Scagel, R.F. *Marine Algae of British Columbia and Northern Washington, Part I: Chlorophyceae (Green Algae)*; National Museum of Canada Bulletin 207; National Museum: Ottawa, ON, Canada, 1966; 257p.
35. Phillips, R.C.; Vadas, R.L. Marine algae of Whidbey Island, Washington. *J. Inst. Res. Ser. A* **1967**, *7*, 2–81.
36. Tanner, C.E. The Taxonomy and Morphological Variation of Distromatic Ulvaceous Algae (Chlorophyta) from the Northeast Pacific. Ph.D. Thesis, Department of Botany, University of British Columbia, Vancouver, BC, Canada, 1979.
37. Nelson, T.A.; Olson, J.; Imhoff, L.; Nelson, A.V. Aerial exposure and desiccation tolerances are correlated to species composition in "green tides" of the Salish Sea (northeastern Pacific). *Bot. Mar.* **2010**, *53*, 103–111. [CrossRef]
38. Van Alstyne, K.L. Seasonal changes in nutrient limitation and nitrate sources in the green macroalga *Ulva lactuca* at sites with and without green tides in a northeastern Pacific embayment. *Mar. Pollut. Bull.* **2016**, *103*, 186–194. [CrossRef]
39. Shelford, V.E.; Weese, A.O.; Rice, L.A.; Rasmussen, D.I.; Maclean, A. Some marine biotic communities of the Pacific coast of North America. Part I. General survey of the communities. *Ecol. Monogr.* **1935**, *5*, 249–329.
40. Hylleberg, J.; Henriksen, K. The central role of bioturbation in sediment mineralization and element re-cycling. *Ophelia Suppl.* **1980**, *1*, 1–16.
41. Nelson, T.A.; Nelson, A.V.; Tjoelker, M. Seasonal and spatial patterns of "green tides" (ulvoid algal blooms) and related water quality parameters in the coastal waters of Washington State, USA. *Bot. Mar.* **2003**, *46*, 263–275. [CrossRef]
42. Nelson, T.A.; Haberlin, K.; Nelson, A.V.; Ribarich, H.; Hotchkiss, R.; Van Alstyne, K.L.; Buckingham, L.; Simunds, D.J.; Fredrickson, K. Ecological and physiological controls of species composition in green macroalgal blooms. *Ecology* **2008**, *89*, 1287–1298. [CrossRef]

43. Nelson, T.A.; Lee, D.J.; Smith, B.C. Are "green tides" harmful algal blooms? Toxic properties of water-soluble extracts from two bloom-forming macroalgae, *Ulva fenestrata* and *Ulvaria obscura* (Ulvophyceae). *J. Phycol.* **2003**, *39*, 874–879. [CrossRef]
44. Hebert, P.D.N.; Cywinska, A.; Ball, S.L.; Dewaard, J.R. Biological identification through DNA barcodes. *Proc. Royal Soc. B* **2003**, *270*, 313–321. [CrossRef]
45. Hebert, P.D.N.; Ratnasingham, S.; Dewaard, J.R. Barcoding animal life: Cytochrome c oxidase subunit 1 divergences among closely related species. *Proc. R. Soc. B* **2003**, *270*, 596–599. [CrossRef]
46. Hollingsworth, M.L.; Clark, A.A.; Forrest, L.L.; Richardson, J.; Pennington, R.T.; Long, D.G.; Cowan, R.; Chase, M.W.; Gaudeul, M.; Hollingsworth, P.M. Selecting barcoding loci for plants: Evaluation of seven candidate loci with species-level sampling in three divergent groups of land plants. *Mol. Ecol. Resour.* **2009**, *9*, 439–457. [CrossRef] [PubMed]
47. McFadden, C.S.; Benayahu, Y.; Pante, E.; Thoma, J.N.; Nevarez, P.A.; France, S.C. Limitations of mitochondrial gene barcoding in Octocorallia. *Mol. Ecol. Resour.* **2011**, *11*, 19–31. [CrossRef] [PubMed]
48. Hall, J.D.; Fucikova, K.; Lo, C.; Lewis, L.A.; Karol, K.G. An assessment of proposed DNA barcodes in freshwater green algae. *Crypt. Algol.* **2010**, *31*, 529–555.
49. Saunders, G.W.; Kucera, H. An evaluation of *rbc*L, *tufA*, UPA, LSU and ITS as DNA barcode markers for the marine green macroalgae. *Crypt. Algol.* **2010**, *31*, 487–528.
50. Taylor, R.L.; Bailey, J.C.; Freshwater, D.W. Systematics of Cladophora spp. (Chlorophyta) from North Carolina, USA, based upon morphology and DNA sequence data with a description of *Cladophora subtilissima* sp. nov. *J. Phycol.* **2017**, *53*, 541–556. [CrossRef]
51. Fama, P.; Wysor, B.; Kooistra, W.; Zuccarello, G.C. Molecular phylogeny of the genus *Caulerpa* (Caulerpales, Chlorophyta) inferred from chloroplast *tufA* gene. *J. Phycol.* **2002**, *38*, 1040–1050. [CrossRef]
52. Freshwater, D.W.; Miller, C.E.; Fankovich, T.A.; Wynne, M.J. DNA sequence analyses reveal two new species of *Caloglossa* (Delesserieaceae, Rhodophyta) from the skin of West Indian Manatees. *J. Mar. Sci. Eng.* **2021**, *9*, 163. [CrossRef]
53. Shimada, S.; Hiraoka, M.; Nabata, S.; Iima, M.; Masuda, M. Molecular phylogenetic analyses of the Japanese *Ulva* and *Enteromorpha* (Ulvales, Ulvophyceae), with special reference to the free-floating *Ulva*. *Phycol. Res.* **2003**, *51*, 99–108. [CrossRef]
54. Edgar, R.C. MUSCLE: Multiple sequence alignment with high accuracy and high throughput. *Nucleic Acids Res.* **2004**, *32*, 1792–1797. [CrossRef]
55. Kumar, S.; Stecher, G.; Tamura, K. MEGA7: Molecular Evolutionary Genetics Analysis Version 7.0 for Bigger Datasets. *Mol. Biol. Evol.* **2016**, *33*, 1870–1874. [CrossRef]
56. Meier, R.; Zhang, G.; Ali, F. The use of mean instead of smallest interspecific distances exaggerates the size of the "barcoding gap" and leads to misidentification. *Syst. Biol.* **2008**, *57*, 809–813. [CrossRef] [PubMed]
57. Kirkendale, L.; Saunders, G.W.; Winberg, P. A molecular survey of *Ulva* (Chlorophyta) in temperate Australia reveals enhanced levels of cosmopolitanism. *J. Phycol.* **2013**, *49*, 69–81. [CrossRef] [PubMed]
58. Karlin, S.; Altschul, S.F. Methods for assessing the statistical significance of molecular sequence features by using general scoring schemes. *Proc. Natl. Acad. Sci. USA* **1990**, *87*, 2264–2268. [CrossRef]
59. Cui, J.; Monotilla, A.P.; Zhu, W.; Takano, Y.; Shimada, S.; Ichihara, K.; Matsui, T.; He, P.; Hiraoka, M. Taxonomic reassessment of *Ulva prolifera* (Ulvophyceae, Chlorophyta) based on specimens from the type locality and Yellow Sea green tides. *Phycologia* **2018**, *57*, 692–704. [CrossRef]
60. Tanner, C.E. Investigations of the taxonomy and morphological variation of *Ulva* (Chlorophyta): *Ulva californica* Wille. *Phycologia* **1986**, *25*, 510–520. [CrossRef]
61. Bloomster, J.; Maggs, C.A.; Stanhope, M.J. Molecular and morphological analysis of *Enteromorpha intestinales* and *E. compressa* (Chlorophyta) in the British Isles. *J. Phycol.* **1998**, *34*, 319–340. [CrossRef]
62. Shimada, S.; Yokoyama, N.; Arai, S.; Hiraoka, M. Phylogeography of the genus *Ulva* (Ulvophyceae, Chlorophyta), with special reference to the Japanese freshwater and brackish taxa. *J. Appl. Phycol.* **2008**, *20*, 979–989. [CrossRef]
63. Phillips, J.A.; Lawton, R.J.; Denys, R.; Paul, N.A.; Carl, C. *Ulva sapora sp. nov.*, an abundant tubular species of *Ulva* (Ulvales) from the tropical Pacific Ocean. *Phycologia* **2016**, *55*, 55–64. [CrossRef]
64. Smith, G.M. *Marine Algae of the Monterey Peninsula California*; Stanford University Press: Stanford, CA, USA, 1944; p. 622.
65. Doty, M.S. The marine algae of Oregon, Part I. Chlorophyta and Phaeophyta. *Farlowia* **1947**, *3*, 1–65. [CrossRef]
66. Abbott, I.A.; Hollenberg, G.J. *Marine Algae of California*; Stanford University Press: Stanford, CA, USA, 1976; 827p.
67. Gabrielson, P.W.; Widdowson, T.B.; Lindstrom, S.C. *Keys to the Seaweeds and Seagrasses of Southeast Alaska, British Columbia, Washington and Oregon*; Phycological Contribution Number 7; PhycoID: Hillsborough, NC, USA, 2006; 209p.
68. Gabrielson, P.W.; Lindstrom, S.C.; O'Kelly, C.J. *Keys to the Seaweeds and Seagrasses of Southeast Alaska, British Columbia, Washington and Oregon*; Phycological Contribution Number 8; Island Blue/Printorium Bookworks: Victoria, BC, Canada, 2012; 192p.
69. Lindstrom, S.C.; Lemay, M.A.; Starko, S.; Hind, K.R.; Martone, P. New and interesting seaweed records from the Hakai area of the central coast of British Columbia, Canada: Chlorophyta. *Bot. Mar.* **2021**, *64*, 343–361. [CrossRef]
70. Setchell, W.A.; Gardner, N.L. The marine algae of the Pacific coast of North America. II. Chlorophyceae. *Univ. Calf. Publ. Bot.* **1920**, *8*, 139–374.
71. O'Kelly, C.J. (University of Hawai'i, Honolulu, HI, USA). Personal communication, 2021.
72. Hiraoka, M.; Ichihara, K.; Zuh, W.; Ma, J.; Shimada, S. Culture and hybridization experiments on an *Ulva* clade including the Qingdao strain blooming in the Yellow Sea. *PLoS ONE* **2011**, *6*, e19371. [CrossRef] [PubMed]

73. Ogawa, T.; Ohki, K.; Kamiya, M. High heterozygosity and phenotypic variation of zoids in apomictic *Ulva prolifera* (Ulvophyceae) from brackish environments. *Aquat. Bot.* **2014**, *120*, 185–192. [CrossRef]
74. Müller, O.F. *Florae Danicae, fasc. 13*; Havniae: Copenhagen, Denmark, 1778; Volume 5, 8p.
75. Dillenius, J.J. *Historia Muscorum in Qua Circiter Sexcentae Species Veteres et Novae ad Sua Genera Relatae Describuntur et Iconibus Genuinis Illustrantur: Cum Appendice et Indice Synonymorum*; e Theatro Sheldoniano: Oxford, UK, 1742; 576p.
76. Gabrielson, P.W. (University of North Carolina at Chapel Hill, Chapel Hill, NC, USA). Personal communication, 2022.
77. O'Kelly, C.J.; Wysor, B.; Bellows, W.K. Gene sequence diversity and the phylogenetic position of algae assigned to the genera Phaeophila and Ochlochaete (Ulvophyceae, Chlorophyta). *J. Phycol.* **2004**, *40*, 789–799. [CrossRef]

Article

The Introduction of the Asian Red Algae *Melanothamnus japonicus* (Harvey) Díaz-Tapia & Maggs in Peru as a Means to Adopt Management Strategies to Reduce Invasive Non-Indigenous Species

Julissa J. Sánchez-Velásquez, Lorenzo E. Reyes-Flores, Carmen Yzásiga-Barrera and Eliana Zelada-Mázmela *

Laboratory of Genetics, Physiology, and Reproduction, Faculty of Sciences, Universidad Nacional del Santa, Chimbote 02801, Peru; julissa.sanchez.velasquez@vub.be (J.J.S.-V.); eduardoreyesf@outlook.com (L.E.R.-F.); cyzasiga@uns.edu.pe (C.Y.-B.)

* Correspondence: ezelada@uns.edu.pe

Abstract: Early detection of non-indigenous species is crucial to reduce, mitigate, and manage their impacts on the ecosystems into which they were introduced. However, assessment frameworks for identifying introduced species on the Pacific Coast of South America are scarce and even non-existent for certain countries. In order to identify species' boundaries and to determine the presence of non-native species, through morphological examinations and the analysis of the plastid ribulose-1,5-bisphosphate carboxylase/oxygenase large subunit (*rbc*L-5P) gene, we investigated the phylogenetic relationships among species of the class Florideophyceae from the coast of Ancash, Peru. The *rbc*L-5P dataset revealed 10 Florideophyceae species distributed in the following four orders: Gigartinales, Ceramiales, Halymeniales, and Corallinales, among which the Asian species, *Melanothamnus japonicus* (Harvey) Díaz-Tapia & Maggs was identified. *M. japonicus* showed a pairwise divergence of 0% with sequences of *M. japonicus* from South Korea, the USA, and Italy, the latter two being countries where *M. japonicus* has been reported as introduced species. Our data indicate a recent introduction event of *M. japonicus* in Peru, and consequently, the extension of its distribution into South America. These findings could help to adopt management strategies for reducing the spread and impact of *M. japonicus* on the Pacific Coast of South America.

Keywords: DNA barcode; Florideophyceae; non-indigenous species; *Melanothamnus japonicus*; *rbc*L

Citation: Sánchez-Velásquez, J.J.; Reyes-Flores, L.E.; Yzásiga-Barrera, C.; Zelada-Mázmela, E. The Introduction of the Asian Red Algae *Melanothamnus japonicus* (Harvey) Díaz-Tapia & Maggs in Peru as a Means to Adopt Management Strategies to Reduce Invasive Non-Indigenous Species. *Diversity* **2021**, *13*, 176. https://doi.org/10.3390/d13050176

Academic Editors: Michael Wink and Manuel Elias-Gutierrez

Received: 10 February 2021
Accepted: 26 March 2021
Published: 21 Apirl 2021

Publisher's Note: MDPI stays neutral with regard to jurisdictional claims in published maps and institutional affiliations.

1. Introduction

The deliberate or accidental introduction of non-indigenous species has been described as one of the four greatest threats to marine ecosystems [1]. Even though diverse international guidelines have been established to reduce the human-mediated exchange of species [2–4], on a global scale, the rate of first records is increasing, with the highest rates being observed in recent years [5]. For marine algae, the annual rate of introduced species has rapidly increased as they are more difficult to regulate and are associated with increasing trade [5–8]. Currently, macroalgae represent approximately 12.5% of the world's introduced species [9], a value that could be underestimated since different studies have shown that many macroalgal introductions go unnoticed due to cryptic introductions, i.e., introduced species that are morphologically indistinguishable but genetically different from native species [10–13]. Common cryptic introductions include those from the phylum Rhodophyta, one of the phyla containing most introduced algae species reported worldwide [11,12,14–19]. Within this phylum, the species *Melanothamnus japonicus* (Harvey) Díaz-Tapia & Maggs causes great concern since it has been successfully established in many non-native areas without being noticed due to its morphological similarity with native *Melanothamnus* species [20–22]. Molecular evidence for the introduction of *M. japonicus* has been shown in Italy [18], the USA, Spain, Australia, and New Zealand [11,22], without reports for South American countries. Even though *M. japonicus* has not been reported on the Pacific and Atlantic coasts of South America, ten *Melanothamnus* species have been identified in these regions, with Peru and Chile being the countries with the highest number of native species. The species reported in Peru and Chile are *M. peruviensis* (D.E.Bustamante, B.Y.Won, M.E.Ramirez & T.O.Cho) Díaz-Tapia & Maggs, *M. sphaerocarpus* (Børgesen) Díaz-Tapia & Maggs, *M. savatieri* (Hariot) Díaz-Tapia & Maggs, *M. ramireziae* (D.E.Bustamante, B.Y.Won & T.O.Cho) Díaz-Tapia & Maggs (only reported for Peru), and *M. unilateralis* (Levring) Díaz-Tapia & Maggs (only reported for Chile) [23–26]. It is noteworthy to mention that the identification of many of these species, along with the newly identified Rhodophyta species for South America [27–31], was performed thanks to the use of molecular tools, highlighting the importance of molecular analysis for revealing the hidden biodiversity in those regions where the phycological studies have recently begun to increase [25].

Even though the introduction of non-indigenous *Melanothamnus* species such as *M. japonicus* can directly affect native species [22], their presence has not been evaluated in South America, a region where the determination of non-indigenous seaweed introductions has been hindered by insufficient knowledge of its endemic diversity [32]. The lack of studies focused on monitoring the presence of non-indigenous species can strongly affect this region since, except for Chile and Guyana, all South American countries show a low proactive capacity to manage alien species. A proactive capacity demonstrates that a country monitors the presence of non-indigenous species and establishes measures to tackle them [33]. Therefore, if a South American country has an undetected non-indigenous species, it is unlikely that it will be able to contain the emerging invasion. Indeed, Barrios et al. [34] showed that the introduction of *Kappaphycus alvarezii* (Doty) L.M.Liao, a native species from the Philippines, to Cubagua Island, Venezuela, was one of the main causes of coral bleaching, an effect that could not be controlled using the current mitigating approaches, such as the manual removal of the invasive algae. Although studies that have evaluated the negative impacts caused by non-indigenous algae species are scarce for South America, globally, different authors have shown that the introduction of non-native seaweed species negatively impacts local communities. For instance, Cebrian et al. [35] showed that *Womersleyella setacea* (Hollenberg) R.E.Norris, an exotic species in the Mediterranean, decreases the survival of coralligenous assemblages significantly. Smith et al. [36] also showed that *Acanthophora spicifera* (M.Vahl) Børgesen, an alga introduced in Hawaii, displaces most native species where it is abundant. Therefore, early detection of non-indigenous species is crucial to control the potential negative impacts on native communities.

In South America, besides a low proactive capacity, Peru is the only country that also shows a low reactive capacity regarding the degree to which a national action plan exists to reduce the impacts of non-indigenous species invasions [33]. Indeed, there is no current Peruvian law regarding management actions to control marine species introductions or prioritizes the studies focused on monitoring possible introduction events. For example, even though the Peruvian port of Chimbote in Ancash is a potential hotspot for introduced species: it receives many international ships [37] and has been the focus of illegal fishing from Asian ships [38], studies to determine the presence of exotic species are non-existent. The presence of Asian ships in Chimbote's Port is especially worrisome since they have been identified as the vectors driving the high degree of species exchange between Asian and South American ports [39], increasing the likelihood of introducing unnoticed species in this region.

To evaluate the presence of non-indigenous species along the coast of Ancash, Peru, we examined species diversity by performing morphological examinations and analyzing the phylogenetic relationships among the species of the class Florideophyceae using the ribulose-1,5-bisphosphate carboxylase/oxygenase large subunit (*rbc*L-5P) genetic marker. We report the introduction of the Asian species *M. japonicus* in Peru for the first time, and consequently, the extension of its distribution into South America. Moreover, we provide further evidence regarding the phylogenetic relationships among native Florideophyceae species, enriching our knowledge of the biodiversity of native aquatic species from this country.

2. Materials and Methods

Forty-seven specimens of the class Florideophyceae collected along the coast of Ancash, Peru (from 8°59′23.64″ S, 78°39′12.60″ W to 10°1′42.96″ S, 78°11′8.88″ W, Table S1), were examined in the present study. The collected organisms were identified within the currently recognized species using identification keys, original description, and species' redescriptions [28,29,40–48]. Organisms that could not be identified at the species level were reported as "*Genus* sp." The microscopic observations were made in manual sections stained with Orange G (Azer Scientific, Morgantown, PA, USA) as follows: The samples were placed in the dye for 3 min, transferred to acid alcohol for 5 s (3% concentrated HCl), and mounted in 100% glycerin. Photomicrographs were taken with a Canon PowerShot G1X camera (Canon USA, Huntington, NY, USA). Voucher specimens were deposited in the Herbarium Truxillense of the Universidad Nacional de Trujillo, Trujillo, Peru. The data of the samples, vouchers, and GenBank accession numbers for the *rbc*L-5P gene sequences are listed in Table S1.

2.1. Molecular Analysis

DNA extractions were performed on silica gel dried tissue. A small amount of tissue (approximately 5 mg) was placed in a microcentrifuge tube containing 800 μL of cetyltrimethylammonium bromide (CTAB) buffer (2% CTAB, 0.1 M Tris–HCl pH 8.0, 1.4 M NaCl, and 20 mM EDTA) and 10 μL of proteinase K (Promega, Madison, WI, USA). The procedure was carried out following the protocol of Zuccarello and Lokhorst [49]. The primers for the amplification and sequencing of the *rbc*L-5P gene were F7-R753 [50]. The PCR master mix reaction consisted of 7.99 μL water PCR grade, 1.88 μL KAPA *Taq* Buffer B (10X), 3.6 μL MgCl2 (25 mM), 0.9 μL dNTPs (2.5 mM), 0.18 μL each primer (50 μM each), 3.6 μL TBT-PAR 5X (1 mg mL^{-1} BSA, 1% Tween 20, 8.5 mM Tris HCl pH 8 [51]), 0.18 μL KAPA *Taq* DNA polymerase (5 U μL^{-1}), and 1.8 μL genomic DNA. The PCR conditions consisted of 4 min of initial denaturation at 95 °C and 28 cycles of denaturation at 94 °C for 30 s, annealing at 50 °C for 30 s, and extension at 72 °C for 45 s. The PCR reaction was finalized with a 5 min final extension at 72 °C. The success in the amplification was determined by electrophoresis in 1% agarose gels. Successfully amplified samples were cleaned using exonuclease I and FastAP Thermosensitive Alkaline Phosphatase (Thermo Fisher Scientific, Waltham, MA, USA). The PCR products were sent to Macrogen, USA for bidirectional sequencing reactions via capillary electrophoresis using the ABI3730XL automated DNA sequencer (Applied Biosystems®, Foster, CA, USA). All forward and reverse nucleotide sequences were edited and aligned using Sequencher 4.1.4 (Gene Codes Corporation, Ann Arbor, MI, USA). The sequences were aligned using the default alignment strategy (L-INS-i) of the multiple alignments program for nucleotide sequences (MAFFT 7.402) [52] and edited manually where required. The alignment was trimmed to 622 bp to reduce missing data and erroneous base calls at the ends of the sequences [53].

To determine the identity of each species, we searched a representative sequence of the *rbc*L-5P gene for each clade in the Barcode of Life Database (BOLD) Systems (https://www.boldsystems.org/, (accessed on 10 February 2021)) and the database of the National Center for Biotechnology Information (NCBI) (https://www.ncbi.nlm.nih.gov/, (accessed on 10 February 2021)), using the BOLD identification system (IDS) and the basic local alignment search tool (BLAST), respectively. The final identification of species was only accepted if they demonstrated more than a 99% similarity regarding the sequences available in both databases. The data presented in this study are openly available in GenBank and the dataset with code "DS-LGFYR" on BOLD systems (dx.doi.org/10.5883/DS-LGFYR, (accessed on 10 February 2021).).

2.2. Phylogenetic Analysis

To determine phylogenic relationships, a dataset of 78 *rbc*L-5P sequences was constructed, including 47 own sequences and 23 sequences with the highest identity obtained through the standard nucleotide BLAST search. For comparative purposes, six specimens representing five species native to the North Pacific were also included. The final alignment included 78 specimens with 622 nucleotide positions. The selection of the best model of nucleotide substitution was based on the Bayesian information criterion (BIC) through jModelTest 2 [54]. The chosen model was the general time-reversible model with a proportion of invariable sites and a gamma distribution (GTR + Γ + G). The statistical support, bootstraps, and subsequent Bayesian probabilities were calculated using maximum likelihood (ML) and Bayesian inference (BI) methods. Phylogenetic inference via ML was performed with RAxML-HPC2 on XSEDE 8.2.12 [55] through the CIPRES Science Gateway (http://www.phylo.org/, (accessed on 10 February 2021)) [56] using the GTRGAMMA model. Branch support was evaluated using bootstrapping with 1000 replicates. BI was performed using MrBayes 3.2.6 [57]. The BI analyses were run using the Metropolis coupled Markov chain Monte Carlo (MCMC) algorithm. Two independent runs were performed with four MCMC chains (three hot and one cold) for 50,000,000 generations. The trees were sampled every 50,000 generations. The convergence of both runs was tested using Tracer 1.6 [58] to see if the executions reached an effective sample size greater than 200. To calculate the potential scale reduction factor and posterior probabilities, we established a burn-in value to discard the first 25% of trees. The ML and BI trees were compared to evaluate the consistency of the results.

3. Results

For accurate species identification, both a detailed morphological examination and the use of molecular tools are necessary. Thus, we performed a preliminary survey to examine species diversity of the class Florideophyceae from the coast of Ancash, Peru, by performing microscopic observations and inferring their phylogenetic relationships through ML and BI. Our morphological identification and *rbc*L-5P phylogenetic analysis showed the presence of the Asian species *M. japonicus* in the port of Chimbote, Peru, for the first time. The morphological observations and phylogenetic analysis are described below.

3.1. Morphological Observations

Epiphytic. Solitary or in aggregates. Thallus 2–6 cm height, reddish-brown to dark brown. Tangled prostrated axes with a delicate texture, fixed to the substrate by rhizoids from which erect axes arise with few side branchlets (Figure 1a). The main erect axes are prominent and composed of four pericentral cells. Branching points occurred at intervals of 5–12 axial cells in the main axes. Scar cells developed between distal terminations of the pericentral cells (Figure 1b). Apices with a prominent apical cell of 8 × 5 µm average size (Figure 1c). Apical cells were divided transversally (Figure 1c). Branches emerged in connection with trichoblasts (Figure 1c,d). Trichoblasts were delicate, few, 20–30 µm long, and emerged from adventitious branches (Figure 1c,d). Tetrasporangia spirally arranged (Figure 1e), slightly bulky, 40–60 µm in diameter, and tripartitely divided (Figure 1f). Female and carpogonial branches were not observed.

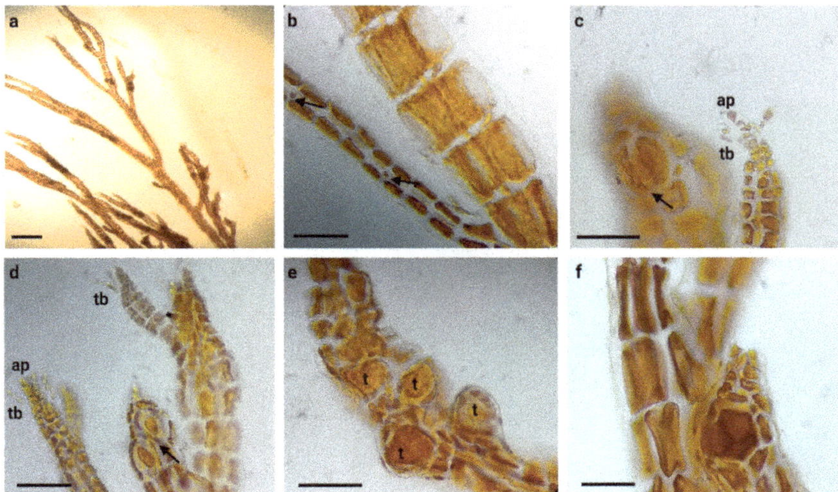

Figure 1. Habit, vegetative, and male structures of *Melanothamnus japonicus*. Microscopic observations were made in manual sections stained with Orange G, *Azer Scientific*. (**a**) Male thallus, scale bar = 1.5 mm; (**b**) Scar cells (arrows) on an erect axis, scale bar = 70 µm; (**c**) Adventitious lateral arising from main axes with short trichoblasts (tb) and apical cell (ap); an apex with young tetrasporangia surrounded by two presporangial cover cells, scale bar = 70 µm; (**d**) Apex of thallus with short trichoblasts and apical cell, an apex with young tetrasporangia (arrow), scale bar = 70 µm; (**e**) Lateral branch with young tetrasporangia (t) in irregular series, scale bar = 70 µm; (**f**) Surface view of a segment with tetraspores tripartitely divided and surrounded by cover cells, scale bar = 70 µm.

3.2. Molecular Phylogeny

To analyze species diversity in the class Florideophyceae from Ancash, Peru, and subsequently, to identify non-indigenous species from this region, we performed phylogenetic analyses using *rbc*L-5P sequences from 78 Florideophyceae specimens. The phylogeny resolved a monophyletic lineage corresponding to the class Florideophyceae with ML bootstraps (MLB) of 100% and a Bayesian posterior probability (BPP) of 1.00 (Figure 2). Within this class, we found the non-indigenous species *M. japonicus*, which showed a pairwise divergence of 0% with sequences from South Korea (the native range of this species), the USA, and Italy. The *rbc*L-5P phylogeny also showed that within the *Melanothamnus* group, the clade containing *M. japonicus* was a sister taxon (100% MLB and 1.00 BPP) to the clade containing native *Melanothamnus* species from Japan and South Korea, such as *M. harlandii* (Harvey) Díaz-Tapia & Maggs, *M. decumbens* (T.Segi) Díaz-Tapia & Maggs, *M. flavimarinus* (M.S.Kim & I.K.Lee) Díaz-Tapia & Maggs, and *M. yendoi* (T.Segi) Díaz-Tapia & Maggs. The sequences of native Peruvian species, such as *M. peruviensis* and *M. ramireziae*, instead, formed a distinct clade (87% MLB and 1.00 BPP), basal to the one containing *M. japonicus*. Therefore, it is highly likely that *M. japonicus* has been introduced to Peru by Northeast Pacific populations.

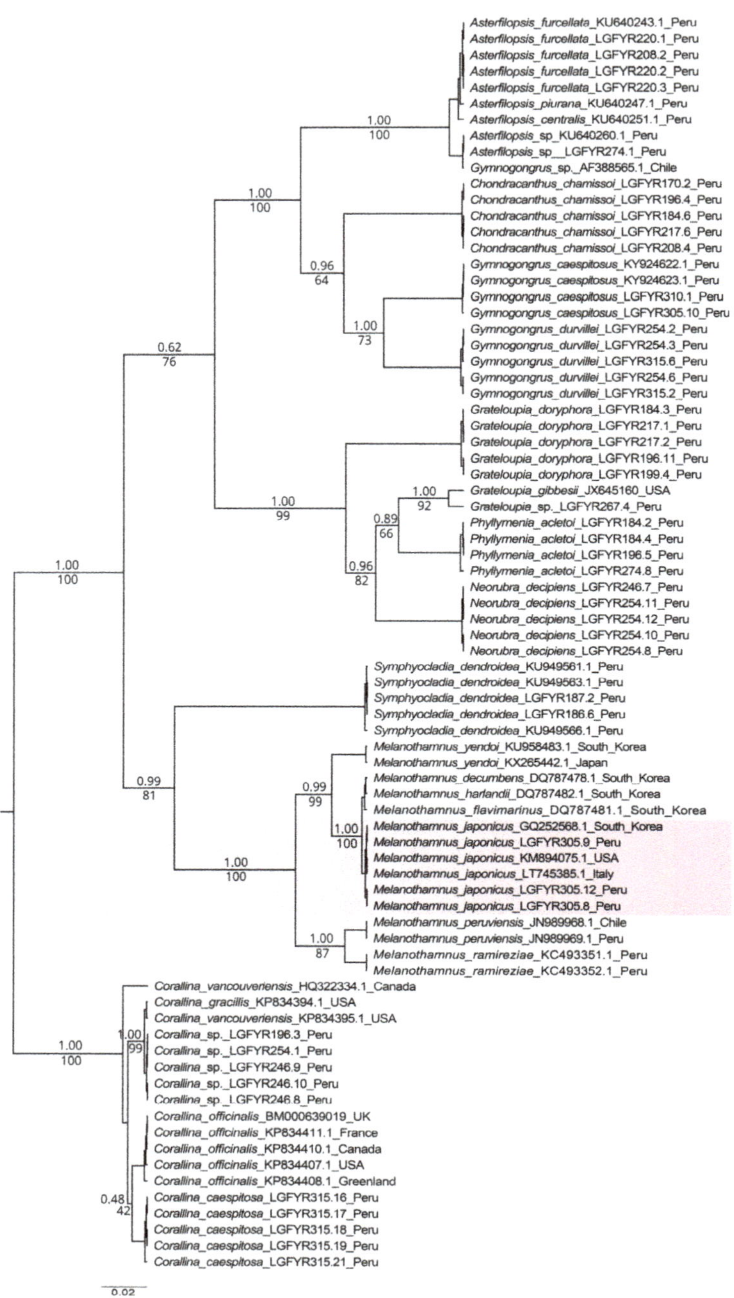

Figure 2. Bayesian inference tree of Florideophyceae species from Peru. The Bayesian inference analysis was performed for 50,000,000 generations using ribulose-1,5-bisphosphate carboxylase/oxygenase large subunit (*rbc*L-5P) sequences of Florideophyceae species collected along the coast of Ancash, Peru, and phylogenetically close species with available *rbc*L-5P sequences. The phylogenetic tree shows the sequences of *Melanothamnus japonicus* from Peru, the USA, and Italy in a strongly supported clade with sequences of *Melanothamnus* species native to Japan and South Korea, the native range of this species. The clade containing *M. japonicus* is shown in red. Bayesian posterior probability (BPP) and maximum likelihood bootstrap (MLB) values are indicated above and below the branches, respectively.

The *rbc*L-5P phylogeny also revealed a non-identified species from the genus *Corallina*, *Corallina* sp. This species showed high sequence similarity with sequences of *C. vancouveriensis* Yendo from the USA (0.3% pairwise divergence). Despite the low sequence divergence between *Corallina* sp. and *C. vancouveriensis*, an assignation to the species level could not be made since *Corallina* sp. also showed 2.5% pairwise divergence with *C. vancouveriensis* from Canada. Even though the sequence from Canada has the same species name as the sequence from the USA, it formed a different clade that was phylogenetically closer to *C. caespitosa* R.H.Walker, J.Brodie & L.M.Irvine from Peru. *Corallina* sp. also showed 1.8% and 2.1% pairwise divergence with sequences of *C. caespitosa* and *C. officinalis* Linnaeus, respectively. It is also important to note that even though *C. officinalis* has been reported for Peru, we could not find DNA sequences of this species from Peru in either BOLD Systems or GenBank databases, highlighting the absence of genetic studies for the identification of *Corallina* species from Peru.

Overall, among the 47 Florideophyceae specimens examined, the morphological and genetic information revealed the first report of a non-indigenous macroalgal introduction in Peru, i.e., *M. japonicus*, which extends the current distribution of *M. japonicus* into the Pacific coasts of South America.

4. Discussion

The morphological and phylogenetic analysis of this study showed the presence of the Asian species *M. japonicus* on the Pacific coast of South America for the first time. The morphological features of *M. japonicus* were in agreement with those described for *P. japonica*, the lectotype of *M. japonicus* characterized by Masuda et al. [59] and Kudo and Masuda [60]. *M. japonicus* was characterized by having four pericentral cells, few trichoblasts being composed of uninucleate cells, with branching points occurring at intervals of 5 to 12 axial cells in the main axes, and tripartitely divided tetrasporangia. The morphological identification was further confirmed by the molecular analysis, which showed a 100% similarity between the sequences of *M. japonicus* from Peru, South Korea, the USA, and Italy; the latter two being regions where *M. japonicus* has been reported as introduced [18,21,22]. It is important to note that to report a species as introduced, it is necessary to verify its genetic variation, which in most cases represents a subset of the genetic variation found in the native range of the species [61]. Even though the *rbc*L-5P sequences of *M. japonicus* collected in Chimbote's Port were identical, stating the presence of this species corresponds to an introduction event, based solely on the absence of intraspecific variation could not be performed, since this study has only three collections for comparison and utilized a molecular marker that, despite its ability to reconstruct phylogenetic relationships, is generally incapable of detecting intraspecific genetic variation [62,63]. A sampling design including other Peruvian coastal regions and the generation of supplementary molecular data using more variable markers, such as the mitochondrial cytochrome oxidase subunit 1 (*COI*-5P) gene [50,64], are still needed in order to determine whether there is intraspecific variation within the *M. japonicus* population present in Peru and whether it represents a subset of the genetic variation found within *M. japonicus* in its native range. Nevertheless, our phylogenetic analysis did show the sequences of *M. japonicus* and species such as *M. harlandii*, *M. decumbens*, *M. flavimarinus*, and *M. yendoi*, native to the Northeast Pacific, in a fully supported clade that diverged from the clade containing *Melanothamnus* species native to Peru. Hence, this study shows the introduction of *M. japonicus* into the Pacific coasts of South America and confirms the rapid settlement of this species outside its native range.

Although additional studies are required to determine the introduction vector for *M. japonicus*, this species was found in an important international port in Ancash, Peru. Several studies have shown that global shipping is the primary source of non-indigenous species [20,65]. For instance, using DNA metabarcoding on samples of ballast water taken from ships arriving at the Bay of Biscay, Ardura et al. [66] showed that about 22% of the total algae found in ballast waters were non-indigenous to that location. Importantly, it has also been shown that among the algal species found in ballast water, the species from the genus *Polysiphonia*, a phylogenetically close genus to *Melanothamnus*, are able to survive prolonged periods in the ships' tanks and even increase in density [67]. Moreover, most algal spores have the ability to germinate when they are exposed to long periods of darkness (e.g., the conditions in the ships' tanks), an ability that appears to be unrelated to a particular taxonomic group or life-history style [68]. Like ballast water, ships' hulls also transport many sessile algal species between ports [69,70]. In fact, compared to ballast water and ship sediments, most non-indigenous species tend to be found on ships' hulls [71]. Because the oceanographic conditions of Chimbote are favorable for industrial vessels to enter, both ballast water and hull fouling are the most likely vectors for the introduction of *M. japonicus*. It is also important to note that using a species flow network based on ballast water exchange, Xu et al. [39] showed that the ships coming from the Pacific cluster (from Asian and Oceanian regions) are the most common vectors for the non-indigenous species invading South American countries. Hence, and considering that several Asian ships have been shown to arrive at the port of Chimbote [37], we hypothesize that *M. japonicus* was likely introduced from ships coming from Asia rather than those coming from the Mediterranean or the USA, regions where *M. japonicus* has also successfully settled.

We should also mention that species imported for aquaculture are known vectors for non-indigenous species [72,73]. Nevertheless, the only imported species native to Asia is the Japanese oyster *Crassostrea gigas* Thunberg, a species imported to Peru not from Japan but Chile, a country that has not reported the presence of *M. japonicus* within their biogeographic limits [74]. Hence, imported species for aquaculture can be excluded as an introduction vector for *M. japonicus*. In light of these observations, ballast waters and hull fouling from Asian ships are the most apparent sources of *M. japonicus*. It is very likely that after a first introduction event, the presence of favorable environmental conditions in Chimbote's Port (a water temperature between 17.2 and 22.8 °C, a temperature range in which *M. japonicus* can grow rapidly [60,75], and the minimal water current activity) contributed to the rapid adaptation and successful settlement of *M. japonicus* in this port.

Although several studies have shown the introduction and successful establishment of *M. japonicus* in different coastal regions [18,22], it is still unknown if its introduction will negatively impact native communities. By simulating the global spreading dynamics of marine non-indigenous species and comparing the predicted and observed species ranges, Seebens et al. [76] observed that *M. harveyi* (Bailey) Díaz-Tapia & Maggs, a species morphologically indistinguishable from *M. japonicus*, was placed within the top ten highly invasive species. Even though not all species from the same genus will necessarily demonstrate invasiveness [77], it has been shown that *M. japonicus* has rapidly spread through the regions into which it was introduced. For instance, Rindi et al. [20] and Sfriso et al. [21] reported that *M. japonicus* is nowadays one of the most common and widespread macroalgae in the Conero Riviera and the Venice Lagoon, respectively, indicating the high invasive potential of *M. japonicus*. Another important consequence of the establishment of *M. japonicus* in Peru is DNA introgression from the introduced species into a native species, a phenomenon already shown between *M. japonicus* and *M. harveyi* [22]. If this occurs, it could give rise to a more vigorous individual with a higher spreading ability, which could displace the parental species [78]. Future studies are still required to determine whether hybridization between *M. japonicus* and native Peruvian species is possible.

Additionally, Dijkstra et al. [7] indicated that the introduction of non-indigenous algal species leads to positive ecological effects, such as the generation of a more biogenetically complex habitat that is able to support twice or three times the richness and abundance of species from lower trophic levels as compared with native seaweed assemblages. Since the introduction of non-indigenous species can have different consequences, Katsanevakis et al. [79] stated that the "native good, alien bad" view is a misconception and that the role of most non-indigenous species in marine ecosystems is rather complex. Because species introductions can result in either good or bad consequences for the native community, it is necessary to establish preventive and mitigating measures to avoid any potential negative impacts. On a global scale, Ficetola et al. [80] showed that climatic and land use information are useful tools for creating environmental models that can delineate areas with the highest risk of invasions. Considering that the databases containing the global distribution of algal species, such as Algaebase [81] and the Global Biodiversity Information Facility [82], are publicly available, and that from there, the climate tolerance range for each species can be inferred, an approach similar to the one implemented by Ficetola et al. could be conducted to identify probable introduction points for invasive algal species. If this were possible, such data could accelerate the implementation of preventive methods in areas with the highest risk of invasion.

In Peru, as a preventive measure, the law requires that all ships renew their ballast waters outside 12 nautical miles from the coast before entering the ports (Supreme Decree N° 009-2014-MINAM). However, this measure was not sufficient to stop the introduction of *M. japonicus* to Peru, which may have been introduced through hull fouling. Thus, it is also necessary to establish mitigating measures that take into account present introductions. This will require both studies focusing on detecting non-indigenous species and studies on the ecology and genetics of current introductions being prioritized. For the Peruvian government, these actions should be of utmost importance considering that in November 2014, the National Peruvian Strategy for Biological Diversity established as one of its goals to increase the regulatory mechanisms of invasive non-indigenous species by 2021 [83]. Considering that there is no previous information related to the study of marine non-indigenous species in Peru, studies concerning the presence of these species are not only needed but are urgent for the effective management of non-indigenous species introductions.

In conclusion, our phylogenetic analyses using the *rbc*L-5P gene dataset revealed the introduction of the Asian species *M. japonicus* into a South American country. Future research needs to focus on revealing the geographic extension of *M. japonicus* and determining the principal risks of its presence on the coasts of Peru, which until 2008 were considered pristine in terms of non-indigenous marine species [84]. Because our study showed the successful settlement of a non-native species likely transported into Chimbote's Port aboard international ships, our data could promote studies prioritizing the biomonitoring of *M. japonicus* along the Pacific coasts of Peru and other South American countries where this species might also have been introduced. Importantly, because the source region and possible vectors for the introduced species can be identified by comparing the introduced genotypes' distribution in the native range [61], the studies focused on biomonitoring exotic species should consider using markers able to detect genetic variation within species. Supplementary molecular data generated from mitochondrial DNA-derived markers provide an ideal approach for this purpose. Their high mutation rate in red algae, approximately four times that of the plastid and nuclear DNA [63,85], makes them the best markers to infer intraspecific genetic variation [50,64,86]. Finally, this study could serve as the starting point to implement preventive and mitigating measures against invasive species, especially in South American countries that adhere to the International Convention for the Control and Management of Ships' Ballast Water and Sediments (BWM) [87], whose primary goal is to eliminate the risk of modifying the oceans' environmental conditions through the transfer of non-native species. This is strongly encouraged since although our survey did not constitute a comprehensive barcoding library of the Peruvian Florideophyceae diversity, an exotic species was revealed, suggesting there are more to be discovered.

Supplementary Materials: The following are available online at https://www.mdpi.com/article/10.3390/d13050176/s1, Table S1: Sample information including locality, collectors name, collection codes, and GenBank accession number. GenBank accession numbers of the sequences generated during the present study are shown in bold.

Author Contributions: Conceptualization, E.Z.-M.; investigation, methodology, formal analysis, and data curation, J.J.S.-V. and L.E.R.-F.; visualization, C.Y.-B.; writing—original draft preparation, J.J.S.-V.; writing—review and editing, J.J.S.-V.; E.Z.-M.; C.Y.-B. and L.E.R.-F.; supervision, L.E.R.-F.; project administration, E.Z.-M. and C.Y.-B. All authors have read and agreed to the published version of the manuscript.

Funding: This research received no external funding.

Institutional Review Board Statement: Not applicable.

Informed Consent Statement: Not applicable.

Data Availability Statement: The data presented in this study are openly available in BOLD Systems with the project code "LGF" and GenBank.

Acknowledgments: We thank the Research Circle "Código de Barras de ADN para el estudio, conservación y uso sostenible de los recursos" for supporting the DNA sequencing costs of this study, and the Laboratory of Genetics, Physiology, and Reproduction working group for all its constructive ideas, and the time and effort dedicated to the collection of samples.

Conflicts of Interest: The authors declare no conflict of interest.

References

1. UNEP. United Nations Environment Programme. Invasive Alien Speceis—A Growing Threat in Regional Seas Alien Species. Available online: https://wedocs.unep.org/bitstream/handle/20.500.11822/13623/invasive_alien_brochure.pdf?sequence=1&%3BisAllowed= (accessed on 18 December 2020).
2. ICES. Code of Practice on the Introductions and Transfers of Marine Organisms 2005. Technical Report. Available online: https://www.nobanis.org/globalassets/ices-code-of-practice.pdf (accessed on 18 November 2020).
3. Genovesi, P.; Shine, C. European strategy on invasive alien species. In *Convention on the Conservation of European Wildlife and Natural Habitats*, 1st ed.; Council of Europe Publishing: Strasbourg, France, 2004; 68p.
4. IUCN. Guidelines for the Prevention of Biodiversity Loss Caused by Alien Invasive Species. Prepared by the IUCN/SSC Invasive Species Specialist Group (ISSG) and Approved by the 51st Meeting of the IUCN Council, Gland, Switzerlan, February 2000. Available online: https://portals.iucn.org/library/efiles/documents/Rep-2000-052.pdf (accessed on 18 November 2020).
5. Seebens, H.; Blackburn, T.M.; Dyer, E.E.; Genovesi, P.; Hulme, P.E.; Jeschke, J.M.; Pagad, S.; Pyšek, P.; Winter, M.; Arianoutsou, M.; et al. No saturation in the accumulation of alien species worldwide. *Nat. Commun.* **2017**, *8*, 14435. [CrossRef]
6. Pierucci, A.; De La Fuente, G.; Cannas, R.; Chiantore, M. A new record of the invasive seaweed Caulerpa cylindracea Sonder in the South Adriatic Sea. *Heliyon* **2019**, *5*, e02449. [CrossRef]
7. Dijkstra, J.A.; Harris, L.G.; Mello, K.; Litterer, A.; Wells, C.; Ware, C. Invasive seaweeds transform habitat structure and increase biodiversity of associated species. *J. Ecol.* **2017**, *105*, 1668–1678. [CrossRef]
8. Kim, M.-S.; Yang, E.C.; Mansilla, A.; Boo, S.M. Recent introduction of Polysiphonia morrowii (Ceramiales, Rhodophyta) to Punta Arenas, Chile. *Bot. Mar.* **2004**, *47*, 389–394. [CrossRef]
9. World Register of Introduced Marine Species. Available online: http://www.marinespecies.org/introduced/ (accessed on 18 November 2020).
10. Manghisi, A.; Armeli-Minicante, S.; Bertuccio, C.; Morabito, M.; Torricelli, P.; Genovese, G. A cryptic alien seaweed spreading in Mediterranean coastal lagoons. *Transit. Waters Bull.* **2011**, *5*, 1–7. [CrossRef]
11. Piñeiro-Corbeira, C.; Verbruggen, H.; Díaz-Tapia, P. Molecular survey of the red algal family Rhodomelaceae (Ceramiales, Rhodophyta) in Australia reveals new introduced species. *Environ. Boil. Fishes* **2020**, *32*, 2535–2547. [CrossRef]
12. Bolton, J.J.; Andreakis, N.; Anderson, R.J. Molecular evidence for three separate cryptic introductions of the red seaweedAsparagopsis(Bonnemaisoniales, Rhodophyta) in South Africa. *Afr. J. Mar. Sci.* **2011**, *33*, 263–271. [CrossRef]
13. Morais, P.; Reichard, M. Cryptic invasions: A review. *Sci. Total. Environ.* **2018**, *613*, 1438–1448. [CrossRef]
14. Williams, S.L.; Smith, J.E. A Global Review of the Distribution, Taxonomy, and Impacts of Introduced Seaweeds. *Annu. Rev. Ecol. Evol. Syst.* **2007**, *38*, 327–359. [CrossRef]
15. McIvor, L.; Maggs, C.A.; Provan, J.; Stanhope, M.J. *rbc*L sequences reveal multiple cryptic introductions of the Japanese red alga Polysiphonia harveyi. *Mol. Ecol.* **2001**, *10*, 911–919. [CrossRef]
16. Galil, B.S.; Occhipinti-Ambrogi, A.; Gollasch, S. Chapter 4. Biodiversity impacts of species introductions via marine vessels. In *Maritime Traffic Effects on Biodiversity in the Mediterranean Sea. Review of Impacts, Priority Areas and Mitigation Measures*, 1st ed.; Abdulla, A., Linden, O., Eds.; IUCN Centre for Mediterranean Cooperation: Malaga, Spain, 2008; Volume 1, pp. 118–150.
17. Mineur, F.; Johnson, M.P.; Maggs, C.A. Macroalgal Introductions by Hull Fouling on Recreational Vessels: Seaweeds and Sailors. *Environ. Manag.* **2008**, *42*, 667–676. [CrossRef] [PubMed]
18. Wolf, M.A.; Buosi, A.; Juhmani, A.-S.F.; Sfriso, A. Shellfish import and hull fouling as vectors for new red algal introductions in the Venice Lagoon. *Estuar. Coast. Shelf Sci.* **2018**, *215*, 30–38. [CrossRef]

19. Cianciola, E.N.; Popolizio, T.R.; Schneider, C.W.; Lane, C.E. Using Molecular-Assisted Alpha Taxonomy to Better Understand Red Algal Biodiversity in Bermuda. *Diversity* **2010**, *2*, 946–958. [CrossRef]
20. Rindi, F.; Gavio, B.; Díaz-Tapia, P.; Di Camillo, C.G.; Romagnoli, T. Long-term changes in the benthic macroalgal flora of a coastal area affected by urban impacts (Conero Riviera, Mediterranean Sea). *Biodivers. Conserv.* **2020**, *29*, 2275–2295. [CrossRef]
21. Sfriso, A. Invasion of alien macroalgae in the Venice Lagoon, a pest or a resource? *Aquat. Invasions* **2020**, *15*, 245–270. [CrossRef]
22. Savoie, A.M.; Saunders, G.W. Evidence for the introduction of the Asian red algaNeosiphonia japonicaand its introgression withNeosiphonia harveyi(Ceramiales, Rhodophyta) in the Northwest Atlantic. *Mol. Ecol.* **2015**, *24*, 5927–5937. [CrossRef]
23. Bustamante, D.E.; Won, B.Y.; Cho, T.O. Neosiphonia ramirezii sp. nov. (Rhodomelaceae, Rhodophyta) from Peru. *ALGAE* **2013**, *28*, 73–82. [CrossRef]
24. Bustamante, D.E.; Won, B.Y.; Ramírez, M.E.; Cho, T.O. Neosiphonia peruviensis sp. nov. (Rhodomelaceae, Rhodophyta) from the Pacific coast of South America. *Bot. Mar.* **2012**, *55*, 359. [CrossRef]
25. Avila-Peltroche, J.; Padilla-Vallejos, J. The seaweed resources of Peru. *Bot. Mar.* **2020**, *63*, 381–394. [CrossRef]
26. Ramírez, M.E.; Santelices, B. *Catálogo de las Algas Marinas Bentónicas de la Costa Temperada del Pacífico de Sudamérica*; Pontificia Universidad Católica de Chile: Santiago, Chile, 1991.
27. Calderon, M.S.; Boo, S.M. A new genus Phyllophorella gen. nov. (Phyllophoraceae, Rhodophyta) from central Peru, including Phyllophorella peruviana comb. nov., *Phyllophorella Humboldtiana* sp. nov., and *Phyllophorella Limaensis* sp. nov. *Bot. Mar.* **2016**, *59*, 339–352. [CrossRef]
28. Calderon, M.S.; Boo, S.M. Phylogeny of Phyllophoraceae (Rhodophyta, Gigartinales) reveals Asterfilopsis gen. nov. from the Southern Hemisphere. *Phycologia* **2016**, *55*, 543–554. [CrossRef]
29. Calderon, M.S.; Boo, S.M. The Phyllophoraceae (Gigartinales, Rhodophyta) from Peru with descriptions of Acletoa tarazonae gen. & sp. nov. and *Gymnogongrus caespitosus* sp. nov. *Phycologia* **2017**, *56*, 686–696. [CrossRef]
30. Bustamante, D.E.; Won, B.Y.; Lindstrom, S.C.; Cho, T.O. The new genusSymphyocladiella gen. nov. (Ceramiales, Rhodophyta) based onS. bartlingiana comb. nov. from the Pacific Ocean. *Phycologia* **2019**, *58*, 9–17. [CrossRef]
31. Ramírez, M.E.; Contreras-Porcia, L.; Guillemin, M.-L.; Brodie, J.; Valdivia, C.; Flores-Molina, M.R.; Núñez, A.; Contador, C.B.; Lovazzano, C. Pyropia orbicularis sp. nov. (Rhodophyta, Bangiaceae) based on a population previously known as Porphyra columbina from the central coast of Chile. *Phytotaxa* **2014**, *158*, 133. [CrossRef]
32. Miloslavich, P.; Klein, E.; Díaz, J.M.; Hernandez, C.E.; Bigatti, G.; Campos, L.; Artigas, F.; Castillo, J.; Penchaszadeh, P.E.; Neill, P.E.; et al. Marine Biodiversity in the Atlantic and Pacific Coasts of South America: Knowledge and Gaps. *PLoS ONE* **2011**, *6*, e14631. [CrossRef]
33. Early, R.; Bradley, B.A.; Dukes, J.S.; Lawler, J.J.; Olden, J.D.; Blumenthal, D.M.; Gonzalez, P.; Grosholz, E.D.; Ibañez, I.; Miller, L.P.; et al. Global threats from invasive alien species in the twenty-first century and national response capacities. *Nat. Commun.* **2016**, *7*, 12485. [CrossRef]
34. Barrios, J.; Bolaños, J.; López, R. Blanqueamiento de arrecifes coralinos por la invasión de Kappaphycus alvarezii (Rhodo-phyta) en Isla Cubagua, Estado Nueva Esparta, Venezuela. *Boletín Inst. Ocean. Venez.* **2007**, *46*, 147–152.
35. Cebrian, E.; Linares, C.; Marschal, C.; Garrabou, J. Exploring the effects of invasive algae on the persistence of gorgonian populations. *Biol. Invasions* **2012**, *14*, 2647–2656. [CrossRef]
36. Smith, J.E.; Hunter, C.L.; Smith, C.M. Distribution and Reproductive Characteristics of Nonindigenous and Invasive Marine Algae in the Hawaiian Islands. *Pac. Sci.* **2002**, *56*, 299–315. [CrossRef]
37. Produce. Information Obtained by Requesting "Access to the Information", an Online Service Implemented by the Min-Isterio de la Producción from the Peruvian Government. Available online: https://www.gob.pe/produce (accessed on 12 December 2020).
38. Aroni, E. Global Fishing Watch Holds Workshops with Authorities in Peru's Three Main Fishing Ports. Available online: https://globalfishingwatch.org/fisheries/workshops-peru-ports/#:~{}:text=GlobalFishingWatchholdsworkshopswithauthoritiesinPeru\T1\textquoterightsthreemainfishingports,-ByEloyAroni&text=GlobalFishingWatchrecentlymet,Paita%2CChimboteandCallao (accessed on 18 November 2020).
39. Xu, J.; Wickramarathne, T.; Grey, E.; Steinhaeuser, K.; Keller, R.; Drake, J.; Chawla, N.; Lodge, D.M. Patterns of ship-borne species spread: A clustering approach for risk assessment and management of non-indigenous species spread. *arXiv* **2014**, arXiv:1401.5407.
40. Kützing, F.T. *Phycologia Generalis: Oder, Anatomie, Physiologie und Systemkunde der Tange/Bearb. von Friedrich Traugott Kützing. Mit 80 Farbig Gedruckten Tafeln, Gezeichnet und Gravirt vom Verfasser*; Milner and Sowerby: London, UK, 1843; p. 143.
41. Yoon, H.Y. A Taxonomic study of genus Polysiphonia (Rhodophyta) from Korea. *Algae* **1986**, *1*, 3–86.
42. Howe, M.A. *The marine Algae of Peru*; Press of the New era Printing Company: New York, NY, USA, 1914; Volume 15, 330p.
43. Riosmena-Rodriguez, R.; Siqueiros-Beltrones, D.A. Morphology and distribution of Corallina vancouverensis (Corallinales, Rhodophyta) in Northwest Mexico. *Ciencias Mar.* **1995**, *21*, 187–199. [CrossRef]
44. Walker, R.H.; Brodie, J.; Russell, S.; Irvine, L.M.; Orfanidis, S. Biodiversity of Coralline Algae in the Northeastern Atlantic Including *Corallina Caespitosa* sp. nov. (corallinoideae, rhodophyta). *J. Phycol.* **2009**, *45*, 287–297. [CrossRef] [PubMed]
45. Calderon, M.; Ramírez, M.E.; Bustamante, D. Notas sobre tres especies de Gigartinaceae (Rhodophyta) del litoral peruano. *Rev. Peru. Biol.* **2011**, *17*, 115–121. [CrossRef]
46. Calderon, M.S.; Boo, G.H.; Boo, S.M. Morphology and phylogeny of *Ramirezia osornoensis* gen. & sp. nov. and *Phyllymenia acletoi* sp. nov. (Halymeniales, Rhodophyta) from South America. *Phycologia* **2014**, *53*, 23–36. [CrossRef]

47. Calderon, M.S.; Boo, G.H.; Boo, S.M. Neorubra decipiens gen. & comb. nov. and Phyllymenia lancifolia comb. nov. (Halymeniales, Rhodophyta) from South America. *Phycologia* **2014**, *53*, 409–422. [CrossRef]
48. Bustamante, D.E.; Won, B.Y.; Cho, T.O. Morphology and phylogeny of Pterosiphonia dendroidea (Rhodomelaceae, Ceramiales) described as Pterosiphonia tanakae from Japan. *Bot. Mar.* **2016**, *59*, 353–361. [CrossRef]
49. Zuccarello, G.C.; Lokhorst, G.M. Molecular phylogeny of the genus Tribonema (Xanthophyceae) using rbcL gene sequence data: Monophyly of morphologically simple algal species. *Phycologia* **2005**, *44*, 384–392. [CrossRef]
50. Tan, J.; Lim, P.-E.; Phang, S.-M.; Hong, D.D.; Sunarpi, H.; Hurtado, A.Q. Assessment of Four Molecular Markers as Potential DNA Barcodes for Red Algae Kappaphycus Doty and Eucheuma J. Agardh (Solieriaceae, Rhodophyta). *PLoS ONE* **2012**, *7*, e52905. [CrossRef]
51. Samarakoon, T.; Wang, S.Y.; Alford, M.H. Enhancing PCR Amplification of DNA from Recalcitrant Plant Specimens Using a Trehalose-Based Additive. *Appl. Plant Sci.* **2013**, *1*, 1200236. [CrossRef] [PubMed]
52. Katoh, K.; Rozewicki, J.; Yamada, K.D. MAFFT online service: Multiple sequence alignment, interactive sequence choice and visualization. *Brief. Bioinform.* **2019**, *20*, 1160–1166. [CrossRef]
53. Stoeckle, M.Y.; Kerr, K.C.R. Frequency Matrix Approach Demonstrates High Sequence Quality in Avian BARCODEs and Highlights Cryptic Pseudogenes. *PLoS ONE* **2012**, *7*, e43992. [CrossRef] [PubMed]
54. Darriba, D.; Taboada, G.L.; Doallo, R.; Posada, D. jModelTest 2: More models, new heuristics and parallel computing. *Nat. Methods* **2012**, *9*, 772. [CrossRef] [PubMed]
55. Stamatakis, A.; Hoover, P.; Rougemont, J. A Rapid Bootstrap Algorithm for the RAxML Web Servers. *Syst. Biol.* **2008**, *57*, 758–771. [CrossRef] [PubMed]
56. Miller, M.A.; Pfeiffer, W.; Schwartz, T. Creating the CIPRES Science Gateway for inference of large phylogenetic trees. In Proceedings of the 2010 Gateway Computing Environments Workshop (GCE), New Orleans, LA, USA, 14 November 2010; pp. 1–8. [CrossRef]
57. Ronquist, F.; Teslenko, M.; Van Der Mark, P.; Ayres, D.L.; Darling, A.; Höhna, S.; Larget, B.; Liu, L.; Suchard, M.A.; Huelsenbeck, J.P. MrBayes 3.2: Efficient Bayesian Phylogenetic Inference and Model Choice Across a Large Model Space. *Syst. Biol.* **2012**, *61*, 539–542. [CrossRef]
58. Rambaut, A.; Suchard, M.A.; Xie, D.; Drummond, A.J. Tracer v1.6. 2014. Available online: http://beast.bio.ed.ac.uk/Tracer (accessed on 12 March 2021).
59. Masuda, M.; Kudo, T.; Kawaguchi, S.; Guiry, M.D. Lectotypification of some marine red algae described by Harvey, W.H. from Japan. *Phycol. Res.* **1995**, *43*, 191–202. [CrossRef]
60. Kudo, T.; Masuda, M. A taxonomic study of Polysiphonia japonica Harvey and P. akkeshiensis Segi (Rhodophyta). *Jpn. J. Phycol.* **1986**, *34*, 293–310.
61. Miura, O. Molecular genetic approaches to elucidate the ecological and evolutionary issues associated with biological invasions. *Ecol. Res.* **2007**, *22*, 876–883. [CrossRef]
62. Smith, D.R. Mutation Rates in Plastid Genomes: They Are Lower than You Might Think. *Genome Biol. Evol.* **2015**, *7*, 1227–1234. [CrossRef] [PubMed]
63. Smith, D.R.; Hua, J.; Lee, R.W.; Keeling, P.J. Relative rates of evolution among the three genetic compartments of the red alga Porphyra differ from those of green plants and do not correlate with genome architecture. *Mol. PhyloGenet. Evol.* **2012**, *65*, 339–344. [CrossRef]
64. Robba, L.; Russell, S.J.; Barker, G.L.; Brodie, J. Assessing the use of the mitochondrial cox1 marker for use in DNA barcoding of red algae (Rhodophyta). *Am. J. Bot.* **2006**, *93*, 1101–1108. [CrossRef] [PubMed]
65. Geburzi, J.C.; McCarthy, M.L. How Do They Do It?—Understanding the Success of Marine Invasive Species. In *Youmares 8—Oceans Across Boundaries: Learning from Each Other*; Springer International Publishing: Berlin/Heidelberg, Germany, 2018; pp. 109–124.
66. Ardura, A.; Borrell, Y.J.; Fernández, S.; Arenales, M.G.; Martínez, J.L.; Garcia-Vazquez, E. Nuisance Algae in Ballast Water Facing International Conventions. Insights from DNA Metabarcoding in Ships Arriving in Bay of Biscay. *Water* **2020**, *12*, 2168. [CrossRef]
67. Zaiko, A.; Martinez, J.L.; Schmidt-Petersen, J.; Ribicic, D.; Samuiloviene, A.; Garcia-Vazquez, E. Metabarcoding approach for the ballast water surveillance—An advantageous solution or an awkward challenge? *Mar. Pollut. Bull.* **2015**, *92*, 25–34. [CrossRef]
68. Santelices, B.; Aedo, D.; Hoffmann, A. Banks of microscopic forms and survival to darkness of propagules and microscopic stages of macroalgae. *Rev. Chil. Hist. Nat.* **2002**, *75*, 547–555. [CrossRef]
69. Godwin, L.S. Hull Fouling of Maritime Vessels as a Pathway for Marine Species Invasions to the Hawaiian Islands. *Biofouling* **2003**, *19*, 123–131. [CrossRef]
70. Mineur, F.; Johnson, M.P.; Maggs, C.A.; Stegenga, H. Hull fouling on commercial ships as a vector of macroalgal introduction. *Mar. Biol.* **2006**, *151*, 1299–1307. [CrossRef]
71. Gollasch, S. The Importance of Ship Hull Fouling as a Vector of Species Introductions into the North Sea. *Biofouling* **2002**, *18*, 105–121. [CrossRef]
72. Mineur, F.; Belsher, T.; Johnson, M.P.; Maggs, C.A.; Verlaque, M. Experimental assessment of oyster transfers as a vector for macroalgal introductions. *Biol. Conserv.* **2007**, *137*, 237–247. [CrossRef]
73. Raffo, M.P.; Geoffroy, A.; Destombe, C.; Schwindt, E. First record of the invasive red alga Polysiphonia morrowii Harvey (Rhodomelaceae, Rhodophyta) on the Patagonian shores of the Southwestern Atlantic. *Bot. Mar.* **2014**, *57*, 21–26. [CrossRef]

74. IMARPE. Adaptación y reproducción de la ostra japonesa Crassostrea gigas en ambiente controlado. Informe Preliminar. 1995. Available online: http://biblioimarpe.imarpe.gob.pe/bitstream/123456789/892/1/IP13.pdf (accessed on 16 November 2020).
75. IMARPE. Informe de las Condiciones Oceanográficas y Biológico-Pesqueras Abril 2020. Available online: http://www.imarpe.pe/imarpe/archivos/informes/imarpe_informe_gti_abril_2020.pdf (accessed on 25 November 2020).
76. Seebens, H.; Schwartz, N.; Schupp, P.J.; Blasius, B. Predicting the spread of marine species introduced by global shipping. *Proc. Natl. Acad. Sci. USA* **2016**, *113*, 5646–5651. [CrossRef]
77. Sakai, A.K.; Allendorf, F.W.; Holt, J.S.; Lodge, D.M.; Molofsky, J.; With, K.A.; Baughman, S.; Cabin, R.J.; Cohen, J.E.; Ellstrand, N.C.; et al. The Population Biology of Invasive Species. *Annu. Rev. Ecol. Syst.* **2001**, *32*, 305–332. [CrossRef]
78. Daehler, C.C.; Strong, D.R. Hybridization between introduced smooth cordgrass (Spartina alterniflora; Poaceae) and native California cordgrass (S. foliosa) in San Francisco Bay, California, USA. *Am. J. Bot.* **1997**, *84*, 607–611. [CrossRef]
79. Katsanevakis, S.; Wallentinus, I.; Zenetos, A.; Leppäkoski, E.; Çinar, M.E.; Oztürk, B.; Grabowski, M.; Golani, D.; Cardoso, A.C. Impacts of invasive alien marine species on ecosystem services and biodiversity: A pan-European review. *Aquat. Invasions* **2014**, *9*, 391–423. [CrossRef]
80. Ficetola, G.F.; Thuiller, W.; Miaud, C. Prediction and validation of the potential global distribution of a problematic alien invasive species—The American bullfrog. *Divers. Distrib.* **2007**, *13*, 476–485. [CrossRef]
81. Guiry, M.D.; Guiry, M.D.; Base, A. World-wide electronic publication, National University of Ireland, Galway. Available online: https://www.algaebase.org/search/genus/detail/?genus_id=37461&-session=abv4:AC1F06401b4682AB03gHD18097D0 (accessed on 17 November 2020).
82. Global Biodiversity Information Facility. Available online: www.gbif.org. (accessed on 3 November 2010).
83. DICAPI. *Dirección General de Capitanías y Guardacostas. Informe para sustentar la adhesión del Perú al "Convenio internacional para el control y la gestión del agua de lastre y los sedimentos de los buques, 2004." Convenio de Agua de Lastre*; Ministry of Foreign Affairs: Lima, Peru, 2015; pp. 30–104.
84. Castilla, J.C.; Neill, P.E. Marine Bioinvasions in the Southeastern Pacific: Status, Ecology, Economic Impacts, Conservation and Management. In *Biogeography of Mycorrhizal Symbiosis*; Springer Science and Business Media LLC: Berlin/Heidelberg, Germany, 2008; Volume 204, pp. 439–457.
85. Smith, D.R.; Arrigo, K.R.; Alderkamp, A.-C.; Allen, A.E. Massive difference in synonymous substitution rates among mitochondrial, plastid, and nuclear genes of Phaeocystis algae. *Mol. Phylogenet. Evol.* **2014**, *71*, 36–40. [CrossRef]
86. Yow, Y.-Y.; Lim, P.-E.; Phang, S.-M. Assessing the use of mitochondrial cox1 gene and cox2-3 spacer for genetic diversity study of Malaysian Gracilaria changii (Gracilariaceae, Rhodophyta) from Peninsular Malaysia. *Environ. Boil. Fishes* **2013**, *25*, 831–838. [CrossRef]
87. International Convention for the Control and Management of Ships' Ballast Water and Sediments (BWM). Available online: https://www.imo.org/en/About/Conventions/Pages/International-Convention-for-the-Control-and-Management-of-Ships%27-Ballast-Water-and-Sediments-(BWM).aspx (accessed on 23 December 2020).

Article

Uncovering Hidden Diversity: Three New Species of the *Keratella* Genus (Rotifera, Monogononta, Brachionidae) of High Altitude Water Systems from Central Mexico †

Alma E. García-Morales [1,*], Omar Domínguez-Domínguez [2] and Manuel Elías-Gutiérrez [1]

1 Laboratorio de Zooplancton, El Colegio de la Frontera Sur, Av. Centenario km 5.5, Chetumal 77014, Quintana Roo, Mexico; melias@ecosur.mx

2 Laboratorio de Biología Acuática, Facultad de Biología, Universidad Michoacana de San Nicolás de Hidalgo, Ciudad Universitaria, Morelia 58000, Michoacan, Mexico; goodeido@yahoo.com.mx

* Correspondence: aegarcia@ecosur.mx; Tel.: +52-(983)-835-0440

† urn:lsid:zoobank.org:act:7F194E23-7A36-46C4-90D9-EA38EB5F4380;urn:lsid:zoobank.org:act:FD30C7A8-C63E-435D-A1C6-99003007789B;urn:lsid:zoobank.org:act:DCA4000E-4B2B-4221-A15A-8CF15EF69B30.

Abstract: The correct identification of species is an essential step before any study on biodiversity, ecology or genetics. *Keratella* is a genus with a predominantly temperate distribution and with several species being endemics or restricted geographically. Its diversity may be underestimated considering the confusing taxonomy of species complexes such as *K. cochlearis*. In this study, we examined genetic diversity and morphology among some *Keratella* populations from Mexico in order to determine if these populations represent different species. We analyzed a dataset of previously published and newly generated sequences of the mitochondrial COI gene and the nuclear ITS1 marker. We conducted phylogenetic analyses and applied three methods of species delimitation (ABGD, PTP and GMYC) to identify evolutionary significant units (ESUs) equivalent to species. Morphological analyses were conducted through scanning electron microscope (SEM) and morphometry under a compound microscope. In the present study, three new species *Keratella cuitzeiensis* sp. nov., *Keratella huapanguensis* sp. nov., and *Keratella albertae* sp. nov., are formally described. These species were collected in high-altitude water bodies located in the Central Plateau of Mexico. Combining DNA results through COI and ITS1 molecular markers and morphology it was possible to confirm the identity of the new species.

Keywords: integrative taxonomy; biodiversity; DNA taxonomy; ABGD; morphology; genetic entities

Citation: García-Morales, A.E.; Domínguez-Domínguez, O.; Elías-Gutiérrez, M. Uncovering Hidden Diversity: Three New Species of the *Keratella* Genus (Rotifera, Monogononta, Brachionidae) of High Altitude Water Systems from Central Mexico. *Diversity* **2021**, *13*, 676. https://doi.org/10.3390/d13120676

Academic Editor: Michael Wink

Received: 17 November 2021
Accepted: 14 December 2021
Published: 17 December 2021

Publisher's Note: MDPI stays neutral with regard to jurisdictional claims in published maps and institutional affiliations.

1. Introduction

Species are the fundamental unit of biodiversity; therefore, any biodiversity analyses, as well as genetic, physiological and ecological studies, rely on the proper delimitation and identification of species [1]. However, estimates of species richness are often hampered by the presence of cryptic species, which are groups of species that are not confidently distinguishable based only on morphology [2].

Rotifera is a phylum of microscopic animals (50–2000 μm) that are globally distributed in aquatic ecosystems [3]. They play an essential role in aquatic food webs by transferring energy to higher trophic levels [4,5]. Rotifera harbors a high level of cryptic diversity [6], and this hidden diversity is expected due to the small size of rotifers, the scarcity of rotifer taxonomists that can identify them reliably, the lack of taxonomically relevant morphological features, little or no morphological variation between species, as well as the high level of phenotypic plasticity present in several species [2,7]. Cryptic species complexes have been described for taxa such as *Brachionus plicatilis* (Müller, 1786) [8,9], *B. calyciflorus* Pallas, 1766 [10], *Epiphanes senta* (Müller, 1773) [11], *Polyarthra dolichoptera* (Idelson, 1925) [12], *Keratella cochlearis* (Gosse, 1851) [13,14], *Limnias melicerta* (Weisse, 1848) and *L. ceratophylli* (Schrank, 1803) [15], and other such as *Platyias quadricornis* (Ehrenberg, 1832) and *Testudinella patina* (Hermann, 1783) [6].

Within Brachionidae, *Keratella* (Bory de St. Vincent, 1822) is the genus with the highest degree of endemicity, mainly in temperate zones (e.g., Palearctic, Nearctic and Australian regions), with some endemics from Neotropical and Oriental regions. The rest of the *Keratella* species are cosmopolitan or widespread in some regions [16]. Currently, there are approximately 53 *Keratella* taxa recognized as valid species [17]. It seems that diversity in *Keratella* is low because of the number of the registered species; nevertheless, diversity within this genus is underestimated due to the presence of cryptic species complexes, for example as *K. cochlearis* s.l. [16]. Morphologically, *Keratella* species bear a stiff lorica, which is split into two plates, one dorsal and a ventral one. This lorica can be rectangular, trapezoidal or ovoid in shape. The dorsal part of the lorica can be smooth or covered by different types of ornamentation such as granules, pustules, spinules or reticulation, whereas the ventral part of the lorica is generally smooth, but may have ornamentation on its anterior part [18].

Besides, the dorsal lorica presents several fields (also named plaques, polygones, panels, facets) with symmetrical and asymmetrical polygonal shapes. The arrangement and shape of these fields of the dorsal plate have taxonomic importance for the identification of the species [18–20]. In *Keratella*, the anterodorsal margin of the lorica presents six curved spines, being the anteromedian pair the most curved inward. Whereas, the posterior margin of the lorica may have two posterolateral spines, a single spine or the posterior spines can be absent [19,20]. The genus can be split roughly into two main groups: (1) the "quadrata" group, which presents a row of median fields over the dorsal plate, and (2) the "cochlearis" group, which has a median ridge over the dorsal plate with the fields arranged on each side of the ridge [18].

On the other hand, *Keratella tropica* (Apstein, 1907) was described from Colombo Lake in Ceylon (now Sri Lanka) by [21]. It is widely distributed in tropical and subtropical regions of the world [22] and has also been reported in Netherlands and Siberia during summer [23]. *K. tropica* is morphologically similar to *K. valga* (Ehrenberg, 1834), but differs from this late by the presence of an additional field on the dorsal lorica: the postero-median remnant [24,25]. In *K. tropica,* the dorsal lorica has a row of five median fields, four of these are hexagonal, and the fifth (the postero-median remnant) is squared [18,19]. In this species, the length of the posterior spines varies widely, as well as the size of the lorica [18]. Sometimes is difficult to observe the dorsal median fields and especially the small remnant, resulting in confusion when examining different specimens within *K. tropica* s.l. Besides the wide variation in the length of the posterior spines and size reported in *K. tropica*, this indicates that some of these variants could be cryptic species.

The development of DNA-based taxonomic tools provides a means to study biodiversity through the analysis of genetic variation in molecular markers to delimit species [26]. Markers such as the mitochondrial cytochrome c oxidase subunit 1 (COI) and the nuclear internal transcribed spacer (ITS) were useful for the study of cryptic speciation, and genetic differentiation in some rotifer taxa [2,27]. In addition, combining molecular, morphological, ecological and crossmating analyses have proved to be a suitable approach to assess cryptic diversity and to delimit species in rotifers [2,26]. This approach falls in the so-called integrative taxonomy [28]. For example, in the well studied *B. plicatilis* complex, morphology and molecular analyses and mating experiments have shown that this taxon contains at least 14 possible cryptic species [2,29]. From these, only *B. paranguensis* Guerrero-Jiménez, Vannucchi, Silva-Briano, Adabache-Ortiz, Rico-Martínez, Roberts, Neilson, Elías-Gutiérrez, 2019 was described [9]. Another study with the widespread *Epiphanes senta* demonstrated that this taxon is a species complex and three new species were formally described based on morphological and genetic evidence [10].

In the present study, we used integrative taxonomy tools to explore the diversity of some *Keratella* populations from seven water bodies of Mexico. Specimens from these populations resemble *K. tropica* morphologically.

2. Materials and Methods

2.1. Sampling

We collected samples in seven high altitude (>1700 m above sea level, masl) water bodies along the Trans-Mexican Volcanic Belt (TMVB): Santa Teresa dam, Cuitzeo lagoon, Huapango dam, Yuriria dam, Tepatitlan-Yahualica pond, Ignacio Ramirez dam and Timilpan pond. One additional sample was taken near Kohunlich in the lowlands of the Yucatan Peninsula (Figure 1, Table S1). TMVB is a morphotectonic province extended from Gulf of Mexico to Pacific Ocean in central Mexico [30]. It is characterized to be a cold region formed by complex mountains and volcanoes with average annual temperatures varying between 12–22 °C and average annual precipitation between 300–2000 mm depending on the zone [31]. We collected samples using a 50 µm mesh size plankton net. Posteriorly, all samples were sieved to extract all water, and fixed with 96% ethanol. Samples were transported on ice to the laboratory, and stored in a freezer.

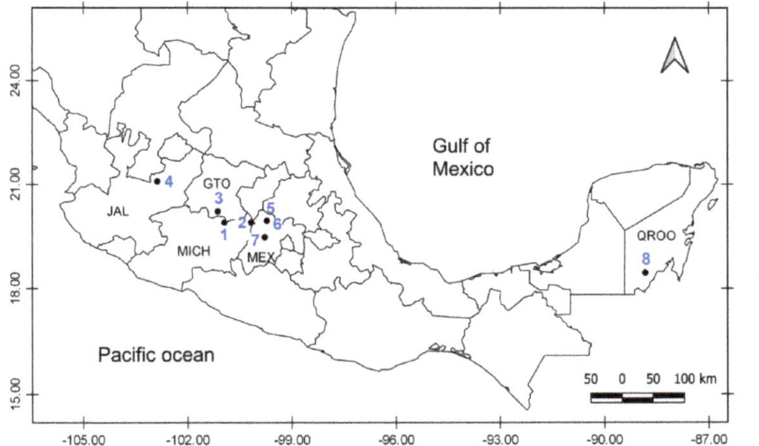

Figure 1. Sample locations of the *Keratella* taxa from Mexico. Numbers on the map correspond to the numbers in Table S1. Locations 1 to 5 were collected in this study. Samples for locations 6 to 8 were obtained from [6]. 1 = Cuitzeo lagoon, 2 = Santa Teresa dam, 3 = Yuriria dam, 4 = Tepatitlan-Yahualica pond, 5 = Huapango dam, 6 = Timilpan pond, 7 = Ignacio Ramirez dam, 8 = Aguada Kohunlich. JAL = Jalisco state, GTO = Guanajuato state, MICH = Michoacan state, MEX = Mexico state, QROO = Quintana Roo state.

2.2. DNA Extraction and Amplification

We sorted *Keratella* specimens from the samples under a stereomicroscope, rinsed them with distilled water to remove debris, and transferred them into PCR tubes for DNA extraction. We conducted the DNA extraction and PCR amplification of the COI gene according to [6], using the primers LCO1490 and HCO2198 [32]. Additionally, the ITS1 nuclear marker was amplified and the amplification profile is provided in Table S2. For ITS1 we used the primers III 5′-CACACCGCCCGTCGCTACTACCGATTG-3′ and VIII 5′-GTGCGTTCGAAGTGTCGATGATCAA-3′ from [33]. We deposited all sequences from COI and ITS markers in GenBank under accession numbers, COI: OL678378-OL678396 and ITS1: OL664525-OL664561.

2.3. Alignment and Phylogenetic Analyses

We downloaded COI (mitochondrial) and ITS1 (nuclear) sequences of *Keratella* taxa from different parts of the world available from GenBank and Barcode of Life Database (BOLD, boldsystems.org) (398 for COI and 157 for ITS1), and included them in this study.

Accession numbers for all the downloaded sequences are available in Table S3. In total, we aligned 417 COI sequences (19 from this study and 398 obtained from GenBank and BOLD), using MEGA 7.0 [34] through ClustalW with default settings. For ITS1, we aligned 194 sequences (37 from this study and 157 from GenBank) through MAFFT v.7 using the Q-INS-I algorithm as the optimal strategy for ribosomal markers [35]. This last aligment was carried out on the MAFFT webserver http://mafft.cbrc.jp/alignment/server/index.html (accessed on 29 March 2021). Both final alignments were subsequently reviewed by us.

First, we ran JMODELTEST v.2.1.1 [36] to identify the model of molecular evolution that best fit the COI (TVM + I + G), and ITS1 (HKY + G) datasets, defined by the Akaike Information Criterion. Posteriorly, all sequences for each dataset were collapsed in haplotypes using DNASP v.5.10 [37]. We used Bayesian inference (BI) and maximum likelihood (ML) analyses to infer phylogenetic relationships among the different *Keratella* samples with mtDNA COI and nuclear ITS1 analyzed separately. We did not carry out concatenated analyses because our sequences of the COI and ITS1 markers were obtained from different individuals, except five individuals from whom we have their COI and ITS1 sequences (JX216635, JX216636, JX216637, JX216638 and JX216639). We conducted The BI and ML analyses through MrBAYES v.3.2.7 [38] and RAXML v.1.5 [39], respectively. The settings for the BI analysis for each molecular dataset were four simultaneous Markov Chain Monte Carlo (MCMC) runs for six and five million generations for mitochondrial and nuclear data, respectively, with trees sampled every 100 generations. We used TRACER v.1.7 [40] to assess convergence between runs and monitor the standard deviation of split frequencies and by using the effective sampling size (ESS) criterion (>200), discarding 25% of generations as burn-in to construct the majority-rule consensus tree. For ML analysis we used a GTR + I + G (mtDNA data), and GTR + G (nuclear data) models and we ran both ML analyses with 5000 bootstrap replications. We used *Plationus patulus* (Müller, 1786) (accession number JX216784 for COI and KC431010 for ITS1) as outgroup for the phylogenetic analyses.

2.4. Species Delimitation

We applied three methods of species delimitation for mtDNA and nrDNA datasets and compared the results. Generalized Mixed Yule-Coalescent model (GMYC) [41], uses a maximum likelihood approach to identifying the shift in the branching patterns between species level (Yule model) and population level (Coalescent model) to delimit independently evolving entities. For the GMYC method, we generated ultrametric trees from the two datasets (COI and ITS1) using BEAST v.2.1.3 [42]. The settings comprised a GTR + G + I (for COI) and HKY + G (for ITS1) substitution model, a relaxed lognormal clock, and a birth–death prior [43]. Because of the absence of a molecular clock specific to Rotifera, we used calibration clocks for COI of 1.76% sequence divergence per Myr [44] and 1.2% per Myr for ITS1 [45,46] tested in aquatic invertebrates. We ran the analyses with 100 million MCMC for COI and 70 million MCMC for ITS1, sampling every 1000 generations. We checked the MCMC runs for convergence in TRACER v.1.7 [40]. We combined trees in TREEANNOTATOR 2.1.2 using a maximum credibility tree, with the first 10% discarded as burn-in. We ran the GMYC model through the GMYC webserver, using the single threshold option (http://species.h-its.org/gmyc/) (accessed on 1 April 2021).

We conducted the Poisson Tree Processes method (PTP) [47] to search for evidence of independently evolving entities considered to be species. This method uses a phylogenetic tree as input, optimizing differences in branching events in terms of number of substitutions, and adding support values to that branching events. We used the ML trees (from COI and ITS1) generated in the phylogenetic analyses. We ran both analyses with 500,000 MCMC generations on the PTP webserver http://species.h-its.org/ (accessed on 2 April 2021) and using the two types of PTP: Maximum likelihood approach (PTP-ML) and Bayesian approach (PTP-B). Before running the PTP analyses we discarded the outgroup. We also applied the Automatic Barcode Gap Discovery method (ABGD) for COI and ITS1 markers. This method clusters sequences based on the genetic distances by detecting the gaps (barcode gap) in the distribution of genetic pairwise distances. Thus, the genetic distance among individuals belonging to the same species is smaller than the distance between individuals from different species [48]. We carried out ABGD analyses through its online webserver https://bioinfo.mnhn.fr/abi/public/abgd/abgdweb.html (accessed on 2 April 2021), using default settings.

For this study purpose, we are going to consider the genetic entities discriminated by all species delimitation methods as evolutionary significant units (ESUs), and we will distinguish between COI-ESUs and ITS1-ESUs accordingly to [49].

2.5. Measurements of Specimens and Morphological Analyses

We identified specimens morphologically following the taxonomic keys of [18–20]. Specimens were identified using the features of the lorica, mainly its overall shape, shape of the lorica plates, length of the caudal spines and shape of the median fields. Several specimens from the seven populations sampled in this study were separated under a stereomicroscope and measured on a compound microscope Olympus BX51 at 40× using a micrometer.

Morphometric parameters considered for the study were: TL (total length), LL (lorica length without considering anterior and posterior spines), LW (lorica width), RPS, LPS (right and left posterior spines) and AMS, AIS, ALS (anteromedian, intermediate and lateral spines), following [50] (See Figure 2). Body dimensions are in micrometers. We also observed the five main median fields of the dorsal plate: FMF (frontomedian field), AMF (anteromedian field), MMF (mesomedian field), PMF (posteromedian field) and PMR (posteromedian remnant) (See Figure 2). For these analyses, some *Keratella* specimens were gold-coated to be observed in the scanning electron microscope (SEM) JEOL-JSM6010 located in El Colegio de la Frontera Sur in Chetumal.

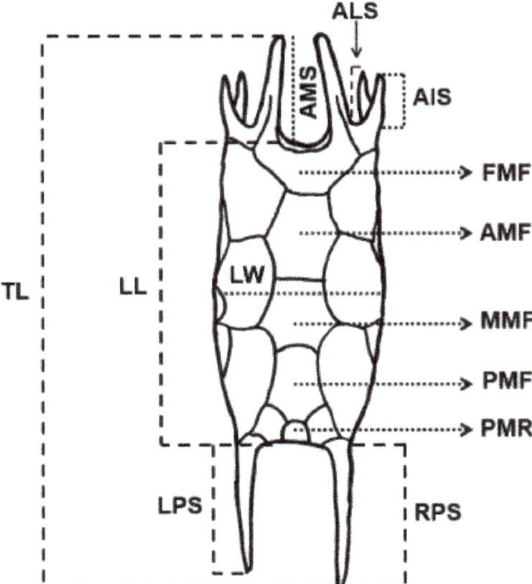

Figure 2. Lorica drawing of *Keratella* sp., with measured parameters. Total length (TL), lorica length excluding anterior and posterior spines (LL), lorica width at its widest part (LW), right posterior spine length (RPS), left posterior spine length (LPS), anteromedian dorsal spine length (AMS), anterointermediate dorsal spine length (AIS), anterolateral dorsal spine length (ALS). Frontomedian field (FMF), anteromedian field (AMF), mesomedian field (MMF), posteromedian field (PMF) and posteromedian remnant (PMR).

Our specimens of *Keratella* from the seven sampling sites were compared with the type material of *K. tropica*, which was deposited by [21] in the "Vermes" collection of the Zoological Museum of Berlin with catalog number 10121, in order to determine if our specimens morphologically correspond or not to the *K. tropica* species.

2.6. Statistical Analysis of Morphological Measurements

Morphometric measures transformed as the square root of a + ā of adult females were examined with a Principal Component Analysis (PCA) performed with the Multi Variate Statistical Package (MVSP 3.21). We performed a PCA analysis to investigate if the populations examined in this study could be distinguished as separated entities based in morphological measurements. To perform the PCA analysis we measured specimens from the seven sampling sites (see above section).

3. Results

3.1. DNA Taxonomy

The COI alignment was 580 bp, defining 80 unique haplotypes from 417 sequences; while, ITS1 alignment was 350 bp, with 57 haplotypes from 194 sequences. The trees produced by the BI and ML methods for both markers retrieved the same topology. We present only the ML trees (Figure 3 for COI and Figure 4 for ITS1). For the COI gene, nine well-defined lineages were discriminated, while for the ITS1 marker, four lineages were discriminated (See Figures 3 and 4). We consider two "Keratella tropica" groups in the COI and ITS1 trees, because specimens grouped in these lineages, morphologically resemble the *K. tropica* species (Figures 3 and 4).

In particular, within the "Keratella tropica 1 and 2" lineages for COI gene the three delimitation methods discriminated the same six ESUs (ESU1-ESU6 in Figure 3). ESU1 corresponds to the Ignacio Ramirez and Timilpan populations and were identified as *K.* cf. *morenoi*. ESU2 corresponds to the Cuitzeo population and it is *Keratella cuitzeiensis* sp. nov., ESU3 is from the Huapango population and was named *Keratella huapanguensis* sp. nov. Whereas ESU4 from the Santa Teresa population and ESU5 from the Kohunlich population, both considered as *Keratella albertae* sp. nov. ESU6 include the Yuriria and Tepatitlan-Yahualica populations herein considered here as *K. tropica* s. str., (Figure 3). For the ITS1 marker, within the "Keratella tropica I and II" lineages the three species delimitation tests delimited the same four ESUs, and we called these ESUI to ESUIV (Figure 4). With this marker individuals from the Huapango population (ESU3 with COI) correspond to the ESUI, and individuals from China and Mexico (ESU6 with COI) were nested forming the ESUII. Cuitzeo, Ignacio Ramirez and Timilpan populations (ESU1 and ESU2 with COI) were nested together within the ESUIII, whereas individuals from Santa Teresa and Kohunlich populations (ESU4 and ESU5 with COI) were nested together within the ESUIV (Figure 4). We must clarify that ITS1 sequences are not from the same individuals, except for some specimens (See Methods Section).

The uncorrected p distances within the six COI-ESUs (ESU1 to ESU6) ranged from 0 to 0.9% (Table S4), whereas distances between these ESUs ranged from 4 to 20% (Table S4). In the four ITS1-ESUs (ESUI to ESUIV), the uncorrected p distances within these ESUs ranged from 0 to 0.5% (Table S5); whereas distances between these ESUs ranged from 3.8 to 10.8% (Table S5). Most of the COI and ITS1 ESUs were formed by only a single haplotype (not a singleton) which were present in a single location, except ESU6 which was formed by four haplotypes as well ESUII and ESUIII with 12 and two haplotypes respectively.

According to the DNA results, and following a conservative approach, we propose the existence of three new *Keratella* species based on the analysis of both markers, supported also with the morphological analyses (see below). These three species are clearly genetically different from the *Keratella tropica* species.

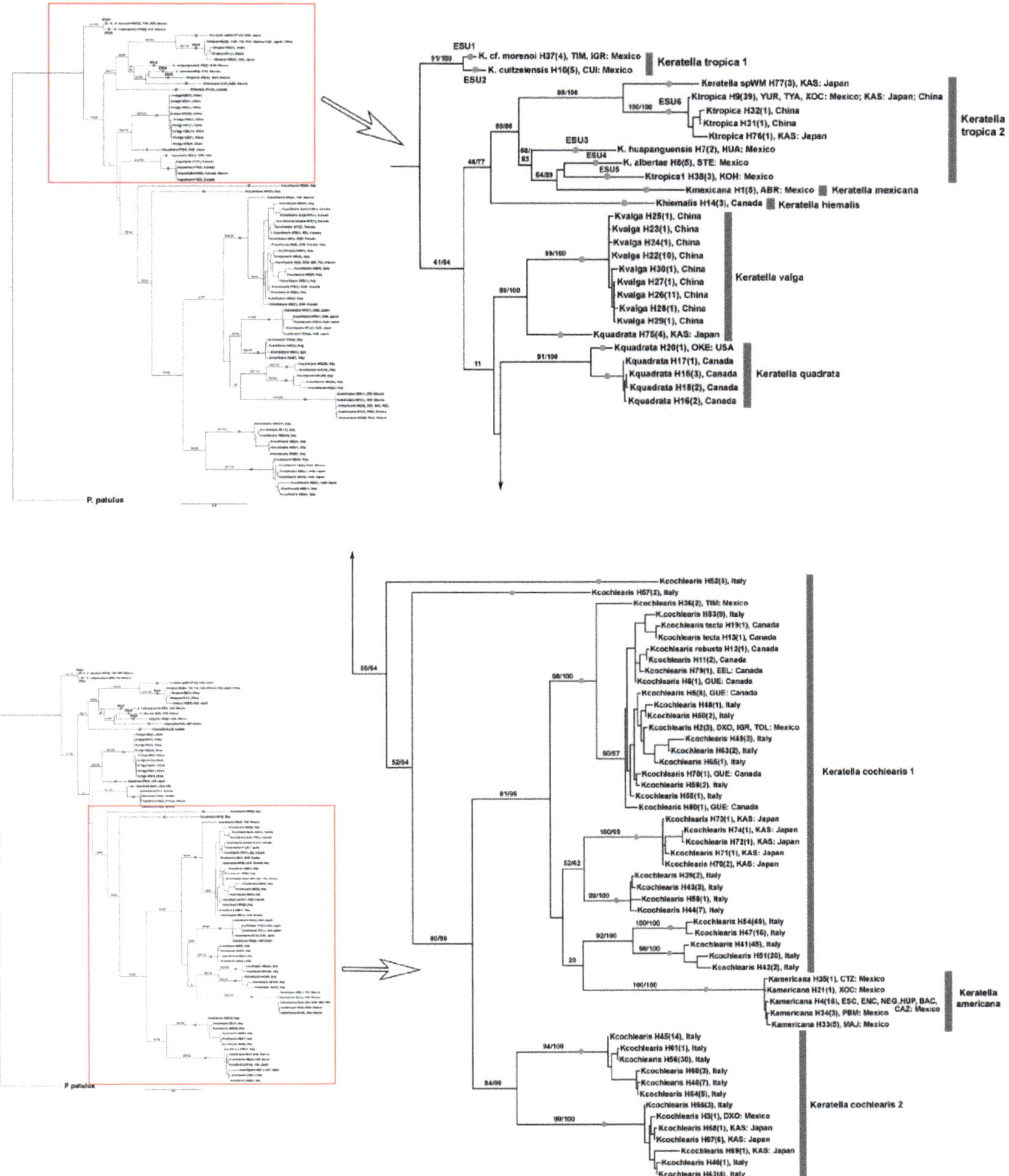

Figure 3. Maximum likelihood phylogram of the COI gene showing the relationship of the *Keratella* taxa. Haplotypes are accompanied by the number of individuals displaying that particular haplotype within parentheses and by the acronym of the water body or country in which they were isolated (when the information is available). Numbers on major branches are the percentages of branch support in the Maximum likelihood (bootstrap) and Bayesian (posterior probability) analyses respectively. Dark bars indicate that haplotypes are part of a *Keratella* group. Dark circles over branches indicate a putative species delimited by all the species delimitation methods.

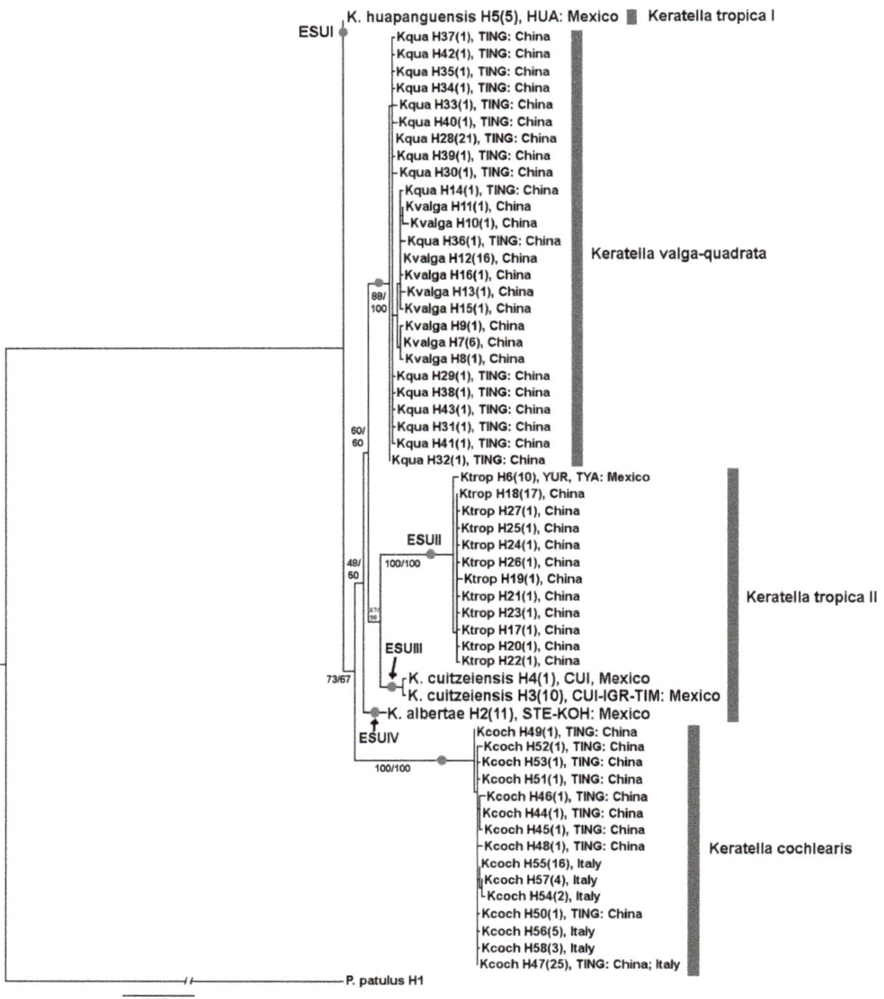

Figure 4. Maximum likelihood phylogram of the ITS1 nuclear marker showing the relationship of the *Keratella* taxa. Haplotypes are accompanied by the number of individuals displaying that particular haplotype within parentheses and by the acronym of the water body or country in which they were isolated (when the information is available). Numbers on major branches are the percentages of branch support in the Maximum likelihood (bootstrap) and Bayesian (posterior probability) analyses respectively. Dark bars indicate that haplotypes are part of a *Keratella* group. Dark circles over branches indicate a putative species delimited by all the species delimitation methods.

3.2. Statistical Analysis of Morphometric Measurements

We obtained lorica measurements from 84 individuals of *Keratella* taxa from seven populations of Mexico (Tables 1 and S6). From the biplot, a gradient was observed on axis 1, representing the morphometric features in the specimens. Axis 1 explained 98% of the variability. Populations formed a gradient without a well-defined grouping, but a subtle separation can be observed in some populations. Specimens from Ignacio Ramirez and Cuitzeo populations (ESU1 and ESU2 with COI, ESUIII with ITS1) formed a group (Figure 5, see Table 1). Specimens from the Huapango population (ESU3 with COI and ESUI with ITS1) formed another group a little more separated from the other populations

(Figure 5, see Table 1). Specimens from Santa Teresa and Kohunlich populations (ESU4 and ESU5 with COI and ESUIV with ITS1) are mixed and together form a group. The specimens in these two populations are similar in size and morphometric measurements, besides several specimens from the Yuriria population (COI-ESU6 and ITS1-ESUII) are grouped with these previous populations because of their similar size (Figure 5, Table 1). Finally, specimens from the Tepatitlan-Yahualica population (COI-ESU6 and ITS1-ESUII) formed another group separated from the other populations because specimens are small in size compared with the other populations (Figure 5, Table 1). Specimens from Yuriria are separated from specimens from Tepatitlan-Yahualica despite belonging to the same ESU, due to the great morphometric variation in this ESU (which corresponds to *K. tropica* s. str.).

Table 1. Length measurements of main lorica features based on 84 specimens of *Keratella* taxa. Measurements were obtained by population, which match with the COI-ESUs. Total length (TL), lorica length excluding anterior and posterior spines (LL), lorica width at its widest part (LW), right posterior spine (RPS), left posterior spine (LPS), anteromedian dorsal spine (AMS), anterointermediate dorsal spine (AIS), anterolateral dorsal spine (ALS). Number of individuals measured is given between brackets.

	TL	LL	LW	RPS	LPS	AMS	AIS	ALS
COI-ESU1								
Mean	312.9	149.6	84.8	130.4	49.3	32.8	25.9	20.4
Median	314	150	86	131	51	32	25	20
Min	297	142	77	115	30	25	25	17
Max	322	157	87	147	57	37	30	25
Population = Ignacio Ramirez (12)								
COI-ESU2								
Mean	293.4	132.3	75.4	134.8	49.7	26.1	17.9	15.8
Median	301	132	75	142	50	26	18	16
Min	260	124	74	100	42	26	17	15
Max	314	140	80	152	62	27	18	16
Population = Cuitzeo (15)								
COI-ESU3								
Mean	201.3	113.3	65.9	56.1	45.9	31.8	19.6	17.6
Median	202	114	66	56	45	32	20	18
Min	194	108	64	52	40	30	19	17
Max	208	117	72	60	58	32	20	18
Population = Huapango (15)								
COI-ESU4								
Mean	265.6	128.4	80.6	101.4	63	35.6	21.8	19.8
Median	264	128	80	100	67	36	22	20
Min	260	126	74	96	48	34	21	19
Max	274	132	84	106	70	36	22	20
Population = Santa Teresa (10)								
COI-ESU5								
Mean	257.7	109.6	74.4	104.7	58.1	43.3	22.5	20.2
Median	259	107	75	105	57	43.5	22	20
Min	240	105	67	95	52	40	20	17
Max	275	125	77	122	65	47	25	22
Population = Kohunlich (12)								
COI-ESU6								
Mean	215.8	103.6	61.4	84.7	31.9	27.3	20.4	19.4
Median	206	99	63	80	30	26	20	20
Min	168	90	48	54	8	20	18	18
Max	264	120	70	118	50	36	24	20
Population = Yuriria (10), Tepatitlan (10)								

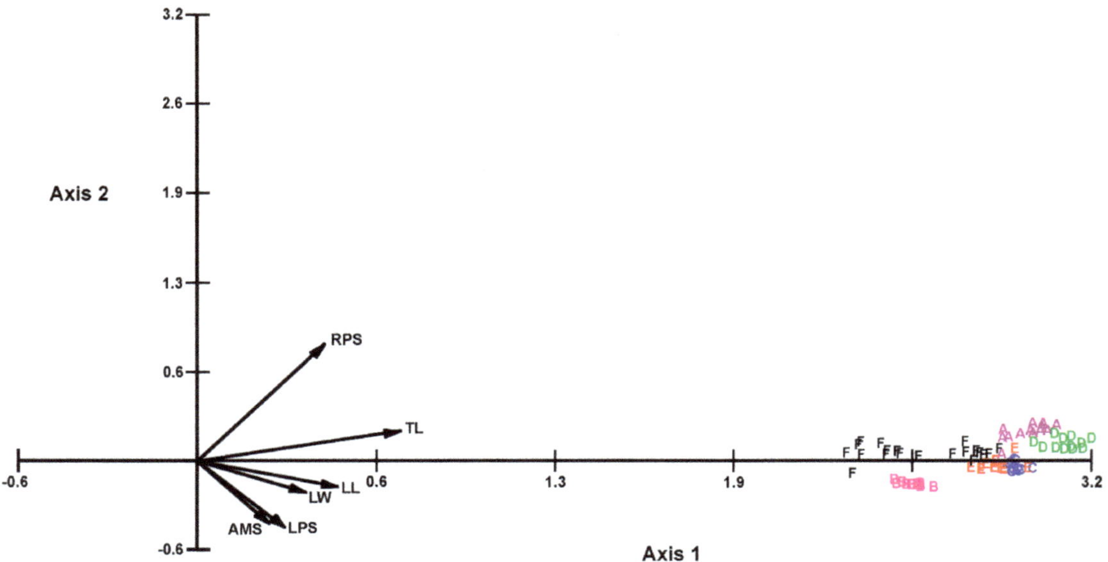

Figure 5. PCA biplot of *Keratella* populations analyzed. Symbols are as follow: A = ESU2, B = ESU3, C = ESU4, **D** = ESU1, E = ESU5 and **F** = ESU6. LT = Total lenght, LL = Lorica lenght, LW = Lorica width, RPS = Right posterior spine, LPS = Left posterior spine, AMS = Antero median spine.

3.3. Taxonomy

Keratella tropica species is diagnosed by a stiff lorica, bloated laterally. With six spines over the anterior dorsal margin. The dorsal lorica has a row of five median fields, four of these fields are hexagonal and the posteromedian remnant is smaller and square. Dorsal lorica is ornamented by a granular pattern. Posterior end of lorica slightly rounded, with two stiff posterior spines. In this species, the length of the posterior spines varies widely, with the right one longer than the left one. Although there are specimens where the left posterior spine is greatly reduced or absent.

The morphology of specimens from Yuriria and Tepatitlan-Yahualica populations compared with the descriptions in taxonomic keys and a careful review of the type specimens mounted in a permanent slide and deposited by [21] in the "Vermes" collection of the Zoological Museum of Berlin with catalog number 10,121 allowed us the conclusion that specimens ITS1-ESUII (ESU6 from COI) from Yuriria and Tepatitlan-Yahualica correspond to *Keratella tropica* s. str (See Figures 6 and 7).

Below, we present the taxonomic description of the three new *Keratella* species. In general these three new species present 19 fields over their dorsal plate: five median fields, four pairs of large polygonal lateral fields and three pairs of triangular marginal fields, all of them delimited by ridges. Moreover, we compared our specimens from these three new species with the type specimens and we confirm that these new species are different from *K. tropica* s. str.

Phylum Rotifera Cuvier, 1817
Class Eurotatoria De Ridder, 1957
Subclass Monogononta Plate, 1889
Order Ploima Hudson and Gosse, 1886
Family Brachionidae Ehrenberg, 1838
Genus *Keratella* Bory de St. Vincent, 1822
Keratella cuitzeiensis sp. nov.
Zoobank ID: urn:lsid:zoobank.org:act:7F194E23-7A36-46C4-90D9-EA38EB5F4380

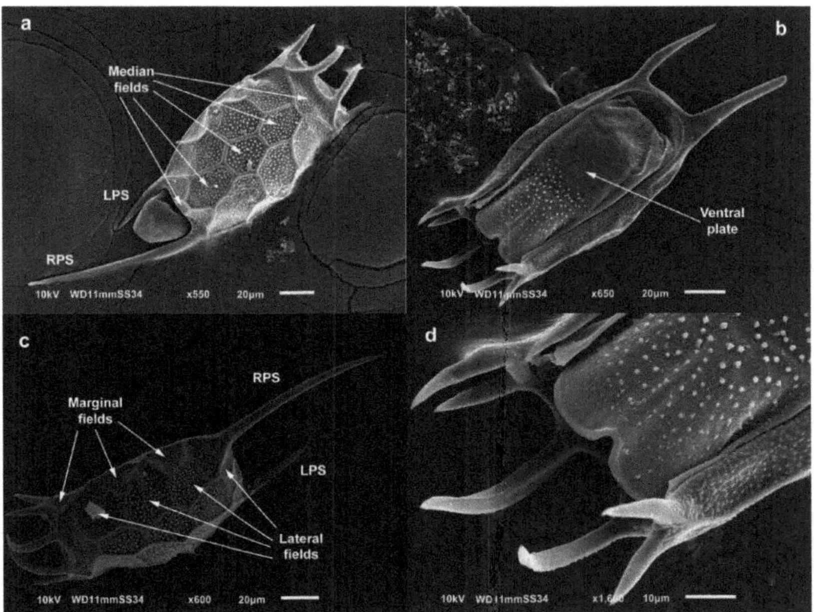

Figure 6. SEM photographs of *Keratella tropica* from Yuriria dam. (**a**). Dorsal view: arrows indicate the median fields, RPS (Right Posterior Spine), LPS (Left Posterior Spine). (**b**). Ventral view showing ornamentation over anterior part of the lorica. (**c**). Dorsal view: arrows indicate the lateral and the marginal fields, RPS (Right Posterior Spine), LPS (Left Posterior Spine). (**d**). Ventral view of the anterior part of the lorica.

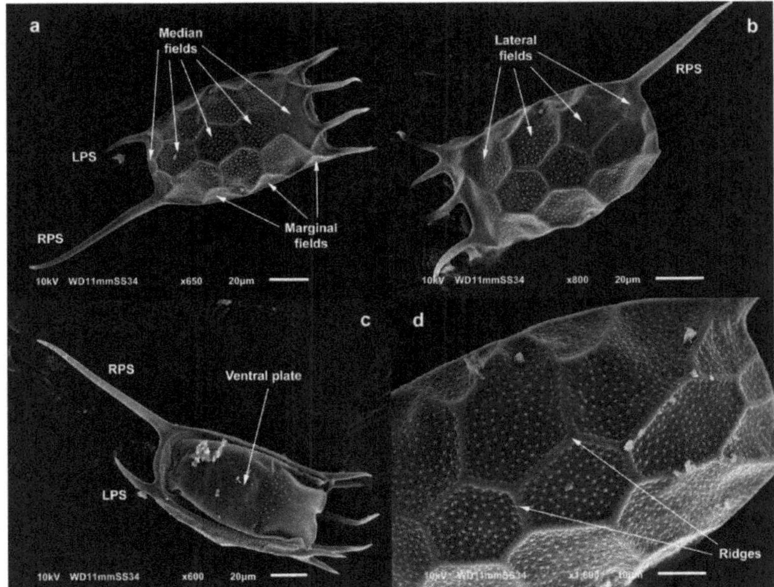

Figure 7. SEM photographs of *Keratella tropica* from Tepatitlan-Yahualica pond. (**a**). Dorsal view: arrows indicate the median and the marginal fields. (**b**). Dorsal view: arrows indicate lateral fields. (**c**). Ventral view, RPS (Right Posterior Spine), LPS (Left Posterior Spine). (**d**). Dorsal view closeup, arrows indicate a ridge.

Figures 8a,b and 9a,b.

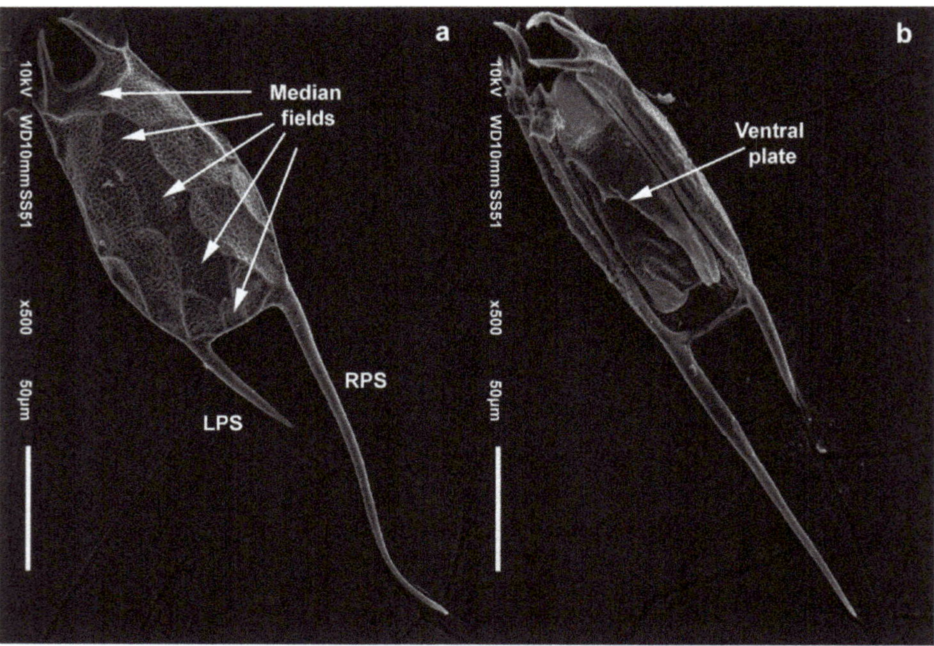

Figure 8. Morphology of *Keratella cuitzeiensis* sp. nov. (**a**). Dorsal view: arrows indicate the five median fields, RPS (Right Posterior Spine), LPS (Left Posterior Spine). (**b**). Ventral view. Females from Cuitzeo lagoon.

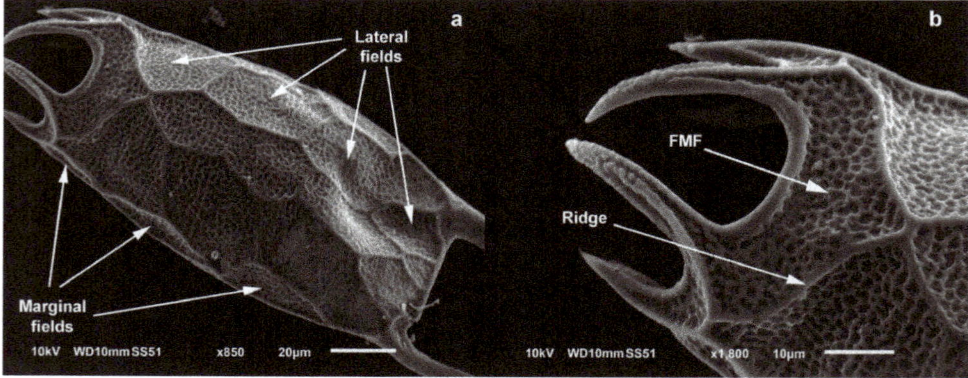

Figure 9. Morphology of *Keratella cuitzeiensis* sp. nov. (**a**). Dorsal view: arrows indicate the lateral and the marginal fields. (**b**). Anterior margin of the lorica in dorsal view, FMF (Frontomedian field), thin arrow indicate a ridge. Female from Cuitzeo lagoon.

Type locality and types. Cuitzeo lagoon is located in Michoacan, Mexico (19.8894 N, −100.9439 W). Sample collected on 25 February 2014. It is a shallow and saline water body, located at 1836 masl. The surface of the lagoon is 42,000 ha, with a maximum depth of 2.2 m.

Holotype: A parthenogenetic female mounted on a permanent slide. Paratypes: 10 females in a tube with ethanol. Holotype and paratypes are deposited in the Zooplankton Reference Collection of El Colegio de la Frontera Sur with accession numbers ECO-CH-Z-10589 and ECO-CH-Z-10590, respectively.

Differential diagnosis: *Keratella cuitzeiensis* sp. nov., most closely resembles the *Keratella tropica* species, *K. huapanguensis* sp. nov., and *K. albertae* sp. nov. It is diagnosed by a lorica that is a little bloated laterally. The anteromedian field is pentagonal. Mesomedian and posteromedian fields are hexagonal and elongated. The posteromedian remnant is conspicuous and slightly elongated. The dorsal plate is ornamented by a reticulate pattern. With two stiff posterior spines, the right one is longer than the left one (Figure 8a,b).

Description: Lorica stiff, the posterior margin of the lorica is slightly wider than the anterior margin. The posterior end of the lorica is almost straight, with two stiff and unequal posterior spines (Figure 8a). With three pairs of anterior spines that are short (Figure 9b). Anteromedian spines are recurved inward. Frontomedian field is an open short pentagon with lateral ridges prolonged into anteromedian spines (Figure 9a,b). The ventral plate is delicate, narrower and shorter than the dorsal plate. The ventral plate is bilobate, smooth without ornamentation in its anterior part (Figure 8b). Average measurements: 293.4 µm of total length, 75.4 µm of lorica width, 132.3 µm of lorica length, 134.8 µm of the right posterior spine, 49.7 µm of the left posterior spine (See Table 1).

Ecology and distribution: In Cuitzeo lagoon the new species coexists with *Brachionus quadridentatus* Hermann, 1783, *Cyclops* (Müller, 178), *Mastigodiaptomus patzcuarensis* (Kiefer, 1938), fishes (e.g., *Chirostoma* Swainson, 1839; *Xenotoca* Hubbs and Turner, 1939; *Zoogoneticus* Meek, 1902) and ostracods (*Potamocypris* Brady, 1870). The lagoon is a turbid environment with an average conductivity of 6595 µS/cm, temperature of 22 °C and pH 8 to 11.5. Cuitzeo is located in a region with a dry climate, with annual precipitation of between 6.0–150 mm. Specimens from this new species were also found in the Ignacio Ramirez dam and Timilpan pond located in the state of Mexico.

Etymology: The species name refers to the type locality where it was collected.

G + C content: ITS1 marker 0.293; COI gene 0.365.

Keratella huapanguensis sp. nov.

Zoobank ID: urn:lsid:zoobank.org:act:FD30C7A8-C63E-435D-A1C6-99003007789B

Figure 10a–d.

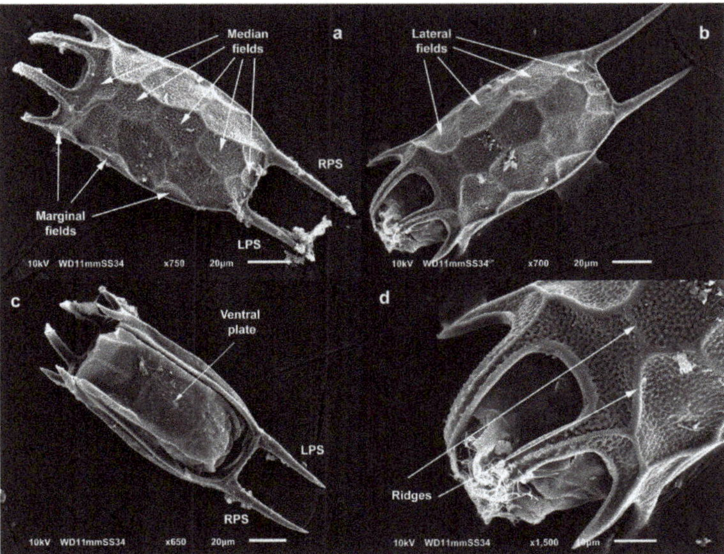

Figure 10. Morphology of *Keratella huapanguensis* sp. nov. (**a**). Dorsal view: arrows indicate the median and the marginal fields, RPS (Right Posterior Spine), LPS (Left Posterior Spine). (**b**). Dorsal view: arrows indicate lateral fields. (**c**). Ventral view, RPS (Right Posterior Spine), LPS (Left Posterior Spine). (**d**). Anterior margin of the lorica in dorsal view, arrows indicate a ridge. Females from Huapango dam.

Type locality and types: Huapango dam is located in the state of Mexico, Mexico (19.9483 N, −99.7144 W). Sample collected on 17 August 2014. It is a freshwater system, located at 2619 masl. The surface of the dam is 1000 ha, with a maximum depth of 14 m. It belongs to the basin of the Lerma River that runs 708 km from the state of Mexico to Jalisco, flowing out into Chapala Lake.

Holotype: A parthenogenetic female mounted on a permanent slide. Paratypes: 10 females in a tube with ethanol. Holotype and paratypes are deposited in the Zooplankton Reference Collection of El Colegio de la Frontera Sur with accession numbers ECO-CH-Z-10591 and ECO-CH-Z-10592, respectively.

Differential diagnosis: *Keratella huapanguensis* sp. nov. Most closely resembles the *Keratella tropica* species, *K. cuitzeiensis* sp. nov., and *K. albertae* sp. nov. It is diagnosed by a lorica almost rectangular in dorsal view. Anteromedian, mesomedian and posteromedian fields are hexagonal with almost the same size (Figure 10a,b). The posteromedian remnant is rounded and small (Figure 10a). It presents two short posterior spines (Figure 10a,c). The dorsal plate is ornamented by a reticular-granular pattern (Figure 10d).

Description: Lorica stiff, the anterior margin of the lorica is slightly wider than the posterior margin. The posterior end of the lorica is almost straight, with two stiff and relatively short posterior spines, the right one slightly longer than the left one (Figure 10a,b). With three pairs of anterior spines, which are elongated (Figure 10d). Anteromedian spines are the longest and recurved. Frontomedian field is an open short hexagon with ridges prolonged into anteromedian spines (Figure 10a,d). The ventral plate is delicate, narrower and shorter than the dorsal plate. The ventral plate is bilobate, smooth with some granules in its anterior part (Figure 10c). Average measurements: 201.3 μm of total length, 65.9 μm of lorica width, 113.3 μm of lorica length, 56.1 μm of the right posterior spine, 45.9 μm of the left posterior spine (See Table 1).

Ecology and distribution: In the Huapango dam the new species coexists with fish, e.g., *Girardinichthys multiradiatus* (Meek, 1904), "ajolote" *Ambystoma granulosum* Taylor, 1944, and "acocil" *Cambarellus montezumae* (Saussure, 1857). The sample was taken in the littoral zone among aquatic vegetation, and the water temperature and depth in that zone were 26 °C and 0.3 m respectively. Huapango is located in a region with a temperate sub-humid climate, with annual precipitation of between 700–1200 mm.

Etymology: The species name refers to the type locality where it was collected.

G + C content: ITS1 marker 0.296; COI gene 0.353.

Keratella albertae sp. nov.

Zoobank ID: urn:lsid:zoobank.org:act:DCA4000E-4B2B-4221-A15A-8CF15EF69B30

Figure 11a–d.

Type locality and types: Santa Teresa is a dam located in the state of Michoacan, Mexico (19.8886 N, −100.1722 W). The sample was collected on 17 August 2014. It is a freshwater system, located at 2307 masl. The surface of the lagoon is 149 ha, with a maximum depth of 49 m.

Holotype: A parthenogenetic female mounted on a permanent slide. Paratypes: 10 females in a tube with ethanol. Holotype and paratypes are deposited in the Zooplankton Reference Collection of El Colegio de la Frontera Sur with accession numbers ECO-CH-Z-10593 and ECO-CH-Z-10594, respectively.

Differential diagnosis: *Keratella albertae* sp. nov. Most closely resembles the *Keratella tropica* species, *K. cuitzeiensis* sp. nov., and *K. huapanguensis* sp. nov. It is diagnosed by a lorica slightly bloated in dorsal view. Anteromedian, mesomedian and posteromedian fields are hexagonal, of which posteromedian field is more elongated (Figure 11a,b). The posteromedian remnant is a conspicuous and elongated field (Figure 11b). The dorsal plate is ornamented by a granular pattern (Figure 11d). With two stiff posterior spines, the right one is longer than the left one (Figure 11a).

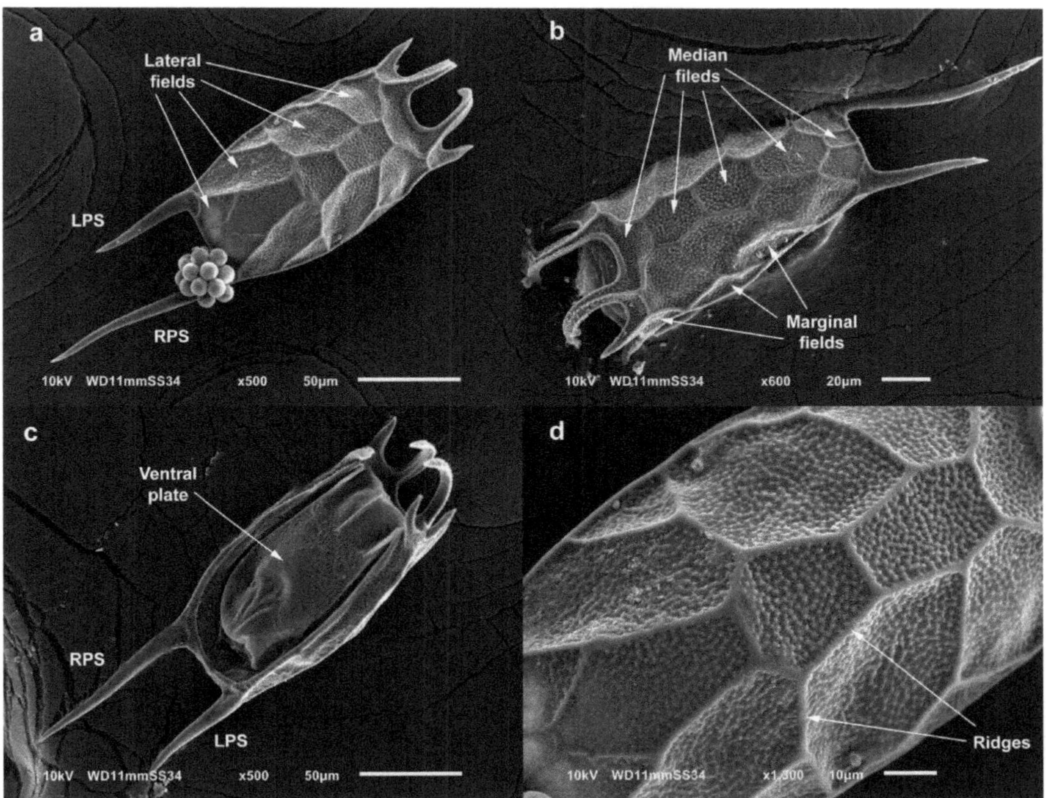

Figure 11. Morphology of *Keratella albertae* sp. nov. (**a**). Dorsal view: arrows indicate the lateral fields, RPS (Right Posterior Spine), LPS (Left Posterior Spine). (**b**). Dorsal view: arrows indicate the median and the marginal fields. (**c**). Ventral view. (**d**). Dorsal view closeup, arrows indicate a ridge. Females from Santa Teresa dam.

Description: Lorica stiff, the anterior margin of the lorica is slightly wider than the posterior margin. The posterior end of the lorica is almost straight, with two stiff and unequal posterior spines, the right one longer than the left one (Figure 11a,c). With three pairs of anterior spines, which are elongated. Anteromedian spines are the longest and recurved (Figure 11a,c). Frontomedian field is an open short hexagon with ridges prolonged into anteromedian spines (Figure 11a,b). The ventral plate is delicate, narrower and shorter than the dorsal plate. The ventral plate is bilobate, smooth with some ornamentation in its anterior part (Figure 11c). Average measurements: 265.6 µm of total length, 80.6 µm of lorica width, 128.4 µm of lorica length, 101.4 µm of the right posterior spine, 63 µm of left posterior spine (See Table 1).

Ecology and distribution: In Santa Teresa dam the new species coexists with the common carp (*Cyprinus carpio* Linnaeus, 1758). The sample was taken in the littoral zone, and the water temperature in that zone was 26 °C. Santa Teresa is located in a region with a temperate sub-humid climate, with annual precipitation of between 700–1000 mm. Specimens from this new species were also found in the Kohunlich pond located in the state of Quintana Roo.

Etymology: The species name refers to the name of the mother of AEGM.

G + C content: ITS1 marker 0.305; COI gene 0.349.

4. Discussion

Cryptic species are by definition, a set of closely related species that share very similar morphological features, and therefore are not readily distinguished [2]. Additionally, the tiny size (50–2000 microns), and translucent body of the rotifers make sometimes difficult to observe the morphological characteristics of the specimens under a compound microscope. For all above the specimens can be confused with morphologically similar species and be mistakenly identified. For example, due to the morphological stasis of the external features that characterize the *B. plicatilis* species complex [8,29], the use of molecular analysis was a fundamental basis to help unravel the cryptic diversity within this group [2]. Our study demonstrates that taxonomy-based only on morphology is not effective at providing an accurate assessment of the diversity of the *Keratella* taxa. Combining morphology with other data as genetics and ecology has demonstrated to be a more reliable approach to study diversity in rotifers and new species have been described as is the case of *Brachionus paranguensis* from central Mexico [9]. In the present study, DNA taxonomy through the use of the two markers COI and ITS1 represented an important tool that helped to confer identity to the new entities, named here as *Keratella cuitzeiensis* sp. nov., *Keratella huapanguensis* sp. nov., and *Keratella albertae* sp. nov.

Our results suggest that the divergence between the COI and ITS1 is ancient. The sequence divergence between the three new species found in both mitochondrial and nuclear markers (4–20% for COI and 3–10% for ITS1) exceeds the values usually found between congeneric species, indicating that each of these species has an independent evolutionary history. These genetic divergences are similar to the values found in species complex as *Brachionus plicatilis* (11–23% for COI and 2.5–22% for ITS1, [2]) and *B. calyciflorus* (9–13% for COI and 3–6% for ITS1, [11]).

However, we found mitonuclear discordance between the mitochondrial COI and the nuclear ITS1, as the nuclear marker revealed four ESUs (ESUI-ESUIV) and mitochondrial six ESUs (ESU1-ESU6). Mitonuclear discordance between COI and ITS1 in rotifers has already been observed in previous studies with *B. calyciflorus* [51], *K. cochlearis* and *Polyarthra dolichoptera* [49], and *B. paranguensis* by [9]. In fact, the discordance observed here is similar to the latter species, where ITS1 did not split the species as COI. Mitonuclear discrepancies were attributed to processes such as hybridization, incomplete lineage sorting and horizontal gene transfer [7,49]. Nevertheless, other factors as concerted evolution of the ribosomal nuclear markers can be considered to explain the incongruence between the mitochondrial and nuclear phylogenies [52].

Concerted evolution is defined as the coordinate evolution of repetitive DNA sequences (such as rDNA) resulting in a sequence similarity of repeating units that are greater within than among species [53]. Therefore, concerted evolution leads to sequence homogeneity within a species, but also a divergence between species [54]. COI gene evolves much faster than the nuclear ITS1 marker [9], therefore the rapid mutational rate of COI explains the high genetic variation found for this gene within the examined *Keratella* taxa. Nevertheless, it is hard to say what process is responsible for the mitonuclear discordance found in our study. We speculate that incomplete lineage sorting can be the most likely cause of the mitonuclear discordance observed in our work. This is because differentiation has not been completed in the nuclear marker. However, other processes as concerted evolution or hybridization can be operating. Further research will be needed to determine which process is responsible for the mitonuclear discordance observed in the *Keratella* taxa examined in the present study.

4.1. Morphology

Keratella tropica is a cosmopolitan species that shows morphological variation in the length of the lorica and posterior spines [55,56]. However, with the morphological and morphometric analyses, we could observe morphological differences between our *K. tropica* specimens and specimens of the three new species that can allow us distinguishing among them. Differences were observed in the shape of the median fields, length of the posterior and anterior spines, dorsal lorica ornamentation, body size and lorica shape. Other differences were observed in the ecological preferences of the species (see next section).

4.2. Distribution and Ecological Comments

Keratella tropica is distributed in tropical and subtropical regions of the world [23,56,57]. In America, it has been recorded in temperate zones from Mexico, USA and Argentina (Patagonia) [58–60]. The three new species are distributed in central Mexico, which is a temperate region with mountains.

Although we did not measure environmental parameters (except temperature), we want to make some remarks about the environment of the water bodies of the three new species based on published literature. This is important because differences in salinity preference among *Brachionus* species, for example, is considering another factor that has allowed discrimination between species [9]. The water bodies where these three new species were found show some differences in their water chemistry [61–64]. Cuitzeo lagoon is the most saline water body displaying salt levels up to 5 g L^{-1} [65], with pH fluctuating between 9.8 and 10.4, and conductivity of 6595 µS cm^{-1} [61]. Santa Teresa dam displays levels of conductivity between 200–280 µS cm^{-1}, and alkalinity of 120 mg L^{-1} [63]. Whereas, Huapango dam displays levels of alkalinity between 24 to 49 mg L^{-1}, with pH fluctuating among 5.7 to 7.6 and dissolved oxygen among 5.6 to 9.4 mg L^{-1} [62]. Therefore, the Cuitzeo lagoon is the system that has the most particular environmental conditions from all systems.

Moreover, Cuitzeo lagoon is also an interesting case, because is an antique lagoon originated during the upper Miocene and experimented several changes during the Pleistocene [66]. Today Cuitzeo is a saline system affected by anthropogenic activities. However, it is an important water body due that harbors some native fish species, for example, *Chirostoma compressum* de Buen, 1940; *C. jordani* Woolman, 1894 and *Xenotoca variata* (Bean, 1887) [67], and recently a copepod species *Mastigodiaptomus patzcuarensis* was found here [68]. Recently, another cryptic species belonging to the *Brachionus quadridentatus* species complex was found in this lagoon [7].

In addition, it was shown that certain environmental parameters as salinity and temperature of the water have important influences over the adaptation of the species and eventually over their genetic differentiation [5,69]. Some sibling species within the *B. plicatilis* species complex possess differential responses to salinity. For example, ref. [70] found that *B. plicatilis* s. str. occurs at low to high salinities (3–45 g L^{-1}), whereas *B. ibericus* Ciros-Pérez, Gómez and Serra, 2001, and *B. rotundiformis* Tschugunoff, 1921 occur in waters with medium to high salinities (8–50 and 10–57g L^{-1} respectively). However, *B. ibericus* occurs at high temperatures (>15 °C) and *B. rotundiformis* at temperatures between 10–30 °C. A laboratory study with *B. manjavacas* Fontaneto, Giordani, Melone and Serra, 2007 presented similar results where the optimal salinity for this species was observed in the range 10–30 g L^{-1} [71]. Whereas [72] reported levels of optimal salinity for *B. asplanchnoides* Charin, 1947 between 3.8–8.5 g L^{-1}. However, *B. paranguensis* a species recently described seems to be adapted to high salinity (>25 g L^{-1}), which allowed the species to colonize the hypersaline volcanic maar lake Rincón de Parangueo [9]. Maar lake Rincón de Parangueo is 66 km away from the Cuitzeo lagoon. According to the above, differences in salinity tolerance can be considered as an additional parameter to discriminate the species in some brachionids.

5. Conclusions

This is the first study conducted on several populations of *Keratella* from Mexico using integrative taxonomy. A formal description was provided for *Keratella cuitzeiensis* sp. nov., *Keratella huapanguensis* sp. nov., and *Keratella albertae* sp. nov., combining morphology and genetics. These three new species are related to *K. tropica* species. Comparison of SEM images and morphometry among the three new species showed differences in body shape, the shape of median fields, length of the posterior and anterior spines, body size, and dorsal lorica ornamentation. Therefore, DNA sequences along with morphological data support the existence of the three new species. Results of genetic variation were different among the two markers used, with a higher genetic divergence in the COI gene (six ESUs) compared to the ITS1 marker (four ESUs), and thus, providing evidence of mitonuclear discordance. This incongruence might be due to differences in mutation rate between markers, probably because of incomplete lineage sorting. Environmental conditions reported for the water systems of the three new species suggest different salinity preferences of the species, with *Keratella cuitzeiensis* sp. nov., adapted to a more saline water body than *Keratella huapanguensis* sp. nov., and *Keratella albertae* sp. nov.

Supplementary Materials: The following are available online at https://www.mdpi.com/article/10.3390/d13120676/s1, Table S1: Geographic coordinates of the sampling sites, Table S2: PCR profile for ITS1 marker, Table S3: GenBank accessions of COI and ITS1 used in this study, Table S4: Percentages of uncorrected genetic distances of COI-ESUs, Table S5: Percentages of uncorrected distances of ITS1-ESUs, Table S6: Measurements of *Keratella* specimens.

Author Contributions: Conceptualization, A.E.G.-M.; validation, A.E.G.-M., O.D.-D. and M.E.-G.; formal analysis, A.E.G.-M., O.D.-D. and M.E.-G.; resources, O.D.-D. and M.E.-G.; data curation, A.E.G.-M.; writing—original draft preparation, A.E.G.-M.; writing—review and editing, A.E.G.-M., O.D.-D. and M.E.-G.; funding acquisition, O.D.-D. All authors have read and agreed to the published version of the manuscript.

Funding: This research was funded in part by grant number CIC-UMSNH-2015.

Institutional Review Board Statement: Not applicable.

Data Availability Statement: The data presented in this study are available in Supplementary Materials; Tables S1–S6.

Acknowledgments: We thank Birger Neuhaus, Scientific Head Collection "Vermes" of the Zoological Museum of Berlin, Germany, who sent us several photographs of the *Keratella tropica* types deposited by Apstein (1907). We thank Adrian Cervantes Martínez, who kindly assisted us with the PCA analysis.

Conflicts of Interest: The authors declare no conflict of interest.

References

1. Fontaneto, D.; Kaya, M.; Herniou, E.A.; Barraclough, T.G. Extreme levels of hidden diversity in microscopic animals (Rotifera) revealed by DNA taxonomy. *Mol. Phylogenet. Evol.* **2009**, *53*, 182–189. [CrossRef]
2. Mills, S.; Alcántara-Rodríguez, J.A.; Ciros-Pérez, J.; Gómez, A.; Hagiwara, A.; Hinson, K.; Jersabek, C.D.; Malekzadeh-Viayeh, R.; Leasi, F.; Lee, J.S.; et al. Fifteen species in one: Deciphering the *Brachionus plicatilis* species complex (Rotifera, Monogononta) through DNA taxonomy. *Hydrobiologia* **2017**, *796*, 39–58. [CrossRef]
3. Wallace, R.L.; Snell, T.W.; Ricci, C.; Nogrady, T. *Rotifera: Biology, Ecology and Systematics. Guides to the Identification of the Microinvertebrates of the Continental Waters of the World,* 2nd ed.; Kenobi Productions and Backhuys Publishers: The Hague, The Netherlands, 2006; pp. 1–20.
4. Brandl, Z. Freshwater copepods and rotifers: Predators and their prey. *Hydrobiologia* **2005**, *546*, 475–489. [CrossRef]
5. Xiang, X.L.; Xi, Y.L.; Zhu, L.Y.; Xu, Q.L. Comparative studies of the population genetic structure of the *Brachionus calyciflorus* species complex from four inland lakes in Wuhu, China. *Biochem. Syst. Ecol.* **2017**, *71*, 69–77. [CrossRef]
6. García-Morales, A.E.; Elías-Gutiérrez, M. DNA barcoding of freshwater Rotifera in Mexico: Evidence of cryptic speciation in common rotifers. *Mol. Ecol. Res.* **2013**, *13*, 1097–1107. [CrossRef] [PubMed]
7. García-Morales, A.E.; Domínguez-Domínguez, O. Cryptic molecular diversity in the morphologically variable rotiferan *Brachionus quadridentatus* (Rotifera: Monognonta). *Rev. Biol. Trop.* **2019**, *67*, 1114–1130. [CrossRef]
8. Gómez, A.; Serra, M.; Carvalho, G.R.; Lunt, D.H. Speciation in ancient cryptic species complex: Evidence from the molecular phylogeny of *Brachionus plicatilis* (Rotifera). *Evolution* **2002**, *56*, 1431–1444. [CrossRef] [PubMed]
9. Guerrero-Jiménez, G.; Vannucchi, P.E.; Silva-Briano, M.; Adabache-Ortiz, A.; Rico-Martínez, R.; Roberts, D.; Neilson, R.; Elías-Gutiérrez, M. *Brachionus paranguensis* sp. nov. (Rotifera, Monogononta), a member of the L group of the *Brachionus plicatilis* complex. *Zookeys* **2019**, *880*, 1–23. [CrossRef]
10. Schröder, T.; Walsh, E.J. Cryptic speciation in the cosmopolitan *Epiphanes senta* complex (Monogononta, Rotifera) with the description of new species. *Hydrobiologia* **2007**, *593*, 129–140. [CrossRef]

11. Gilbert, J.J.; Walsh, E.J. *Brachionus calyciflorus* is a species complex: Mating behavior and genetic differentiation among four geographically isolated strains. *Hydrobiologia* **2005**, *546*, 257–265. [CrossRef]
12. Obertegger, U.; Flaim, G.; Fontaneto, D. Cryptic diversity within the rotifer *Polyarthra dolichoptera* along an altitudinal gradient. *Freshw. Biol.* **2014**, *59*, 2413–2427. [CrossRef]
13. Cieplinski, A.; Weisse, T.; Obertegger, U. High diversity in *Keratella cochlearis* (Rotifera, Monogononta): Morphological and genetic evidence. *Hydrobiologia* **2017**, *796*, 145–159. [CrossRef]
14. Derry, A.M.; Hebert, P.D.N.; Prepas, E.E. Evolution of rotifers in saline and subsaline lakes: A molecular phylogenetic approach. *Limnol. Oceanogr.* **2003**, *48*, 675–685. [CrossRef]
15. Kordbacheh, A.; Wallace, R.L.; Walsh, E.J. Evidence supporting cryptic species within two sessile microinvertebrates, *Limnias melicerta* and *L. ceratophylli* (Rotifera, Gnesiotrocha). *PLoS ONE* **2018**, *13*, e0205203. [CrossRef] [PubMed]
16. Segers, H. Global diversity of rotifers (Rotifera) in freshwater. *Hydrobiologia* **2008**, *595*, 49–59. [CrossRef]
17. Segers, H.; De Smet, W.H. Diversity and endemism in Rotifera: A review, and *Keratella* Bory de St Vincent. *Biodivers. Conserv.* **2008**, *17*, 303–316. [CrossRef]
18. Koste, W.; Shiel, R.J. Rotifera from Australian inland waters II Ephiphanidae and Brachionidae (Rotifera: Monogononta). *Invertebr. Taxon.* **1987**, *1*, 949–1021. [CrossRef]
19. Koste, W. *Rotatoria. Die Rädertiere Mitteleuropas. Ein Bestimmungswerk, begründet von Max Voigt. Überordnung Monogononta. I. Textband*; Gebrüder Borntraeger: Stuttgart, Germany, 1978.
20. Koste, W. *Rotatoria. Die Rädertiere Mitteleuropas. Ein Bestimmungswerk, begründet von Max Voigt. Überordnung Monogononta. II. Tafelband*; Gebrüder Borntraeger: Stuttgart, Germany, 1978; pp. 16–25.
21. Apstein, C. Das plankton im Colombo-See auf Ceylon. *Zool. Jahrbücher Abt. Syst. Geogr. Und Biol. Tiere* **1907**, *25*, 201–244.
22. Segers, H. Annotated checklist of the rotifers (Phylum Rotifera), with notes on nomenclature, taxonomy and distribution. *Zootaxa* **2007**, *1564*, 1–104. [CrossRef]
23. Yermolaeva, N.I.; Kirillov, V.V. First record of *Keratella tropica* (Apstein, 1907) (Rotifera: Brachionidae) in western Siberia. *Russ. J. Biol. Invasions* **2018**, *9*, 38–43. [CrossRef]
24. Berzins, B. Taxonomie und Verbreitung von *Keratella valga* und verwandten Formen. *Arkiv Zoologi.* **1955**, *8*, 549–559.
25. Xi, Y.L.; Xu, D.D.; Ma, J.; Ge, Y.L.; Wen, X.L. Differences in life table parameters between *Keratella tropica* and *Keratella valga* (Rotatoria) from subtropical shallow lakes. *J. Freshw. Ecol.* **2013**, *28*, 539–545. [CrossRef]
26. Fontaneto, D.; Flot, J.F.; Tang, C.Q. Guidelines from DNA taxonomy with focus on the meiofauna. *Mar. Biodivers.* **2015**, *45*, 433–451. [CrossRef]
27. Walsh, E.J.; Schröder, T.; Wallace, R.L.; Rico-Martinez, R. Cryptic speciation in *Lecane bulla* (Monogononta: Rotifera) in Chihuahuan Desert waters. *Int. Ver. Theor. Angew. Limnol. Verhandlungen* **2009**, *30*, 1046–1050.
28. Dayrat, B. Towards integrative taxonomy. *Biol. J. Linn. Soc.* **2005**, *85*, 407–415. [CrossRef]
29. Suatoni, E.; Vicario, S.; Rice, S.; Snell, T.; Caccone, A. An analysis of species boundaries and biogeographic patterns in a cryptic species complex: The rotifer *Brachionus plicatilis*. *Mol. Phylogenet. Evol.* **2006**, *41*, 86–98. [CrossRef]
30. Ferrusquía-Villafranca, I. Ensayo sobre la caracterización y significación biológica. In *Biodiversidad de la Faja Volcánica Transmexicana*; Luna, I., Morrone, J.J., Espinosa, D., Eds.; Jiménez Editores e Impresores S.A de C.V.: Ciudad de México, México, 2007; pp. 7–23.
31. Hernández, M.E.; Carrasco, G. Rasgos climáticos más importantes. In *Biodiversidad de la Faja Volcánica Transmexicana*; Luna, I., Morrone, J.J., Espinosa, D., Eds.; Jiménez Editores e Impresores S.A de C.V.: Ciudad de México, México, 2007; pp. 57–72.
32. Folmer, O.; Black, M.; Hoeh, W.; Lutz, R.; Vrijenhoek, R. DNA primers for amplification of mitochondrial cytochrome c oxidase subunit I from diverse metazoan invertebrates. *Mol. Mar. Biol. Biotech.* **1994**, *3*, 294–297.
33. Palumbi, S.R. The polymerase chain reaction. In *Molecular Systematics*; Hillis, D.M., Moritz, C., Marble, K., Eds.; Sinauer Associates: Sunderland, MA, USA, 1996; pp. 205–247.
34. Kumar, S.; Stecher, G.; Tamura, K. MEGA7: Molecular evolutionary genetic analysis version 7.0 for bigger datasets. *Mol. Biol. Evol.* **2016**, *33*, 1870–1874. [CrossRef]
35. Katoh, K.; Asimenos, G.; Toh, H. Multiple alignment of DNA sequences with MAFFT. *Methods. Mol. Biol.* **2009**, *537*, 39–64. [PubMed]
36. Darriba, D.; Taboada, G.L.; Doallo, R.; Posada, D. jModelTest 2: More models, new heuristics and parallel computing. *Nat. Methods.* **2012**, *9*, 772. [CrossRef]
37. Librado, P.; Rozas, J. DnaSP v5: A software for comprehensive analysis of DNA polymorphism data. *Bioinformatics* **2009**, *25*, 1451–1452. [CrossRef]
38. Ronquist, F.; Teslenko, M.; van der Mark, P.; Ayres, D.; Darling, A.; Höhna, S.; Larget, B.; Liu, L.; Suchard, M.A.; Huelsenbeck, J.P. Mrbayes 3.2: Efficient Bayesian phylogenetic inference and model choice across a large model space. *Syst. Biol.* **2012**, *61*, 539–542. [CrossRef]
39. Stamatakis, A.; Hoover, P.; Rougemont, J. A rapid bootstrap algorithm for the RAxML web servers. *Syst. Biol.* **2008**, *57*, 758–771. [CrossRef] [PubMed]
40. Rambaut, A.; Drummond, A.J.; Xie, D.; Baele, G.; Suchard, M.A. Posterior summarization in Bayesian phylogenetics using Tracer 1.7. *Syst. Biol.* **2018**, *67*, 901–904. [CrossRef] [PubMed]
41. Pons, J.; Barraclough, T.G.; Gómez-Zurita, J.; Cardoso, A.; Duran, D.P.; Hazell, S.; Kamoun, S.; Sumlin, W.D.; Vogler, A. Sequence based species delimitation for the DNA taxonomy of undescribed insects. *Syst. Biol.* **2006**, *55*, 595–609. [CrossRef]

42. Bouckaert, R.; Heled, J.; Kuehnert, D.; Vaughan, T.; Wu, C.H.; Xie, D.; Suchard, M.; Rambaut, A.; Drummond, A.J. BEAST 2: A software platform for Bayesian evolutionary analysis. *PLoS Comput. Biol.* **2014**, *10*, e1003537. [CrossRef]
43. Tang, C.Q.; Obertegger, U.; Fontaneto, D.; Barraclough, T.G. Sexual species are separated by larger genetic gaps than asexual species in rotifers. *Evolution* **2014**, *68*, 2901–2916. [CrossRef]
44. Wilke, T.; Schultheiß, R.; Albrecht, C. As time goes by: A simple fool´s guide to molecular clock approaches in invertebrates. *Am. Malacol. Bull.* **2009**, *27*, 25–45. [CrossRef]
45. DeJong, R.J.; Morgan, J.A.; Lobato, W.; Pointier, J.P.; Amarista, M.; Ayeh-Kumi, P.F.; Babiker, A.; Barbosa, C.S.; Brémond, P.; Canese, A.P.; et al. Evolutionary relationships and biogeography of *Biomphalaria* (Gastropoda: Planorbidae) with implications regarding its role as host of the human bloodfluke, *Schistosoma mansoni*. *Mol. Biol. Evol.* **2001**, *18*, 2225–2239. [CrossRef]
46. Schlötterer, C.; Hauser, M.; von Haeseler, A.; Tautz, D. Comparative evolutionary analysis of rDNA ITS regions in *Drosophila*. *Mol. Biol. Evol.* **1994**, *11*, 513–522. [PubMed]
47. Zhang, J.; Kapli, P.; Pavlidis, P.; Stamatakis, A. A general species delimitation method with applications to phylogenetic placements. *Bioinformatics* **2013**, *29*, 2869–2876. [CrossRef] [PubMed]
48. Puillandre, N.; Lambert, A.; Brouillet, S.; Achaz, G. ABGD, Automatic Barcode Gap Discovery for primary species delimitation. *Mol. Ecol.* **2012**, *21*, 1864–1877. [CrossRef]
49. Obertegger, U.; Cieplinski, A.; Fontaneto, D.; Papakostas, S. Mitonuclear discordance as a confounding factor in the DNA taxonomy of monogonont rotifers. *Zool. Scr.* **2018**, *47*, 122–132. [CrossRef]
50. Modenutti, B.E.; Diéguez, M.C.; Segers, H. A new *Keratella* from Patagonia. *Hydrobiologia* **1998**, *389*, 1–5. [CrossRef]
51. Papakostas, S.; Michaloudi, E.; Proios, K.; Brehm, M.; Verhage, L.; Rota, J.; Peña, C.; Stamou, G.; Pritchard, V.L.; Fontaneto, D.; et al. Integrative taxonomy recognizes evolutionary units despite widespread mitonuclear discordance: Evidence from a rotifer cryptic species complex. *Syst. Biol.* **2016**, *65*, 508–524. [CrossRef]
52. Liao, D. Concerted evolution. In *Encyclopedia of the Human Genome*; Macmillan Publishers Ltd.: New York, NY, USA; Nature Publishing Group: London, UK, 2003; Available online: www.ehgonline.net (accessed on 6 October 2021).
53. Elder, J.F.; Turner, B.J. Concerted evolution of repetitive DNA sequences in eukaryotes. *Q. Rev. Biol.* **1995**, *70*, 297–320. [CrossRef] [PubMed]
54. Gong, L.; Shi, W.; Yang, M.; Kong, X.Y. Characterization of 18S-ITS1-5.8S rDNA in eleven species in Soleidae: Implications for phylogenetic analysis. *Hydrobiologia* **2018**, *819*, 161–175. [CrossRef]
55. Egborge, A.B.M.; Ogbekene, L. Cyclomorphosis in *Keratella tropica* (Apstein) of Lake Asejire, Nigeria. *Hydrobiologia* **1986**, *135*, 179–191. [CrossRef]
56. Marinone, M.C.; Zagarese, H.E. A field and laboratory study on factors affecting polymorphism in the rotifer *Keratella tropica*. *Oecologia* **1991**, *86*, 372–377. [CrossRef]
57. Green, J. Asymmetry and variation in *Keratella tropica*. *Hydrobiologia* **1980**, *73*, 241–248. [CrossRef]
58. Dieguez, M.C.; Modenutti, B.E. *Keratella* distribution in North Patagonian lakes (Argentina). *Hydrobiologia* **1996**, *321*, 1–6. [CrossRef]
59. Garza, G.; Silva-Briano, M.; Nandini, S.; Sarma, S.S.S.; Castellanos-Páez, M.E. Morphological and morphometrical variations of selected rotifer species in response to predation: A seasonal study of selected brachionid species from Lake Xochimilco (Mexico). *Hydrobiologia* **2005**, *546*, 169–179. [CrossRef]
60. Tausz, C.; Beaver, J.R.; Renicker, T.R.; Klepach, J.A.; Pollard, A.I.; Mitchell, R.M. Biogeography and co-occurrence of 16 planktonic species of *Keratella* Bory de St. Vincent, 1822 (Rotifera, Ploima, Brachionidae) in lakes and reservoirs of the United States. *Zootaxa* **2019**, *4624*, 337–350. [CrossRef]
61. Alcocer, J.; Bernal-Brooks, F.W. Limnology in Mexico. *Hydrobiologia* **2010**, *644*, 15–68. [CrossRef]
62. Arriaga, L.; Aguilar, V.; Alcocer, J. *Aguas Continentales y Diversidad Biológica de México*; Comisión Nacional para el Conocimiento y Uso de la Biodiversidad: Ciudad de México, México, 2000; pp. 153–167.
63. COMPESCA. *Atlas Pesquero y Acuícola de Michoacán*; Gobierno del Estado de Michoacán: Morelia, Mexico, 2013; pp. 165–170.
64. García, M.; Herrera, F. La cuenca hidrosocial presa Huapango, México: Un análisis de la gestión integrada de los recursos hídricos y la gobernanza en cuerpos de agua compartidos. *Agua y Territorio* **2019**, *14*, 69–84. [CrossRef]
65. Sarma, S.S.S.; Nandini, S.; Morales-Ventura, J.; Delgado- Martínez, I.; González-Valverde, L. Effects of NaCl salinity on the population dynamics of freshwater zooplankton (rotifers and cladocerans). *Aquatic Ecol.* **2006**, *40*, 349–360. [CrossRef]
66. Israde, I.; Velázquez-Durán, R.; Lozano, M.S.; Bischoff, J.; Domínguez, G.; Garduño, V.H. Evolución paleolimnológica del lago Cuitzeo, Michoacán durante el Pleistoceno-Holoceno. *Bol. Soc. Geol. Mex.* **2010**, *62*, 345–357. [CrossRef]
67. Chacón, A.; Rosas, C.; Alvarado, J. El lago de Cuitzeo. In *Las Aguas Interiores de México: Conceptos y Casos*; De la Lanza, G., Hernandez, S., Eds.; AGT Editor S.A: Ciudad de México, México, 2007; pp. 305–338.
68. Gutiérrez-Aguirre, M.A.; Cervantes-Martínez, A.; Elías-Gutiérrez, M. An example of how barcodes can clarify cryptic species: The case of the calanoid copepod *Mastigodiaptomus albuquerquensis* (Herrick). *PLoS ONE* **2014**, *9*, e85019. [CrossRef]
69. Campillo, S.; García-Roger, E.M.; Carmona, M.J.; Serra, M. Local adaptation in rotifer populations. *Evol. Ecol.* **2011**, *25*, 933–947. [CrossRef]
70. Ciros-Pérez, J.; Gómez, A.; Serra, M. On the taxonomy of three sympatric sibling species of the *Brachionus plicatilis* (Rotifera) complex from Spain, with the description of *B. ibericus* n. sp. *J. Plankton. Res.* **2001**, *23*, 1311–1328. [CrossRef]

71. Montero-Pau, J.; Ramos-Rodríguez, E.; Serra, M.; Gómez, A. Long-term coexistence of rotifer cryptic species. *PLoS ONE* **2011**, *6*, e21530. [CrossRef] [PubMed]
72. Papakostas, S.; Michaloudi, E.; Triantafyllidis, A.; Kappas, I.; Abatzopoulos, T.J. Allochronic divergence and clonal succession: Two microevolutionary processes sculpturing population structure of *Brachionus* rotifers. *Hydrobiologia* **2013**, *700*, 33–45. [CrossRef]

Article

Integrative Taxonomy of Two Peruvian Strains of *Brachionus plicatilis* Complex with Potential in Aquaculture

Pedro Pablo Alonso Sánchez-Dávila [1,*], Giovanna Sotil [2], Araceli Adabache-Ortiz [3], Deivis Cueva [2] and Marcelo Silva-Briano [3]

1 Instituto del Mar del Perú—IMARPE, DGIA, AFIA, BGOA, Esquina Gamarra y Gral. Valle s/n, Callao 07021, Peru
2 Laboratorio de Genética Molecular, Instituto del Mar del Perú—IMARPE, DGIA, Esquina Gamarra y Gral. Valle s/n, Callao 07021, Peru; gsotil@imarpe.gob.pe (G.S.); dcueva@imarpe.gob.pe (D.C.)
3 Laboratorio No. 1, de Ecología, Centro de Ciencias Básicas, Edificio 202, Departamento de Biología, Universidad Autónoma de Aguascalientes, Av. Universidad No. 940, Ciudad Universitaria, Aguascalientes 20100, Mexico; aadaba@correo.uaa.mx (A.A.-O.); msilva@correo.uaa.mx (M.S.-B.)
* Correspondence: psanchez@imarpe.gob.pe

Abstract: Two Peruvian strains of the genus Brachionus were isolated from impacted coastal wetlands. With an integrative taxonomic view, we described their taxonomic status, morphological characters, productive parameters, and phylogenetic position. In the case of both strains, the relationship between biometrics and productive parameters obtained with Principal Components Analysis indicated that the lorica length was associated with longevity, progeny, egg production, and reproductive age, while the lorica width and aperture were associated with the maximum number of eggs carried. Maximum Likelihood and Bayesian Inference analysis carried out with mtDNA COI gene and rDNA ITS1 region showed that both strains were clustered in two clades with distinct phylogenetic positioning from what is currently known for *Brachionus plicatilis s.l.* One of the strains, Z010-VL, is proposed to be a subspecies of L4 (*B. paranguensis*), and the other strain, Z018-SD, is proposed as a sub species of SM2 (*B. koreanus*). In addition, 33 and 31 aquaculture production lineages are proposed, delimited by COI and concatenated COI+ITS1 sequences, respectively. Finally, this study provides new tools that enhance the traceability of the origin of each sub-species throughout the world.

Keywords: *Brachionus plicatilis* complex; integrative taxonomy; biometrics; lifespan; strain; aquaculture production lineages

Citation: Sánchez-Dávila, P.P.A.; Sotil, G.; Adabache-Ortiz, A.; Cueva, D.; Silva-Briano, M. Integrative Taxonomy of Two Peruvian Strains of *Brachionus plicatilis* Complex with Potential in Aquaculture. *Diversity* **2021**, *13*, 671. https://doi.org/10.3390/d13120671

Academic Editor: Chang-Bae Kim

Received: 9 November 2021
Accepted: 10 December 2021
Published: 15 December 2021

1. Introduction

The taxonomy of the *Brachionus plicatilis* species complex began with the first description performed by Müller in 1786, and then other reports went on to describe the species as potentially cosmopolitan [1,2]. One of the first insights was the dominance of morphotypes according to seasonal changes, postulated by Serra and Miracle [3]. Fifteen different species have currently been postulated, but only seven formal delimited species have been described [4,5], taking into consideration the analysis of the nuclear rDNA ITS1 region. Thus, the current complex group (*B. plicatilis sensu lato*) is formed by *B. plicatilis sensu stricto* (L1), *Brachionus manjavacas* (L2), *Brachionus asplanchnoidis* (L3), *Brachionus paranguensis* (L4), *Brachionus ibericus* (SM1), *Brachionus koreanus* (SM2), and *Brachionus rotundiformis* (SS).

In Peru, studies on the taxonomy of *B. plicatilis s.l.* are scarce. Species records, ecological and morphological descriptions [6,7], as well as applied aspects [8–12] have been reported. Although there are important efforts to begin formal descriptions and to provide useful information for those wishing to experiment with multidisciplinary research with local strains, especially considering their high utility in the local aquaculture [13–16], the current taxonomic status of Peruvian strains of *B. plicatilis* is still unclear.

Biometrics is currently an instrument considered in the monitoring of resources such as fish and mollusks, which guides towards selection criteria to improve production and management. Therefore, the goal of this study is to investigate the possibility of its functional application with rotifers. The rotifers of the *B. plicatilis* species complex are the most widely cultured zooplankton used in aquaculture worldwide [17]. Hence, a taxonomic record in addition to their individual productive characteristics could prove to be a new tool to better use its potential.

In this sense, we present the results of an integrative taxonomic study of the lineage of two Peruvian strains IMP-BG Z010-VL and IMP-BG Z018-SD, from species L4 and SM2, respectively, in order to contribute to the description of the taxonomy and diversity of the species. In addition, we evaluated the correlation between productivity and biometric parameters of these *Brachionus* isolated from impacted coastal wetlands, drawing attention to this environmental issue based on individual descriptions of the life cycle of two strains and the relationship of their production parameters against three morphometric measurements.

Finally, we strengthen the hypothesis of the existence of delimited lineages in productive terms, looking to improve their traceability back to the designation of origin. Therefore, we suggest a subspecific classification to formally delimit them as aquaculture production lineages, based on morphological and molecular (mitochondrial and nuclear DNA markers) evidence.

2. Material and Methods

2.1. Sample Collection

Organisms were collected from 2 impacted coastal wetlands surrounded by desert, located in the central south of Peru (Figure 1), using a 10 μm phytoplankton net. The 2 isolated strains were coded correlatively. Strain IMP-BG Z010-VL was collected in 2009 from the Municipal wetland of the Ventanilla district, Callao (WGS84 11°52′16.11″ S; 77°08′17.88″ W), which was used as a rubbish heap by a nearby human settlement, causing it to be full of litter; while the strain IMP-BG Z018-SD was collected in 2014 from a very shallow relictual wetland near a private club in Santo Domingo, located in the Paracas district in Ica (13°51′25″ S; 76°15′11″ W). The areas of the sampling stations were registered using a Garmin GPS, model GPSMAP 60CSx (Shijr, New Taipei City, Taiwan), and the datum used in the mapping was WGS1984, displayed with the ArcGIS Desktop program, version 10.5.

The isolation of the rotifers was carried out in the laboratory to avoid undesirable protozoan. A modified pipetting technique was used for successive washes of organisms. Washes among drops of filtered and sterile seawater were performed on a glass slide, according to the protocol (unpublished) of the Instituto del Mar del Peru (IMARPE), based on Andersen [18]. Both strains are conserved ex-situ in the Germplasm Bank of Aquatic Organisms of IMARPE (http://www.imarpe.pe/imarpe/index2.php?id_seccion=I0170050400000000000000, accessed on 10 December 2021), in batch culture. Samples from different batches and from different years were selected for morphological and molecular characterization.

2.2. Culture Conditions, Morphometry, and Parameter Evaluation

The isolated organisms were cultured in 100 mL beakers at a density of 5 rot/mL with seawater filtered at 0.22 μm and sterilized in an autoclave. The experiments were carried out in 2 plastic 48-well culture plates, with one rotifer of each strain per well. Then, the F1 obtained the day after placing the F0 in each well was used to start the lifespan experiment. Both cultures were performed under controlled conditions in a Torrey climate chamber model R-14AI, with 14:10 h photoperiod, a temperature of 24 ± 1 °C, with water changes every 7 to 10 days, dissolved oxygen from 6.2 to 3.5 mg/L and pH from 8 to 6. The salinity was adapted in order to provide the best fit for each strain, which was observed a priori, and hence was considered 35‰ for IMP-BG Z010-VL and 25‰ for IMP-BG Z018-SD. Organisms were fed with *Nannochloris* sp.

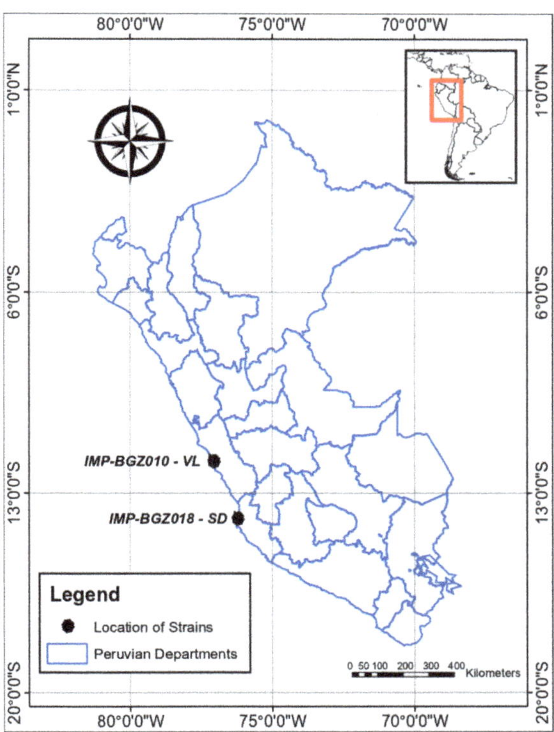

Figure 1. Map indicating (with black dots) the origin of the Peruvian strains, collected from Ventanilla-Callao (IMP-BG Z010-VL) and Santo Domingo-Ica (IMP-BG18-SD).

Currently, worldwide, there are no taxonomic keys to describe all the species from this species complex; consequently, 10 parthenogenetic females of each native strain were obtained from cultures and fixed in 2% formalin. These organisms were then observed under an optical microscope to describe the morphological characters used to compare and classify the *Brachionus* complex, such as the presence of gastric glands, the shape of the dorsal sinus, pores, lateral antennas, form of the eye, and the shape of the upper part of the lorica. A group of organisms from each strain was sent to the University of Aguascalientes in Mexico for the surface electronic microscopic (SEM) analysis. For this, trophi were removed according to the methodology of Segers et al. [19] and mounted on glass slides. Observations were made with SEM JEOL 5900 LV, and habit images were taken according to Silva-Briano et al. [20].

Nine morphometric character measurements of the cultivated rotifers were performed in the lapse of 3 years, using a microscope Leica DM1000 LED (Wetzlar, Hesse, Germany), with a 3 Mpxs image capture CMOS DFC290 HD (Wetzlar, Hesse, Germany) and the Leica Application Suite v. 4.10.0 software. Measurements of lorica length, distance between lateral spines, lorica width, the distance between central spines and dorsal sinus depth, the distance between central and medial spines, and medial spine length (indicated in Figure 2, from "a" to "g", respectively) were selected on the basis of Fu et al. [21]. Additionally, the characters: head aperture (h), and lateral spine length (i), were selected according to Ciros-Pérez et al. [22].

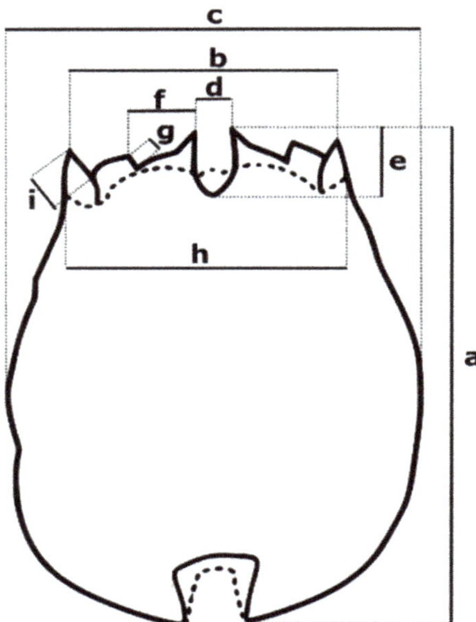

Figure 2. Morphometric measurements considered for the Peruvian strain descriptions. (a) lorica length, (b) distance between lateral spines—DbLS or lorica aperture, (c) lorica width, (d) distance between central spines, (e) dorsal sinus depth, (f) distance between central and medial spines, (g) medial spine length, (h) head aperture, (i) lateral spine length. Characters from (a–g) were selected based on Fu et al. [21], while (h,i) were selected considering Ciros-Pérez et al. [22] recommendations.

Initially, 5 morphometric characters ("a", "b", "c", "d" and "e") of 48 parthenogenetic females were measured at the end of each rotifer's lifespan. Moreover, from the small-scale culture, the individual production parameters from each rotifer were registered by removing the newly hatched rotifer from the plate with a small, very fine-tipped, heat-treated Pasteur pipette and replenishing the removed aliquot. We registered longevity (number of days until death), progeny (number of offspring), egg production (number of viable and non-viable eggs), maximum load of eggs (the maximum number of eggs carried by a female during her lifetime), pre-reproductive, reproductive, and post-reproductive age (periods before, during and after the reproductive stage, respectively). This production data were associated with the biometric parameters "a", "b," and "c", by the Principal Component Analysis (PCA) using the PRIMER-e version 5.0 (PRIMER-E Ltd., Plymouth, UK; https://www.primer-e.com/, accessed on 10 December 2021) software. To describe reliable size confidence intervals, we measured 5 morphometric characters (from "a" to "e") in 120 parthenogenetic females (including the previous 48 rotifers considered for PCA) of IMP-BG Z010-VL (up to 48 measures for "d" and "e"), and in 125 rotifers of IMP-BG Z018-SD.

In addition, the 9 parameters were randomly measured, regardless of age, in other 60 parthenogenetic females of both strains (IMP-BG Z010-VL and IMP-BG Z018-SD) isolated in this study and compared against 60 rotifers of 2 reference strains maintained in culture *B. plicatilis* s.s. L-size and *B. rotundiformis* SS-size. These last 240 measurements were used to statistically compare their morphology by Canonical Discriminant Analysis (CDA) using the IBM SPSS Statistics software for Windows Inc. version 22 (IBM Corp. Released, 2013 (Armonk, NY, USA); https://www.ibm.com/analytics/spss-statistics-software, accessed on 10 December 2021). The functions were obtained by stepwise discriminant analysis performed on each lorica measurement from 4 strains, and the data were transformed considering the natural logarithm (Ln).

2.3. Molecular Analysis

For both strains, organisms from parthenogenetic-originated monocultures were collected, and DNA extraction was conducted from each individual, as well as in pools, in 0.2 mL microtubes, using the HotSHOT method based on Montero-Pau et al. [23], with slight modifications in alkaline volume (25 μL) and neutralizing (25 μL) solutions, and with incubation on an Eppendorf MixMate microplate shaker at 800 rpm, 95 °C for 40 min. A small number of males was obtained randomly during the measurements of the females and also considered for the molecular analysis.

For species identification, the mtDNA COI gene and the nuclear rDNA ITS1 region were analyzed. The COI gene was amplified using primers ZplankF1_t1/ZplankR1_t1, while ITS1 with primers III/VIII, indicated in Table 1. PCR reactions were performed using the HotStartTaq Plus Master Mix kit (QIAGEN), with final concentrations of 0.2 μM of each primer, 2 mM $MgCl_2$, and 1.5–3 μL of template DNA, in 10 μL of final reaction volume. The thermal cycling conditions considered were an initial denaturation of 95 °C for 5 min, followed by 36 cycles of 95 °C for 40 s, 45 °C (COI) or 54 °C (ITS1) for 50 s, and 72 °C for 1 min (COI) or 45 s (ITS1), with a final extension of 72 °C for 7 min. All reactions were evaluated by electrophoresis in 1% agarose gels, amplified products with the expected size were purified with the AccuPrepPCR Purification kit (Bioneer), and bidirectionally sequenced in an ABI 3500 genetic analyzer (Applied Biosystems Inc. Foster City, CA, USA). Electropherograms were manually edited using Chromas 2.6.6 (Technelysium Pty Ltd., South Brisbane, QLD, Australia), sequences were aligned using MUSCLE [24] from MEGA 7.026 [25] and trimmed to 683 bp for COI sequences and 557 bp (VL 543 bp plus indels) for ITS1. All consensus sequences (n = 9 for each marker and strain) obtained in this study were registered in GenBank, with accession numbers for strains IMP-BG-Z018-SD (COI: MK534737 and MZ662909-MZ662916; ITS1: MZ569584 and MZ695037-MZ695044); and IMP-BG-Z010-VL (COI: MK534738 and MZ662901-MZ662908; ITS1: MZ569507 and MZ695046-MZ695053) detailed in Table S1. In addition, some isolates of reference strains used in this study were selected, and the COI gene was sequenced for the confirmation of species *B. plicatilis* s.s. L1-size (accession numbers OL700039–OL700040) and *B. rotundiformis* SS-size (accession numbers OL700041-OL700042).

Table 1. Primers used for the amplification of mtDNA COI gene and rDNA ITS1 region of two Peruvian strains (IMP-BG-Z018-SD and IMP-BG-Z018-VL) of genus *Brachionus* isolated from impacted coastal wetlands. (*) indicates primers used for COI gene sequencing.

Marker	Primer	Sequence (5′-3′)	Size (pb)	Reference
COI	ZplankF1_t1	TGTAAAACGACGGCCAGTTCTASWAATCATAARGATATTGG	~700	[26]
	ZplankR1_t1	CAGGAAACAGCTATGACTTCAGGRTGRCCRAARAATCA		
	* M13F	TGTAAAACGACGGCCAGT		[27]
	* M13R	CAGGAAACAGCTATGAC		
ITS1	III	CACACCGCCCGTCGCTACTACCGATTG	~560	[28]
	VIII	GTGCGTTCGAAGTGTCGATGATCAA		

For comparative purposes, all COI and ITS1 sequences of *Brachionus plicatilis s.l.*, available in GenBank and BOLD databases were retrieved (avoiding the selection of misidentified sequences, with a small number of bp, a high number of gaps, or ambiguous nucleotide specifications) and aligned with those obtained in this study. A total of 256 of COI (569 bp), 197 of ITS1 (370 bp) and 197 concatenated COI+ITS1 sequences (948 bp). were analyzed. Each concatenated sequence was derived from the same organism. Parameters, including nucleotide composition, the conserved, variable, and parsimony informative (PI) sites, were calculated in MEGA 7.026 [25]. In addition, p-distances were calculated for all pairwise comparisons of taxa, with 1000 bootstrap replicates.

The phylogenetic relationship was reconstructed using the Maximum Likelihood (ML) and Bayesian Inference (BI) methods, with COI and COI+ITS1 sequences. *Brachionus calyciflorus* (KC431011 for COI, and KC431009 for ITS1) from Brazil was included as an outgroup.

For ML, the best substitution models were determined using the SMS algorithm [29] with optimized frequency balance and 1000 permutations. The ML analysis was performed with PHYML 3.0 [30], considering GTR + G (1.402) + I (0.555) for COI, and GTR + G (0.656) + I (0.465) for COI+ITS1. The BI analysis was carried out with MrBayes v3.2.6 x86 [31]; the Markov Chain Monte Carlo (MCMC) algorithm was run for COI, 16 million generations and 10 million for COI+ITS, and trees were sampled at intervals of 1000 generations. The first 25% of the generations (4000 trees for COI and 2500 for COI+ITS) were discarded as the burn-in phase, and a consensus tree was constructed summarizing the branching patterns of the remaining trees (12,000 for COI and 7500 for COI+ITS) and a majority rule equal to 50%. The convergence diagnostic showed all effective sample size (ESS) values greater than 1000 and the average potential scale reduction factor (PSRF+) parameters equal to 1. For the COI-based tree, the average standard deviation of the split frequencies was 0.005477, and the maximum was 0.063693. For the COI+ITS-based tree, the average standard deviation of the split frequencies was 0.005832 and the maximum was 0.052696.

3. Results

3.1. Taxonomy

- Strain IMP-BG Z010-VL
- Class Eurotatoria De Ridder, 1957
- Subclass Monogononta Plate, 1889
- Superorder Pseudotrocha Kutikova, 1970
- Order Ploima Hudson & Gosse, 1886
- Family Brachionidae Ehrenberg, 1838
- Genus *Brachionus* Pallas, 1766
- Species *Brachionus paranguensis* Guerrero-Jiménez, 2019
- Sub species *Brachionus paranguensis ventanillensis* subsp. nov.

- Strain IMP-BG Z018-SD
- Class Eurotatoria De Ridder, 1957
- Subclass Monogononta Plate, 1889
- Superorder Pseudotrocha Kutikova, 1970
- Order Ploima Hudson & Gosse, 1886
- Family Brachionidae Ehrenberg, 1838
- Genus *Brachionus* Pallas, 1766
- Species *Brachionus koreanus* Hwang, 2013
- Sub species *Brachionus koreanus santodomingensis* subsp. nov.

3.2. Etymology

Based on morphological and molecular analysis, the species under study have been nominated as subspecies; although it is a new category and not yet valid, it is necessary for aquaculture traceability. Strains were *Brachionus paranguensis ventanillensis* subsp. nov. (with holotype and paratypes record MUSM-PL 2021-0031) and *Brachionus koreanus santodomingensis* subsp. nov. (with holotype and paratypes record MUSM-PL 2021-0032). Both were proposed to highlight their origins since the urban sprawl growth is increasing, and these water bodies may disappear in the next few decades. The holotypes and paratypes were deposited in the Natural History Museum of the Universidad Nacional Mayor de San Marcos.

3.3. Morphological Differences of Peruvian Strains

A total of 180 parthenogenetic females from IMP-BG Z010-VL and IMP-BG Z018-SD, were examined and measured to describe their morphology.

3.3.1. Strain IMP-BG Z010-VL

Description

Strain with morphology similar to *B. paranguensis* L4, according to Guerrero-Jiménez et al. [5]. Organisms with ventral and dorsal plates fused dorsally and laterally, one triangular reddish-brown cerebroid ocellus (Figure 3b), long sensory bristles, corona with two typical concentric rings of cilia, first on the upper part trochus, formed by four bands surrounded by cirrus and cingulum on the lower part near the dorsal antenna (Figure 3c). Dorsal margin of the lorica with three pairs of spines framing the U-shaped sinus (Figure 3a), two gastric glands (Figure 3b), and two emerging lateral antennas from a lateral pore, each one placed near the widest part of the lorica (Figure 3d). Biometric results (expressed in μm) are detailed in Tables S2–S4; while productive parameters are in Table S5.

Type Locality

The parthenogenetic females were collected from the Ventanilla municipal wetland in Callao, Perú (S. 11°52′16.11″; W. 77°08′17.88″).

Differential Diagnosis

Lorica were similar to *B. plicatilis* s.s. "L1", however, statistically, the differences were enough to place it in another group. Although it is genetically related to *B. paranguensis*, one difference was found by its inconspicuous orange peel-like surface of the lorica. The anterior ventral margin of the lorica with two pairs of rounded lobules was located on both sides of the wide sinus (Figure 3a). The inner lobes with a wider base than the outer one. The toes (spurs/pedal glands) emerged completely from the foot terminal (apical) part (Figure 3e), which explains why these organisms were usually found attached to the walls of beakers, particularly during any stress due to vessel changes. The egg presented a diagonal incision on the bottom (Figure 3f). In contrast to other species such as *B. plicatilis* s.s., *B. rotundiformis* and *B. koreanus* maintained in culture for morphometric measurements in this essay, Z010-VL swam very superficially until many individuals perished, trapped by the surface tension of the water.

Trophi

The basic archetype, also called "Mallei" or "Maleate type", corresponded to the genus *Brachionus*. Hollow fulcrum, short, and truncated pyramid shape. Satellites form an irregular quadrilateral and anterior processes with rough edges in ventral view (Figure 3g). Rami with two posterior asymmetrical projections, the left one ending in a point and narrower than the right one. Basifenestras were circular and similar. Unci with four long teeth and brush-shaped subuncus. Flattened manubrium in the form of hollowed planes, each with projections towards the center and three tunnel-like cavities directed at the distally and dorsally folded ends, articulation presented its processes with a half-elliptical shape. The joints of the manubrium presented, in dorsal view, rougher ornamentation with more marked lobular bifurcation than the Z018-SD strain (Figure 3g,h).

3.3.2. Strain IMP-BG Z018-SD

Description

Strain with morphology similar to *B. koreanus* SM2 [32]. Organisms with ventral and dorsal plates fused dorsally and laterally, one brown cerebroid ocellus, two gastric glands (Figure 4b), sensory bristles, corona with two typical concentric rings of cilia, the first one being in the upper part trochus, formed by four bands surrounded by cirrus and cingulum on the lower part near the dorsal antenna. Pear-shaped and smoothed surface of the lorica, three pairs of spines that were triangular and dissimilar in length and width; the lateral and central spines are longer than the medial one, lateral spines have sigmoid outer margins (Figure 4a). U-shaped sinus was narrower than strain IMP-BG Z010-VL (Figure 4c). Biometric results (in μm) and productive parameters are reported in Tables S2–S5.

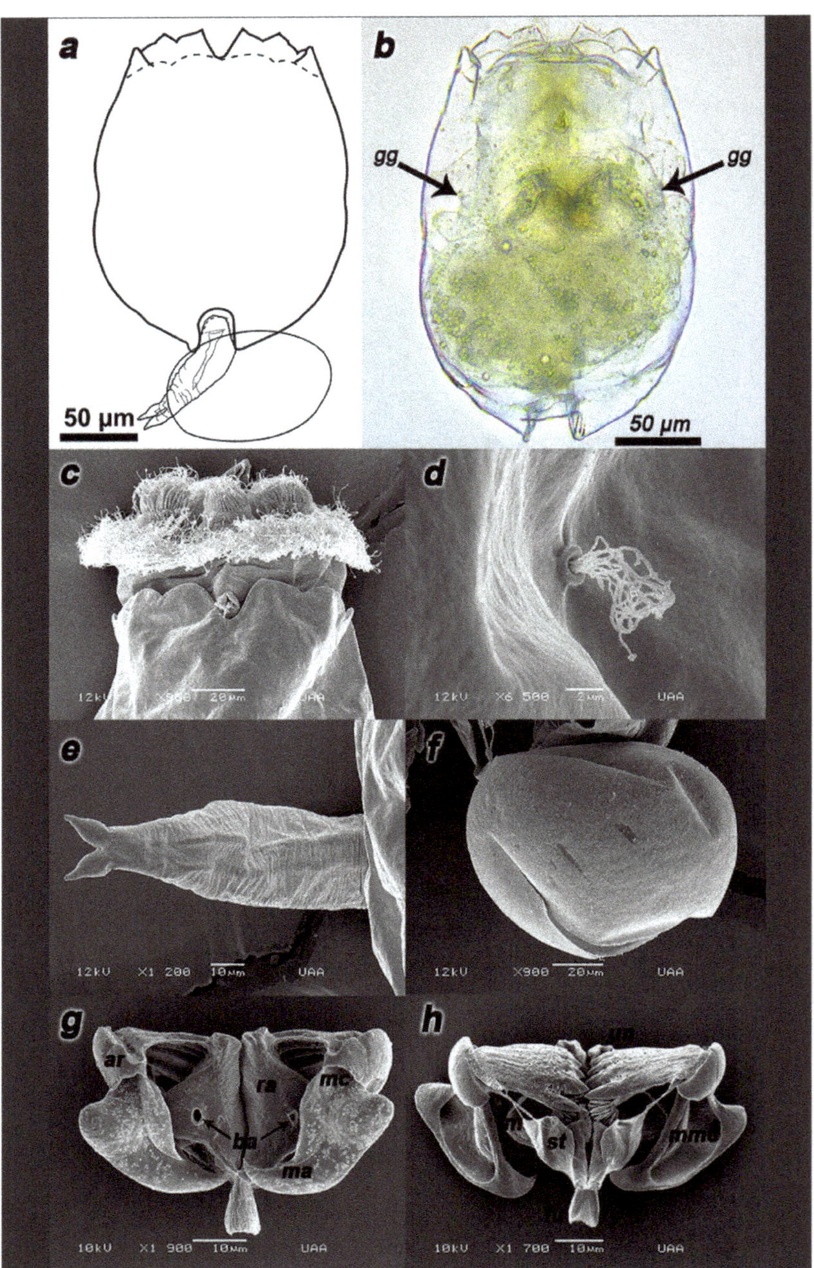

Figure 3. Description of the parthenogenetic female strain IMP-BG Z010-VL. Illustration of the strain (**a**) and its habitus (**b**). SEM microphotographs of the dorsal spines (**c**), pore with lateral antenna (**d**), foot (**e**), parthenogenetic egg (**f**), trophi dorsal view (**g**) and the trophi ventral view (**h**). Arrows indicate gastric gland (gg) structures. Components of trophi are indicated in cursive letters: membrane (*m*), manubrium (*ma*), manubrium middle crest (*mmc*), satellites (*st*), uncus (*un*), articulation of manubrium (*ar*), basifenestras (*ba*); fulcrum (*fu*), manubrium cavities (*mc*) and rami (*ra*). *n* = 20 rotifers.

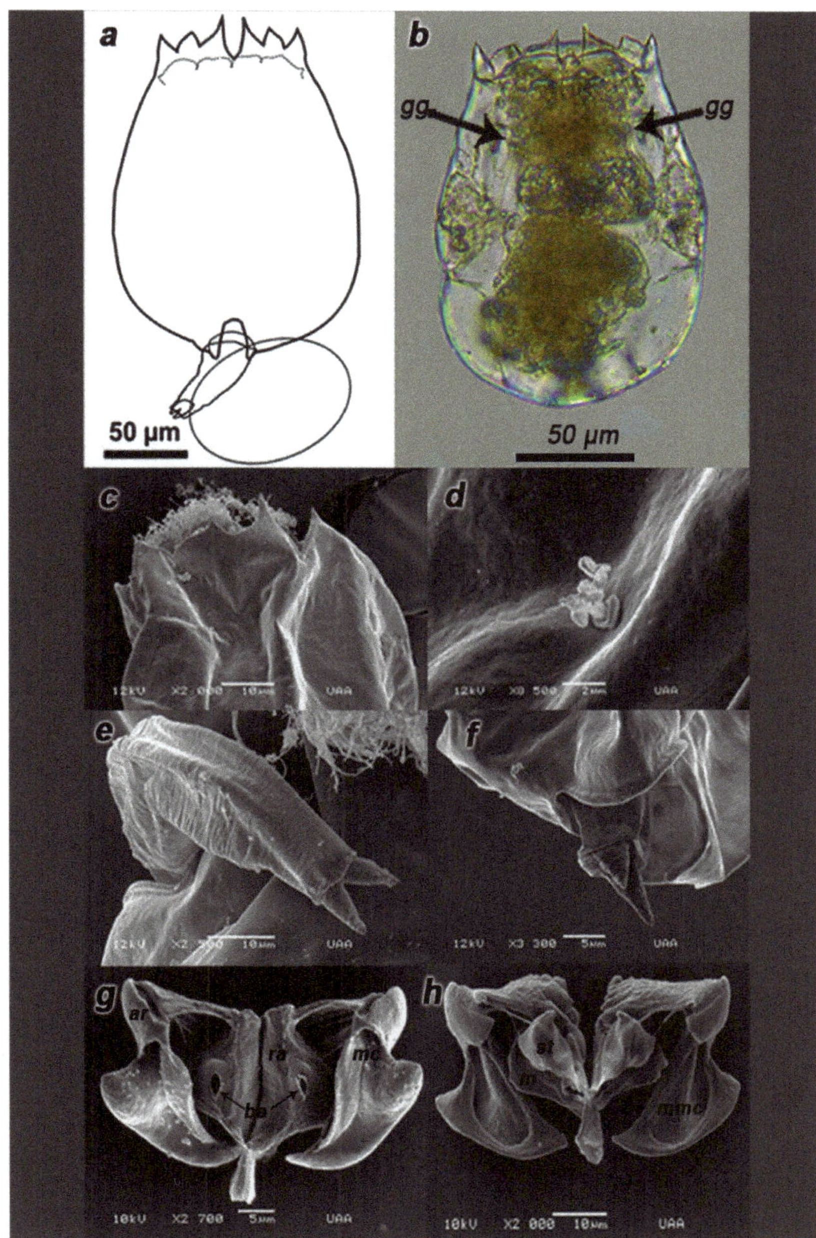

Figure 4. Description of the parthenogenetic female strain IMP-BG Z018-SD. Illustration of the strain (**a**), and its habitus (**b**). SEM microphotographs of the dorsal spines (**c**), pore with lateral antenna (**d**), foot (**e**), foot aperture (**f**), trophi dorsal view (**g**), and the trophi ventral view (**h**). Arrows indicate gastric gland (gg) structures. Components of trophi are indicated in cursive letters: membrane (*m*), manubrium (*ma*), manubrium middle crest (*mmc*), satellites (*st*), uncus (*un*), articulation of manubrium (*ar*), basifenestras (*ba*); fulcrum (*fu*), manubrium cavities (*mc*) and rami (*ra*). *n* = 20 rotifers.

Type Locality

The parthenogenetic females were collected from the Santo Domingo relict wetland, in Paracas, Ica, Peru (S 13°51′25″; W 76°15′11″).

Differential Diagnosis

The anterior ventral margin of the lorica with two pairs of rounded lobules was located on both sides of the slender sinus. Inner lobes showed a narrower base than the outer ones. Two emerging lateral antennas from a lateral pore, each placed near the widest part of the lorica (Figure 4d). The toes (spurs/pedal glands) did not emerge completely from the foot terminal part, which explains why these organisms were not usually found stuck to the wall of the beaker (Figure 4e,f).

Trophi

Trophi with Mallei archetype presented. Hollow fulcrum, short, and truncated pyramid-shape. Satellites form irregular quadrilateral, anterior processes with rough edges in ventral view. (Figure 4g). Rami with two posterior asymmetrical projections, the left one ending in a point to the right one, this structure was similar but thinner as opposed to IMP-BG Z010-VL. Basifenestras were irregular and oval. Uncus with four long teeth and brush-shaped subuncus. Flattened manubrium in the form of hollowed planes, each with projections towards the center and three tunnel-like cavities directed at the distally and dorsally folded ends, articulation presented its processes with an irregular triangular shape. The joints of the manubrium had, in dorsal view, smoother ornamentation with softer lobular bifurcation than the IMP-BG Z010-VL strain (Figure 4g,h).

3.4. Relationship between Biometric and Production Parameters

The first two principal components (PC1 and PC2) explained the 88.3% and 94.6% accumulated variation for strains IMP-BG Z010-VL and IMP-BG Z018-SD, respectively. Both PCA determined that the lorica length was positively associated with longevity, progeny, egg production, and reproductive age, while the width and aperture of the lorica (DbLS) were slightly associated with the maximum load of eggs (Figures 5 and 6).

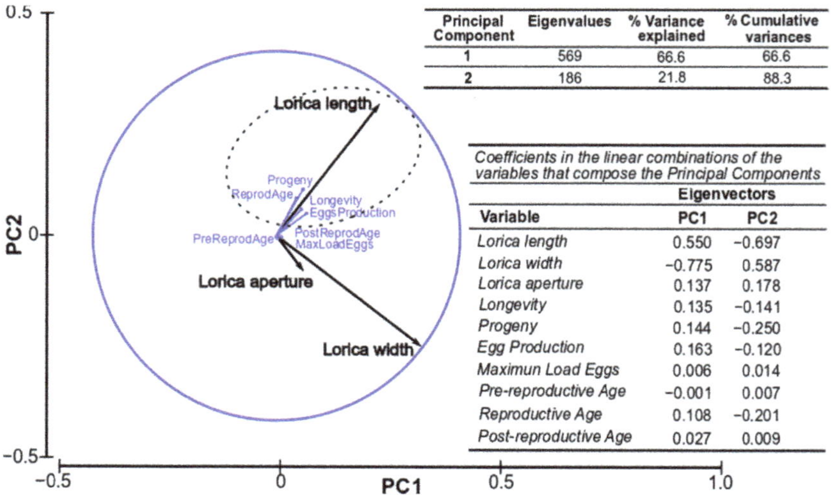

Figure 5. Representation of the studied production and biometric parameters of the Ventanilla Strain (Z010-VL) along the first two axes (PC1 and PC2) of the PCA. (n = 48 rotifers).

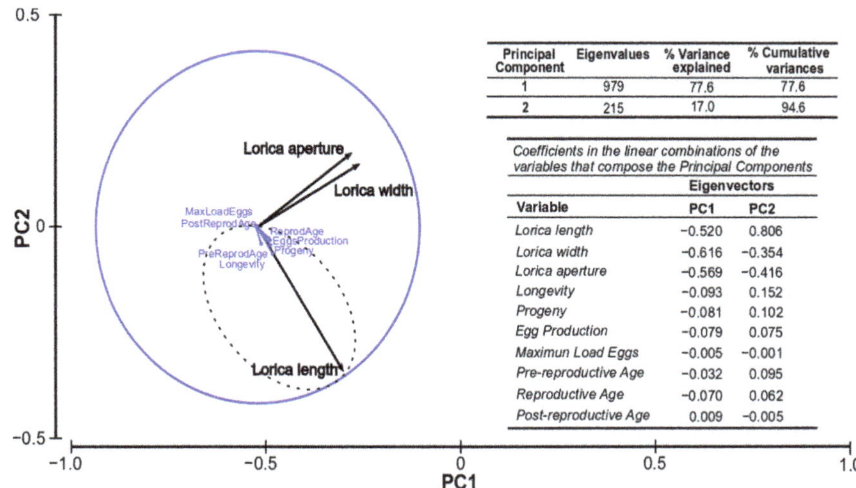

Figure 6. Representation of the studied production and biometric parameters of the Santo Domingo Strain (Z018-SD) along the first two axes (PC1 and PC2) of the PCA. (n = 48 rotifers).

3.5. Canonical Discriminant Analysis

According to the 96.8% (Table 2) accumulated variance from the first two canonical functions, statistical differences were reported on the four strains (*B. rotundiformis* SS-size, *B. plicatilis* s.s. L1-size, and the two strains from this study) analyzed. Only one value of *B. plicatilis* s.s. L1-size strain corresponded to IMP-BG Z010-VL. For this reason, 99.6% of original grouped cases were correctly classified. Furthermore, it is interesting to mention that strain Z010-VL was discriminated from *B. plicatilis* s.s. L1 when functions 1 and 2 were considered (Figure 7). Thus, the Brachionus strains were clearly grouped statistically in four different clusters. The higher within-group correlation coefficient in function 1 was lorica length (a), approximately duplicating the correlation values of the distance between lateral spines (b) and the head aperture (h); the same proportion of correlations gradually decreasing with dorsal sinus depth (e), lorica width (c), and lateral spine length (i), meanwhile within function 2 the distance between central spines (d), dorsal sinus depth (e), the distance between central and medial spines (f), and medial spine length (g) had the highest correlations (Table 3).

Table 2. Eigenvalues and variances of each function. The first two canonical discriminant functions were used in the analysis.

Function	Eigenvalues	Variance (%)	Accumulated Variance (%)
1	28.259	90.6	90.6
2	1.916	6.1	96.8
3	1.008	3.2	100.0

3.6. Molecular Taxonomy

The COI sequences obtained in this study showed a nucleotide frequency of T (42.2%), C (17.7%), A (22.3%) and G (17.9%), while for ITS1 region was T (33.2), C (16.3%), A (30.2%) and G (20.4%). Nucleotide frequencies of L4 and SM2 from different origins are detailed in Tables S6 and S7.

Considering 61 mtDNA COI sequences (570 bp) of L4 group organisms from 3 lineages of 8 countries (including those from the Peruvian strain Z010-VL), 108 PI sites were registered, and no singletons identified. In particular, 11 PI were recorded when we compared sequences from Peru and Chile (Table S4), most being due to differences between Chile sequences. In addition, when we compared 30 sequences from the SM2 group (8 countries, including those from the Peruvian strain Z018-SD), we registered 105 variable sites (89 PI). Sequences from VL and SD Peruvian strains showed 103 variable sites (PI) and no differences within each group.

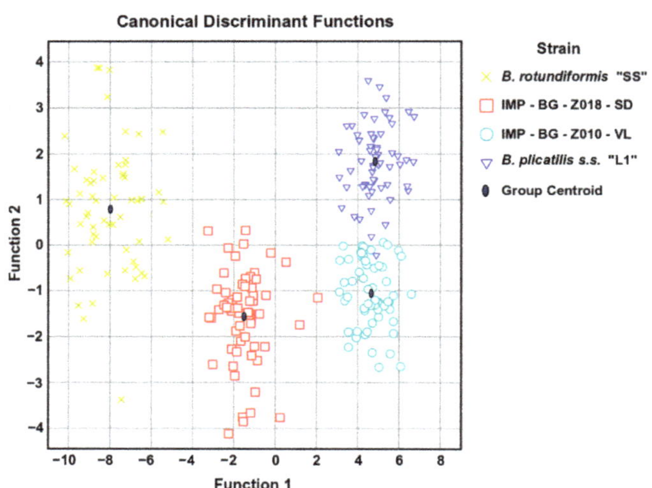

Figure 7. Scatter plot of each measured *Brachionus* defined by the first two canonical discriminant functions.

Table 3. Stepwise discriminant analyses of the body measurements for the first two canonical functions (Function 1 and Function 2). Coefficient: represents the standardized coefficient for the canonical discriminant function. Correlation: represents the pooled within-group correlation coefficient between the body measurements and the canonical discriminant function. No lorica measurement was excluded in the stepwise analyses. Measurements are presented in natural log (Ln).

Measurement (in Ln)	Function 1		Function 2	
	Coefficient	Correlation	Coefficient	Correlation
(a) Lorica length	1.236	0.857	0.092	0.195
(b) Lorica aperture	−0.054	0.43	0.01	−0.08
(c) Lorica width	−0.64	0.252	−0.452	0.079
(d) Distance between central spines	0.054	0.022	0.564	0.69
(e) Dorsal sinus depth	−0.131	0.266	0.255	0.542
(f) Distance between central and medial spines	0.002	−0.015	−0.232	0.498
(g) Medial spine length	−0.26	0.112	0.58	0.46
(h) Head aperture	−0.122	0.432	0.28	−0.034
(i) Lateral spine length	0.539	0.237	−0.038	0.059

On the other hand, less variability was observed when we compared ITS1 sequences from L4 and SM2 groups of different countries (along 334 positions including indels), Along 44 sequences of L4 group (from 6 countries), only one PI site (in position 328 nt, from Chile) was observed, while along 30 sequences of SM2 group (8 countries), we recorded 7 variable sites (5 PI). In addition, 51 PI sites were observed when we compared all sequences from VL and SD from L4 and SM2, respectively; and no differences were observed comparing sequences within each Peruvian strain.

Generally, higher values of uncorrected *p*-distances were registered using COI gene sequences than with ITS1. Genetic distances between COI sequences of IMP-BG Z010-VL and the L4 complex ranged from 0.9 to 13.1%, while no differences were observed with ITS1. The lowest distances were observed when comparing this Peruvian strain with sequences from Chile (0.9–1.1%), which also were grouped in the same clade after phylogenetic reconstruction; and above 10% distance was observed compared with North America, Australasia, and Europe. IMP-BG Z010-VL and the Mexican holotype *B. paranguensis* showed 10.3% of genetic distance (Table 4). On the other hand, comparing IMP-BG Z018-SD against the other countries of complex SM2, *p*-distances ranged from 4.5 to 14.6% with COI, and from 0.3 to 1.2% with ITS1. Less genetic distance was observed comparing COI sequences of this Peruvian strain with Spain and USA (4.5–6%); whereas higher *p*-distances (up to 12–14%) were observed with Asia, the Caribbean, and Europe. After phylogenetic reconstruction, IMP-BG Z018-SD sequences showed high similarity with those from Spain but ordered in an independent clade and disrupting with North America, Asia, the Caribbean, and Europe. The IMP-BG Z018-SD strain and the Korean holotype *B. koreanus* showed 6.2–12.2% (COI) and 0.9% (ITS1) of genetic distances (Table 5).

Table 4. Uncorrected *p*-distance between strain IMP-BG Z010-VL and other L4 haplotypes from different countries, calculated using COI (569 bp) and ITS1 (371 bp) regions. * Corresponds to the holotype *B. paranguensis*.

Marker	Origin of Haplotypes					
	(This Study)	Chile	* Mexico	China–USA	Australia–Japan–USA	France
COI	Peru (Z010-VL)	0.009–0.011	0.103	0.103	0.131	0.131
ITS1	Peru (Z010-VL)	0	0	0	0	0

Table 5. Uncorrected *p*-distance between strain IMP-BG Z018-VL and other SM2 haplotypes from different countries, calculated using COI (569 bp) and ITS1 (371 bp) regions. * Corresponds to the holotype *B. koreanus*.

Marker	Origin of Haplotypes						
	(This Study)	Spain	USA	Philippines–* Korea	Cayman Islands	Turkey	Italy
COI	Peru (Z018-SD)	0.045–0.056	0.056–0.060	0.062–0.122	0.060–0.124	0.124	0.146
ITS1	Peru (Z018-SD)	0.003–0.009	0.009	0–0.003	0.009	0.003–0.009	0.012

It is important to mention that, based only on the mtDNA COI gene, one new lineage was observed for *B. rotundiformis* (SS) from Hawaii, and another new one for *Brachionus* sp. (SM3) from Iran; while with ML analysis, lineages were not clear in L1 for *Brachionus plicatilis s.s.* (Figure 8). The BI tree (Figure 9) showed the presence of two additional aquaculture production lineages, discriminating a total of 33 lineages.

The ML and BI phylogenetic analysis based on concatenated COI+ITS1 clearly showed the presence of three clades related to the morphotypes L (L1 to L4), SS and SM (SM1 to SM7). It was also possible to discriminate the presence of 31 new aquaculture production lineages with a clear clustering related to origin in most cases. In this sense, L1 was integrated by six lineages proposed for *B. plicatilis* s.s., L2 by three for *B. manjavacas*, L3 by four lineages for *B. asplanchnoidis*, L4 by three for *B. paranguensis*, SS by four lineages for *B. rotundiformis*, SM1 by one for *B. ibericus*, SM2 by four for *B. koreanus*, SM3 by two for *Brachionus* sp., SM4 by one for *Brachionus* sp., SM5 by one for *Brachionus* sp.; SM6 by one for *Brachionus* sp. and SM7 integrated by one lineage for *Brachionus* sp. (Figures 10 and 11).

The Peruvian strain IMP-BG-Z010-VL *Brachionus paranguensis ventanillensis* subsp. nov. was clustered in the L4 group, integrated by lineage from Peru-Chile (n = 36 sequences). In addition, another lineage of L4 was formed by Mexico, USA, and China (n = 11), and a third collapsed group by France, Australia, Japan, and the USA (n = 14). Meanwhile the strain IMP-BG-Z018–SD *Brachionus koreanus santodomingensis* subsp. nov. was clustered in SM2 forming one lineage, discriminated from the other three lineages from SM2 formed by Spain (n = 3), South Korea-Turkey-Italy (n = 5), and USA-South Korea-Spain-Philippines-Cayman Islands (n = 14).

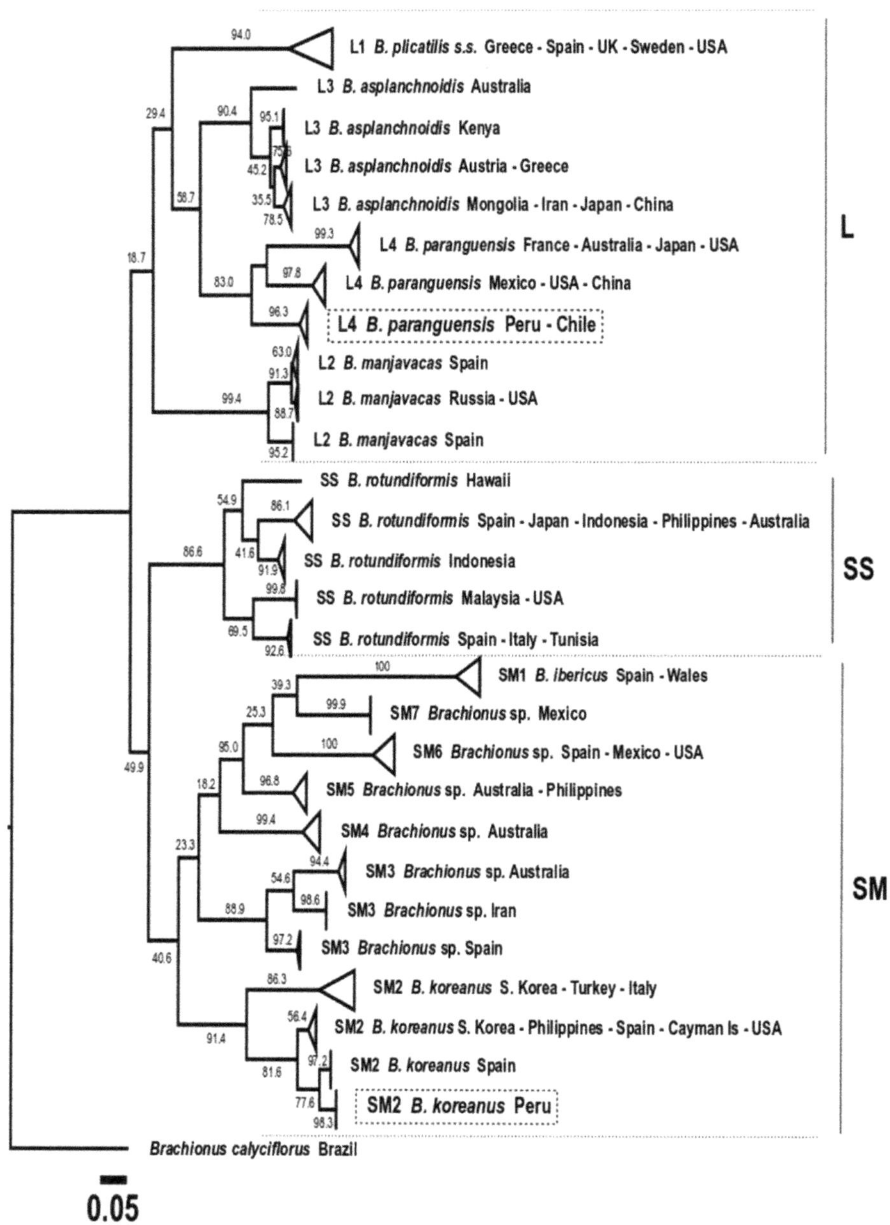

Figure 8. Phylogenetic tree obtained with the Maximum likelihood method based on 256 COI gene sequences (569 bp). A total of 28 production lineages were discriminated. The three morphotypes (L, SS, and SM) are indicated. Boxes in dotted lines indicate the Peruvian strain sequences obtained in this study.

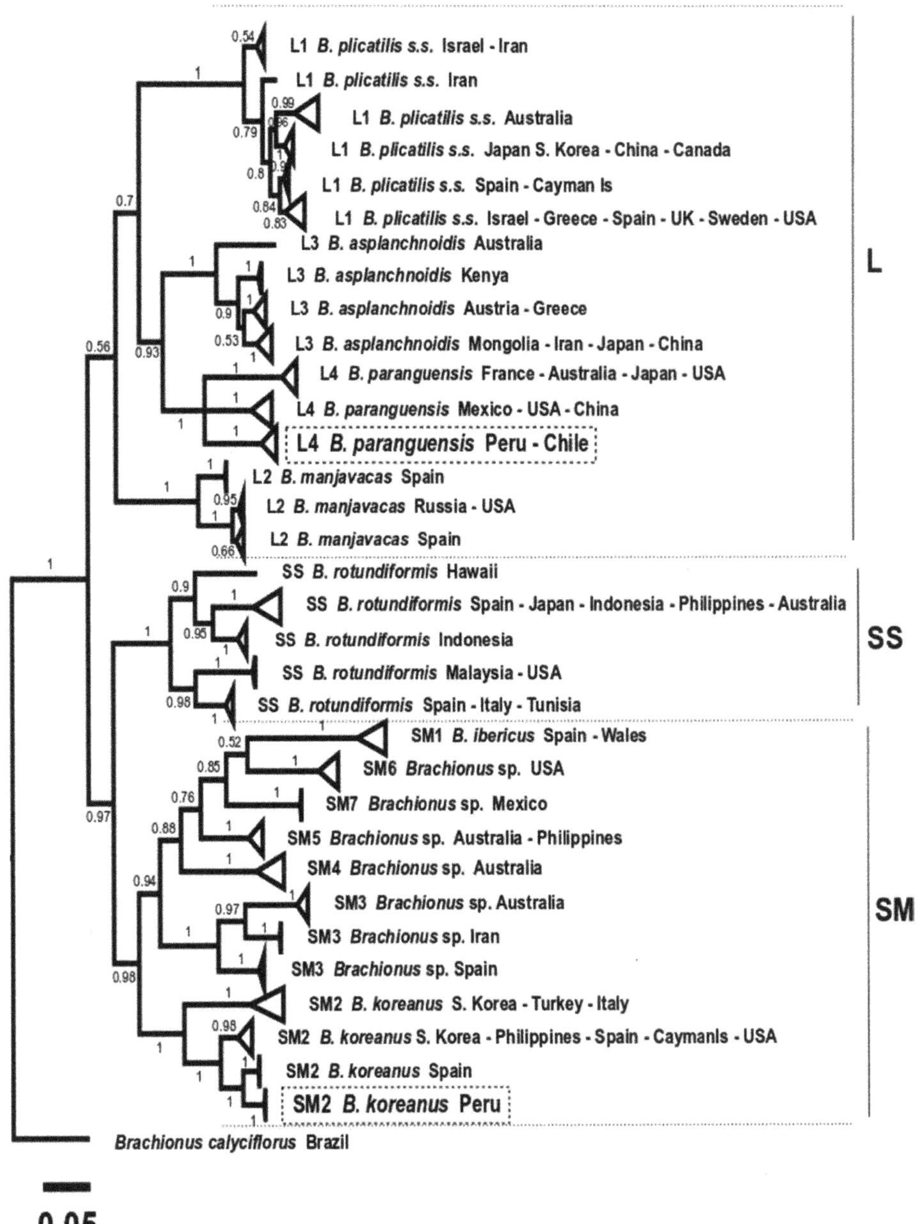

Figure 9. Phylogenetic tree obtained with the Bayesian Inference method based on 256 COI gene sequences (569 bp). A total of 33 production lineages are observed. The three morphotypes (L, SS, and SM) are indicated. Boxes in dotted lines indicate the Peruvian strain sequences obtained in this study.

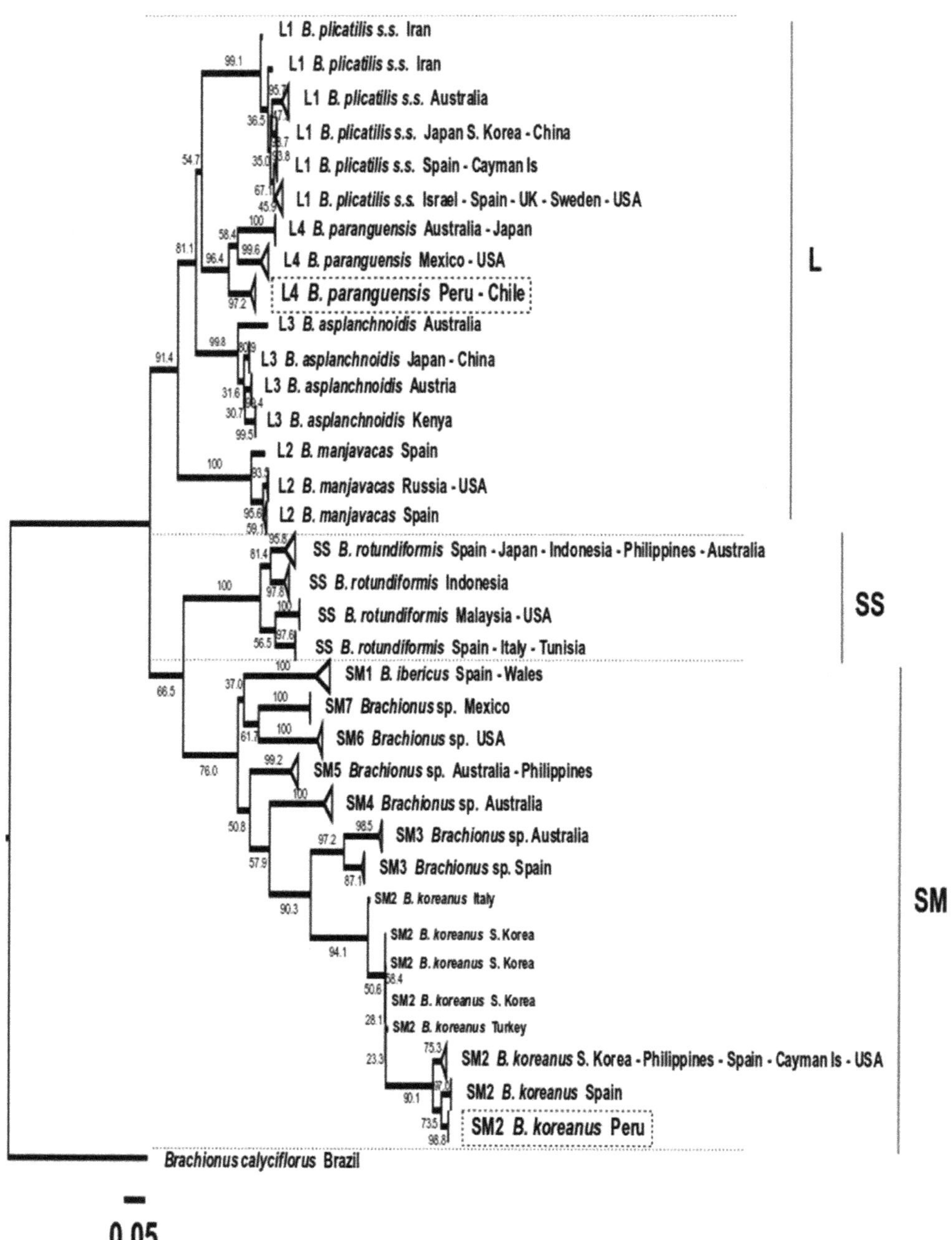

Figure 10. Phylogenetic tree obtained with the Maximum likelihood method based on 197 concatenated COI + ITS1 sequences, observing the discrimination of 31 production lineages. The three morphotypes (L, SS and SM) are indicated. Boxes in dotted lines indicate the Peruvian strain sequences obtained in this study.

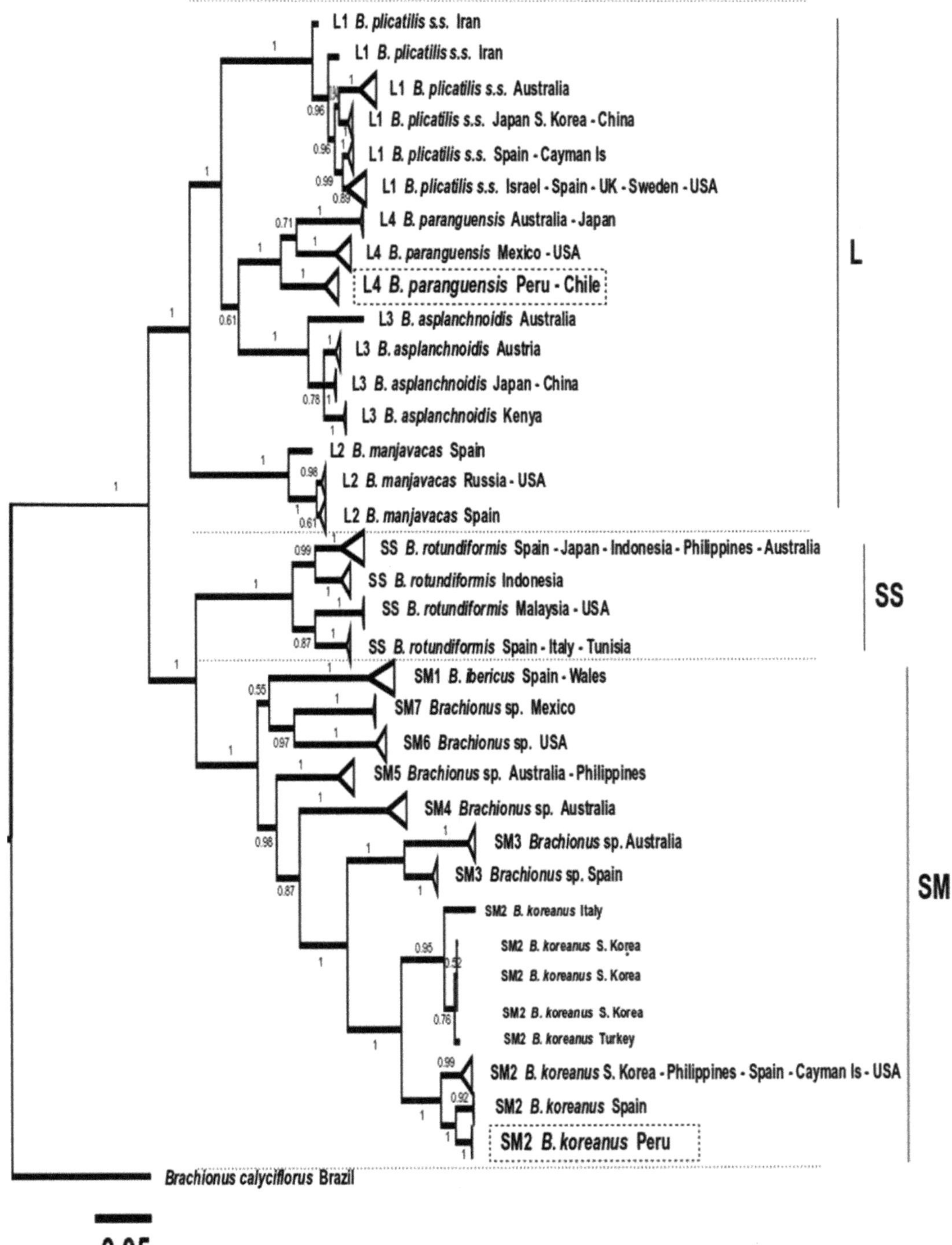

Figure 11. Phylogenetic tree obtained with the Bayesian Inference method based on 197 concatenated COI + ITS1 sequences (948 bp), with 31 production lineages discriminated. The three morphotypes (L, SS, and SM) are indicated. Boxes in dotted lines indicate the Peruvian strain sequences obtained in this study.

A disadvantage when analyzing lineages using concatenated COI + ITS1 was that 59 were less used than when the analysis was conducted with COI sequences, and thus samples from Hawaii (SS) and Iran (SM3) were absent, consequently showing a not well-discriminated clade. However, six subspecies groups were observed for the species *B. plicatilis* s.s.-L1.

4. Discussion

This study shows a morphological and molecular description of two Peruvian strains proposed here as two new subspecies, *Brachionus paranguensis ventanillensis* subsp. nov. and *B. koreanus santodomingensis* subsp. nov. These strains are considered with a potential in aquaculture, particularly in larviculture, due to the production parameters and the resilience in live food handling, also considering the habitat from which they were isolated since desert environments tend to have high levels of contamination and desiccation. In effect, taking into account a relationship between productivity and biometrics fosters value-added economic prospects to local strains that could be promoted, as well as the conservation of lineages in terms of productivity, traceability, and their intrinsic potential as new resources. Thus, the hypothesis of a delimitation subspecies is complemented with molecular (COI and COI + ITS1 regions) marker analysis. In this sense, it is important to avoid the generalization of productive parameters for all the species or genus is avoided [33] because those may be an exclusive evolutionary product by the parapatric divergence of organisms isolated from remote water bodies [34].

Morphological variability throughout ontogenetic changes is important to be considered for the characterization and selection of parameters related to productivity. Thus, contrary to different reported 72 h essays with measurements of juvenile forms in some groups [5,22,32] and even in some Asian species where they reached their maximum size after 40 h [35], here, following nine measurements recorded in four different strains of *Brachionus*, it was possible to include descriptions. In addition, with ontogenetic changes from gravid females, and also considered their standard deviation to obtain a 99% confidence interval, bearing in mind that strain Z018-SD becomes gravid the fourth or fifth day onwards (Table S5).

Due to the results obtained in this study, we propose the use of the lorica measurements ("d" to "g") as excellent parameters for successful morphometry discrimination within the genus *Brachionus*. Thus, the selected measurements (among the nine evaluated) statistically discerned four different groups (99.6% of these cases), mainly observing a taxonomic discriminant value between L1 and L4. Although the data were previously transformed to a natural logarithm to minimize the dispersion, similar differences were found without any transformation (data not shown). In addition, another important aspect to mention is that statistical differences were registered as the number of females measured randomly (up to 60) increased. Some authors reported that the somatic maintenance of females is independent of reproduction under food limitations (chronic caloric restrictions). Despite the fact that individual growth may be altered, biometrics should not be biased towards the first days of life in order to describe species but should ideally consider all measurements throughout the lifespan [36].

Phylogenetic analysis demonstrated the existence of the two new putative Peruvian subspecies, mainly based on the mtDNA COI gene analysis (p-distances > 3%), whilst rDNA ITS1 (p-distances < 1.2%) rigidly grouped them together within SM2 or L4 species. These results were consistent with previous studies, where a greater number of terminal taxa were resolved with COI-based trees compared to ITS1-based within *B. plicatilis s.l.* [4,37,38] and *B. quadridentatus* [39] and others monogonont rotifer species [40]. It has even been suggested that COI would not be appropriate for species delimitation due to the risk of an over-splitting [4], but in an intraspecific perspective, the greater variability hosted in COI sequences could reveal evolutionarily important lineages, also supported by other data and regarding economic perspectives. In fact, the biometric analysis conducted on the Peruvian strains supported the presence of statistically discrete units due to the association of three morphological variables with production parameters. Thus, this confirmed the existence of the COI-based lineages with productive importance.

It is also noteworthy that tree topology was certainly influenced and modified depending on the molecular markers (COI or ITS1) used for the phylogenetic analysis of genus *Brachionus*, due to the low variability that ITS1 showed (variable and PI sites) compared to COI gene diversity across different *Brachionus* strains analyzed from different countries. Thus, for example, *Brachionus rotundiformis* was associated with SM group according to the COI gene analysis but was associated with L group when based on COI+ITS1. In addition, differences in the relationship between L4 and L3 were observed, COI tree topology showing them more related than with L1; while with COI+ITS1 were more related L4 and L1 than with L3. This discrepancy has also been previously reported in the phylogenetic reconstruction with ML [5], where, as in our study, higher node support values were consistently observed in the COI and COI+ITS1 tree. Conversely, Mills et al. [4] found a concordant topology between markers, supporting a closer relationship between L4 and L3 than between any of these with L1. The incongruence both in topology and level of estimated divergence between COI and ITS1, termed mitonuclear discordance, is a pervasive challenge when it is required to reconcile phylogenies singly obtained from mitochondrial and nuclear marker data sets [41]. Among the causes that generate these discrepancies are mitochondrial introgression, sex-biased dispersal, natural selection, Wolbachia-mediated genetic sweeps, incomplete lineage sorting, and unresolved phylogenetic polytomies [40,42,43], the two latter being suggested as the likely drivers of mitonuclear conflicts in some monogonont rotifer species [40].

As we mentioned previously, using only the ITS1 region may not be effective for phylogeographic studies of genus *Brachionus*, due to its remarkably conservative nature and evidence of a possible concerted evolution of these genes within the species complex [44]. Concerted evolution implies that homogenizing forces have promoted a high similarity between ITS1 sequences within the species, thus hindering the opportunity to recover bifurcation events in phylogenetic reconstructions [45,46].

Brachionus paranguensis ventanillensis subsp. nov.—L4 showed to have been structured in three clades (more detectable in the Bayesian than ML tree). One clade conformed by isolates from the central coast from Peru and the northern coast in Antofagasta, Chile [47], where the old migration relationship from South America to North America is located [34], continue with a volcanic area from Guanajuato, Mexico [5] and the state of Nevada, USA [4,48–50], as well as a further relationship, shown with Australasia, from Japan [4,48,51,52] and Australia [4]. We cannot explain the relationship of Chinese strains within the clade of North America or the strains from the USA within the clade of Australasia. The case of the French strains is very relevant, as well as subject to any reservations because these were also reported in Norwegian and Greek hatcheries, and their origin was mentioned by the countries where the hatcheries were located [4,53–55]. Papakostas et al. [53] investigated the species/biotype composition of *Brachionus* strains used in Europe to improve its culture conditions. One of his conclusions was that Norwegian strain "SINTEF" was the same as the Spanish strain "PL". However, our results showed these strains become a clone strain. The Norwegian strain "SINTEF" was a donation by IFREMER (France) in 1984 (Retain K.I., com.pers). They were first collected by Dr. Pourriot in 1974 in the south of Camargue [56] and were first mentioned in the work of Blanchot and Pourriot [57], naming it "*Brachionus plicatilis* GS74". Therefore, with this evidence, we considered France the country of origin for the strains SINTEF (DQ314559 and DQ314558), BEARC015, and BEARC016 (KU299273 and KU299274), and the Greek strain "K" (AM180752).

Meanwhile, *Brachionus koreanus santodomingensis* subsp. nov.—SM2 showed to have been structured in four different clades, without a clear phylogeographic relationship. One group conformed by Eurasia, Korea [32], Italy [4] and Turkey [58]; a second group was formed by Asia plus America and Europe, Korea [32], Philippines [4], Cayman

Islands [4,48], Spain [4] and USA [48]; a third group Europe, from Spain [4] and the fourth group from South America, in which Peru was included.

For *Brachionus* morphotypes analyzed in this study, we confirmed the possibility of differentiating clades related to geographic origin observed in this study based on COI and concatenated data, and we suggest including information on production parameters in the study of these groups as a future aspect for traceability of exchanges among hatcheries. Previous studies highlighted the importance of the intraspecific component in understanding the geographic patterns of biodiversity in microorganisms [59], but disentangling such geographic patterns below the species level, this sublevel could also be considered for an economic perspective when information on production parameters is added. For example, knowing the origin place of the production of lineages would improve the management of local strains by allowing traceability during their commercialization. Therefore, information on production parameters is an important aspect to be completed for the other worldwide production lineages complementing those two presented in this study in order to allow business traceability among hatcheries.

This new subspecific delimitation proposed using COI and COI+ITS1 markers could be understood, based on the economic delimitation of aquaculture production lineages, that despite a few morphological differences, this could be supported with production parameters. For example, it has been suggested to supplement taxonomy with the life cycle and the highest reproductive rate at the best salinity and temperature [60]. Those parameters must be the result of their adaptability to divergent habitats with different food, salinities, and temperature as limiting factors [61]. Kutikova [62] remarks behavioral patterns to suggest phylogeny within rotifers. However, the family *Brachionidae* preserves differences in food capture and types of body movements. Snell [63] conducted a complete revision about rotifer lifespan and reproduction, interpreting some production relationships with dietary restrictions and genetic expression.

Considering the above, a common pattern in this essay was observed in both local strains, where lorica length is associated with positive production parameters. However, this inference could be better represented with more robust statistical tests, such as CoInertia analysis (Table S8), also including more strains or subspecies. Apparently, this could happen because longer rotifers may be faster than shorter ones. This swimming speed range could be interpreted as a better chance to find food (i.e., motile algae) as well as better fitness to spread their eggs further, therefore increasing the likelihood of their progeny's survival. Whereas association of lorica width and aperture (DbLS) with a maximum load of eggs could be explained because all rotifers were measured after their complete lifespan. Thus, their bodies were influenced in their final form by these two width parameters, as carrying more eggs requires more strength to pull them, slowing them down. Korstad et al. [64] reported an inverse relationship between swimming speed and density (rotifers number per ml). In particular, swimming speed became slower in the stationary phase. In this way, we suggest that the corona needs to be stronger, and thus the trochus and cingulum must fill more space in its lorica. We have also observed that rotifers with more than two eggs had lesser hatching thereof, and these eggs were often carried joined downside the lorica near the foot. In this study, the unique case of six eggs carried by a female of *B. paranguensis ventanillensis* subsp. nov. had a single offspring, which was male and lived for 3 days.

We must also mention that both Peruvian strains showed different stress tolerance during cultures, this being another parameter for discrimination. The management of each strain must be considered separately. For instance, IMP-BG Z010-VL in a batch or clonal culture [65] tends to stick onto the beaker, and it is common to find them dead and floating on the surface due to this behavior of agglomerating very close to the waterline, causing them to become trapped by the surface tension. For this reason, it is highly recommended to gently shake the beaker occasionally to avoid the loss of this strain. The case of Z018-SD is possibly simpler, considering they tend to swim homogeneously in the water column. However, this strain is more delicate to water changes and is easily stressed due to any other changes such as food, temperature, oxygen, or salinity. Both cases highlight first drafts of new behavioral differences among local and regional lineages, particularly in fresh tropical waters [51].

The incorporation of biometric data for the other two subspecies of *B. paranguensis* and the other three subspecies of *B. koreanus* not considered in this study is recognized as an important aspect that should be included in subsequent studies in order to strengthen this hypothesis. In addition, the lack of updated morphological taxonomic keys for the *Brachionus plicatilis* species complex is a relevant issue to new researchers and students who wish to deepen future investigations, which is why more research must be performed in order to integrate them all in one complete morphological key, backed up with molecular support, at least for the 15 putative species reported [4], also considering a need for preventative practices in uncontrolled international exchanges of rotifer strains among institutions or common sales, as they are settled with unnamed strains or without registration of origin traceability, which is unfortunately regarded as a common informal practice.

Finally, it is important to call attention to Peruvian ecological issues. The Ventanilla wetlands in the Province of Callao have serious environmental plights [66–68], as there is also a growing human settlement nearby that contributes to the large amounts of litter found in the area. In addition, other wetlands nearby, called the Regional wetlands, have been re-designed from their natural state for landscape tourism reasons. An example of this is the small colorful fish that have been introduced, with authorities further linking part of those scarce bodies of water with canals and building wooden bridges over them. The Santo Domingo wetland in the Paracas district is a complex of very shallow and little brackish water bodies (<30 cm depth $\times$ 2 m length), with a few trees and lower vegetation in the middle of a wasteland near the highway and a residential area called Santo Domingo, from which its name originates. In spite of Monogononta dormancy strategies to resist long periods without water [69,70], it is very likely that these two Peruvian coastal wetlands will disappear in the next few years, as others less known did, and in consequence, local fauna such as birds, reptiles and above all plankton and benthos invertebrates too, unless authorities start to better organize urban growth regarding the surrounding wetlands [71].

5. Conclusions

Morphological and molecular differences were reported against four *Brachionus* species. We point out that the COI marker is suitable for classifying hydrobiological economic resources, which can be ordered as subspecies, while the ITS marker can delimit species complex.

The IMP-BG Z010-VL strain presented resilience to daily manipulation, showing potential in aquaculture applications, and corresponded to *Brachionus paranguensis*—L4. Meanwhile, the Z018-SD strain was less resistant to manipulation, getting stressed easily, and corresponded to *Brachionus koreanus*—SM2.

Brachionus paranguensis—L4 showed three clades geographically differentiated (Group 1: North America; Group 2: Australasia in addition to an old formal registry from France; and Group 3: Peru and Chile); while *Brachionus koreanus*—SM2 showed four clades that were not geographically well defined (Group 1: Eurasia; Group 2: Asia, Europe and America; Group 3: Spain; and Group 4: Peru).

In both cases, the biometrical relationship with production parameters determined that the length of the lorica is positively associated with longevity, progeny, egg production, and reproductive age, while the width and lorica aperture are slightly associated with the maximum egg load.

The presence of local subspecies of *Brachionus plicatlis* complex from impacted and vulnerable environments was evaluated. We recommend further studies to support taxonomy beyond biodiversity inventories and suggest a more multidisciplinary and applied approach must be performed in order to offer more decision-making tools, such as economic added value as an instrument to support conservation.

Supplementary Materials: The following are available online at https://www.mdpi.com/article/10.3390/d13120671/s1, Table S1: List of GenBank accession numbers for COI and ITS1 sequences obtained in this study, from organisms of strains isolated from Peru, Table S2: Measures from parthenogenetic females of the strains IMP-BG Z010-VL and Z018-SD, Table S3: Measures from parthenogenetic females of the strains IMP-BG Z010-VL and Z018-SD used for statistical comparison (n = 60), Table S4: Measures from parthenogenetic females of the reference strains L (*B. plicatilis* s.s.) and SS (*B. rotundiformis*) used for statistical comparison (n = 60), Table S5: Record of production parameter of both parthenogenetic females of the strains IMP-BG Z010-VL and Z018-SD. Min = minimum, Max = maximum, SD = Standard deviation. The units of each parameter are indicated in parenthesis. Data obtained from the F1 with less than 24 h after hatching. (n = 48), Table S6: Nucleotide composition registered in mtDNA COI sequences (570 bp) of L4 and SM2 organisms from different origins, considered for phylogenetic analysis, Table S7: Nucleotide composition registered in rDNA ITS1 sequences (331 bp) of L4 and SM2 organisms from different origins, considered for phylogenetic analysis, Table S8: Summary of the outputs of the R-Studio programme. They showed low correlation expressed by Monte-Carlo Test on the sum of eigenvalues of a co-inertia analysis (RV).

Author Contributions: P.P.A.S.-D. conceived the idea for the project and registered morphological data. M.S.-B. and A.A.-O. performed the SEM and morphological description. G.S. performed the DNA extraction, amplification, and sequencing. P.P.A.S.-D., G.S. and D.C. analyzed the molecular data. All authors have read and agreed to the published version of the manuscript.

Funding: Research was partially funded by the Ministry of Production—IMARPE-DGIA (PpR Budget program 0094 "Ordenamiento y Desarrollo de la Acuicultura"). The specimens analyzed are property of the Germplasm Bank of Aquatic Organisms Germplasm Bank (BGOA for its acronym in Spanish) of the Instituto del Mar del Peru.

Institutional Review Board Statement: Ethical review and approval were waived for this study, due to the institutional management of ex-situ conservation of native strains with aquaculture potential. Besides, regarding Peruvian law N° 30407. Law for the protection and welfare of animals. Article 25. Prohibitions and exceptions for the use of animals in acts of experimentation, research and teaching. Clause 3.

Data Availability Statement: The data presented in this study are available on request from the corresponding author. The data are not publicly available because it belongs to the Instituto del Mar del Peru.

Acknowledgments: The authors thank the BGOA team and the Biologist Elder Fernandez for plotting the map, and also thank Gerardo Guerrero Jiménez for his help preparing the trophi of the specimens.

Conflicts of Interest: The authors declare no conflict of interest.

References

1. Jennings, H.S. Rotatoria of the United States, with especial reference to those of the great lakes. *Bull. US Fish Comm.* **1900**, *19*, 14–22.
2. Harring, H.K.; Myers, F.J. The rotifer fauna of Wisconsin. IV. The Dicranophorinae. *Trans. Wis. Acad. Sci. Arts Lett.* **1928**, *23*, 667–808.
3. Serra, M.; Miracle, M.R. Biometric analysis of *Brachionus plicatilis* ecotypes from Spanish lagoons. *Hydrobiologia* **1983**, *104*, 279–291. [CrossRef]
4. Mills, S.; Alcántara-Rodríguez, J.A.; Ciros-Pérez, J.; Gómez, A.; Hagiwara, A.; Galindo, K.H.; Jersabek, C.D.; Malekzadeh-Viayeh, R.; Leasi, F.; Lee, J.-S.; et al. Fifteen species in one: Deciphering the *Brachionus plicatilis* species complex (Rotifera, Monogononta) through DNA taxonomy. *Hydrobiologia* **2017**, *796*, 39–58. [CrossRef]
5. Guerrero-Jiménez, G.; Vannucchi, P.E.; Silva-Briano, M.; Ortiz, A.A.; Rico-Martínez, R.; Roberts, D.; Neilson, R.; Elías-Gutiérrez, M. *Brachionus paranguensis* sp. nov. (Rotifera, Monogononta), a member of the L group of the *Brachionus plicatilis* complex. *ZooKeys* **2019**, *880*, 1–23. [CrossRef]
6. Romero, G.L. Caracterización Morfométrica y Aspectos Filogenéticos de Cepas de Rotíferos del Grupo *Brachionus plicatilis* (Rotifera: Brachionidae) Utilizados en la Acuicultura Peruana. Master's Dissertation, Universidad Nacional Mayor de San Marcos, Lima, Peru, 2008.
7. Toscano, E.; Severino, R. Brachionidae (Rotifera: Monogononta) de la albufera El Paraíso y el reporte de *Brachionus ibericus* en el Perú. *Rev. Peru. Biol.* **2013**, *20*, 177–180. [CrossRef]
8. Chinchayán, M. Cultivo de la Microalga *Nannochloropsis oculata* y su Consumo por el Rotífero *Brachionus plicalitis* (línea S). Master's Dissertation, Universidad Nacional Agraria La Molina, Lima, Peru, 1996.

9. Alayo, M.; Iannacone, J. Population growth of the rotífer eurihalin *Brachionus plicatilis* hepatotomus fed on the microalga *nannochloris* sp. *Bol. Lima* **2001**, *23*, 87–93.
10. Alayo, M.; Iannacone, J. Ensayos ecotoxicológicos con petróleo crudo, diesel 2 y diesel 6 con dos subespecies de *Brachionus plicatilis* müller 1786 (rotifera: Monogononta). *Gayana* **2002**, *66*, 45–58. [CrossRef]
11. Cisneros, R. Rendimiento poblacional del rotífero nativo *Brachionus* sp. "Cayman", utilizando diferentes enriquecedores. *Ecol. Apl.* **2011**, *10*, 99–105. [CrossRef]
12. Rosales-Barrantes, R. Efecto de la Temperatura, la Salinidad y sus Interacciones Sobre el Crecimiento Poblacional del Rotífero Nativo *Brachionus* sp. Cayman, Cepa Chilca, Perú. Master's Dissertation, Universidad Ricardo Palma, Lima, Peru, 2012.
13. Alcedo-Durán, N.; Córdova-Calle, J.C. Efecto de Tres Concentraciones de Harina de Ensilado de Vísceras de Gallus gallus Domesticus "pollo" en el Crecimiento Poblacional y Contenido de Proteínas y Lípidos de *Brachionus* sp. en Laboratorio. Master's Dissertation, Universidad Nacional del Santa, Nuevo Chimbote, Peru, 2014.
14. Murrieta-Morey, G.A.; Nájar, J.; Alcantara-Bocanegra, F. Producción experimental de rotíferos en bolsas de plástico utilizando harina de pescado como fuente de nutrientes. *Folia Amaz.* **2015**, *24*, 15. [CrossRef]
15. Huanacuni-Pilco, J.I.; Espinoza-Ramos, L.A. Producción de alimento vivo para la investigación en acuicultura de peces marinos en la UNJBG, Tacna. *Cienc. Desarro.* **2019**, *17*, 82–86. [CrossRef]
16. Aguilar-Samanamud, C.P.; Gaspar, W.; Inga, G.; Flores, L.; Sanchez, P.; Hernandez-Acevedo, H.; Borda-Soares, R.; Olivera-Galvez, A. Variables that intervene in the weight of rotifer biomass and fatty acids. *J. World Aquac. Soc.* **2021**, 1–13. [CrossRef]
17. Lubzens, E. Raising rotifers for use in aquaculture. In *Rotifer Symposium IV*; May, L., Wallace, R., Herzig, A., Eds.; Springer: Dordrecht, The Netherlands, 1987; pp. 245–255, ISBN 9789401083027/9789400940598.
18. Andersen, R.A. *Algal Culturing Techniques*; Elsevier: Amsterdam, The Netherlands, 2005; p. 589.
19. Segers, H.; Murugan, G.; Dumont, H.J. On the taxonomy of the Brachionidae: Description of *Plationus* n. gen. (Rotifera, Monogononta). *Hydrobiologia* **1993**, *268*, 1–8. [CrossRef]
20. Silva-Briano, M.; Adabache-Ortiz, A.; Guerrero-Jiménez, G.; Rico-Martínez, R.; Zavala-Padilla, G. Ultrastructural and Morphological Description of the Three Major Groups of Freshwater Zooplankton (Rotifera, Cladocera, and Copepoda) from the State of Aguascalientes, Mexico. In *The Transmission Electron Microscope: Theory and Applications*; INTECH Open Science/Open Minds: Rijeka, Croatia, 2015; pp. 307–325.
21. Fu, Y.; Hirayama, K.; Natsukari, Y. Morphological differences between two types of the rotifer *Brachionus plicatilis* O.F. Müller. *J. Exp. Mar. Biol. Ecol.* **1991**, *151*, 29–41. [CrossRef]
22. Ciros-Pérez, J.; Gómez, A.; Serra, M. On the taxonomy of three sympatric sibling species of the *Brachionus plicatilis* (Rotifera) complex from Spain, with the description of *B. ibericus* n. sp. *J. Plankton Res.* **2001**, *23*, 1311–1328. [CrossRef]
23. Montero-Pau, J.; Gomez, A.; Munoz, J. Application of an inexpensive and high-throughput genomic DNA extraction method for the molecular ecology of zooplanktonic diapausing eggs. *Limnol. Oceanogr. Methods ASLO* **2008**, *6*, 218–222. [CrossRef]
24. Edgar, R.C. MUSCLE: A multiple sequence alignment method with reduced time and space complexity. *BMC Bioinform.* **2004**, *5*, 113. [CrossRef]
25. Kumar, S.; Stecher, G.; Tamura, K. MEGA7: Molecular Evolutionary Genetics Analysis Version 7.0 for Bigger Datasets. *Mol. Biol. Evol.* **2016**, *33*, 1870–1874. [CrossRef]
26. Prosser, S.; Martínez-Arce, A.; Elías-Gutiérrez, M. A new set of primers for COI amplification from freshwater microcrustaceans. *Mol. Ecol. Resour.* **2013**, *13*, 1151–1155. [CrossRef]
27. Messing, J. New M13 vectors for cloning. In *Methods in Enzymology*; Recombinant DNA Part C; Academic Press: Cambridge, MA, USA, 1983; Volume 101, pp. 20–78.
28. Palumbi, S.R. Nucleic acids II: The polymerase chain reaction. In *Molecular Systematics*, 2nd ed.; Hillis, D., Moritz, C., Mable, B., Eds.; Sinauer Associates: Sunderland, MA, USA, 1996; pp. 205–247, ISBN 0878932828 9780878932825.
29. Lefort, V.; Longueville, J.-E.; Gascuel, O. SMS: Smart Model Selection in PhyML. *Mol. Biol. Evol.* **2017**, *34*, 2422–2424. [CrossRef]
30. Guindon, S.; Dufayard, J.-F.; Lefort, V.; Anisimova, M.; Hordijk, W.; Gascuel, O. New Algorithms and Methods to Estimate Maximum-Likelihood Phylogenies: Assessing the Performance of PhyML 3.0. *Syst. Biol.* **2010**, *59*, 307–321. [CrossRef]
31. Huelsenbeck, J.P.; Ronquist, F. MRBAYES: Bayesian inference of phylogenetic trees. *Bioinformatics* **2001**, *17*, 754–755. [CrossRef] [PubMed]
32. Hwang, D.-S.; Dahms, H.-U.; Park, H.G.; Lee, J.-S. A new intertidal Brachionus and intrageneric phylogenetic relationships among Brachionus as revealed by allometry and CO1-ITS1 gene analysis. *Zool. Stud.* **2013**, *52*, 13. [CrossRef]
33. Campaña-Torres, A.; Martinez-Cordova, L.R.; Villarreal-Colmenares, H.; Hernández-López, J.; Ezquerra-Brauer, J.M.; Cortés-Jacinto, E. Efecto de la adición del rotífero *Brachionus rotundiformis* (Tschugunoff, 1921) sobre la calidad del agua y la producción, en cultivos super-intensivos de camarón blanco del Pacífico *Litopenaeus vannamei* (Boone, 1931). *Rev. Biol. Mar. Oceanogr.* **2009**, *44*, 335–342. [CrossRef]
34. Dumont, H.J. Biogeography of rotifers. In Proceedings of the Biology of Rotifers; Pejler, B., Starkweather, R., Nogrady, T., Eds.; Springer: Dordrecht, The Netherlands, 1983; pp. 19–30.
35. Castellanos Páez, M.E.; Garza Mouriño, G.; Marañón Herrera, S. *Aislamiento, Caracterización, Biología y Cultivo del Rotífero Brachionus plicatilis (O.F. Muller)*; Universidad Autónoma Metropolitana, Unidad Xochimilco: Mexico City, Mexico, 1999.

36. Gribble, K.; Welch, D.B.M. Life-Span Extension by Caloric Restriction Is Determined by Type and Level of Food Reduction and by Reproductive Mode in *Brachionus manjavacas* (Rotifera). *J. Gerontol. Ser. A Boil. Sci. Med. Sci.* **2013**, *68*, 349–358. [CrossRef] [PubMed]
37. Fontaneto, D.; Kaya, M.; Herniou, E.A.; Barraclough, T.G. Extreme levels of hidden diversity in microscopic animals (Rotifera) revealed by DNA taxonomy. *Mol. Phylogenet. Evol.* **2009**, *53*, 182–189. [CrossRef]
38. Malekzadeh-Viayeh, R.; Pak-Tarmani, R.; Rostamkhani, N.; Fontaneto, D. Diversity of the rotifer *Brachionus plicatilis* species complex (Rotifera: Monogononta) in Iran through integrative taxonomy. *Zool. J. Linn. Soc.* **2014**, *170*, 233–244. [CrossRef]
39. García-Morales, A.E.; Elías-Gutiérrez, M. DNA barcoding of freshwater Rotifera in Mexico: Evidence of cryptic speciation in common rotifers. *Mol. Ecol. Resour.* **2013**, *13*, 1097–1107. [CrossRef]
40. Obertegger, U.; Cieplinski, A.; Fontaneto, D.; Papakostas, S. Mitonuclear discordance as a confounding factor in the DNA taxonomy of monogonont rotifers. *Zool. Scr.* **2018**, *47*, 122–132. [CrossRef]
41. Rokas, A.; Williams, B.L.; King, N.; Carroll, S.B. Genome-scale approaches to resolving incongruence in molecular phylogenies. *Nature* **2003**, *425*, 798–804. [CrossRef]
42. Toews, D.P.L.; Brelsford, A. The biogeography of mitochondrial and nuclear discordance in animals. *Mol. Ecol.* **2012**, *21*, 3907–3930. [CrossRef]
43. Després, L. One, two or more species? Mitonuclear discordance and species delimitation. *Mol. Ecol.* **2019**, *28*, 3845–3847. [CrossRef]
44. Snell, T.W.; Shearer, T.L.; Smith, H.A.; Kubanek, J.; Gribble, K.E.; Welch, D.B.M. Genetic determinants of mate recognition in *Brachionus manjavacas* (Rotifera). *BMC Biol.* **2009**, *7*, 60. [CrossRef]
45. Wendel, J.F.; Doyle, J.J. Phylogenetic Incongruence: Window into Genome History and Molecular Evolution. In *Molecular Systematics of Plants II: DNA Sequencing*; Soltis, D.E., Soltis, P.S., Doyle, J.J., Eds.; Springer: Boston, MA, USA, 1998; pp. 265–296.
46. Álvarez, I.; Wendel, J.F. Ribosomal ITS sequences and plant phylogenetic inference. *Mol. Phylogenet. Evol.* **2003**, *29*, 417–434. [CrossRef]
47. Aránguiz-Acuña, A.; Pérez-Portilla, P.; De La Fuente, A.; Fontaneto, D. Life-history strategies in zooplankton promote coexistence of competitors in extreme environments with high metal content. *Sci. Rep.* **2018**, *8*, 11060. [CrossRef]
48. Suatoni, E.; Vicario, S.; Rice, S.; Snell, T.; Caccone, A. An analysis of species boundaries and biogeographic patterns in a cryptic species complex: The rotifer—*Brachionus plicatilis*. *Mol. Phylogenet. Evol.* **2006**, *41*, 86–98. [CrossRef] [PubMed]
49. Stelzer, C.-P.; Riss, S.; Stadler, P. Genome size evolution at the speciation level: The cryptic species complex *Brachionus plicatilis* (Rotifera). *BMC Evol. Biol.* **2011**, *11*, 90. [CrossRef] [PubMed]
50. Michaloudi, E.; Mills, S.; Papakostas, S.; Stelzer, C.-P.; Triantafyllidis, A.; Kappas, I.; Vasileiadou, K.; Proios, K.; Abatzopoulos, T.J. Morphological and taxonomic demarcation of *Brachionus asplanchnoidis* Charin within the *Brachionus plicatilis* cryptic species complex (Rotifera, Monogononta). *Hydrobiologia* **2017**, *796*, 19–37. [CrossRef]
51. Yoshinaga, T.; Minegishi, Y.; Rumengan, I.; Kaneko, G.; Furukawa, S.; Yanagawa, Y.; Tsukamoto, K.; Watabe, S. Molecular phylogeny of the rotifers with two Indonesian Brachionus lineages. *Coast. Mar. Sci.* **2004**, *29*, 45–56.
52. Yasuike, M.; Tezuka, N.; Sekino, M.; Katoh, M. Species identification of the four rotifer genetic resources preserved at the FRA Genebank Project based on ITS1 and COI sequences. Unpublished work. 2017.
53. Papakostas, S.; Dooms, S.; Triantafyllidis, A.; Deloof, D.; Kappas, I.; Dierckens, K.; De Wolf, T.; Bossier, P.; Vadstein, O.; Kui, S.; et al. Evaluation of DNA methodologies in identifying Brachionus species used in European hatcheries. *Aquaculture* **2006**, *255*, 557–564. [CrossRef]
54. Kostopoulou, V.; Miliou, H.; Katis, G.; Verriopoulos, G. Changes in the Population Structure of the Lineage 'Nevada' Belonging to the *Brachionus plicatilis* Species Complex, Batch-Cultured under Different Feeding Regimes. *Aquac. Int.* **2006**, *14*, 451–466. [CrossRef]
55. Kostopoulou, V.; Miliou, H.; Krontira, Y.; Verriopoulos, G. Mixis in rotifers of the lineage 'Nevada', belonging to the *Brachionus plicatilis* species complex, under different feeding regimes. *Aquac. Res.* **2007**, *38*, 1093–1105. [CrossRef]
56. Pourriot, R.; Rougier, C. Taux de reproduction en fonction de la concentration en nourriture et de la température chez trois espèces du genre Brachionus (Rotifères). *Ann. Limnol. Int. J. Limnol.* **1997**, *33*, 23–31. [CrossRef]
57. Blanchot, J.; Pourriot, R. Influence de trois facteurs de l'environnement, lumiere, temperature et salinite sur l'eclosion des oeufs de duree d'un clone de *Brachionus plicatilis* (O. F. Müller) rotifere. *CR Acad. Sci. Paris* **1982**, *295*, 243–246.
58. Gómez, A.; Serra, M.; Carvalho, G.R.; Lunt, D.H. Speciation in ancient cryptic species complexes: Evidence from the molecular phylogeny of *Brachionus plicatilis* (Rotifera). *Evolution* **2002**, *56*, 1431–1444. [CrossRef]
59. Mills, S.; Lunt, D.H.; Gómez, A. Global isolation by distance despite strong regional phylogeography in a small metazoan. *BMC Evol. Biol.* **2007**, *7*, 225. [CrossRef] [PubMed]
60. Anitha, P.; George, R. The taxonomy of *Brachionus plicatilis* species complex (Rotifera: Monogononta) from the Southern Kerala (India) with a note on their repro- ductive preferences. *J. Mar. Biol. Assoc. India* **2006**, *48*, 6–13.
61. Derry, A.M.; Hebert, P.D.N.; Prepas, E.E. Evolution of rotifers in saline and subsaline lakes: A molecular phylogenetic approach. *Limnol. Oceanogr.* **2003**, *48*, 675–685. [CrossRef]
62. Kutikova, L.A. Parallelism in the evolution of rotifers. *Hydrobiologia* **1983**, *104*, 3–7. [CrossRef]
63. Snell, T.W. Rotifers as models for the biology of aging. *Int. Rev. Hydrobiol.* **2014**, *99*, 84–95. [CrossRef]

64. Korstad, J.; Neyts, A.; Danielsen, T.; Overrein, I.; Olsen, Y. Use of swimming speed and egg ratio as predictors of the status of rotifer cultures in aquaculture. *Hydrobiologia* **1995**, *313–314*, 395–398. [CrossRef]
65. Gilbert, J.J. Monoxenic cultivation of the rotifer *Brachionus calyciflorus* in a defined medium. *Oecologia* **1970**, *4*, 89–101. [CrossRef] [PubMed]
66. Vizcardo, C.; Gil-Kodaka, P. Estructura de las comunidades Macrozoobentónicas de los Humedales de Ventanilla, Callao, Perú. *An. Cient.* **2015**, *76*, 1–11. [CrossRef]
67. Aponte, H.; Ramírez, D.W.; Vargas, R. First stages of the post-fire natural regeneration of vegetation in the ventanilla wetlands (Lima-Peru). *Ecol. Apl.* **2017**, *16*, 23–30. [CrossRef]
68. Rodríguez, R.; Retamozo-Chavez, R.; Aponte, H.; Valdivia, E. Evaluación microbiológica de un cuerpo de agua del ACR Humedales de Ventanilla (Callao, Perú) y su importancia para la salud pública local. *Ecol. Apl.* **2017**, *16*, 15–21. [CrossRef]
69. Gilbert, J.J. Dormancy in Rotifers. *Trans. Am. Microsc. Soc.* **1974**, *93*, 490–513. [CrossRef]
70. Ricci, C. Dormancy patterns in rotifers. *Hydrobiologia* **2001**, *446*, 1–11. [CrossRef]
71. WWT Consulting. *Good Practices Handbook for Integrating Urban Development and Wetland Conservation*; WWT Consulting: Slimbridge, UK, 2018; pp. 1–49.

Article

Application of COI Primers 30F/885R in Rotifers to Regional Species Diversity in (Sub)Tropical China

Ya-Nan Zhang [1], Shao-Lin Xu [1], Qi Huang [1], Ping Liu [1] and Bo-Ping Han [1,2,*]

1 Department of Ecology and Institute of Hydrobiology, Jinan University, Guangzhou 510632, China; zhangyanan@stu2018.jnu.edu.cn (Y.-N.Z.); shaolin@stu2018.jnu.edu.cn (S.-L.X.); hq2416@stu2018.jnu.edu.cn (Q.H.); liuping329098@163.com (P.L.)
2 Engineering Research Center for Tropical and Subtropical Aquatic Ecological Engineering, Ministry of Education, Jinan University, Guangzhou 510632, China
* Correspondence: tbphan@jnu.edu.cn

Abstract: Rotifers are the most diverse group in freshwater zooplankton and play an important role in food webs and ecosystems. DNA barcoding has become a useful approach to investigate species diversity at local and regional scales, but its application is still limited by efficient primers for the group. To test a pair of primers 30F/885R recently designed for rotifers, we applied them to investigating regional species diversity in the freshwater of South China. We sequenced the cytochrome c oxidase subunit I (COI) gene of rotifers collected from the investigated 23 reservoirs in a large river basin and obtained 145 COI sequences from 33 species in 14 genera. The mean PCR success rate for all tested species was 50%. The 145 sequenced mtCOI in this study covered 33 of 64 identified morphological taxa, including most of the common species in the basin. The intraspecific genetic distance was calculated with a K2P model for 24 rotifer species occurring in the quantitative samples, in which 15 rotifers, such as *Keratella cochlearis* and *Brachionus calyciflorus*, had a genetic distance higher than 5%. The high intraspecific genetic differentiation indicates that cryptic species are probably common in (sub)tropical China.

Keywords: rotifers; cryptic species; freshwater zooplankton; reservoir; species diversity; tropics

Citation: Zhang, Y.-N.; Xu, S.-L.; Huang, Q.; Liu, P.; Han, B.-P. Application of COI Primers 30F/885R in Rotifers to Regional Species Diversity in (Sub)Tropical China. *Diversity* **2021**, *13*, 390. https://doi.org/10.3390/d13080390

Academic Editors: Manuel Elias-Gutierrez and Bert W. Hoeksema

Received: 14 May 2021
Accepted: 18 August 2021
Published: 19 August 2021

1. Introduction

Rotifers are a group of zooplankton with high species richness in freshwater ecosystems [1]. There are more than 2000 species of rotifers described worldwide, including 1571 species from Monogononta and 461 species from Bdelloidea [2,3]. This group plays a critical role in the flow of energy and the cycling of matter in freshwater ecosystems [4]. Most species in the group graze or feed mainly on algae or bacteria and serve as food for small invertebrates and fish [5,6]. Due to a short lifespan and high reproduction, rotifers are highly dynamic in natural waters and sensitive to environmental change. Knowing their species richness and species composition is of great significance for understanding ecosystem functions and environmental monitoring [2,7].

The morphological taxonomy of rotifers is based on external shapes and internal structures. Their ciliated corona and lorica are important to species identification [8]. However, their small body size and complicated morphology make morphological identification difficult. In addition, environmental conditions, such as temperature and food concentration, can induce morphological changes in many species [9], due to the phenotypic plasticity, especially in monogononts [10]. Relying only on morphological features may lead to faulty identification, especially for species with high phenotypic plasticity. Up to now, more than 40 species complexes have been discovered in *Keratella cochlearis* (Gosse, 1851), *B. calyciflorus*, *Philodina flaviceps* Bryce, 1906, and *Lecane bulla*, (Gosse, 1886) [11–15]. High genetic variation can occur within local populations despite insignificant morphological differentiation [11,16]. Molecular classification has already been extensively applied to rotifers [12]. Earlier studies relied on allozyme electrophoresis [17–19], but today, DNA barcoding has become an essential technique to identify species in monogonont and bdelloid rotifers [15,20–22].

Molecular classification provides critical supplementary information for morphological taxonomy. The mitochondrial cytochrome oxidase subunit I (COI) gene is the most widely used sequence segment in DNA barcode classification and has proved to help detect cryptic species, intraspecific variation, and phylogeographical patterns [17,23]. At present, Folmer's universal primers are commonly used for amplifying COI [24,25]. Meyer et al. [26] modified the Folmer primers to obtain the primers dgLCO/dgHCO and amplified the COI sequence of the *B. plicatilis* complex [24,27]. Wilts et al. [28] developed primers COI-F/COI-R to amplify the COI of *Proales daphnicola* (Thompson, 1892) [25,27,29]. Elías-Gutiérrez et al. [30] used the Zplank primers to amply rotifer COI and obtained 11 BINs (Barcode Index Numbers) of rotifers, with a sequencing success rate of 100%. Recently, Zhang et al. [31] used a metagenomics method to assemble nine mitochondrial genomes from *Brachionus* and *Keratella*, with which they designed a new pair of primers just for rotifer COI: 30F/885R. The pair of primers performed efficiently (86%), much higher than dgLCO/dgHCO (32%) and Folmer primers (59%). Despite that, the newly designed primers need further testing in rotifers from different water bodies.

Here, we test the primers 30F/885R and apply them to investigating rotifer species diversity and composition of rotifers at a regional scale. Hanjiang River Basin is located in Guangdong Province, South China, where rotifers dominate zooplankton in most drinking water [32,33]. We conducted both morphological identification and COI sequence amplification on rotifers from the 23 investigated reservoirs in the basin and tested further the primers 30F/885R and their usefulness in the assessment of species diversity in (sub)tropical regions.

2. Materials and Methods

2.1. Collection, Identification, and Counting of Rotifers

Rotifers were sampled from 23 reservoirs (Figure 1) in the Hanjiang River Basin, Guangdong Province, in southern China from November 2019 to January 2020. For quantitative samples used for the assessment of species diversity in rotifer communities, a 5 L water sampler was used to vertically collect 50 L water from the surface to the bottom evenly. The sample was filtered and concentrated with a plankton net with a mesh size of 38 μm and fixed with 5% formalin. For the qualitative samples used for DNA extraction of rotifers, a plankton net with a mesh size of 64 μm was trawled horizontally and vertically. The obtained zooplankton was immediately fixed with BBI's DNA-EZ Reagents F DNA-Be-Locked A and stored in a refrigerator at 4 °C in the field.

All rotifer species in our samples were first identified based on external shape and internal structure [8]. For rotifer species that could not be easily identified by morphological characteristics, individuals were picked out to check the shape of their lorica. After adding 10% glycerol and 5–10% sodium hypochlorite, the shape of the lorica was further observed under a microscope (400×) for morphological identification. All species were identified, measured, and counted under a microscope (Olympus BX41, Tokyo, Japan). Their individual body volume was calculated with approximate geometric volume formulae, and the density of 1 g/cm^3 was set to estimate the bodyweight [34,35]. If a species contributes at least 2% of the total abundance, it is considered dominant in that reservoir (Table S1) [36].

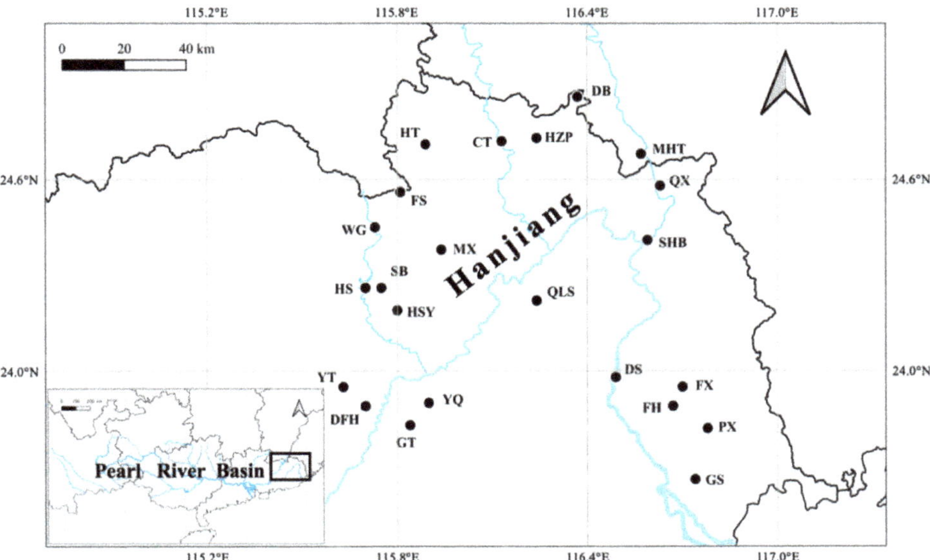

Figure 1. Localities of 23 investigated reservoirs in the Hanjiang River Basin, South China (Abbreviations of reservoir names are listed in Table S2.

2.2. DNA Extraction, PCR Amplification, and Sequencing

Before DNA extraction, rotifer specimens were washed with MilliQ water, and three or four individuals from one species were put into a 0.2 mL tube. Three microliters Proteinase K and 30 μL Chelex resin (BioRad, Hemel Hempstead, UK) were added into the tube for DNA extraction. The tube was centrifuged at 8000 rpm for 1 min and finally put into the PCR instrument. The DNA samples were incubated at 56 °C for 60 min at 99 °C for 10 min and stored at 12 °C. All DNA samples were stored at 4 °C, and/or at −20 °C for long-term storage. Finally, the samples were centrifuged at 10,000 rpm for 2 min, and the supernatant was directly used in each PCR reaction.

DNA from a single species was used as a template. A 760-bp segment of COI was amplified using the primers 30F and 885R [31]. The total amplification volume of primers 30F/885R was 30 μL, including 15 μL 2 × HieffTM PCR Master Mix (With Dye), 11 μL ddH_2O, 0.5 μL of forward and reverse primers (100 μM), 3 μL DNA, respectively. The amplification started with initial denaturation 2 min at 98 °C, then six cycles of (95 °C for 30 s, 54 °C for 40 s (−0.5 °C/each cycle), 72 °C for 30 s), and 36 cycles of (95 °C for 30 s, 51 °C for 40 s, 72 °C for 30 s) and final extension of 72 °C for 2 min [31].

The PCR products were detected in 1.0% agarose gel. The amplified products with clear and bright target bands were selected and sent to Tianyi Huiyuan Gene Technology Company for purification and sequencing. All DNA samples were paired-end sequenced. After that, all the chromatograms of forward and reserve sequences were checked with Finch TV1.5.0, and poor-quality sequences and repeated sequences were discarded (Geospiza Inc. https://www.digitalworldbiology.com/FinchTV (accessed on 1 December 2019)). The forward and reverse sequences from each sample were assembled into one sequence with Geneious v10.22, and all sequences were aligned with MAFFT and MACSE [37–39]. Poor-quality flanking regions of the sequences were discarded. We calculated the coverage of DNA barcodes for rotifers in the 23 reservoirs. The coverage of DNA barcodes is defined as the percentage of species with successfully obtained COI to the number of species identified morphologically in the quantitative sample.

2.3. Species Identification and Analysis Based on Molecular Methods

Each amplified sequence was submitted to NCBI for BLAST [40]. We obtained sequences for 33 species, among which six species, including *Pompholyx sulcata* (Hudson, 1885), *Ploesoma truncatum* (Levander, 1894), *Filinia opoliensis* (Zacharias, 1898), *Filinia camasecla cambodgensis* (Bērzinš, 1973), *Trichotria pocillum* (Müller, 1776), *Trichotria tetractis similis* (Stenroos, 1898), had no COI sequences deposited to NCBI. The interspecific distance, intraspecific genetic difference (K2P), and a NJ tree were calculated or constructed in MEGA 10.1.8 [41].

3. Results

A total of 64 rotifer species were identified from both quantitative samples and qualitative samples. Forty-seven species (Table S3) were morphologically identified from the quantitative samples that were fixed with formaldehyde. The dominant species included *Keratella tropica* (Apstein, 1907), *Keratella tect* (Gosse, 1851), *Trichocerca similis* (Wierzejski, 1893), *Synchaeta stylata* (Wierzejski, 1893), *Anuraeopsis fissa* (Gosse, 1851), *Polyarthra dolichoptera* (Idelson, 1925). There were 14 rare species, including *Brachionus quadridentatus* (Hermann, 1783), *Anuraeopsis coelata* (de Beauchamp, 1932), *T. tetractis similis*, *Lecane lunaris crenata* (Harring, 1913), *Lecane flexilis* (Gosse, 1886), *Lecane galeata* (Bryce, 1892), *Lecane pyriformis* (Daday, 1905), *Lecane arcuata* (Bryce, 1891), *Lecane thailandensis* (Segers et Sanoamuang, 1994), *Gastropus stylifer* (Imhof, 1891), *Trichocerca longiseta* (Schrank, 1802) *Trichocerca vargai* (Wulfert, 1961), *Filinia saltator* (Gosse, 1886), and *Filinia terminalis* (Plate, 1886).

In the qualitative samples fixed with DNA-EZ Reagents F DNA-Be-Locked A, 45 species were picked and sequenced (Table S4). A total of 145 COI sequences were obtained from 33 species in 14 genera (Table 1). The remaining 12 species failed to be amplified and sequenced. Seven of these are rare species in the samples. To test the 30F/885R primers, we amplified more than two times for the remaining five common species = (i.e., *Conochilus unicornis* (Rousselet, 1892) was done for nine times).

Among 47 species in the quantitative samples fixed with formaldehyde, only 23 species were covered at least by a COI sequence from the qualitative samples (Figure 2). Among the remaining 24 morphological species without COI sequences, 12 rare and 11 common species had too low abundance for PCR amplification and sequencing (i.e., <3 individuals), while one dominant species (*A. fissa*) failed to be amplified and sequenced. With morphological identification, the identified species number was between 6 and 20 for a single investigated reservoir, while with the molecular classification of COI sequences, the identified species number was between 0 and 13. The barcode recovery rate for a single reservoir was between 0% and 67%, with an average of 29% (Figure 3).

Among 33 species with COI sequences, we calculated the intraspecific K2P (Kimura two-parameter) genetic distance for 24 species that had at least two sequences. The average intraspecific genetic distance was from 0.00 to 0.32 (Table 2), with an averaged distance of 0.08. Many rotifers had high intraspecific genetic distances at the regional scale (Figure 4). Fifteen species had intraspecific genetic distance above 0.05: *K. cochlearis*, *Keratella tecta* (Gosse, 1851), *K. tropica*, *Brachionus leydigi* (Cohn, 1862), *B. calyciflorus*, *Plationus patulus* (Müller, 1786), *Asplanchna brightwelli* (Gosse, 1850), *Polyarthra vulgaris* (Carlin, 1943), *P. dolichoptera*, *T. similis*, *Ascomorpha ovalis* (Bergendahl, 1892), *Trichocerca dixonnuttalli* (Jennings, 1903), *Synchaeta oblonga* (Ehrenberg, 1831), and *P. sulcata*, *Hexarthra mira* (Hudson, 1871).

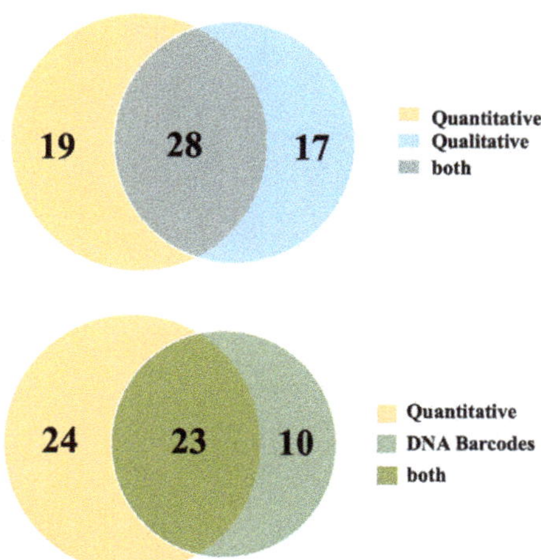

Figure 2. Venn diagrams, "Quantitative": the numbers of species identified with quantitative samples; "Qualitative": the numbers of species identified with qualitative samples; "DNA Barcodes": the number of species obtained mtCOI barcodes.

Table 1. PCR success rates for species that had at least one successful sequence.

Species	Number of Specimens	Number of Sequences	PCR Success Rate (%)
Keratella cochlearis	24	17	71
Keratella tropica	15	8	53
Keratella tecta	24	10	42
Brachionus diversicornis	8	6	67
Brachionus calyciflorus	5	4	80
Brachionus caudatus	2	1	50
Brachionus urceolaris	2	1	50
Brachionus angularis	3	2	67
Brachionus quadridentatus	3	1	33
Brachionus forficula	3	2	67
Brachionus leydigi	2	2	100
Brachionus budapestinensis	1	1	100
Asplanchna priodonta	9	6	67
Asplanchna brightwelli	2	2	100
Polyarthra vulgaris	14	4	29
Polyarthra dolichoptera	9	9	100
Ploesoma hudsoni	7	5	71
Ploesoma truncatum	1	1	100
Pompholyx sulcata	5	4	80
Trichocerca dixonnuttalli	4	4	100
Trichocerca capucina	23	11	48
Trichocerca cylindrica	18	4	22
Trichocerca similis	16	10	63
Filinia opoliensis	1	1	100
Filinia camaseclacambodgensis	2	2	100
Trichotria pocillum	1	1	100
Trichotria tetractis similis	3	1	33
Synchaeta oblonga	3	2	67
Synchaeta stylata	10	9	90
Lecane bulla	3	1	33
Ascomorpha ovalis	11	9	82
Hxarthra mira	8	2	25
Plationus patulus	3	2	67

Note: PCR success rate for a given species denotes the percentage of the obtained mtCOI sequences in the total number of tested specimens.

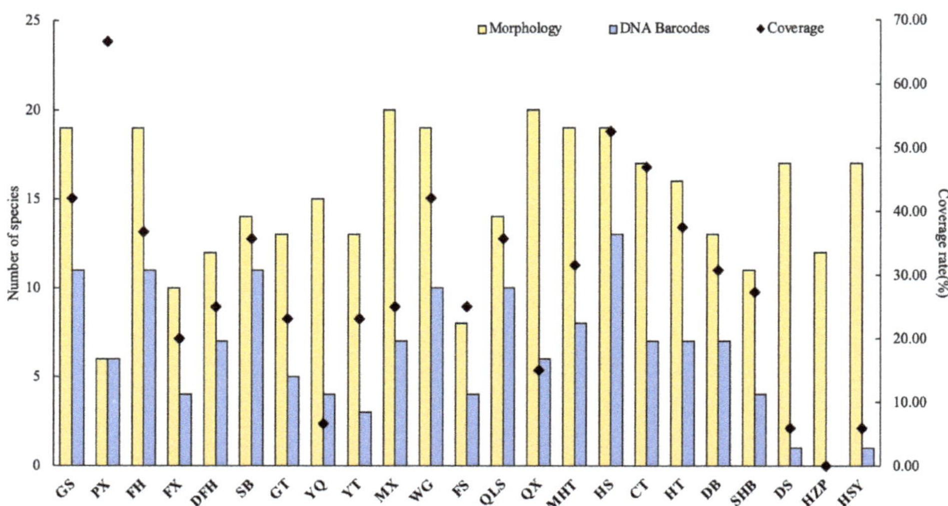

Figure 3. The number of rotifer species identified morphologically in the quantitative samples and number of species identified with COI barcodes from in the 23 reservoirs, and the coverage rate (%) of COI barcodes in the 23 investigated reservoirs (The coverage of DNA barcodes is the percentage of the number of species successfully obtained COI to the number of species identified morphologically in the quantitative sample.).

Table 2. Intraspecific genetic distance (K2P) of 24 rotifer species.

Species	Numbers of Sequences	Min Distance	Max Distance	Mean Distance
Keratella cochlearis	17	0.00	0.19	0.11
Keratella tecta	10	0.00	0.18	0.04
Keratella tropica	8	0.00	0.12	0.06
Brachionus diversicornis	6	0.00	0.00	0.00
Brachionus leydigi	2	N/A	N/A	0.15
Brachionus forficula	4	0.00	0.01	0.01
Brachionus angularis	2	N/A	N/A	0.01
Brachionus calyciflorus	4	0.04	0.18	0.14
Plationus patulus	2	N/A	N/A	0.10
Asplanchna priodonta	7	0.00	0.03	0.02
Asplanchna brightwelli	2	N/A	N/A	0.31
Polyarthra vulgaris	4	0.06	0.25	0.07
Polyarthra dolichoptera	9	0.00	0.25	0.19
Ascomorpha ovalis	11	0.00	0.14	0.08
Trichocercasimilis	10	0.00	0.32	0.17
Trichocerca cylindrica	5	0.00	0.07	0.04
Trichocerca dixon-nuttalli	4	0.00	0.18	0.09
Trichocerca capucina	13	0.00	0.01	0.00
Synchaeta stylata	9	0.00	0.06	0.02
Synchaeta oblonga	2	N/A	N/A	0.13
Ploesoma hudsoni	5	0.02	0.00	0.01
Pompholyx sulcata	4	0.00	0.12	0.06
Filiniacamasecla cambodgensis	2	N/A	N/A	0.00
Hexarthramira	2	N/A	N/A	0.07

Note: "N/A" means missing value.

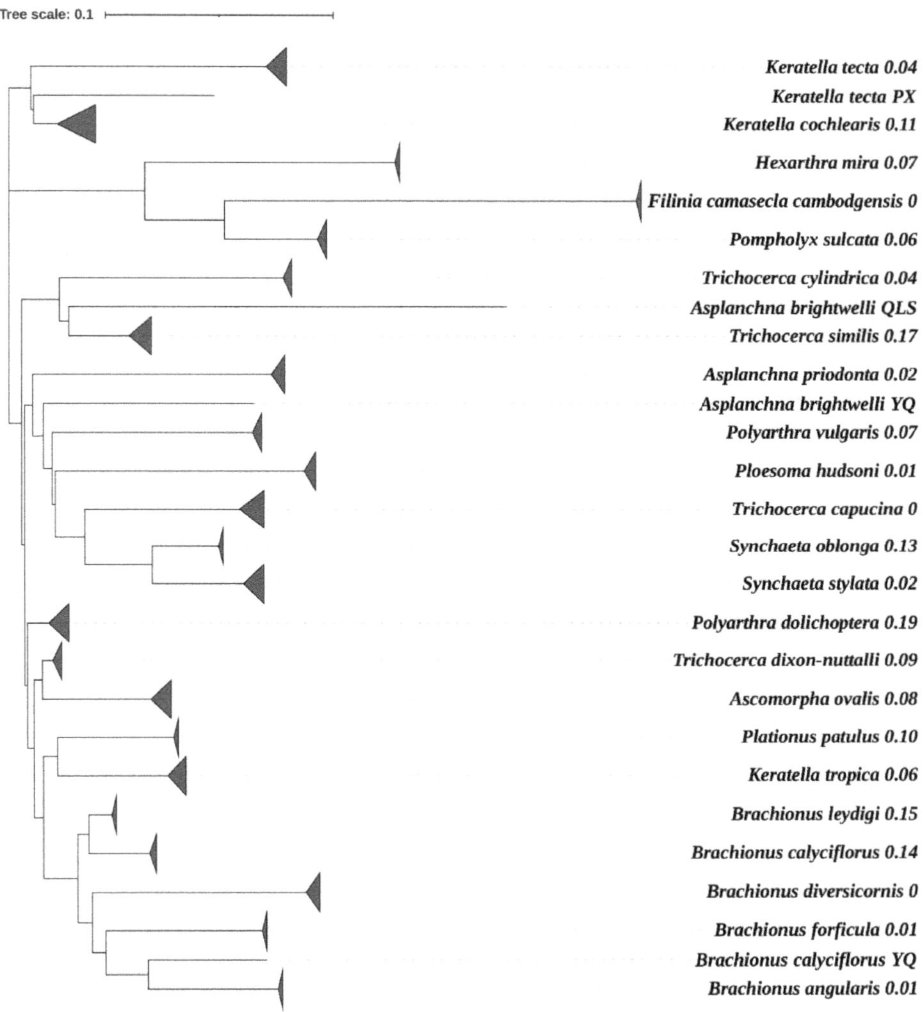

Figure 4. A simplified NJ tree with K2P distance for 24 morphological taxa. The triangle size represents the number of sequences, and the number after the species name is the mean K2P distance. The abbreviations after a species with a single sequence is the acronym of the reservoir name.

4. Discussion

The present study expanded the testing of a pair of recently reported DNA barcoding primers (30F/885R) and assessed the species diversity of rotifers at a regional scale. The average amplification success rate for all tested specimens (245) in 45 species was up to 50%. Among 45 species from our qualitative samples fixed for DNA sequencing, 12 species failed to be amplified and sequenced. Seven of 12 species are rare in this basin. The abundances for seven of 12 species might be too low for PCR amplification and sequencing. More individuals need to be collected to test these rare species for the amplification of COI. Surprisingly, *A. fissa* is one of the remaining five common species of the 12 failed species. It is dominant in the basin but failed to be sequenced. Rotifer species vary largely in body weight; the minimum weight of *A. fissa* is only 0.009 μg, while the individual weight of *Asplanchna girodi* (de Guerne, 1888) is up to 31.85 μg (Figure S1). In general, body weight determines the DNA amount for extraction, particularly mtDNA, which might affect subsequent amplification and sequencing [42]. However, smaller species, such as *P. sulcata*, *K. tecta*, and *K. cochlearis*, were successfully amplified and sequenced. Having a small body size might not be the main reason for the failure of amplification and sequencing for *A. fissa*. More likely, the COI sequences among different species in rotifers varied greatly [24]. We suspect that the primer incompatibility might be the main reason for the failure of amplification and sequencing of the five common species.

The DNA barcoding in the present study covered most species (33) of the rotifers in the investigated basin, and the obtained barcode library (145 COI sequences) will benefit the future survey of rotifers in similar regions. First, they can provide references for validating species identification. DNA barcode libraries can be used as a standard for species identification and improve the accuracy of rotifer morphological classification. Second, the barcoding is suitable for all life stages of rotifers, including resting eggs [43,44]. In addition, the barcode library constructed in this basin could be used as a reference in high-throughput-based monitoring techniques, such as eDNA metabarcoding and mitochondrial metagenomics [45–49].

As previously reported, using COI sequences can efficiently identify most species in rotifers, with the divergence among conspecific individuals being less than 1% [24]. *K. cochlearis* is a widely distributed species with phenotypic diversity. The COI nucleotide sequence divergence of 4.4% was detected between spined and unspined forms. As a result, the species was split into different species [50]. In the present study, the intraspecific genetic distance for *K. cochlearis* was between 0 and 0.19 for pairs of 17 individuals from all the reservoirs, with an average equal to 0.11, indicating a high hidden diversity. *B. calyciflorus* is a widely distributed species that shows a significant morphological difference with multiple subspecies and varieties [51]. Xiang et al. [52] collected eight geographical groups of *B. calyciflorus* from eastern China and concluded that this complex was composed of three cryptic species. In the present study, *B. calyciflorus* was found in four reservoirs, and the average intraspecific genetic distance was high up to 0.14, indicating high genetic diversity. As reported in other studies [24,47], high genetic distance also occurred within *B. leydigi*, *A. ovalis*, and *S. oblonga*, and based on the intraspecific genetic distances estimated here, species complexes, such as *K. cochlearis*, *B. calyciflorus*, *B. leydigi*, *A. ovalis*, and *S. oblonga*, might co-exist in the Hanjiang River Basin. Therefore, further investigation of cryptic species in this basin is recommended.

In conclusion, our study showed that the COI primers (30F/885R) utilized in this study can be used to investigate the regional diversity of rotifers and that the 145 mtCOI sequences obtained will be helpful to uncover rotifer species diversity in South China. Intraspecific genetic variation is high in some species in our study, especially within some "cosmopolitan" species or species complexes, such as *B. calyciflorus*. Therefore, detailed sampling and in-depth analysis for detecting cryptic species are necessary for uncovering the full regional diversity of rotifers.

Supplementary Materials: The following are available online at https://www.mdpi.com/article/10.3390/d13080390/s1, Figure S1: Average body weight of individual rotifers of 44 species, Table S1: The abundances of rotifers in 23 reservoirs; Table S2: The information of 23 investigated reservoirs in the Hanjiang River Basin; Table S3: Morphological identification of rotifers from both quantitative and qualitative samples in the 23 investigated reservoirs and species identified by COI sequencing; Table S4: Amplification of COI sequence in 23 reservoir rotifers.

Author Contributions: Conceptualization, Y.-N.Z. and S.-L.X.; methodology, Y.-N.Z. and S.-L.X.; software, Q.H. and P.L.; validation, Y.-N.Z., S.-L.X., Q.H., P.L., and B.-P.H.; formal analysis, Y.-N.Z., S.-L.X., and B-P.H.; investigation, Y.-N.Z.; resources, Y.-N.Z., S.-L.X., Q.H., P.L., and B.-P.H.; data curation, Y.-N.Z.; writing—original draft preparation, Y.-N.Z. and B.-P.H.; writing—review and editing, Y.-N.Z., S.-L.X., Q.H., P.L., and B.-P.H.; visualization, Y.-N.Z. and S.-L.X.; supervision, Y.-N.Z. and B.-P.H.; project administration, S.-L.X., Q.H., and P.L.; funding acquisition, B.-P.H. All authors have read and agreed to the published version of the manuscript.

Funding: This research was funded by grants from NSF of China, grant numbers 31901098 and 32171538.

Institutional Review Board Statement: Not applicable.

Informed Consent Statement: Not applicable.

Data Availability Statement: Newly sequenced COI sequence in this study is openly available in NCBI with GenBank number: MZ438131-MZ438183, MZ438185-MZ438227, MZ461497-MZ461518, and MZ461520-MZ461546. The collection for the specimens and their sequences are available in the project file "The COI of Rotifers from 23 reservoirs in the Hanjiang River Basin in southern China", code RSC on the Barcode of Life Data System (BOLD) at http://www.boldsystems.org (accessed on 16 July 2021).

Acknowledgments: The authors are grateful to Henri Dumont (Belgium) and Eric Zeus Rizo (Philippines) for reading and commenting. The authors acknowledge the National Natural Science Foundation of China, grant numbers 31901098 and 32171538. The authors thank the help of the colleagues from the Department of Ecology and Institute of Hydrobiology at the Jinan University during the course of this work.

Conflicts of Interest: The authors declare no conflict of interest.

References

1. Duggan, I.C. The ecology of periphytic rotifers. *Rotifera IX* **2001**, *153*, 139–148.
2. Wallace, R.L.; Snell, T.W.; Ricci, C.; Nogrady, T. Guides to the identification of the microinvertebrates of the continental waters of the world. In *Rotifera Part 1: Biology, Ecology and Systematics*; Dumont, H.J.F., Ed.; Kenobi Production & Backhuys Publishers: Hague, The Netherlands, 2006; Volume 23, p. 299.
3. Segers, H. Global diversity of rotifers (Rotifera) in freshwater. *Hydrobiologia* **2007**, *595*, 49–59. [CrossRef]
4. Wallace, R.L. Rotifers: Exquisite metazoans. *Integr. Comp. Biol.* **2002**, *42*, 660–667. [CrossRef] [PubMed]
5. Arndt, H. Rotifers as predators on components of the microbial web (bacteria, heterotrophic flagellates, ciliates)—A review. *Hydrobiologia* **1993**, *255–256*, 231–246. [CrossRef]
6. Schmid-Araya, J.M.; Schmid, P.E. Trophic relationships: Integrating meiofauna into a realistic benthic food web. *Freshw. Biol.* **2000**, *44*, 149–163. [CrossRef]
7. Rao, T.R.; Sarma, S.S.S. Demographic parameters of *Brachionus patulus* Muller (Rotifera) exposed to sublethal DDT concentrations at low and high food levels. *Hydrobiologia* **1986**, *139*, 193–200.
8. Koste, W. *Rotatoria. Die Rädertiere Mitteleuropas, Bd. II.*; Gebrüder: Berlin, Germany, 1978.
9. Pavón-Meza, E.L.; Sarma, S.S.S.; Nandini, S. Combined effects of temperature, food (*Chlorella vulgaris*) concentration and predation (*Asplanchna girodi*) on the morphology of *Brachionus havanaensis* (Rotifera). *Hydrobiologia* **2007**, *593*, 95–101. [CrossRef]
10. Segers, H.; De Smet, W.H. Diversity and endemism in Rotifera: A review, and *Keratella* Bory de St Vincent. *Biodivers. Conserv.* **2007**, *8*, 69–82.
11. Fontaneto, D.; Boschetti, C.; Ricci, C. Cryptic diversification in ancient asexuals: Evidence from the bdelloid rotifer *Philodina flaviceps*. *J. Evol. Biol.* **2008**, *21*, 580–587. [CrossRef]
12. Fontaneto, D. Molecular phylogenies as a tool to understand diversity in rotifers. *Int. Rev. Hydrobiol.* **2014**, *99*, 178–187. [CrossRef]
13. Mills, S.; Alcántara-Rodríguez, J.A.; Ciros-Pérez, J.; Gómez, A.; Hagiwara, A.; Galindo, K.H.; Jersabek, C.D.; Malekzadeh-Viayeh, R.; Leasi, F.; Lee, J.-S.; et al. Fifteen species in one: Deciphering the *Brachionus plicatilis* species complex (Rotifera, Monogononta) through DNA taxonomy. *Hydrobiologia* **2017**, *796*, 39–58. [CrossRef]
14. Michaloudi, E.; Papakostas, S.; Stamou, G.; Neděla, V.; Tihlaříková, E.; Zhang, W.; Declerck, S.A.J. Reverse taxonomy applied to the *Brachionus calyciflorus* cryptic species complex: Morphometric analysis confirms species delimitations revealed by molecular phylogenetic analysis and allows the (re) description of four species. *PLoS ONE* **2018**, *13*, e0203168. [CrossRef]
15. García-Morales, A.E.; Domínguez-Domínguez, O. Cryptic species within the rotifer *Lecane bulla* (Rotifera: Monogononta: Lecanidae) from North America based on molecular species delimitation. *Rev. Mex. Biodiv.* **2020**, *91*, 1–12. [CrossRef]
16. John, G.; Elizabeth, W. *Brachionus calyciflorus* is a species complex: Mating behavior and genetic differentiation among four geographically isolated strains. *Hydrobiologia* **2005**, *546*, 257–265.
17. Gómez, A.; Temprano, M.; Serra, M. Ecological genetics of a cyclical parthenogen in temporary habitats. *J. Evol. Biol.* **1995**, *8*, 601–622. [CrossRef]
18. Ortells, R.; Snell, T.W.; Gómez, A.; Serra, M. Patterns of genetic differentiation in resting egg banks of a rotifer species complex in Spain. *Hydrobiologia* **2000**, *149*, 529–551. [CrossRef]
19. Gómez, A. Allozyme electrophoresis: Its application to rotifers. *Hydrobiologia* **1998**, *387/388*, 385–393. [CrossRef]
20. Kordbacheh, A.; Wallace, R.L.; Walsh, E.J. Evidence supporting cryptic species within two sessile microinvertebrates, *Limnias melicerta* and *L. ceratophylli* (Rotifera, Gnesiotrocha). *PLoS ONE* **2018**, *13*, e0205203. [CrossRef]
21. Fontaneto, D.; Jondelius, U. Broad taxonomic sampling of mitochondrial cytochrome c oxidase subunit I does not solve the relationships between Rotifera and Acanthocephala. *Zool. Anz.* **2011**, *250*, 80–85. [CrossRef]

22. Hwang, D.S.; Dahms, H.U.; Park, H.G.; Lee, J.S. A new intertidal *Brachionus* and intrageneric phylogenetic relationships among *Brachionus* as revealed by allometry and CO1-ITS1 gene analysis. *Zool. Stud.* **2013**, *52*, 1–10. [CrossRef]
23. Birky, C.W.; Wolf, C.; Maughan, H.; Herbertson, L.; Henry, E. Speciation and selection without sex. *Hydrobiologia* **2005**, *546*, 29–45. [CrossRef]
24. García-Morales, A.E.; Elías-Gutiérrez, M. DNA barcoding of freshwater Rotifera in Mexico: Evidence of cryptic speciation in common rotifers. *Mol. Ecol. Resour.* **2013**, *13*, 1097–1107. [CrossRef] [PubMed]
25. Folmer, O.; Black, M.; Hoeh, W.; Lutz, R.; Vrijenhoek, R. DNA primers for amplification of mitochondrial cytochrome c oxidase subunit I from diverse metazoan invertebrates. *Mol. Mar. Biol. Biotechnol.* **1994**, *3*, 294–299.
26. Meyer, C.P.; Geller, J.B.; Paulay, G. Fine scale endemism on coral reefs: Archipelagic differentiation in turbinid gastropods. *Evolution* **2005**, *59*, 113–125. [CrossRef]
27. Rico-Martínez, R.; Snell, T.W.; Shearer, T.L. Synergistic toxicity of Macondo crude oil and dispersant Corexit 9500A® to the *Brachionus plicatilis* species complex (Rotifera). *Environ. Pollut.* **2013**, *173*, 5–10. [CrossRef]
28. Wilts, E.F.; Bruns, D.; Fontaneto, D.; Ahlrich, W.H. Phylogenetic study on *Proales daphnicola* Thompson, 1892 (Proalidae) and its relocation to Epiphanes (Rotifera: Epiphanidae). *Zool. Anz.* **2012**, *251*, 180–196. [CrossRef]
29. Prosser, S.; Martínez-Arce, A.; Elías-Gutiérrez, M. A new set of primers for COI amplification from freshwater microcrustaceans. *Mol. Ecol. Resour.* **2013**, *13*, 1151–1155. [CrossRef]
30. Elías-Gutiérrez, M.; Valdez-Moreno, M.; Topan, J.; Young, M.R.; Cohuo-Colli, J.A. Improved protocols to accelerate the assembly of DNA barcode reference libraries for freshwater zooplankton. *Ecol. Evol.* **2018**, *8*, 3002–3018. [CrossRef] [PubMed]
31. Zhang, Y.N.; Xu, S.L.; Sun, C.H.; Dumont, H.; Han, B.P. A new set of highly efficient primers for COI amplification in rotifers. *Mitochondrial DNA Part B* **2021**, *6*, 636–640. [CrossRef]
32. Lan, Z.H.; Du, L.M. Studies on zooplankton in the Hanjiang River. *J. Hanshan Norm. Univ.* **1996**, 97–104.
33. Lin, Q.Q.; Zhao, S.Y.; Han, B.P. Rotifer distribution in tropical reservoirs, Guangdong Province, China. *Acta Ecol. Sin.* **2005**, *25*, 1123–1131.
34. Zhang, Z.X.; Huang, X.F. *Freshwater Plankton Research Method*; Science Press: Beijing, China, 1991.
35. Hillebrand, H.; Dürselen, C.D.; Kirschtel, D. Biovolume calculation for pelagic and benthic microalgae. *J. Phycol.* **1999**, *35*, 403–424. [CrossRef]
36. Xu, Z.L.; Chen, Y.Q. Aggregated intensity of dominant species of zooplankton in autumn in the East China Sea and Yellow Sea. *Chin. J. Ecol.* **1989**, *8*, 13–15.
37. Kearse, M.; Moir, R.; Wilson, A.; Stones-Havas, S.; Cheung, M.; Sturrock, S.; Buxton, S.; Cooper, A.; Markowitz, S.; Duran, C.; et al. Geneious Basic: An integrated and extendable desktop software platform for the organization and analysis of sequence data. *Bioinformatics* **2012**, *28*, 1647–1649. [CrossRef]
38. Katoh, K.; Standley, D.M. MAFFT multiple sequence alignment software version 7: Improvements in performance and usability. *Mol. Biol. Evol.* **2013**, *30*, 772–780. [CrossRef]
39. Ranwez, V.; Douzery, E.J.P.; Cambon, C.; Chantret, N.; Delsuc, F. MACSE v2: Toolkit for the alignment of coding sequences accounting for frameshifts and stop codons. *Mol. Biol. Evol.* **2018**, *35*, 2582–2584. [CrossRef]
40. Henzinger, T.A.; Jhala, R.; Majumdar, R.; Sutre, G. Software verification with BLAST. In *International SPIN Workshop on Model Checking of Software*; Ball, T., Rajamani, S.R., Eds.; Springer: Berlin/Heidelberg, Germany, 2003; pp. 235–239.
41. Kumar, S.; Stecher, G.; Li, M.; Knyaz, C.; Tamura, K. MEGA X: Molecular evolutionary genetics analysis across computing platforms. *Mol. Biol. Evol.* **2018**, *35*, 1547–1549. [CrossRef]
42. Leutbecher, C. A routine method of DNA-extraction from extremely small metazoans, eg single rotifer specimens for RAPD-PCR analyses. *Hydrobiologia* **2000**, *437*, 133–137. [CrossRef]
43. Makino, W.; Maruoka, N.; Nakagawa, M.; Takamura, N. DNA barcoding of freshwater zooplankton in Lake Kasumigaura, Japan. *Ecol. Res.* **2017**, *32*, 481–493. [CrossRef]
44. Hebert, P.D.N.; Cywinska, A.; Ball, S.L.; Dewaard, J.R. Biological identifications through DNA barcodes. *Proc. R. Soc. B* **2003**, *270*, 313–321. [CrossRef]
45. Leray, M.; Yang, J.Y.; Meyer, C.P.; Mills, S.C.; Agudelo, N.; Ranwez, V.; Boehm, J.T.; Machida, R.J. A new versatile primer set targeting a short fragment of the mitochondrial COI region for metabarcoding metazoan diversity: Application for characterizing coral reef fish gut contents. *Front. Zool.* **2013**, *10*, 34. [CrossRef]
46. Lim, N.K.M.; Tay, Y.C.; Srivathsan, A.; Yeo, D.C.J. Next-generation freshwater bioassessment: eDNA metabarcoding with a conserved metazoan primer reveals species-rich and reservoir-specific communities. *R. Soc. Open Sci.* **2016**, *3*, 160635. [CrossRef]
47. Yang, J.H.; Zhang, X.W.; Zhang, W.W.; Sun, J.Y.; Xie, Y.W.; Zhang, Y.M.; Burton, G.A.; Yu, H.X. Indigenous species barcode database improves the identification of zooplankton. *PLoS ONE* **2017**, *12*, e0185697. [CrossRef]
48. Crampton-Platt, A.; Timmermans, M.J.T.N.; Gimmel, M.L.; Kutty, S.N.; Cockerill, T.D.; Khen, C.V.; Vogler, A.P. Soup to tree: The phylogeny of beetles inferred by mitochondrial metagenomics of a Bornean rainforest sample. *Mol. Biol. Evol.* **2015**, *32*, 2302–2316. [CrossRef]
49. Gómez-Rodríguez, C.; Crampton-Platt, A.; Timmermans, M.J.T.N.; Baselga, A.; Vogler, A.P. Validating the power of mitochondrial metagenomics for community ecology and phylogenetics of complex assemblages. *Methods Ecol. Evol.* **2015**, *6*, 883–894. [CrossRef]
50. Derry, A.M.N.; Hebert, P.D.; Prepas, E.E. Evolution of rotifers in saline and subsaline lakes: A molecular phylogenetic approach. *Limnol. Oceanogr.* **2003**, *48*, 675–685. [CrossRef]

51. Koste, W.; Shiel, R.J. Rotifera from Australian inland waters. II. Epiphanidae and Brachionidae (Rotifera: Monogononta). *Invertebr. Syst.* **1987**, *1*, 949–1021. [CrossRef]
52. Xiang, X.L.; Xi, Y.L.; Wen, X.L.; Zhang, J.Y.; Ma, Q. Spatial patterns of genetic differentiation in *Brachionus calyciflorus* species complex collected from East China in summer. *Hydrobiologia* **2010**, *638*, 67–83. [CrossRef]

Article

Checklist of Arrenurids (Acari: Hydrachnidia: Arrenuridae) of Mexico, with New Records from the Yucatan Peninsula, and the Description of Five New Species of the Subgenera *Megaluracarus* and *Dadayella* †

Lucia Montes-Ortiz [1], Manuel Elías-Gutiérrez [1,*] and Marcia María Ramírez-Sánchez [2]

[1] Unidad Chetumal, El Colegio de la Frontera Sur, Avenida Centenario km 5.5, Quintana Roo, Chetumal 77014, Mexico; lumontes@ecosur.edu.mx
[2] Facultad de Ciencias, Coyoacán Campus, Universidad Nacional Autónoma de México, Ciudad de Mexico 04510, Mexico; ramirezsanchezmarciam@gmail.com
* Correspondence: melias@ecosur.mx
† urn:lsid:zoobank.org:pub:96EC5085-0EE0-47C3-9B64-E0437BA71EA6.

Abstract: A checklist of arrenurids of Mexico is presented, including three new records from the Yucatan Peninsula. We provide updated descriptions of *Arrenurus mexicanus*, *A.* (*Megaluracarus*) *colitus*, and *A.* (*Megaluracarus*) *marshalli*. Additionally, four new species of the subgenus *Megaluracarus* and one of *Dadayella* are described by using integrative taxonomy: *Arrenurus* (*Megaluracarus*) *eduardoi* n. sp., characterized by a large, thorn-shaped hump in the middle dorsal shield; *Arrenurus* (*Megaluracaurus*) *federicoi* n. sp., with large pores in the body, including the idiosoma; *Arrenurus* (*Megaluracarus*) *ecosur* n. sp., with a peculiar pattern of setation in the legs; *Arrenurus* (*Megaluracarus*) *beatrizae* n. sp., with a short cauda with two pairs of lateral notches, and *Arrenurrus* (*Dadayella*) *cristinae* n. sp., characterized by a male cauda with two falcate setae. Non-destructive methods allowed the taking of scanning electron microscope images and DNA sequencing of the designed type material. All new species have a divergence using the DNA mitochondrial gene COI from 21.1% to 28.6% within them. With these records and descriptions, the number of *Arrenurus* registered for Mexico increases to 42, most of them from a single locality.

Keywords: taxonomy; morphology; DNA barcodes; COI; karstic; *Arrenurus*

Citation: Montes-Ortiz, L.; Elías-Gutiérrez, M.; Ramírez-Sánchez, M.M. Checklist of Arrenurids (Acari: Hydrachnidia: Arrenuridae) of Mexico, with New Records from the Yucatan Peninsula, and the Description of Five New Species of the Subgenera *Megaluracarus* and *Dadayella*. *Diversity* **2022**, *14*, 276. https://doi.org/10.3390/d14040276

Academic Editor: Michael Wink

Received: 19 March 2022
Accepted: 2 April 2022
Published: 7 April 2022

1. Introduction

Arrenurus Dugés, 1834 is the most species-rich water mite genus, with approximately 1000 species described worldwide, and is currently divided into 11 accepted subgenera: *Arrenurus*, *Arrhenuropsides*, *Arrhenuropsis*, *Brevicaudaturus*, *Dadayella*, *Dividuracarus*, *Megaluracarus*, *Micruracarus*, *Rhinophoracarus*, *Truncaturus*, and *Stygarrenurus* [1,2]. In Mexico, 37 species are reported, divided into five subgenera: *Arrenurus*, *Arrhenuropsis*, *Dadayella*, *Megaluracarus*, and *Truncaturus*. Six of the total species recorded are only known from the Yucatan Peninsula [3–7]. Likely, this number does not represent the total species number for the genus in this region, as Mexico is one of the countries with the greatest biological diversity in the world due to its complex topography, the variety of climates, and the convergence of the two main biogeographic zones of the Americas: The Nearctic and the Neotropics [8]. In particular, the Yucatan Peninsula, one of the worlds' largest karstic aquifer systems, has a great diversity of aquatic ecosystems with unique geohydrological characteristics [9,10].

Recently, Montes-Ortiz and Elías-Gutiérrez [11] studied the water mites' diversity from 24 sites in the Yucatan Peninsula using the sequences of the mitochondrial cytochrome subunit I (COI). Their main results indicated the presence of 77 genetic groups or putative species represented through a barcode index number (BIN), and 17 of them corresponded to the genus *Arrenurus*. This result illustrates the potential water mite diversity for this region since only six species are described in this area.

Megaluracarus Viets, 1911 can be considered the most complex subgenus of *Arrenurus* in terms of diversity and range of morphological characteristics [6]. Another subgenus, *Dadayella*, is difficult in taxonomy, since some of its descriptions have been based only on females [1,4]. This study supplies a checklist of arrenurids from Mexico, providing three new records and describing four new species of the subgenus *Megaluracarus*: *A. eduardoi* n. sp., *A. federicoi* n. sp., *A. ecosur* n. sp., and *A. beatrizae* n. sp. and one from the *Dadayella* subgenus, *A. cristinae* n. sp., using morphological and molecular data.

2. Materials and Methods

The specimens were collected in five different karst systems from the southern Yucatan Peninsula (Figure 1) during a sampling survey in April and August 2019 [11] using light traps and a hand net with a mesh size of 50 μm. The mites collected with the light trap were sieved, washed, and fixed in 96% cold ethanol [12]. Specimens collected with a hand net were sorted in the field from the samples using a pipette and fixed in ethanol 96%. All specimens were stored at −18 °C for at least seven days [13].

Figure 1. Sampling sites. (**A**) Bacalar lagoon, Cenote Cocalitos (**front**) and Cenote Azul (**back**); (**B**) Silvituc lagoon; (**C**) Ramonal wetland; (**D**) Acapulquito stream.

The arrenurids were separated under a stereomicroscope; representative morphospecies were photographed using a Zeiss Discovery stereomicroscope with an attached Eos Rebel T3i camera. Five individuals (when this was possible) from every morphospecies were used for DNA analyses, using a non-destructive extraction method [14]. After the process, most of the specimens were recovered, and the selected type was dissected and mounted in glycerin jelly. In the case of new species from subgenus *Megaluracarus*, after the DNA extraction, detailed images were obtained with a low vacuum and a freezing platina to −31 °C attached to a Jeol JSM-6010 scanning electron microscope (SEM) at the Chetumal Unit of El Colegio de la Frontera Sur. This non-destructive method allows the recovery of the studied specimens, as it does not need the critical drying point and gold coating. Subsequently, whole specimens and the dissected parts were examined and measured under a compound microscope, LW Scientific. The drawings were made using a graphic digital tablet on Inkscape V. 0.92.4 (www.inkscape.org, accessed on 18 March 2022) [15].

All specimen preparations recovered were deposited in the Reference Collection of Zooplankton (ECO-CH-Z) at El Colegio de la Frontera Sur (ECOSUR, Chetumal, Mexico), except for the two paratypes of *Arrenurus* (*Megaluracarus*) *beatrizae* n. sp. deposited in the water mites collection of the Aquatic Zoology Laboratory (AAL) at Facultad de Ciencias, Universidad Nacional Autónoma de México.

Molecular analysis. DNA extraction was performed using a standard glass fiber method [16] modified, following Porco et al. [14]. For voucher recovery, specimens were recovered after the lysis step from the glass fiber filter plates or the 96-well original plates and preserved in Koenike's fluid. For the PCR process, see [11,13,17]. PCR products were visualized on 2% agarose gels (E-Gel 96 Invitrogen), and positive PCR products were selected for sequencing bidirectionally at Eurofins Scientific in Louisville, Kentucky.

All sequences were edited using Codon Code v. 3.0.1 and uploaded to the Barcode of Life Database (BOLD: www.boldsystems.org, accessed on 18 March 2022) and are in the public dataset DS_ARRENURI; DOI: XXXX. The sequences of the new species of *Megaluracarus* were included in a maximum likelihood (ML) tree generated with 1000 replicates using MEGA version X [18]. Two sequences of the *Krendowskia* genus were used to root the three GENWM130-16 and GENWM138-16 (Table 1).

Finally, a total of 1111 good-quality public sequences of the genus *Arrenurus* from the BOLD database were used to build a neighbor-joining (NJ) tree for a general comparison with all sequenced specimens from the globe.

We provide the consensus sequence for each species described in this study as an additional character. The resulting tree is included as Supplementary File S1.

All measurements are given in μm. Terminology and abbreviations in the descriptions of the new species follow [1,6,19].

Abbreviations used: BIN = barcode index number; Cxgl-1 = coxoglandularia 1; Cxgl-2 = coxoglandularia 2; Cxgl-4 = coxoglandularia 4; Cx-I–IV = first to fourth coxae; Dgl-1–4 = first to fourth dorsoglandularia; L = length; IV-Leg-1–6 = first to sixth segments of the fourth leg; P1–P5 = first to fifth palp segments; W = width.

Nomenclatural Acts

This published work and the nomenclatural acts were registered in ZooBank, the online registration system for the ICZN. The ZooBank Life Science Identifiers (LSIDs) were resolved, and the associated information can be viewed through any standard web browser by appending the LSID to the prefix http://zoobank.org/, accessed on 18 March 2022. The online version of this work is archived and available from the following digital repositories: Diversity, Basel.

Table 1. Sequences used in the descriptions for this study.

Species	Type Locality *	Accession Number of the Type Material	ID in BOLD	Barcode Index Number
Arrenurus (*Megaluracarus*) *eduardoi* n. sp.	Acapulquito, Riviera del Río Hondo, Quintana Roo (Mexico) *	ECO-CH_000XXXXX	YUCWM195-20 YUCWM087-19 YUCWM085-19 YUCWM084-19	AEA7844
Arrenurus (*Megaluracarus*) *federicoi* n. sp.	Acapulquito, Riviera del Río Hondo, Quintana Roo (Mexico) *	ECO-CH_000XXXXX	YUCWM198-20 YUCWM196-20 YUCWM197-20	AEB7095

Table 1. *Cont.*

Species	Type Locality *	Accession Number of the Type Material	ID in BOLD	Barcode Index Number
Arrenurus (*Megaluracarus*) *ecosur* n. sp.	Bacalar lagoon *, Cenote Cocalitos, Chichancanab, Muyil lagoon 1, Cenote Azul, Cenote Chancah, Cenote Sijil Noh Ha, Cenote del Padre, Quintana Roo (Mexico)	ECO-CH_000XXXXX	BACWM287-16 BACWM016-15 BACWM014-15 BACWM007-15 BACWM005-15 BACWM003-15 BACWM002-15 BACWM244-15 BACWM243-15 BACWM193-15 BACWM133-15 BACWM127-15 BACWM100-15 BACWM083-15 BACWM082-15 BACWM078-15 BACWM074-15 BACWM073-15 BACWM059-15 YUCWM103-19 YUCWM036-19 YUCWM035-19 YUCWM047-19 YUCWM040-19 YUCWM039-19 YUCWM037-19 YUCWM034-19 YUCWM032-19 YUCWM031-19 CAZUL452-17 SKAAN-079-19 SKAAN-019-19 SKAAN-370-19 BACWM047-15 BACWM046-15 BACWM045-15 BACWM043-15 BACZP2234-16 SKAAN-160-19	ACX8463
Arrenurus marshallae	Silvituc lagoon *, Escarcega, Campeche (Mexico)	ECO-CH_000XXXXX	EXD479-20 EXD493-20 EXD510-20 EXD567-21	ACL2521
Arrenurus (*Dadayella*) *cristinae* n. sp.	Ramonal, Quintana Roo * (Mexico)	ECO-CH_000XXXXX	YUCWM012-19 YUCWM017-19	AEA7842

3. Results

Before our study, there were 37 *Arrenurus* species registered for Mexico, grouped into five subgenera: *Megaluracarus*, *Arrenurus*, *Dadayella*, *Truncaturus*, and *Arrhenuropsis* and one species represented by a female, without a subgenus assigned, *A.* (?) *nayaritensis* (Table 2). Of these, six species are distributed in the Yucatan Peninsula; with our new records and species descriptions (Figure 2), the total number increases to 42 arrenurids registered for the country, and 11 of them are found in the Yucatan Peninsula (Table 2).

Table 2. List of *Arrenurus* species (Acari: Hydrachnidia: Arrenuridae) known from Mexico.

Genus	Subgenus	Specie	Author	Distribution	Habitat
Arrenurus	*Arrenurus*	*dentipetiolatus*	Marshall, 1908	United States of America, Mexico (Oaxaca/Guanajuato)	Pond
Arrenurus	*Arrenurus*	*valencius*	Marshall, 1919	Venezuela, Cuba, Haití, Guatemala, Mexico (Campeche/Tabasco)	Water-filled roadside
Arrenurus	*Arrenurus*	*munovus*	Cook, 1980	Mexico (Chiapas)	Stream
Arrenurus	*Arrenurus*	*wucabus*	Cook, 1980	Mexico (Oaxaca)	Pond
Arrenurus	*Arrenurus*	*tamaulipensis*	Cramer and Cook, 1992	Mexico (Tamaulipas)	Lake
Arrenurus	*Arrenurus*	*xochimilcoensis*	Cramer and Cook, 1992	Mexico (Mexico City)	Lake
Arrenurus	*Megaluracarus*	*manubriator*	Marshall, 1903	Marshall, 1903	Standing waters
Arrenurus	*Megaluracarus*	*birgei*	Marshall, 1903	United States of America, Mexico (Tabasco)	Pond
Arrenurus	*Megaluracarus*	*marshallae*	Piersig 1904	United Sates of America, Canada, Mexico (Campeche)	Lagoon
Arrenurus	*Megaluracarus*	*gricalus*	Cook, 1980	Mexico (Campeche)	Water-filled ditch
Arrenurus	*Megaluracarus*	*hartesus*	Cook, 1980	Mexico (Veracruz)	Pond
Arrenurus	*Megaluracarus*	*neoexpansus*	Cook, 1980	Mexico (Tabasco)	Pond
Arrenurus	*Megaluracarus*	*tabascoensis*	Cook, 1980	Mexico (Tabasco)	Pond
Arrenurus	*Megaluracarus*	*trassamus*	Cook, 1980	Mexico (Campeche)	Water-filled ditch
Arrenurus	*Megaluracarus*	*zitavus*	Cook, 1980	Mexico (Tabasco)	Pond
Arrenurus	*Megaluracarus*	*campechensis*	Cook, 1980	Mexico (Campeche)	Water-filled ditch
Arrenurus	*Megaluracarus*	*wolardus*	Cook, 1980	Mexico (Campeche)	Water-filled ditch
Arrenurus	*Megaluracarus*	*costeroae*	Cramer and Cook, 1992	Mexico (Veracruz, Colima)	Pond
Arrenurus	*Megaluracarus*	*alloexpansus*	Cramer and Cook, 1992	Mexico (Tamaulipas)	Lake
Arrenurus	*Megaluracarus*	*apizanus*	Cramer and Cook, 1992	Mexico (Colima)	Not specified
Arrenurus	*Megaluracarus*	*catoi*	Cramer and Cook, 1992	Mexico (Tamaulipas)	Lake
Arrenurus	*Megaluracarus*	*champayanus*	Cramer and Cook, 1992	Mexico (Tamaulipas)	Lake
Arrenurus	*Megaluracarus*	*colitus*	Cramer and Cook, 1992	Mexico (Tamaulipas)	Lake
Arrenurus	*Megaluracarus*	*anae*	Cramer and Cook, 1998	Mexico (Tamaulipas)	Lake
Arrenurus	*Megaluracarus*	*anitahoffmannae*	Ramírez-Sánchez and Rivas, 2013	Mexico (Tabasco)	Lake, pond, canal
Arrenurus	*Megaluracarus*	*olmeca*	Ramírez-Sánchez and Rivas, 2013	Mexico (Tabasco)	Lake, pond, canal
Arrenurus	*Megaluracarus*	*maya*	Ramírez-Sánchez and Rivas, 2013	Mexico (Yucatan/Quintana Roo)	Cenote
Arrenurus	*Megaluracarus*	*urbanus*	Ramírez-Sánchez and Rivas, 2013	Mexico (Mexico City)	Canal
Arrenurus	*Megaluracarus*	*eduardoi* n. sp.	Montes-Ortiz et al., 2022	Mexico (Quintana Roo)	Pool (in a stream)
Arrenurus	*Megaluracarus*	*federicoi* n. sp.	Montes-Ortiz et al., 2022	Mexico (Quintana Roo)	Pool (in a stream)
Arrenurus	*Megaluracarus*	*ecosur* n. sp.	Montes-Ortizet al., 2022	Mexico (Quintana Roo)	Cenote, lagoon, wetlands
Arrenurus	*Megaluracarus*	*beatrizae* n. sp.	Montes-Ortiz et al., 2022	Mexico (Quintana Roo/Tabasco)	Wetland, lagoon
Arrenurus	*Dadayella*	*zempoala*	Cook, 1980	Mexico (Mexico state)	Small stream
Arrenurus	*Dadayella*	*adrianae*	Cramer and Cook, 1992	Mexico (Colima/Michoacán)	Wetland, lagoon
Arrenurus	*Dadayella*	*veracruzensis*	Cramer and Cook, 1992	Mexico (Veracruz)	Pond
Arrenurus	*Dadayella*	*aztecus*	Cramer and Cook, 1992	Mexico (Veracruz)	Wetland, lagoon
Arrenurus	*Dadayella*	*colimensis*	Cramer and Cook, 1992	Mexico (Colima)	Wetland, lagoon
Arrenurus	*Dadayella*	*cristinae* n. sp.	Montes-Ortizet al., 2022	Mexico (Quintana Roo)	Wetland
Arrenurus	*Truncaturus*	*plevamus*	Cook, 1980	Costa Rica, Mexico (Guerrero)	Small stream
Arrenurus	*Truncaturus*	*zukovus*	Cook, 1980	Mexico (Chiapas)	Gravel-bottom stream
Arrenurus	*Truncaturus*	*teoceloensis*	Rivas and Cramer, 1998	Mexico (Veracruz)	Stream
Arrenurus	*Arrhenuropsis*	*mexicanus*	Cramer and Cook, 1992	Mexico (Tamaulipas/Colima)	Lagoon
	?	*nayaritensis*	Cook, 1980	Mexico (Nayarit)	Small stream

We obtained four sequences for *Arrenurus* (*Megaluracarus*) *marshallae*, and 14 more are public in the BOLD database with the associated BIN ACL2521. In the case of *A.* (? *Arrhenuropsis*) *mexicanus* Cramer and Cook, 1992, we could not obtain the genetic information. However, we provide morphological notes. *A.* (*Megaluracarus*) *colitus* (Cramer and Cook, 1992) was represented by one sequence and the BIN AEA8234. Some measurements and notes are provided for these three species to achieve a more complete record. For *A.* (*Megaluracarus*) *eduardoi* n. sp., we obtained four sequences, and the BIN AEA7844 was assigned. *A.* (*Megaluracarus*) *federicoi* n. sp. has three sequences and the BIN AEB7095. *A.* (*Megaluracarus*) *ecosur* n. sp. is represented by 39 sequences and the BIN ACX8463. In the case of *A.* (*Megaluracarus*) *beatrizae* n. sp., we were unable to obtain the genetic information. Nonetheless, all the morphological data are given. Finally, for *A.* (*Dadayella*) *cristinae* n. sp., we obtained two sequences, and the BIN assigned was AEA 7842.

In the NJ tree comparing our material with all worldwide, sequenced arrenurids (Supplementary File S1), the 1111 specimens represented 148 BINs, of which only 50 have a taxonomical identification. The BINs reported for Mexico (including those used for descriptions or new records in this study) are separated from those reported for other world regions, except for *A. marshallae*, BIN ACL2521 and BIN ACL2418, which are found in Canada as well.

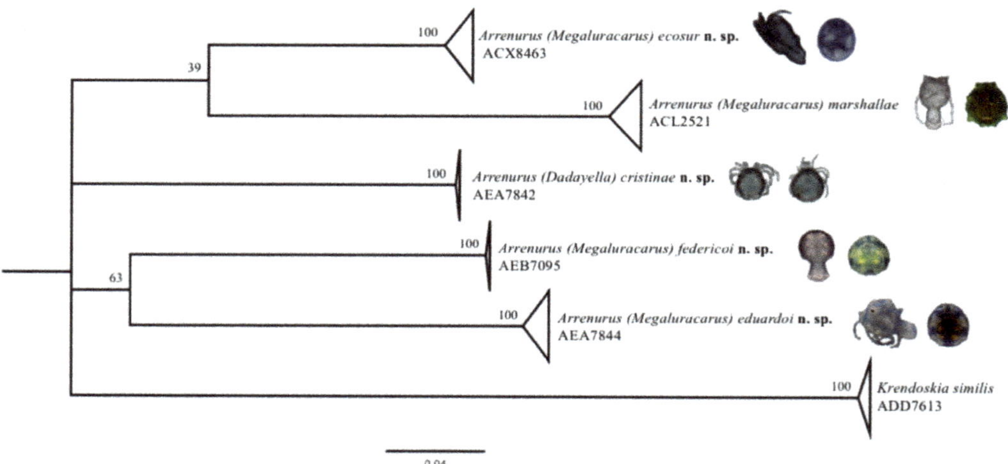

Figure 2. Maximum likelihood tree, based on COI sequences. Bootstrap support values were generated after 1000 replicates. The name is followed by the barcode index number and corresponding photograph of male and female. *Krendoskia similis* was used as an outgroup.

Systematic Part

Family Arrenuridae (Thor, 1900)
Genus *Arrenurus* (Dugés, 1834)
Subgenus *Arrhenuropsis* (Viets, 1954)
Arrenurus (? *Arrhenuropsis*) *mexicanus* (Cramer and Cook, 1992), (Figure 3).

Material examined: One male from Ramonal pond (access number: ECO-CH-Z-10608), Quintana Roo, 19°23′31″ N, −82°37′27″ W; emergent vegetation, 14 April 2019.

Description. MALE: Idiosoma blue-greenish with white areas in the Dgl 1–4 regions, 799 L without petiole and 493 W. Dorsal shield small, oval, and located in the anterior part of the dorsum, 296 L and 345 W (Figure 3A). Genital field 394 W, gonopore 69 L and 48 W (Figure 3B). Dorsal L of palpal segments L: P1: 27; P2: 74; P3: 29; P4: 84; P5: 84. Dorsal L of fourth leg segments: IV-Leg-1: 32; IV-Leg-2: 104; IV-Leg-3: 101; IV-Leg-4: 148; IV-Leg-5: 151; IV-Leg-6: 109.

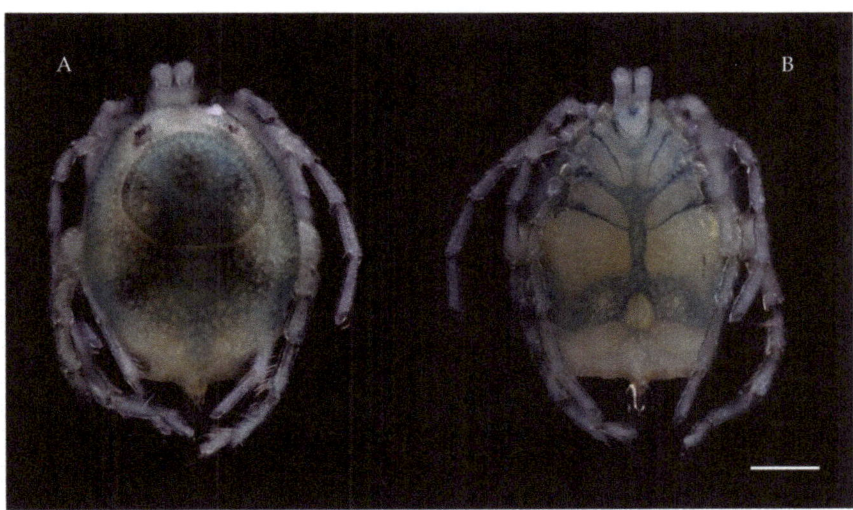

Figure 3. *Arrenurus* (? *Arrhenuropsis) mexicanus* (Cramer and Cook), male. (**A**) Habitus, dorsal view; (**B**) Habitus, ventral view. Scale bar: 200 µm.

Remarks. Male and female were described by Cramer and Cook [4]. Therefore, we only give some diagnostic measurements. This record represents the second of this species for the country.

Distribution. Previously known from the Champayan lagoon, Altamira, Tamaulipas state, 22°22′49″ N, −97°58′34″ W (Mexico).

Subgenus *Megaluracarus* (Viets, 1911).

Arrenurus (Megaluracarus) colitus (Cramer and Cook, 1992), (Figure 4).

Material examined: One female from Ramonal pond (access number: ECO-CH-Z-10609), Quintana Roo state, 19°23′31″ N, −82°37′27″ W; emergent vegetation, 14 April 2019.

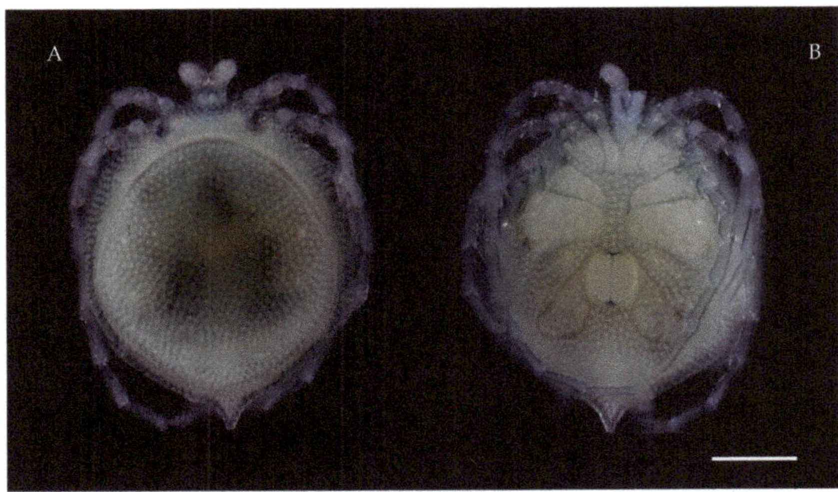

Figure 4. *Arrenurus* (*Megaluracarus*) *colitus* (Cramer and Cook), female. (**A**) Dorsal view; (**B**) ventral view. Scale bar: 200 µm. The sequence of this specimen, recovered after DNA extraction, is represented by the BIN AEA8234.

Description. FEMALE: Idiosoma bluish-green with white areas in the Dgl 1–4 regions, 680 L and 552 W; dorsal furrow complete, dorsal shield 512 L and 483 W (Figure 4A). Genital field 305 W, gonopore 99 L and 116 W (Figure 4B). Dorsal L of fourth leg segments: IV-Leg-1: 74; IV-Leg-2: 101; IV-Leg-3: 99; IV-Leg-4: 119; IV-Leg-5: 106; IV-Leg-6: 116.

Sequence: ATCTATGATACCATTGGAACAGCCCATGCTTTGATTATAATTTTCTTT-ATAGTCATACCCATCATAATTGGAGGATTTGGAAATTGACTCGTTCCCCTCATGTT-AGAAGCTCCAGATATAGCATTCCCACGAATAAATAACATAAGATTTTGATTACTTC-CACCCCCCTTAACACTCCTTCTATCAAGATCATTAACTTCTTCAGGAGCAGGAAC-TGGATGAACAGTTTACCCTCCTTTATCAAGAAATATCGCCCATGGAGGACCTTCA-GTAGACCTAGCAATTTTCTCCCTACACCTTGCAGGTGTGTCCTCAATTTTAGGAGC-AATTAACTTCTTGGCTACATTTATAAACATAAAACCTAAACATATAAAATATGACC-GAATCCCCCTATTTGTAATTTCTATTTTTATCACCGTAATCCTCCTTCTTCTTTCCCT-CCCCGTATTAGCTGGAGCCATTACTATACTTCTTACTGATCGAAATTTTAATACTTC-ATTTTTTGACCCGGCGGGGGAGGAGATCCCATCCTTTACCAACATCTATTT.

Remarks. Female and male were described by Cramer and Cook [4]. We provide some additional measurement data. The chaetotaxy of the palp and IV-Leg-5–6, as well as the position of ventral and dorsal glandularia, agree with the original description. The only noticeable difference is that the first and second coxae tips extend slightly beyond the body proper in our specimen. The associated sequence was obtained, representing a unique BIN (BOLD: AEA8234).

Distribution. Previously known from the Champayan lagoon, Altamira, Tamaulipas state (Mexico). This record represents the second of this species for the country.

Arrenurus (*Megaluracarus*) *marshalli* (Piersig, 1904)

Syn. A. globator (*err*) (Marshall, 1903); A. marshallae (Viets, 1914); *A. marshallae* (Marshall, 1940), (Figure 5).

Material examined: One male, one female, and one nymph from Silvituc lagoon (access number: ECO-CH-Z-10610-10611), Escarcega municipality, Campeche state, 18°37′26″ N, −90°17′5.9″ W, 18 March 2020.

Description. MALE: Idiosoma light-bluish, 962 L (including cauda) and 560 W; dorsal shield 776 L (including cauda) and 422 W (Figure 5A). Genital field 281 W, gonopore 47 L and 61 W (Figure 5B). Dorsal L of palpal segments: P1: 34; P2: 63; P3: 33; P4: 66; P5: 47. Dorsal L of fourth leg segments: IV-Leg-1: 86; IV-Leg-2: 128; IV-Leg-3: 151; IV-Leg-4: 178; IV-Leg-5: 165; IV-Leg-6: 138.

FEMALE: Idiosoma light bluish, 986 L and 907 W; dorsal furrow complete, dorsal shield 719 L and 680 W. Genital field 454 W, gonopore 138 L and 140 W.

Consensus sequence: ACATTATACTTCGCATTCGGAGCTTGATCGGGTATAG-TAGGAGCAAGACTTAGAAGTCTAATCCGACTAGAATTAGGGCAACCAGGAAGAC-TTTTAGGAAATGATCAAATTTACAACACCATTGTTACAGCGCATGCTTTCATTATA-ATCTTCTTTATAGTTATACCAATTATAATCGGAGGATTCGGAAACTGATTAGTACCC-CTAATACTAGCCGCCCCTGATATGGCATTCCCACGAATAAATAATATAAGATTCTG-ACTTCTACCGCCAGCCTTAACACTTCTTTTATCAAGATCGTTAACTTCAGTAGGAG-CAGGAACCGGATGAACAGTCTACCCTCCCCTATCCAGAAACATTGCACATGGTG-GACCTTCAGTAGATATAGCTATCTTCTCATTACATTTAGCAGGAGTCTCCTCAATTT-TAGGAGCTATCAATTTTCTAGCTACAATTTTAAATATAAAGCCTAAACATATAAAAT-ATGACAGAATTCCATTATTTGTAGTTTCAATTTTTATTACAGTAATTCTTCTTTTACTT-TCACTGCCTGTATTAGCAGGAGCTATTACTATACTTCTTACAGATCGAAATTTTAAC-ACCTCTTTCTTCGATCCAGCTGGAGGAGGAGATCCTATTTTATACCAA.

Remarks. Our specimens agree with the descriptions given by Marshall (1903) and Wilson (1961). According to Cook [20], the status of *A. marshallae* is complex because the species is a member of a closely related group characterized by the possession of a long cauda and horn-like projections over the eyes. It can be separated from other species (*A. megalurus megalurus*, *A. megalurus intermedius*) by the slightly indented posterior end of the cauda.

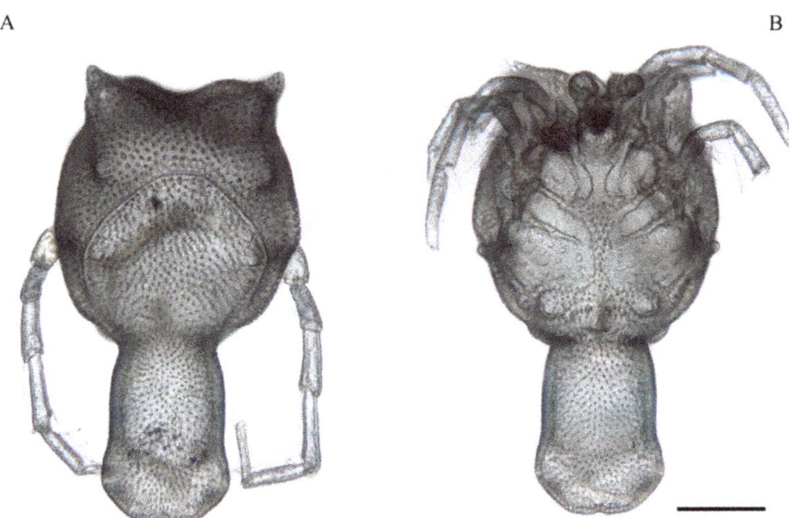

Figure 5. *Arrenurus* (*Megaluracarus*) *marshallae* (Piersig), male. (**A**) Dorsal view; (**B**) ventral view. Scale bar: 200 μm. The sequence of this specimen, recovered after DNA extraction, is represented by the BIN ACL2521.

The sequences obtained in this study with the BIN ACL2521 agree with another 14 public sequences from *A. marshallae*. Some of these were identified morphologically by Bruce Smith. Morphological and molecular identification agree (Figure 2). These public sequences in the BOLD database integrated with the morphology make it possible to verify the records of putative *A. marshallae* in other localities and other members of this complex group.

Distribution. Previously known from the United States and Canada. This record constitutes the first for Mexico.

Arrenurus (*Megaluracarus*) *eduardoi* n. sp., (Figures 6 and 7).

Holotype: Male from Acapulquito stream, Rivera del Río hondo, Othon P. Blanco municipality, Quintana Roo state (access number: ECO-CH-Z-10612), 18° 25′ 55″ N, 88° 31′51″ W; emergent vegetation and submerged roots, 11 April 2019, coll. L. Montes.

Paratypes: Two females and two males, same data as holotype (access number: ECO-CH-Z-10613-10614).

Diagnosis. Male with a large, thorn-shaped hump in the middle of the dorsal shield (Figure 6C,D), falcate setae on Dgl-2 and Dgl-3 (Figure 7A); three pinnate setae on P2 (two in anterolateral position and one in anteromedial position) and one falcate seta on medial position on P3. Bipectinate setae on all lateral IV-Leg-3 segment and serrate setae in the anterolateral position of IV-Leg-2 segment (Figure 7G).

Description. MALE: Idiosoma bluish with yellow spots in the Dgl-1–4 regions (Figure 7A), 1178 L and 785 W; anterior part of idiosoma very wide (Figures 6 and 7A). Dgl-2 and Dgl-3 setae falcate. Dorsal shield 1000 L (cauda included dorsal portion), 571 W. Cauda long, representing almost half of the total body length, 470 L and 478 W, small humps in Lgl-4 region. Dorsal furrow complete, passing ventrally at base of cauda and continuing immediately posterior to the acetabular plates. In lateral view, there is a large, thorn-shaped hump centrally on the dorsum (212 height) (Figure 6C,D). Anterior and posterior coxal groups separated, Cxgl-1 between Cx-II and Cx-III, Cx-IV laterally slightly extending beyond the idiosoma, posterior region concave. Cxgl-2 is located between Cx-IV and the acetabular plates (Figure 7C). Genital field 457 W, gonopore 113 L and 102 W, acetabular plate extending laterally from the gonopore region with two setae posterior to each plate (Figure 7C). Dorsal L of palpal segments: P1: 21; P2: 73; P3: 47; P4: 79;

P5: 47; P3 with a long, falcate seta on anterolateral position (Figure 6B D). Dorsal L of fourth leg segments: IV-Leg-1 120; IV-Leg-2: 155; IV-Leg-3: 196; IV-Leg-4: 210; IV-Leg-5: 189; IV-Leg-6: 172; IV-Leg-2 with three serrate setae in anterolateral position and IV-Leg-3 with ten bipectinate setae on IV-Leg-4 and IV-Leg-5 with 10 and 11 small, pinnate setae, respectively (Figure 7G). IV-Leg-2–6 bear numerous swimming setae.

FEMALE: Idiosoma 1000 L and 948 W, dorsal shield 800 L and 680 W, bears the postocularia and four pairs of glandularia. Anterior idiosoma margin rounded with distinctive posterolateral projections (Figures 6E and 7F). Acetabular plates wing-shaped, laterally directed, narrow and slightly bowed. Genital field 514 W, gonopore 182 L and 187 W. Anterior and posterior coxal group separated, Cx-I and Cx-II extending beyond the anterior margin of idiosoma (Figures 6 and 7F). Idiosoma and legs are bluish with yellow areas on Dgl-1–4 regions (Figure 7E).

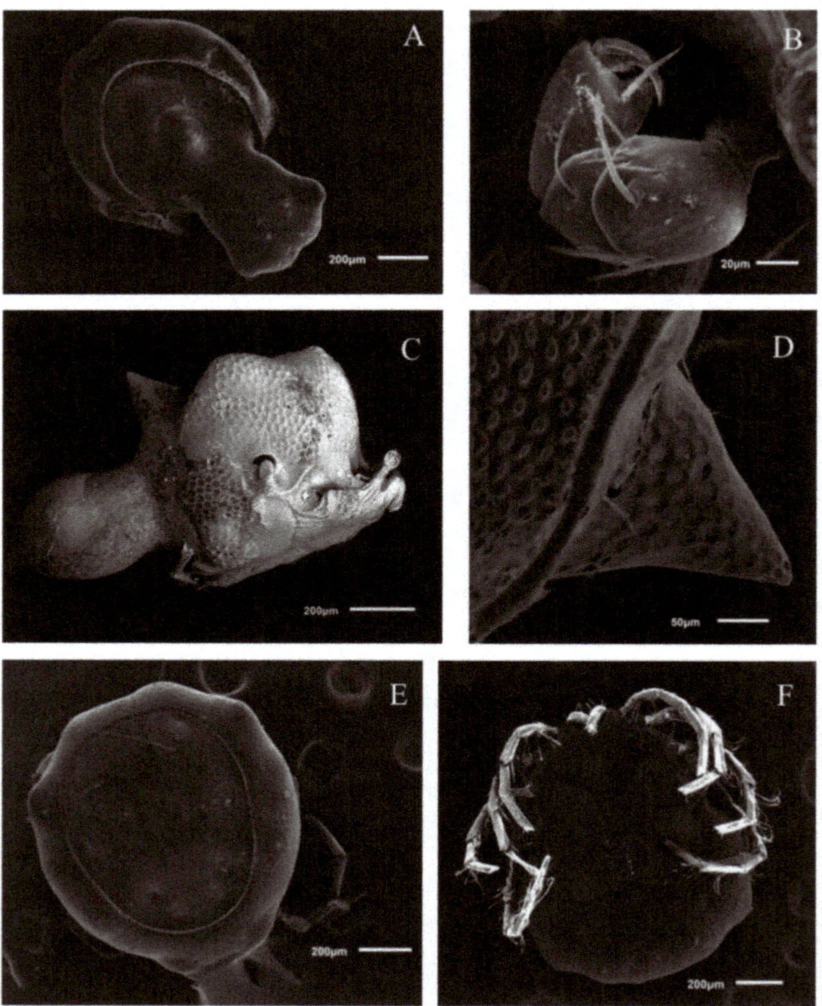

Figure 6. SEM micrograph at a low vacuum of *Arrenurus* (*Megaluracarus*) *eduardoi* n. sp. Male. (**A**) Dorsal view; (**B**) palp, medial view; (**C**) lateral view; (**D**) detail on the thorn-shaped hump on the dorsal shield. Female. (**E**) Dorsal view; (**F**) ventral view. The sequence of this specimen, recovered after DNA extraction, is represented by the BIN AEA7844.

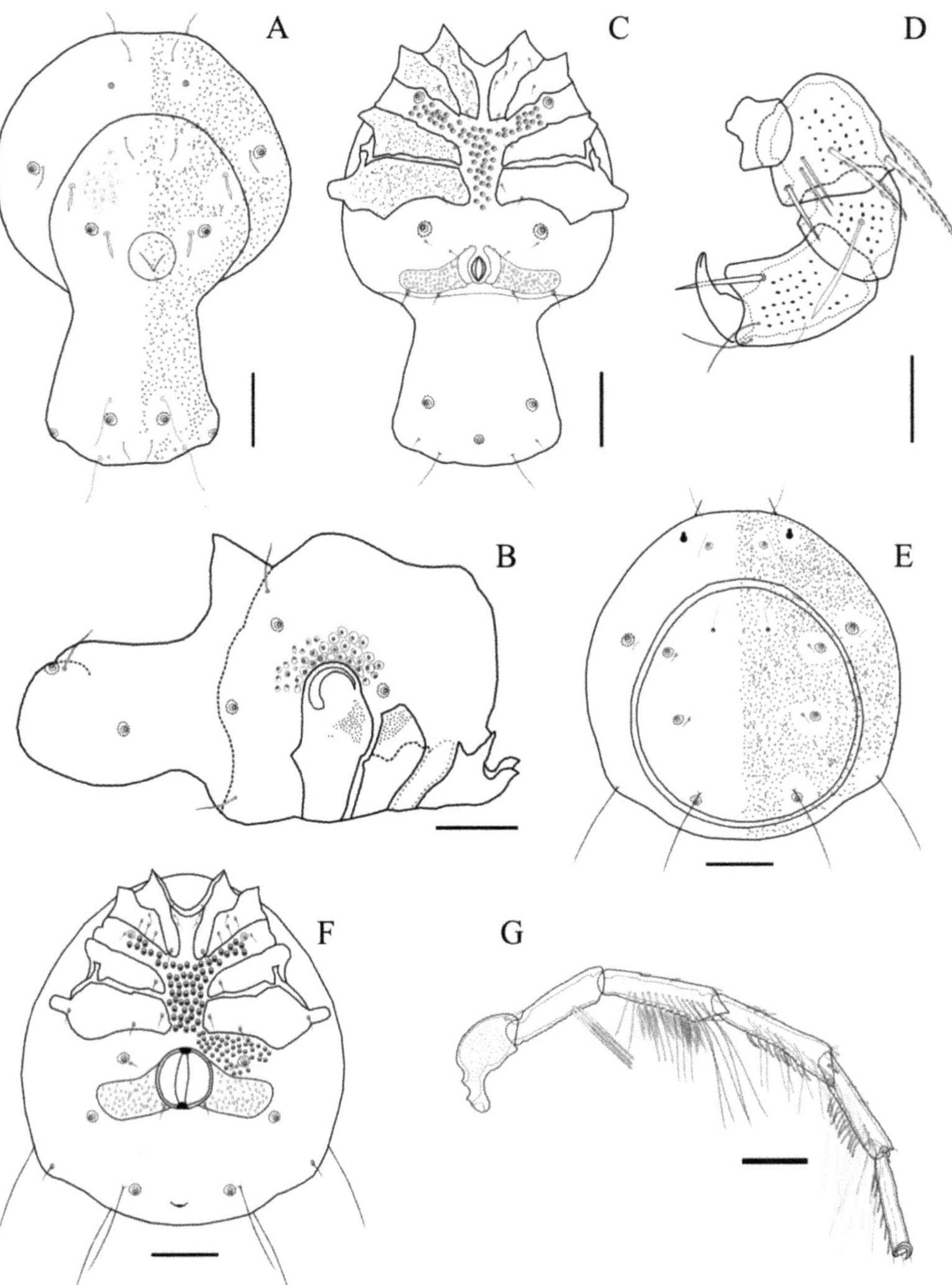

Figure 7. *Arrenurus* (*Megaluracarus*) *eduardoi* n. sp. Male. (**A**) Dorsal view; (**B**) lateral view; (**C**) ventral view; (**D**) palp, medial view; (**G**) Leg IV, lateral view; Female. (**E**) Dorsal view; (**F**) ventral view. Scale bars: (**A–C**,**E**,**F**) = 200 μm, (**D**) = 50 μm, (**G**) = 100 μm.

Consensus sequence: ACTCTATACTTCGCTTTTGGCGCTTGATCAGGCATAATC-GGAGCTAGCCTTAGAAGTCTTATCCGTTTAGAACTTGGACAACCTGGTAATCTTTT-

AGGAAACGATCAAATATACAACACAATTGTCACTGCCCACGCATTTGTTATAATCT-TTTTCATAGTTATGCCAATCATAATCGGAGGATTCGGAAACTGATTAGTTCCTATTA-TACTAGCAGCCCCAGATATAGCTTTCCCACGAATAAATAATATAAGATTTTGACTT-CTTCCCCCCGCTTTAACTCTTCTACTNTCAAGATCTCTATCTTCTTCAGGAGCGGG-GACTGGCTGAACAGTTTACCCCCCTCTTTCNAGTAATATCGCTCACGGAGGACCA-TCTGTCGATATAGCTATTTTCTCACTTCACTTAGCAGGAGTTTCGTCTATTCTAGGG-GCAATTAACTTCTTAGCCACAACTATAAACATAAAGCCAAAATATATAAAATATGA-CCGAATCCCCTTATTTGTAGTCTCAATTTTCATCACAGTCATTCTCCTCCTCTTATC-ATTACCAGTCTTAGCTGGAGCTATCACAATACTATTAACTGATCGAAACTTTAACA-CATCATTCTTTGACCCTGCCGGAGGGGGAGACCCAATTCTTTACCAA.

Etymology. This species is named after Eduardo Montes, brother of the first author, for his empathy, solidarity, and for the lovingly provided support.

Remarks. *A.* (*Megaluracarus*) *eduardoi* n. sp. is similar to *A. campechensis* (Cook, 1980) and *A. maya* (Ramírez-Sánchez and Rivas, 2013) in terms of overall shape and sturdy idiosoma. However, males of *A. eduardoi* n. sp. present a distinctive, large, thorn-shaped hump in the middle of the dorsal shield (in lateral view) that easily separates this species from the latter two. This hump resembles *A. gibberifer* (Viets, 1933), originally described from Uruguay. Nevertheless, the shape of both species is quite different, especially in the dorsal view of cauda; *A. eduardoi* n. sp. presents a trapezoidal shape (Figure 6A), while *A. gibberifer* has a quadrangular shape. Additionally, the reported size for *A. gibberifer* is much smaller (742 L and 528 W) than that registered for *A. eduardoi* n. sp. Furthermore, the palp chaetotaxy of these two species is distinct. The BOLD database assigned the unique BIN AEA7844 (Table 1), used to pair the sexes. The result of the ML tree (Figure 2) and the NJ tree (Supplementary File S1) separates *A. eduardoi* n. sp. from the others registered in the database and strongly supports the status of these new species.

Distribution. So far only known from the type locality, Acapulquito stream, Riviera del Río Hondo, Quintana Roo (Mexico).

Arrenurus (*Megaluracarus*) *federicoi* n. sp., (Figures 8 and 9).

Type material. **Holotype**: Male from Acapulquito stream, Riviera del Río Hondo, Othon P. Blanco, Quintana Roo (access number: ECO-CH-Z-10615), 18°25′55″ N, 88°31′51″ W; emergent vegetation and submerged roots, 11 April 2019, coll. L. Montes.

Paratypes: Two females and one male, same data as holotype (access number: ECO-CH-Z-10616-10617).

Diagnosis. Pores huge (on the idiosoma as well as the legs and palps), Dgl-1 and Cxgl-2 on distinct humps in males. Numerous setae surround the acetabular field in both sexes.

Description. MALE: Idiosoma 1037 L and 693 W, uniformly bluish with large pores. The anterior part of the idiosoma is wide, with noticeable humps in the Dgl-1 area which are visible in the lateral view (Figures 8C and 9B). Dorsal shield 718 L (cauda included dorsal portion) and 436 W. Cauda of medium length, representing a third of the total length of the body, 365 L and 394 W (Figures 8 and 9A), with lobes posterolaterally directed and Lgl-4 on small humps. Dorsal furrow complete, passing ventrally at base of cauda and continuing immediately posterior to the acetabular plates. In lateral view, a big hump is visible in the anterior part of the idiosoma in the Dgl-1 region (Figures 8C and 9B). Coxae with a porous surface, anterior and posterior coxal groups separated, Cxgl-1 located in the middle of Cx-II and Cx-III; Cx-II and Cx-IV slightly extending beyond the anterolateral margin of the idiosoma; Cx-III slightly overlapping Cx-IV (Figures 8B and 9C). Cxgl-2 is located between Cx-IV and the acetabular plates. Genital field 403 W, gonopore 102 L and 75 W. Acetabular plates extending laterally from the gonopore and surrounded by numerous setae (anterior ones small, 24 L, posterior ones longer, 82 L) (Figures 8D and 9C). Dorsal L of palpal segments: P1: 37, P2: 63, P3: 41, P4: 63, P5: 38 (Figure 9D). Dorsal L of fourth leg segments: IV-Leg-1: 125, IV-Leg-2: 165, IV-Leg-3: 209, IV-Leg-4: 159, IV-Leg-5: 193, IV-Leg-6: 165: IV-Leg-5 bears six swimming setae, IV-Leg-4 distal process bears nine short swimming setae, IV-Leg-3 bears 12 swimming setae, both IV-Leg-2 and IV-Leg-3 bear three tiny, spine-like setae on lateral surface (Figure 9G).

FEMALE: Idiosoma bearing huge pores, bluish with yellow spots on the region of Dgl-1–4 and eyes (Figure 9E), 1170 L and 1066 W, dorsal shield 714 L and 790 W, bears the postocularia and three pairs of glandularia. Idiosoma rounded in the anterior margin and with posterolateral lobes (Figures 8F and 9E). Acetabular plates curved and anterolaterally directed, narrow in telation to the gonopore length. Genital field 499 W surrounded by small setae (38–52 L), gonopore 190 L and 204 W. (Figure 8E). The anterior and posterior coxal groups separated, Cx-II and Cx-IV extending slightly beyond the margin of the idiosoma (Figures 8 and 9G).

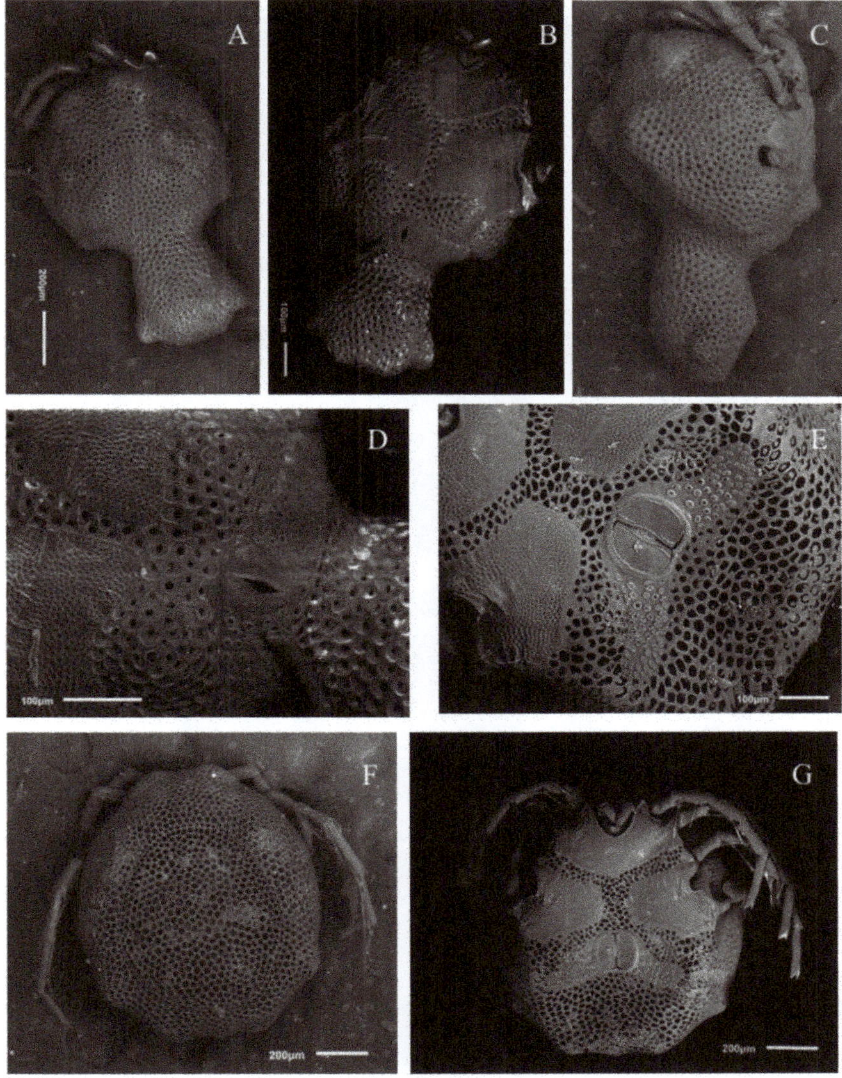

Figure 8. SEM micrograph at a low vacuum of *Arrenurus* (*Megaluracarus*) *federicoi* n. sp. Male. (**A**) Dorsal view; (**B**) ventral view; (**C**) lateral view; (**D**) detail of genital field. Female. (**E**) Detail of genital field; (**F**) dorsal view; (**G**) ventral view. The sequence of this specimen, recovered after DNA extraction, is represented by the BIN AEB7095.

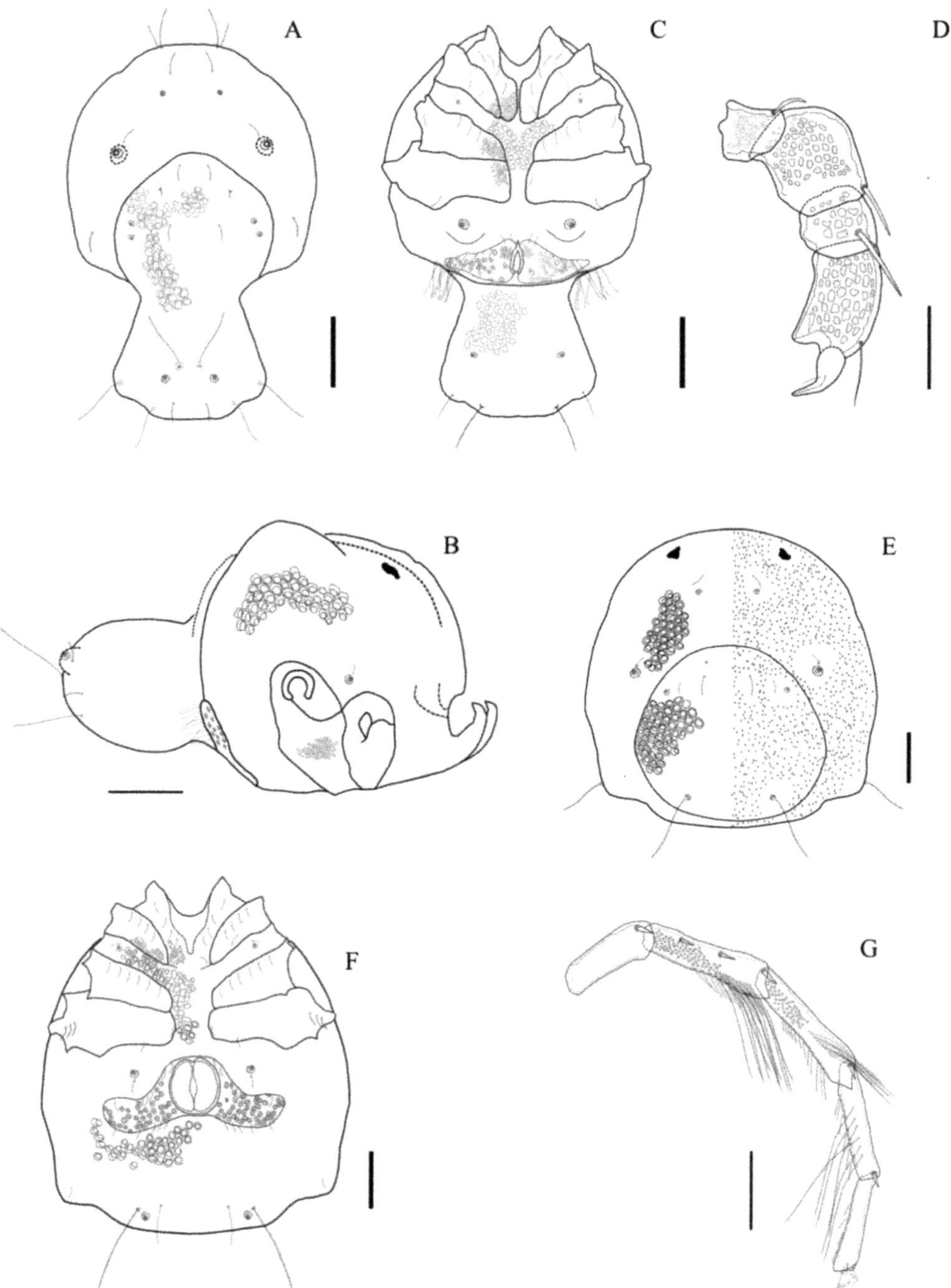

Figure 9. *Arrenurus* (*Megaluracarus*) *federicoi* n. sp. Male. (**A**) Dorsal view; (**B**) lateral view; (**C**) ventral view; (**D**) palp, medial view; (**G**) IV-Leg-2-6, lateral view. Female. (**E**) Dorsal view; (**F**) ventral view. Scale bars: (**A–C**,**E**,**F**) = 200 µm, (**D**) = 50 µm, (**G**) = 100 µm.

Consensus sequence: CTCTATTTCGCCTTTGGAGCATGATCAGGAATAATTGGAGCAAGATTAAGAAGTCTAATTCGCTTAGAACTAGGACAACCAGGAAGACTATTAGGAAACGATCAAATTTATAACACTATTGTTACAGCTCATGCATTCATCATAATTTTCTTCATAGTAATACCTATCATAATCGGAGGATTTGGAAATTGATTAGTACCATTAATACTAGCTGCTCCAGATATAGCTTTTCCTCGAATAAATAATATAAGATTTTGACTATTACCCCCAGCATTATCCCTTCTACTAGCAAGCTCCCTTTCTTCATCAGGAGCAGGAACAGGATGAACAGTTTACCCCCCATTATCAAGAAATATCGCACACGGAGGACCTTCAGTTGATATAGCTATCTTCTCACTCCACCTAGCAGGAGTATCTTCAATTCTAGGAGCCATCAATTTTCTAGCAACAATCATAAATATAAAACCTAAATACATAAAATATGATCGAATCCCTTTATTTGTTATCTCTATCTTTATCACAGTAATCTTACTCTTATTATCCTTACCAGTTTTAGCTGGAGCTATCACTATACTATTAACAGATCGAAACTTTAATACATCATTCTTCGACCCAGCAGGAGGAGGAGATCCTATCCTATACCAACAT.

Etymology. This species is named after Federico Montes, father of the first author, in the form of gratitude for bringing her closer to science since childhood.

Remarks. *Arrenurus (Megaluracarus) federicoi* n. sp. is similar to *A. maya* (Ramírez-Sánchez and Rivas, 2013), described from a cenote in Yucatan, in the shape of idiosoma, the pattern of dorsoglandularia position, and in the presence of setae surrounding the genital field. The significant difference is in the palp chaetotaxy; *A. maya* presents three long, thickened setae while *A. federicoi* n. sp. does not. In IV-Leg-6, *A. federicoi* n. sp. presents four spine-like setae while *A. maya* presents ten. Furthermore, *A. maya* has very small, Dgl-4 associated setae, while *A. federicoi* n. sp. Dgl-4 associated setae are at least four times longer than in *A. maya* (Figure 9A). Additionally, the cauda in *A. maya* is more slender than in *A. federicoi* **n. sp.** Both *A. catoi* (Cramer and Cook, 1992) and *A. campechensis* (Cook, 1980) are similar to the new species in the shape of the anterior idiosoma in dorsal view and Dgl-1 over humps. However, *Arrenurus federicoi* n. sp. can be separated from both latter species by the chaetotaxy of the palp, IV-Leg, the distinctive shape of cauda in dorsal view, and especially the integument with large pores. The BOLD database assigned the unique BIN AEB7095 (Table 1), used to pair the sexes. The ML tree (Figure 2) and the NJ tree (Supplementary Material) separate *A. federicoi* **n. sp.** from the others registered in the database and strongly support the status of these new species.

Distribution. So far only known from the type locality, Acapulquito stream, Riviera del Río Hondo, Quintana Roo (Mexico).

Arrenurus (Megaluracarus) ecosur n. sp., (Figures 10 and 11).

Holotype: Male from Mis Casas, Bacalar lagoon, Bacalar, Quintana Roo (access number: ECO-CH-Z-10618), 18°25′55 N, 88°31′51 W; littoral, emergent vegetation, 14 April 2019, coll. L. Montes.

Paratypes: Three males and one female, with same data as the holotype. Six females and one male from Chichancanab lagoon, José María Morelos, Quintana Roo (access number: ECO-CH-Z-10619-10622), 19°55′26 N, 88°36′14 W.

Diagnosis. Male with cauda of moderate length (330) with Dgl-3 and Dgl-4 on distinct humps. P2 with three long, pinnate setae laterally and three medial, short, spine-like setae in the posterior margin; P3 with one thin and long, pinnate seta lateromedially situated; IV-Leg-3 with three pilose setae lateromedially situated. Considerably long setae of Cxgl-2.

Description. MALE: Idiosoma 864 L and 483 W, light blue, some specimens with purple legs. Dorsal shield 729 L (including cauda) and 374 W. Dorsal furrow complete, continuing posterior to genital field. The non-caudal portion of the dorsal shield bearing two pairs of glandularia, Dgl-3 on a hump, each one. (Figures 10C and 11B). Cauda is relatively short, representing one-third of the total length of idiosoma, with a rounded posterior margin. Dgl-4 on small humps. In lateral view, the base of the cauda is thicker than the anterior idiosoma (Figures 10C and 11B). Anterior and posterior coxal groups separated. Cx-I and Cx-II extend slightly beyond the idiosoma margin. Cxgl-2 between Cx-IV and the acetabular plates, with the associated setae considerably long (146 L) (Figure 11C). Genital field 293 W, gonopore 58 L and 56 W. Acetabular plates extending laterally from the gonopore region with numerous long (50 L) setae along their posterior margin (Figure 11C).

Dorsal L of palpal segments: P1: 29; P2: 58; P3: 31; P4: 62; P5: 25 (Figures 10B and 11D). Dorsal L of fourth leg segments: IV-Leg-1: 151, IV-Leg-2: 119, IV-Leg-3: 112, IV-Leg-4: 135, IV-Leg-5: 154, IV-Leg-6: 109. IV-Leg-3 bears eight swimming setae, three small, pilose setae, and six medium-length, swimming setae on the dorsal surface.

Female: Idiosoma oval, uniformly bluish, 655 L and 590 W, with the postocularia and four pairs of glandularia, dorsal shield 773 L and 716 W (Figures 10D and 11E). Acetabular plates wing-shaped, laterally directed, narrow in relation to gonopore length. Genital field 378 W, gonopore 138 L and 141 W. Anterior and posterior coxal groups separated, Cx-I slightly reaching the margin of the ventral shield (Figures 10E and 11F).

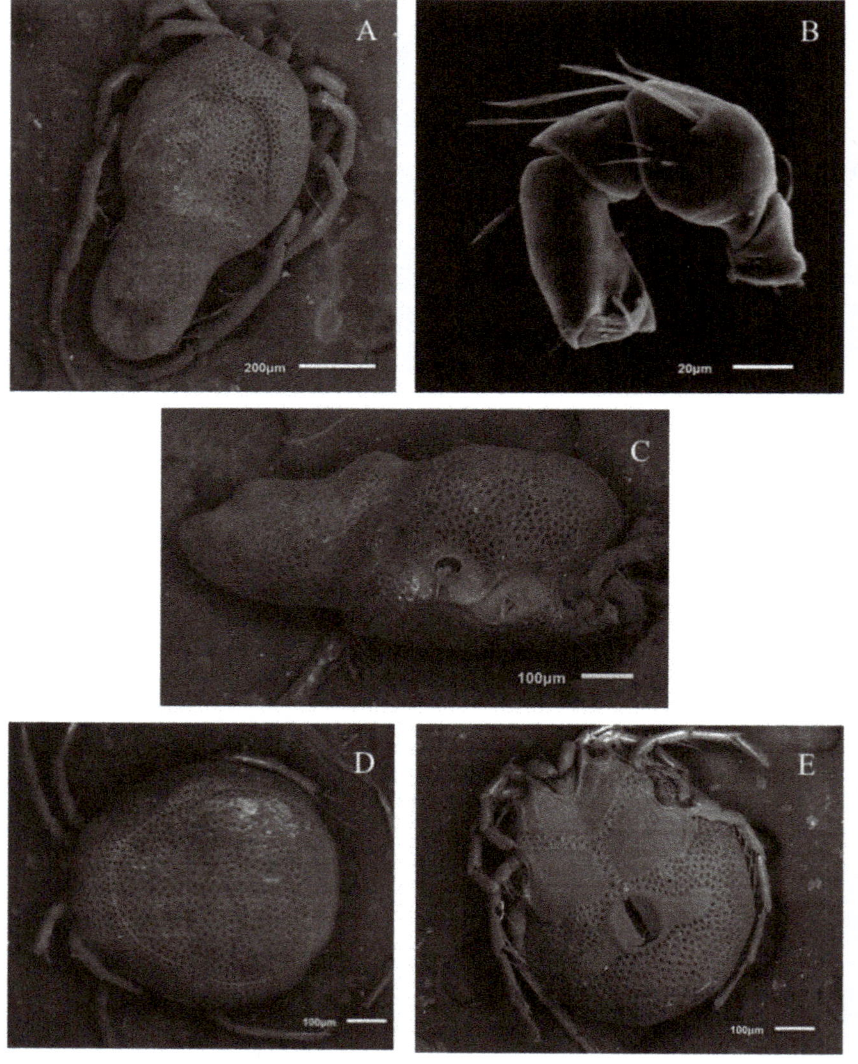

Figure 10. *Arrenurus* (*Megaluracarus*) *ecosur* n. sp. SEM micrograph of n. sp. Male. (**A**) Dorsal view; (**B**) palp; (**C**) lateral view; Female. (**D**) Dorsal view; (**E**) ventral view. The sequence of this specimen, recovered after DNA extraction, is represented by the BIN ACX8463.

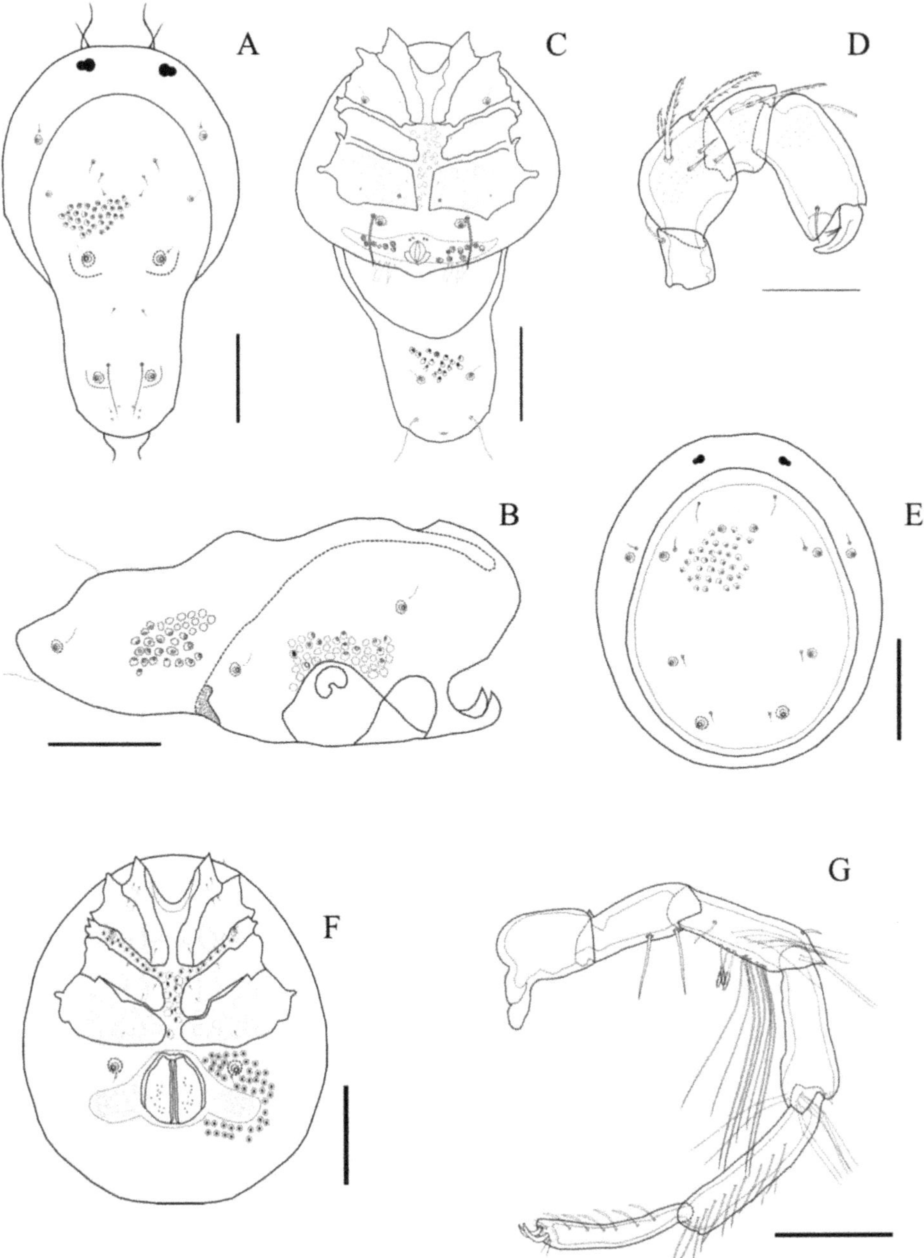

Figure 11. *Arrenurus* (*Megaluracarus*) *ecosur* n. sp. Male. (**A**) Dorsal view; (**B**) lateral view; (**C**) ventral view; (**D**) palp, medial view; (**G**) IV-Leg, lateral view. Female. (**E**) Dorsal view; (**F**) ventral view. Scale bars: (**A–C**,**E**,**F**) = 200 μm, (**D**) = 50 μm, (**G**) = 100 μm.

Consensus sequence: ACACTTTATTTTGCATTTGGAGCTTGATCAGGTATAGTA-GGAGCTAGACTAAGAAGTCTAATTCGCCTAGAACTAGGACAACCAGGAAATCTT-TTAGGAAACGATCAAATTTACAACACAATTGTAACAGCTCACGCTTTTATTATAAT-CTTTTTCATAGTTATACCAATCATAATCGGAGGATTCGGAAACTGACTAGTTCCATT-

AATACTAGCAGCCCCAGACATAGCGTTCCCACGAATAAACAATATAAGATTCTGA-CTTTTACCACCTGCCCTTACACTCCTACTATCTAGATCACTATCATCCACTGGAGC-AGGAACAGGGTGAACTGTTTATCCACCCCTTTCAAGAAACATTGCCCATGGAGG-ACCGTCAGTAGACATAGCAATCTTCTCACTACACTTAGCAGGTGTGTCATCAATTT-TAGGAGCTATCAACTTTTTAGCCACAATCATAAACATAAAACCTAAACACATAAA-ATACGATCGAATTCCCCTTTTTGTTGTATCAATTTTTATTACTGTTATCCTACTTCTTC-TCTCACTTCCAGTTTTAGCAGGAGCTATTACAATGCTACTAACAGATCGAAATTTC-AATACATCATTCTTTGACCCAGCCGGGGGGGGAGACCCTATCTTATACCAA.

Etymology. This species is named in honor of El Colegio de la Frontera Sur (ECOSUR), the research center where the first author completed her graduate studies.

Remarks. *Arrenurus ecosur* n. sp. is similar to *A. tabascoensis* (Cook, 1980) and *A. birgei* (Marshall, 1903), both known from Tabasco (Mexico), mainly in the distinct hump in the area of Dgl-3 (when viewed laterally). However, the cauda of the new species is slightly tapering, contrary to *A. tabascoenis*. Moreover, the posterior margin of the cauda is convex in *A. ecosur* **n. sp**. and straight in *A. tabascoensis*. Aditionally the principal difference among these species is the chaetotaxy of the palps. The new species presents three distinct, pinnate setae on P2. *Arrenurus ecosur* n. sp. is also similar to *A. urbanus* (Ramírez-Sánchez and Rivas, 2013) in the overall shape of the idiosoma in the lateral and dorsal view. Nevertheless, the cauda of the new species is longer and thinner.

Additionally, *A. urbanus* possesses a characteristic patch of two types of seta medially on P2, which are absent in *A. ecosur* **n. sp**. The BOLD database assigned the BIN ACX8463 (Table 1), which was used to pair the sexes. The result of the ML tree (Figure 2) and the NJ tree (Supplementary Material) separate *A. eduardoi* **n. sp**. from the others registered in the database and support the status of this new species.

Distribution. Wide regional distribution in the Yucatan Peninsula: Bacalar lagoon, Chichancanab lagoon, Muyil lagoon, Cenote Azul, Cenote Chancah Veracruz, Cenote Sijil Noh Ha, and Cenote del Padre, Quintana Roo (Table 1).

Arrenurus (*Megaluracarus*) *beatrizae* n. sp., (Figures 12 and 13).

Holotype: One male from Ramonal wetland, Quintana Roo (access number: ECO-CH-Z-10623), 19°23′31″ N, –82°37′27″ W, emergent vegetation, 14 April 2019. Coll. L. Montes and T. Goldschmidt.

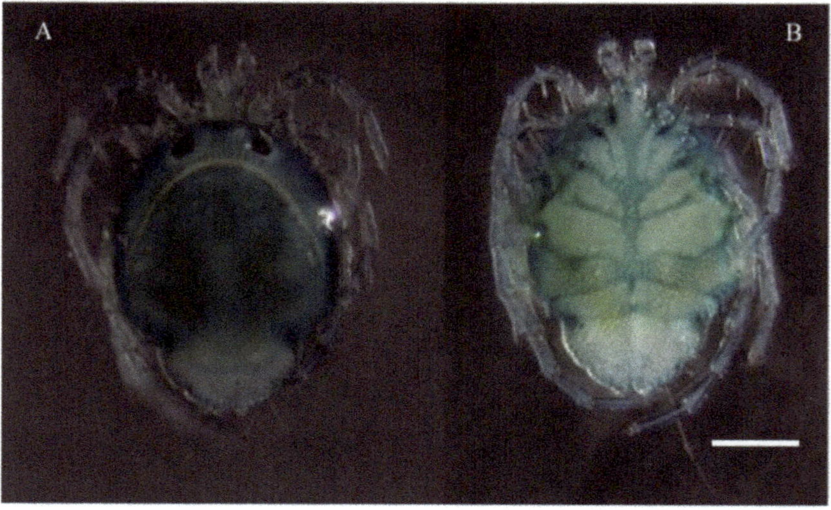

Figure 12. *Arrenurus* (*Megaluracarus*) *beatrizae* n. sp. Male. (**A**) Dorsal view; (**B**) lateral view. Scale bar = 200 µm.

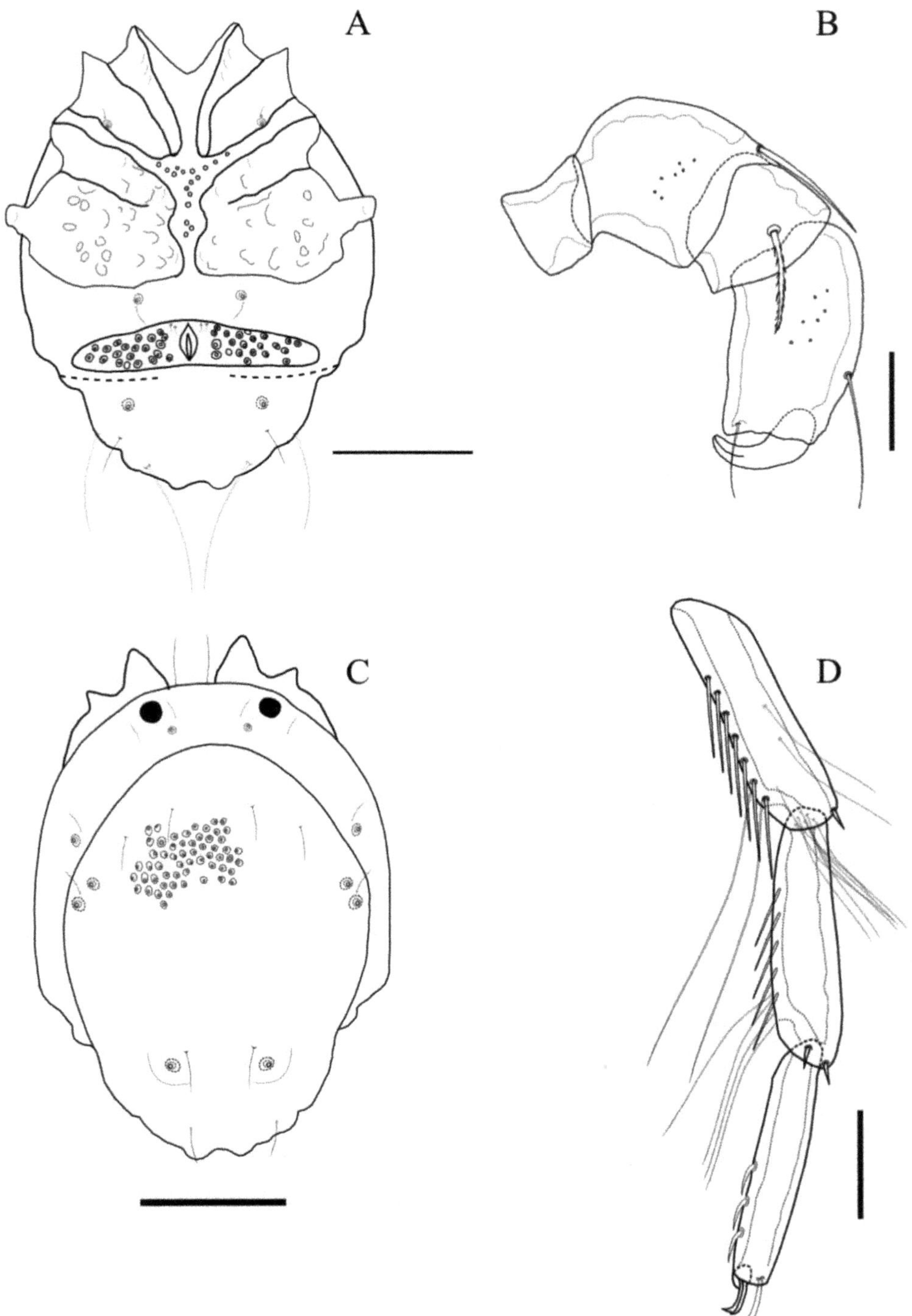

Figure 13. *Arrenurus* (*Megaluracarus*) *beatrizae* n. sp. Male. (**A**) Ventral view; (**B**) palp, medial view; (**C**) dorsal view; (**D**) IV-Leg-3-6, distal segments. Scale bars: (**A**,**C**) = 200 μm, (**B**) = 30 μm, (**D**) = 50 μm.

Paratypes: Three males, one with the same data as the holotype (access number: ECO-CH-Z-10624), the other two from San Pedrito lagoon, Pantanos de Centla, Tabasco (access number: AAL00273, AAL00274), 18°21′58.7″ N, −92°36′03.6″ W, 6 February 2002. Coll. M. Ramírez-Sánchez.

Diagnosis. Characteristic short cauda with two pairs of lateral notches, tips of Cx-II significantly protruding beyond the anterior margin of the idiosoma, P3 presents a long, pinnate seta located medially.

Description. MALE: Idiosoma 640 L, 512 W, dark blue with whitish cauda. Dorsal shield 581 L, 423 W. Dorsal furrow incomplete but continuing posterior to genital field. Cauda is short, 187 L and 285 W, bearing one medial and two pairs of lateral notches (Figures 12A and 13C). The anterior part of the idiosoma is wide with a slight constriction at the base of the cauda. Dgl-2 and Dgl-3 are close to each other. Dgl-4 is located at the end of the cauda on small humps. The anterior coxal group with complete suture lines, Cx-III and Cx-IV, separated with an incomplete suture line. Tips of Cx-II significantly protrude beyond the idiosoma's anterior margin (Figure 13A). Cxgl-1 is located posteromedially in the margin of Cx-I. Apodemes of Cx-IV protrude slightly beyond the lateral part of the idiosoma. Cxgl-2 with an associated seta posteriorly to Cx-IV (Figure 13A). Genital field 315 W, gonopore 69 L and 27 W. Dorsal L of palpal segments: P1: 27; P2: 47; P3: 41; P4: 58; P5: 33, P3 with a long, pinnate seta located medially (Figure 13B). Dorsal L of fourth leg segments: IV-Leg-3: 104, IV-Leg-4: 126, IV-Leg-5: 119, IV-Leg-6: 116, IV-Leg-4-5, with numerous swimming setae and lateral, spine-like setae (seven on IV-Leg-4 and five on IV-Leg-5) (Figure 13D). FEMALE: Unknown.

Etymology. This species is named after Beatriz Rosso de Ferradás for her invaluable contributions to water mite acarology in South America.

Remarks. This species belongs to the subgenus *Megaluracarus*. However, the cauda is relatively short compared with other members of the subgenus. The short cauda is a particular characteristic only shared by *A. olmeca* (Ramírez-Sánchez and Rivas, 2013) from Mexico and *A. amazonicus* (Viets, 1954) from Brazil. However, both *A. olmeca* and *A. amazonicus* have a patch of spatulate setae on the medial side of P2, while *A. beatrizae* exhibits only one long, pinnate seta. Additionally, the cauda posterior margin in both *A. olmeca* and *A. amazonicus* is not indented. Finally, the number of swimming setae on IV-Leg-4 is reduced in *A. olmeca* compared with *A. beatrizae* n. sp.

Distribution. So far only known from el Ramonal, Quintana Roo and San Pedrito lagoon, Tabasco.

Subgenus *Dadayella* (Koenike, 1907)

Arrenurus (*Dadayella*) *cristinae* n. sp., (Figures 14–16).

Holotype: Male from Ramonal wetland, Quintana Roo (access number: ECO-CH-Z-10625), 19°23′31″ N, −82°37′27″ W; emergent vegetation, 14 April 2019. Coll. L Montes and T. Goldschmidt.

Paratypes: One male and two females. Same data as holotype (access number: ECO-CH-Z-10626-10627).

Diagnosis. Male cauda with two falcate setae located posterolaterally, P2 medially with three simple setae, and one pinnate seta on the anterolateral part.

Description. MALE: Idiosoma 364 L and 295 W, uniformly dark blue (Figure 14A). Dorsal furrow incomplete. Dorsal shield 305 L and 207 W, short and relatively square cauda, 49 L. Dgl-4 anteriorly located on the cauda with the associated setae located on small humps and posteriorly in the idiosoma, with two small falcate setae on the posterolateral part of the cauda (Figure 16A). Coxae are occupying two-thirds of the ventral region, suture lines complete. Suture lines of Cx-I–III are diagonally elongated. Cxgl-2 between Cx-II and Cx-IV. Posteriorly to Cx-IV, is the Cxgl-2 located (Figure 16B). Genital field, 246 W, elongated almost reaching the sides of the ventral area, gonopore 59 L and 14 W. Dorsal L of palpal segments L: P1: 30; P2: 58; P3: 38; P4: 63; P5: 30. P2 with three simple setae medially located and one pinnate seta on the anterolateral, P4 rotated (Figure 16C). L of fourth leg segments: IV-Leg-3: 63, IV-Leg-4: 73, IV-Leg-5: 100, IV-Leg-6: 101, IV-Leg-5 with one pinnate seta posteromedially located and four spine-like setae along the dorsal medially surface (Figure 16E).

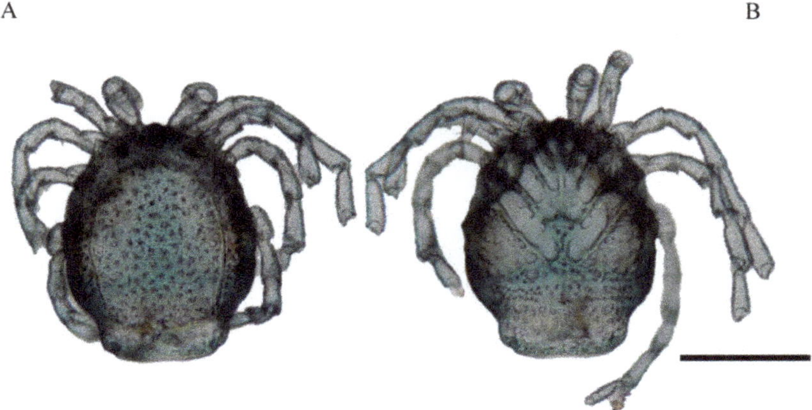

Figure 14. *Arrenurus* (*Dadayella*) *cristinae* n. sp. Male. (**A**) Dorsal view; (**B**) ventral view. Scale bar = 200 µm. The sequence of this specimen, recovered after DNA extraction, is represented by the BIN AEA7842.

FEMALE: Idiosoma 522 L and 483 W, uniformly dark blue (Figure 15A), dorsal shield oval, 463 L and 384 W (dorsal furrow complete), four dorsal pairs of glandularias present. Dgl-2 (on the ventral plate) is close to Dgl-3 (on the dorsal plate). Dgl-3 setae are located posteriorly and separated from their respective glandularia (Figure 16D). With complete suture lines, coxae occupy half of the ventral area, Cx-I, and Cx-II, elongated and extended diagonally. Cx-III and Cx-IV separated, Cx-III elongated and diagonally located, suture lines of Cx-III–IV sloping, Cx-IV triangular without medial margin. Cxgl-1 is located between Cx-II and Cx-III. Genital field 335 W, straight and with numerous associated acetabula, gonopore 118 L and 112 W. Cxgl-2 between genital area and Cx-IV (Figure 16F).

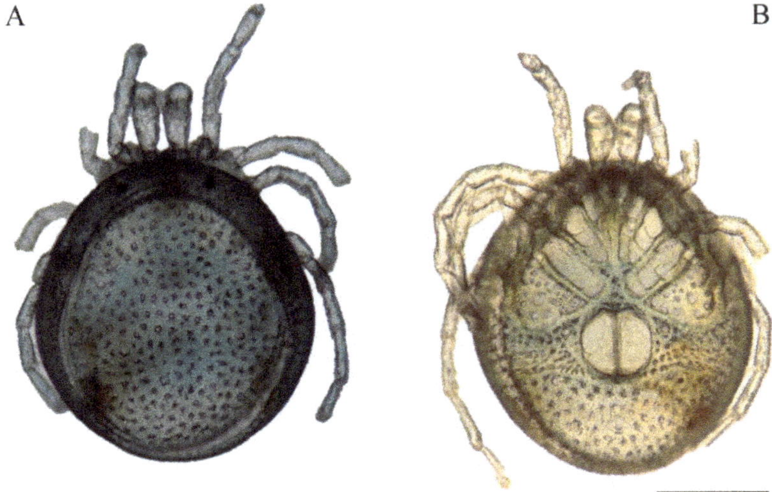

Figure 15. *Arrenurus* (*Dadayella*) *cristinae* n. sp. Female. (**A**) Dorsal view; (**B**) ventral view. Scale bar = 200 µm. The sequence of this specimen, recovered after DNA extraction, is represented by the BIN AEA7842. The difference in color is due to the DNA extraction process.

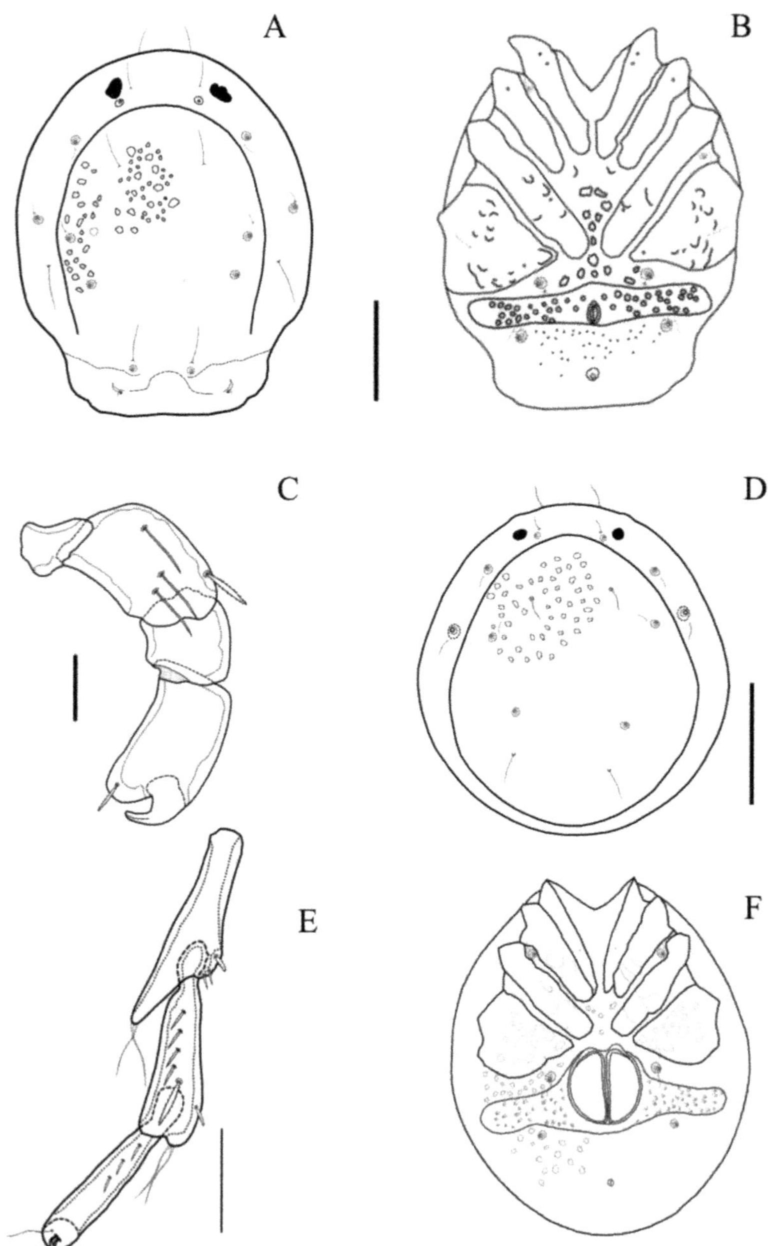

Figure 16. *Arrenurus* (*Dadayella*) *cristinae* n. sp. Male. (**A**) Dorsal view; (**B**) ventral view; (**C**) palp; (**E**) IV-Leg-3-6. Female. (**D**) Dorsal view; (**F**) ventral view. Scale bars: (**A**,**B**) = 100 μm, (**C**) = 30 μm, (**E**) = 50 μm, (**D**,**F**) = 150 μm.

Consensus sequence: ACTCTTTATTTTGCCTTTGGATTCTGATCAGGTATGGTA-GGTGCAAGATTAAGAAGACTAATTCGCTTAGAATTAGGACAACCAGGGAGACTCT-TAGGGAGAGACCAAATTTACAACACAATCGTAACAGCTCATGCTTTTATCATAAT-CTTTTTTATAGTTATACCTATTATAATTGGAGGTTTCGGAAACTGACTAGTTCCTCTT-ATACTAGCAGCTCCAGATATGGCATTCCCACGAATAAACAATATAAGATTTTGAC-

TTCTTCCCCCAGCTTTAATTCTCCTTCTATCTAGATCTCTCTCCTCAACAGGAGCAG-GAACAGGGTGAACAGTATATCCNCCACTTTCAAGTAACATTGCACATGGAGGAC-CTTCAGTTGACATAGCAATCTTTTCCCTCCATTTAGCAGGAGTCTCATCAATTCTA-GGTGCTATCAATTTCTTAGCTACAACCATCAATATAAAACCAAAATATATAAAATA-TGATCGTATTCCTCTATTTGTCATTTCAATTTTCATCACAGTTATTCTCCTTCTCCTA-TCTTTACCAGTCTTAGCAGGAGCCATTACCATACTTTTAACTGACCGAAACTTTAA-TACATCATTTTTTGATCCAGCTGGAGGAGGAGACCCAATTCTATATCAA.

Etymology. This species is named after Cristina Cramer Hemkes for her invaluable contributions to water mite acarology in Mexico.

Remarks. The present species belong to the *Dadayella* subgenus, characterized by males with a small or undifferentiated cauda with an incomplete dorsal furrow and P2 with a simple chaetotaxy [4]. *Arrenurus* (*Dadayella*) *cristinae* n. sp. is similar to *A. veracruzensis* (Cramer and Cook, 1992) in the shape and size of the idiosoma, particularly in the quadrangular silhouette of the cauda. The female of *A. veracruzensis* is similar to the new species. However, the dorsal shield of the A. veracruzensis female has three pairs of glandularia, while *A. cristinae* n. sp. has two. Furthermore, the chaetotaxy of P2 is quite different; *A. cristinae* n. sp. presents three simple, medial setae and a little, pinnate seta in the anterior-lateral part while *A. veracruzensis* presents four medial, spine-like setae. Additionally, *A. cristinae* n. sp. presents two falcate setae on the posterolateral part of the cauda, which are absent in *A. veracruzensis*. Most of the *Dadayella* species described are known from females, making comparisons difficult due to their scarce morphological variation. It was possible for *A. cristinae* n. sp. to obtain the DNA barcode with the BIN AEA7842. Therefore, we could, undoubtedly, assign the female to the respective male (Figure 2). These data represent the first sequences obtained for this subgenus.

Distribution. So far only known from the type locality (Ramonal, Quintana Roo).

4. General Remarks

With these new records and species descriptions, the list of arrenurids from Mexico increases from 37 to 42. The subgenus *Megaluracarus* is the richest in species, with 26 known species (as well as four of the new species described in the present paper). This figure is followed by subgenera *Arrenurus* and *Dadayella*, with six species each. The subgenera with fewer representatives are *Truncaturus* and *Arrhenuropsis*, with only three and one species, respectively. The case of *Arrenurus* (?) *nayaritensis* is particular, and the relationships of this species will not be known until the male is described [3]. According to the checklist (Table 2), only five species have a continuous distribution between the USA and Mexico, one between Costa Rica and Mexico, and one with a more extensive range of distribution in the Neotropics and the Caribbean islands: *Arrenurus valencius*, known from Venezuela, Cuba, Haití, Guatemala, and Mexico.

The new record of *Arrenurus marshallae* from Mexico is shared with Canada and the USA. The remaining species exhibit a restricted distribution to one or two localities (at the present stage of knowledge), and the new records of *Arrenurus colitus* and *A.* (? *Arrhenuropsis*) *mexicanus*, previously known from Tamaulipas state, are now extending the known distribution of these species to Quintana Roo state.

The available molecular information also supports the species diagnoses. Comparing all available sequences of genus *Arrenurus* from the BOLD database (1111 sequences, see Supplementary Information) the discriminated sequences from Mexico indicate a restricted distribution as only two putative species are shared with Canada. This pattern is repeated in the rest of the tree, where other putative species are recorded in only one country or a maximum of two. However, these inferences are strongly biased due to the few sequences and countries with molecular information available. However, this comparison supports our previous conclusion about the new species presented here.

All the arrenurids currently known from Mexico have been reported for 14 of the 32 states in the country. From these, Tamaulipas heads the listing with six species, while Mexico state, Michoacán, and Yucatan have only one species recorded. For 18 entities, particularly in the north, there is no information. As stated in the introduction, due to the geographical position of Mexico and its great variety of ecosystems (many unique in the world, e.g., Bacalar lagoon in the tropics and Cuatrocienegas in the semi-desert), a great diversity of water mites should be expected.

Once we know the diversity of mites, we can make progress to understand their ecological significance and value as water quality indicators.

Supplementary Materials: The following are available online at: https://www.mdpi.com/article/10.3390/d14040276/s1. Figure S1: N.J. compressed tree based on worldwide COI sequences of *Arrenurus* (In total 1111 sequences, representing 148 putative species).

Author Contributions: Conceptualization, L.M.-O.; methodology, M.E.-G., L.M.-O.; software, M.E.-G., L.M.-O.; validation, M.E.-G. and M.M.R.-S.; formal analysis, L.M.-O.; investigation, L.M.-O.; resources, M.E.-G.; data curation, L.M.-O.; writing—original draft preparation, L.M.-O.; writing—review and editing, M.E.-G., M.M.R.-S. and L.M.-O.; visualization, M.E.-G., M.M.R.-S. and L.M.-O.; supervision, M.E.-G. and M.M.R.-S.; project administration, M.E.-G.; funding acquisition, M.E.-G. All authors have read and agreed to the published version of the manuscript.

Funding: This study was partially financed by the Global Environment Fund through the United Nations Development Programme (UNDP, Mexico), Comisión Nacional para el Conocimiento y Uso de a Biodiversidad (CONABIO) and Comision Nacional de Áreas Naturales Protegidas (CONANP) as part of the investigation called: Programa de detección temprana piloto de especies acuáticas invasoras a través de los métodos de código de barras de la vida y análisis de ADN ambiental en la Reserva de la Biosfera Sian Ka'an within Project 00089333 "Aumentar las capacidades de México para manejar especies exóticas invasoras a través de a implementación de la Estrategia Nacional de Especies Invasoras" granted to Martha Valdez Moreno, who kindly shared the samples from her project with us.

Institutional Review Board Statement: Not applicable.

Informed Consent Statement: Not applicable.

Data Availability Statement: Not applicable.

Acknowledgments: The results presented here are part of the first author's doctoral research, being conducted in El Colegio de la Frontera Sur, supported with a fellowship from the National Council of Science and Technology (CONACYT). We thank Alma Estrella Morales García from the Chetumal node of MEXBOL, who assisted with molecular analysis. We are indebted to Margarita Ojeda Carrasco, who performed measurements of some specimens, Bruce Smith, who facilitated literature to the revision of *A. marshallae*, and Tom Goldschmidt, who accompanied and guided LMO during the field collection and made valuable comments that significantly improved this manuscript.

Conflicts of Interest: The authors declare no conflict of interest.

References

1. Smit, H. New records of the water mite genus Arrenurus Dugès, 1834 from South America (Acari: Hydrachnidia: Arrenuridae), with the description of five new species and one new subspecies. *Acarologia* **2020**, *60*, 371–389. [CrossRef]
2. Smit, H. Water mites of the world, with keys to the families, subfamilies, genera and subgenera (Acari: Hydrachnidia). *Monogr. Nederl. Entomol. Ver.* **2020**, *12*, 1–774.
3. Cook, D.R. *Studies on Neotropical Water Mites*; The American Entomological Institute: Ann Arbor, MI, USA, 1980.
4. Cramer, C.; Cook, D.R. New species of Arrenurus (*Dadayella*) (Acari: Arrenuridae) from Mexico, with a discussion of the latter's relationships. *Int. J. Acarol.* **1992**, *18*, 221–229. [CrossRef]
5. Cramer, C.; Cook, D.R. A new species of Stygarrenurus (Acari: Hungarohydracaridae) and a discussion of its systematic position. *Int. J. Acarol.* **1996**, *22*, 29–32. [CrossRef]
6. Ramírez-Sánchez, M.M.; Rivas, G. New species of subgenus *Megaluracarus* (Acari: Hydrachnidiae: Arrenuridae: Arrenurus) from Mexico. *Zootaxa* **2013**, *3718*, 317–330. [CrossRef] [PubMed]
7. Cramer, C.; Cook, D. New species of *Arrenurus* (Acari: Arrenuridae) from Mexican lakes. *Acarologia* **1992**, *33*, 349–366.
8. Morrone, J.J. Biogeographic regionalization and biotic evolution of Mexico: Biodiversity's crossroads of the New World. *Rev. Mex. Biodivers.* **2019**, *90*, e902980. [CrossRef]
9. Perry, E.; Velazquez-Oliman, G.; Marin, L. The hydrogeochemistry of the karst aquifer system of the northern Yucatan peninsula, Mexico. *Int. Geol. Rev.* **2002**, *44*, 191–221. [CrossRef]

10. Gondwe, B.R.N.; Lerer, S.; Stisen, S.; Marín, L.; Rebolledo-Vieyra, M.; Merediz-Alonso, G.; Bauer-Gottwein, P. Hydrogeology of the south-eastern Yucatan Peninsula: New insights from water level measurements, geochemistry, geophysics and remote sensing. *J. Hydrol.* **2010**, *389*, 1–17. [CrossRef]
11. Montes-Ortiz, L.; Elías-Gutiérrez, M. Water Mite Diversity (Acariformes: Prostigmata: Parasitengonina: Hydrachnidiae) from Karst Ecosystems in Southern of Mexico: A Barcoding Approach. *Diversity* **2020**, *12*, 329. [CrossRef]
12. Montes-Ortiz, L.; Elías-Gutiérrez, M. Faunistic survey of the zooplankton community in an oligotrophic sinkhole, cenote azul (Quintana roo, Mexico), using different sampling methods, and documented with DNA barcodes. *J. Limnol.* **2018**, *77*, 428–440. [CrossRef]
13. Elías-Gutiérrez, M.; Valdez-Moreno, M.; Topan, J.; Young, M.R.; Cohuo-Colli, J.A. Improved protocols to accelerate the assembly of DNA barcode reference libraries for freshwater zooplankton. *Ecol. Evol.* **2018**, *8*, 3002–3018. [CrossRef] [PubMed]
14. Porco, D.; Rougerie, R.; Deharveng, L.; Hebert, P. Coupling non-destructive DNA extraction and voucher retrieval for small soft-bodied Arthropods in a high-throughput context: The example of Collembola. *Mol. Ecol. Resour.* **2010**, *10*, 942–945. [CrossRef] [PubMed]
15. Fisher, J.R.; Dowling, A.P.G. Modern methods and technology for doing classical taxonomy. *Acarologia* **2010**, *50*, 395–409. [CrossRef]
16. Ivanova, N.V.; Dewaard, J.R.; Hebert, P.D.N. An inexpensive, automation-friendly protocol for recovering high-quality DNA. *Mol. Ecol. Notes* **2006**, *6*, 998–1002. [CrossRef]
17. Montes-Ortiz, L.; Goldschmidt, T.; Vásquez-Yeomans, L.; Elías-Gutiérrez, M. A new species of Litarachna Walter, 1925 (Acari: Hydrachnidia: Pontarachnidae) from Corozal Bay (Belize), described upon morphology and DNA barcodes. *Acarologia* **2021**, *61*, 602–613. [CrossRef]
18. Kumar, S.; Stecher, G.; Li, M.; Knyaz, C.; Tamura, K. MEGA X: Molecular evolutionary genetics analysis across computing platforms. *Mol. Biol. Evol.* **2018**, *35*, 1547–1549. [CrossRef] [PubMed]
19. Gerecke, R.; Gledhill, T.; Pešić, V.; Smit, H. *Süßwasserfauna von Mitteleuropa, Bd. 7/2-3 Chelicerata*; Springer: Berlin/Heidelberg, Germany, 2016; ISBN 9783827426895.
20. Cook, D. Preliminary List of the Arrenuri of Michigan Part II. The Subgenus *Megaluracarus*. *Trans. Am. Microsc. Soc.* **1954**, *73*, 367–380. [CrossRef]

Article

Improved Chironomid Barcode Database Enhances Identification of Water Mite Dietary Content

Adrian A. Vasquez [1,2,*,†], Brittany L. Bonnici [1,2,†], Safia Haniya Yusuf [1], Janiel I. Cruz [3], Patrick L. Hudson [4] and Jeffrey L. Ram [1]

1 Department of Physiology, School of Medicine, Wayne State University, Detroit, MI 48201, USA; eh5829@wayne.edu (B.L.B.); safiahaniya@wayne.edu (S.H.Y.); jeffram@med.wayne.edu (J.L.R.)
2 Healthy Urban Waters, Department of Civil and Environmental Engineering, Wayne State University, Detroit, MI 48202, USA
3 Prism Education Center, Fayetteville, AR 72703, USA; nature.rsch@outlook.com
4 US Geological Survey, Great Lakes Science Center, Ann Arbor, MI 48105, USA; patrick_hudson@sbcglobal.net or phudson@usgs.gov
* Correspondence: avasquez@wayne.edu
† These authors contributed equally to this work.

Abstract: Chironomids are one of the most biodiverse and abundant members of freshwater ecosystems. They are a food source for many organisms, including fish and water mites. The accurate identification of chironomids is essential for many applications in ecological research, including determining which chironomid species are present in the diets of diverse predators. Larval and adult chironomids from diverse habitats, including lakes, rivers, inland gardens, coastal vegetation, and nearshore habitats of the Great Lakes, were collected from 2012 to 2019. After morphological identification of chironomids, DNA was extracted and cytochrome oxidase I (COI) barcodes were PCR amplified and sequenced. Here we describe an analysis of biodiverse adult and larval chironomids in the Great Lakes region of North America based on new collections to improve chironomid identification by curating a chironomid DNA barcode database, thereby expanding the diversity and taxonomic specificity of DNA reference libraries for the Chironomidae family. In addition to reporting many novel chironomid DNA barcodes, we demonstrate here the use of this chironomid COI barcode database to improve the identification of DNA barcodes of prey in the liquefied diets of water mites. The species identifications of the COI barcodes of chironomids ingested by *Lebertia davidcooki* and *L. quinquemaculosa* are more diverse for *L. davidcooki* and include *Parachironomus abortivus*, *Cryptochironomus ponderosus*. *Parachironomus tenuicaudatus*, *Glyptotendipes senilis*, *Dicrotendipes modestus*, *Chironomus riparius*, *Chironomus entis/plumosus*, *Chironomus maturus*, *Chironomus crassicaudatus*, *Endochironomus subtendens*, *Cricotopus sylvestris*, *Cricotopus festivellus*, *Orthocladius obumbratus*, *Tanypus punctipennis*, *Rheotanytarsus exiguus* gr., and *Paratanytarsus nr. bituberculatus*.

Keywords: non-biting midge; barcode gap; food web; *Lebertia*; Laurentian Great Lakes

Citation: Vasquez, A.A.; Bonnici, B.L.; Yusuf, S.H.; Cruz, J.I.; Hudson, P.L.; Ram, J.L. Improved Chironomid Barcode Database Enhances Identification of Water Mite Dietary Content. *Diversity* **2022**, *14*, 65. https://doi.org/10.3390/d14020065

Academic Editor: Manuel Elias-Gutierrez

Received: 8 December 2021
Accepted: 7 January 2022
Published: 19 January 2022

1. Introduction

Understanding trophic cascades of freshwater ecosystems can be extremely useful for managing aquatic habitats. Freshwater habitats are among the most threatened, and more research into their biodiversity and understanding the ecological interactions of organisms have been recommended [1]. Knowledge of prey is important to construct food web pathways of aquatic systems. We focus here on the chironomid prey of water mites.

Chironomidae (commonly referred to as chironomids, nonbiting flies, midges, or bloodworms) is an insect family whose aquatic larvae are an important constituent of freshwater systems. All stages of chironomid development, including eggs, larvae and adult flies, are used as food sources for various organisms [2]. The biomass of chironomids is so great that, at times, they may be considered pests [3]. Chironomids have been used as biological indicators of aquatic health [4,5] and cultured as fish food [6].

Water mites are true aquatic arachnids that are ubiquitous and are considered the most biodiverse arachnid class [7]. Water mites belong to the suborder Parasitengona, and, as the name suggests, most water mites are parasitic as larvae [8]. They have been observed as parasitizing a wide array of aquatic hosts, and most of these associations are still not well understood [9]. Despite being "neglected" in freshwater research, water mites are also important predators with potentially significant predation effects on the variety of prey they consume, including crustaceans, ostracods, nematodes and aquatic Dipteran larvae, including chironomids and mosquitoes [10]. Water mite predation of chironomids can significantly reduce the standing crop of chironomids [11]. Since water mites digest their prey extra-orally [12,13], analysis of their diets cannot be accomplished by dissection and visualization of gut contents under a microscope. However, the DNA of ingested organisms, like chironomids, remains sufficiently intact so that their DNA sequences can be detected up to 24 h after ingestion [14]. Application of next-generation sequencing (NGS) to analyze DNA fragments of ingested prey in water mites freshly collected from the field revealed many chironomid taxa as prey items [15]. The identification of the species level of many of these prey organisms was difficult because many reference sequences in barcode databases were not identified for chironomids beyond generic or family-level classification. To resolve this difficulty and improve identifications of organisms in water mite diets, we used morphological identification and DNA barcodes to generate a more specific and broader curated database of chironomids that improved identifications.

Cytochrome oxidase I (COI) DNA barcodes are useful for characterizing biodiversity and studying diet [16,17]. Our previous work on chironomid COI barcodes resulted in the identification of several taxa of chironomids from the Lake Erie region [18]. DNA barcodes have been assisting with taxonomy since the development of metazoan primers called Folmer primers [19]. A combination of classical taxonomy and COI DNA barcodes helped us previously in multiple projects on biodiversity, invasive species detection, and clarification of cryptic species [18,20–22].

In this paper, we present an expanded, curated database of identified chironomid COI barcode sequences, including many novel chironomid DNA barcodes. In addition to exploring chironomid "barcode gaps" and the possible presence of cryptic species, we further demonstrate its application to improve the specificity of prey identification in water mites.

2. Materials and Methods

2.1. Sampling of Chironomids and Water Mites

We collected Chironomidae larvae and adults from sediment and aerial collections in the Laurentian Great Lakes region, focusing mainlyon Western Lake Erie, Southeast Michigan and several locations outside the Great Lakes watershed (Figure 1). Collection methods for samples from Lake Erie are described in Failla, Vasquez, Hudson, Fujimoto and Ram [18]. We collected sediments by Ponar grab, washed on a 500 µm sieve, and then stored in ethanol. Chironomid larvae from other sites were collected using either a Ponar grab or a circular 250 µm collecting net and then washed through a 250 µm sieve. Adult chironomid flies were collected from bushes and other structures—such as spider webs, surfaces of cars, boats, leaves and buildings—with 250 µm mesh sweep nets directly into vials containing either isopropanol or ethanol. We sampled for water mites from Blue Heron Lagoon, Detroit, MI, using a 250 µm circular net followed by washing on a 250 µm sieve and preservation in ethanol, as described in Vasquez, Mohiddin, Li, Bonnici, Gurdziel and Ram [15]. Specimens were transported to the laboratory for morphological identification.

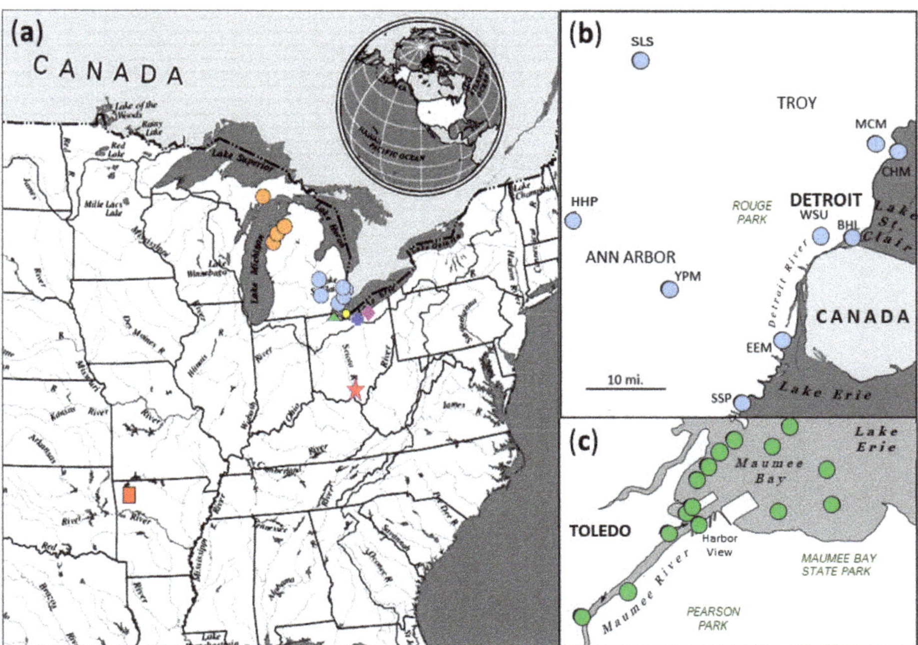

Figure 1. (**a**) Map depicting collection sites of chironomids across the Laurentian Great Lakes and North American rivers. (**b**) Inset showing detailed locations of collection sites throughout Southeastern Michigan. (**c**) Inset showing detailed locations of collection sites near Toledo Harbor, Ohio. Collection site latitudes and longitudes for the chironomids in the curated set are included in their GenBank accession annotations.

2.2. Morphological Identification of Chironomids and Water Mites

Taxonomic methods for the identification of larval and adult chironomids to species were the same as previously described [18]. The keys of Townes [23], Saether [24], Saether [25], Epler [26], Dendy and Sublette [27], Cranston et al. [28], Roback [29], and Heyn [30] were used. Water mite genera studied for this work were *Lebertia*, *Limnesia* and *Arrenurus* and were the same specimens described in Vasquez, Mohiddin, Li, Bonnici, Gurdziel and Ram [15].

2.3. DNA Extraction, Amplification, Sequencing

Adult and larval chironomid tissue samples from morphologically identified specimens were used for DNA extraction. Chironomid tissue was either sequenced by the Canadian Center for DNA Barcoding (CCDB; Biodiversity Institute of Ontario, University of Guelph, Ontario, Canada) as described in Failla, Vasquez, Hudson, Fujimoto and Ram [18] or was extracted for subsequent amplification and sequencing. DNA extraction employed the Qiagen DNeasy blood and tissue kit (Hilden, Germany) as described in Failla, Vasquez, Hudson, Fujimoto and Ram [18]. Tissues were put in lysis buffer in a 1.5 mL centrifuge tube, homogenized with a hand-held fitted pestle and treated with proteinase K enzyme for 3 h at 56 °C. Spin columns were then used to concentrate and purify DNA for subsequent PCR and sequencing. For water mites, a sterile minutien pin was used to puncture each water mite. The mites were then transferred to lysis buffer, where DNA was extracted using the Qiagen DNeasy kits similarly to the chironomid extraction—with the exception that the water mites were lysed overnight rather than homogenized—to more completely extract the DNA from the gut of the punctured water mite, as described in Vasquez, Mohiddin, Li, Bonnici, Gurdziel and Ram [15]. For chironomid specimens, DNA barcodes were generated by PCR as reported in Failla, Vasquez, Hudson, Fujimoto and Ram [18], amplifying the COI gene (658 bases in length) with the Folmer primers [19]. The PCR products were sequenced using Sanger sequencing by Genewiz company (South Plainfield, NJ, USA). For water mites, COI barcodes were generated using Folmer primers to verify the mite identity and a second set of primers (modified mLEP and Folmer LCOI primers, which amplify insect sequences but not arachnids (and hence are able to amplify chironomid sequences but not the DNA from the water mite host)) [15]. The mLEP:FolmerLCOI primer set amplifies a somewhat shorter (332 bases) region of the COI gene than the Folmer primers and were modified with adapters for next-generation sequencing on a MiSeq V2 Illumina platform at the Michigan State University RTSF Genomics Core, as described by Vasquez, Mohiddin, Li, Bonnici, Gurdziel and Ram [15].

2.4. Bioinformatics of Chironomid Sequences

COI barcode sequences from chironomid specimens were assembled bidirectionally and trimmed to remove primer sequences using DNABaser (Heracle BioSoft SRL, Mioveni, Romania) and MEGA X, respectively [31]. DNABaser was used to determine sequence quality as described in Vasquez, Hudson, Fujimoto, Keeler, Armenio and Ram [20]. For graphical assistance in identifying clusters or unique branches of chironomid sequences, 631 sequences were displayed in a neighbor-joining tree using MEGA X, which generated the tree using the maximum composite likelihood method. We subsequently selected representative sequences from each branch to generate a curated set of sequences from branches that differed in sequence by no more than 3.5%. The branch distance of 3.5% was based on the previously described "barcode gap", below which chironomid sequences that differed by a smaller amount were always the same species when species identification was known [18]. Sequences within 3.5% were numbered sequentially for labelling and referencing purposes. Selected sequences were chosen to represent each branch, prioritizing sequence length, most specific taxon identification, and consensus in identification with the other members of the cluster. Generally, when a sequence from an identified adult was available, that sequence was chosen to represent the branch in the curated set, as identification to species level is usually more reliably accomplished in adults than in larval chironomids (see Appendix A for summarized methods).

2.5. Curation of Chironomid Sequences

Twenty-one specimens accounting for ~3% of the sequences and affecting approximately 20% of the branches (see Table S1) with different morphospecies identifications were present in a single cluster. The curation process involved making a rules-based decision as to whether a specific sequence should be excluded. The selection of which taxon would represent the branch was based on consensus among the other sequences in the cluster (e.g., a cluster that had three sequences identified as one species and one as a related species was represented as the first) or by comparison with GenBank or the Barcode of Life Database (BOLD) (i.e., if other sequences identified by reliable taxonomists were available and agreed with either identification, the consensus identification was used to represent the branch in the final version of the curated database). Discrepancies between database matches and morphological identifications were reviewed and decided in consultation with taxonomic evaluation and the taxonomic literature (see Table S1). In the case of one branch (*Chironomus entis/plumosus*), both identifications have been applied to the branch. Another branch in which two species appeared included *Dicrotendipes lucifer* and *D. simpsoni*. These two species have been described as members of a *D. lucifer* complex [32], and the branch has been given the name of the complex (*Dicrotendipes lucifer* agg.). The cause of these ambiguities could be several, including difficulty in determining morphospecies characters in closely related species, variability in the species, possible errors in labeling, sequencing, etc. Careful chain of custody methods were used. While errors or mistakes affecting approximately 3% of the sequences cannot be ruled out, other explanations, such as the presence of hybrids having the morphological characters of one species but mitochondria that are maternally inherited from another, are also possible.

Following this selection process, representative sequences were compiled and aligned. A "curated neighbor-joining tree" of the database of representative sequences was made in MEGA X. Sequence alignment was performed with CLUSTALW. The best-fit DNA substitution model was determined using the maximum composite likelihood method. The resulting phylogenetic trees were chosen from a heuristic search with a bootstrap value of 200 replicate iterations. The pairwise patristic distances between sequences used for the heuristic search were estimated with the Tamura–Nei model using the Neighbor-Join and BioNJ algorithms, with a discrete gamma distribution rate (5 categories (+G, parameter = 0.83) and invariable sites [31]. Representative sequences of each branch will be uploaded to GenBank [accession IDs will be provided upon acceptance of the manuscript].

Pairwise distance analysis matrices generated by MEGA X were used to generate histograms of pairwise distances among the 631 curated chironomid sequences. These histograms were examined for the presence of "barcode gaps" that might identify the distances which most reliably identified species and genera among chironomids (see Appendix A for the summarized method).

2.6. Identification of Water Mite Prey Using the Curated Chironomid Database

Chironomid sequences from high-throughput sequencing of water mite molecular gut contents from 16 *Lebertia quinquemaculosa*, 21 *Lebertia davidcooki*, 2 *Lebertia* sp., 1 *Arrenurus* sp., and 1 *Limnesia* sp. Specimens were compared to the sequences in the curated chironomid database. Chironomid sequences amplified by the mLEP:LCOI primer pair from each water mite were combined with 160 sequences representing all branches in the curated chironomid dataset and analyzed with MEGA X. For each of these combined datasets, MEGA X was used to generate a neighbor-joining tree that allowed for graphical taxa identification comparisons in which water mite diet sequences clustered together with sequences from the curated chironomid database. Pairwise distance matrices were generated for each of these combined datasets. Mite diet sequences that were <3.5% different from an identified database branch were putatively identified as having that taxonomic identity—i.e., to the species level if the matching branch of the curated database provided species-level identification, or to the genus level if the matching branch taxon was only identified to the genus level. Water mite diet sequences for which the closest curated sequence was >3.5% distant but <9.5% distant were identified only to the genus level even if the nearest pairwise match was at the species level. Subsequent reconsideration of these barcode gap boundaries in the Results indicates that these pairwise differences are reasonable for assigning genus and species to chironomid sequences. Previously, these mite diet chironomid sequences had been identified by family, genus, or species level only in relation to the existing GenBank chironomid sequences [15]. In the current paper, we, therefore, summarize quantitatively the improvements in taxonomic identification (from genus to species or from family to genus or species) by application of the new chironomid database, in comparison to what was previously available in GenBank. We also checked for additional identifications in the Barcode of Life Database.

3. Results

3.1. Chironomid Biodiversity and Barcode Gap Revealed by Morphology and DNA Barcodes

A total of 99 identified sequences were selected from the 631 identified chironomid sequences to represent each <3.5% similarity group for the curated database. Figure 2 shows the maximum composite likelihood tree constructed from the consensus set. Due to its large size of 99 major branches, this curated consensus tree is shown in 3 connected figures (Figure 2A–C). A total of 73 branch clusters were identified to species, while 26 were identified only to genus. A total of 42 branches were based on the sequence of a single identified specimen. Furthermore, 70 branches contained at least 1 sequence from a morphologically identified adult chironomid, and 29 branches were based on 2 or more morphologically identified adults.

As previously noted for the much smaller tree described in Failla, Vasquez, Hudson, Fujimoto and Ram [18], the clades of the curated database tree mostly show excellent congruence with previous morphological taxonomic classification at the family, subfamily, or tribe levels. Thus, Figure 2A comprises all Tanytarsini tribe (Chironominae) specimens (*Cladotanytarsus, Paratanytarsus, Tanytarsus, Rheotanytarsus,* and *Stempellina*). Greater than 90% of the branches in Figure 2B represent specimens of the Chironomini tribe of the Chironominae subfamily (*Axarus, Benthalia, Chironomus, Cladopelma, Cryptochironomus, Cryptotendipes, Glyptotendipes, Harnischia, Kiefferulus, Lobochironomus, Microchironumus, Parachironomus, Paracladopelma, Robackia*). Figure 2B also has several branches of *Dicrotendipes* (Chironominae) and all the Pseudochironomini tribe specimens (*Pseudochironomus*), as well as several remaining Chironomini tribe specimens (*Endochironomus, Polypedilum, Stictochironomus,* and *Tribelos*). Figure 2C has all the representatives of the subfamily Tanypodinae (*Ablabesmyia, Clinotanypus, Coelotanypus, Procladius,* and *Tanypus*) and >85% of the subfamily Orthocladinae (*Cricotopus, Eukieferriella, Hydrobaenus, Nanocladius, Orthocladius, Parakiefferella, Smittia,* and *Stilocladius*).

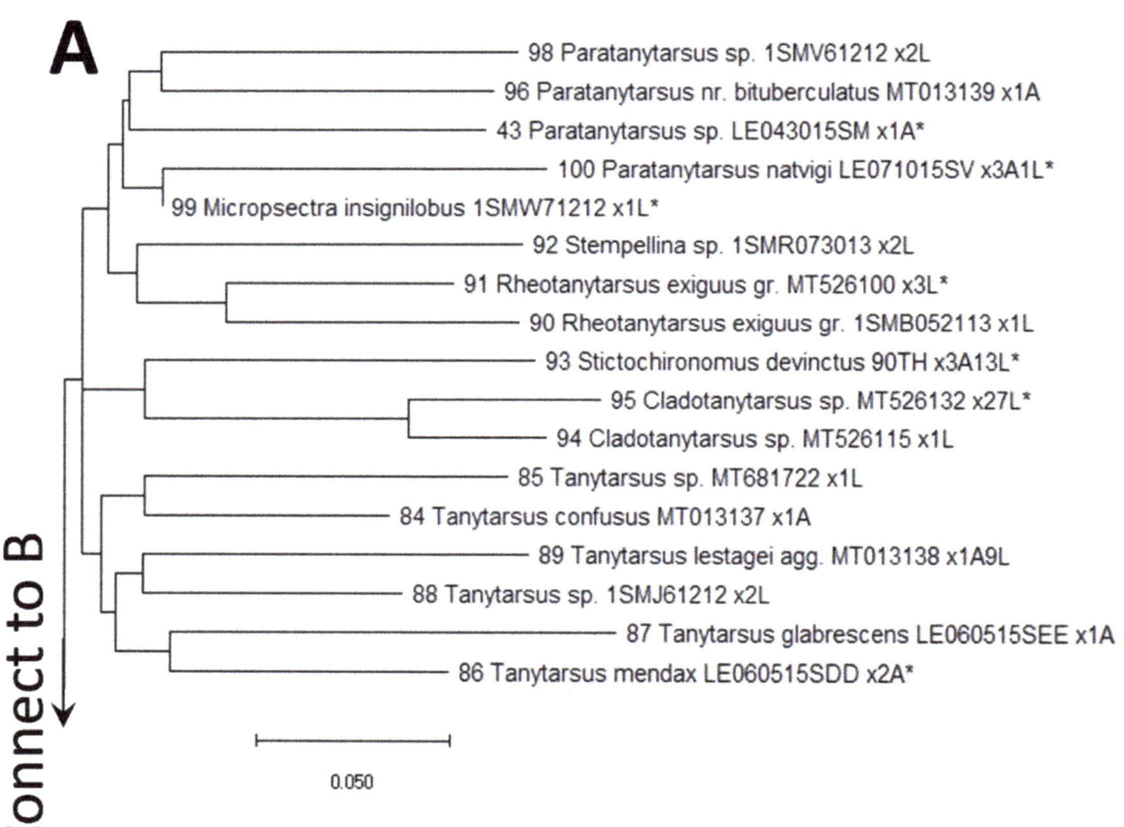

Figure 2. *Cont.*

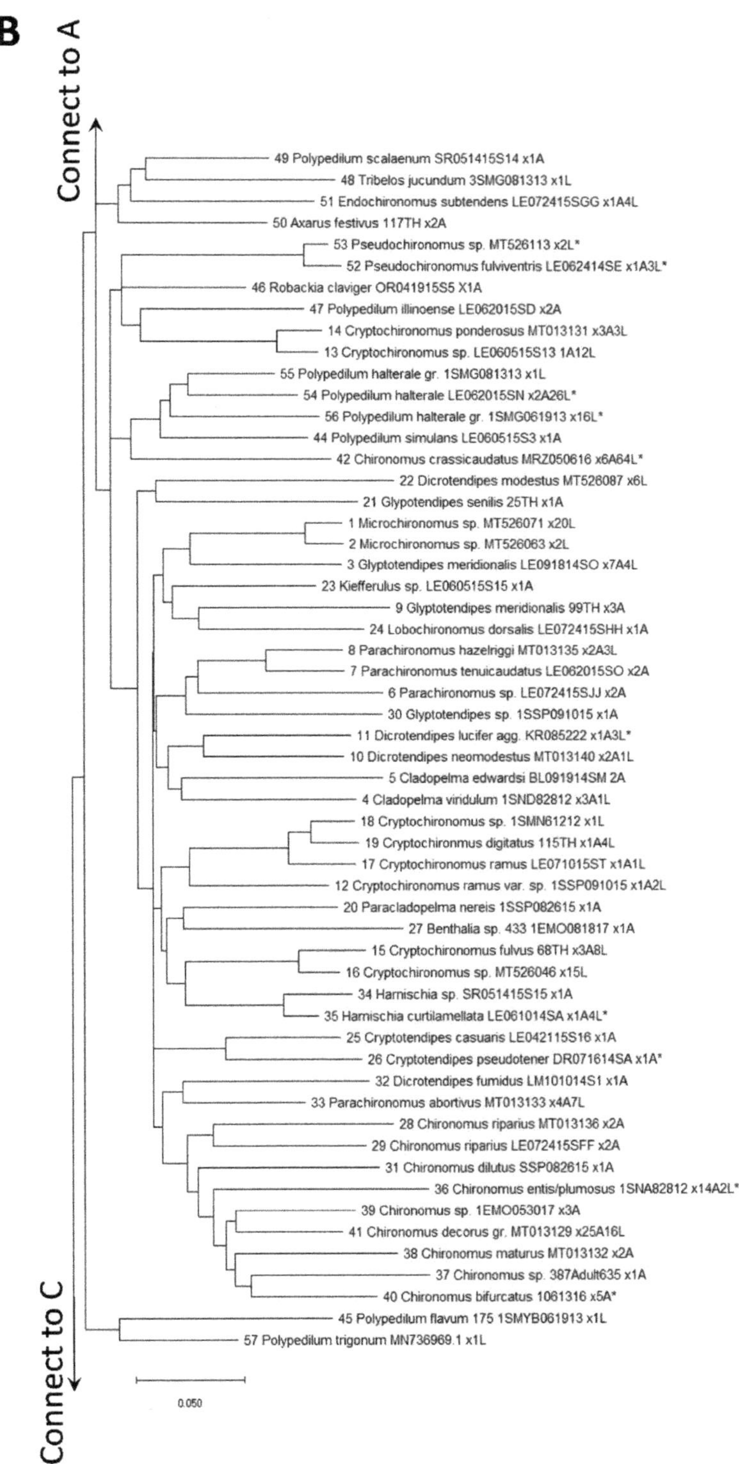

Figure 2. *Cont.*

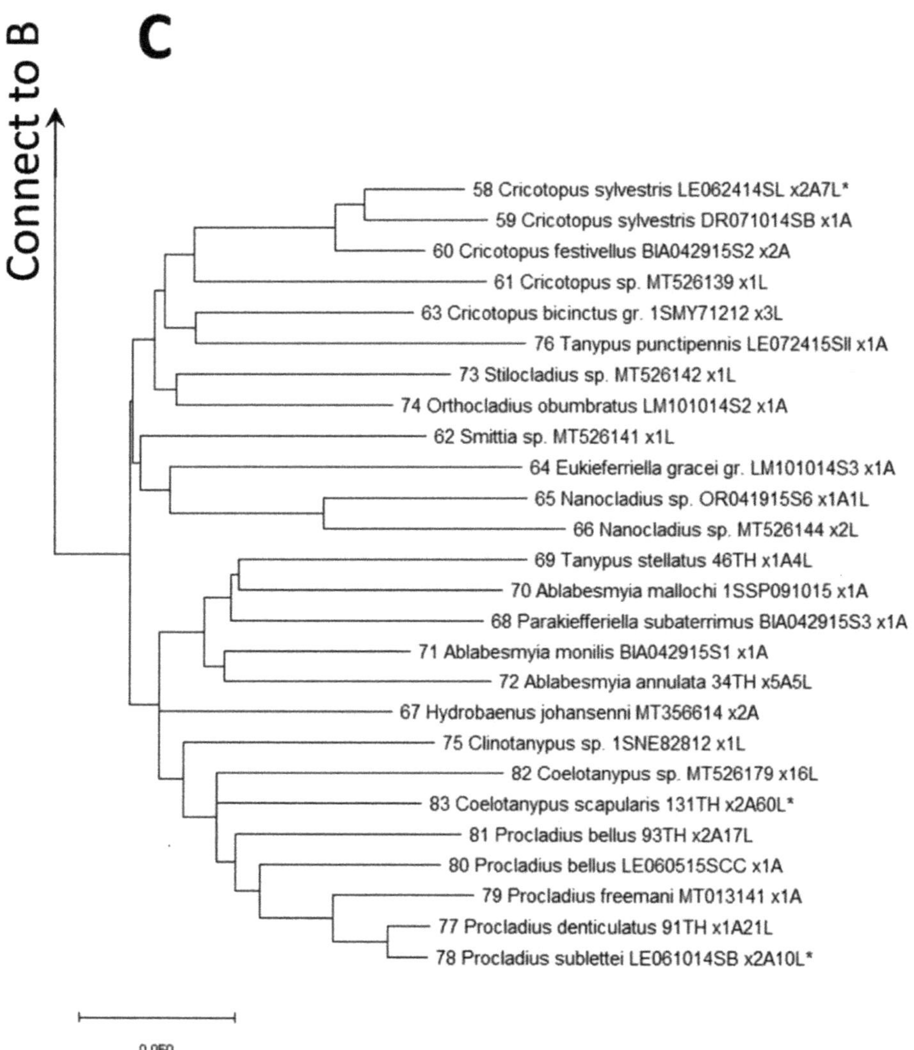

Figure 2. Curated consensus chironomid reference tree. The names on each branch represent the consensus of clusters of up to 71 specimens, taking into account criteria of sequence length, most specific taxon identification, and consistency with identification with the other members of the cluster. Due to the large size of the tree, the full tree is shown in three parts (**A**–**C**), of which the major constituents are (**A**) tribe Tanytarsini, (**B**) tribes Chironomini and Pseudochironomini, plus several *Dicrotendipes* spp. and (**C**) subfamilies Tanypodinae and Orthocladinae. See Figure S1 to for larger view of full tree seen in (**A**). Naming convention for each branch: Lab ID number, genus (XXX sp.) or species (XXX yyy), GenBank accession ID or RamLab ID (if not already uploaded), x number of adults (**A**) and number of larvae (L) used for branch consensus, * shown if a specimen has been removed or name revised due to non-consensus identification or other comment about the branch (see Table S1). The lines in the image at the left of A shows how the entire tree was split into three parts.

3.2. Pairwise Analysis of Distances between Curated Chironomid Sequences

A histogram of pairwise differences among the 631 identified curated sequences (i.e., after the removal of ~3% of non-consensus sequences) is shown in Figure 3. The analysis starts with a high number of pairs having small pairwise differences belonging to the same species. The number of pairwise differences decreases to a relative minimum (i.e., not quite a definitive barcode "gap") at about 3.5% and rises to a small peak in pairs at around 4% that falls to a low level above 6%. This is followed by several peaks in the pairwise differences at around 8% and 10% pairwise differences before the differences arise in a continuum, more or less, with a broad peak at about 20–25% difference.

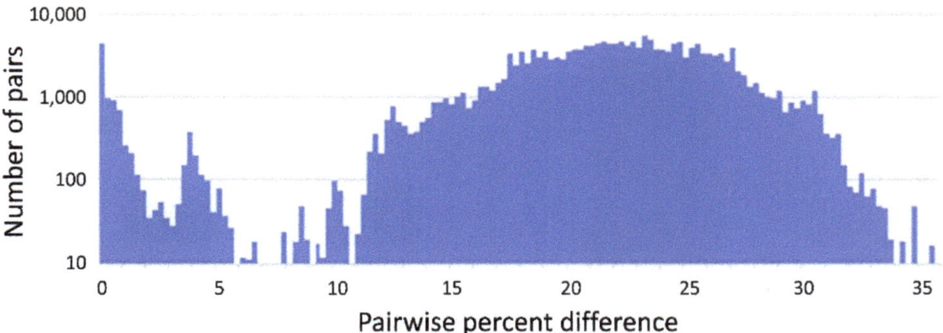

Figure 3. Histogram of pairwise similarities among the 631 chironomid sequences from the curated database (all specimens, minus 21 sequences known to be in error, listed in Table S1). The vertical axis is the number of pairwise observations, converted to percent pairwise difference and plotted on a semi-log scale for each 0.25% bin. The horizontal axis is the percent range of the matches.

With the exception of one, all pairs below 3.5% difference agreed in species when species was known (and, in any case, they always agreed in genus if identified only to sp.). The exceptional pair of species with less than a 3.5% difference is *Procladius denticulatus* and *P. sublettei*, which are separated in the curated tree by only 2.6%. The next closest pairwise difference between specimens identified to the species level was the pair *Cryptochironomus ramus* and *Cryptochironomus digitatus*, which differ in sequence by 5.8%.

In the range of 3.5% to 11%, all pairs agreed on the genus, with some pairs agreeing on the species as well. The lowest percent difference at which a pairwise difference occurred for 2 specimens differing in genus was at 11.0%; this occurred for the pairwise distance between *Parachironomus abortivus* and *Chironomus decorus*. Another example of a pair of genera with approximately this difference is *Coelotanypus scapularis* and *Procladius denticulatus*, differing by 11.1%.

Some pairs with identical morphospecies identification differed in sequence by more than 10% and could potentially represent cryptic species. These pairs include the following: *Glyptotendipes meridionalis* (represented by branches 3 and 9), differing by 15.2%; *Polypedilum halterale* (branches 54 and 56), 11.4%; *Rheotanytarsus exiguus* (branches 90 and 91), 11.8%; *Procladius bellus* (branches 80 and 81), 11.9%; and *Chironomus riparius* (branches 28 and 29), 13.5%.

3.3. Improved Identification of Water Mite Prey Using the Curated Chironomid Sequences Database

The curated chironomid sequence database was used in the present study both to confirm previous identifications of sequences in the water mite gut and, in many cases, to improve its specificity. An example is illustrated in Figure 4, in which the branches of a previously published neighbor-joining tree of dietary sequences in a specimen of *L. davidcooki* is paired with closely related sequences in the chironomid database [15]. In the original tree, half of the chironomid sequences were identified only to the genus level, and numerous other branches were identified only to family (Chironomidae) or subfamily (Orthocladinae, or Chironominae subfamilies) [15]. The species identities of some branches were confirmed (e.g., *Chironomus riparius* identities were supported by curated *Chironomus riparius* sequences that were 97–98% identical), while the genus of other branches was confirmed by a match better than 90.5% identity (e.g., KM995443.1 *Cricotopus* sp. at 91.3%

identity to *Cricotopus festivellus*). For other branches, the closest database sequence improved the identification from family to species (e.g., KR278055.1 Chironominae improved to *Endochironomus subtendens*, 97.6% identity match, highlighted in blue) or genus to species (e.g., KR276527.1 *Paratanytarsus* sp. was improved to *Paratanytarsus natvigi*, with 97.9% identity). One of the branches of KR955123.1 Chironominae was identified to genus by a 90.2% match to *Robackia claviger* (0.3% less than our usual "genus rule" of 90.5%).

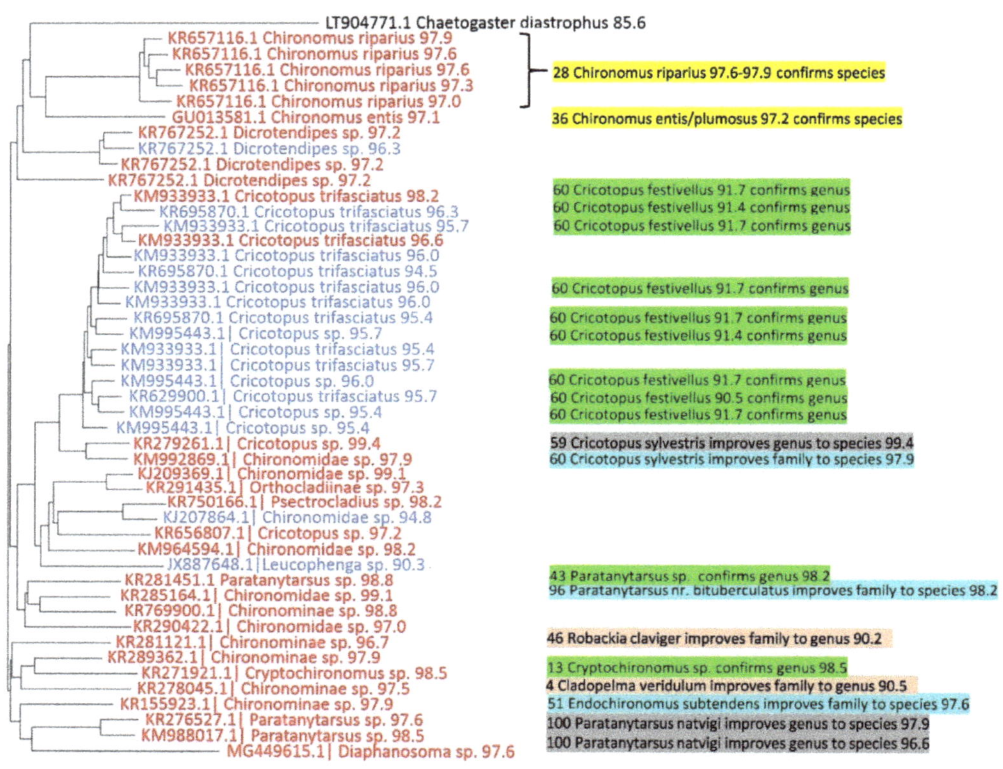

Figure 4. A representative example of confirmed and improved identification of water mite dietary content using the curated chironomid barcode database. The neighbor-joining tree (reproduced from Vasquez, Mohiddin, Li, Bonnici, Gurdziel and Ram [15]) aligns with the confirmations of identification highlighted in yellow for species and green for genus. Additionally, improvements are highlighted: grey for genus to species, blue for family to species and brown for family to genus. The curated sequence names are listed in the following format: Lab ID branch number, genus or species, percent similarity to the mite diet sequence.

In other water mites, similar improvements (family to species, genus to species, and family to genus) were observed, as well as various confirmations of genus or species. Figure 5 summarizes the number of members of the curated chironomid database that have confirmed or improved a previous identification of a dietary sequence in the set of 40 water mites whose diets were analyzed in this study. A total of 11 of the 99 branches of the curated chironomid database improved identifications to species that had previously been identified only to family (6, Table 1) or only to genus (5, Table 2). In addition, 13 members of the curated database improved family identifications in water mite diets to genus (Table 3).

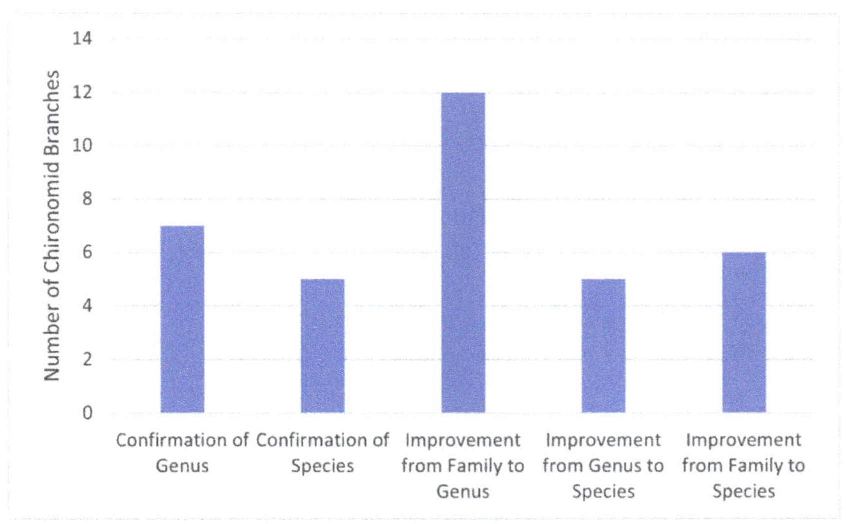

Figure 5. Summary of numbers of members of the curated chironomid database that either confirmed the identity of mite diet sequences or improved them from family to either genus or species or from genus to species.

Table 1. Summary of family to species improvements.

Previous GenBank Identification in Mite Diet	Improved Identification
Chironomidae	*Dicrotendipes modestus*
Chironomidae	*Parachironomus abortivus*
Chironominae	*Endochironomus subtendens*
Chironomidae	*Cricotopus festivellus*
Chironomidae	*Tanypus punctipennis*
Chironomidae	*Paratanytarsus nr. bituberculatus*

Table 2. Summary of genus to species Improvements.

Previous GenBank Identification in Mite Diet	Improved Identification
Parachironomus sp.	*Parachironomus tenuicaudatus*
Cricotopus sp. [1]	*Cricotopus sylvestris*
Cricotopus sp. [1]	*Cricotopus sylvestris*
Orthocladius sp.	*Orthocladius obumbratus*
Rheotanytarsus sp.	*Rheotanytarsus exiguus* gr.

[1] Two different branches identified as *Cricotopus* sp. in the mite diet correspond to two different branches in the curated tree identified as *C. sylvestris*, respectively.

Figure 6 summarizes that 38 of the 41 water mites had improvements in the identification of their dietary sequences. While a few water mites experienced improvements only of family to genus, 18 water mites experienced all 3 types of improvements. A total of 27 water mites had dietary constituents with improvements from family level identifications to species. Table 4 summarizes the dietary differences observed in the various water mite species that were the subject of this study, taking into account all of the improvements in the identification of sequences provided by the chironomid database. Among the specimens analyzed, the 21 specimens of *L. davidcooki* had by far the more diverse chironomid diet (33 different chironomid barcodes in its diet) compared to the 16 specimens of *L. quinquemaculosa* (17 chironomid barcodes) and the other species of water mites in the table. Only one *Arrenurus*, two *Limnesia* and two unidentified specimens of *Lebertia* were analyzed.

Table 3. Summary of family to genus improvements.

Previous GenBank. Identification in Mite Diet	Improved Identification [1]
Chironominae	*Cladopelma veridulum*
Chironomidae	*Polypedilum simulans*
Chironominae	*Robackia claviger*
Chironomidae	*Ablabesmyia mallochi*
Chironomidae	*Ablabesmyia annulata*
Chironomidae	*Tanytarsus glabrescens*
Chironominae	*Chironomus* sp.
Chironomidae	*Cricotopus bicinctus* gr.
Chironomidae	*Polypedilum* cf. *halterale*
Chironomidae	*Polypedilum halterale* gr.
Chironominae	*Polypedilum trigonum*
Chironomidae	*Polypedilum scalaenum*

[1] Although the database provides species-level identification, the examples given are considered improvements only to genus because the distance was more than 3.5% and below 9.5%. Alternatively, if the database sequence was only identified to "sp.", the improvement was also only to genus.

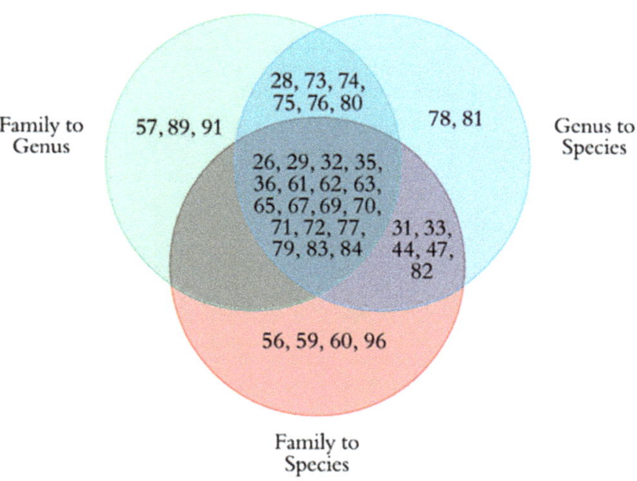

Figure 6. Water mite chironomid diet improvements from curated chironomid database. Improvements from family to genus (green circle), genus to species (blue circle) and family to species (red circle) are represented. The numbers in the diagram refer to specific mites in the water mite database for this project.

Table 4. Chironomids that were observed as prey in *Lebertia davidcooki*, *Lebertia quinquemaculosa*, *Lebertia* sp., *Limnesia* sp. and *Arrenurus* sp. indicated by check marks. Identifications that are between 3.5% and 9.5% pairwise difference from the mite diet sequence or that are identified as no better than to genus are marked with an asterisk.

Chironomids		Water Mite Species				
Chironomid Name	OTU Number	*Lebertia davidcooki*	*Lebertia quinquemaculosa*	*Lebertia* sp.	*Limnesia* sp.	*Arrenurus* sp.
Glyptotendipes meridionalis *	3	✓				

Table 4. *Cont.*

Chironomids		Water Mite Species				
Chironomid Name	**OTU Number**	***Lebertia davidcooki***	***Lebertia quinquemaculosa***	***Lebertia* sp.**	***Limnesia* sp.**	***Arrenurus* sp.**
Cladopelma veridulum *	4	✓	✓	✓	✓	✓
Parachironomus sp. *	6	✓	✓	✓		
Parachironomus tenuicaudatus	7	✓				
Parachironomus hazelriggi *	8	✓				
Cryptochironomus sp. *	13	✓	✓	✓		
Cryptochironomus ponderosus *	14	✓		✓		
Glypotendipes senilis	21	✓				
Dicrotendipes modestus	22	✓	✓			
Chironomus riparius	28	✓	✓	✓		✓
Chironomus riparius *	29	✓				
Parachironomus abortivus	33		✓			
Chironomus entis/plumosus	36	✓		✓		
Chironomus maturus	38	✓				✓
Chironomus sp. *	39	✓				
Chironomus crassicaudatus	42	✓	✓	✓		
Paratanytarsus sp. *	43	✓	✓	✓	✓	
Polypedilum simulans *	44	✓		✓		
Robackia claviger *	46	✓		✓		
Polypedilum illinoense *	47	✓				
Polypedilum scaleneum *	49	✓				
Endochironomus subtendens	51	✓	✓	✓		
*Polypedilum halterale**	54			✓		
Polypedilum halterale gr. *	55			✓		
Polypedilum trigonum *	57	✓	✓			
Cricotopus sylvestris	58	✓	✓	✓	✓	
Cricotopus sylvestris	59	✓	✓	✓	✓	
Cricotopus festivellus	60	✓	✓	✓	✓	
Cricotopus bicinctus gr. *	63	✓	✓			
Ablabesmyia mallochi *	70			✓		
Ablabesmyia annulate *	72			✓		
Orthocladius obumbratus	74	✓		✓		
Tanypus punctipennis	76	✓				
Coelotanypus sp. *	82	✓				
Tanytarsus glabrescens *	87	✓	✓			
Rheotanytarsus exiguous gr.	91	✓				
Paratanytarsus nr. *bituberculatus*	96	✓	✓	✓		
Paratanytarsus natvigi *	100	✓	✓	✓		✓

* Check marks that indicate chironomid taxa with asterisk were only identified to genus.

4. Discussion

Chironomids are a speciose dipteran found in many diverse aquatic habitats and are an important food source for multiple organisms, including water mites. A previous study on water mite prey DNA revealed that for *Lebertia* water mites, chironomids make up more than 80% of the diet content, but definitive identification of the majority of chironomid prey seen in the water mite diets was lacking [15]. Our current work developed an expanded chironomid reference sequence database by an intense multi-year sampling for chironomids to bioinformatically improve the identification of chironomids in water mite diets and to improve the study of chironomids in the environment in general. This work now contributes: (1) several new DNA COI barcodes for North American chironomid taxa, (2) insights into COI barcode gap parameters for future chironomid DNA barcoding work, (3) improved identification of chironomids found in the molecular gut contents of water mites, and (4) increased knowledge of chironomid genetic diversity in part of the Laurentian Great Lakes watershed.

4.1. New Barcodes

To generate the expanded chironomid database, we sampled further at our previously published chironomid collection sites: our paper by Failla et al. [18] was based on 2012 collections; the current paper includes a comparable number of specimens collected in Toledo Harbor in 2013. We also collected at additional locations in Michigan, especially in the Detroit metropolitan area (Figure 1). These additional collections enabled the identification of specimens with novel DNA barcodes that had not previously been identified to the species level (and in many cases not appearing at all, even at genus identification) in either GenBank or BOLD databases. The novel species barcodes in this work include *Parachironomus abortivus, Endochironomus subtendens, Robackia claviger, Orthocladius obumbratus, Cryptochironomus ponderosus, Parachironomus hazelriggi, Paracladopelma nereis, Cryptochironomus ramus, Dicrotendipes fumidus, Cryptotendipes casuarius, Cryptotendipes pseudotener, Dicrotendipes neomodestus, Tribelos jucundum, Tanytarsus confuses, Pseudochironomus fulviventris, Polypedilum illinoense, Polypedilum trigonum, Rheotanytarsus exiguous, Eukiefferiella gracei, Parakiefferiella subaterrimus, Hydrobaenus johansenni, Procladius freeman,* and *Procladius sublettei. Harnischia curtilamellata, Tanypus punctipennis, Paratanytarsus nr. bituberculatus* and *Polypedilum flavum* were the first barcodes reported in North America for those species.

Among these new barcodes, we provide here additional information about *Robackia claviger*. This sequence is the first barcode for this genus and species in Genbank; however, several sequences of the same genus (*Robackia demeijerei*) are present in BOLD. Compared to the *R. demereijerei* sequences in BOLD, the *R. claviger* sequence differs by 6.5%, fully within the range expected to be considered belonging to the "same genus". A sequence in the diet of a specimen of *Lebertia davidcooki* that was originally identified from GenBank as Chironominae (i.e., only to family) was 90.2% identical to the *Robackia claviger* in our curated database, just slightly more distant than our usual requirement of 90.5% identity for assigning genus (however, see the discussion of barcode gaps below). *Robackia* has at least four recognized species as erected by Saether [33]. In the present study, *Robackia claviger* was sampled from the Ohio River at Shawnee State Park and has previously been found in lotic habitats in the southeastern United States [34]. In the Great Lakes region, *R. demereijerei* larvae were found in coarse sediments in Lake Michigan, and it was thought that their narrow head and tough outer body integuments allowed them to inhabit this embenthic habitat [35]. The *R. demereijerei* specimens in BOLD are from a lake in Sweden and Chequamegon Bay in Lake Superior. Due to the >3.5% distance from both *R. claviger* and *R. demereijerei*, we speculate that the species in the water mite diet may be either *R. pilicauda, R. aculeate*, or a new species since new species are still being described, such as *R. parallela sp. n.* from China [36].

4.2. Insights into "Barcode Gaps" in Chironomids

Figure 2 of Failla et al. [18] shows that pairwise differences >3.5% and <11% (but having very few pairs between 6% and 11%) were always of the same genus. Although the pairwise distance analysis in this paper of 631 chironomid sequences shows only a relative minimum at 3.5% and not a distinct "gap", that difference still seems to be a good delineation, at least among chironomids, of how far a pairwise distance could be and still be used reliably for species identification. We had only one exceptional species set below the 3.5% difference that did not follow this 3.5% parameter: two *Procladius* species (*denticulatus* and *sublettei*) were separated in the curated chironomid tree by only 2.6%. The next closest pairwise distance between specimens identified to the species level was the pair *Cryptochironomus ramus* and *Cryptochironomus digitatus,* which differed in sequence by 5.8%. The peak at 4% in the histogram of pairwise distances may also contain species differences. However, all of the pairs represented in this peak had at least one specimen that was identified only to genus (e.g., the distance between 13 *Cryptochironomus* sp. and 14 *Cryptochironomus ponderosus* is 3.8%, and the distance of 34 *Harnischia* sp. and 35 *Harnischia curtilamellata* is 4.6%). While these branches are clearly separate on the basis of sequence, it is unknown whether they represent different species. Additional species-level identifications of specimens that differ in sequence by 3.5–5.8% will be necessary to resolve whether differences beyond 3.5% (and how far?) are usually of the same species or not.

In the present study, we were conservative in assigning genus identifications based on sequence alone. We limited such assignments to differences of no greater than 9.5% except in the case of *Robackia*, noted above, in which we assigned a tentative identification with a 9.8% pairwise distance. In fact, analysis of the pairwise data in the present study indicates that reliable genus identification extends out to greater distances. All of the pairs in the small pairwise peaks centered around 8% and 10% were between sequences of the same genus. We found that reliable genus identifications could be made up to 11% differences, beyond which pairs of different genera begin to be detected.

We also observed instances when specimens with identical morphospecies identifications differed greatly (i.e., defined as differences of >11%) from one another. These include *Glyptotendipes meridionalis*, *Polypedilum halterale*, *Rheotanytarsus exiguous*, *Procladius bellus*, and *Chironomus riparius*. Further study might reveal them as cryptic species. A review of the morphospecies characters of representative specimens of each cluster may reveal some new differentiating character by which animals in the two clusters can be distinguished.

From the above considerations and data, we, therefore, conclude that (a) assignment of sequences of operational taxonomic units to species can be done reliably up to a 3.5% pairwise difference, (b) assignment of genus can be done with confidence up to at least 9.5% and possibly more, and (c) identical morphospecies designations with greater than 11% difference in their sequence indicate the possible presence of cryptic species represented by one or both branches being compared.

4.3. Improved Water Mite Diet Identifications

Improving the identity of molecular water mite diet sequences enables us to better understand the diversity of chironomids in the diet contents and the trophic interactions of aquatic food webs in which water mites are embedded. For example, our curated chironomid database allowed us to identify the following species of chironomids as prey for water mites: *Parachironomus tenuicaudatus*, *Glyptotendipes senilis*, *Dicrotendipes modestus*, *Parachironomus abortivus*, *Chironomus riparius*, *Chironomus entis/plumosus*, *Chironomus maturus*, *Chironomus crassicaudatus*, *Endochironomus subtendens*, *Cricotopus sylvestris*, *Cricotopus festivellus*, *Orthocladius obumbratus*, *Tanypus punctipennis*, *Rheotanytarsus exiguous* gr., and *Paratanytarsus nr. bituberculatus*. Many of the barcode sequences of these prey species were previously known at best only to family.

Table 4, which lists the genera and species of chironomids that were found in the guts of *Lebertia quinquemaculosa*, *L. davidcooki*, *Lebertia* sp., *Limnesia* and *Arrenurus*, is expected to be a reliable list of taxa that the water mites have been ingesting. The greater richness of chironomid prey for *L. davidcooki* may indicate a dietary difference that is a result of niche partitioning. The dietary difference of the two species might also or alternatively be related to seasonal variation between collection dates for the two species, as described by Vasquez, Mohiddin, Li, Bonnici, Gurdziel and Ram [15].

In some cases, a dietary sequence of a water mite could be identified only to the genus, as in Table 3. This occurred in two ways: (1) BLAST comparisons to GenBank or to the curated database returned high identity (>96.5%) to sequences that were identified only to the genus in the database; or (2) the best match to the reference database was a <96.5% match to a known species or genus. Both types of genus identifications indicate inadequate species coverage in reference databases of the chironomids that *Lebertia* are ingesting.

While inadequate sampling effort may be part of the reason for incomplete species coverage of water mite diet sequences, another possibility is that water mites may be able to access chironomid habitats that our collecting and taxonomic methods have not yet encountered. Traditional sampling methods to capture chironomid larvae and adults are limited since chironomid larvae may inhabit unusual habitats, such as mined substrates like submerged wood [37], that may not be picked up by a ponar dredge and other collection methods used in this study. Chironomid larva of the genera *Cricotopus*, *Endochironomus*, *Glyptotendipes*, and *Parachironomus* are miners of substrates such as macrophytes, bryozoans and sponges [37] and were, nevertheless, found in the diets of the water mites studied. Water mites may be active predators seeking out and digging out chironomids from unusual habitats that human collectors may miss, as pointed out by Hudson many years ago [37].

In this regard, water mites seem to function as "DNA detectives", sometimes ingesting the DNA of rare or difficult to collect benthic microinvertebrates. This was also true for the DNA of oligochaetes that water mites ingested: the numerous oligochaete sequences associated with *L. quinquemaculosa* rarely matched any previously barcoded species or genus within 10% [15], suggesting that water mites are "discovering" species of organisms that collectors have not yet encountered or at least not yet bar-coded and submitted to a public database.

Specific Taxa Found in Water Mite Diets

Among the taxa that this study has newly identified in mite diets are *Tanypus punctipennis* and *Ablabesmyia*. *Tanypus punctipennis* is an example of a prey organism for which identification was improved from family to species, enabling a more specific analysis of trophic relationships. *Tanypus punctipennis* is a midge with distribution in all world regions except Australia [38]. *T. punctipennis* has a wide Palearctic distribution in temperate climates, consistent with its possible presence in the Great Lakes [39]. The late instar larvae of *T. punctipennis* are relatively small [40] compared to larger genera such as *Chironomus* and *Ablabesmyia annulata*. We speculate that the small size of *T. punctipennis* makes it a more suitable prey for the smaller *Lebertia davidcooki* water mite.

Some sequences in the *Lebertia* diet that were previously identified only to family (Chironomidae) were similar enough to *Ablabesmyia* (approximately 9% pairwise distance) to be identified with that genus. *Ablabesmyia* is a genus found worldwide, with over 90 identified species [41]. In the Americas, they are primarily found in the Nearctic region, including the Laurentian Great Lakes. *Ablabesmyia* is typically found on substrate in shallow littoral zones [42,43]. *A. monilis* has been reported from Northern Michigan and is found in muddy-bottomed lakes [44]. This is consistent with the Blue Heron Lagoon (Detroit, MI collection site) habitat, where the water mites studied in this work were obtained. *Ablabesmyia annulata* has the largest head capsule among 30 species of North American chironomids in which third instar larvae were compared [45]. *Ablabesmyia* is known to have a symbiotic relationship with freshwater mussels [46] and are predators of *Tubifex tubifex*, their own early instars, and other benthic macroinvertebrates [29,41,47,48]. *Ablabesmyia mallochi* is known to be more difficult to morphologically identify, so having an available barcode may assist in its identification in the future. Adults of *Ablabesmyia annulate*, on the other hand, are easily identifiable as they have anterior and posterior parapodia that are not darkened, three palpal segments, a long procercus, and a more quadrate head [49,50]. The larvae are just as easily differentiable from other species of *Ablabesmyia* [51]. *Ablabesmyia* DNA sequence was detected in only one water mite, suggesting that these genera of midges are rare in this habitat or that the species of water mites studied do not prefer *Ablabesmyia* sp. as prey potentially due to their larger larval size compared to other chironomids.

4.4. Need for Further Improvements of Knowledge of Chironomid Diversity

In our previous work on molecular barcodes of chironomids [18], we reported a tree with 45 larval operational taxonomic units and an additional adult barcode sequence not yet observed in larvae. That publication improved the species identification of the hitherto mostly genus-level identifications from 15.5% of the operational taxonomic units (OTUs) to more than 40% of the OTUs and reported sequences for 22 chironomid genera and 19 species. The present work expands the number of distinct chironomid OTUs from 46 to 99 and now includes barcodes from 39 genera and 60 species, a significant increase from previous studies.

However, in the curated database, several genera are identified only to genus level, either because they are for larval specimens for which species-level keys are not available or because the specimens were insufficiently intact to determine species. Of the 99 members of the curated chironomid barcode database, the following genera lack even one specimen identified to species: *Cladotanytarsus*, *Stempellina*, *Microchironomus*, *Kiefferulus*, *Benthalia*, *Smittia*, and *Clinotanypus*. We originally included *Nanocladius* in this list; however, a sequence of *N. distinctus* from Sweden, updated on BOLD on 16 November 2021, is a 99.8% match to *Nanocladius* sp. MT526144. The other *Nanocladius* sp. sequence (65 *Nanocladius* sp. OR041915S6) in our curated database has no species-level match in either GenBank or BOLD; the closest species match on BOLD is *N. distinctus* at a pairwise distance of 11% and therefore is likely a different species. The recency of the *N. distinctus* record could suggest that this is an ongoing activity in several laboratories. The lack of species identifications of some OTUs of these genera emphasizes the need for more collecting, morphological identification and barcoding to attain more complete coverage of chironomids.

5. Conclusions and Future Considerations

The use of DNA molecular barcoding on chironomids has yielded significant advances in assisting with differentiating among multiple species of chironomids since specialized taxonomy in these aquatic organisms is not readily available [52]. DNA barcoding studies on chironomids combined with next-generation sequencing (NGS) techniques have been suggested as an efficient method to assess biodiversity and future environmental monitoring of these important aquatic invertebrates [53]. In addition to water mites, DNA barcoding has also been shown to be a useful tool in identifying the diet of other important aquatic organisms, such as fish [17]. Since chironomids are a significant part of fish diets, this curated database should assist diet analyses of fish as well.

Improvements in the species-level identifications of chironomids may enable investigations of the determinants of prey choice by water mites. Why, for example, is *Cladopelma* preyed upon by all five types of water mites studied and the various species of *Cricotopus* by four of the five types of water mites, in contrast to others that were detected as prey only in *Lebertia davidcooki* (notably, *Glyptotendipes*, *Tanypus*, *Coelotanypus*, and *Rheotanytarsus*)? Considering that the water mite diets matched five *Polypedilum* species only to genus level, what other species of *Polypedilum* might *Lebertia* be ingesting? Are there comparative features of the larvae of the four species of *Chironomus* (and two identified only to genus) that the various species of water mites ingested that would explain the different patterns of occurrence among the different water mite species? Identifying these organisms to the species level may enable future studies of differences in predator-prey dynamics in the laboratory and the field.

The methodology of our work does not differentiate whether the predator is feeding on the eggs, larvae, pupae or even the emerging adult stage of the prey. Some water mites feed on the larval stage of chironomids, although some other species of water mites are known to feed on dipteran eggs, including chironomids [10]. Follow-up experiments in the laboratory may be able to determine on which life stage(s) the predator is feeding.

Curation of barcoding data, as we have done here, is a critical step for using DNA barcodes in the future [54]. This paper demonstrates the use of DNA barcoding beyond simply biodiversity and biomonitoring analysis. The improved water mite diet information also sets the stage for future studies looking deeper into trophic interactions that require molecular analysis where morphological and observational data are not sufficient. While the advent of DNA barcoding and other genetic identification tools has brought advancement to identifying species, many more barcode sequences accompanied by careful morphospecies analyses are needed so that the sequences can be more usefully applied. Many taxonomic identifications were made decades ago, but with new technology and data, inconsistencies should be reviewed and updated (i.e., curated, as we have done). Expert taxonomists for aquatic invertebrate organisms are few and overburdened with material [55]; therefore, future advancements will require increased investment in the education and research of taxonomists who can combine morphological and molecular approaches to taxonomy whenever possible.

Supplementary Materials: The following are available online at https://www.mdpi.com/article/10.3390/d14020065/s1, Table S1: Curation notes regarding the chironomid database: comments, corrections, and omitted sequences. Figure S1: 99 sequences in the Chironomid curated database, shown in a neighbor-joining tree (samedata as Figure 2 in the main text).

Author Contributions: Conceptualization, A.A.V. and J.L.R.; methodology, P.L.H., B.L.B., J.L.R., and A.A.V.; validation, A.A.V., S.H.Y., P.L.H. and J.L.R.; formal analysis, A.A.V., B.L.B., S.H.Y., P.L.H. and J.L.R.; investigation, A.A.V., B.L.B., J.I.C., P.L.H. and J.L.R.; resources, A.A.V., J.I.C., P.L.H. and J.L.R.; data curation, A.A.V., B.L.B., S.H.Y., J.I.C., P.L.H. and J.L.R.; writing—original draft preparation, A.A.V. and B.L.B.; writing—review and editing, A.A.V., B.L.B., S.H.Y., P.L.H. and J.L.R.; visualization, J.L.R., B.L.B., J.I.C., S.H.Y. and A.A.V.; supervision, J.L.R. and A.A.V.; project administration, J.L.R. and A.A.V.; funding acquisition, J.L.R. and A.A.V. All authors have read and agreed to the published version of the manuscript.

Funding: This research was funded by a National Institutes of Health grant from the National Institutes of Health Common Fund and Office of Scientific Workforce Diversity under three linked awards, RL5GM118981, TL4GM118983, and 1UL1GM118982, administered by the National Institute of General Medical Sciences and by the Fred A. and Barbara M. Erb Family Foundation and the Sharon L. Ram Aquatic Sciences Fund to A.A.V.; collection of chironomids by J.L.R. was funded by a grant from the Environmental Protection Agency (Grant number GL00E00808-0); and B.L.B. was supported by Healthy Urban Waters with support from the Fred A. and Barbara M. Erb Family Foundation. J.L.R. covered the cost of open-access publication of this manuscript.

Data Availability Statement: Sequence data generated from this work are openly available in the GenBank repository for nucleotide sequences (https://www.ncbi.nlm.nih.gov/genbank/ accessed on 10 January 2022).

Acknowledgments: We are grateful for the work of many taxonomists who have painstakingly studied chironomids and other aquatic organisms, including Mike Sergeant, who identified several species of *Polypedilum*. We are grateful for many student volunteers who assisted us over the years by collecting chironomids from different locations in the aquatic habitats surrounding the Laurentian Great Lakes and also assisted with generating molecular barcodes. Thanks are extended to Armin Namayandeh who provided comments on an earlier version of this manuscript and to two anonymous reviewers who also helped to improve this work.

Conflicts of Interest: The authors declare no conflict of interest.

Appendix A

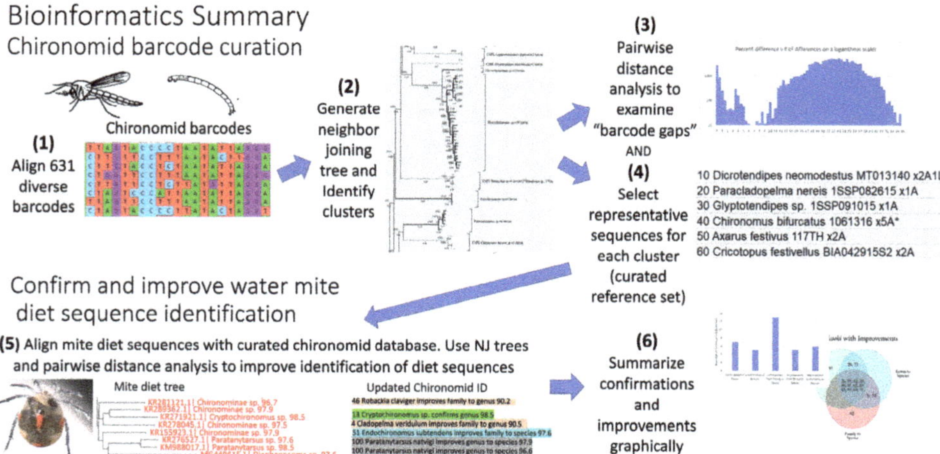

Figure A1. Graphical summary of methodology.

References

1. Albert, J.S.; Destouni, G.; Duke-Sylvester, S.M.; Magurran, A.E.; Oberdorff, T.; Reis, R.E.; Winemiller, K.O.; Ripple, W.J. Scientists' warning to humanity on the freshwater biodiversity crisis. *Ambio* **2020**, *50*, 85–94. [CrossRef]
2. Rapp, T.; Shuman, D.A.; Graeb, B.D.S.; Chipps, S.R.; Peters, E.J. Diet composition and feeding patterns of adult shovelnose sturgeon (Scaphirhynchus platorynchus) in the lower Platte River, Nebraska, USA. *J. Appl. Ichthyol.* **2011**, *27*, 351–355. [CrossRef]
3. Failla, A.; Vasquez, A.; Fujimoto, M.; Ram, J. The ecological, economic and public health impacts of nuisance chironomids and their potential as aquatic invaders. *Aquat. Invasions* **2015**, *10*, 1–15. [CrossRef]
4. Mantilla, J.G.; Gomes, L.; Cristancho, M. The differential expression of Chironomus spp genes as useful tools in the search for pollution biomarkers in freshwater ecosystems. *Briefings Funct. Genom.* **2017**, *17*, 151–156. [CrossRef] [PubMed]
5. Koperski, P. Taxonomic, phylogenetic and functional diversity of leeches (Hirudinea) and their suitability in biological assessment of environmental quality. *Knowl. Manag. Aquat. Ecosyst.* **2017**, *418*, 49. [CrossRef]
6. Hamidoghli, A.; Falahatkar, B.; Khoshkholgh, M.; Sahragard, A. Production and Enrichment of Chironomid Larva with Different Levels of Vitamin C and Effects on Performance of Persian Sturgeon Larvae. *North Am. J. Aquac.* **2014**, *76*, 289–295. [CrossRef]
7. Vasquez, A.A.; Kabalan, B.A.; Ram, J.L.; Miller, C.J. The Biodiversity of Water Mites That Prey on and Parasitize Mosquitoes. *Diversity* **2020**, *12*, 226. [CrossRef]
8. Smith, B.P. Host-parasite interaction and impact of larval water mites on insects. *Annu. Rev. Entomol.* **1988**, *33*, 487–507. [CrossRef]
9. Smith, I.M.; Oliver, D. Review of parasitic associations of larval water mites (acari: Parasitengona: Hydrachnida) with insect hosts. *Can. Èntomol.* **1986**, *118*, 407–472. [CrossRef]
10. Proctor, H.; Pritchard, G. Neglected predators-water mites (Acari, Parasitengona, Hydrachnellae) in fresh-water communities. *J. N. Am. Benthol. Soc.* **1989**, *8*, 100–111. [CrossRef]
11. Winkel, E.H.T.; Davids, C.; De Nobel, J. Food and Feeding Strategies of Water Mites of the Genus Hygrobates and the Impact of Their Predation on the Larval Population of the Chironomid Cladotanytarsus Mancus (Walker) in Lake Maarsseveen. *Neth. J. Zool.* **1988**, *39*, 246–263. [CrossRef]
12. Shatrov, A.B. Anatomy and ultrastructure of the salivary (prosomal) glands in unfed water mite larvae Piona carnea (C.L. Koch, 1836) (Acariformes: Pionidae). *Zool. Anz.-A J. Comp. Zool.* **2012**, *251*, 279–287. [CrossRef]
13. Cohen, A.C. Solid-to-Liquid Feeding: The Inside(s) Story of Extra-Oral Digestion in Predaceous Arthropoda. *Am. Èntomol.* **1998**, *44*, 103–117. [CrossRef]
14. Martin, P.; Koester, M.; Schynawa, L.; Gergs, R. First detection of prey DNA in Hygrobates fluviatilis (Hydrachnidia, Acari): A new approach for determining predator-prey relationships in water mites. *Exp. Appl. Acarol.* **2015**, *67*, 373–380. [CrossRef]
15. Vasquez, A.A.; Mohiddin, O.; Li, Z.; Bonnici, B.L.; Gurdziel, K.; Ram, J.L. Molecular diet studies of water mites reveal prey biodiversity. *PLoS ONE* **2021**, *16*, e0254598. [CrossRef] [PubMed]
16. Hebert, P.D.N.; Cywinska, A.; Ball, S.L.; DeWaard, J.R. Biological identifications through DNA barcodes. *Proc. R. Soc. B Biol. Sci.* **2003**, *270*, 313–321. [CrossRef] [PubMed]
17. Leray, M.; Yang, J.Y.; Meyer, C.P.; Mills, S.C.; Agudelo, N.; Ranwez, V.; Boehm, J.T.; Machida, R.J. A new versatile primer set targeting a short fragment of the mitochondrial COI region for metabarcoding metazoan diversity: Application for characterizing coral reef fish gut contents. *Front. Zool.* **2013**, *10*, 34. [CrossRef] [PubMed]
18. Failla, A.; Vasquez, A.; Hudson, P.; Fujimoto, M.; Ram, J. Morphological identification and COI barcodes of adult flies help determine species identities of chironomid larvae (Diptera, Chironomidae). *Bull. Èntomol. Res.* **2015**, *106*, 34–46. [CrossRef]
19. Folmer, O.; Black, M.; Hoeh, W.; Lutz, R.; Vrijenhoek, R. DNA primers for amplification of mitochondrial cytochrome c oxidase subunit I from diverse metazoan invertebrates. *Mol. Mar. Biol. Biotechnol.* **1994**, *3*, 294–299. [PubMed]
20. Vasquez, A.A.; Hudson, P.L.; Fujimoto, M.; Keeler, K.; Armenio, P.M.; Ram, J.L. Eurytemora carolleeae in the Laurentian Great Lakes revealed by phylogenetic and morphological analysis. *J. Great Lakes Res.* **2016**, *42*, 802–811. [CrossRef] [PubMed]
21. Vasquez, A.A.; Qazazi, M.S.; Fisher, J.R.; Failla, A.J.; Rama, S.; Ram, J.L. New molecular barcodes of water mites (Trombidiformes: Hydrachnidiae) from the Toledo Harbor region of Western Lake Erie, USA, with first barcodes forKrendowskia(Krendowskiidae) andKoenikea(Unionicolidae). *Int. J. Acarol.* **2017**, *43*, 494–498. [CrossRef]
22. Vasquez, A.A.; Carmona-Galindo, V.; Qazazi, M.S.; Walker, X.N.; Ram, J.L. Water mite assemblages reveal diverse genera, novel DNA barcodes and transitional periods of intermediate disturbance. *Exp. Appl. Acarol.* **2020**, *80*, 491–507. [CrossRef]
23. Townes, H.K. The nearctic species of Tendipedini-Diptera, Tendipedidae (=chironomidae). *Am. Midl. Nat.* **1945**, *34*, 1–206. [CrossRef]
24. Saether, O.A. Glyptotendipes Kieffer and Demeijerea Kruseman from Lake Winnipeg, Manitoba, Canada, with the description of four new species (Diptera: Chironomidae). *Zootaxa* **2011**, *2760*, 39–52. [CrossRef]
25. Saether, O.A. Cryptochironomus Kieffer from Lake Winnipeg, Canada, with a review of Nearctic species (Diptera: Chironomidae). *Zootaxa* **2009**, *2208*, 1–24. [CrossRef]
26. Epler, J.H. Biosystematics of the genus Dicrotendipes Kieffer, 1913 (Diptera: Chironomidae) of the world. *Mem. Am. Entomol. Soc.* **1988**, *36*, 1–214.
27. Dendy, J.S.; Sublette, J.E. The Chironomidae (=Tendipedidae: Diptera) of Alabama with Descriptions of Six New Species1. *Ann. Èntomol. Soc. Am.* **1959**, *52*, 506–519. [CrossRef]
28. Cranston, P.S.; Dillon, M.E.; Pinder, L.C.V.; Reiss, F. The adult males of Chironominae (Diptera, Chironomidae) of the holarctic region-keys and diagnoses. *Entomol. Scand.* **1989**, *34*, 353–502.

29. Roback, S.S. Monograph 17 the academy of natural sciences of Philadelphia the adults of the subfamily Tanypodinae equals pelopiinae in North America Diptera chironomidae. *Monogr. Acad. Nat. Sci. Phila.* **1971**, *17*, 410.
30. Heyn, M.W. A review of the systematic position of the North American species of the genusGlyptotendipes. *Aquat. Ecol.* **1992**, *26*, 129–137. [CrossRef]
31. Kumar, S.; Stecher, G.; Li, M.; Knyaz, C.; Tamura, K. MEGA X: Molecular Evolutionary Genetics Analysis across Computing Platforms. *Mol. Biol. Evol.* **2018**, *35*, 1547–1549. [CrossRef] [PubMed]
32. Epler, J.H. Revision of the Nearctic Dicrotendipes Kieffer, 1913 (Diptera: Chironomidae). *Evol. Monogr.* **1987**, *9*, 1–102.
33. Saether, O.A. Taxonomic studies on Chironomidae Nanocladius pseudochironomus and the Harnischia complex. *Bull. Fish. Res. Board Can.* **1977**, *196*, 1–141.
34. Hudson, P.L.; Lenat, D.R.; Caldwell, B.A.; Smith, D. *Chironomidae of the Southeastern United States: A Checklist of Species and Notes on Biology, Distribution, and Habitat;* U S Fish and Wildlife Service Fish and Wildlife Research: Washington, DC, USA, 1990; pp. 1–46.
35. Winnell, M.H.; Jude, D.J. Associations among Chironomidae and Sandy Substrates in Nearshore Lake Michigan. *Can. J. Fish. Aquat. Sci.* **1984**, *41*, 174–179. [CrossRef]
36. Yan, C.; Wang, X. Robackia Saether from China (Diptera: Chironomidae). *Zootaxa* **2006**, *1361*, 53–59. [CrossRef]
37. Hudson, P.L. Unusual larval habitats and life-history of Chironomid (Diptera) genera. *Entomol. Scand.* **1987**, *29*, 369–373.
38. Roback, S. *Adults of the Subfamily Tanypodinae (-Pelopinae) in North America (Diptera: Chironomidae)*; Academy of Natural Sciences: Hinckley, MN, USA, 2007; Volume 17, p. 410.
39. Aydın, G.B. The growth of Tanypus punctipennis meigen (Diptera, Chironomidae) larvae in laboratory conditions and the effects of water temperature and ph. *Trak. Univ. J. Nat. Sci.* **2018**, *19*, 101–105. [CrossRef]
40. Specziár, A. Life history patterns of Procladius choreus, Tanypus punctipennis and Chironomus balatonicus in Lake Balaton. *Ann. de Limnol.-Int. J. Limnol.* **2008**, *44*, 181–188. [CrossRef]
41. Stur, E.; da Silva, F.L.; Ekrem, T. Back from the Past: DNA Barcodes and Morphology Support Ablabesmyia americana Fittkau as a Valid Species (Diptera: Chironomidae). *Diversity* **2019**, *11*, 173. [CrossRef]
42. Int Panis, L.; Boudewijn, G.; Lieven, B.; Verheyen, R.F. Ablabesmyia longistyla Fittkau, 1962 (Diptera: Chironomidae), new for the Belgian fauna. *Bull. Annls Soc. R. Belg. Ent.* **1992**, *128*, 316–318.
43. Oliver, D.R.; Roussel, M.E. *The Genera of Larval Midges of Canada Diptera: Chironomidae*; Insectes et Arachnides du Canada; Research Branch, Agriculture Canada: Ottawa, ON, Canada, 1983; pp. 1–263.
44. Egan, A.T.; Ferrington, L.C., Jr. Chironomidae (Diptera) in Freshwater Coastal Rock Pools at Isle Royale, Michigan. *Trans. Am. Èntomol. Soc.* **2015**, *141*, 1–25. [CrossRef]
45. Hudson, P.L.; Adams, J.V. Sieve efficiency in benthic sampling as related to chironomid head capsule width. *J. Kans. Entomol. Soc.* **1998**, *71*, 456–468.
46. Roback, S.S.; Bereza, D.J.; Vidrine, M.F. Description of an Ablabesmyia [Diptera: Chironomidae: Tanypodinae] Symbiont of Unionid Fresh-Water Mussels [Mollusca:Bivalvia:Unionacea], with Notes on Its Biology and Zoogeography. *Trans. Am. Entomol. Soc.* **1979**, *105*, 577–620.
47. Kaster, J.L.; Bushnell, J.H. Occurrence of Tests and Their Possible Significance in the Worm, Tubifex tubifex (Oligochaeta). *Southwest. Nat.* **1981**, *26*, 307. [CrossRef]
48. Oliveira, C.S.N.; Fonseca-Gessner, A.A.; Silva, M.A.N. The immature stages of Ablabesmyia (Sartaia) metica Roback, 1983 (Diptera: Chironomidae) with keys to subgenera. *Zootaxa* **2008**, *1808*, 61–68. [CrossRef]
49. Roback, S.S. Ablabesmyia (Sartaia) metica, a New Subgenus and Species (Diptera: Chironomidae: Tanypodinae). *Proc. Acad. Nat. Sci. Phila.* **1983**, *135*, 236–240.
50. Beck, W.M. Biology of the larval chironomids. *State Fla. Dep. Environ. Regul.* **1976**, *2*, 58.
51. Boesel, M.W. The early stages of Ablabesmyia annulata (Say) (Diptera, Chironomidae). *Ohio J. Sci.* **1972**, *72*, 3.
52. Pfenninger, M.; Nowak, C.; Kley, C.; Steinke, D.; Streit, B. Utility of DNA taxonomy and barcoding for the inference of larval community structure in morphologically cryptic Chironomus (Diptera) species. *Mol. Ecol.* **2007**, *16*, 1957–1968. [CrossRef] [PubMed]
53. Brodin, Y.; Ejdung, G.; Strandberg, J.; Lyrholm, T. Improving environmental and biodiversity monitoring in the Baltic Sea using DNA barcoding of Chironomidae (Diptera). *Mol. Ecol. Resour.* **2012**, *13*, 996–1004. [CrossRef] [PubMed]
54. Grant, D.; Brodnicke, O.; Evankow, A.; Ferreira, A.; Fontes, J.; Hansen, A.; Jensen, M.; Kalaycı, T.; Leeper, A.; Patil, S.; et al. The Future of DNA Barcoding: Reflections from Early Career Researchers. *Diversity* **2021**, *13*, 313. [CrossRef]
55. Elías-Gutiérrez, M.; Jerónimo, F.M.; Ivanova, N.V.; Valdez-Moreno, M.; Hebert, P.D.N. DNA barcodes for Cladocera and Copepoda from Mexico and Guatemala, highlights and new discoveries. *Zootaxa* **2008**, *1839*, 1–42. [CrossRef]

 *diversity* 

Article

Diversity and Life-Cycle Analysis of Pacific Ocean Zooplankton by Videomicroscopy and DNA Barcoding: Gastropods

Peter J. Bryant [1,*] and Timothy Arehart [2]

[1] Department of Developmental and Cell Biology, University of California Irvine, Irvine, CA 92697, USA
[2] Crystal Cove Conservancy, Newport Coast, CA 92657, USA
* Correspondence: pjbryant@uci.edu

Abstract: The life cycles and biodiversity of Pacific coast gastropods were analyzed by videomicroscopy and DNA barcoding of individuals collected from tide pools and in plankton nets from a variety of shore stations. In many species (Families Calyptraeidae, Cerithiopsidae, Strombidae, Vermetidae, Columbellidae, Nassariidae, Olivellidae, Hermaeidae, Onchidorididae, Gastropteridae, Haminoeidae), the free-swimming veligers were recovered from plankton collections; in *Roperia poulsoni* (family Muricidae) veligers were usually recovered from egg sacs where they had been retained although some escapees were found in plankton collections; in *Pteropurpura festiva* (family Muricidae) free-living veligers were also found; and in *Atlanta californiensis* (family Atlantidae) both veligers and adults were obtained from plankton collections making this a holoplanktonic species. The results confirm that DNA barcoding based on COI gene sequencing is a useful strategy to match life-cycle stages within species as well as to identify species and to document the level of biodiversity within the gastropods.

Keywords: Mollusks; gastropods; Zooplankton; plankton; COI mitochondrial gene; Pacific Ocean; larvae; DNA barcoding

Citation: Bryant, P.J.; Arehart, T. Diversity and Life-Cycle Analysis of Pacific Ocean Zooplankton by Videomicroscopy and DNA Barcoding: Gastropods. *Diversity* **2022**, *14*, 912. https://doi.org/10.3390/d14110912

Academic Editor: Manuel Elias-Gutierrez

Received: 18 September 2022
Accepted: 14 October 2022
Published: 27 October 2022

1. Introduction

Gastropods (snails and slugs) occur in saltwater, freshwater, and terrestrial environments. They are the most diverse class in the phylum Mollusca, containing about 476 families with 65,000–80,000 living species. The class is thought to be second only to the insects in overall species number [1].

The fertilized egg of gastropods hatches directly into a spherical or pear-shaped free-swimming larval stage called the trochophore, carrying a ring of cilia [2]. The ciliary girdle then expands into large, heavily ciliated lobes, giving rise to a larval stage called the veliger. Later the larva undergoes torsion, a 180° twisting that brings the posterior part of the body to an anterior position behind the head. Torsion is unique to the gastropods.

The veliger has a shell (secreted by the dorsal shell gland), a foot, and a velum, which is a lobed, ciliated structure used for swimming and feeding [2]. In most cases, the veliger eventually sinks to the seabed, loses its velum, and completes its metamorphosis into a juvenile or adult with typical snail-like morphology (a heteroplanktonic life cycle). In some cases, the adult is also planktonic, making the life cycle holoplanktonic

Analyzing the development of gastropod larvae can be done by rearing and documenting individuals, but here we show that a simpler and more efficient method is to gather individuals and identify them by sequencing their DNA barcode, which is a portion of the cytochrome *c* oxidase I (COI or COX1) gene, found in mitochondrial DNA [3,4]. We have previously used this approach to document the life cycles of cnidarians [5] and crustaceans [6].

The class Gastropoda includes both shelled and unshelled species. The marine shelled species include whelks, abalone, conches, periwinkles, turbans, cowries, limpets, chitons, and others. In most cases the one-piece shell is coiled in both larval and adult stages, but in the limpets it is coiled only in the larval stage. Gastropods are distinguished by their asymmetrical anatomy, in which most of the organs are more developed on one side of the body than the other. They typically have a distinct head carrying two or four sensory tentacles bearing eyes. Their ventral foot is the basis for the name gastropod ("stomach-foot"). Many species have an operculum which allows closing of the shell. In the following we use the taxonomy established by the World Register of Marine Species [7].

2. Materials and Methods

Zooplankton was collected under Scientific Collecting Permit SC-12162 from the California Department of Fish and Wildlife. Collections were made from 16 sites between Newport Beach and Dana Point, Orange County California, as well as one off Santa Barbara and two from Baja California (Table 1 and Figure 1).

Table 1. Collection locations.

Locality	Latitude	Longitude	Collecting Site #
Balboa Island at Coral Street	33 D 60′ N	117 D 89′ W	1
Via Lido and Genoa, Lido Island	33 D 62′ N	117 D 92′ W	2
Crew Dock, Back Bay Science Center	33 D 62′ N	117 D 89′ W	3
Newport Harbor	33 D 61′ N	117 D 91′ W	4
Newport Harbor Entrance	33 D 59′ N	117 D 88′ W	5
Newport Pier	33 D 61′ N	117 D 93′ W	6
Little Treasure Cove, Crystal Cove St Park	33 D 58′ N	117 D 86′ W	7
Ocean off Crystal Cove State Park	33 D 57′ N	117 D 85′ W	8
Pelican Point, Crystal Cove State Park	33 D 58′ N	117 D 85′ W	9
Reef Point, Crystal Cove State Park	33 D 57′ N	117 D 85′ W	10
Crescent Bay, Laguna Beach	33 D 55′ N	117 D 80′ W	11
Shaw's Cove, Laguna Beach	33 D 54′ N	117 D 80′ W	12
Rocky Bight, Crystal Cove State Park	33 D 57′ N	117 D 83′ W	13
Twin Points, Crystal Cove State Park	33 D 33′ N	117 D 48′ W	14
Ocean off Dana Point from a plankton tow aboard R/V Sea Explorer to 800 ft depth. 4/18/15, 9 p.m.	33 D 46′ N	117 D 71′ W	15
Rock Field, Dana Point	33 D 46′ N	117 D 71′ W	16
Naples Reef, off Santa Barbara, CA	34 D 42′ N	119 D 93′ W	
Boca near whale shark, Baja California	28 D 95′ N	113 D 55′ W	
La Profunda, Baja California	28 D 96′ N	113 D 55′ W	

Shore-based collections were made with a 150 μm mesh net (aperture 30 cm) attached to a rope, with a 50 mL collection tube at the base. They were made from public docks using repeated horizontal sweeps near the surface and diagonal sweeps down to about 5 m depth. About 5–10 sweeps of a total of about 35 m usually yielded sufficient specimens, but no attempt was made to monitor collections quantitatively.

Ocean collection #15 was made with a 250 μm. mesh net attached to a 35 m rope. The net (aperture 30 cm) was towed behind the vessel, just below the surface, for a period of 7 min at the slowest possible speed. Deployment and retrieval extended the total tow period to 10 min.

Field work. In the following account, only the localities outside of Orange County are identified specifically.

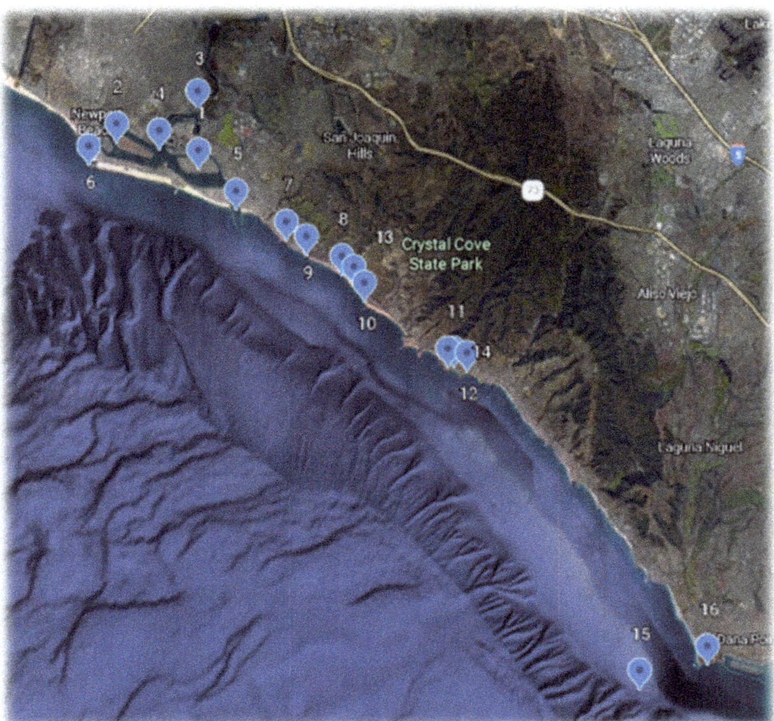

Figure 1. Map of collecting sites listed in Table 1.

Laboratory analysis. Plankton collections were brought to the laboratory at the University of California, Irvine and examined under a dissecting microscope with lateral light and a dark background. Each specimen of interest was removed using a Pasteur Pipette, transferred to a depression slide, and recorded by video microscopy using a Zeiss microscope with a dark-field condenser, fitted with a phototube attached to a Nikon D5100 single-lens reflex camera. The most informative frames were taken from the videos and used in the figures for this paper. Each plankton specimen was preserved in 90% ethanol in a well of a 96-well microplate.

Each adult specimen was photographed in place, removed physically from its location, brought to the laboratory, and examined under the dissecting microscope. Using the microwave method [8] live adults were quickly heated sufficient to kill the animals and firm up their tissues; the steam from inside the shell forcing the body from the shell for easy removal. Multiple tissue samples were then removed using dissecting tools and transferred to the microplates. If available, typically three individuals per species were sacrificed, though occasionally more were used due to our inability to positively identify some species using field characteristics. Many of the species that also have a benthic stage in the life cycle have already been listed by the Southern California Association of Marine Invertebrate Taxonomists (SCAMIT) [9].

Filled plates were sent to the Canadian Centre for DNA Barcoding at the University of Guelph, 50 Stone Road East, Guelph, ON, N1G2W1, Canada for DNA extraction using an in-house protocol (http://ccdb.ca/resources/ accessed on 1 October 2022), and bidirectional sequencing of the standard 648-bp "DNA barcode" [3,4] in the COI mitochondrial gene. All samples except three were run with cocktail primers C_GasF1_t1 + GasR1_t1; Gast14_A01, Gast14_A02, and Gast14_A03 were run with both C_GasF1_t1 + GasR1_t1 and ZplankF1_t1 + ZplankR1_t1.

The procedure usually produced a DNA barcode of 658 nucleotides, and only those containing > 300 nucleotides were included in the sequence analysis. Groups of specimens with identical or almost identical DNA barcodes were assigned BIN numbers. They were compared with all barcode records on BOLD (10,580,183 Sequences) including the Public Record Barcode Database (2,529,561 Sequences/153,565 Species/66,474 Interim Species) using the Bold Aligner (Amino Acid Based HMM). The identification system on BOLD delivers a species identification if the query sequence shows less than 1% divergence to a reference sequence.

Images of all specimens as well as the DNA barcode sequences are in the public domain under code GASSC at the Canadian Centre for DNA Barcoding (http://www.boldsystems.org/index.php/MAS_Management_DataConsole?codes=GASSC accessed on 1 October 2022). In this paper, we concentrate on those species where we have found pelagic larval stages.

Conceptualization, Project administration, Data curation, Formal analysis: Peter Bryant; Investigation, Methodology: Timothy Arehart, Peter Bryant; Writing—original draft, Peter Bryant; Writing—review and editing, Peter Bryant and Timothy Arehart. Collections were made and analyzed with the assistance of Undergraduate students Taylor Sais, Alicia Navarro, Debbie Chung, Lesly Ortiz, and Bita Rostam.

3. Results

This project GASSC (Gastropoda of Southern California) included 1238 specimens of which 1235 provided images. 680 specimens provided a COI-5P sequence and 589 of these were Barcode-compliant, falling into 143 BINs containing 127 species (See Supplementary Materials). Our data (Figure 2) show a much larger range of interspecific divergences (seen in the graph of "Distance to Nearest Neighbour") compared to intraspecific divergences in this DNA sequence for gastropods suggesting the existence of a "DNA barcode gap". Within species in this set of samples, the mean% divergence in sequence was 0.66 +/− 0.0 S.E. Within the largest set of conspecifics (*Crepidula onyx*, $n = 54$) the mean% divergence in sequence was 0.69 +/− 0.0 S.E., maximum 2.49%, minimum 0.0%.

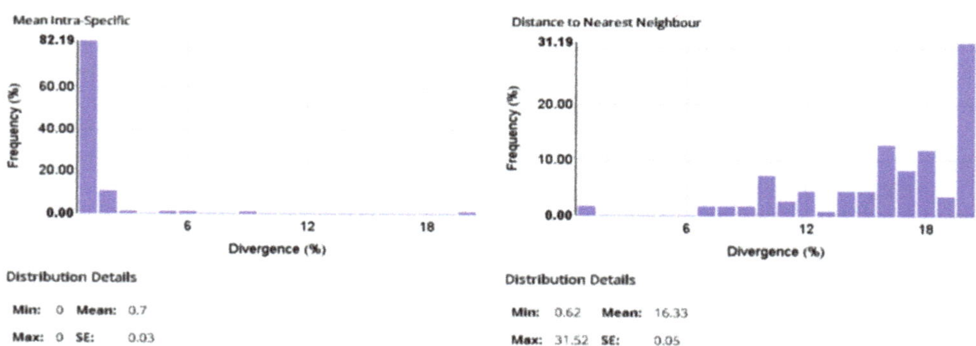

Figure 2. Sequence divergence in the COI Barcode for intraspecific and interspecific comparisons using the data included in this publication. Distance Model: Kimura 2 Parameter; Alignment: BOLD Aligner; Length filter >/= 300 bp; Excluded: contaminants and misidentifications, records with stop codons.

Subclass caenogastropoda

Order Littorinimorpha: Sea Snails

Family Atlantidae Wiegmann and Ruthe, 1832

A family of microscopic (<1 cm shell diameter), holoplanktonic gastropods [1]. They have a transparent, coiled shell into which their bodies can be retracted, and an operculum that is used to close off the opening. The larval stage is a veliger in which the velum is initially small and bilobed, but with growth it develops three lobes (Figure 3). The larval shell and operculum are retained in the adult.

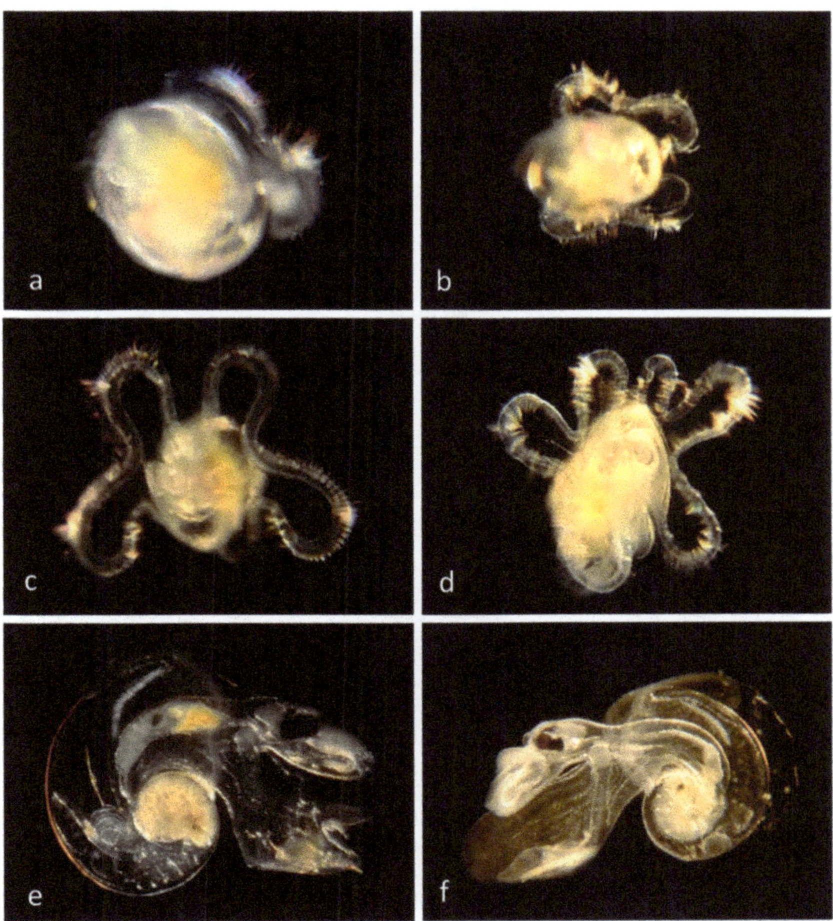

Figure 3. Family Atlantidae: Sea snail, *Atlanta californiensis* (BINACV0973; *n* = 19). (**a**) Early veliger BIOUG01205-B08 from off Dana Point. (**b**) Later veliger with bilobed velum CCDB-24236-G01 from off Newport Pier. (**c**) Veliger with bilobed velum CCDB 31597 E12 from Newport Harbor entrance. (**d**) Later veliger with 6-lobed velum CCDB-24236-H05 from Newport Harbor. (**e**) Adult CCDB 31597 E09 from Newport Harbor entrance. (**f**) Adult CCDB-24722-D06 from La Profunda, Baja California.

Family Calyptraeidae Lamarck, 1809, the Slipper Shells

The shell is quite flat and has an internal shelf-like half shell, so that it resembles a traditional bedroom slipper; hence the name "slipper shell" or "slipper limpet" (although these are not true limpets). During mating they pile on top of each other to form a tower called a mating chain, in which all individuals start as males but the basal member transforms into a female. Females produce eggs that are fertilized internally, and developing embryos are held beneath the mother's shell until they hatch into microscopic trochophore larvae before developing into swimming larvae (veligers) carrying shells, thousands of which disperse and later metamorphose into juveniles on the ocean floor.We have examined individuals of *Crepidula onyx* (Figure 4) *Crepidula naticarum* (Figure 5) and *Crepidula huerta* (Figure 6). In some species (e.g., Figures 6a and 7b) the vela are decorated with yellow spots, the nature and function of which is unknown.

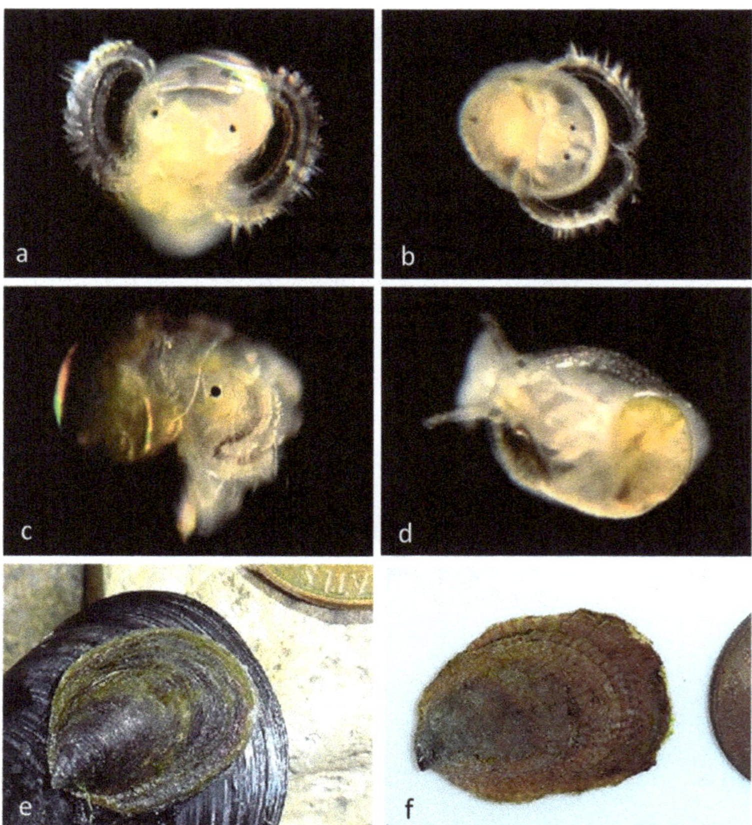

Figure 4. Family Calyptraeidae. Onyx slipper snail, *Crepidula onyx*. (**a**) Veliger BIOUG01229-H01 from off Newport Pier. (**b**) Veliger BIOUG01229-B01 from off Newport Pier. (**c**) Late veliger CCDB 31597 F09 from Newport Harbor entrance. (**d**) Juvenile BIOUG01205-D02 from off Newport Pier. (**e**) Adult CCDB-32336 D06 from Balboa at Coral. (**f**) Adult CCDB-32336 E12 from Dana Point south shore.

Our adult specimens of *Crepidula huertae*, (Figure 6) found inside the empty shell of a Bubble Snail (*Bulla gouldiana*) from Balboa at Coral (Newport Harbor), supported by the presence of DNA-sequenced veligers at the Harbor Entrance, extend the host and geographic range of this species, which was otherwise known from hermit crab shells at Naples Reef, near Santa Barbara, Santa Barbara County, California [10].

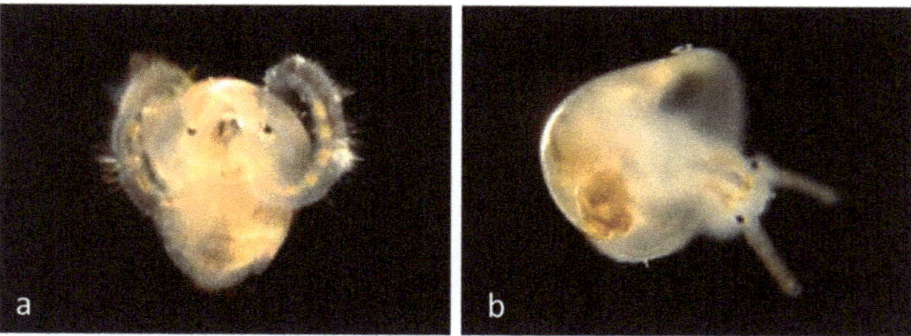

Figure 5. Family Calyptraeidae. Slipper snail, *Crepidula naticarum*. (**a**) Veliger CCDB-24722-H04 from Newport Beach Harbor entrance. (**b**) Juvenile CCDB-24722-H09 from Newport Beach Harbor entrance.

Figure 6. Family Calyptraeidae. Slipper snail, *Crepidula huertae* (=*Crepidula* cf. *perforans*). (**a**) Veliger CCDB 31597 G03 from Newport Harbor entrance. (**b**) Adult CCDB-32336 D10 from inside the empty shell of a Bubble Snail (*Bulla gouldiana*) from Balboa at Coral. (**c**) Adult CCDB-32336 E01, 2&3 inside the empty shell of the same Bubble Snail (*Bulla gouldiana*). (**d**) Adult CCDB-32336 D10 and CCDB-32336 E01, 2&3 ventral views, live animals (no DNA sequences obtained). (**e**) CCDB-32336 D10 shell—ventral view. (**f**) CCDB-32336 E01, 2&3 shell—ventral view. (**g**,**h**), adult FMNH 282243. Subtidal in hermit crab shells, from Naples Reef, near Santa Barbara, Santa Barbara County, California. Length 30 mm [10]. We have found many stages of a different species of Slipper snail, *Crepipatella lingulata*. (Figure 6) which also has yellow spots on the vela.

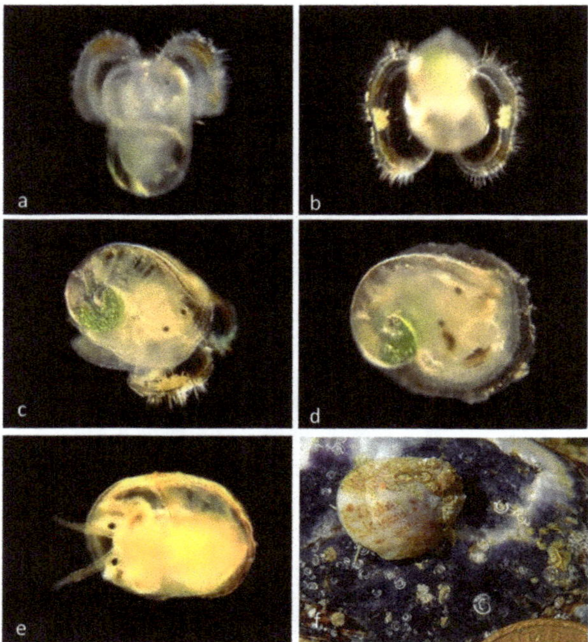

Figure 7. Family Calyptraeidae. Slipper snail, *Crepipatella lingulata.* (BIN AAF2329; n = 27 veligers, 4 pre-adults, 2 adults) (**a**) Veliger CCDB-24236-E04 from Little Treasure Cove, Crystal Cove State Park. (**b**) Veliger CCDB 31597 F05 from Newport Beach harbor entrance. (**c**) Late veliger BIOUG01229-F06 from Newport Beach harbor entrance. (**d**) Juvenile CCDB-24722-H11 from Newport Beach harbor entrance. (**e**) Juvenile BIOUG01229-F02 from Newport Beach harbor entrance, (**f**) Adult CCDB-32336 G10, dorsal, from Shaw's Cove south, Laguna Beach.

Family Cerithiopsidae H. Adams & A. Adams, 1853: The Cerithiopsids

A family of very small gastropods (Figure 8) with high spires and multiple whorls.

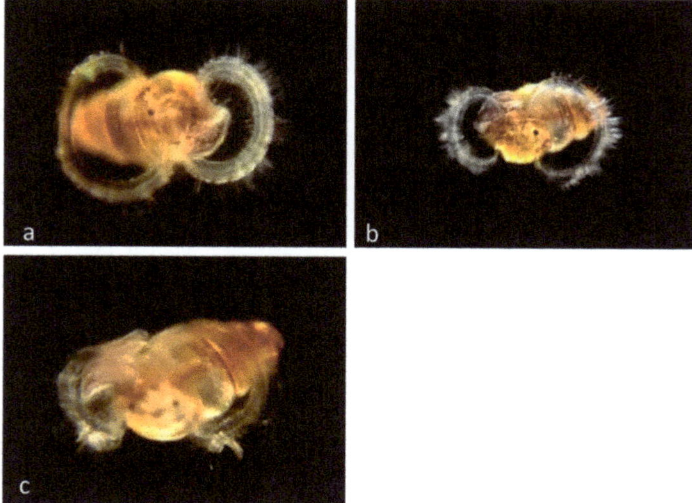

Figure 8. Family Cerithiopsidae. *Cerithiopsis* sp.? (**a**) Veliger BIOUG01205-H03 from off Newport Pier. (**b**) Veliger CCDB-24236-B10 from off Newport Pier. (**c**) Veliger CCDB-24722-D04 from La Profunda, Baja California.

Family Strombidae Rafinesque, 1815: The true Conchs

Medium to very large snails (Figure 9), with eyes on long stalks. The shell has a long, narrow aperture with an anterior indentation that accommodates one of the eye stalks.

Figure 9. Family Strombidae. Eastern Pacific fighting conch, *Strombus gracilior*. (**a**) Veliger, CCDB-24722-E09 from Bahia de Los Angeles, Baja California. (**b**) Adult, from Hardy's Internet Guide to Marine Gastropods (This species is not recorded from Orange County).

Family Vermetidae Rafinesque, 1815: Worm Snails

Small to medium-sized snails (Figure 10), with very irregular elongated tubular shells often forming large clumps. Some species have opercula at the ends of the tubes, while others do not.

Figure 10. Family Vermitidae. Scaly Worm Shell, ***Thylacodes squamigerus***. (**a**) Veliger, BIOUG01205-B01. Off Crystal Cove State Park, Laguna Beach, Orange County, CA. (**b**) Veliger, BIOUG01229-D05. Off Newport Pier, Orange County, CA. (**c**) Adult CCDB 31729 A11 from Twin Points, Laguna Beach. (**d**) Adult CCDB-24002-C03 from Shaw's Cove, Laguna Beach.

Subclass heterobranchia

The veliger has a shell, but this is lost during metamorphosis into the adult.

Family Hermaeidae H. Adams & A. Adams 1854

Small sea slugs (Figure 11) with cerata containing branches of the digestive gland, and rhinophores with a recessed tip.

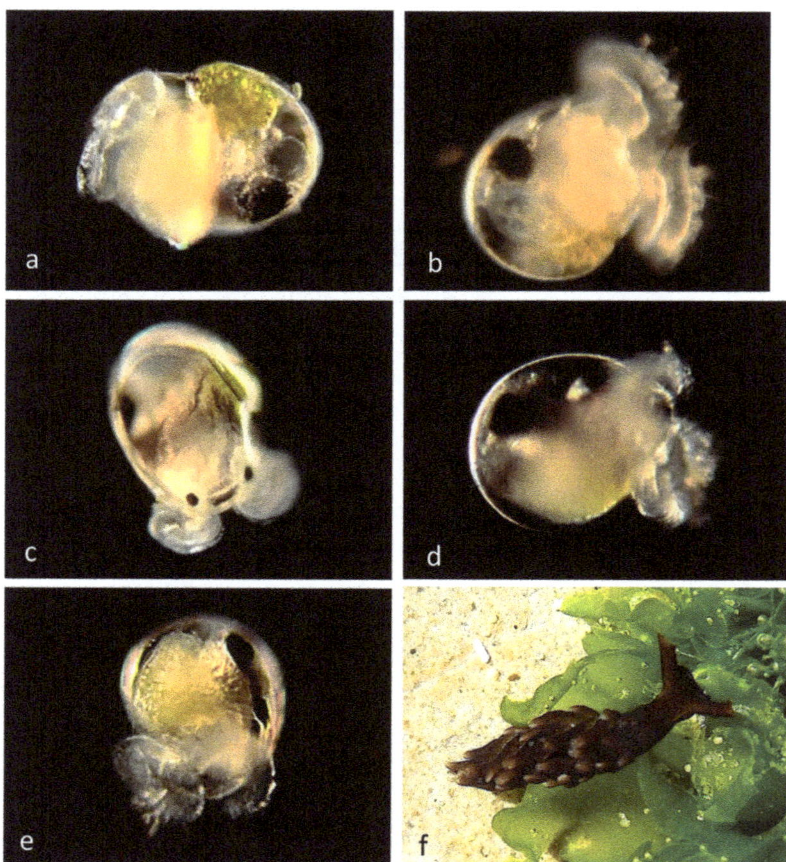

Figure 11. Family Hermaeidae. ***Aplysiopsis enteromorphae*****.** (**a**) Veliger larva BIOUG01229-H05 from Newport Pier. (**b**) Veliger larva CCDB-32329 H01 from Balboa at Coral. (**c**) Veliger larva BIOUG01205-H12 from Via Lido, Lido Island. (**d**) Late veliger BIOUG01229-H03 from Balboa at Coral. (**e**) Late veliger BIOUG01229-H04 from off Newport Pier. (**f**) Adult (no DNA barcode) from Rocky Bight, Crystal Cove State Park, Orange County, CA.

Family Onchidorididae Gray, 1827

Dorid nudibranchs (Figure 12), of which we have found only one pelagic larva.

Order Neogastropoda: Sea Snails

Characterized by a long incurrent siphon and accompanying siphonal structure on the base of the shell.

Family Columbellidae Swainson, 1840: The Dove Snails

Minute to small snails (Figure 13) with a thick shell with a narrow opening. The foot is narrow, and the siphon is very long.

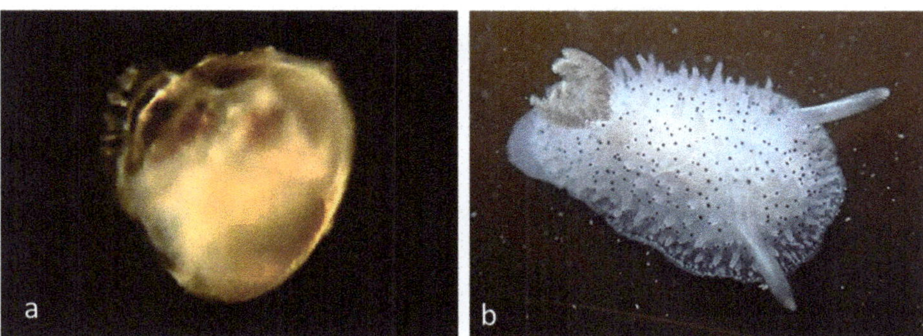

Figure 12. Family Onchidorididae. Black-tipped Spiny Doris, *Acanthodoris rhodoceras*. (**a**) Larva CCDB 31597 D11 from Balboa at Coral. (**b**) Juvenile from Cabrillo Beach, Los Angeles County, CA. (http://nathistoc.bio.uci.edu/Molluscs/Acanthodoris%20rhodoceras/index.html accessed on 1 October 2022).

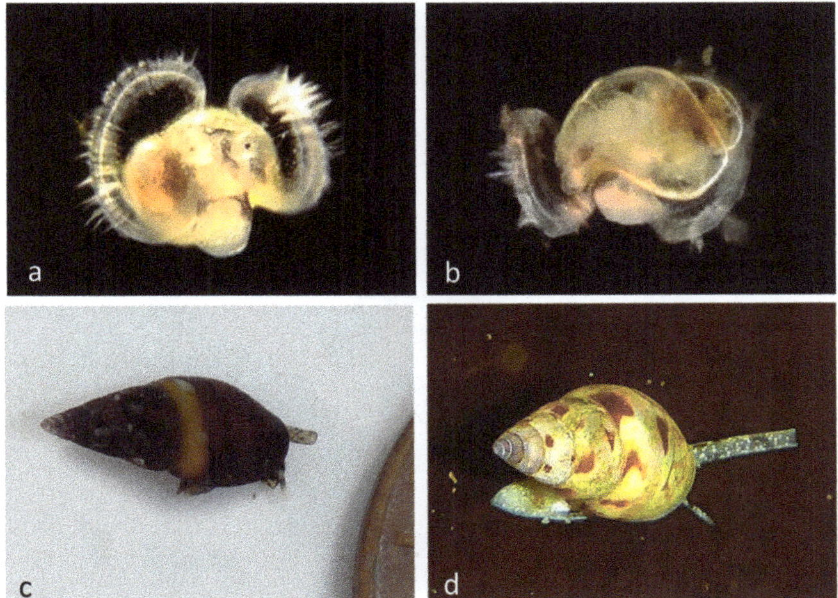

Figure 13. Family Columbellidae. Carinate dove shell, *Alia carinata* (BIN AAZ4577). (**a**) Veliger BIOUG01229-G06 from Newport Beach, Harbor entrance. (**b**) Late veliger BIOUG01229-C06 from Pacific Ocean from Crystal Cove State Park. (**c**) Adult CCDB-24002-H02 from Dana Point rock field. (**d**) Adult CCDB 31729 D02 from Twin Points, Crystal Cove State Park.

Family Nassariidae Iredale, 1916 (1835): The Dog Whelks

Snails with rounded shells (Figure 14), a high spire, an oval aperture, and a siphonal notch.

Family Muricidae Rafinesque, 1815

Within the family Muricidae, our collections include 27 adults but no larvae of *Nucella ostrina* (BIN AAA4209), 26 adults but no larvae of *Acanthinucella spirata*, 18 adults but no larvae of *Ceratostoma nuttalli* and 2 adults but no larvae of *Mexacanthina lugubris*. This is consistent with the finding that in these species of Muricidae the equivalent of the veliger stage occurs within the egg capsule, and the individuals hatch as juveniles which are not represented in our collections because of the limitations of our collection methods. In other family members Poulson's Dwarf Triton, *Roperia poulsoni*, and Festive Murex, *Pteropurpura festiva*, individuals escape as veligers (Figures 15 and 16).

Figure 14. Family Nassariidae. Western mud nassa, *Nassarius tiarula.* (BIN ACV0694; $n = 1$ veliger, 7 adults). (**a**) Veliger BIOUG01205-G06 from Via Lido and Genoa, Lido Island. (**b**) Adult CCDB32329 G01 from Balboa at Coral (mud grab). (**c**) Adult CCDB-32329 F07 from Crew Dock, Back Bay Science Center (mud grab). (**d**) Adult CCDB-32329 F06 from Crew Dock, Back Bay Science (mud grab).

Figure 15. Family Muricidae. Poulson's Dwarf Triton, *Roperia poulsoni* (BIN ADG1012, $n = 6$ adults, 2 veligers, 1 eggs). (**a**) Eggs CCDB 31597 C05 from Ocean off Dana Point. (**b**) Egg cases containing veligers, Gast14_A01—A03. Reef Point, Crystal Cove State Beach, Orange County, CA. 5/17/2021. (**c**) escaped veligers. (**d**) Veliger BIOUG01229-H10 from Newport Harbor entrance. (**e**) Adult CCDB 31729 B12 from Crescent Beach. (**f**) Adult CCDB 31729 B08 from Crescent Beach.

Figure 16. Family Muricidae. Festive Murex, *Pteropurpura festiva* (BIN ADF9780, *n* = 6 veligers, 3 adults). (**a**) Veliger BIOUG01229-H09 from Newport Beach, Harbor entrance; (**b**) Adult (lateral) CCDB-36354 G07 from Reef Point, Crystal Cove State Park; (**c**) Adult (lateral) CCDB-36354 G09 from Pelican Point, Crystal Cove State Park.

Family Olivellidae Troschel, 1869: The Dwarf Olives

A family of small predatory snails (Figure 17) with smooth, shiny, elongated shells.

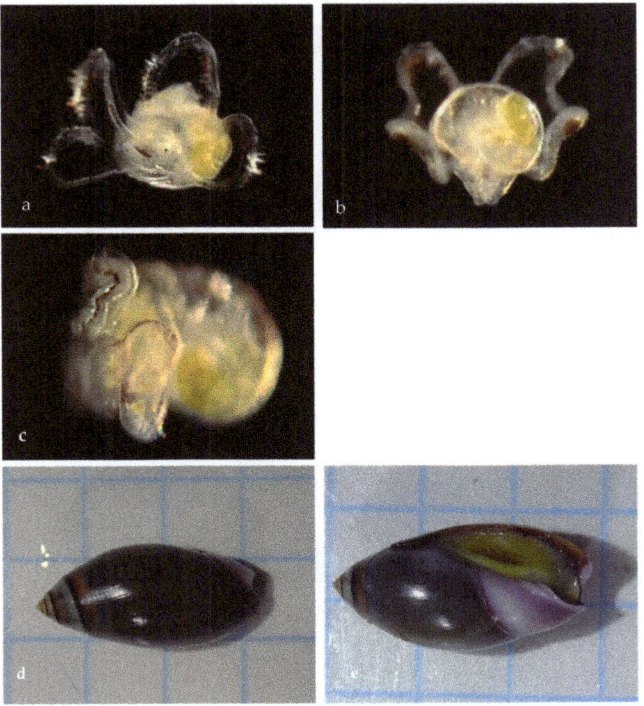

Figure 17. Family Olivellidae. Purple dwarf olive, *Callianax biplicata* (BINACB8152; *n* = 8 veligers + 2 adults pending). (**a**) Veliger CCDB-24722-H05 from Newport Beach Harbor entrance. (**b**) Veliger CCDB-24722-H07 from Newport Beach Harbor entrance. (**c**) Veliger CCDB 31597 F10 from Ocean off Newport Beach. (**d**) Adult, dorsal CCDB-32329 G10 from Balboa at Coral. (**e**) Adult, ventral CCDB-32329 G09 from Balboa at Coral. Background ¼ inch squares.

Order Cephalaspidea: Sea Slugs and Bubble Snails

Family Gastropteridae Swainson, 1840 [1] Bat-Winged Slugs

Adults have no shell, or an internal reduced shell (Figure 18). They have outgrowths from the mantle wall called parapodia, used in swimming.

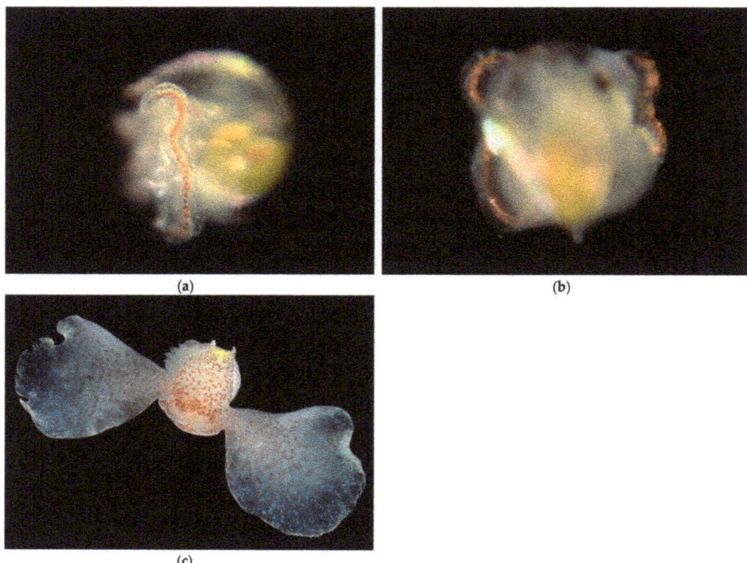

Figure 18. Family Gastropteridae. Pacific Batwing Seaslug *Gastropteron pacificum*. (**a**,**b**) Veliger CCDB 31729 A07 from Newport Beach Harbor entrance. (**a**) lateral, (**b**) dorsal. (**c**) adult © Gustav Paulay. https://cdn.floridamuseum.ufl.edu/IZ/04ff1938-4872-41fd-9a4a-270ef0c5cc7e/ (accessed on 1 January 2022).

Family Haminoeidae Pilsbry, 1895

The shell (Figure 19) is partially or completely enfolded by lateral fleshy parapodial lobes. Represented in our collection by the Blister Glassy-bubble, *Haminoea virescens* (Figure 19) and the White Bubble snail, *Haminoea vesicula* (Figure 20).

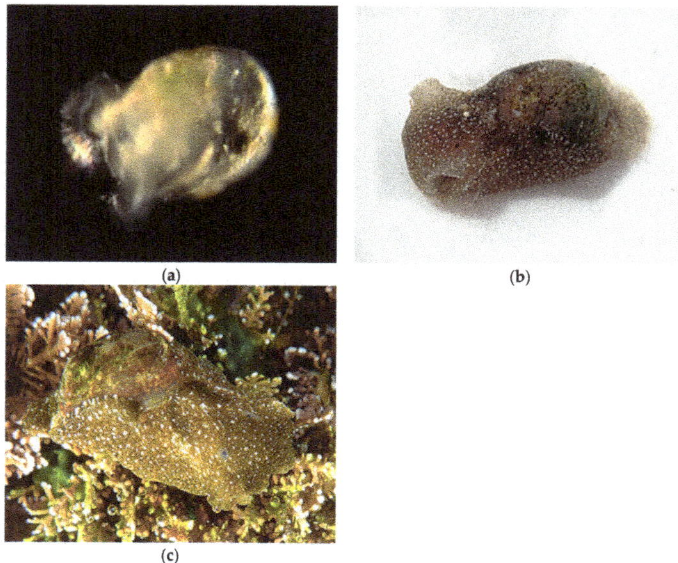

Figure 19. Family Haminoeidae. Blister Glassy-bubble, *Haminoea virescens*. (**a**), Veliger CCDB-24722-G10 from Balboa at Coral. (**b**), Adult CCDB-24002-D05 from Shaw's Cove south. (**c**), Adult CCDB-31729 B02 from Twin Points, Crystal Cove State Park.

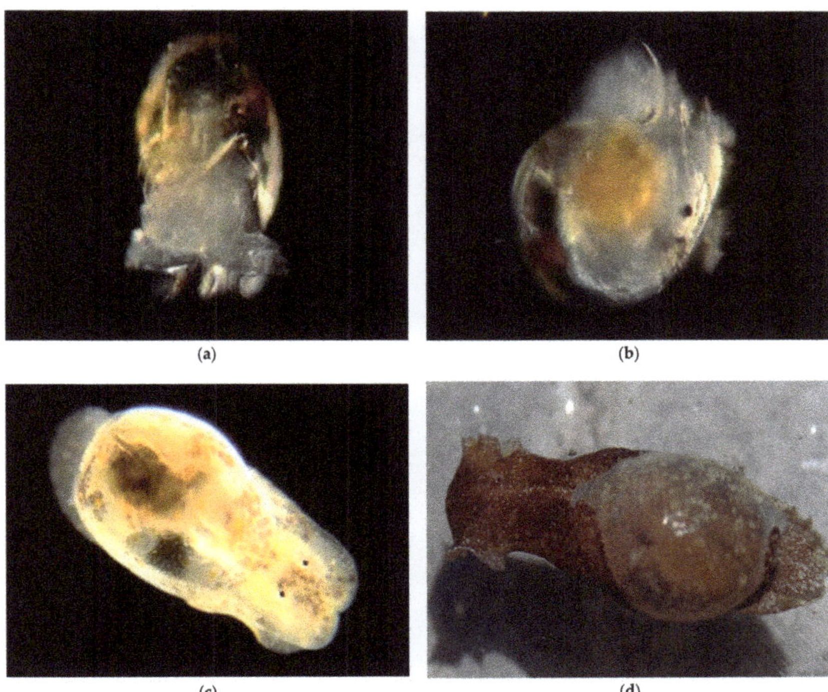

(a) (b) (c) (d)

Figure 20. Family Haminoeidae. White Bubble snail, *Haminoea vesicula*. (**a**), Veliger CCDB-24722-G12 from off Newport Pier. (**b**) Veliger CCDB-24002-A06 from Balboa at Coral. (**c**) Juvenile BIOUG01205-H09 from Balboa at Coral. (**d**) Adult from Bodega Head, CA, photo courtesy of Jackie Sones.

4. Discussion

Our results confirm that DNA barcoding using the COI barcode [3,4,11] is a useful strategy to match life-cycle stages within species as well as to identify species and to document the level of biodiversity within a taxon, in this case the gastropods. Our data also show the expected "DNA barcode gap"; i.e., a much larger range of interspecific divergences versus intraspecific divergences in this DNA sequence for gastropods (Figure 1). DNA barcoding has often revealed unexpected species diversity in many taxa [11,12] and the present study leads to the same conclusion for gastropods. The data will contribute to the development of a DNA barcode reference library, which will allow the rapid and convenient identification of individual gastropod individuals and parts collected at any developmental stage.

In this study we have begun to compile a collection of images of veligers, which are identified to the species level by matching of their DNA barcodes to those of morphologically recognized adults. These images confirm that the respective species have free-swimming veligers, and show that there is considerable morphological diversity, with some veligers having two velar lobes, some having four velar lobes, and some having yellow- or red-spotted vela. However, as expected from studies of larval stages in general, it is difficult to identify many species from the morphology of veligers.

The DNA sequence differences in the COI barcode are, of course, not responsible for the morphological differences we have observed between specimens in separate taxa. However, the DNA barcode differences that have evolved between morphologically distinct organisms can be used to examine the degree of relatedness between them. When the DNA sequence data are organized into a taxonomic tree (see Taxon Tree in Supplementary Materials), the results are generally consistent with the taxonomic tree according to conventional morphological methods. This can be explored by cladistic analysis, in which the taxonomic tree is examined for "DNA clades"—groups of species that are uniquely and exclusively related by DNA sequence. According to this analysis, all four species of *Lottia*, three species of *Crepidula*, two of *Cerithidea*, two of *Nassarius*, three of *Littorina*, two of *Nuttalina*, two of *Epitonium*, four of *Tegula*, three of *Doriopsilla*, two of *Ancula*, two of *Felimare*, two of *Corambe*, and two of *Aplysia* form clades, although the *Lottia* and *Tegula* clades include some different species.

DNA barcode reference libraries can lead to the development of more global sequencing strategies including metabarcoding and parallel sequencing of complex bulk samples including "environmental DNA", which are being developed for monitoring ecosystem health. For example, in a recent study of plankton communities in the Baltic Sea, five nonindigenous species were discovered, and four of these were identified exclusively by metabarcoding [13]. Our work illustrates the need for more larval/adult matching to build sequence libraries specifically for meta- and eDNA barcoding.

Supplementary Materials: The following supporting information can be downloaded at: http://www.boldsystems.org/index.php/MAS_Management_DataConsole?codes=GASSC (accessed on 1 January 2022) (1238 records selected) Distance Model: Kimura 2 Parameter Marker: COI-5P Labels: Taxon, Sample ID, Life Stage, Barcode Cluster (BIN) Colorization: Barcode Cluster (BIN) Alignment: BOLD Aligner (Amino Acid based HMM) Filters Applied: ≥300 bp only, exclude records flagged as misidentifications, records with stop codons, contaminants. Sequence Count: 647 sequences; 104 Species; 84 Genera; 59 Families; 76 Unidentified; 143 BINs. For specimens that were large enough to allow separation of multiple samples (usually three samples to allow for sequencing failures) all of the sequences obtained are included in the tree; when full-length barcodes were obtained they always matched perfectly.

Author Contributions: Conceptualization, Project administration, Data curation, Formal analysis: P.J.B.; Investigation, Methodology: T.A., P.J.B.; Writing—original draft, P.J.B.; Writing—review and editing, P.J.B. and T.A. All authors have read and agreed to the published version of the manuscript.

Funding: This research received no external funding.

Institutional Review Board Statement: Not applicable.

Data Availability Statement: The DNA barcodes generated during and/or analyzed during the current study are available at the Centre for Biodiversity Genomics, Biodiversity Institute of Ontario under the project GASSC: Gastropods of Southern California: http://www.boldsystems.org/index.php/MAS_Management_DataConsole?codes=GASSC (accessed on 1 October 2022).

Acknowledgments: Collections were made and analyzed with the assistance of Undergraduate students Taylor Sais, Alicia Navarro, Debbie Chung, Lesly Ortiz, and Bita Rostam.

Conflicts of Interest: The authors declare no conflict of interest.

References

1. Bouchet, P.; Rocroi, J.P.; Hausdorf, B.; Kaim, A.; Kano, Y.; Nützel, A.; Parkhaev, P.; Schrödl, M.; Strong, E.E. Revised Classification, Nomenclator and Typification of Gastropod and Monoplacophoran Families. *Malacologia* **2017**, *61*, 1–526. [CrossRef]
2. Boss, K.J. SHELLFISH | Characteristics of Molluscs. In *Encyclopedia of Food Sciences and Nutrition*; Academic Press: Cambridge, MA, USA, 2003; pp. 5216–5221.
3. Hebert, P.D.N.; Cywinska, A.; Ball, S.L.; DeWaard, J.R. Biological identifications through DNA barcodes. *Proc. R. Soc. B Biol. Sci.* **2003**, *270*, 313–321. [CrossRef] [PubMed]
4. Hubert, N.; Hanner, R. DNA Barcoding, species delineation and taxonomy: A historical perspective. *DNA Barcodes* **2016**, *3*, 44–58. [CrossRef]
5. Bryant, P.J.; Arehart, T.E. Diversity and life-cycle analysis of Pacific Ocean zooplankton by videomicroscopy and DNA barcoding: Hydrozoa. *PLoS ONE* **2019**, *14*, e0218848. [CrossRef] [PubMed]
6. Bryant, P.; Arehart, T. Diversity and life-cycle analysis of Pacific Ocean zooplankton by video microscopy and DNA barcoding: Crustacea. *J. Aquac. Mar. Biol.* **2021**, *10*, 108–136.
7. WoRMS Editorial Board World Register of Marine Species. Available online: https://www.marinespeces.org (accessed on 27 August 2022).
8. Galindo, L.A.; Puillandre, N.; Strong, E.E.; Bouchet, P. Using microwaves to prepare gastropods for DNA barcoding. *Mol. Ecol. Resour.* **2014**, *14*, 700–705. [CrossRef] [PubMed]

9. Barwick, K.; Cadien, D.B.; Lovell, L.L. (Eds.). *A Taxonomic Listing of Benthic Macro-and Megainvertebrates*, 12th ed.; The Southern California Association of Marine Invertebrate Taxonomists: Los Angeles, CA, USA, 2018.
10. Collin, R. Calyptraeidae from the northeast Pacific (Gastropoda: Caenogastropoda). *Zoosymposia* **2019**, *13*, 107–130. [CrossRef]
11. Bucklin, A.; Steinke, D.; Blanco-Bercial, L. DNA Barcoding of Marine Metazoa. *Ann. Rev. Mar. Sci.* **2011**, *3*, 471–508. [CrossRef] [PubMed]
12. Waugh, J. DNA barcoding in animal species: Progress, potential and pitfalls. *Bioessays* **2007**, *29*, 188–197. [CrossRef] [PubMed]
13. Zaiko, A.; Samuiloviene, A.; Ardura, A.; Garcia-Vazquez, E. Metabarcoding approach for nonindigenous species surveillance in marine coastal waters. *Mar. Pollut. Bull.* **2015**, *100*, 53–59. [CrossRef] [PubMed]

Article

DNA Barcoding and Distribution of Gastropods and Malacostracans in the Lower Danube Region

Selma Menabit [1,2], Tatiana Begun [1], Adrian Teacă [1], Mihaela Mureşan [1], Paris Lavin [3] and Cristina Purcarea [2,*]

1 National Institute for Research and Development on Marine Geology and Geoecology-GeoEcoMar, 024053 Bucharest, Romania; selma.menabit@geoecomar.ro (S.M.); tbegun@geoecomar.ro (T.B.); ateaca@geoecomar.ro (A.T.); mmuresan@geoecomar.ro (M.M.)
2 Department of Microbiology, Institute of Biology Bucharest of the Romanian Academy, 060031 Bucharest, Romania
3 Facultad de Ciencias del Mar y Recursos Biológicos, Departamento de Biotecnología, Universidad de Antofagasta, Antofagasta 1240000, Chile; paris.lavin@uantof.cl
* Correspondence: cristina.purcarea@ibiol.ro

Abstract: This survey reports the spatial distribution of gastropods belonging to Caenogastropoda, Architaenioglossa, Littorinimorpha, Cycloneritida and Hygrophila orders, and malacostracans from Amphipoda and Mysida orders in the lower sector of the Danube River, Romania, using DNA barcoding based on the cytochrome C oxidase I (COI) gene sequence. Sampling was performed for eight locations of Danube Delta branches and Bechet area during three consecutive years (2019–2021). Molecular identification of sixteen gastropods and twelve crustacean individuals was confirmed to the species level, providing the first molecular identification of gastropods from the Lower Danube sector. Phylogenetic analysis showed that species of gastropods and crustaceans clustered in monophyletic groups. Among gastropods, *Microcolpia daudebartii acicularis, Viviparus viviparus, Bithynia tentaculata, Physa fontinalis, Ampullaceana lagotis* and *Planorbarius corneus* were identified in Chilia and Sulina branches; and the Bechet area was populated by *Holandriana holandrii, Theodoxus transversalis* and *Gyraulus parvus*. The amphipods and mysids were present along the three main Danube branches. The calculated density of these species revealed an abundant community of crustacean *Chelicorophium robustum* on Sulina branch, and *Dikerogammarus haemobaphes* and *D. villosus* in extended areas of the Danube Delta. The presence of these invertebrates along Danube River was reported in relation to the sediment type and water depth.

Keywords: DNA barcoding; gastropoda; amphipoda; mysidae; Danube River; distribution

Citation: Menabit, S.; Begun, T.; Teacă, A.; Mureşan, M.; Lavin, P.; Purcarea, C. DNA Barcoding and Distribution of Gastropods and Malacostracans in the Lower Danube Region. *Diversity* **2022**, *14*, 533. https://doi.org/10.3390/d14070533

Academic Editor: Manuel Elias-Gutierrez

Received: 31 May 2022
Accepted: 28 June 2022
Published: 30 June 2022

1. Introduction

Danube is one of the most important inland waterways and the second-largest river in Europe. It has a length of 2857 km from the source (Black Forest, Germany) to the delta and the Black Sea, Romania [1]. The lower Danube course, between Baziaş to its mouth at the Black Sea, with a length of 1075 km, represents Romania's natural borders with Serbia, Bulgaria, Ukraine and the Republic of Moldova [2]. Over a third of the river's length is in Romania, covering almost a third of the surface area of the Basin [3]. With a hydrographic basin of 816,028 km^2, covering 11% of the European continent [4], the river discharges into the Black Sea in a characteristic delta formed by three main branches.

The Danube Delta represents one of the continent's most valuable habitats for wetland and is the largest remaining natural wetland. Its unique ecosystems consist of a labyrinthine network of river channels, shallow bays and hundreds of lakes. The three main channels flowing through the delta are represented by the Chilia branch, which carries 63% of the total flow; the Sulina branch, which accounts for 16%; and the St. George branch, which carries the remainder [3,5]. As the largest delta in the European Union covering about 5640 km^2 (including the outer lagoons areas), of which 4400 km^2 is in Romanian territory, the Danube Delta acts as a natural filter for about 7 to 10% of the total water, sediment and pollutant discharges of the river into the sea [6].

Benthic invertebrates are an important component of freshwater ecosystems; they contributing to accelerating detrital decomposition [7,8], material circulation and energy flow and supply food for both aquatic and terrestrial vertebrate consumers [9].

Gastropods are one of the largest benthic groups with regard to the number of species and their relative abundances in large lowland rivers [10,11]. They have a substantial function in riverine systems, controlling the growth of algal communities and grazed systems, resulting in decreased algal biomass [12], and provide an important food source for some fish species [13]. Danube's gastropod fauna belongs to the richest in Europe, encompassing species with a wide European distribution, but also with unique Danubian and Ponto-Caspian elements [14–16].

Malacostracan crustacean groups represented by amphipods and mysids play key roles in water quality assessment and ecological [17] and ecotoxicological studies, being sensitive to some chemical contaminants at environmentally relevant concentrations [18]. Taking into consideration their large distribution, their ecological role in the food chain and their susceptibility to pollutants, these organisms are frequently used as bioindicators [19,20] and contribute to nutrient recycling and water purification, representing an important food source for a variety of animals [17]. Ponto–Caspian amphipods, isopods, mysids and cumaceans represent some of the most successful groups of aquatic invaders, comprising several high-impact species, such as *Chelicorophium robustum*, *Dikerogammarus villosus* and *D. haemobaphes* [21].

Owing to their sensitivity to water quality, hydrology and sediment conditions, benthic invertebrates are the most commonly used organisms for biological monitoring of freshwater ecosystems worldwide; they are frequently used in environmental assessment studies and as indicators of functional change [22,23]. However, monitoring functions depend, to a large extent, on the accuracy of the species identification [24,25].

For the last few decades, DNA barcoding based on mitochondrial cytochrome c oxidase 1 (COI) gene sequencing [26] was extensively used for efficient and accurate species identification, facilitating the discovery of cryptic and new species [27]. To date, this method has been successfully applied for the identification of gastropods [28], amphipods [29] and mysid crustaceans [30] to overcome the limitations of specimen identification based on morphological characters [31,32].

Several studies carried out in the Lower Danube region based on morphological identification aimed to assess the distribution and ecology of macroinvertebrates [33], including gastropod fauna [34–37]. Only limited data on Ponto-Caspian amphipods and mysids from this region have been provided [38–42]. Moreover, molecular identification of amphipods [43–45] and mysids [30,46] from the Lower Danube sector and Danube Delta targeted only a few species. Additionally, [30,45,46] conducted studies on specimens collected from unspecified locations of the Danube Delta. Meanwhile, no such investigations based on molecular identification were carried out so far regarding gastropod fauna.

In this context, the current report based on molecular identification by DNA barcoding provided new data on the distributions of several gastropod and Ponto-Caspian malacostracan amphipods and mysids species along the Lower Danube River sector in relation to the depth and substrate type of their habitat.

2. Materials and Methods

2.1. Study Area, Sampling and Sample Preparation

Sediment samples were collected from 8 sites along the Lower Danube sector in 2019, 2020 and 2021 during 3 field trip sessions in late spring (May–June) periods (Figure 1, Table 1). Among these, the sites P01 and P01A were located within the Ceatal Izmail area, the apex of the Danube Delta where the splitting of the river in Chilia and Tulcea distributaries occurs—P06 on Chilia branch, P12 and P13 on Sulina branch, P20 and P24 on the St. George branch and D20 in the Bechet area (km 676), respectively (Figure 1). Sampling was carried out at various depths, corresponding to different substrate composition (Table 1). The substrate from the Danube branches site was represented by mixed, sandy and muddy sediments, and in the Bechet area by submerged vegetation and a solid substrate. The water depth in the D20 station located in Bechet area was 3.7 m on average and ranged from 4.7 to 24 m along the Danube Delta river branches.

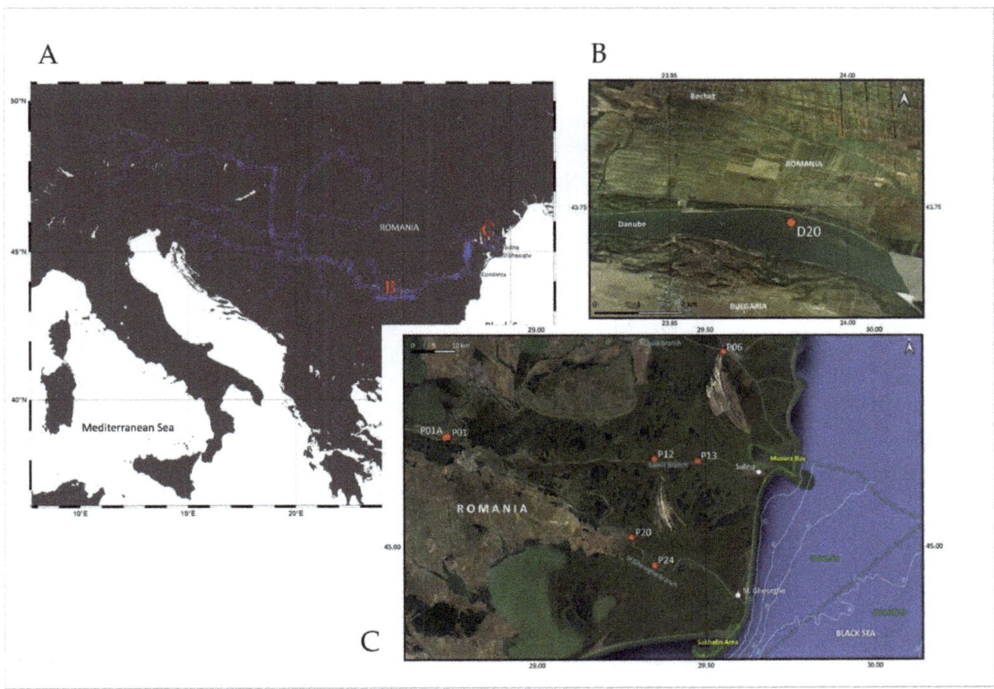

Figure 1. Sampling sites of the Lower Danube River sector. (**A**) Overview map of the Danube River; (**B**) sampling location in Bechet area; (**C**) sampling locations along the Danube Delta river branches.

Table 1. Study area. Sampling sites of the lower sector of the Danube River (Figure 1), year of collection, water depth and substrate type. MS: mixed sediments; MSM: mixed sediments dominated by mud with aquatic vegetation; SM: sandy mud with aquatic vegetation; S sand; M: mud; AV: aquatic vegetation with solid substrate.

Station	Year of Sampling	Coordinates		Depth (m)	Substrate Type
		Lat. (α)	Long. (λ)		
P01	2019, 2020, 2021	45°13′36.23″	28°43′57.49″	24.0	MS
P01A	2019, 2020, 2021	45°13′36.23″	28°43′57.49″	24.0	MS
P06	2019, 2020, 2021	45°24′19.16″	29°33′12.71″	6.8	SM
P12	2019, 2020, 2021	45°10′53.65″	29.20′47.84″	4.7	MSM
P13	2019, 2020, 2021	45°10′35.34″	29°28′28.66″	5.1	MSM
P20	2019, 2020, 2021	45°01′10.2″	29°16′36.64″	13.8	S
P24	2019, 2020, 2021	44°57′33.80″	29°20′48.80″	19.5	M
D21	2019, 2021	43°44′13.0″	23°59′4.3″	3.7	AV

Sediments were collected using a Van Veen grab with a surface of 0.1 m^2 and a limnological net. Samples were washed immediately after collection using 250 and 125 μm mesh sieves to remove excess sediment particles and preserve macrofauna. For collecting the phytophilous organisms, the vegetation was swept by using a limnological net with 125 μm mesh size. Each specimen selected for genetic analyses was washed with sterile water and placed in 200 μL Tris-EDTA pH 8 buffer at −20 °C [47]. For samples collected in 2021 with the Van Veen grab, the quantitative distribution of each species was evaluated by counting all individuals and calculating their theoretical density per unit surface (1 m^2) using a multiplication factor of 10 [48]. For the samples collected with the limnological net, the abundance was expressed as the total number of individuals collected.

2.2. Morphological Identification

In a first attempt, all collected species were morphologically assigned according to the identification keys for gastropods [49], amphipods [50], and mysids [51], and further submitted to DNA barcoding analysis.

2.3. DNA Extraction, PCR Amplification and COI Gene Sequencing

Total genomic DNA was extracted using the DNeasy Blood and Tissue Kit (Qiagen, Hilden, Germany), following an optimized protocol including an initial stage of cell disruption [52]. The specimens were introduced into Tris-EDTA pH 8 buffer and homogenized at 20 °C, 50 Hz, for 12 min, in a SpeedMill PLUS Cell Homogenizer (Analytik, Jena, Germany) in the presence of 5 ZR BashingBead lysis matrix 0.2 mm (Zymo Research, Irvine, CA, USA), and further processed following the manufacturer's protocol. A partial region of mitochondrial COI gene was amplified using metazoan universal primers (CO1490 (5′-GGTCAACAAATCAAA-GATATTGG-3′) and HCO2198 (5′-TAAACTTCAGGGTGAC-CAAAAAATCA-3′)) [53]. PCR amplification was carried out in a 50 μL reaction volume containing 1 unit ofTaq DNA polymerase (ThermoFisher Scientific, Waltham, MA, USA), 1 μL genomic DNA, 1 μL each of LCO1490 and HCO2198 primer, 0.1 mM of dNTP (ThermoFisher Scientific), 1 × BSA (New England Biolab, BiolabIpswich, MA, USA) and 1 × Taq buffer containing 2.5 mM $MgCl_2$ (ThermoFisher Scientific). The COI fragment was amplified after an initial incubation at 95 °C for 2 min, followed by 5 cycles of incubation at 94 °C for 30 s, annealing at 45 °C for 1.5 min and extension at 72 °C for 1 min; and 35 cycles of 94 °C for 30 s, 50 °C for 1.5 min and 72 °C for 1 min, with a final extension step of 5 min at 72 °C. The size and integrity of the amplified DNA were analyzed by electrophoresis in 1% agarose gel (Cleaver Scientific, Ltd., England). The amplicons were further purified using a QIAquick PCR Purification Kit (Qiagen) and sequenced on both strands using the amplification primers (Macrogen, Amsterdam, The Netherlands).

The resulting COI nucleotide sequences were edited using Sequence Assembly and Alignment—CodonCode Aligner Software (CodonCode Corporation 2003). Sequence identification was performed using the BLAST-NCBI platform [54]. Molecular identification of isolated gastropods and crustaceans collected from the Danube branches was based on the sequence identity of the COI amplicons using a combination of approaches that included the use of the R package INSECT with the database classifier version [55,56] and a 97% threshold for BLAST sequence screening of the NCBI GenBank database [26].

The COI sequence of all identified gastropod and crustacean specimens from the Lower Danube sector were deposited in GenBank (Supplementary Table S1).

2.4. Phylogenetic Analysis and Calculation of Intra- and Interspecific Genetic Distance

The alignment of gastropods and crustaceans' COI sequences retrieved from NCBI GenBank (Supplementary Table S1) was performed using MUSCLE with default parameters [57]. Phylogenetic analysis for both gastropod and crustacean species based on COI sequences was performed via maximum likelihood statistical method using the IQ-TREE web server (Available online: http://iqtree.cibiv.univie.ac.at/ (accessed on 20 June 2022)). We used the ModelFinder to find the best substitution model according to BIC (K3Pu + F + I + G4), with ultrafast bootstrapping (1000 iterations), single branch test SH-aLRT (1000 iterations) and the Approximate Bayes test [58]. The resulting tree was visualized using Interactive Tree of Life (Available online: https://itol.embl.de/ (accessed on 20 June 2022)) [59]. *Spongilla lacustris* COI gene (HQ379431) was used as an outgroup for both gastropods and crustacean phylogenetic trees. Pairwise intraspecific and interspecific genetic distances were calculated using the Kimura two-parameter (K2P) model using the Molecular Evolutionary Genetics Analyses (MEGA) platform version 11 [60,61]. A 3% molecular threshold was taken into account as the most used cut-off value for species delimitation [26].

3. Results

3.1. Molecular Identification and Phylogeny of Gastropods and Crustaceans

Following an initial morphological screening of the invertebrates collected from the eight stations of the Lower Danube sector (Figure 1), all specimens were successfully identified by DNA barcoding, including the 16 gastropods and 12 crustacean individuals collected (Supplementary Table S1). The taxonomic assignment of these new species using the R package INSECT analysis confirmed their affiliation (Supplementary Table S2).

The COI amplicons' sizes varied between 513 and 654 bp with an average was 602 bp for gastropods, and between 539 and 651 bp with an average of 600 bp for crustaceans.

The nine gastropod species from Caenogastropoda, Architaenioglossa, Littorinimorpha, Cycloneritida and Hygrophila orders belonged to eight families (Amphimelaniidae, Melanopsidae, Viviparidae, Bithyniidae, Neritidae, Lymnaidae, Physidae and Planorbidae), and the five identified crustaceans from Amphipoda and Mysida orders were classified into three different families (Corophiidae, Gammaridae and Misidae) (Table 2).

A phylogenetic tree for gastropod species was constructed based on 28 individuals' DNA barcode sequences, of which 16 were from the current study and 10 additional sequences were retrieved from the NCBI GenBank database (Figure 2). All individuals assigned to the same species belonged to monophyletic clusters, and all individuals of the same species formed a branch, each with high bootstrap support values (<90%) (Figure 2).

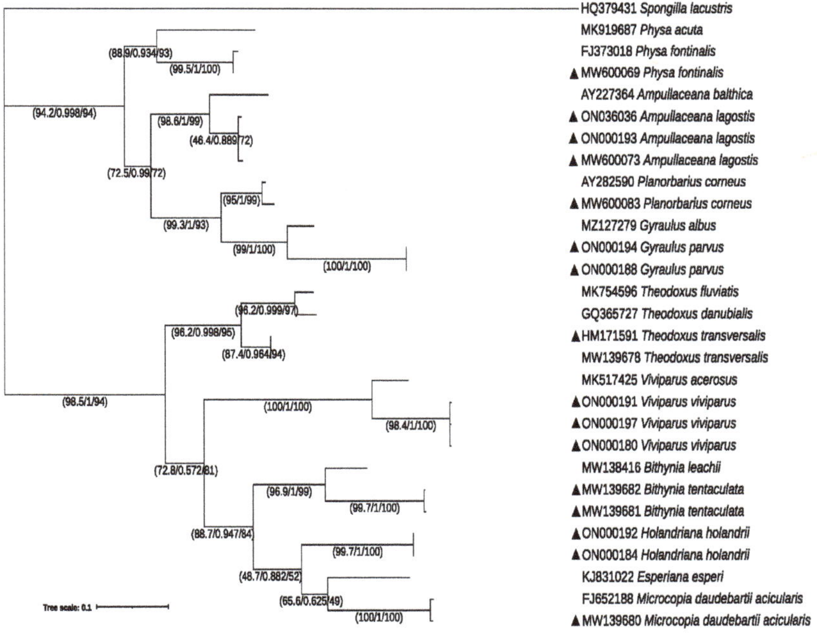

Figure 2. Phylogenetic tree of gastropod species based on cytochrome oxidase subunit I (COI). ▲: sequences from the current study; *Spongilla lacustris* COI (HQ379431) was used as an outgroup; numbers in parentheses are SH-aLRT support (%)/aBayes support/ultrafast bootstrap support (%).

Table 2. Occurrences of gastropods and malacostracans according to the year of sampling, location, water depth and type of substrate. Sampling sites P01, P01A, P06, P12, P13, P20, P24, D20 (Figure 1).

Species/Orders	Sampling Sites								Depth (m)	Type of Substrate
	P01	P01A	P06	P12	P13	P20	P24	D20		
Gastropoda										
Caenogastropoda										
Holandriana holandrii								2021	3.7	AV
Microcolpia daudebartii acicularis	2019 2021			2019 2021					4.7 24	MS MSM
Architaenioglossa										
Viviparus viviparus	2020 2021			2020 2021					4.7 24	MS MSM
Littorinimorpha										
Bithynia tentaculata	2020 2021				2020 2021				5.1 24	MS MSM
Cycloneritida										
Theodoxus transversalis								2019 2021	3.7	AV
Hygrophila										
Ampullaceana lagotis			2020 2021						6.8	SM
Physa fontinalis					2020 2021				5.1	MSM
Planorbarius corneus				2020 2021					4.7	MSM
Gyraulus parvus								2021	3.7	AV
Malacostraca										
Amphipoda										
Chelicorophium robustum	2019 2021	2019 2021	2019 2021	2019 2021					4.7 6.8 24	MS MSM
Dikerogammarus haemobaphes	2019 2020 2021	2019 2021	2019 2020 2021	2019 2020 2021	2019 2020 2021				4.7 5.1 6.8 24	MS MSM
Dikerogammarus villosus	2020 2021	2021		2020 2021	2020 2021	2020 2021	2020 2021		4.7 5.1 13.8 19.5 24	MS MSM
Mysida										
Limnomysis benedeni					2020 2021				5.1	MSM
Paramysis (Mesomysis) lacustris				2020 2021	2020 2021				4.7 5.1	MSM

The intraspecific and interspecific distances were measured for the specimens assigned to the same species and same family, in order to validate the existence of the 3% molecular threshold. For gastropods, the intraspecific distance calculation was conducted for four species which were represented by more than one individual, as follows: *H. holandrii*, *V. viviparus*., *B. tentaculata* and *A. lagotis*. Intraspecific T3P distances of the COI sequences within species ranged from 0% to 0.7%, the highest distance being found in *A. lagotis*. Interspecific distances for gastropods ranged from 17.2% to 37.8%. In the case of Planorbidae being represented by two species, the average genetic distance within this family was 16.9% (Table 3).

Table 3. Intraspecific and interspecific K2P genetic pairwise distances for gastropod species. The values calculated for Danube River specimens are represented in bold.

	1.	2.	3.	4.	5.	6.	7.	8.	9.	10.	11.	12.	13.	14.	15.	16.
1. ON000184 *H. holandrii*	-															
2. ON000192 *H. holandrii*	**0.000**	-														
3.MW139680 *M. daudebartii acicularis*	0.205	0.207	-													
4. ON000180 *V. viviparus*	0.281	0.282	0.241	-												
5. ON000191 *V. viviparus*	0.275	0.276	0.241	**0.003**	-											
6. ON000197 *V. viviparus*	0.277	0.278	0.240	**0.003**	**0.003**	-										
7. MW139681 *B. tentaculata*	0.235	0.236	0.240	0.274	0.268	0.271	-									
8. MW139682 *B. tentaculata*	0.242	0.241	0.243	0.278	0.268	0.278	**0.003**	-								
9. MW139678 *T. transversalis*	0.240	0.238	0.211	0.263	0.259	0.261	0.224	0.225	-							
10. ON000193 *A. lagotis*	0.327	0.317	0.327	0.347	0.337	0.345	0.319	0.321	0.277	-						
11. ON036036 *A. lagotis*	0.299	0.294	0.307	0.332	0.329	0.329	0.307	0.307	0.257	**0.004**	-					
12. MW600073 *A. lagotis*	0.315	0.305	0.306	0.333	0.330	0.332	0.308	0.306	0.258	**0.005**	**0.007**	-				
13. MW600069 *P. fontinalis*	0.365	0.352	0.355	0.315	0.312	0.312	0.298	0.298	0.272	0.180	0.180	0.185	-			
14. MW600083 *P. corneus*	0.345	0.329	0.302	0.305	0.302	0.304	0.321	0.318	0.273	0.173	0.172	0.175	0.202	**-**		
15. ON000188 *G. parvus*	0.362	0.345	0.333	0.356	0.347	0.353	0.299	0.307	0.295	0.215	0.207	0.208	0.218	**0.164**	**-**	
16. ON000194 G. parvus	0.362	0.362	0.350	0.378	0.372	0.378	0.316	0.320	0.316	0.232	0.219	0.226	0.230	**0.175**	**0.000**	-

For crustacea, the phylogenetic reconstruction was based on 16 COI sequences, of which 12 were from this study and 4 were retrieved from GenBank (Figure 3). All individuals belonging to the same species analyzed in the present study formed distinct clusters in the tree, with bootstrap support values <90% (Figure 3).

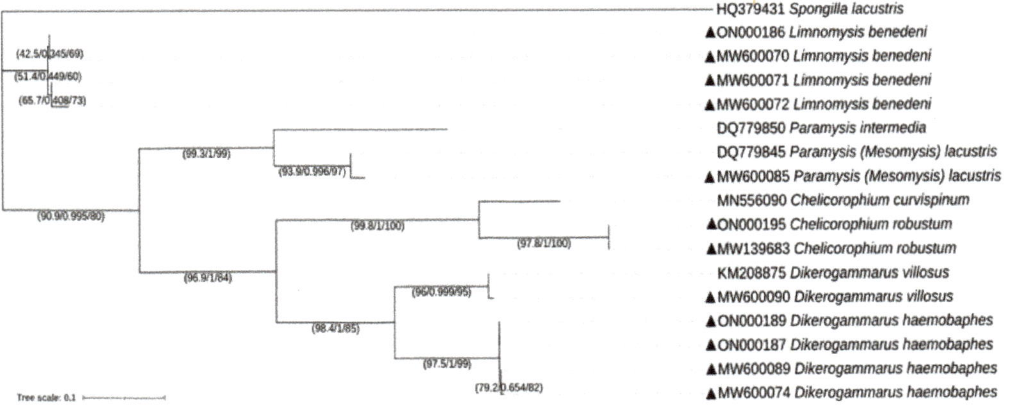

Figure 3. Phylogenetic tree of amphipod and mysid species based on cytochrome oxidase subunit I (COI); (▲): sequences from the current study; *Spongilla lacustris* (HQ379431) was used as an outgroup; numbers in parentheses are SH-aLRT support (%)/aBayes support/ultrafast bootstrap support (%).

The calculated intraspecific distances for *C. robustum*, *D. haemobaphes* and *L. benedeni* showed the highest genetic distance (2.7%) for *L. benedeni*, in addition to the identified 3% threshold for different species. The lowest interspecific distance of 20.4% was obtained between D. *haemobaphes* and *D. villosus*, and the highest value was 38.9% between *C. robustum* and *P. lacustris*, and the same between *D. haemobaphes* and *L. benedeni*. Additionally, the interspecific distances between species belonging to the same family varied in the case of Gammaridae. For the genus *Dikerogammarus*, the interval was 19.9–20.5% and the average value was 20.2%. The values for the two Misidae species ranged from 30.5% to 33.8%, and the average value was 31.4% (Table 4).

Table 4. Intra- and interspecific K2P pairwise distances for crustacean species. The values calculated for Danube River specimens are represented in bold.

	1.	2.	3.	4.	5.	6	7.	8.	9.	10.	11.	12.
1. MW139683 *C.robustum*	-											
2. ON000195 *C. robustum*	**0.000**	-										
3. MW600074 *D. haemobaphes*	0.332	0.342	-									
4. MW600089 *D. haemobaphes*	0.331	0.336	**0.003**	-								
5. ON000187 *D. haemobaphes*	0.337	0.332	**0.004**	**0.002**	-							
6. ON000189 *D. haemobaphes*	0.338	0.331	**0.002**	**0.002**	**0.000**	-						
7. MW600090 *D. villosus*	0.322	0.332	0.204	**0.199**	**0.205**	**0.203**	-					
8. MW600070 *L. benedeni*	0.350	0.373	0.350	0.347	0.358	0.349	0.326	-				
9. MW600071 *L. benedeni*	0.348	0.370	0.353	0.350	0.361	0.353	0.333	**0.008**	-			
10. MW600072 *L. benedeni*	0.374	0.400	0.380	0.383	0.389	0.381	0.355	**0.025**	**0.020**	-		
11. ON000186 *L. benedeni*	0.350	0.370	0.348	0.344	0.355	0.346	0.333	**0.002**	**0.006**	**0.027**	-	
12. MW600085 *P. lacustris*	0.366	0.389	0.336	0.333	0.347	0.348	0.346	**0.310**	**0.306**	**0.338**	**0.305**	-

Intra- and interspecific pairwise distances for crustacean species COI analyses showed that the obtained intraspecific and interspecific genetic distances between individuals of both gastropod and crustacean species do not overlap, further supporting the species identification.

3.2. Distribution and Ecology of Gastropod and Crustacean Species along the Lower Danube Region

Out of the eight investigated sites, gastropod species were identified in four locations along the Danube branches and in the Bechet area site (Figure 4) during different field trips. The crustacean species were found in all seven sampling sites located along the Danube branches (Figure 5), during the whole time interval (Table 2).

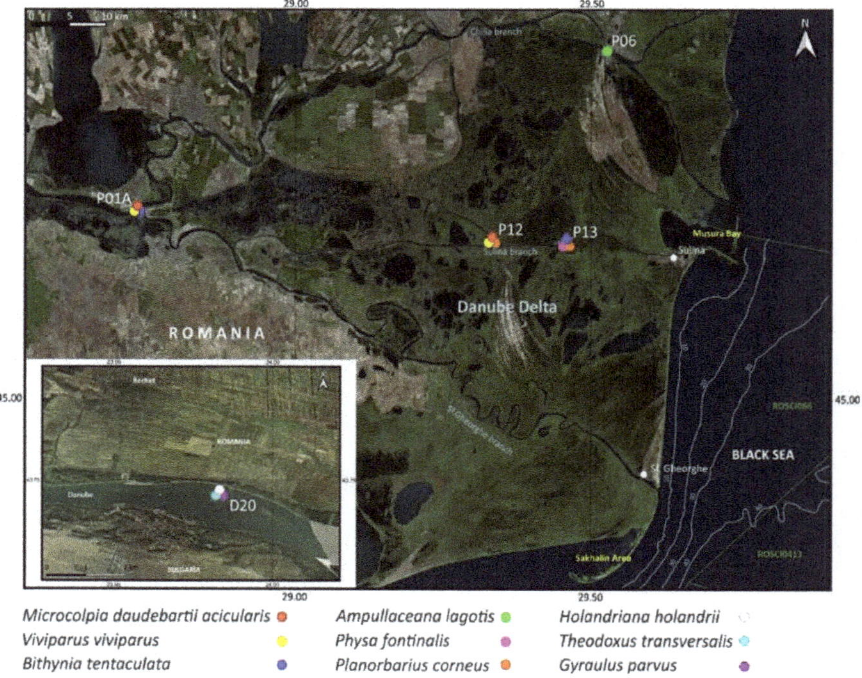

Figure 4. Distribution of gastropod species along Danube branches and in the Bechet area (inset).

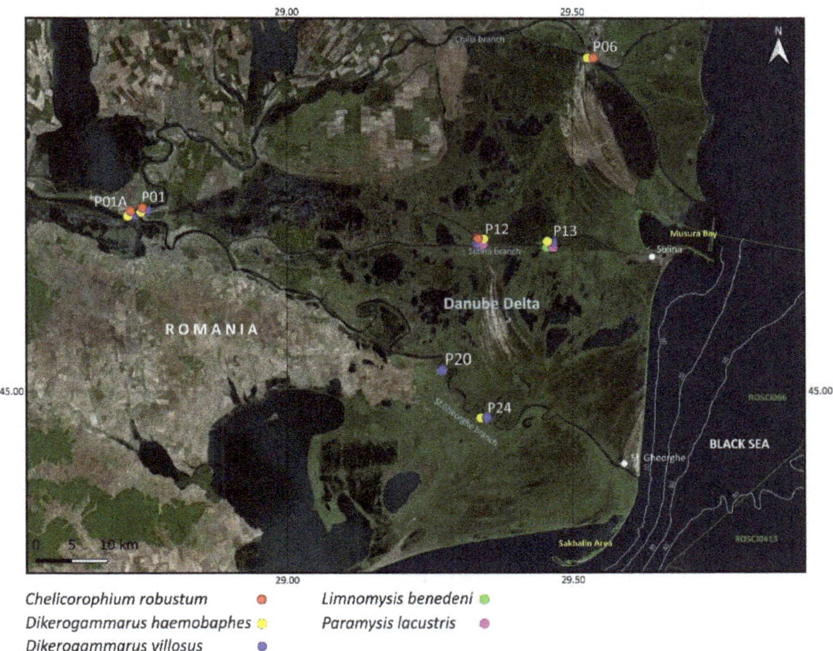

Figure 5. Distribution of crustacean species along Danube branches.

The gastropod species identified along the Danube branches (*M. daudebartii acicularis*, *V. viviparus*, *B. tentaculata*, *P. fontinalis* and *P. corneus*) were present only on Chilia and Sulina Danube branches (Figure 4). Meanwhile, only three phytophilous gastropods were retrieved from the Bechet area, *H. holandrii*, *T. transversalis* and *G. parvus* (Figure 4). These specimens were retrieved from different water depths ranging from 3.7 to 24 m (Table 2). The river-bed substrate below the isolated species was also variable. In this respect, *M. daudebartii acicularis*, *V. viviparus*, *B. tentaculata*, *P. fontinalis* and *P. corneus* species were associated with mixed sediments, whereas *A. lagotis* was detected in areas dominated by sandy and muddy sediments rich in submerged vegetation (Table 2).

The calculated density of gastropod species was relatively low, ranging from 10 to 20 individuals/m^2 interval. There was a high presence (30 ind/m^2) of *P. corneus* in site P12 (Sulina branch) (Table 5).

The amphipods and mysids were identified from the investigated sites located along the three main Danube channels (Chilia, Sulina and St. George). *D. haemobaphes* was detected in all three branches, *D. villosus* only in Sulina and St. George areas and *C. robustum* in Chilia and Sulina; the mysids were identified only in the Sulina branch (Figure 5). Their habitat was characterized by a variable water depth ranging from 4.7 to 25 m (Table 2). All investigated species were detected in substrates characterized by mixed sediments, but only representatives of Amphipoda order were encountered in sandy and muddy sediments (Table 2).

Overall, the amphipods recorded higher calculated density as compared to mysid species. The highest values were found for *C. robustum* and *D. haemobaphes*, reaching 420 and 300 ind./m2 at P12 and P13 stations, respectively. The two mysid species registered relatively low densities, the highest being recorded by *P. lacustris* with 30 ind./m2 in P13 station. (Table 5).

Table 5. Calculated densities of gastropod and crustacean species for the sampling sites of the Lower Danube River. Locations of sampling sites P01, P01A, P06, P12, P13, P20, P24 and D20 are indicated in Figure 1.

Species	Density (Individuals/m^2)							
	P01 *	**P01A ***	**P06 ***	**P12 ***	**P13 ***	**P20 ***	**P24 ***	**D20 ****
				Gastropoda				
M. daudebartii acicularis		20		20				
V. viviparus		20		20				
B. tentaculata		10			10			
A. lagotis			10					
P. fontinalis					10			
P. corneus				30				
H. holandrii								2
T. transversalis								3
G. parvus								2
				Malacostraca				
C. robustum	10	30	20	420				
D. haemobaphes	10	130	210	20	300		30	
D. villosus	40	110		10	30	10	20	
L. benedeni					10			
P. lacustris				10	30			

* Samples collected with VV grab (density: expressed as individuals/m^2, ** samples collected with the limnological net (abundance: total number of individuals collected).

4. Discussion

For the last decades, the molecular approach based on DNA barcoding has become an important tool for biodiversity assessment worldwide, being suitable for the identification of species from different life stages and species with sexual dimorphism, or for putative cryptic species, from both fresh and preserved materials [62]. In the current study, DNA barcoding was proven to be an effective instrument for identifying gastropods and crustaceans.

Traditionally, the method for validating presumptive species using DNA barcoding analysis is based on the comparison between intraspecific and interspecific genetic nucleotide divergence enabling the inference of a molecular threshold to help taxonomic identification [26,63]. There are many debates in the scientific literature about the most appropriate similarity threshold. It can vary in the 2–4% interval depending on the taxonomic group of macroinvertebrate species [64]. The variation between species needs to exceed the variation within species, which allows clear genetic differentiation of species by the existence of the barcoding gap [65]. Here, for performing pairwise genetic distances, the 3% molecular threshold was used. The calculated intraspecific and interspecific divergence values for gastropods (Table 3) were comparable to those reported for species retrieved from the Portuguese coast, Vaal River and Adriatic Sea, varying in the 8.44–74.67% and 0–2.9% intervals, respectively [66–68]; and for amphipods and mysids collected from the Pacific coast of Canada, the Black Sea, the Caspian Sea, Danube River and Don and Rhine river systems, these values were in 0%–4.3% and 4.92–34.2% intervals [69–71]. These findings support the efficacy of DNA barcoding based on COI gene sequencing in species delineation. Moreover, our results showed no overlap between intra- and interspecific genetic divergence for both gastropod and crustacean taxa.

COI represent a better target, having major advantages: the universal LCO1490 and HCO2198 primers for this gene are very robust, allowing the recovery of the 5′ end from the majority of the representatives of animal phyla [53,72]. The evolution of this gene was fast enough to enable the discrimination of closely related species and phylogeographic groups within a single species [73,74]. For both gastropod and crustacean taxa, the ML tree showed distinctness of all the studied species. For instance, although sequences belonging to the specimens identified as *V. viviparus* grouped closely with the GenBank retrieved sequences belonging to *V. acerosus*, the species formed distinct clusters in the COI phylogenetic reconstruction. Our results are consistent with those other studies showing that COI-based phylogeny could confirm the genetic differentiation between *Viviparus* species [75]. Moreover, [76] reported morphological similarities between the aforementioned species, but a later revision and molecular analyses confirmed their delimitation. Another example is given by the representatives of the Planorbidae family. The phylogenetic reconstruction showed that the species were grouped in distinct branches and sequences belonging to the same species clustered together. Recently, phylogenetic analyses based on mitochondrial and nuclear DNA sequences [77] showed that *G. laevis and G. parvus* are in fact part of the same species-level clade, the latter having nomenclatural priority. Our data indicated similar results for both amphipods and mysids collected from the lower sector of the Danube River. While members of the same family (e.g., Gammaridae, Misidae) or the same genus (*Dikerogammarus*) did not cluster together, sequences of same species grouped together, suggesting the efficacy of COI sequences in species delineation. Previous debates related to the taxonomic status of *Dikerogammarus* species considered that *D. villosus* and *D. bispinosus* are synonyms of *D. haemobaphes* [78]. However, the taxonomical revision performed by [79] and the analyses based on mitochondrial genomes performed by [80] revealed genetic distinctions among these taxa. Furthermore, a COI gene analysis performed on *Chelicorophium* revealed that the specimens clustered in two separate groups corresponding to *C. curvispinum* and to *C. robustum* [81], which is consistent with our data obtained in the case of the Lower Danube specimens.

Mollusca represent the most abundant organisms of the Danube River in terms of biomass. Owing to their size, Bivalvia make up to 80% of the total biomass, followed by Gastropoda, covering 10% to 35% of the community [82].

The types of substrate associated with invertebrates could also vary. Along the Danube branches, the mixed sediments dominated by mud with aquatic vegetation were populated by all species detected in this area, except for *A. lagotis*, which was identified in sandy mud. The mixed sediments were associated with *M. daudebartii acicularis*, *V. viviparus* and *B. tentaculata*. A recent investigation [83] reported that *V. viviparus*, *B. tentaculata*, *M. daudebartii acicularis* and *P. fontinalis* could also inhabit several types of substrate, such as gravelly, muddy and sandy river bottoms, and areas with aquatic vegetation. Although gastropods can populate areas with sandy and muddy river bottoms, these organisms are frequently associated with solid substrata (boulders, stones, plant parts) [84]. Along the Danube branches, the substrate was represented by mixed sediments, mud and sand, explaining the low density of the species identified in this area. Both *A. lagotis* and *P. corneus* were identified in only one substrate each, in mixed sediments and sandy mud, respectively. Previous studies reported that these species are pelophilous and phytophilous, characteristic of stagnant waters [85,86]; this may explain their absence from the majority of the investigated samples. *B. tentaculata* and *M. daudebartii acicularis* were previously reported in the Danube Delta area, along the St. George branch [42]. In addition, our study showed that these species also populated Ceatal Izmail and Sulina branch, whereas no individuals were found in the St. George sites. *H. holandrii*, *T. trasversalis* and *G. albus* were identified only in the Bechet area, which is characterized by the presence of submerged vegetation and a solid substrate. The Ponto-Caspian snail *T. transversalis*, listed as Endangered in the IUCN Red List, is nowadays found in the Danube River in a very restricted area only in the lower stretch [82,87,88], and *H. holandrii* is known as one of the Balkanian fauna of the Danube River [82,89]. Both species are known to populate river bottoms with hard substrates [82,87–89], and our data confirm these ecological preferences of the species.

In terms of abundance, the fauna of the Danube River was dominated by crustaceans. Amphipoda was reported to be the dominant group in al Danube branches, representing up to 75% of the total abundance [82]. While previous reports indicated that *D. villosus*, *D. haemobaphes* and *C. robustum* are associated with gravelly substrates [90], the current data revealed the presence of the amphipod species in several types of substrates. Although the two representatives of *Dikerogammarus* showed a strong preference for large cobble and artificial substrate, the species are adapted to various ecological conditions [91–93], as confirmed by the current study where these species were identified in mixed sediments, sand and mud. A previous report on the macroinvertebrate communities from the Danube Delta [42] also showed the occurrence of *D. haemobaphes* and *D. villosus* in several locations along the St. George branch, similarly to our data, in support of the resilience of these species for more than 5 years. Furthermore, *C. robustum*, which is reported to inhabit gravelly and muddy substrates [94,95], was also found in the current survey to populate areas dominated by mixed sediments and sandy mud. Our data revealed the association of both *L. benedeni* and *P. lacustris* with mixed sediments dominated by mud, in accordance with initial reports regarding their preferred habitat being characterized by fine sediments (sand and mud) with standing water or slow to moderate currents [51]. *L. benedeni* is often found in great densities on the shore at depths of only 0–0.5 m, although they can occur at a depth of 6 m [51]. *D. haemobaphes* was identified in the littoral zone at 50–70 cm depth [96].

5. Conclusions

The current findings regarding the distributions of several gastropod and Ponto-Caspian amphipods and mysids populating the Lower Danube region extended the knowledge on the presence and density of these benthic invertebrates based on molecular identification by DNA barcoding using COI gene sequencing, and complementary meta-data regarding their habitats (substrate type and river depth), thereby adding to the ecological profile of these fauna populating the Danube Delta sector. The accuracy of species identification by this method was highlighted in the cases of several specimens belonging to same species of gastropods or crustaceans clustered together in monophyletic groups. Moreover, this survey contributed to the first gastropod barcode dataset for the Romanian Danube sector.

Supplementary Materials: The following is available online at https://www.mdpi.com/article/10.3390/d14070533/s1. Table S1. Specimen taxonomy and accession number, sampling period and location, best match COI gene sequence for gastropod, amphipod and mysid species isolated from the Lower Danube sector and Table S2. Taxonomy assignation by the INSECT R package for all COI sequences used on the present research (including sequences recovered from NCBI).

Author Contributions: S.M. and C.P. wrote the manuscript; S.M. isolated the DNA, performed DNA barcoding, performed phylogenetic analysis and contributed to experimental design and data interpretation; T.B. performed the statistical analyses; A.T. contributed to map construction and species identification; M.M. contributed to map construction and sample collection; P.L. performed the phylogenetic tree construction and R package INSECT analyses; C.P. performed the experimental design and coordinated the study. All authors have read and agreed to the published version of the manuscript.

Funding: The study was financially supported by the Romanian Ministry of Research as part of the CORE Programme projects PN19200401, PN19200204 and PN19200302; by the Romanian Academy RO1567-IBB05/2021 project; by the European Union's Horizon 2020 DOORS project, grant agreement number 101000518; and by the Program Development of the National R&D system—AMBI AQUA—No. 23PFE. T.

Institutional Review Board Statement: Not applicable.

Data Availability Statement: All data supporting the conclusions of this article are included in the manuscript.

Acknowledgments: The authors thank Ana Bianca Pavel from GeoEcoMar for assisting in collecting and sorting the benthic material, and Lavinia Iancu for helping with the COI barcoding experiment.

Conflicts of Interest: The authors declare no conflict of interest.

References

1. Helmer, R.; Hespanhol, I. *Water Pollution Control—A Guide to the Use of Water Quality Management Principles*; United Nations Environment Programme, the Water Supply & Sanitation Collaborative Council and the World Health Organization WHO/UNEP: New York, NY, USA, 1997. Available online: https://www.who.int/water_sanitation_health/resourcesquality/wpcbegin.pdf (accessed on 25 April 2022).
2. Gâştescu, P.; Ţuchiu, E. The Danube River in the Pontic Sector—Hydrologycal Regime. Water Resources and Wetlands. In Proceedings of the 1st Water and Wetlands Resources Conference, Tulcea, Romania, 14–16 September 2012; Gâştescu, P., Lewis, W., Jr., Breţcan, P., Eds.; Riscuri si Catastrofe: Cluj-Napoca, Romania, 2012; Volume 12, pp. 13–26, ISBN 978-606-605-038-8.
3. ICPDR. *Danube River Basin District Management Plan*; ICPDR: Vienna, Austria, 2009.
4. Oaie, G.; Secrieru, D.; Bondar, C.; Szobotka, Ş.; Duţu, L.; Manta, T. Lower Danube River: Characterization of sediments and pollutants. *GeoEcoMarina* **2015**, *21*, 19–34.
5. Hanganu, J.; Grigoras, I.; Stefan, N.; Sarbu, I.; Zhmud, E.; Dubyna, D.; Menke, U.; Drost, H. *Transboundary Vegetation Map of the Biosphere Reserve Danube Delta*; Danube Delta Institute DDI, Institute for Inland Water Management and Waste Water Treatment RIZA, Danube Biosphere Reserve of the National Academy of Sciences of Ukraine DPA, I&KIB: Tulcea, Romania, 2002.
6. Mee, L.D. The Black Sea in crisis: A need for concerted international action. *Ambio* **1992**, *21*, 278–286.
7. Hutchinson, G.E. A Treatise on Limnology. In *The Zoobenthos*; John Wiley & Sons: New York, NY, USA, 1993; Volume 4, p. 944.
8. Wallace, J.B.; Webster, J.R. The role of macroinvertebrates in stream ecosystem function. *Annu. Rev. Entomol.* **1996**, *41*, 115–139. [CrossRef]
9. Clarke, K.D.; Knoechel, R.; Ryan, P.M. Influence of trophic role and life-cycle duration on timing and magnitude of benthic macroinvertebrate response to whole-lake enrichment. *Can. J. Fish. Aquat. Sci.* **1997**, *54*, 89–95. [CrossRef]
10. Gomes, M.A.; Novelli, R.; Zalmon, I.R.; Souza, C.M. Malacological assemblages in sediments of eastern Brazilian continental shelf, coordinates 108 and 208 S, between Bahia and Espírito Santo State. *Bios* **2004**, *12*, 11–24.
11. Tubić, B.; Simić, V.; Zorić, K.; Gačić, Z.; Atanacković, Z.; Csányi, B.; Paunović, M. Stream section types of the Danube River in Serbia according to the distribution of macroinvertebrates. *Biologia* **2013**, *68*, 294–302. [CrossRef]
12. Rosemond, A.D.; Mullholland, P.J.; Ellwood, J.W. Top-down and bottom-up control of stream periphyton: Effects of nutrients and herbivores. *Ecology* **1993**, *74*, 1264–1280. [CrossRef]
13. Brown, K.M.; Alexander, J.E.; Thorp, J.H. Differences in the ecology and distribution of lotic pulmonate and prosobranch gastropods. *Am. Malacol. Bull.* **1998**, *14*, 91–101.
14. Frank, C.; Jungbluth, J.; Richnovszky, A. *Die Mollusken der Donau vom Schwarzwald bis zum Schwarzen Meer*; Akaprint: Budapest, Hungary, 1990; pp. 1–142.
15. Moog, O.; Humpesch, U.H.; Konar, M. The distribution of benthic invertebrates along the Austrian stretch of the River Danube and its relevance as an indicator of zoogeographical and water quality patterns—Part 1. *Arch. Hydroboil. Robiol.* **1995**, *9* Suppl. 101, 121–213. [CrossRef]
16. Wesselingh, F.P.; Neubauer, T.A.; Anistratenko, V.V.; Vinarski, M.V.; Yanina, T.; ter Poorten, J.J.; Kijashko, P.; Albrecht, C.; Anistratenko, O.Y.; D'Hont, A.; et al. Mollusc species from the Pontocaspian region—An expert opinion list. *ZooKeys* **2019**, *827*, 31–124. [CrossRef]
17. Glazier, D.S. *Amphipoda. Reference Module in Earth Systems and Environmental Sciences*; Elsevier: Amsterdam, The Netherlands, 2014; p. 49. [CrossRef]
18. Roast, S.D.; Thompson, R.S.; Widdows, J.; Jones, M.B. Mysids and environmental monitoring: A case for their use in estuaries. *Mar. Freshw. Res.* **1998**, *49*, 827–832. [CrossRef]
19. Alonso, Á.; De Lange, H.J.; Peeters, E.T.H.M. Development of a feeding behavioural bioassay using the freshwater amphipod Gammarus pulex and the Multispecies Freshwater Biomonitor. *Chemosphere* **2009**, *75*, 341–346. [CrossRef] [PubMed]
20. Grabowski, M.; Bacela-Spychalska, K.; Pešić, V. Reproductive traits and conservation needs of the endemic gammarid *Laurogammarus scutarensis* (Schäferna, 1922) from the Skadar Lake system, Balkan Peninsula. *Limnologica* **2014**, *47*, 44–51. [CrossRef]
21. Borza, P.; Huber, T.; Leitner, P.; Remund, N.; Graf, W. Success factors and future prospects of Ponto-Caspian peracarid (Crustacea: Malacostraca) invasions: Is "the worst over"? *Biol. Invasions* **2017**, *19*, 1517–1532. [CrossRef]
22. Heino, J.; Louhi, P.; Muotka, T. Identifying the scales of variability in stream macroinvertebrate abundance, functional composition and assemblage structure. *Freshw. Biol.* **2004**, *49*, 1230–1239. [CrossRef]
23. Seymour, M.; Deiner, K.; Altermatt, F. Scale and scope matter when explaining varying patterns of community diversity in riverine metacommunities. *Basic Appl. Ecol.* **2016**, *17*, 134–144. [CrossRef]
24. Frézal, L.; Leblois, R. Four years of DNA barcoding: Current advances and prospects. *Infect. Genet. Evol.* **2008**, *8*, 727–736. [CrossRef]
25. Macher, J.N.; Salis, R.K.; Blakemore, K.S.; Tollrian, R.; Matthaei, C.D.; Leese, F. Multiple-stressor effects on stream invertebrates: DNA barcoding reveals contrasting responses of cryptic mayfly species. *Ecol. Indic.* **2016**, *61*, 159–169. [CrossRef]
26. Hebert, P.D.N.; Cywinska, A.; Ball, S.L. Biological identifications through DNA barcodes. *Proc. R. Soc. Lond. B* **2003**, *270*, 313–321. [CrossRef]
27. Puillandre, N.; Cruaud, C.; Kantor, Y.I. Cryptic species in Gemmuloborsonia (Gastropoda: Conoidea). *J. Mollus. Stud.* **2009**, *76*, 11–23. [CrossRef]

28. Sands, A.F.; Glöer, P.; Gürlek, M.E.; Albrecht, C.; Neubauer, T.A. A revision of the extant species of Theodoxus (Gastropoda, Neritidae) in Asia, with the description of three new species. *Zoosyst. Evol.* **2020**, *96*, 25–66. [CrossRef]
29. Lipinskaya, T.; Radulovici, A.E. DNA barcoding of alien Ponto-Caspian amphipods from the Belarusian part of the Central European invasion corridor. *Genome* **2017**, *60*, 963–964. [CrossRef]
30. Audzijonyte, A.; Daneliya, M.; Väinölä, R. Comparative phylogeography of Ponto-Caspian mysid crustaceans: Isolation and exchange among dynamic inland sea basins. *Mol. Ecol.* **2006**, *15*, 2969–2984. [CrossRef] [PubMed]
31. Barco, A.; Claremont, M.; Reid, D.G.; Houart, R.; Bouchet, P.; Williams, S.T.; Cruaud, C.; Couloux, A.; Oliverio, M. A molecular phylogenetic framework for the Muricidae, a diverse family of carnivorous gastropods. *Mol. Phylogenet. Evol.* **2010**, *56*, 1025–1039. [CrossRef] [PubMed]
32. Layton, K.K.; Martel, A.L.; Hebert, P.D.N. Patterns of DNA Barcode Variation in Canadian Marine Molluscs. *PLoS ONE* **2014**, *9*, e95003. [CrossRef]
33. Graf, W.; Csányi, B.; Leitner, P.; Paunović, M.; Huber, T.; Szekeres, J.; Nagy, C.; Borza, P. *Joint Danube Survey 3, Full Report on Macroinvertebrates*; ICPDR, International Commission for the Protection of the Danube River: Sofia, Bulgaria, 2014; p. 87. Available online: www.icpdr.org (accessed on 20 May 2022).
34. Glöer, P.; Sîrbu, I. New freshwater molluscs species found in the Romanian fauna. *Heldia* **2005**, *6*, 229–238.
35. Gomoiu, M.T.; Begun, T.; Opreanu, P.; Teaca, A. Present state of benthic ecosystem in Razelm-Sinoie Lagoon Complex (RSLC). In Proceedings of the 37th IAD Conference, The Danube River Basin in a Changing World, Chişinău, Moldova, 29 October–1 November 2008; pp. 108–112.
36. Paraschiv, G.M.; Begun, T.; Teaca, A.; Bucur, M.; Tofan, L. New data about benthal populations of the Golovita and Zmeica lakes. *J. Environ. Prot. Ecol.* **2010**, *11*, 253–260.
37. Begun, T.; Teacă, A.; Mureşan, M.; Pavel, A.B. Current state of the mollusc populations in the Razim-Sinoe Lagoon System. *Anim. Sci.* **2020**, *63*, 553–561.
38. Begun, T. Complex Ecological Studies of Crustaceans Populations (Cumaceans and Mysidaceans) of the North Western Black Sea Coast. Ph.D. Thesis, University "Ovidius" Constanta, Constanța, Romania, 2006; p. 276.
39. Paraschiv, G.M.; Schroder, V.; Samargiu, M.D.; Sava, D. Ecological study of zoobenthos communities from the Matita and Merhei lakes (Danube Delta). *Res. J. Agric. Sci.* **2007**, *39*, 489–498.
40. Graf, W.; Csányi, B.; Leitner, P.; Paunović, M.; Janeček, B.; Šporka, F.; Chiriac, G.; Stubauer, I.; Ofenböck, T. *Joint Danube Survey 2, Report on Macroinvertebrates*; ICPDR International Commission for the Protection of the Danube River: Vienna, Austria, 2008; p. 87.
41. Borza, P.; Csányi, B.; Paunović, M. Corophiids (Amphipoda, Corophioidea) of the river Danube—The results of a longitudinal urvey. *Crustaceana* **2010**, *83*, 839–849.
42. Stoica, C.; Gheorghe, S.; Petre, J.; Lucaciu, I.; Nita-Lazar, M. Tools for assessing Danube Delta systems with macro invertebrates. *Environ. Eng. Manag. J.* **2014**, *13*, 2243–2252. [CrossRef]
43. Cristescu, M.E.; Hebert, P.D.; Onciu, T.M. Phylogeography of Ponto-Caspian crustaceans: A benthic–planktonic comparison. *Mol. Ecol.* **2003**, *12*, 985–996. [CrossRef] [PubMed]
44. Rewicz, T.; Wattier, R.; Grabowski, M.; Rigaud, T.; Bącela-Spychalska, K. Out of the Black Sea: Phylogeography of the invasive killer shrimp *Dikerogammarus villosus* across Europe. *PLoS ONE* **2015**, *10*, e0118121. [CrossRef] [PubMed]
45. Jażdżewska, A.M.; Rewicz, T.; Mamos, T.; Wattier, R.; Bącela-Spychalska, K.; Grabowski, M. Cryptic diversity and mtDNA phylogeography of the invasive demon shrimp, *Dikerogammarus haemobaphes* (Eichwald, 1841), in Europe. *NeoBiota* **2020**, *57*, 53–86. [CrossRef]
46. Cristescu, M.E.A.; Hebert, P.D.N. The 'Crustacean Seas'—An evolutionary perspective on the Ponto-Caspian peracarids. *Can. J. Fish. Aquat. Sci.* **2005**, *62*, 505–551. [CrossRef]
47. Ross, K.S.; Haites, N.E.; Kelly, K.F. Repeated freezing and thawing of peripheral blood and DNA in suspension: Effects on DNA yield and integrity. *J. Med. Genet.* **1990**, *27*, 569–570. [CrossRef]
48. *SR EN ISO 10870*; Water Quality—Guidelines for the Selection of Sampling Methods and Devices for Benthic Macroinvertebrates in Fresh Waters. ISO: Geneva, Switzerland, 2012.
49. Grossu, A.V. *Mollusca—Gastropoda Prosobranchia si Opistobranchia*; Fauna Republicii Populare Române: Bucharest, Romana, 1956; Volume 3.
50. Cărăuşu, S.; Dobreanu, E.; Manolache, C. *Fauna Republicii Populare Romîne. Crustacea. Volumul IV. Fascicula 4. Amphipoda. Forme Salmastre şi de apă Dulce*; Editura Acedemiei Republicii Populare Romîne: Bucharest, Romana, 1955; pp. 1–401.
51. Băcescu, M. *Crustacea: Mysidacea*; Editura Academiei Republicii Populare Romîne: Bucureşti, Romania, 1954; Volume 126, p. 52.
52. Iancu, L.; Carter, D.O.; Junkins, E.N.; Purcarea, C. Using bacterial and necrophagous insect dynamics for postmortem interval estimation during cold season: Novel case study in Romania. *Forensic Sci. Int.* **2015**, *254*, 106–117. [CrossRef]
53. Folmer, O.; Black, M.; Hoeh, W.; Lutz, R.; Vrijenhoek, R. DNA primers for amplification of mitochondrial cytochrome c oxidase subunit I from diverse metazoan invertebrates. *Mol. Mar. Biol. Biotechnol.* **1994**, *3*, 294–299.
54. Available online: https://blast.ncbi.nlm.nih.gov (accessed on 28 April 2022).
55. Leray, M.; Yang, J.Y.; Meyer, C.P.; Mills, S.C.; Agudelo, N.; Ranwez, V.; Machida, R.J. A new versatile primer set targeting a short fragment of the mitochondrial COI region for metabarcoding metazoan diversity: Application for characterizing coral reef fish gut contents. *Front. Zool.* **2013**, *10*, 34. [CrossRef]

56. Wilkinson, S.P.; Davy, S.K.; Bunce, M.; Stat, M. Taxonomic identification of environmental DNA with informatic sequence classification trees. *PeerJ Prepr.* **2018**, *6*, e26812v1. [CrossRef]
57. Edgar, R.C. MUSCLE: Multiple sequence alignment with high accuracy and high throughput. *Nucleic Acids Res.* **2004**, *32*, 1792–1797. [CrossRef] [PubMed]
58. Trifinopoulos, J.; Nguyen, L.T.; von Haeseler, A.; Minh, B.Q. W-IQ-TREE: A fast online phylogenetic tool for maximum likelihood analysis. *Nucleic Acids Res.* **2016**, *44*, W232–W235. [CrossRef] [PubMed]
59. Letunic, I.; Bork, P. Interactive Tree of Life (iTOL) v5: An online tool for phylogenetic tree display and annotation. *Nucleic Acids Res.* **2021**, *49*, W293–W296. [CrossRef] [PubMed]
60. Kimura, M. A simple method for estimating evolutionary rates of base substitutions through comparative studies of nucleotide sequences. *J. Mol. Evol.* **1980**, *16*, 111–120. [CrossRef]
61. Tamura, K.; Stecher, G.; Kumar, S. MEGA 11: Molecular Evolutionary Genetics Analysis Version 11. *Mol. Biol. Evol.* **2021**, *38*, 3022–3027. [CrossRef]
62. Radulovici, A.E.; Archambault, P.; Dufresne, F. DNA Barcodes for Marine Biodiversity: Moving Fast Forward? *Diversity* **2010**, *2*, 450–472. [CrossRef]
63. Hebert, P.D.N.; Penton, E.H.; Burns, J.M.; Janzen, D.H.; Hallwachs, W. Ten Species in One: DNA Barcoding Reveals Cryptic Species in the Neotropical Skipper Butterfly *Astraptes fulgerator*. *Proc. Natl. Acad. Sci. USA* **2004**, *101*, 14812–14817. [CrossRef]
64. Sweeney, B.W.; Battle, J.M.; Jackson, J.K.; Dapkey, T. Can DNA Barcodes of Stream Macroinvertebrates Improve Descriptions of Community Structure and Water Quality? *J. N. Am. Benthol. Soc.* **2011**, *30*, 195–216. [CrossRef]
65. Hebert, P.D.N.; Stoeckle, M.Y.; Zemlak, T.S.; Francis, C.M. Identification of Birds through DNA Barcodes. *PLoS Biol.* **2004**, *2*, e312. [CrossRef]
66. Borges, L.; Hollatz, C.; Lobo, J.; Cunha, A.M.; Vilela, A.P.; Calado, G.; Coelho, R.; Costa, A.C.; Ferreira, M.S.G.; Costa, M.H.; et al. With a little help from DNA barcoding: Investigating the diversity of Gastropoda from the Portuguese coast. *Sci. Rep.* **2016**, *6*, 20226. [CrossRef]
67. Lawton, S.P.; Allan, F.; Hayes, P.M.; Smit, N.J. DNA barcoding of the medically important freshwater snail *Physa acuta* reveals multiple invasion events into Africa. *Acta Trop.* **2018**, *188*, 86–92. [CrossRef] [PubMed]
68. Buršić, M.; Iveša, L.; Jaklin, A.; Arko Pijevac, M.; Kučinić, M.; Štifanić, M.; Neal, L.; Bruvo Mađarić, B. DNA Barcoding of Marine Mollusks Associated with Corallina officinalis Turfs in Southern Istria (Adriatic Sea). *Diversity* **2021**, *13*, 196. [CrossRef]
69. Costa, F.; deWaard, J.R.; Boutillier, J.; Ratnasingham, S.; Dooh, R.T.; Hajibabaei, M.; Hebert, P.D.N. Biological identifications through DNA barcodes: The case of the Crustacea. *Can. J. Fish. Aquat. Sci.* **2007**, *64*, 272–295. [CrossRef]
70. Costa, F.; Henzler, C.; Lunt, D.; Whiteley, N.M.; Rock, J. Probing marine Gammarus (Amphipoda) taxonomy with DNA barcodes. *Syst. Biodivers.* **2009**, *7*, 365–379. [CrossRef]
71. Audzijonyte, A.; Wittmann, K.J.; Ovcarenko, I.; Väinölä, R. Invasion phylogeography of the Ponto-Caspian crustacean Limnomysis benedeni dispersing across Europe. *Divers. Distrib.* **2009**, *15*, 346–355. [CrossRef]
72. Zhang, D.-X.; Hewitt, G.M. Assessment of the universality and utility of a set of conserved mitochondrial primers in insects. *Insect Mol. Biol.* **1997**, *6*, 143–150. [CrossRef]
73. Cox, A.J.; Hebert, P.D.N. Colonization, extinction and phylogeographic patterning in a freshwater crustacean. *Mol. Ecol.* **2001**, *10*, 371–386. [CrossRef]
74. Wares, J.P.; Cunningham, C.W. Phylogeography and historical ecology of the North Atlantic intertidal. *Evolution* **2001**, *12*, 2455–2469. [CrossRef]
75. Rysiewska, A.; Hofman, S.; Osikowski, A.; Beran, L.; Pešić, V.; Falniowski, A. *Viviparus mamillatus* (Küster, 1852), and partial congruence between the morphology-, allozyme- and DNA-based phylogeny in European Viviparidae (Caenogastropoda: Architaenioglossa). *Folia Malacol.* **2019**, *27*, 43–51. [CrossRef]
76. Beran, L.; Horsák, M.; Hofman, S. First records of *Viviparus acerosus* (Bourguignat, 1862) (Gastropoda: Viviparidae) from the Czech Republic outside its native range. *Folia Malacol.* **2019**, *27*, 223–229. [CrossRef]
77. Lorencová, E.; Beran, L.; Nováková, M.; Horsáková, V.; Rowson, B.; Jaroslav, Č.; Hlaváč, J.Č.; Nekola, J.C.; Horsák, M. Invasion at the population level: A story of the freshwater snails *Gyraulus parvus* and *G. laevis*. *Hydrobiol.* **2021**, *848*, 4661–4671. [CrossRef]
78. Pjatakova, G.M.; Tarasov, A.G. Caspian Sea amphipods: Biodiversity, systematic position and ecological peculiarities of some species. *Int. J. Salt Lake Res.* **1996**, *5*, 63–79. [CrossRef]
79. Müller, J.C.; Schramm, S.; Seitz, A. Genetic and morphological differentiation of *Dikerogammarus* invaders and their invasion history in Central Europe. *Freshw. Biol.* **2002**, *47*, 2039–2048. [CrossRef]
80. Mamos, T.; Grabowski, M.; Rewicz, T.; Bojko, J.; Strapagiel, D.; Burzyński, A. Mitochondrial Genomes, Phylogenetic Associations, and SNP Recovery for the Key Invasive Ponto-Caspian Amphipods in Europe. *Int. J. Mol. Sci.* **2021**, *22*, 10300. [CrossRef] [PubMed]
81. Marescaux, J.; Latli, A.; Lorquet, J.; Virgo, J.; Van Doninck, K.; Beisel, J.N. Benthic macro-invertebrate fauna associated with Dreissena mussels in the Meuse River: From incapacitating relationships to facilitation. *Aquat. Ecol.* **2016**, *50*, 15–28. [CrossRef]
82. ICPDR. *Joint Danube Survey 2. Final Scientific Report*; ICPDR: Vienna, Austria, 2008.
83. Martinovic-Vitanovic, V.; Raković, M.; Popović, N.; Kalafatic, V. Qualitative study of Mollusca communities in the Serbian Danube stretch (river km 1260–863.4). *Biologia* **2013**, *68*, 112–130. [CrossRef]
84. ICPDR. *Joint Danube Survey 1. Ecological Status Characterization. Macrozoobenthos*; ICPDR: Vienna, Austria, 2002.

85. Surugiu, V.; Mustaţă, G.; Hârţăscu, M. Contributions to the qualitative and quantitative study of the macrozoobenthos from the Danube—Black Sea Canal. *Stud. Şi Cercet. Biol. Univ. Din Bacău* **2004**, *9*, 75–80.
86. Georgiev, D.; Hubenov, Z. Freshwater snails (Mollusca: Gastropoda) of Bulgaria: An updated annotated checklist. *Folia Malacol.* **2013**, *21*, 237–263. [CrossRef]
87. Sîrbu, I.; Sîrbu, M.; Benedek, A. The freshwater molluskfauna from Banat (Romania). *Trav. du Mus. Natl. d'Hist. Nat. Grigore Antipa* **2010**, *53*, 21–43. [CrossRef]
88. Solymos, P.; Feher, Z. *Theodoxus transversalis* (C. Pfeiffer1828). In *IUCN 2011. IUCN Red List of Threatened Species. Version 2011.* 2011. Available online: www.iucnredlist.org (accessed on 18 May 2022).
89. Novakovic, B.; Markovic, V.; Tomović, J. Distribution of the snail *Amphimelania holandrii* Pfeiffer, 1828 (Melanopsidae; Gastropoda) in Serbia in the 2009–2012 period. *Water Res. Manag. J. Belgrade* **2013**, *3*, 21–26.
90. ICPDR. *Joint Danube Survey 3. Full Report on Macroinvertebrates*; ICPDR: Vienna, Austria, 2014.
91. Muskó, I.B. The life history of *Dikerogammarus haemobaphes* (Eichw.) (Crustacea: Amphipoda) living on macrophytes in Lake Balaton (Hungary). *Arch. Für Hydrobiol.* **1993**, *127*, 227–238. [CrossRef]
92. Boets, P.; Lock, K.; Messiaen, M.; Goethals, P.L.M. Combining data-driven methods and lab studies to analyse the ecology of *Dikerogammarus villosus*. *Ecol. Inform.* **2010**, *5*, 133–139. [CrossRef]
93. Clinton, K.E.; Mathers, K.L.; Constable, D.; Gerrard, C.; Wood, P.J. Substrate preferences of coexisting invasive amphipods, *Dikerogammarus villosus* and *Dikerogammarus haemobap-hes*, under field and laboratory conditions. *Biol. Invasions* **2018**, *20*, 2187–2196. [CrossRef]
94. Lipták, B.; Šporka, F.; Necpálová, K.; Stloukal, E. First record of Ponto-Caspian amphipod *Corophium curvispinum* in Slovakside of the Danube River. *Folia Faun. Slovaca* **2012**, *17*, 183–186.
95. Lipták, B. Non-indigenous invasivefreshwater crustaceans (Crustacea: Malacostraca) in Slovakia. *Water Res. Manag.* **2013**, *3*, 21–31.
96. Lipinskaya, T.; Makaranka, A.; Razlutskij, V.; Semenchenko, V. First records of the alien amphipod *Dikerogammarus haemobaphes* (Eichwald, 1841) in the Neman River basin (Belarus). *BioInvasions Rec.* **2021**, *10*, 319–325. [CrossRef]

MDPI

Article

DNA Barcodes Applied to a Rapid Baseline Construction in Biodiversity Monitoring for the Conservation of Aquatic Ecosystems in the Sian Ka'an Reserve (Mexico) and Adjacent Areas

Martha Valdez-Moreno [1], Manuel Mendoza-Carranza [2], Eduardo Rendón-Hernández [3], Erika Alarcón-Chavira [3] and Manuel Elías-Gutiérrez [1,*]

1 Departamento de Sistemática y Ecología Acuática, El Colegio de la Frontera Sur, Av. Centenario Km 5.5, Chetumal 77014, Quintana Roo, Mexico; mvaldez@ecosur.mx
2 Departamento de Ciencias de la Sustentabilidad, El Colegio de la Frontera Sur, Carretera a Reforma Km 15.5, Villahermosa 86280, Tabasco, Mexico; mcarranza@ecosur.mx
3 Comisión Nacional de Áreas Naturales Protegidas, Av. Ejército Nacional 223, Miguel Hidalgo, Ciudad de México 11320, Mexico; erendon@sinergiaplus.org (E.R.-H.); erika.alarcon@conanp.gob.mx (E.A.-C.)
* Correspondence: melias@ecosur.mx

Abstract: This study is focused on the aquatic environments of the Sian Ka'an reserve, a World Heritage Site. We applied recently developed protocols for the rapid assessment of most animal taxa inhabiting any freshwater system using light traps and DNA barcodes, represented by the mitochondrial gene Cytochrome Oxidase I (COI). We DNA barcoded 1037 specimens comprising mites, crustaceans, insects, and fish larvae from 13 aquatic environments close or inside the reserve, with a success rate of 99.8%. In total, 167 barcode index numbers (BINs) were detected. From them, we identified 43 species. All others remain as a BIN. Besides, we applied the non-invasive method of environmental DNA (eDNA) to analyze the adult fish communities and identified the sequences obtained with the Barcode of Life Database (BOLD). All round, we found 25 fish species and other terrestrial vertebrates from this region. No alien species was found. After comparing the BINs from all systems, we found that each water body was unique with respect to the communities observed. The reference library presented here represents the first step for future programs to detect any change in these ecosystems, including invasive species, and to improve the knowledge of freshwater zooplankton, enhancing the task of compiling the species barcodes not yet stored in databases (such as BOLD or GenBank).

Keywords: biodiversity; sinkhole; Chironomidae; Copepoda; Trombidiformes; cladocera; Ostracoda; Yucatan Peninsula

Citation: Valdez-Moreno, M.; Mendoza-Carranza, M.; Rendón-Hernández, E.; Alarcón-Chavira, E.; Elías-Gutiérrez, M. DNA Barcodes Applied to a Rapid Baseline Construction in Biodiversity Monitoring for the Conservation of Aquatic Ecosystems in the Sian Ka'an Reserve (Mexico) and Adjacent Areas. *Diversity* **2021**, *13*, 292. https://doi.org/10.3390/d13070292

Academic Editor: Bert W. Hoeksema

Received: 10 May 2021
Accepted: 22 June 2021
Published: 28 June 2021

1. Introduction

Freshwater is the most valued resource globally, representing around 1% of all water in the biosphere. However, pollution, excess use, and alterations due to global warming and dry seasons are changing its availability, affecting all life forms that depend on it for their survival. As a result, all the animal communities dwelling in these freshwater ecosystems are highly vulnerable, especially to the invasions of alien species. Mexico is not an exception [1] to this situation. Hence, we chose to study an important system and its surroundings, located on the east coast of the Yucatan Peninsula: the Sian Ka'an biosphere reserve (Figure 1).

First and foremost, it is essential to point out that the inventories of freshwater species are still in an incipient phase of knowledge since few groups of specialists in the matter are scattered throughout the world. None can identify all individuals present in any particular system to the species level, making it even more relevant to join efforts. For example, back in 2005, an estimation of the diversity of branchiopods, a crustacean group dominated by freshwater species, was approximately 1180 species in the world, and this figure was also reported to be at least two times higher elsewhere [2]. Since then, about 156 species of branchiopods have been described formally (Web of Science, 3 May 2021; Search: "new species" and "branchiopod"). Adamowicz and Purvis concluded that the neotropics are one of the regions where more species remain to be discovered [2], suggesting the importance of its study. Some recent developments have helped to understand the biodiversity in these particular environments, e.g., through DNA barcoding, a powerful tool to study aquatic organisms, from invertebrates [3] to vertebrates [4]. Therefore, the construction of databases with this information is fundamental. The Barcode of Life Database (BOLD, boldsystems.org) is one of the best to use because it focuses on biodiversity exclusively [5].

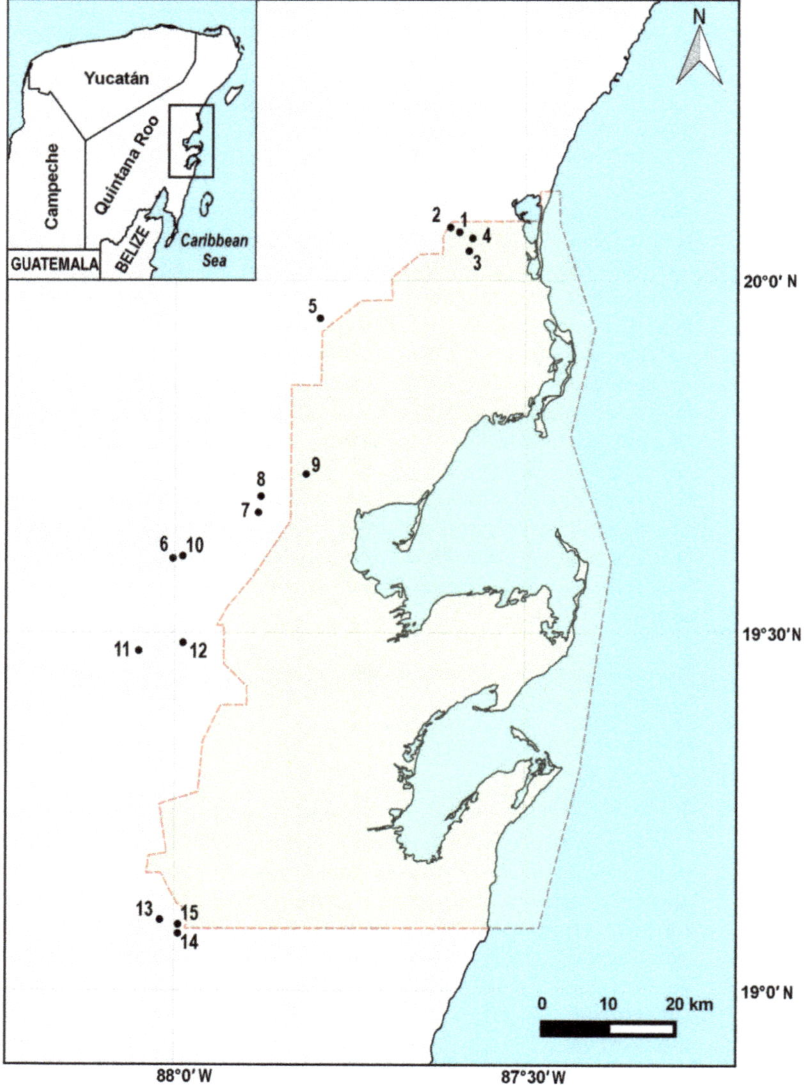

Figure 1. Map of the Yucatan Peninsula's east coast with localities sampled. Dotted lines show the limits of the Sian Ka'an reserve. Numbers of localities are the same as Table 1.

With the popularity of metabarcoding after the development of new generation sequencers, many labs remain focused on searching targeted species in aquatic environments, as aliens [6] or endangered [7]. Others focused on a whole community, like fish, but these latter studies rely upon a public database of DNA barcodes [8]. The relevance of the baseline uploaded in public databases has been highlighted for many groups, as the insects, before the metabarcoding, even though the detailed curation of names is incomplete [9]. Hence, the main problem for any aquatic community study is the lack of knowledge of the species (the so-called taxonomic impediment) and the consequently limited information on the DNA analysis.

For several years now, our group has been working on constructing a public database with the most accurate information for DNA barcodes. This effort has been an example for others [10]. Nonetheless, a proper assessment of the biodiversity within Mexico remains distant, mainly because of the global taxonomy crisis associated with the fact that many new barcodes have not been matched with their corresponding species. To overcome this taxonomic impediment has particular relevance in the neotropics, the most species-rich region in the world [11]. Consequently, we consider the best practice is to document the specimens studied with vouchers deposited in museums. If their DNA is barcoded, there are several approaches to establish them as molecular taxonomic units (MOTUs) that can be a hypothesis to be tested against a real species identification [12]. BOLD provides a useful calculation of these MOTUs, and the barcode index numbers (BINs) [13] were created as part of a rapid solution to this taxonomic impediment.

In the case of the Mexican continental territory, the Yucatan Peninsula, set in the neotropical region, has a complex system of underground freshwater that emerges to the surface as sinkholes (locally known as "cenotes"), lagoons, and "aguadas," mostly shallow rounded water bodies [14]. Mexico's east coast hosts the second largest coral reef in the world, known as the Mesoamerican Reef. Unfortunately, tourism has been growing with no order along this coastline, from Cancun to Chetumal city (near the border with Belize), including the Sian Ka'an biosphere reserve. This expansion shows no respect for natural resources.

As a result, the freshwater systems in the region present pollution of hydrocarbons going from Cancun (in the north) to Bacalar Lake and Milagros lagoons, located 10 km from Chetumal city [15]. This situation is relevant because the physical and biological interactions between mainland freshwater and the sea are not well understood due to the lack of studies. For example, just recently, larvae and juveniles of the fish *Cyprinodon artifrons* (Hubbs, 1936), whose adult stages are found around the mesoamerican reef, were discovered to have migrated to breed into the freshwater system of Bacalar Lake, located more than 70 km from the ocean [16]. We do not know how these larvae reached the lake because adults have not been found here.

Sian Ka'an, which means "Origin of the Sky" in the Mayan language, is a Natural Protected Area with nearly 4000 km^2 of land surface and continental waters (Figure 1). It has an outstanding conservation status on its hydrology and ecosystems, including tropical forests, mangroves, wetlands, beaches, sinkholes, and marshes (Figure 2). It was designated as a biosphere reserve of international importance by UNESCO in 1986 and recognized as a World Heritage Site in 1987 [17]. In 2003, it was recognized as a wetland of importance by the Ramsar Convention [18]. Nevertheless, notwithstanding its importance, there is surprisingly limited information regarding the freshwater species inhabiting the aquatic habitats here. The existing records include more than 40 fish species, with only two alien species: the Mozambique tilapia, *Oreochromis mossambicus*, reported long ago from an "aguada" near the reserve limits [19] and the Nile tilapia, *O. niloticus*. Among aquatic invertebrates, recent surveys in these environments allowed the description of one calanoid copepod, *Mastigodiaptomus siankaanensis* [20], and the harpacticoid *Remaneicaris siankaan* [21]. Nevertheless, after these studies, there is no other formal description of the aquatic biota from Sian Ka'an reserve, except non-peer-reviewed lists provided by Comisión Nacional para el Conocimiento y Uso de la Biodiversidad (CONABIO) [22] and Naturalista [23]. There is another unverified list published by Comisión Nacional de Áreas Naturales Protegidas (CONANP, Mexico) [24]. Finally, a report from United Nations Development Programme [25] listed 80 operational taxonomic units (OTUs), but the authors warned that this report was preliminary.

Recently, Elías-Gutiérrez et al. [16] and Montes-Ortiz & Elías-Gutiérrez [26] showed the usefulness of light traps combined with DNA barcoding to recognize the zooplankton biodiversity despite the little taxonomical knowledge in some temperate and tropical water systems on the Yucatan Peninsula.

Here, we present how to establish a new and rapid baseline of all possible BINs or potential freshwater species found, including invasive alien species, if present, around Sian Ka'an reserve and nearby aquatic systems. This will be based on the sequencing of a fragment of the COI gene (for zooplankton) and metabarcoding methodologies [8], using the eDNA for fish as an alternative to collecting the specimens directly. Finally, we will discuss how both results could be used in the future as a biomonitoring tool.

Table 1. Locations where the sampling was made. The buffer zone is systems within the reserve but near the limits of it. Influence area means less than 20 km from the limits of the polygon of the reserve.

Number	Name	Coordinates	Zone	Location in Sian Ka'an	Municipality
		Latitude N	Longitude W		
1	Laguna Muyil 1	20.0686	87.5944	Buffer zone	Felipe Carrillo Puerto
2	Laguna Muyil 2	20.0753	87.6073		
3	Chunyaxché 1	20.0422	87.5807	Buffer zone	Felipe Carrillo Puerto
4	Chunyaxché 2	20.0601	87.5757	Buffer zone	
5	Km 48	19.9431	97.7938	Influence area	Felipe Carrillo Puerto
6	Del Padre sinkhole	19.6038	88.0028	Influence area	Felipe Carrillo Puerto
7	Tres Reyes 1 sinkhole	19.6682	87.8812	Influence area	Felipe Carrillo Puerto
8	Tres Reyes sinkhole 2	19.6916	87.8774	Influence area	Felipe Carrillo Puerto
9	Santa Teresa sinkhole	19.7240	87.8130	Buffer zone	Felipe Carrillo Puerto
10	Minicenote sinkhole	19.6070	87.9887	Influence area	Felipe Carrillo Puerto
11	Sijil Noh Há sinkhole	19.4746	88.0516	Influence area	Felipe Carrillo Puerto
12	Chancah Veracruz sinkhole	19.4855	87.9879	Influence area	Felipe Carrillo Puerto
13	El Toro sinkhole	19.0981	88.0206	Influence area	Bacalar
14	Pucté Cafetal sinkhole	19.0788	87.9943	Influence area	Bacalar
15	Pucté 2 sinkhole	19.0915	87.9942	Influence area	Bacalar

Figure 2. Aerial photographs of some of the studied systems. (**A**) Laguna Muyil. At the bottom is the Chunyaxché lagoon. The arrow points to the ancient channel that communicates both systems. (**B**) Km 48, surrounded by dried wetlands. (**C**) Del Padre sinkhole. Notice the surrounding vegetation. Arrow points to a car for scale. (**D**) Sijil Noh Há sinkhole inside of a lagoon with the same name. (**E**) Tres Reyes 1 sinkhole. This system is the only eutrophic water body studied, as the color of the water indicates. Arrow points to parked cars. Notice the influence area and the surrounding vegetation. All photographs were taken by Manuel Elías-Gutiérrez.

2. Materials and Methods

2.1. Sampling

Samples were collected from 13 water bodies near and around the buffer zone of the Sian Ka'an reserve (Table 1, Figure 1) on 21–26 August 2019. We decided to work within the reserve, including other adjacent water bodies, because we know that all these systems are not isolated. They can be connected by subterranean waters or on the surface after floods, a common phenomenon in this region.

Laguna Muyil and Chunyaxché were the only systems sampled in two sites. Light traps with 50 µm mesh [26] were deployed overnight in the limnetic and (or only) littoral zones (the latter designed with a number with the same name of the locality), depending on the size of the water body. In 2019, it was uncommonly dry during August (supposedly part of the rain season, occurring from May to October). This phenomenon possibly triggered forest fires. Consequently, these fires limited the possibility of visiting more systems (also unusually dry) [26].

After collection, water from all the samples was extracted with a 50 µm sieve by washing them with cold (4 °C) 96% ethanol. Specimens were then transferred into jars with approximately 1/3 of the sample and 2/3 of ethanol [16]. Next, the sample jars were placed in a container with ice before being transferred to the laboratory, where they were stored in a freezer at −18 °C for at least one week. After this period, samples were kept at room temperature.

For metabarcoding from each system, we collected 1 L of water from the sub-surface in a sterilized CIVEQ bottle as suggested for the nearby Bacalar Lake [20]. Water samples were filtered in a field lab in Carrillo Puerto town to minimize eDNA degradation. We filtered at least 250 mL of water from each point in 0.22 µm filters. All filters were stored in cold gels till further analysis (see Section 2.4. for more details).

2.2. DNA Extraction and Amplification of Individuals

All specimens collected were sorted out under a stereomicroscope from the zooplankton samples. Representatives of each morphologically distinct taxon were then photographed using the same microscope. After being photographed, specimens were again placed in ethanol and stored at room temperature.

Small specimens (less than one mm) were destructively analyzed. DNA from larger specimens was obtained from an antenna, eggs or embryos, or a piece of the abdomen, preserving the remaining parts, like the head and the end of the abdomen (e.g., chironomid pupae). In the case of water mites, voucher specimens were recovered, as suggested for Collembola [27], and preserved in 96% ethanol with a drop of glycerol. Finally, for fish larvae, one eye of the right side or a piece of muscle was used. The vouchers (specimens not lost during extraction) were deposited in the Reference Collection at El Colegio de la Frontera Sur (ECOSUR), Unidad Chetumal. All specimens were identified to the finest possible taxonomical level by using general identification keys [28] or detailed descriptions, as *M. siankaanensis*, for example [20], or by comparison with previous DNA barcodes in the Barcode of Life Database (BOLD, bodsystems.org).

For each individual, DNA was extracted using a standard glass fiber method [29]. After DNA extraction, polymerase chain reaction (PCR) was performed as follows. First, 2 µL of each DNA extract were added to a PCR mixture consisting of 2 µL of Milli-Q water (Merck), 6.25 µL of 10% D-(+)-trehalose dihydrate (Fluka Analytical), 1.25 µL of 10× Platinum Taq buffer (Invitrogen), 0.625 µL of 50 µM $MgCl_2$ (Invitrogen), 0.0625 µL of 10 µM dNTP (KAPA Biosystems), 0.125 µL of each 10 µM Zplank primer [29], and 0.06 µL of PlatinumTaq (Invitrogen). The reactions were cycled at 95 °C for 1 min, followed by 5 cycles of (94 °C for 40 s, 45 °C for 40 s, 72 °C for 1 min), then 35 cycles of (94 °C for 40 s, 51 °C for 40 s, 72 °C for 1 min), and a final extension of 72 °C for 5 min. PCR products were visualized on a 2% agarose gel using an E-Gel 95 well Pre-cast Agarose Electrophoresis System (Invitrogen), and those showing a PCR product were selected for sequencing.

2.3. Sequencing and Data Analysis

PCR products were cycle sequenced using a modified [30] BigDye© Terminator v.3.1 Cycle Sequencing Kit (Applied Biosystems, Inc., San Francisco, CA, USA) and sequenced bi-directionally in a sequencing facility (Eurofins, Louisville, KY, USA) using M13F and M13R primers. Sequences were edited using CodonCode v.9.0.1 (CodonCode Corporation,

Dedham, MA, USA), uploaded to BOLD, and are available in the dataset Baseline Sian Ka'an I (DS-BASKAAN; dx.doi.org/10.5883/DS-BASKAAN). All data were analyzed with the quality tools on BOLD, and all sequences were examined for the presence of stop codons and indels as a check against NUMTS.

A neighbor-joining (NJ) tree was calculated by using the Maximum Likelihood method based on the Kimura two-parameter (K2P) distance model [31] with 500 Bootstrap replications including all taxa found in the four major groups: Arachnida, Crustacea, Insecta, and Actinopterygii found in all systems, using the MEGA 7 software [32]. Each tree was simplified with the Compress feature of Mega 7. Additionally, we prepared a general ID tree with all specimens, with the analytical tools provided by BOLD (Figure S1). We selected this method because it allows the rapid analysis of large data sets of specimens [33]. MOTUs as a proxy to species were delimited with the barcode index number (BIN) [13] that has proven more than 80% of effectiveness [16] and has been widely accepted [9]. Based on the BINs (= MOTUs), we prepared a list for the finest possible identification of all species found. Mexico has one of the most extensive datasets in public databases of freshwater species in the world [10].

2.4. Metabarcoding and eDNA

Water filters from each sampling point were sent to be processed in the Centre for Biodiversity Genomics in the University of Guelph (Canada). The interval between filtration and DNA extraction was less than 48 hrs.

Each sample tube was kept at −20 °C before processing. Before DNA extraction, all lab surfaces and pipettors were sterilized using 10% bleach, followed by 70% ethanol [8]. Prior to extraction, each filter was placed in DNEasy PowerWater Bead Tube (forceps were sterilized with 20% bleach, followed by 100% ethanol and triple-flame sterilization between samples). DNA was extracted with minor modifications to previous publications [34]: a volume of 900 μL of ILB buffer with 100 μL Proteinase K was added to each tube, tubes were incubated at 56 °C for 30 min and vortexed on Genie 2 vortex at max speed for 5 min, followed by 1.5-h incubation at 56 °C. Tubes were centrifuged at 2000× *g* for 2 min and ~700 μL of lysate transferred to a clean tube containing 1.4 mL of 5 M GuSCN buffer; all resulting volume was applied to Epoch Biolabs column in three subsequent centrifugation steps at 6000 g (each 700 μL transfer). Silica membrane was washed 1× with 400 μL of PWB, 2× with 700 μL of WB, and dried at 56 °C for 20 min. DNA was eluted in 100 μL of EB buffer at 11,000× *g*. DNA was transferred into a 96-well plate in 3 replicates per sample. After extraction, we followed a two-step PCR approach with the first round employing conventional primers, while the diluted PCR product served as a template for a second round of PCR with fusion primers containing sequencing adapters and UMI-tags. A 184–187 bp segment of the barcode region of COI was amplified with two primer sets (AquaF2/C_FishR1, AquaF3/C_FishR1). The PCR reactions employed the master mix described previously [35] and Platinum Taq. The first round of PCR employed the following thermal regime: 94 °C for 2 min, 40 cycles of 94 °C for 40 s, annealing at 51 °C for AquaF2 or 50 °C for AquaF3 for 1 min, and 72 °C for 1 min, with a final extension at 72 °C for 5 min. PCR resulting products were diluted 2× and used for second round PCR with fusion primers for 20 cycles to create IonXpress MID-tag labeled libraries. The PCR regime for the second round consisted of 94 °C for 2 min, 20 cycles of 94 °C for 40 s, annealing at 51 °C for 1 min, and 72 °C for 1 min, with a final extension at 72 °C for 5 min. PCR products were visualized on an E-Gel96 (Invitrogen, Thermo Fisher Scientific, Waltham, MA, USA).

The library for each unique primer pair was pooled without normalization and purified using magnetic beads with a 0.5:1 bead to product ratio for the uppercut and 0.8:1 ratio for the lower cut. Each library was diluted to 26 pM and mixed in a 1:1 ratio for the S5 run using Ion 510/520/530 chef kit. Each DNA replicate for each primer combination was processed under a separate IonXpress MID-tag.

The following procedure to process the raw NGS reads: Cutadapt (v1.8.1) was used to trim primer sequences; Sickle (v1.33) was used for size filtering (sequences from 150–250 bp were retained), while Uclust (v1.2.22q) served to recognize OTUs based on the >98% identity and a minimum read depth of 2 thresholds. The Local Blast 2.2.29+ algorithm was then used to compare each OTU to the reference sequences in five datasets: public fish data from BOLD filtered to genus and species ID level (130,357 sequences), and public BOLD data for Amphibia, Aves, Mammalia, and Reptilia represented by the following datasets: DS-EBACAMPH (11,018 sequences), DS-EBACAVES (28,914 sequences), DS-EBACMAMM (39,890 sequences), DS-EBACREPT (5424 sequences). Raw Blast output results were parsed using custom-built Python scripts. Processed results in tab-delimited format were imported to MS Excel, then filtered by a minimum score of 250 and 97–100% percent identity range. Blast search results were parsed and concatenated using custom-built Python scripts, exported to Excel, and visualized using Tableau software.

2.5. Statistical Analyses

Since data are absence–presence, variation in species composition between sampling places was performed using the Sørensen index of dissimilarity [36–38]. All analyses were performed with Betapart Package in R software, version 1.5.4 [39].

3. Results and Discussion

All aquatic ecosystems studied here, except Tres Reyes 1, are of transparent blue water (see Figure 2). Accordingly, most of the aquatic systems in the Yucatan Peninsula, which are oligotrophic, show an average Secchi disk of 7.6 m [40] and exhibit a predominant presence of diatoms [41].

3.1. DNA Barcoding Baseline

Briefly, 1037 specimens were processed in total for DNA barcodes. We had success for 1035, corresponding to 99.8%. These results are remarkable because we used for all groups presented here only a single pair of primers, Zplank forward and reverse [42]. We followed all recommendations by Elías-Gutiérrez et al. [16] concerning the fixation of the material with cold ethanol and preservation in cold storage the first week. We consider these procedures contributed to the high success obtained here. All sequences and chromatograms passed the quality controls of BOLD, and none was contaminated, or with stop codons, only some few sequences were shorter than 500 bp (see dataset DS_BASKAAN, Baseline Sian Ka'an dx.doi.org/10.5883/DS-BASKAAN in BOLD).

A total of 43 species were identified to species level, and they are coincident with the BINs assigned by BOLD (Figures 3–6), with less than 2% of intraspecific divergence. All others remained as a MOTU (= BIN), a putative species.

The list of species and potential species, represented by the BINs presented here, is the largest published freshwater biota of Sian Ka'an reserve with 167 (Supplementary File 1). It includes many groups not considered as "true" zooplankton, such as the water mites or chironomids. However, Montes-Ortiz & Elías-Gutiérrez [25] discussed the presence of these groups in the zooplankton community.

We found that each system has a unique assemblage of species, and Muyil and Chunyaxché have a unique distribution of the species, which will be discussed later.

Arachnida were represented by water mites, with 209 specimens comprising 11 families and two orders, with 40 BINs.

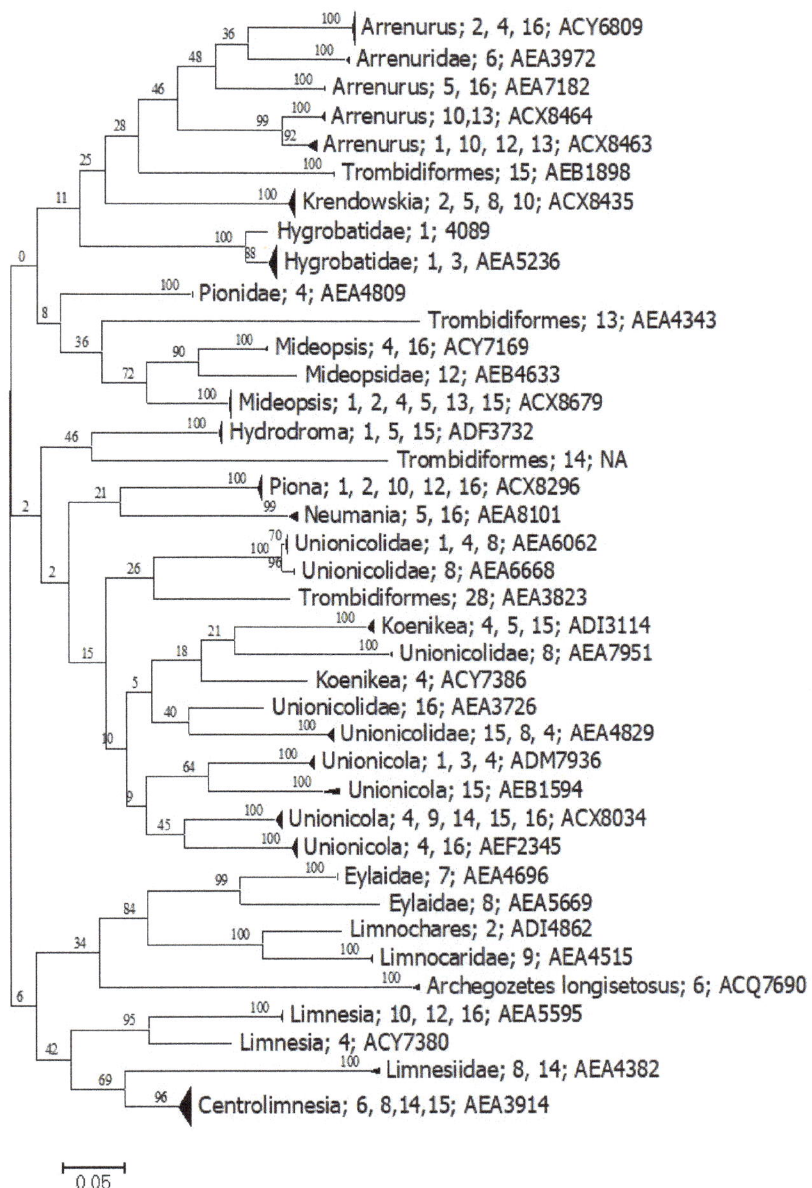

Figure 3. Simplified ID tree for Arachnida represented in the samples. All of them belong to Trombidiformes. Some of the sequences presented here are public [43]. Numbers represent the aquatic system, as presented in Table 1. The last number corresponds to the BIN assigned in BOLD. Support bootstrap values are on each branch.

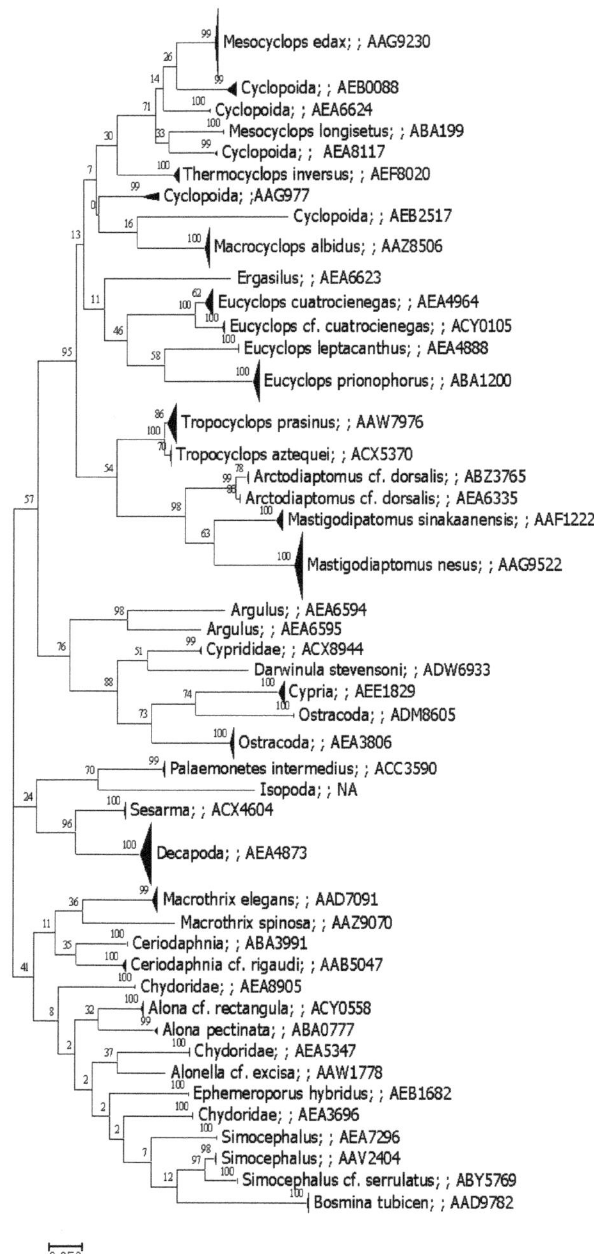

Figure 4. Simplified tree for Crustacea present in the samples. Numbers represent the aquatic system as presented in Table 1. The last number corresponds to the BIN assigned in BOLD. Bootstrap support values are shown in each branch.

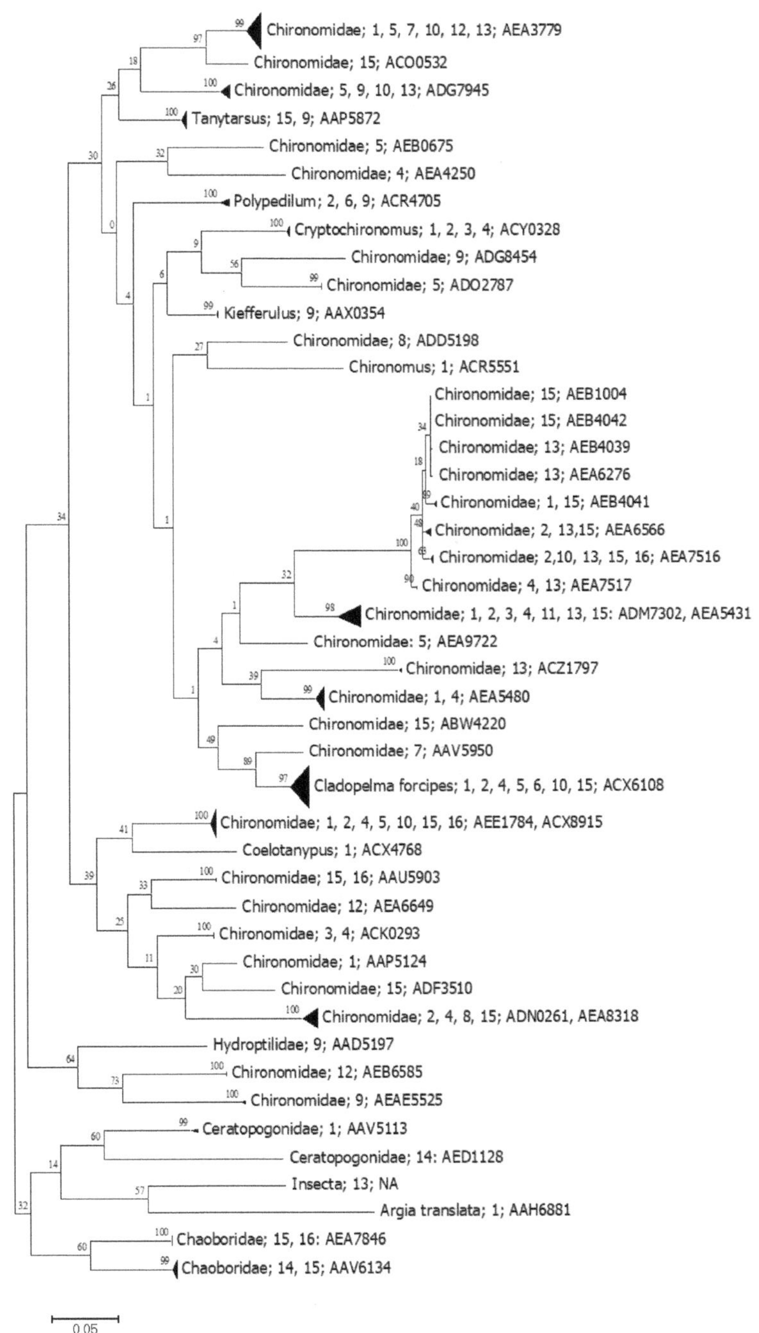

Figure 5. Simplified tree for Insecta. Numbers represent the aquatic system as presented in Table 1. The last number corresponds to the BIN assigned in BOLD. Bootstrap support is shown in each branch.

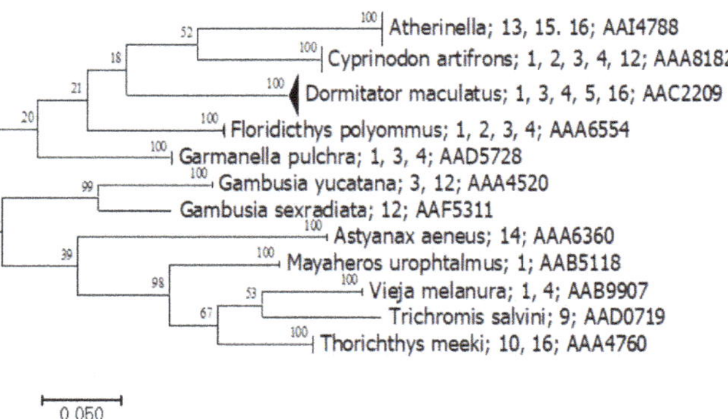

Figure 6. Simplified tree for Actinopterygii. Numbers represent the aquatic system, as presented in Table 1. The last number corresponds to the BIN assigned in BOLD. Bootstrap support is shown in each branch.

The high diversity of mites in this region was noticed in two previous studies [16,43]. Their relevance as indicators of water quality is well known [40,44]. Their taxonomy remains unresolved, and they have a limited distribution in many cases, with possible endemics. Some of them as representatives of *Limnesia* (ACY7380) and *Unionicola* (AEB1594), or some Pionidae (AEA4808) and *Eylaidae* (AEA5669 and AEA4696) were present in only one system (Figure 3) [43].

Crustaceans were represented by 436 specimens comprising five classes, five orders, and 13 families, corresponding to 62 BINs, 20 of them allowing identification to species. Another seven were previously identified as possible comparable species but possibly different (cf.). Finally, seven more were identified to genus, and all remaining can be considered putative species after the BIN.

The Cladocerans were represented by 30% of the total crustaceans and usually were not dominant in the samples. Although scarce, Sminov and Elías-Gutiérrez [41] found eight cladoceran genera after a survey in sediments from 25 systems in Yucatan. For example, recent studies on 56 systems from Mexico and Central America [45,46] found only three taxa in five out of 19 systems located in the Yucatan Peninsula. These numbers indicate, as Smirnov & Elías-Gutiérrez [41] suggested, that cladocerans´ preference for littoral areas and their rapid decomposition in the bottom mud impedes preservation. From the 15 branchiopod BINs found, only four could be identified with certainty, confirming the need to improve the taxonomy for this group [3,47].

It was surprising to find that the sinkhole Tres Reyes 2, with blue and clear water (number 8 in Figure 4), is highly dominated by a species of the daphniid *Ceriodaphnia*, with more than 90% of the total zooplankton. Named here *Ceriodaphnia* cf. *rigaudi*, it is possibly an undescribed species with previous records in Yucatan [3]. All the other systems were dominated by copepods (calanoids and cyclopoids). Tres Reyes 2 is quite remarkable in a region were cladocera seem to have vanished as the dominant group. We cannot explain the reason for this phenomenon, but the system where *Ceriodaphnia* cf. *rigaudi* dominates seems to be a deep oligotrophic sinkhole with vertical walls. This particular system requires further studies, but access to it is difficult.

The ostracods were found in ten systems (Figure 4). They did not appear in all systems, probably because sampling was conducted only with light traps. It is necessary to compare results using different devices such as plankton tows to clarify if this group was underrepresented here. Although recently several publications included this group [48–50], we could identify only *Darwinula stevensoni* to species level. Ostracods are an important group in the Yucatan Peninsula, more abundant in sediments than cladocerans [41], with great value as paleobioindicators [49]. Although the study of the diversity of this group is just starting here, a new genus was described recently [48].

The most common and dominant members of the planktonic community in most systems were the copepods, represented by two groups: Calanoida and Cyclopoida. Among the identified species, only the copepod *Mesocyclops edax* is distributed from Yucatan to Canada. We cannot establish if it was introduced or not. Previous Mexican records of this species are only from Yucatan [28]. Even though some of the taxa registered here are restricted to Yucatan or the south of Mexico, others have a more global range, such as *Tropocyclops prasinus* and *Macrocyclops albidus,* which are found from the southern lowlands to highlands of the central plateau of Mexico [28]. However, these two species seem to be a complex of a sibling or translocated species [28,51], requiring a detailed study.

The analyses of these animals in Yucatan allowed the discovery of new species, possibly endemic in sinkholes [52] where cyclopoids have been barcoded since 2008 [3], allowing the identification of ten species of the 24 BINs found. In the case of calanoids, we identified the four BINs found to species. However, two could be cryptic, named *Arctodiaptomus* cf. *dorsalis* and *A.* cf. *dorsalis* 1 [3]. Calanoids were absent in five systems (Km 48, Tres Reyes 2, Santa Teresa, El Toro, and Pucté 2), but cyclopoids were present in all. When calanoids were present, they were usually the most common group in the samples.

From the six BINs representing Decapoda, we identified only *Palaemonetes intermedius.* This caridean shrimp that can penetrate from the sea to freshwater systems, widely distributed on the Western Atlantic coast of the continent [53]. Most decapods, mainly represented by zoea larvae, were found in specific sites of Muyil and Chunyaxché lagoons, both with direct connection to the Caribbean, located 10 km away in a straight line, through a series of channels.

In the case of insects, from the 53 BINs found, Chironomids were the most important group present in all samples (Figure 5). We also found three other three orders and four families. In total, 44 BINs were chironomids, and all of them need further studies since most of this fauna is not well known in the study area. Three BINS were shared in three clusters, but the number of sequenced specimens is still low (two of them are singletons), and this phenomenon has been noticed before in aquatic insects [54]. Chironomids were represented basically by pupae, but some adults emerged inside the trap. Recently, in a study about subfossil Chironomidae present in sediments from Yucatan Peninsula, the maximum resolution was to groups of species, and the authors concluded that due to poor sedimentation and preservation of remains, cenotes have limited potential for palaeolimnological studies [40]. This conclusion highlights the urgency to bio-assess this community in the present day. Our material was sequenced, but taxonomical parts (head and tail) are preserved in the zooplankton reference collection at ECOSUR Chetumal, allowing future identification of the BINs. The only detailed recent study on this fauna is the description of six new species from several systems of Yucatan [55]. Previously, the same author pointed out that this region remains unknown regarding chironomid fauna [56]. In the list presented by these latter authors, they identified only 13 species (most of them represented by adults) from a total list of 86 potential taxa. Another small report lists 42 genera from Calakmul reserve in the south of the Yucatan Peninsula [57], with no comments. Other insects less common but important in some systems were the larvae of Chaoboridae, mostly restricted to the three southernmost areas (Figure 5). These were represented by two BINs. It is remarkable the presence of two Ceratopogonidae BINs because most of the larvae for these tiny biting midges are known to be associated with wetlands but have not been reported as part of the "true" zooplankton from lakes or sinkholes. In this case, the larvae swim into the trap, attracted by the light, although they do not seem to be common. Ceratopogonids were only found at Muyil Lake and El Toro sinkhole. From the whole group, only two BINs could be identified to species level: the common chironomid *Cladopelma forcipes* and the odonate *Argia translata*. All others need further studies to assign the correct species name.

Finally, chordates collected with light traps were represented by 94 fish (Figure 6), mostly larvae. They included six orders, seven families, 11 genera, and 12 species with a total of 12 BINs. All fish species collected with the light trap were also detected with metabarcoding. The latter method added 15 more species (Table 2).

Table 2. Primers used in the eDNA process. All listed primers are 5′ to 3′. The explanation is in the text.

Primer Name	Direction	Primer Sequence	Reference
AquaF2_t1	F	[M13F]ATCACRACCATCATYAAYATRAARCC	[34]
AquaF3_t1	F	[M13F]CCAGCCATTTCNCARTACCARACRCC	[20]
C_FishR1 cocktail:		Cocktail primers (FR1d: FishR2; 1:1)	[35]
FR1d_t1	R	[M13R]ACCTCAGGGTGTCCGAARAAYCARAA	
FishR2_t1	R	[M13R]ACTTCAGGGTGACCGAAGAATCAGAA	
M13-tails			
M13F	F	TGTAAAACGACGGCCAGT	[58]
M13R	R	CAGGAAACAGCTATGAC	[58]
NGS-fusion			
IonA-M13F-ion1-96	F	CCATCTCATCCCTGCGTGTCTCC[GACT] [IonExpress-MID][M13F]	Ion Torrent, Thermo Fisher Scientific
trP1-M13R	R	CCTCTCTATGGGCAGTCGGTGAT [M13R]	Ion Torrent, Thermo Fisher Scientific

The fish have a complete taxonomy work, and construction began on the database for them in Yucatan Peninsula in 2005 [4]. All BINs from the light traps could be identified to species. Only one of them, with a clear, unique haplotype (*Atherinella*), still needs clarification about its identity. This species was found in three sinkholes (Pucté Cafetal 1, Pucté 2, and Sijil Noh Há), but records from BOLD indicate a distribution of this BIN (AAI4788) in the southeast of Mexico. All other species are well known in this region. The rest of the fish species we found have been previously reported in Yucatan Peninsula.

3.2. Metabarcoding and eDNA

Metabarcoding was focused only on vertebrates, particularly fish. Negative controls did not produce any visible PCR products and meaningful data. Human DNA was detected and filtered from the final results. Only high-quality reads assigned to correct Ion Express MID tags were used for the analysis. In total, 172,244 valid sequences were obtained from a total of 17,944,836 reads. They are summarized in Table 3. In total, 25 fish species, two reptiles, five mammals, and three birds were found. All of them matched our previous baseline. From the total, one slider (*Trachemys* sp.) and two fishes (*Bramocharax-Astyanax* and *Cyprinodon beltrani-simus* complexes) had a low interspecific resolution that has been highlighted previously [8]. The site with more species was Muyil 1 with 12 species. Meanwhile, Sijil Noh Há had the lowest number (two species). All systems were positive for eDNA, and all these species had previous records here or in the nearby systems [59]. Not any alien species were found among the fish. A previous old record of the alien tilapia (*Oreochromis mossambicus*) [19] in Chancah Veracruz (= Yodzonot in the old record) was not found. However, this species was found in a previous eDNA survey of Bacalar Lake, located about 70 km to the south of the Sian Ka'an reserve [8]. All species of larvae found in the light traps were confirmed with metabarcoding.

In a broader scenario, the Muyil, Pucté, Cafetal sinkholes and Chunyaxché lagoon had the higher number of OTUs with 54 (40 invertebrates, 14 fish), 44 (33 invertebrates, 11 fish), 40 (30 invertebrates, 10 fish), and 40 (30 invertebrates, 10 fish), respectively.

Table 3. Summary of identification results after metabarcoding, filtered by min score 250 and genus/species identity filter. Genera (species complexes) with low interspecific resolution are marked with *. Numbers correspond to sequences matched. Muyil 2 and Chunyaxché 2 were not sampled for this analysis. + A possible new species.

Identification	Chunyaxché 2	Muyil 1	Km 48	Santa Teresa	Tres Reyes 2	Tres Reyes 1	Minicenote	Del Padre	El Toro	Sijil Noh Há	Pucté 2	Chacah Veracruz	Pucté Cafetal 1
Aramides cajaneus						13							
Megaceryle torquata												1887	
Meleagris gallopavo				10									
Artibeus lituratus				5									
Lampronycteris brachyotis				47									
Lonchorhina aurita				95									
Oryzomys couesi										5063			
Pteronotus parnellii											3		
Kinosternon acutum				3									
Trachemys sp.					610								
Atherinella sp. +											290		77
Belonesox belizanus						287			3				
Bramocharax-Astyanax * +	321	2				2		2			3	4678	
Cribroheros robertsoni	23									130			83
Cryptoheros chetumalensis		2									3033		13265
Cyprinodon beltrani-simus * +		969											
Dormitator maculatus +		66											
Gambusia sexradiata			4			2840				1563		816	
Gambusia yucatana +		2	31		4				9		2		
Gerres cinereus	71	381											
Gobiomorus dormitor												9	
Hyphessobrycon compressus													1332
Garmanella pulchra +		411											
Lutjanus griseus				108									
Mayaheros urophthalmus +			69	10	16		18		29		1723		
Ophisternon													233
Parachromis friedrichsthalii											25		
Petenia splendida							919				3927		
Poecilia mexicana		9	2		4		28		6		2	528	2
Rhamdia quelen	2	8				2	15	4	4		156	6	
Rocio octofasciata						42							
Thorichthys helleri								15					
Thorichthys meeki +		4							7				
Trichromis salvini +	6	26	2	3	39			6	20		2780		
Vieja melanura +		2									4		2196
Aves													
Mammalia													
Reptilia													
Actinopterygii													

3.3. Species Composition Comparison

The Sørensen index of dissimilarity indicates high differences among the BINs assemblages of sampled places. They reached a value of 0.87 ± 0.006 SD. Based on the pair-to-pair comparison, the cluster analyses reflect a high dissimilarity in the BINs assemblages among sampled sites (Figure 7). The Sørensen dissimilarity plot indicates lower dissimilarities between Chunyaxché 1 and Chunyaxché 2 (0.43), between Chunyaxché 2 and Muyil 2 (0.49), and between Muyil 1 and Chunyaxché 1 (0.55). The remaining dissimilarity values between pairs are higher than 0.60 (Figure 7).

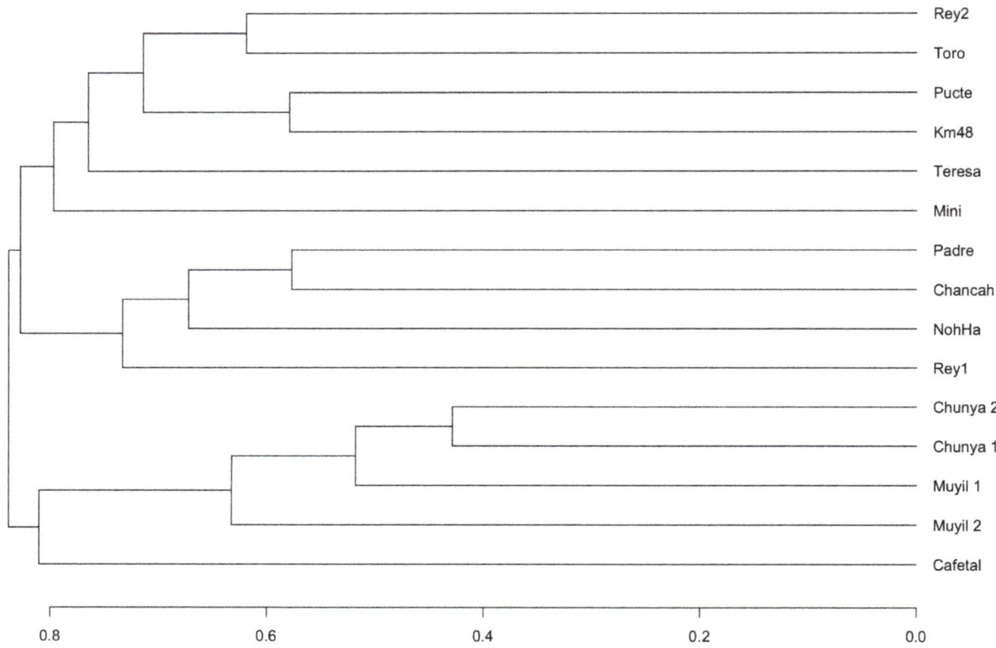

Figure 7. Similarity dendrogram of the localities sampled. Abbreviations: Tres Reyes sinkhole 2, Rey2; Chunyaxché 2, Chunya 2; Pucté 2 sinkhole, Pucté; Chancah Veracruz, Chancah; Santa Teresa sinkhole, Teresa; Laguna Muyil 1, Muyil 1; Pucté Cafetal sinkhole, Cafetal; Tres Reyes sinkhole 1, Rey1; Minicenote sinkhole, Mini; Sijil Noh Há sinkhole, NohHa; Chunyaxché 1, Chunya1; Km 48, Km48; El Toro sinkhole, Toro; Laguna Muyil 2, Muyil 2.

Chunyaxché and El Muyil have well-differentiated zones, with a different community in the eastern localities (Muyil 1 and Chunyaxché 2, see Figure 8). However, these two systems are permanently connected by an ancient channel possibly built by the Mayan culture. Mayans also erected an important ceremony center near the shore of the lagoon with the same name currently. Then the two lagoons keep a similar assemblage of species, as we can observe after Sørensen (Figure 7).

Minicenote, in the southern limit of the reserve, hosts a unique assemblage of species. The peculiarity of this system [14] allowed describing a particular pattern of migration of the *Mastigodiaptomus nesus* population here to the walls [60]. A later, more detailed study of Minicenote concluded it as an oligotrophic system inhabited by 18 taxa, 13 of them rotifers and five crustaceans, and not any chironomids [61]. Our study confirmed the presence of a bosminid in this sinkhole, named by them *Bosmina hagmanni*, and that DNA barcodes allowed to distinguish it as a confusing morpho of *Bosmina tubicen* present in several systems of the Yucatan Peninsula [47]. The presence of *M. nesus* and *Thermocyclops inversus* was also confirmed, but not any rotifer. It seems most rotifers are not attracted by the light of the traps, although some have been reported [16].

Though several BIN's are shared among all systems, the Sørensen analysis shows that each system has a unique assemblage of species. This singularity for each aquatic system makes evident the fragility of these ecosystems in the whole region and confirms previous analyses with mites [43]. We do not know why each system has a unique assemblage of species if they eventually can be connected in the surface after floods or through the complex underground system of rivers. For example, *M. siankaanensis* was found only in two sinkholes (Tres Reyes 2, co-existing with the dominant *C.* cf. *rigaudi* and Pucté Cafetal). Meanwhile, *Mastigodiaptomus nesus* was more widespread and common in the samples. The original localities of the description for *M. siankanensis* near Vigia Chico (Aguada Vigia Chico, Savannah 2 and Aguada Limite de la Reserva, the type locality) [20] were dry during our survey, as was previously mentioned. We do not know anything about the biology of any zooplankter in this region. However, an apparent genetic structure in the haplotypes of *M. siankaanensis* was recently detected from the center to the south of Yucatan [62], with some different haplotypes and other possibly in the process of differentiation.

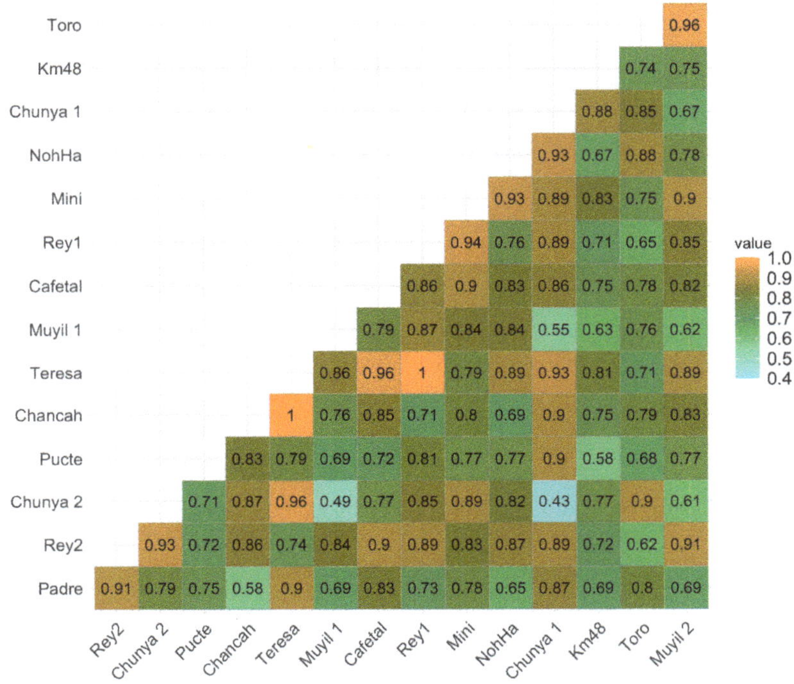

Figure 8. Sorensen dissimilarities among sampling aquatic systems. Abbreviations are the same as Figure 7.

From the origin of the water and possible water flows derived from remote sensing and fault structure, it is suggested that there is a fault from NE-SW communicating systems from Sian Ka'an to the Hondo River (the border between Mexico and Belize) [63,64]. Nevertheless, the biological communities of zooplankton seem to respond more to local variables that could be predation or morphometry of the systems [14] and other unknown factors. This uniqueness of each system or group of systems, e.g., Muyil and Chunyaxché, highlights their vulnerability to environmental issues and the urgency of a permanent bio-monitoring system based on the knowledge of most species dwelling in each water body. This study could be considered a basis to understand any posterior change in these water systems.

3.4. General Remarks and Future Biomonitoring

With the baseline provided here, we have a reliable source of information to apply next-generation techniques connected to eDNA. We demonstrated the usefulness of the DNA barcodes database to identify all species with non-invasive methods with the fish.

In the different freshwater zooplankton groups, there is no laboratory with the ability to identify all the species found in any given aquatic ecosystem. This situation is even worse in megadiverse countries such as Mexico that is the fourth country with the highest biological diversity in the world [65]. Although most of the BINs found with this study have no formal name to species, all material was deposited and photographed. With time, some specialists will collaborate to identify this material.

Though the work to document Mexican freshwater zooplankton with DNA barcodes has been an example for several years [10], most species remain poorly known. The use of new collecting methods such as the light trapping dramatically increased dramatically the species found in neotropical systems of Mexico [16,26]. With the DNA barcode baselines, we can apply the metabarcoding immediately. These methods will allow for the detection of any change in the aquatic community if a periodic biomonitoring system is established. These methods should have the same efficiency with bulk samples of the zooplankton (Elías-Gutiérrez, unpublished).

The subsequent phases should be continuing with the baseline to expand it to more water systems and seasons of the year. Knowledge of the aquatic diversity will help calibrate metabarcoding methodologies to detect any environmental threat from alien species introduction (carrying changes in predation, for example) to pollution. It has been demonstrated that zooplankton responds rapidly to these stressors [66], making it an excellent element to assess any change in the environment [67].

4. Conclusions

We can currently construct a fast baseline of the species in any neotropical aquatic environment despite the taxonomic impediment. If the species are unnamed, they can be kept in a scientific collection. Meanwhile, the genetic data for their identification should be stored in a public database (as BOLD). With time, these BINs will receive a name if the institutions where the collections are deposited maintain a policy to continue the taxonomical studies and encourage the curation of specimens.

This previous step will allow the implementation and continuity of non-invasive biomonitoring with all new tools for eDNA currently available. However, this second step will depend on society being committed to conservation and sustainable development. On the other side, it will depend on the institutions with the equipment and knowledge to perform these complex but efficient techniques.

The economy of Quintana Roo state in the Yucatan Peninsula depends almost entirely on tourism. All the visitors are looking for the natural attractions that this region offers [68]. The recent presence of sargassum in the shoreline of the Mexican Caribbean caused many visitors to move to inland freshwater sinkholes and archaeological sites. As a result, Bacalar (located south of Sian Ka'an) doubled the number of visitors from 2017 to 2018. Currently, there is a noticeable effect of these activities on the microbialites from this site that are unique in the world [69].

Even though Sian Ka'an reserve is a natural protected area, the water systems that support the equilibria in this region are vulnerable to the indirect or direct effects of pollution of the underground or surface water, the introduction of exotic species, and all effects derived from mismanaged tourism. We consider it urgent to consider a proposal to monitor, in a fast and reliable way, all future changes in this environment, as suggested here.

Supplementary Materials: The following are available online at https://www.mdpi.com/article/10.3390/d13070292/s1, Figure S1: ID tree of all specimens found in the dataset used for this paper.

Author Contributions: Conceptualization and methodology, M.E.-G. and M.V.-M.; software, M.M.-C.; validation, M.E.-G., M.V.-M., M.M.-C., E.R.-H. and E.A.-C.; formal analysis, M.E.-G., M.V.-M., M.M.-C., E.R.-H. and E.A.-C.; investigation, M.E.-G. and M.V.-M.; resources, M.E.-G., M.V.-M., M.M.-C., E.R.-H. and E.A.-C.; data curation, M.E.-G. and M.V.-M.; writing—original draft preparation, M.E.-G., M.V.-M., M.M.-C., E.R.-H. and E.A.-C.; writing—review and editing, M.E.-G., M.V.-M., M.M.-C., E.R.-H. and E.A.-C.; visualization, M.E.-G., M.V.-M., M.M.-C., E.R.-H. and E.A.-C.; supervision, M.E.-G. and M.V.-M.; project administration, M.E.-G. and M.V.-M.; funding acquisition, M.E.-G. and M.V.-M. All authors have read and agreed to the published version of the manuscript.

Funding: This study was financed by the Global Environment Fund through the United Nations Development Programme (UNDP, Mexico), Comisión Nacional para el Conocimiento y Uso de la Biodiversidad (CONABIO) and Comisión Nacional de Áreas Naturales Protegidas (CONANP) as part of the investigation called: *Programa de detección temprana piloto de especies acuáticas invasoras a* *través de los métodos de código de barras de la vida y análisis de ADN ambiental en la Reserva de la Biosfera Sian Ka'an* within Project 00089333 "Aumentar las capacidades de México para manejar especies exóticas invasoras a través de la implementación de la Estrategia Nacional de Especies Invasoras".

Institutional Review Board Statement: Ethical review and approval were waived for this study due to it involved material collected under the permit issued by Dirección General de Ordenamiento Pesquero y Acuícola No P2F/DGOPA-063/20.

Data Availability Statement: All data from this study are available in dataset Baseline Sian Ka'an I (DS-BASKAAN; dx.doi.org/10.5883/DS-BASKAAN). For eDNA, we consulted the public fish data from BOLD filtered to genus and species ID level (130,357 sequences), and public BOLD data for Arachnida, Amphibia, Aves, Mammalia, and Reptilia represented by the following datasets: DS-YUCWM (607 sequences), DS-EBACAMPH (11,018 sequences), DS-EBACAVES (28,914 sequences), DS-EBACMAMM (39,890 sequences), DS-EBACREPT (5424 sequences).

Acknowledgments: Special thanks to José Angel Cohuo Colli, Adrian Emmanuel Uh Navarrete, Ivan Canul Palma, and Georgina Alexandra Prisco Pastrana for the field assistance. Alma Estrella García Morales from the Mexican Barcode of Life (MEXBOL) node Chetumal assisted with DNA extraction, PCR reactions, and sequence edition of all material presented here. We appreciate to Felipe Ángel Omar Ortiz Moreno, Director of the Sian Ka'an reserve, his support and facilities to realize this study. Jorge Vidal Quijada, Heliseo Torres Otero, Gilberto Paredes Pat, Isaías Tun Santiago, Alejandro Tuyub Tuyub from Limones town gave us guidance and advice to move inside and the surroundings of the Sian Ka'an Reserve. Edilberto Sosa Martín from the reserve´s main station supported as well our sampling effort. All ejidal commissars supported our study in their communities. Sarah Dolynskyj and Natalia Ivanova from the Biodiversity Institute of Ontario performed the laboratory processes for metabarcoding. Finally, two anonymous referees gave us valuable advice to improve this manuscript.

Conflicts of Interest: The authors declare no conflict of interest.

References

1. Olivares, E.A.O.; Torres, S.S.; Jimenez, S.I.B.; Enriquez, J.O.C.; Zignol, F.; Reygadas, Y.; Tiefenbacher, J.P. Climate Change, Land Use/Land Cover Change, and Population Growth as Drivers of Groundwater Depletion in the Central Valleys, Oaxaca, Mexico. *Remote Sens.* **2019**, *11*, 1290. [CrossRef]
2. Adamowicz, S.J.; Purvis, A. How Many Branchiopod Crustacean Species Are There? Quantifying the Components of Underestimation. *Glob. Ecol. Biogeogr.* **2005**, *14*, 455–468. [CrossRef]
3. Elías-Gutiérrez, M.; Martínez-Jerónimo, F.; Ivanova, N.V.; Valdez-Moreno, M. DNA Barcodes for Cladocera and Copepoda from Mexico and Guatemala, Highlights and New Discoveries. *Zootaxa* **2008**, *1849*, 1–42. [CrossRef]
4. Valdez-Moreno, M.; Ivanova, N.V.; Elías-Gutiérrez, M.; Contreras-Balderas, S.; Hebert, P.D.N. Probing Diversity in Freshwater Fishes from Mexico and Guatemala with DNA Barcodes. *J. Fish. Biol.* **2009**, *74*, 377–402. [CrossRef] [PubMed]
5. Ratnasingham, S.; Hebert, P.D.N. Bold: The Barcode of Life Data System (www.Barcodinglife.Org). *Mol. Ecol. Notes* **2007**, *7*, 355–364. [CrossRef]
6. Ardura, A.; Zaiko, A.; Borrell, Y.J.; Samuiloviene, A.; Garcia-Vazquez, E. Novel Tools for Early Detection of a Global Aquatic Invasive, the Zebra Mussel Dreissena Polymorpha. *Aquat. Conserv. Mar. Freshw. Ecosyst.* **2017**, *27*, 165–176. [CrossRef]
7. Eva, B.; Harmony, P.; Thomas, G.; Francois, G.; Alice, V.; Claude, M.; Tony, D. Trails of River Monsters: Detecting Critically Endangered Mekong Giant Catfish Pangasianodon Gigas Using Environmental DNA. *Glob. Ecol. Conserv.* **2016**, *7*, 148–156. [CrossRef]
8. Valdez-Moreno, M.; Ivanova, N.V.; Elías-Gutiérrez, M.; Pedersen, S.L.; Bessonov, K.; Hebert, P.D.N. Using Edna to Biomonitor the Fish Community in a Tropical Oligotrophic Lake. *PLoS ONE* **2019**, *14*, e0215505. [CrossRef]
9. Moriniere, J.; Balke, M.; Doczkal, D.; Geiger, M.F.; Hardulak, L.A.; Haszprunar, G.; Hausmann, A.; Hendrich, L.; Regalado, L.; Rulik, B.; et al. A DNA Barcode Library for 5200 German Flies and Midges (Insecta: Diptera) and Its Implications for Metabarcoding-Based Biomonitoring. *Mol. Ecol. Resour.* **2019**, *19*, 900–928. [CrossRef] [PubMed]
10. Makino, W.; Maruoka, N.; Nakagawa, M.; Takamura, N. DNA Barcoding of Freshwater Zooplankton in Lake Kasumigaura, Japan. *Ecol. Res.* **2017**, *32*, 481–493. [CrossRef]
11. Lagomarsino, L.P.; Frost, L.A. The Central Role of Taxonomy in the Study of Neotropical Biodiversity. *Ann. Mo. Bot. Gard.* **2020**, *105*, 405–421. [CrossRef]
12. Montoliu-Elena, L.; Elias-Gutierrez, M.; Silva-Briano, M. *Moina Macrocopa* (Straus, 1820): A Species Complex of a Common Cladocera, Highlighted by Morphology and DNA Barcodes. *Limnetica* **2019**, *38*, 253–277.
13. Ratnasingham, S.; Hebert, P.D. A DNA-Based Registry for All Animal Species: The Barcode Index Number (Bin) System. *PLoS ONE* **2013**, *8*, e66213. [CrossRef]

14. Cervantes-Martinez, A.; Elías-Gutiérrez, M.; Suarez-Morales, E. Limnological and Morphometrical Data of Eight Karstic Systems 'Cenotes' of the Yucatan Peninsula, Mexico, During the Dry Season (February–May, 2001). *Hydrobiologia* **2002**, *482*, 167–177. [CrossRef]
15. Medina-Moreno, S.; Jimenez-Gonzalez, A.; Gutiérrez-Rojas, M.; Lizardi-Jimenez, M. Hydrocarbon Pollution Studies of Underwater Sinkholes Along Quintana Roo as a Function of Tourism Development in the Mexican Caribbean. *Revista Mexicana de Ingenieria Quimica* **2014**, *13*, 509–516.
16. Elías-Gutiérrez, M.; Valdez-Moreno, M.; Topan, J.; Young, M.R.; Cohuo-Colli, J.A. Improved Protocols to Accelerate the Assembly of DNA Barcode Reference Libraries for Freshwater Zooplankton. *Ecol. Evol.* **2018**, *8*, 3002–3018. [CrossRef]
17. UNESCO. Sian Ka'an. UNESCO Wolrd Heritage Convention. Available online: https://whc.unesco.org/en/list/410/ (accessed on 7 April 2021).
18. RAMSAR. Sian Ka'an. Ramsar Sites Information Service. Available online: https://rsis.ramsar.org/ris/1329 (accessed on 27 April 2021).
19. Schmitter-Soto, J.J.; Caro, C.I. Distribution of Tilapia, *Oreochromis mossambicus* (Perciformes: Cichlidae), and Water Body Characteristics in Quintana Roo, Mexico. *Rev. Biol. Trop.* **1997**, *45*, 1257–1261.
20. Mercado-Salas, N.F.; Khodami, S.; Kihara, T.C.; Elías-Gutiérrez, M.; Arbizu, P.M. Genetic Structure and Distributional Patterns of the Genus *Mastigodiaptomus* (Copepoda) in Mexico, with the Description of a New Species from the Yucatan Peninsula. *Arthropod Syst. Phylogeny* **2018**, *76*, 487–507.
21. Corgosinho, P.H.C.; Mercado-Salas, N.F.; Arbizu, P.M.; Silva, E.N.D.; Kihara, T.C. Revision of the *Remaneicaris argentina*-Group (Copepoda, Harpacticoida, Parastenocarididae): Supplementary Description of Species, and Description of the First Semi-Terrestrial *Remaneicaris* from the Tropical Forest of Southeast Mexico. *Zootaxa* **2017**, *4238*, 499–530. [CrossRef]
22. CONABIO, Comisión Nacional Para el Uso y Manejo de la Biodiversidad. 108. Sian Ka'an. CONABIO. Available online: http://www.conabio.gob.mx/conocimiento/regionalizacion/doctos/rhp_108.html (accessed on 27 April 2021).
23. CONABIO, Comisión Nacional Para el Conocimiento y Uso de la Biodiversidad. Naturalista. CONABIO. Available online: https://www.naturalista.mx/observations?captive=any&place_id=54190&project_id=1387&taxon_id=47178&verifiable=any&view=species (accessed on 27 April 2021).
24. CONANP, Comisión Nacional de Áreas Protegidas. Sian Ka'an. Gobierno de México. Available online: https://simec.conanp.gob.mx/ficha.php?anp=97®=9 (accessed on 27 April 2021).
25. Valdez-Moreno, M.; Elías-Gutiérrez, M. *Programa De Detección Temprana Piloto De Especies Acuáticas Invasoras a Través De Los Métodos De Código De Barras De La Vida Y Análisis De ADN Ambiental En La Reserva De La Biosfera Sian Ka ´An. Proyecto 00089333 Aumentar Las Capacidades De México Para Manejar Especies Exóticas Invasoras a Través De La Implementación De La Estrategia Nacional De Especies Invasoras*; PNUD México (Programa de Naciones Unidas para el Desarrollo), Ed.; El Colegio de la Frontera Sur: Chetumal, Mexico, 2019; p. 65.
26. Montes-Ortiz, L.; Elías-Gutiérrez, M. Faunistic Survey of the Zooplankton Community in an Oligotrophic Sinkhole, Cenote Azul (Quintana Roo, Mexico), Using Different Sampling Methods, and Documented with DNA Barcodes. *J. Limnol.* **2018**, *77*, 428–440. [CrossRef]
27. Porco, D.; Rougerie, R.; Deharveng, L.; Hebert, P. Coupling Non-Destructive DNA Extraction and Voucher Retrieval for Small Soft-Bodied Arthropods in a High-Throughput Context: The Example of Collembola. *Mol. Ecol. Resour.* **2010**, *10*, 942–945. [CrossRef] [PubMed]
28. Elías-Gutiérrez, M.; Suárez-Morales, E.; Gutiérrez-Aguirre, M.; Silva-Briano, M.; Granados-Ramirez, J.G.; Garfias-Espejo, T. *Guía Ilustrada De Los Microcrustáceos (Cladocera Y Copepoda) De Las Aguas Continentales De México*; Universidad Nacional Autónoma de México: Mexico City, Mexico, 2008.
29. Ivanova, N.V.; DeWaard, J.R.; Hebert, P.D.N. An Inexpensive, Automation-Friendly Protocol for Recovering High-Quality DNA. *Mol. Ecol. Notes* **2006**, *6*, 998–1002. [CrossRef]
30. Hajibabaei, M.; DeWaard, J.R.; Ivanova, N.V.; Ratnasingham, S.; Dooh, R.; Kirk, S.L.; Mackie, P.M.; Hebert, P.D.N. Critical Factors for Assembling a High Volume of DNA Barcodes. *Philos. Trans. R. Soc. Lond. Ser. B Biol. Sci.* **2005**, *360*, 1959–1967. [CrossRef]
31. Kimura, M. A Simple Method of Estimating Evolutionary Rate of Base Substitutions through Comparative Studies. *J. Mol. Evol.* **1980**, *16*, 111–120. [CrossRef] [PubMed]
32. Kumar, S.; Stecher, G.; Tamura, K. Mega7: Molecular Evolutionary Genetics Analysis Version 7.0 for Bigger Datasets. *Mol. Biol. Evol.* **2016**, *33*, 1870–1874. [CrossRef] [PubMed]
33. Mutanen, M.; Kivela, S.M.; Vos, R.A.; Doorenweerd, C.; Ratnasingham, S.; Hausmann, A.; Huemer, P.; Dinca, V.; van Nieukerken, E.J.; Lopez-Vaamonde, C.; et al. Species-Level Para- and Polyphyly in DNA Barcode Gene Trees: Strong Operational Bias in European Lepidoptera. *Syst. Biol.* **2016**, *65*, 1024–1040. [CrossRef] [PubMed]
34. Ivanova, N.V.; Fazekas, A.J.; Hebert, P.D.N. Semi-Automated, Membrane-Based Protocol for DNA Isolation from Plants. *Plant. Mol. Biol. Report.* **2008**, *26*, 186–198. [CrossRef]
35. Ivanova, N.V.; Clare, E.L.; Borisenko, A.V. DNA Barcoding in Mammals. *Methods Mol. Biol.* **2012**, *858*, 153–182.
36. Whittaker, R.H. Vegetation of the Siskiyou Mountains, Oregon and California. *Ecol. Monogr.* **1960**, *30*, 279–338. [CrossRef]
37. Jost, L. Independence of Alpha and Beta Diversities. *Ecology* **2010**, *91*, 1969–1974. [CrossRef]
38. Schroeder, P.J.; Jenkins, D.G. How Robust Are Popular Beta Diversity Indices to Sampling Error? *Ecosphere* **2018**, *9*, e02100. [CrossRef]

39. Baselga, A.; Orme, C.D.L. Betapart: An R Package for the Study of Beta Diversity. *Methods Ecol. Evol.* **2012**, *3*, 808–812. [CrossRef]
40. Hamerlik, L.; Wojewodka, M.; Zawisza, E.; Duran, S.C.; Macario-Gonzalez, L.; Perez, L.; Szeroczynska, K. Subfossil Chironomidae (Diptera) in Surface Sediments of the Sinkholes (Cenotes) of the Yucatan Peninsula: Diversity and Distribution. *J. Limnol.* **2018**, *77*, 213–219. [CrossRef]
41. Smirnov, N.N.; Elías-Gutiérrez, M. Biocenotic Characteristics of Some Yucatan Lentic Water Bodies Based on Invertebrate Remains in Sediments. *Inland Water Biol.* **2011**, *4*, 211–217. [CrossRef]
42. Prosser, S.; Martínez-Arce, A.; Elías-Gutiérrez, M. A New Set of Primers for Coi Amplification from Freshwater Microcrustaceans. *Mol. Ecol. Resour.* **2013**, *13*, 1151–1155. [CrossRef]
43. Montes-Ortiz, L.; Elías-Gutiérrez, M. Water Mite Diversity (Acariformes: Prostigmata: Parasitengonina: Hydrachnidiae) from Karst Ecosystems in Southern of Mexico: A Barcoding Approach. *Diversity* **2020**, *12*, 329. [CrossRef]
44. Goldschmidt, T.; Helson, J.E.; Williams, D.D. Ecology of Water Mite Assemblages in Panama—First Data on Water Mites (Acari, Hydrachnidia) as Bioindicators in the Assessment of Biological Integrity of Neotropical Streams. *Limnologica* **2016**, *59*, 63–77. [CrossRef]
45. Wojewodka, M.; Sinev, A.Y.; Zawisza, E. A Guide to the Identification of Subfossil Non-Chydorid Cladocera (Crustacea: Branchiopoda) from Lake Sediments of Central America and the Yucatan Peninsula, Mexico: Part I. *J. Paleolimnol.* **2020**, *63*, 269–282. [CrossRef]
46. Wojewodka, M.; Sinev, A.Y.; Zawisza, E.; Stanczak, J. A Guide to the Identification of Subfossil Chydorid Cladocera (Crustacea: Branchiopoda) from Lake Sediments of Central America and the Yucatan Peninsula, Mexico: Part II. *J. Paleolimnol.* **2020**, *63*, 37–64. [CrossRef]
47. Elías-Gutiérrez, M.; Montes-Ortiz, L. Present Day Kwnoledge on Diversity of Freshwater Zooplancton (Invertebrates) of the Yucatan Peninsula, Using Integrated Taxonomy. *Teoria Y Praxis* **2018**, *14*, 31–48.
48. Yoo, H.; Cohuo, S.; Macario-Gonzalez, L.; Karanovic, I. A New Freshwater Ostracod Genus from the Northern Neotropical Region and Its Phylogenetic Position in the Family Cyprididae (Podocopida). *Zool. Anz.* **2017**, *266*, 196–215. [CrossRef]
49. Macario-Gonzalez, L.; Cohuo, S.; Elías-Gutiérrez, M.; Vences, M.; Perez, L.; Schwalb, A. Integrative Taxonomy of Freshwater Ostracodes (Crustacea: Ostracoda) of the Yucatan Peninsula, Implications for Paleoenvironmental Reconstructions in the Northern Neotropical Region. *Zool. Anz.* **2018**, *275*, 20–36. [CrossRef]
50. Cohuo-Durán, S.; Elías-Gutiérrez, M.; Karanovic, I. On Three New Species of Cypretta Vávra, 1895 (Crustacea: Ostracoda) from the Yucatan Peninsula, Mexico. *Zootaxa* **2013**, *3636*, 501–524. [CrossRef]
51. Karanovic, T.; Krajicek, M. When Anthropogenic Translocation Meets Cryptic Speciation Globalized Bouillon Originates; Molecular Variability of the Cosmopolitan Freshwater Cyclopoid Macrocyclops albidus (Crustacea: Copepoda). *Ann. Limnol. Int. J. Limnol.* **2012**, *48*, 63–80. [CrossRef]
52. Fiers, F.; Ghenne, V.; Suárez-Morales, E. New Species of Continental Cyclopoid Copepods (Crustacea, Cyclopoida) from the Yucatan Peninsula, Mexico. *Stud. Neotrop. Fauna Environ.* **2000**, *35*, 209–251. [CrossRef]
53. Barba Macias, E. Faunistic Analysis of the Caridean Shrimps Inhabiting Seagrasses Along the Nw Coast of the Gulf of Mexico and Caribbean Sea. *Rev. Biol. Trop.* **2012**, *60*, 1161–1175. [CrossRef]
54. Moriniere, J.; Hendrich, L.; Balke, M.; Beermann, A.J.; Konig, T.; Hess, M.; Koch, S.; Muller, R.; Leese, F.; Hebert, P.D.N.; et al. A DNA Barcode Library for Germanys Mayflies, Stoneflies and Caddisflies (Ephemeroptera, Plecoptera and Trichoptera). *Mol. Ecol. Resour.* **2017**, *17*, 1293–1307. [CrossRef]
55. Vinogradova, E.M. Six New Species of Polypedilum Kieffer, 1912, from the Yucatan Peninsula (Insecta, Diptera, Chironomidae). *Spixiana* **2008**, *31*, 277–288.
56. Vinogradova, E.M.; Riss, W. Chironomids of the Yucatán Peninsula. *Chironomus J. Chironomidae Res.* **2007**, *20*, 32–35. [CrossRef]
57. Contreras-Ramos, A.; Andersen, T. A Survey of the Chironomidae (Diptera) of Calakmul Biosphere Reserve. *Mexico. Chironomus* **1999**, *12*, 3–5.
58. Messing, J. New M13 Vectors for Cloning. *Methods Enzymol.* **1983**, *101*, 20–78.
59. Schmitter-Soto, J.J. *Catálogo De Los Peces Continentales De Quintana Roo*; El Colegio de la Frontera Sur: San Cistóbal de las Casas, Chiapas, Mexico, 1998.
60. Cervantes-Martínez, A.; Elías-Gutiérrez, M.; Gutiérrez-Aguirre, M.A.; Kotov, A.A. Ecological Remarks on Mastigodiaptomus nesus Bowman, 1986 (Copepoda: Calanoida) in a Mexican Karstic Sinkhole. *Hydrobiologia* **2005**, *542*, 95–102. [CrossRef]
61. Cervantes-Martinez, A.; Gutiérrez-Aguirre, M.A. Physicochemistry and Zooplankton of Two Karstic Sinkholes in the Yucatan Peninsula, Mexico. *J. Limnol.* **2015**, *74*, 382–393. [CrossRef]
62. Gutiérrez-Aguirre, M.A.; Cervantes-Martinez, A.; Elías-Gutiérrez, M.; Lugo-Vazquez, A. Remarks on Mastigodiaptomus (Calanoida: Diaptomidae) from Mexico Using Integrative Taxonomy, with a Key of Identification and Three New Species. *PeerJ* **2020**, *8*, e8416. [CrossRef]
63. Bauer-Gottwein, P.; Gondwe, B.R.N.; Charvet, G.; Marín, L.E.; Rebolledo-Vieyra, M.; Merediz-Alonso, G. Review: The Yucatán Peninsula Karst Aquifer, Mexico. *Hydrogeol. J.* **2011**, *19*, 507–524. [CrossRef]
64. Gondwe, B.R.N.; Hong, S.H.; Wdowinski, S.; Bauer-Gottwein, P. Hydrologic Dynamics of the Ground-Water-Dependent Sian Ka'an Wetlands, Mexico, Derived from Insar and Sar Data. *Wetlands* **2010**, *30*, 1–13. [CrossRef]
65. CONABIO. *Capital Natural De México, Vol I: Conocimiento Actual De La Biodiversidad*; Comisión Nacional para el Conocimiento y Uso de la Biodiversidad, México: Mexico City, Mexico, 2008; Volume I.

66. Gutiérrez, M.F.; Molina, F.R.; Frau, D.; Mayora, G.; Battauz, Y. Interactive Effects of Fish Predation and Sublethal Insecticide Concentrations on Freshwater Zooplankton Communities. *Ecotoxicol. Environ. Saf.* **2020**, *196*, 110497. [CrossRef]
67. Almeida, R.; Formigo, N.E.; Sousa-Pinto, I.; Antunes, S.C. Contribution of Zooplankton as a Biological Element in the Assessment of Reservoir Water Quality. *Limnetica* **2020**, *39*, 245–261.
68. Dixon, J.; Hamilton, K.; Pagiola, S.; Segnestam, L. Tourism and the Environment in the Caribbean: An Economic Framework. In *Environmental Economic Series*; World Bank: Washington, DC, USA, 2001; p. 68.
69. Yanez-Montalvo, A.; Gomez-Acata, S.; Aguila, B.; Hernandez-Arana, H.; Falcon, L.I. The Microbiome of Modern Microbialites in Bacalar Lagoon, Mexico. *PLoS ONE* **2020**, *15*, e0230071. [CrossRef]

Article

Phenetic and Genetic Variability of Continental and Island Populations of the Freshwater Copepod *Mastigodiaptomus ha* Cervantes, 2020 (Copepoda): A Case of Dispersal?

Adrián Cervantes-Martínez [1,*], Martha Angélica Gutiérrez-Aguirre [1], Eduardo Suárez-Morales [2] and Sarahi Jaime [1]

1 Unidad Académica Cozumel, Universidad de Quintana Roo, Av. Andrés Quintana Roo, Calle 11 con calle 110 sur s/n, C.P. 77600 Cozumel, Mexico; margutierrez@uqroo.edu.mx (M.A.G.-A.); 1518305@uqroo.mx (S.J.)
2 Unidad Académica Chetumal, El Colegio de la Frontera Sur, Av. Centenario km 5.5, C.P. 77014 Chetumal, Mexico; esuarez@ecosur.mx
* Correspondence: adcervantes@uqroo.edu.mx

Abstract: The diversity of freshwater zooplankton is still little known in Mexico, particularly in reference to insular zooplankton communities. Diaptomid copepods (Crustacea: Copepoda: Calanoida) are a widespread group worldwide, and Mexico harbours high diaptomid diversity. Based on a recent sampling of freshwater zooplankton on a Caribbean Island of Mexico, we present the first record of a diaptomid copepod from an island freshwater ecosystem. It shows the well-known tendency of Neotropical diaptomids to have restricted distributional patterns and high levels of endemism. The species recorded, *Mastigodiaptomus ha* (Cervantes-Martínez, 2020) appears to have a restricted distribution in the Yucatan Peninsula (YP), and the island as well. In order to explore potential differences between the island and continental populations of this species, its phenetic and genetic diversity was analysed by performing morphological comparisons and also by exploring differences of the habitat conditions and genetic sequences (CO1 gene). Our analysis revealed a low (average = 0.33%) genetic divergence between both populations; likewise, both the morphology and habitat conditions closely resemble each other in these two populations. The low genetic divergence between the continental and island populations of *M. ha* suggests an early common origin of the species in the geological history of the YP.

Keywords: barcoding; Calanoida; diaptomids; freshwater; insular water bodies; new record

Citation: Cervantes-Martínez, A.; Gutiérrez-Aguirre, M.A.; Suárez-Morales, E.; Jaime, S. Phenetic and Genetic Variability of Continental and Island Populations of the Freshwater Copepod *Mastigodiaptomus ha* Cervantes, 2020 (Copepoda): A Case of Dispersal? *Diversity* **2021**, *13*, 279. https://doi.org/10.3390/d13060279

Academic Editor: Bert W. Hoeksema

Received: 12 May 2021
Accepted: 13 June 2021
Published: 21 June 2021

1. Introduction

The diverse zooplankton community inhabiting the epicontinental and underground freshwater ecosystems of the Yucatan Peninsula (YP) can be largely constituted by calanoid copepods belonging to the most successful freshwater group; the family Diaptomidae. Diaptomids tend to have restricted distributional patterns, with many endemic species in the Neotropical region [1].

Mastigodiaptomus is one of the most diverse genera in Mexico, currently including 13 species. The genus is widely distributed in the Neotropical region, including the Caribbean islands, Central America, and areas of the Southern United States [2,3].

Recently, Gutiérrez-Aguirre et al. [3] described three new species of the genus from Mexico; *Mastigodiaptomus cihuatlan* (Gutiérrez-Aguirre, 2020), *M. alexei* (Elías Gutiérrez, 2020), and *M. ha* (Cervantes-Martínez, 2020). The latter was found in sinkholes (locally known as cenotes) in the northeastern continental zone region of the YP.

After 15 years of basic studies on the freshwater and anchialine zooplankton in Cozumel Island [4–6], this is the first report of a diaptomid copepod on a Mexican island. Previously, *M. ha* has been recorded in continental freshwater systems in the north-northeastern region of the YP [3]. In this study we analysed the phenetic and genetic distances between the island and continental populations, and specimens from the type locality of *M. ha*, adding new molecular barcodes revealing that, despite the fact that these two populations were isolated for over 8000–6000 years, they are phenotypically and genetically similar.

This work confirms that YP copepod fauna provides the best-known Mexican region for harbouring the greatest diversity of the *Mastigodiaptomus* species in Central America and the Caribbean region [2,3,7].

2. Materials and Methods

2.1. Study Sites and Sampling Methods

Cozumel Island is located off the northeastern coast of the YP, in the northwestern region of the Caribbean Sea (Figure 1) and, because of its area (~500 km^2), it is the third largest island in Mexico and the most populated. An 18 km wide channel separates it from the continental YP.

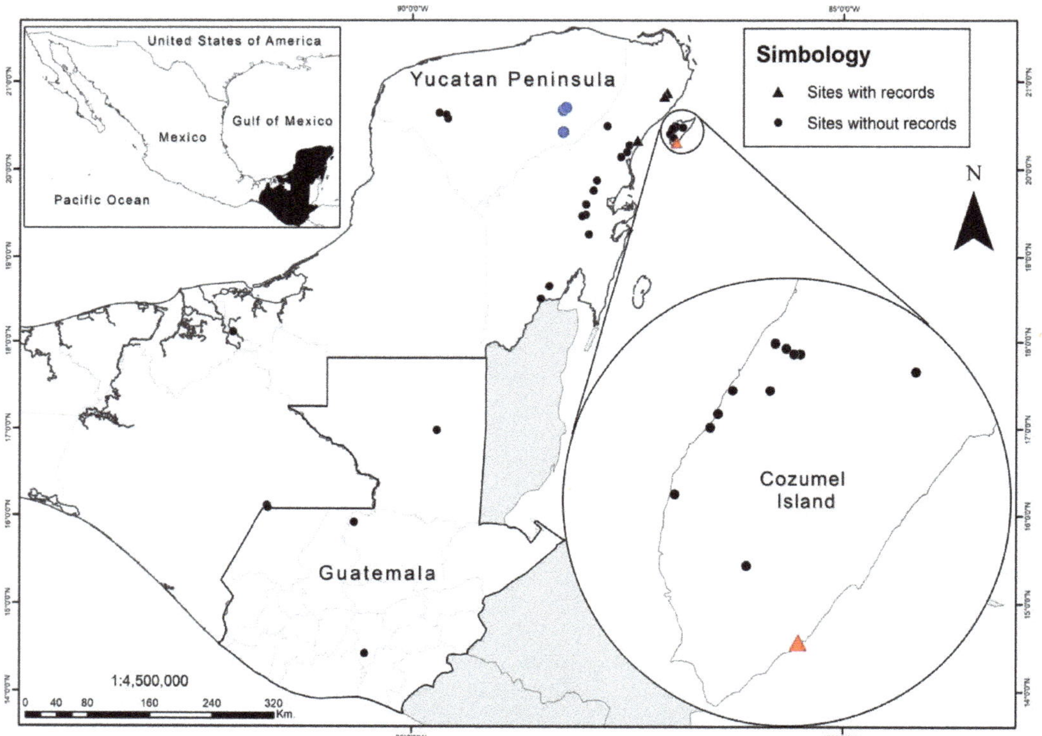

Figure 1. Collection sites for zooplankton samples in the YP, Mexico (2005–2020), where diaptomids are present. Black circles are negative records of *Mastigodiaptomus ha*; triangles are positive sites for records of *M. ha*. The red triangle is the place where the species was founded on Cozumel Island. Blue circles indicate probable presence of this species.

The YP is considered one of the largest karstic aquifers in Mexico and the world [8]; a system that includes Cozumel Island, which was once part of the continental YP geologic plate of the YP (~25 Mya). Like all karstic aquifers, underground aquatic systems are extended (freshwater or anchialine); sinkholes and some superficial lagoons are the surface features of karstic environments [9,10]. Cozumel island has some similar aquatic systems to those mentioned before, but they differ in area, size, and depth [11,12].

2.2. Zooplankton Analysis

During the last 15 years, a systematic sampling of freshwater and anchialine zooplankton has been performed on the YP and Cozumel Island (Figure 1), with high zooplankton diversity and crustacean endemicity [4–6,13]. Water samples for biotic and abiotic variables were collected on the southern part of the island, after the atypical influence of tropical storms in September–October 2020 on Cozumel Island [14].

The species classification of Copepoda was performed according to the methods outlined by Elías-Gutiérrez et al. [15] and Suárez-Morales et al. [16], several specimens were compared with the recently described *M. ha*, inventoried in the north-northeastern region of the YP continental plate [3].

Then, the phenetic variability between the continental vs. island populations of *M. ha* was determined, considering the morphological, environmental, and geological features of the locations inhabited by the copepods. Chemical environmental variables, related to the ionic content, and temperatures were measured in situ with a professional-plus YS Datasonde® (Xylem Inc., Yellow Springs, OH, USA). The trophic status of these systems was determined by the chlorophyll *a* concentration [9]. The sequence of the COI gene was used as evidence of the genetic variability between continental and insular populations of *M. ha*.

2.2.1. Continental Populations Analysed

1. Adult female holotype, dissected on semi-permanent slide (ECOCH-Z-10319), adult male allotype, dissected, mounted on semi-permanent slide (ECOCH-Z-10320), and 20 adult females and 20 males preserved in 96% ethanol and one drop of glycerine (ECOCH-Z-10321). Cenote 7 Bocas, Quintana Roo, Mexico, 20°52′36″ N; 87°02′37″ W, the type locality of *Mastigodiaptomus ha*.

2. Specimens obtained from Verde Lucero (n = 30, ECOCH-Z-10327; 20°52′09.57″ N, 87°04′37.52″ W) and Boca del Puma (n = 6, ECOCH-Z-10326; 20°52′179″ N, 87°03′18″ W) in Quintana Roo, Mexico were also analysed.

2.2.2. Cozumel Island Population

Twenty-two adult females and 36 adult males from 25 Horas Lagoon (20°18′39.7″ N, 86°56′14.2″ W) (ECO-CH-Z-10539). The acronym ECOCH-CH-Z refers to the Zooplankton Reference Collection held at El Colegio de la Frontera Sur, Chetumal, Mexico.

The morphological variability between the two groups of populations was analysed with dissected and whole specimens using a compound Nikon Eclipse 50i microscope. Light microscopic images of variable features were captured with a Lumenera Infinity1 Y-IDT camera (Teledyne Lumenera, Ottawa, Canada) and arranged in Adobe Photoshop V. 6.0 following the current taxonomic standards for the genus *Mastigodiaptomus* [2,17,18].

The DNA extractions of the COI gene, PCR products, and sequence alignment between populations were conducted following the methods of Ivanova et al. [19], Hebert et al. [20], Prosser et al. [21], and Elías-Gutiérrez et al. [22], in accordance with the protocols of the Barcoding Laboratory of Life (ECOSUR, Chetumal, Mexico). Two specimens from the type area, three from the Verde Lucero, and two from the Boca del Puma were considered the continental populations in the genetic analysis, and 28 specimens from the Cozumel population were all included.

Cluster analyses of these sequences were performed to obtain a graphic representation of the divergences among the specimens by using Molecular Evolutionary Genetics Analysis (MEGA X) software (MEGA Freeware, University of Pensilvania, Philadelphia, PA, USA).

Sequences were aligned to 600 base pairs (pb) with CLUSTAL W, and the Kimura 2 parameter (KP2) distance model was used to calculate the sequence divergences. Neighbour-joining (NJ) clusters were created with the gamma distribution model.

All sequences > 500 pb were added to the public data dataset named *Mastigodiaptomus*, created in the Barcode of Life Data Systems portal (BOLDSYSTEMS, http://boldsystems.org/index.php, accessed on 15 April 2021). In the dataset DOI: 10.5883/DS-MMASTIGO (The Barcode of Live Data System, University of Guelph, Guelph, Canada, accessed on

15 April 2021), the individual sequences, trace files, collection data, and primer details are available.

3. Results

3.1. Variability of Environmental, Genetic and Morphological Features

Mastigodiaptomus ha is a freshwater, free-living diaptomid copepod, that is apparently endemic to the north and northeastern zones of the YP. The aquatic ecosystems inhabited by this species showed the following environmental features: low elevations (13.2 ± 6.0 masl), oligo-mesotrophic conditions (≤0.70 mg/m^3 of chlorophyll *a*), low oxygen concentrations (1.5 ± 1.8 mg/L), tropical climates (26.8 ± 1.9 °C), freshwater (0.7 ± 0.2 ppt), low conductivities (1396 ± 559.2 µS/cm^3) (Table 1), and physically the sinkholes where *M. ha* was recorded are type 2, 3, and 4 sinkholes according to Hall [23] (Table 1).

Table 1. Limnological characteristics of the aquatic habitats inhabited by *Mastigodiaptomus ha* (Quintana Roo, Mexico). Elev = elevation (masl), T = water temperature, O_2 = dissolved oxygen (mg/L), EC= electrical conductivity (µS/cm^3), Sal = salinity (ppt), TS = trophic state.

Place Name	Coordinates	Sinkhole Classification [23]	Elev	T	O_2	EC	Sal	TS
Chemuyil	20°21′38.7″ N 87°23′98.1″ W	Vertical walls with wide aperture (as a glass)	10	29.2	4.6	2274	1.1	Oligotrophic
Verde Lucero	20°52′08.7″ N 87°04′37.5″ W	Vertical walls with wide aperture (as a glass)	18	24.7	1.8	1414	0.8	Mesotrophic
7 Bocas	20°52′35.8″ N 87°02′37.5″ W	Cavern, with lateral entrance leading to a chamber with water	16	25.6	0.3	1365	0.7	Oligotrophic
Boca del Puma	20°52′17.9″ N 87°03′18″ W	Cavern, with lateral entrance leading to a chamber with water	18	26.3	0.7	1170	0.6	Oligotrophic
25 Horas	20°18′39.7″ N 86°56′14.2″ W	Superficial lagoon (“aguada”)	4	28.6	0.2	742	0.4	Oligotrophic

This kind of aquatic ecosystem is common on the northern and the eastern fringe zones of the YP, where the most recent, highly permeable sediments (i.e., Miocene, Pliocene, and Quaternary) are widespread [8,24]. These regions are therefore dominated by underground currents or superficial lagoons that formed recently in geological time [8].

3.2. Genetic Variability

The consensus tree with the highest log likelihood is shown in Figure 2. The bootstrap percentages of trees, in which the associated taxa were clustered together, are shown next to the branches. The extended dendrogram displays that the four continental populations and one island population analysed showed one group with low genetic divergence: the BOLDsystems generated the BIN AAU1038 (The Barcode of Live Data System, University of Guelph, Guelph, Canada, accessed on 15 April 2021). The genetic divergence between populations was 0.33% on average (Table 2).

Table 2. Sequence divergence distribution at each taxonomic level examined. To avoid confusion, in this analysis continental vs. island specimens were labelled as two different species, *Mastigodiaptomus ha* vs. *Mastigodiaptomus* sp. (= *M. ha*-island population), to perform a comparison between the populations.

Label	N	Taxa	Comparisons	Min Dist.	Mean Dist.	Max Dist.	SE Dist.
Within species	36	2	416	0.00	0.33	2.22	0.00
Within genus	37	1	179	0.34	4.64	20.15	0.04

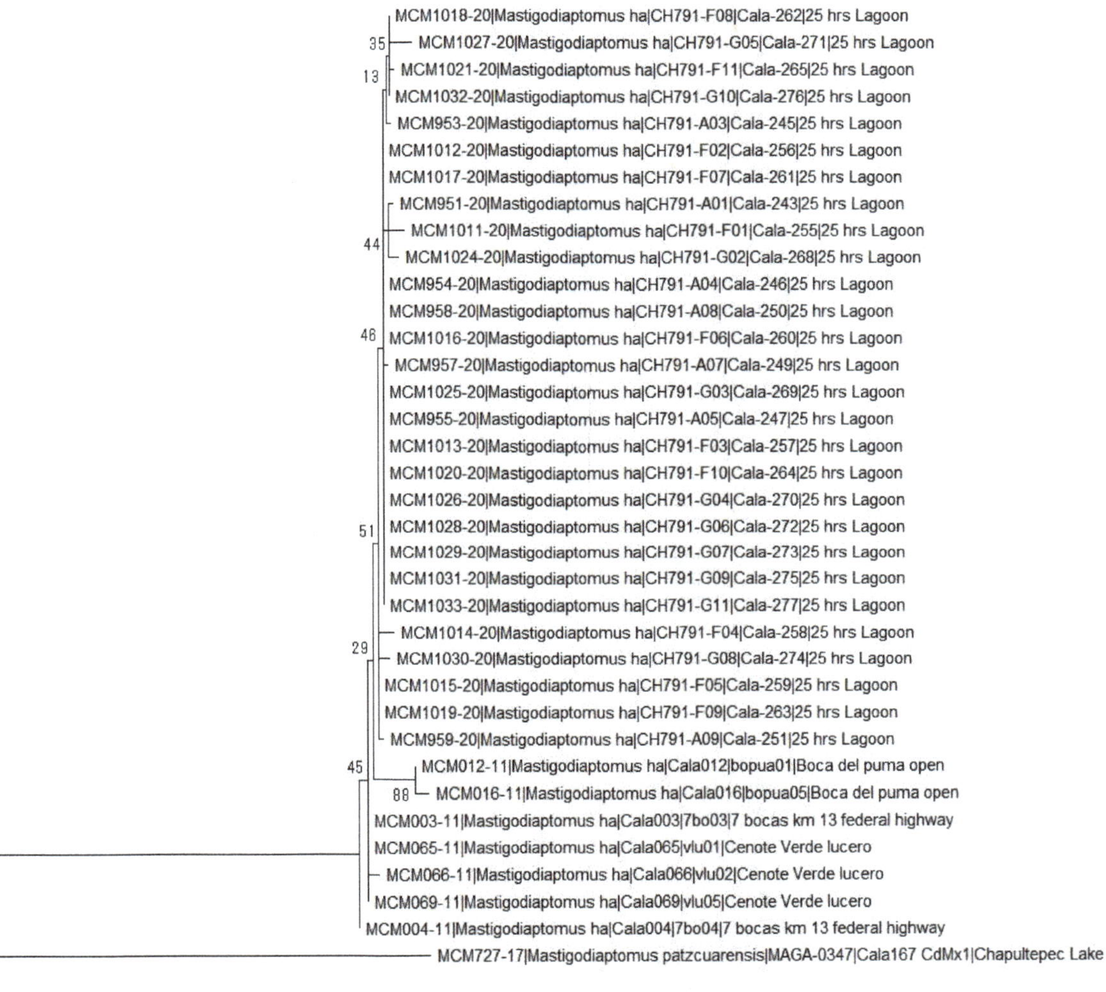

Figure 2. Analysis for Mastigodiaptomus ha (Quintana Roo, Mexico) with the maximum likelihood method to observe genetic relationships between continental and insular populations (1000 replicates), based on the COI gene. The scale length represents the percentage of genetic distance between branches.

The body sizes of specimens of the island population of M. ha were larger than those of the continental population specimens; in the continental populations, the total body lengths were similar between females and males, ranging from 1.2 to 1.3 mm; whereas the females of Cozumel island are larger (i.e., 1.6 to 1.78 mm, n = 28); males are also larger in the island population (1.45 to 1.5 mm, n = 35).

The rostral points shape was also variable in all the analysed continental and island populations: short with rounded points, medium-sized with rounded points or even, large and acute in females (Figure 3A–B). Whereas only 20% of the continental females bear a dorsal keel-like process on the fifth prosomite, all the examined females of the island population (~60 specimens) possess one dorsal, triangle-shaped process (Figure 3C). Twenty percent of the surveyed island females bear large egg sacs with 23–30 eggs (Figure 3D), whereas continental females did not bear egg sacs.

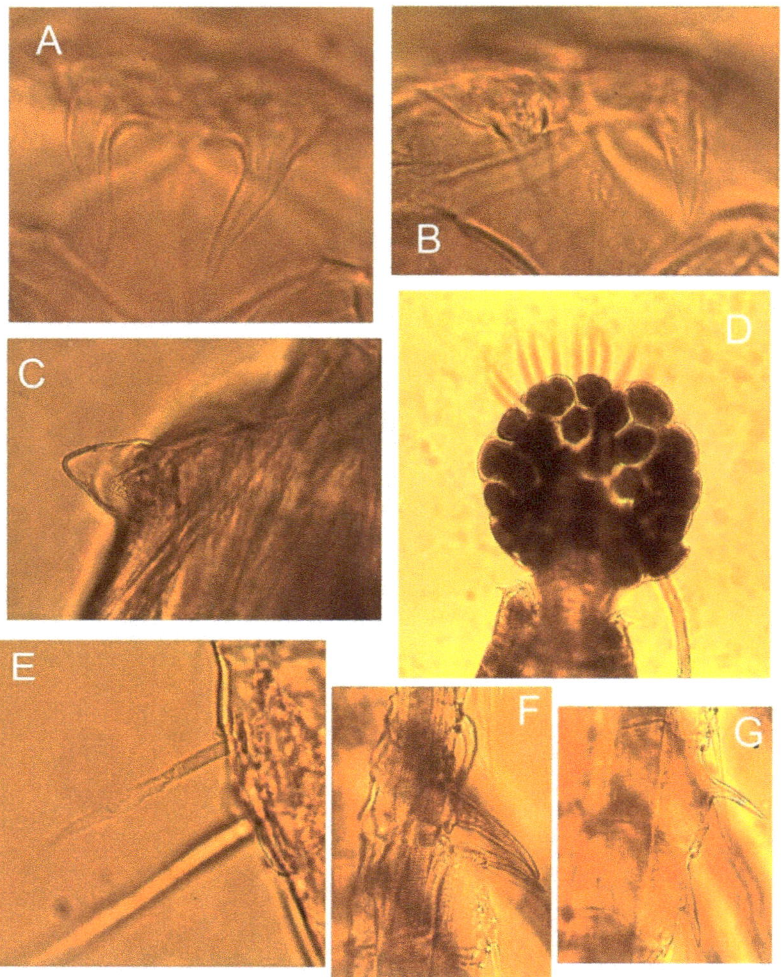

Figure 3. *Mastigodiaptomus ha*, adult female from 25 Horas Lagoon (island population): (**A**) large curved rostral spines; *M. ha*, adult female from Verde Lucero Lagoon (continental population): (**B**) short acute rostral spines; *M. ha*, adult female from 25 Horas Lagoon: (**C**) dorsal process, lateral view; (**D**) egg sac, dorsal view; (**E**) modified setae on antennulary segment 11. *M. ha*, adult male from 25 Horas Lagoon: (**F**) right antennule segments 13, 14; (**G**) same, segments 15, 16.

In M. ha, the presence of modified setae on both female and male antennules was notable as a distinctive feature; these setae are large, flattened, and distally expanded (Figure 3E). In the island population, the spines of male right antennulary segments 13–16 are proportionally larger than those of all the continental populations (Figure 3F–G).

4. Discussion

After systematic surveys of freshwater and anchialine zooplankton were conducted for approximately 15 years in the YP, including on Cozumel Island, this is the first finding on record of a freshwater, free-living calanoid diaptomid on a Mexican island. In the southern region of Cozumel Island where *M. ha* was collected, the pools are flooded in the rainy period (July to October) and the 25 Horas Lagoon suffers a reduced catchment area during the dry season (March to May), with a proportional decrease in the copepod's population.

Therefore, it is hypothesised that *M. ha* is an endemic form of some freshwater systems located in the north-northeastern regions of the YP, and is probably present on other Caribbean islands with geological histories that are homologous to that of Cozumel Island.

The physiography of the YP has been influenced by successive marine regression and transgression processes: during the Cretaceous Period (~145 Mya) and the Pleistocene Yarmouth interglacial stage (220–170 Kya, during which the sea level rose 30 m above the current level), the YP was submerged [7,25]. The YP and Cozumel have been separate since the early Cenozoic Era (~65 MYa) [26]. Allopatric speciation has been advanced as a possible explanation for the divergence of two species of anchialine remipedes: *Xibalbanus cozumelensis* (island species) and *X. tulumensis* (continental). A weak gene flow between these two populations was deemed probable because the marine regressions likely gave place to intermittent communication between these populations, which is not the case for epigean freshwater crustaceans.

The eastern coast of the YP remained submerged for about 33 MY, between the Paleocene and Oligocene periods. During these periods of marine regression, freshwater forms could have recolonised the emerged lands. According to Suárez-Morales [7], one of these forms was an ancestor of the genus *Mastigodiaptomus*, probably related to *M. albuquerquensis*, currently deemed to represent a species complex. This ancestral form successfully radiated across the Yucatan Peninsula, including its central core and eastern coast [7]. The occurrence records of *M. ha* more or less seem to fit the area flooded by seas in the Pleistocene interglacial period, which suggests that this freshwater species could be relatively young (<220–170 Kya).

The most recent soils on the coastal fringe of the peninsula were formed during the Pleistocene Epoch (~1.8 Mya) and the Holocene Epoch (11 Kya) with predominant calcites and dolomites as the most common mineral (of organic origin) [27]. The current physiography of the YP (including its islands and its complex and extensive hypogean landscapes) was present approximately 8–6 Kya [7,27,28]. The specific natures (freshwater, anchialine, or strictly marine) of the cavernous ecosystems in the region were influenced by the intermittent drought/cold periods that took place between 8 and 7, and 6 and 5 Kya [7,28], called stadial and interstadial periods during the Holocene period (~9 Kya). These dry and cold conditions in the YP during the Holocene period [29], the high reproductive rates in copepods, and phoresis of micro-crustaceans by birds [30], probably represented adequate conditions promoting the dispersal of epigean freshwater copepods through glacial refuges in emerged (probably cold lands), including the presence of *M. ha* in Cozumel Island and the eastern coast of the YP.

Additionally, the north and northeastern zones of the YP are considered to have low elevational contrasts, with plains lower than 20 masl, and these regions are apparently separated from the geologically older Meridional Peninsula by hills with elevation levels of 50 to 300 masl [31]. The groundwater ecosystems that span these plains are home to the greatest diversity of species and the highest number of endemic crustaceans in the YP [32]. In accordance with the distribution of other freshwater fauna, such as fish [33], the distribution of freshwater crustaceans that inhabit the plain region can apparently be explained by these geographic barriers.

The presence of *M. ha* individuals in this region supports the hypothesis that the YP is the area with the greatest known diversity and with the highest endemism of the *Mastigodiaptomus* species in the Neotropical region, probably resulting from its habitat diversity and complex geologic history [2,7]. In the YP, seven diaptomid species have been recorded: *Arctodiaptomus dorsalis* (Marsh, 1907); *Leptodiaptomus novamexicanus* (Herrick 1895); *M. nesus* (Bowman, 1986); *M. reidae* (Suárez-Morales and Elías-Gutiérrez, 2000); *M. maya* (Suárez-Morales and Elías-Gutiérrez, 2000); *M. siankaanensis* (Mercado-Salas, Khodami, Kihara, Elías-Gutiérrez and Martínez-Arbizu, 2018), and *M. ha* (Cervantes-Martínez, 2020). The latter four are considered YP endemics [3,7,34]. The late Eocene marine regression exposed most of the YP [7]. This combination of geological, climatological, and ecological conditions probably promoted the isolation conditions for local crustacean speciation and its current distributional patterns in the area [7].

Because low genetic divergence was revealed in and between the analysed continental and island populations of *M. ha*, it is hypothesised that the island population did not result from a very recent invasion process by humans, who have occupied the Cozumel island since 300 BC [35]. Instead, it is suggested that both the continental and island populations have an early common origin, and these populations were isolated after the Pleistocene Yarmouth interglacial stage; the slight morphological divergences featured, found within and between the examined populations of *M. ha*, support the previous statement, as well as a lack of evidence that the distribution of the *Mastigodiaptomus* species is promoted through resistance (or diapause) structures.

The genetic divergence found between populations was remarkably low, ranging between 0 and 2.22%: among the *Mastigodiaptomus* species, the greatest differences recorded within populations reach 2.76% on average [2]. The relatively low genetic, inter- and intra-population divergences found in *M. ha* might indicate the young age of the species. These low divergences have been found in freshwater fish whose current distributional ranges were glaciated during the Pleistocene period in America [36], and are probably related to the reduced population size in glacial refuges causing bottlenecks [37].

Morphological variabilities were described among and within the continental populations of *M. ha* in their original description [3]. In this study, when the continental populations were compared with the island population, a morphological variation was observed in body size (with striking differences between females and males), the rostrum shape, and the size of the spiniform process of the male right antennulary segments 13–16.

In addition, the presence/absence of a dorsal process in females is a variable feature between and within populations of *Mastigodiaptomus texensis*, *M. amatitlanensis*, *M. albuquerquensis*, *M. montezumae*, and *M. ha* [38]. We speculate here that the variability of this character appears to be related to the presence of fish co-occurring with *Mastigodiaptomus*: the dorsal process is absent or very reduced in presence of predator fish, but mostly present and well-developed when fish predation is absent or negligible (unpublished data).

This is the third report of a *Mastigodiaptomus* species inhabiting Caribbean islands with an elevation below 100 masl: *M. nesus* was recorded on San Salvador Island, the Cayman Islands, the Bahamas, and Cuba; *M. purpureus* in Cuba [39], and *M. ha* in Cozumel (this report). Five species of this genus have been recorded at altitudes $\geq$ 1000 masl: *M. montezumae* [40], *M. patzcuarensis*, *M. albuquerquensis* [3,41], *M. suarezmoralesi* [38], and *M. amatitlanensis* [3]. It is likely that the ability of diaptomid genera to dwell in wide elevation and latitudinal ranges is a factor promoting the high regional diversity of *Mastigodiaptomus*.

Individuals of *Mastigodiaptomus* that inhabit island populations appear to be larger on average than their continental counterparts; for instance *M. purpureus*, exclusively recorded in Cuba, is one of the largest species (females = 2.5 mm long, males = 2.2 mm) [42,43]. The length of *M. nesus* in Cuban populations is 1.48 mm in females, and 1.34 in males [39] vs. 0.9–1.0 mm in both sexes in the continental populations [3,44]; the length of *M. ha* from Cozumel is 1.73 mm in females and 1.5 mm in males vs. 1.2–1.3 mm in both sexes in the YP continental populations [3]. As discussed above, it is likely that the presence of fish is related to decreasing copepod size, because in all mentioned cases in which larger island specimens occur, fish are absent.

5. Conclusions

After almost two decades of basic study on freshwater and anchialine zooplankton on Cozumel Island, this is the first report on *Mastigodiaptomus ha* (Cervantes-Martínez, 2020), a freshwater, free-living diaptomid copepod on a Mexican island.

This copepod was previously recorded in freshwater continental ecosystems on the north and northeastern region of the continental plate of the YP; this region is historically and geologically similar to Cozumel Island, and slight morphologic and genetic differences were found between the continental and the island populations of *M. ha*.

Because low genetic divergences were observed between the analysed continental and island *M. ha* populations, both populations of the species probably resulted from the same founder effect.

The YP is the region in which the greatest number of copepod inventories has been documented and the greatest known diversity of the *Mastigodiaptomus* species has been confirmed among the central and northern regions of Mexico, Central America and the Caribbean.

Finally, analyses combining morphological and genetic characteristics in the continental populations of freshwater copepods (and other zooplankton groups) are still scarce. These kinds of studies are even more rare for insular, freshwater organisms. Our study is the first work in Mexico and the Caribbean region, which combined the genetic and morphological tools in research on the freshwater copepod diversity in insular freshwater bodies.

Author Contributions: Conceptualisation, A.C.-M.; methodology, A.C.-M., M.A.G.-A., E.S.-M. and S.J.; software, S.J.; validation, A.C.-M. and E.S.-M.; formal analysis, A.C.-M., M.A.G.-A., E.S.-M. and S.J.; investigation, A.C.-M., M.A.G.-A., E.S.-M. and S.J.; resources, A.C.-M., M.A.G.-A., E.S.-M. and S.J.; data curation, A.C.-M., M.A.G.-A., E.S.-M. and S.J.; writing—original draft preparation, A.C.-M. and M.A.G.-A.; writing—review and editing, M.A.G.-A.; visualisation, A.C.-M., M.A.G.-A., E.S.-M. and S.J.; supervision, A.C.-M.; project administration, A.C.-M.; funding acquisition, A.C.-M. All authors have read and agreed to the published version of the manuscript.

Funding: This research received no external funding.

Institutional Review Board Statement: We collected from several freshwater ecosystems in Mexico. Zooplankton is not under any protection by Mexican laws; thus, no specific permits for this type of field studies are needed.

Informed Consent Statement: Not applicable.

Data Availability Statement: All sequences the public dataset with the name Mastigodiaptomus was created in BOLD database and is avaiable in: DOI: 10.5883/DS-MMASTIGO.

Acknowledgments: The specimens deposited in the Collection of Zooplankton at El Colegio de la Frontera Sur (ECOSUR) were catalogued by R.M. Hernández and J.A. Cohuó-Colli. Mexican Barcode of Life (MEXBOL) provided support for genetic analysis. Facilities to genetic analysis were provided by Barcoding Laboratory of Life, and A. García-Morales (ECOSUR). We are grateful for the academic language editing provided by native English speakers, as well as to the Academic Group Vulnerabilidad y Biodiversidad de Sistemas Acuáticos Continentales y Costeros at the University of Quintana Roo. Three anonymous reviewers improved the manuscript.

Conflicts of Interest: The authors declare no conflict of interest.

References

1. Marsh, C.D. A revision of the north American species of *Diaptomus*. *Trans. Wis. Acad. Sci. Arts Lett.* **1907**, *15*, 381–516.
2. Mercado-Salas, N.F.; Khodami, S.; Kihara, T.C.; Elías-Gutiérrez, M.; Martínez-Arbizu, T. Genetic structure and distributional patterns of the genus *Mastigodiaptomus* (Copepoda) in Mexico, with the description of a new species from Yucatan Peninsula. *Arthropod Syst. Phylogeny* **2018**, *76*, 487–507.
3. Gutiérrez-Aguirre, M.A.; Cervantes-Martínez, A.; Elías-Gutiérrez, M.; Lugo-Vázquez, A. Remarks on *Mastigodiaptomus* (Calanoida: Diaptomidae) from Mexico using integrative taxonomy, with a key of identification and three new species. *PeerJ* **2020**, *8*, e8416. [CrossRef]
4. Cervantes-Martínez, A.; Gutiérrez-Aguirre, M.A.; Delgado-Blas, V.H.; Ruiz-Ramírez, J.D. *Especies del Zooplancton Dulceacuícola de Cozumel*, 2nd ed.; Universidad de Quintana Roo: Quintana Roo, Mexico, 2018; pp. 1–64.
5. Suárez-Morales, E.; Cervantes-Martínez, A.; Gutiérrez-Aguirre, M.A.; Iliffe, T.M. A new *Speleophria* (Copepoda, Misophrioida) from an anchialine cave of the Yucatan Peninsula with comments on the biogeography of the genus. *Bull. Mar. Sci.* **2017**, *93*, 863–878. [CrossRef]

6. Suárez-Morales, E.; Gutiérrez-Aguirre, M.A.; Cervantes-Martínez, A.; Iliffe, T.M. A new anchialine *Stephos* Scott from the Yucatan Peninsula with notes on the biogeography and diversity of the genus (Copepoda, Calanoida, Stephidae). *Zookeys* **2017**, *67*, 1–17. [CrossRef]
7. Suárez-Morales, E. Historical biogeography and distribution of the freshwater calanoid copepods (Crustacea: Copepoda) of the Yucatan Peninsula, Mexico. *J. Biogeogr.* **2003**, *30*, 1851–1859. [CrossRef]
8. Perry, E.; Velazquez-Oliman, G.; Marin, L. The Hydrogeochemistry of the Karts Aquifer System of the Northern Yucatan Peninsula, Mexico. *Int. Geol. Rev.* **2002**, *44*, 191–221. [CrossRef]
9. Cervantes-Martinez, A.; Elías-Gutiérrez, M.; Suárez-Morales, E. Limnological and morphometrical data of eight karstic systems "cenotes" of the Yucatan Peninsula, Mexico, during dry season (February–May, 2001). *Hidrobiologia* **2002**, *482*, 167–177. [CrossRef]
10. Cervantes-Martínez, A. El balance hídrico en cuerpos de agua cársticos de la Península de Yucatán. *Teor. Prax.* **2007**, *3*, 143–152. [CrossRef]
11. Cervantes-Martínez, A. Estudios limnológicos en sistemas cársticos (cenotes). In *Biodiversidad Acuática de la Isla de Cozumel*; Mejía-Ortíz, L., Ed.; Plaza y Valdéz: Ciudad de México, Mexico, 2008; pp. 349–358.
12. Yáñez-Mendoza, G.; Zarza-González, E.; Mejía-Ortiz, L.M. Sistemas anquialinos. In *Biodiversidad Acuática de la Isla de Cozumel*; Mejía-Ortíz, L., Ed.; Plaza y Valdéz: Ciudad de México, Mexico, 2008; pp. 49–69.
13. Suárez-Morales, E.; Reid, J.W. An updated checklist of the continental copepod fauna of the Yucatan Peninsula, Mexico, with notes on its regional associations. *Crustaceana* **2003**, *76*, 977–991. [CrossRef]
14. CONAGUA, Comisión Nacional del Agua. Available online: https://smn.conagua.gob.mx/files/pdfs/comunicados-de-prensa/Comunicado444-20.pdf (accessed on 7 May 2021).
15. Elías-Gutiérrez, M.; Suárez-Morales, E.; Gutiérrez-Aguirre, M.A.; Silva-Briano, M.; Granados-Ramírez, J.G.; Garfias-Espejo, T. *Cladocera y Copepoda de Las Aguas Continentales de México. Guía Ilustrada*; CONABIO, UNAM: Ciudad de México, Mexico, 2008; pp. 1–322.
16. Suarez-Morales, E.; Gutiérrez-Aguirre, M.A.; Gómez, S.; Perbiche-Neves, G.; Previattelli, D.; dos Santos-Silva, N.; da Rocha, C.E.F.; Mercado-Salas, N.F.; Marques, T.M.; Cruz-Quintana, Y.; et al. Class Copepoda. In *Thorp and Covich's Freshwater Invertebrates: Keys to Neotropical and Antarctic Fauna*; Damborenea, C., Rogers, C.D., James, T., Eds.; Elsevier Science Publishing Co., Inc.: San Diego, CA, USA, 2020; Volume 5, pp. 1–1046.
17. Suárez-Morales, E.; Elías Gutiérrez, M. Two new *Mastigodiaptomus* (Copepoda: Diaptomidae) from southeastern Mexico, with a key for the identification of the known species of the genus. *J. Nat. Hist.* **2000**, *34*, 693–708. [CrossRef]
18. Gutiérrez-Aguirre, M.; Cervantes-Martínez, A. A new species of *Mastigodiaptomus* Light 1939 from Mexico, with notes of species diversity of the genus (Copepoda, Calanoida, Diaptomidae). *Zookeys* **2016**, *631*, 61–79. [CrossRef]
19. Ivanova, N.V.; Dewaard, J.R.; Hebert, P.D.N. An inexpensive automation-friendly protocol for recovering high-quality DNA. *Mol. Ecol. Notes* **2006**, *6*, 998–1002. [CrossRef]
20. Hebert, P.D.; Cywisnka, A.; Ball, S.L.; DeWard, J.R. Biological identification through DNA barcodes. *Proc. Biol. Sci.* **2003**, *270*, 313–321. [CrossRef]
21. Prosser, S.; Martínez-Arce, A.; Elías-Gutiérrez, M. A new set of primers for COI amplification from freshwater microcrustaceans. *Mol. Ecol. Res.* **2006**, *13*, 693–708. [CrossRef]
22. Elías-Gutiérrez, M.; Valdéz-Moreno, M.; Topan, J.; Young, M.R.; Cohuo-Colli, J.A. Improved protocols to accelerate the assembly of DNA barcode reference libraries for freshwater zooplankton. *Ecol. Evol.* **2018**, *8*, 3002–3018. [CrossRef]
23. Hall, F.G. Physical and chemical survey of cenotes of Yucatan. In *The Cenotes of Yucatan: A Zoological and Hydrographics Survey*; Pearse, A.S., Creaser, E.P., Hall, F.G., Eds.; Carnegie Inst.: Washington, DC, USA, 1936; p. 304.
24. Gondwe, B.R.N.; Lerer, S.; Stisen, S.; Marín, L.; Rebolledo-Vieyra, M.; Merediz-Alonso, G.; Bayer-Gottwein, P. Hydrogeology of the south-eastern Yucatan Peninsula: New insights from water level measurements, geochemistry, geophysics and remote sensing. *J. Hydrol.* **2010**, *389*, 1–17. [CrossRef]
25. Fiers, F.; Reid, J.W.; Iliffe, T.M.; Suárez-Morales, E. New hypogean cyclopoid copepods (Crustacea) from the Yucatan Peninsula, Mexico. *Contrib. Zool.* **1996**, *66*, 65–102. [CrossRef]
26. Olesen, J.; Meland, K.; Glenner, H.; Van Hengstum, P.J.; Iliffe, T.M. *Xibalbanus cozumelensis*, a new species of Remipedia (Crustacea) from Cozumel, Mexico, and a molecular phylogeny of *Xibalbanus* on the Yucatan Peninsula. *Eur. J. Taxon.* **2017**, *316*, 1–27. [CrossRef]
27. Lugo-Hubp, J.J.; Aceves-Quesada, F.; Espinasa-Pereña, R. Rasgos geomorfológicos mayores de la península de Yucatán. *Rev. Mex. Cienc. Geol.* **1992**, *10*, 143–150.
28. Weide, A.E. Geology of the Yucatan Platform. In *Geology and Hydrogeology of the Yucatán and Quaternary Geology of Northeastern Yucatan Peninsula*; Ward, W.C., Weidie, A.E., Back, W., Eds.; Geological Society: New Orleans, LA, USA, 1985; pp. 1–19.
29. Leyden, B.W.; Brenner, M.; Hodell, D.A.; Curtis, J.H. Late Pleistocene climatic in the Central American lowlands. In *Climatic Change in Continental Isotopic Records*; Swart, P.K., Lohmann, K.C., Mckenzie, J., Savin, S., Eds.; Geophysical Monograph Series; AGU: Washington, DC, USA, 1993; pp. 165–178. [CrossRef]
30. Cortés-Ramírez, G.; Gordillo-Martínez, A.; Navarro-Sigüenza, G. Patrones biogeográficos de las aves de la península de Yucatán. *Rev. Mex. Biodivers.* **2012**, *83*, 530–542. [CrossRef]
31. López Ramos, E. Estudio geológico de la Península de Yucatán. *Assoc. Mex. Geol. Pet. Bol.* **1973**, *25*, 23–76.

32. Álvarez, F.; Iliffe, T.; Benítez, S.; Brankovits, D.; Villalobos, J.L. New records of anchialine fauna from the Yucatan Peninsula, Mexico. *Check List* **2015**, *11*, 1–10. [CrossRef]
33. Schmitter-Soto, J. Ichthyogeography of Yucatan, Mexico. In *Libro Jubilar en Honor al Dr. Salvador Contreras Balderas*; Lozano-Vilano, M.L., Ed.; Universidad Autónoma de Nuevo León: Monterrey, Mexico, 2004; pp. 103–116.
34. Brandorff, G.-O. Distribution of some Calanoida (Crustacea: Copepoda) from the Yucatán Peninsula, Belize and Guatemala. *Rev. Biol. Trop.* **2012**, *60*, 187–202. [CrossRef]
35. Cuarón, A.D. Cozumel. In *Encyclopedia of Islands*; Gillespie, R., Clague, D., Eds.; University of California Press: Berkeley, CA, USA, 2009; pp. 203–206.
36. Avise, J.C.; Bermingham, E.; Kessler, L.G.; Saunders, C. Characterization of mitochondrial DNA variability in a hybrid swam between subspecies of bluegill sunfish (*Lepornis macrochirus*). *Evolution* **1984**, *38*, 931–941.
37. Billington, N.; Hebert, P.D.N. Mitochondrial DNA diversity in fishes and its implications for introductions. *Can. J. Fish. Aquat. Sci.* **1991**, *48*, 80–94. [CrossRef]
38. Gutiérrez-Aguirre, M.A.; Cervantes-Martínez, A. Diversity of freshwater copepods (Maxillopoda: Copepoda: Calanoida, Cyclopoida) from Chiapas, Mexico with a description of *Mastigodiaptomus suarezmoralesi* sp. nov. *J. Nat. Hist.* **2013**, *47*, 479–498. [CrossRef]
39. Bowman, T.E. Freshwater calanoid copepods of the West Indies. *Syllogeus* **1986**, *58*, 237–246.
40. dos Santos-Silva, E.N.; Elías-Gutiérrez, M.; Silva-Briano, M. Redescription and distribution of *Mastigodiaptomus montezumae* (Copepoda, Calanoida, Diaptomidae) in Mexico. *Hydrobiologia* **1996**, *328*, 207–213. [CrossRef]
41. Gutiérrez-Aguirre, M.A.; Cervantes-Martínez, A.; Elías-Gutiérrez, M. An example of how barcodes can clarify cryptic species: The case of the calanoid copepod *Mastigodiaptomus albuquerquensis* (Herrick). *PLoS ONE* **2014**, *9*, e85019. [CrossRef] [PubMed]
42. Wilson, M.S.; Yeatman, H.C. Free-living Copepoda. In *Ward's & Whipple's Freshwater Biology*; Edmonson, W.T., Ed.; John Wiley & Sons, Inc.: New York, NY, USA, 1959; pp. 735–861.
43. Mitch, K.; Fernando, C.H. The freshwater calanoid and cyclopoid copepod Crustacea of Cuba. *Can. J. Zool.* **1978**, *56*, 2015–2023.
44. Cervantes-Martínez, A.; Elías-Gutiérrez, M.; Gutiérrez-Aguirre, M.A.; Kotov, A. Ecological remarks on *Mastigodiaptomus nesus* Bowman, 1986 (Copepoda: Calanoida) in a Mexican karstic sinkhole. *Hydrobiology* **2005**, *542*, 95–102. [CrossRef]

Article

Record of *Caromiobenella* (Copepoda, Monstrilloida) in Brazil and Discovery of the Male of *C. brasiliensis*: Morphological and Molecular Evidence

Judson da Cruz Lopes da Rosa [1,*], Cristina de Oliveira Dias [2], Eduardo Suárez-Morales [3,*], Laura Isabel Weber [4] and Luciano Gomes Fischer [1,4]

1 Programa de Pós-Graduação em Ciências Ambientais e Conservação (PPG-CiAC), Universidade Federal do Rio de Janeiro (UFRJ), Macaé, Rio de Janeiro, RJ 27965-045, Brazil; luciano.fischer@gmail.com
2 Laboratório Integrado de Zooplâncton e Ictioplâncton or Programa de Engenharia Ambiental-PEA, Departa-mento de Zoologia, Instituto de Biologia, Escola Politécnica, Universidade Federal do Rio de Janeiro (UFRJ), Cidade Universitária, Rio de Janeiro, RJ 21941-590, Brazil; crcldias@hotmail.com
3 El Colegio de la Frontera Sur (ECOSUR), Av. Centenario Km. 5.5, Chetumal, Quintana Roo 77014, Mexico
4 Instituto de Biodiversidade e Sustentabilidade (NUPEM), Universidade Federal do Rio de Janeiro (UFRJ), Macaé, Rio de Janeiro, RJ 27965-045, Brazil; lauraweberufrj20@gmail.com
* Correspondence: judsoncruz@yahoo.com.br (J.C.L.R.); esuarez@ecosur.mx (E.S.-M.)

Abstract: Monstrilloid copepods are protelean parasites with a complex life cycle that includes an endoparasitic juvenile phase and free-living early naupliar and adult phases. The monstrilloid copepod genus *Caromiobenella* Jeon, Lee and Soh, 2018 is known to contain nine species, each one with a limited distribution; except for two species, members of this widespread genus are known exclusively from males. Hitherto, members of *Caromiobenella* have not been recorded from tropical waters of the South Western Atlantic (SWA). The nominal species *Monstrilla brasiliensis* Dias and Suárez-Morales, 2000 was originally described from female specimens collected in coastal waters of Espírito Santo and Rio de Janeiro (Brazil), but the male remained unknown. The failure to reliably link both sexes of monstrilloid species is one of the main problems in the current taxonomy of the group, thus leading to a separate treatment for each sex. New zooplankton collections in coastal waters and intertidal rocky pools of the SWA yielded several male and female monstrilloid copepods tentatively identified as *Monstrilla brasiliensis*. Our results of both morphologic and molecular (mtCOI) analyses allowed us to confirm that these males and females were conspecific. We also found evidence suggesting that *Caromiobenella* is not a monophyletic taxon. Our male specimens are morphologically assignable to *Caromiobenella*, therefore, females of the nominal species *Monstrilla brasiliensis*, are matched here with the aforementioned males and, thus, the species should be known as *C. brasiliensis* comb. nov. (Dias and Suárez-Morales, 2000). This finding represents the third documented discovery of a female of *Caromiobenella*, the first record of the genus in the Southwestern Atlantic, and the first documented record of monstrilloids from coastal tidepools. With the addition of *C. brasiliensis*, *Caromiobenella* now includes 10 valid species worldwide. This work represents the second successful use of molecular methods to link both sexes of a monstrilloid copepod. The male of *C. brasiliensis* is herein described, and a key to the known species of *Caromiobenella* and data on the habitat and local abundance of *C. brasiliensis* are also provided.

Keywords: Brazil; integrative taxonomy; monstrilloid copepods; new record; parasitic copepods tropical zooplankton

Citation: Rosa, J.C.L.; Dias, C.O.; Suárez-Morales, E.; Weber, L.I.; Fischer, L.G. Record of *Caromiobenella* (Copepoda, Monstrilloida) in Brazil and Discovery of the Male of *C. brasiliensis*: Morphological and Molecular Evidence. *Diversity* **2021**, *13*, 241. https://doi.org/10.3390/d13060241

Academic Editor: Michael Wink

Received: 22 April 2021
Accepted: 20 May 2021
Published: 31 May 2021

1. Introduction

Monstrilloid copepods are protelean parasites of benthic invertebrates, including polychaetes, molluscs, and sponges [1–3]; most juvenile stages are endoparasitic and free-living adult individuals lacking mouthparts are non-feeding reproductive forms that briefly become part of the zooplankton community [1]. As endoparasites they can cause strong inflammatory processes to their hosts [4]. Because of their rarity in the plankton and taxonomic and nomenclatural complexity [5,6], there are large geographic areas in which the monstrilloid copepod fauna remain largely unknown [2,7]. According to a previous analysis of their known diversity and distribution [2], the regions with the highest number of monstrilloid records are the Northeastern Atlantic (32 species), the northwestern Tropical Atlantic (including the Caribbean Sea and the Gulf of Mexico) (24), and the Indonesia-Malaysia-Philippines, and Japan Seas region (23+). The Brazilian-Argentinean coasts are known to harbor a relatively low monstrilloid diversity (16 species).

The Brazilian monstrilloid copepods have been studied for more than 20 years [4,8–12]. Several new species of Monstrilloida have been described from Brazilian coastal waters, some of them from females only (i.e., *Monstrilla pustulata* Suarez-Morales and Dias, 2001, *M. satchmoi* Suárez-Morales and Dias, 2001, *M. careli* Suarez-Morales and Dias, 2000, *M. brasiliensis* Dias and Suárez-Morales, 2000, and *M. bahiana* Suarez-Morales and Dias, 2001), others from males (i.e., *Monstrillopsis fosshageni* Suárez-Morales and Dias, 2001, *Cymbasoma rochai* Suárez-Morales and Dias, 2001). These taxonomic works [8–10] also include a new geographic record of *M. brevicornis* Isaac, 1974 in the Atlantic Ocean [9,10], ecological aspects of a species of *Cymbasoma* [11], and the discovery and description of the female of *C. rochai* [12]. The male specimens collected in the surveyed area were tentatively identified as *Monstrilla* and presumed to belong to one of the Brazilian species previously known only from females. A more detailed analysis allowed us to recognize our males as members of *Caromiobenella*. Both morphologic and molecular analyses were performed on our male and female specimens to reveal if they are conspecific.

Matching both sexes of monstrilloid species has been raised as one of the main obstacles to determine the true diversity of the group [2,13]. Individuals of both sexes are mixed with those of other species in the plankton samples. Reliable methods to link the sexes of a species include: (1) particular autapomorphies shared by males and females of a species, (2) finding them emerging from the same host and mating, and (3) the use of molecular tools.

The goals of this survey were to: (1) to reveal the taxonomic identity of the male specimens collected from the Rio de Janeiro area, and reliably link them to the female through morphological and molecular analyses; (2) present data about their habitat and local abundance; (3) provide an updated identification key to the known species of *Caromiobenella*.

2. Materials and Methods

2.1. Sampling

Zooplankton samples were collected monthly between August 2017 and December 2018 in marine coastal waters and rocky tidepools at five localities in the State of Rio de Janeiro, Southeastern Brazilian coast (Figure 1). Some of these samples contained male and female monstrilloid copepods that were taxonomically examined. Monstrilloids were firstly found in three rocky tidepools of the municipality of Rio das Ostras (Areias Negras beach: 22°31′47.60″ S, 41°55′32.22″ W, Remanso beach: 22°31′40.57″ S, 41°55′21″ W) and Armação de Búzios (Rasa beach: 22°43′59.6″ S, 41°57′26.2″ W), and subsequently in a shallow costal area in the municipality of Rio das Ostras (Cemitério beach: 22°31′52.21″ S, 41°56′32.18″ W) (Figure 1, Table 1).

The water in the rocky tidepools was drained with an electric bilge pump (12 V) and the zooplankton was retained in a 100 μm mesh filter attached to the end of the water pipe. The drained water volume was recorded to calculate the zooplankton densities (individuals/m^3). The water temperature and salinity were recorded with a thermosalinometer (YSI Yellow Spring Pro 2030). The samples used in taxonomic examination and descriptions were fixed in 4% formaldehyde, analyzed in a stereomicroscope, and quantified in a Dollfus chamber.

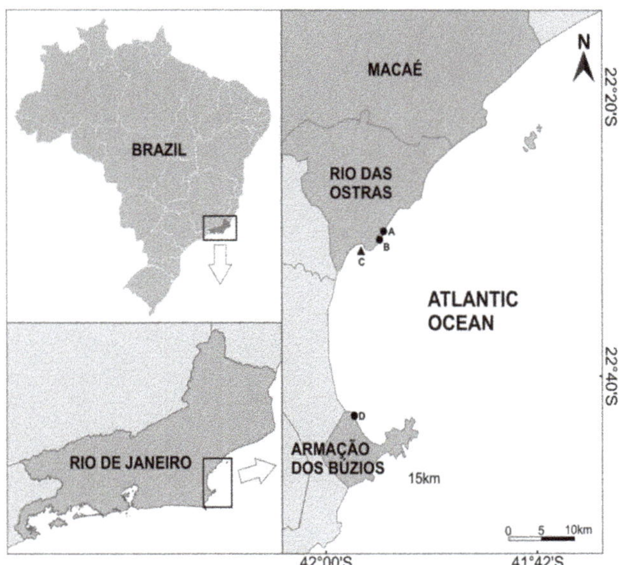

Figure 1. Map showing zooplankton sampling sites in the Rio de Janeiro coast, SW Atlantic, with the coastal municipalities where monstrilloid copepods were collected. Dots: collection in tidepools of rocky shores. Solid triangle: collection in shallow coastal zone (<5 m depth). (**A**) Remanso beach, (**B**) Areias Negras Beach, (**C**) Cemiterio beach, (**D**) Rasa beach.

To complete the genetic analysis, five additional collections were carried out between September and December 2018 (Table 1) in Cemiterio beach, coastal region beyond the surf zone, in an area without rocky shores. These collections were made with a 100 μm mesh plankton net towed by a kayak without using a flowmeter, thus without density estimates. These samples were preserved in 92.8° GL ethanol and monstrilloids were sorted and taxonomically examined in the laboratory.

2.2. Morphologic Analysis

A few male specimens, including those presented here, were tentatively assigned to the genus *Monstrilla* Dana, 1849. A reexamination of these male individuals allowed us to recognize them as a species of *Caromiobenella*, largely known from males and with distinctive morphological characters [3]. One of these specimens (MNRJ30136) was selected to be described. Our description of this male followed the current upgraded descriptive standards in monstrilloid taxonomy [1,13,14] and included the distinctive characters of *Caromiobenella* [3,15]. The morphologic terminology follows Huys and Boxshall (1991) [16]. The Brazilian male specimens were deposited in the collections of the Museu Nacional-Federal University of Rio de Janeiro (MNRJ) and Invertebrate Collection of the Instituto de Biodiversidade e Sustentabilidade (NUPEM/UFRJ) (CIN-NPM), where they are available for inspection.

2.3. Molecular Analysis

Total DNA was extracted from four females and two male copepods by the Chelex resin protocol [17,18] as follows. Ethanol-preserved copepods were picked up with plastic pipettes under a stereomicroscope and then transferred to 0.6 mL Eppendorf vials. The excess of ethanol was then removed and left to dry at room temperature for at least 1 h. Afterwards, 10 μL proteinase-k (10 mg/μL) was added directly over the copepod body, followed by 75 μL lysis buffer 2x (0.1 mM Tris, 0.01 mM EDTA, pH 8.0), 75 μL 12% chelating resin (Chelex 100), SIGMA, shaken in vortex, and incubated overnight at 55 °C.

Table 1. Collection sites data. TD = Tidepool drainage, PN = Plankton net in shallow coastal area, M = male, F = female, temp = water surface temperature (°C), sal = surface water salinity. NA = no data available. Collections at Cemiterio beach were not made in tidepools, but in the coastal zone without using a flowmeter, thus without volume and density estimates.

Date	Site	Collection Type	*C. brasiliensis*		Zoopl. Density (ind. m^3)	Tidepool Volume (L)	Drained Volume (L)	Catalog Number	Temp	Sal
			N°, Sex	Density (ind. m^3)						
08.08.17	A.Negras	TD	3M	15.0	155	345	200	NPM00020 (1M *), (2M * lost%)	24.2	30.0
21.09.17	Remanso	TD	2F	5.0	206	1900	400	NPM000021 (1F), (1F lost%)	20.4	34.3
16.10.17	Remanso	TD	1 M	5.0	991	1900	200	1M (dissected)	21.5	33.2
16.10.17	Remanso	TD	2M	10.0	335	600	200	NPM00022 (1M *), MNRJ30136 (1M *)	21.2	33.6
31.10.17	Rasa	TD	1F	2.5	77	1440	400	NPM00023 (1F)	22.6	28.1
03.02.18	Remanso	TD	1M	2.5	870	1900	400	1M * (dissected)	28.8	32.0
19.09.18	Cemiterio	PN	1F/2M	NA	NA	NA	NA	MNRJ28990 (1M), 1M #, 1F $ *	-	-
20.10.18	Cemiterio	PN	1F/1M	NA	NA	NA	NA	1M #, 2F $	-	-
21.10.18	Cemiterio	PN	2F/1F	NA	NA	NA	NA	2F #, 1F $ *	-	-
28.10.18	Cemiterio	PN	1F	NA	NA	NA	NA	1F # *	-	-
24.11.18	Cemiterio	PN	2F	NA	NA	NA	NA	MNRJ28991 (1F), 1F #	-	-
25.11.18	Cemiterio	PN	1F	NA	NA	NA	NA	NPM00024 (1F)	-	-
02.12.18	Cemiterio	PN	1F	NA	NA	NA	NA	NPM00025 (1F)	-	-

* used in morphometrics, # used in DNA analysis with success; % lost in handling.

The anterior region of the mitochondrial cytochrome c-oxidase, subunit I (COI) was amplified by the polymerase chain reaction (PCR) from 5 and 10 µL of extracted DNA, using Folmer et al. (1994) [19] primers, HCO2198 (TAAACTTCAGGGT GACCAAAAAATCA) and LCO1490 (GGTCAACAAATCATAAAGATATTGG) in a 25 µL final reaction volume containing 1x reaction buffer, 3 mM MgCl2, 0.24 µM of each dNTP, 0.12% Triton-X-100, 0.4 µM of each primer, 2 U of DNA polymerase. PCR was performed in a thermocycler (Gradient Mastercycler, Eppendorf) as follows: 1 cycle at 94 °C for 4 min; 35 cycles at 94 °C, 48 °C and 72 °C for 1 min each; and a final cycle at 72 °C for 7 min. DNA fragments were visualized after electrophoresis over ultraviolet (UV) light using the fluorescent stain UniSafe dye (Uniscience) and PUC19 ladder for fragment size determination. Amplicons were sequenced by the automation system of capillary electrophoresis sequencing (ABI 3730xl System; Sanger method). Sequences were edited using the Chromas Pro, v. 2.1.8 software and then aligned with Clustal W online tool. Edited and revised sequences were analyzed by the Basic Local Alignment Search Tool (BLAST) to find similar sequences deposited in GenBank and verify their proximity to the expected taxonomic group. Multiple alignment and gap insertions were performed retaining conserved coding regions at the same positions.

Pair-wise *p*-distances and Tamura-Nei distances were obtained for the Brazilian group. For comparisons with species from different genera, Tamura-Nei and Kimura-2 parameters distances were used. Maximum likelihood (ML) trees were obtained using Tamura-Nei distance and using 1000 bootstrap iterations for the branch confidence. A parsimony tree was also obtained. Phylogenetic and molecular evolutionary were conducted using MEGA version 6 [20,21]. Sequences of copepod species of other genera obtained from GenBank were used for comparisons and to verify the position of sequences of local copepods. Some of the sequences of other monstrilloid genera available in GenBank were shorter than those found in the present study for the Brazilian copepods. Therefore, to include a greater number of species in the comparison, it was necessary to work with a smaller region, but present in all of them. These comparisons were only possible at over 582 bp DNA fragment of the Brazilian copepods. It was not possible to include *Caromiobenella ohtsukai* (MH638358) and *Monstrillopsis longilobata* (MF447160; MF447163) because they were shorter at the 5′ end.

3. Results

3.1. Taxonomy

Subclass Copepoda Milne-Edwards, 1840
Order Monstrilloida Thorell, 1859

Family Monstrillidae Dana, 1849
Genus Caromiobenella Jeon, Lee and Soh, 2018
Caromiobenella brasiliensis (Dias and Suárez-Morales, 2000) comb. nov.

3.1.1. Material Examined

Eight adult males: 08/08/2017 (1♂NPM-00020, 2♂lost in shipping), 10/16/2017 (1♂NPM-00022, 1♂MNRJ30136, and 1♂dissected), 02/03/2018 (1♂dissected), and 09/19/2018 (1♂MNRJ28990). Specimens NPM-00020, NPM-00022, and MNRJ28990 fixed in 4% formaldehyde and 1 ♂undissected, mounted on semi-permanent slide with glycerin, sealed with acrylic varnish.

3.1.2. Type Locality

Camburi, Baía do Espírito Santo (20°16.383′ S, 40°15.900′ W) [8].

3.1.3. Diagnosis (Female and Male)

Female *Monstrilla* with medium-sized, robust body, cephalothorax with dorsal and ventral scattered fields of striae. Urosome 4-segmented, ovigerous spines short: fifth legs elongate, bilobed, divergent; outer lobe long, slender, with three setae, inner lobe short, cylindrical, with single seta. Antennule 4-segmented, first segment partially fused to cephalothorax, segments 3–4 fused; combined third segment with outer surface produced, forming proximal rounded process furnished with field of coarse cuticular ridges. Caudal rami with 6 setae. Male: *Monstrilla*-like, medium-sized, body segmentation as in *Caromiobenella* (body length ~0.98 mm), cephalothorax representing ~49% of total body length. Pedigerous somites 2–4 representing 39% of total body length. Oral papilla at 34% of way back along ventral surface of cephalothorax. Cephalothorax with dorsal and ventral scattered pores and fields of cuticular striae. Antennules 5-segmented, representing ~35.6% of total body length, geniculated between segments 4–5; first segment partially fused to cephalothorax; second segment with outer margin produced into rounded process ornamented with field of deep transverse wrinkles. Distal antennulary segment with usual armature of genus. Fifth pedigerous somite with reduced fifth legs represented by pair of knob-like processes. Preanal somite with medial pair of small keel-like acute processes on postero-ventral surface. Genital complex of type I, represented by short robust shaft with short, thick distal lappets, branches separated by deep longitudinal slit. Caudal rami armed with 6 setae including short, slender biserially plumose innermost seta.

3.1.4. Description of Adult Male

Body shape and tagmosis as in *Caromiobenella* (see Jeon et al., 2018) (Figure 2A–C). Body robust, total body length of examined individual = 0.98 mm, measured from anterior end of cephalothorax to posterior margin of anal somite. Additional measurements in Table 2. Body short, robust, cephalothorax incorporating first pedigerous somite representing ~49% of total body length. Succeeding pedigerous somites 2–4 each bearing pair of biramous swimming legs; pedigerous somites 2–4 combined accounting for 39% of total body length in dorsal view. Dorsal surface of pedigerous somites 2 and 3 each with pair of "crater-like" cuticular processes; those on third somite being smaller (Figure 2A,C). Cephalic region of cephalothorax wide, smooth, bilaterally protuberant in dorsal view, slightly narrower than cephalothorax; outer margin of cephalic protuberances weakly corrugate. Pair of small dorsal pit setae between antennulary bases. Forehead moderately produced, weakly rounded, with transverse striation fields on dorsal anterior and lateral surfaces; no other cephalic ornamentation was observed on dorsal anterior surface (Figure 2A). Cephalothorax robust, 0.36 mm long, representing almost 37% of total body length; dorsal surface with scattered dorsal pores (Figure 2A). Anterior ventral surface with rounded preoral projection (Figure 3C) between antennule bases and oral papilla, visible in lateral view. Oral papilla at 34% of way back along ventral surface of cephalothorax, with adjacent field of transverse cuticular striae (Figure 3C). Cephalothorax with eyes consisting of relatively small, unpigmented paired lateral cups separated medially by length of less than one eye diameter plus medial cup slightly larger than lateral cups. One pair of relatively small nipple-like cuticular processes present on anterior ventral surface between antennule bases and oral papilla (arrowed in Figure 2B); nipple-like cuticular processes surrounded by field of wrinkles (Figure 2A).

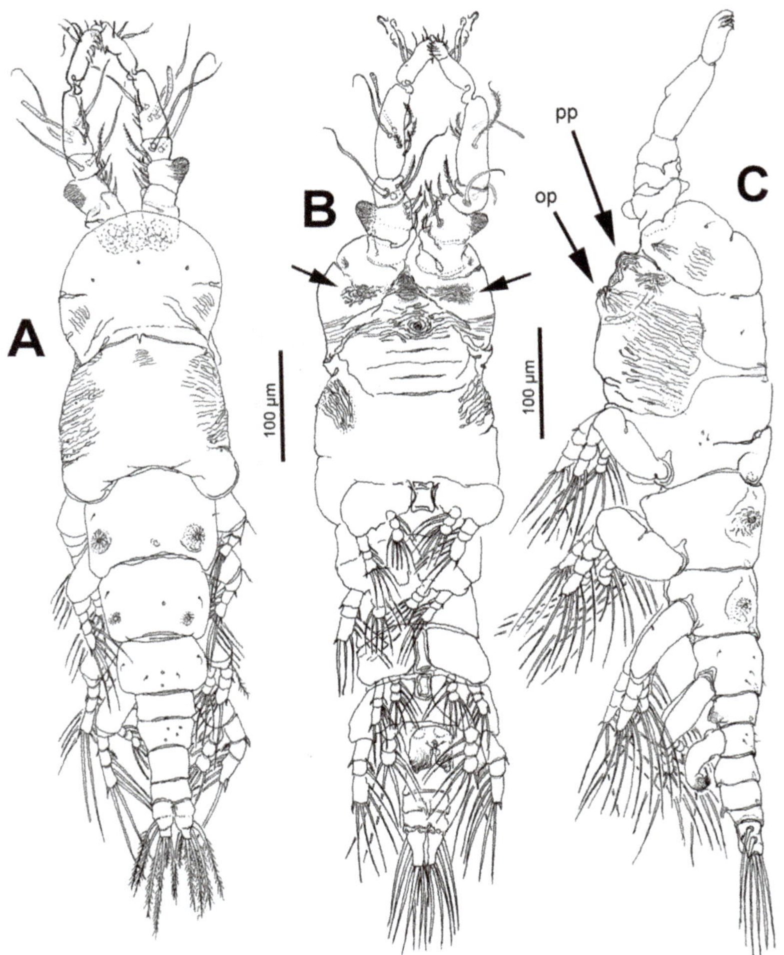

Figure 2. *Caromiobenella brasiliensis* (Dias and Suárez-Morales, 2000) comb. nov., from Rio de Janeiro area, adult male (MNRJ30136). (**A**) Habitus, dorsal view; (**B**) Habitus, lateral view; (**C**) Same, ventral view. Arrows indicate antero-ventral nipple-like cuticular processes. Legend: pp = preoral process, op = oral papilla.

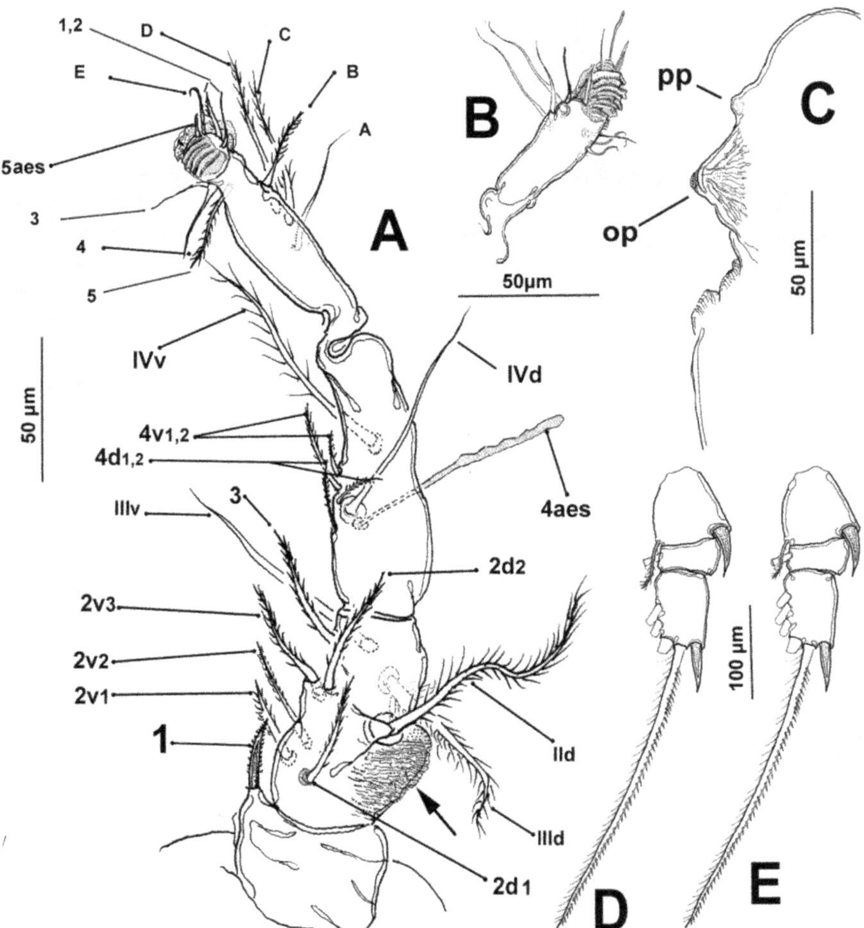

Figure 3. *Caromiobenella brasiliensis* (Dias and Suárez-Morales, 2000) comb. nov., from Rio de Janeiro area, adult male (MNRJ30136). (**A**) Right antennule, dorsal view showing setal elements (segments 1–4 [14], segment 5 [1]; (**B**) Left fifth antennulary segment; (**C**) Anterior part of cephalothorax, lateral view (pp = preoral process, op = oral papilla); (**D**) Leg 1 exopod; (**E**) Leg 3 exopod. Arrow indicates wrinkled outer process on second segment. Legend: pp = preoral process, op = oral papilla.

Table 2. Additional measurements of male specimens of *Caromiobenella brasiliensis* observed in this study (1♂NPM-00020, 1♂NPM-00022, 1♂MNRJ30136, 1♂03/02/2018 (dissected).

Measure	Min (mm)	Max (mm)	Mean (mm)
Total length	0.816	1.170	1.004
Antennule length	0.187	0.373	0.323
Cephalothorax height	0.186	0.269	0.240
Cephalothorax length	0.339	0.475	0.413
Cephalothorax width	0.283	0.456	0.381
Metasome length	0.172	0.239	0.211
Urosome length	0.816	1.170	1.004
Incorporated pediger	0.180	0.256	0.225
Free pediger 1	0.211	0.294	0.266
Free pediger 2	0.164	0.220	0.200
Free pediger 3	0.131	0.203	0.178
Genital somite	0.082	0.810	0.273
Preanal somite	0.065	0.660	0.216
Caudal ramus length	0.053	0.066	0.059
Caudal ramus width	0.036	0.058	0.046

Antennule length = 0.35 mm, representing ~35.6% of total body length. Antennule relatively short, 5-segmented, type III [16] representing 36% of total body length, and 73% of cephalothorax length; antennules indistinctly 5-segmented, segments 1–4 separated by incomplete sutures. First antennulary segment subrectangular, partially fused to cephalothorax proximally and to second segment distally. Second segment with bulging lateral process on outer proximal half; process furnished with deep transverse cuticular wrinkles (arrow in Figure 3A). Fourth segment being longest, representing 37% of total antennulary length. Geniculation between segments 4 and 5 (Figure 3A). In terms of the pattern described by Grygier and Ohtsuka (1995) [14] for antennular armature of segments 1–4 and complemented with Huys et al.'s (2007) [1] nomenclature for elements on the male fifth antennule segment, element 1 present on first segment; element spiniform, lightly pinnate, relatively short, barely reaching midlength of succeeding second segment. Second segment armed with long, lightly plumose elements $2d_{1,2}$, $2v_{1-3}$, and slender seta IId. Third segment partially fused with second, subquadrate, armed with setiform elements 3, IIId, and IIIv. Setal element 3 lightly pinnate, reaching proximal 1/3 of succeeding fourth segment. Fourth segment subrectangular, elongate, about 3.5 times as long as wide, bearing normally developed elements $4d_{1,2}$ and $4v_{1,2}$ as well as long setae IVd and IVv; long, slender aesthetasc 4aes on mid-ventral position. Elements of group 4v short, setiform, lightly pinnate; element $4v_3$ not discernible in the specimen examined. Distal segment with conspicuous proximal geniculation, armed with 11 setal elements (sensu Huys et al., 2007) including elements 1–5 and A–E plus apical aesthetasc 5aes.

Intercoxal sclerites of legs 1–4 subrectangular, smooth. As in other members of *Caromiobenella* [3,15], basis with inner margins produced, forming rounded expansions (arrows in Figure 4A–D); outer margin of basipods of legs 1–4 with lightly setulose basipodal seta; on leg 3, basipodal seta about twice as long as and slightly thicker than in other legs (Figure 4C). Endopodites and exopodites of legs 1–4 triarticulated, outer margins of exopodites smooth. Third exopodal segment of legs 1–4 with 6 setal elements (Figure 3E) except for leg 1 with only 5 elements (Figure 3D). Ramus setae all biserially plumose except for robust apical spiniform seta on exopodal segment 3 (Figure 3D,E); spiniform apical seta on legs 1–4 long, with inner margin spinulose and outer margin lightly setulose (Figure 3D,E). Intercoxal plates of legs 1–4 rectangular, smooth. Armature of legs 1–4 as in Table 3.

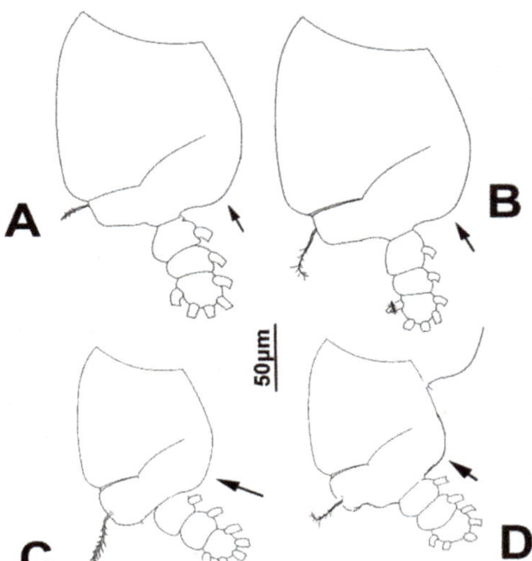

Figure 4. *Caromiobenella brasiliensis* (Dias and Suárez-Morales, 2000) comb. nov., from Rio de Janeiro area, adult male (MNRJ30136). (**A**) Leg 1; (**B**) Leg 2; (**C**) Leg 3; (**D**) Leg 4. Arrows indicate expanded inner margins of basipod.

Table 3. Armature of legs 1–4 including basis, exopods and endopods. Roman numerals indicate spiniform elements, Arabic numbers indicate setiform elements [13,14].

Leg	Basis	Endopod	Exopod
Leg 1	1–0	I-1; 0–1; 2, 2, 1	I-1; 0–1; 2, 2, 1, I
Legs 2–4	1–0	0–1; 0–1; 2, 2, 1	I-1; 0–1; 2, 2, 1, I

Urosome relatively short, representing about 19% of total body length, consisting of fifth pedigerous somite, genital somite (carrying genital complex), and two short free postgenital somites (Figure 5A–C). Anal somite short, with straight outer margins. Fifth pedigerous somite ventrally produced, carrying reduced fifth legs represented by pair of knob-like processes (Figure 5B,C) and proximal half (Figure 5C). Genital somite slightly shorter than preceding fifth pedigerous somite; genital complex of type I [15,20], represented by short, robust, ventrally expanded shaft; genital complex with short, medially conjoined lappets, both tapering distally into apical subtriangular opercular process (Figure 5C–E). Lappets with rugose anterior surface and coarsely striated distal half, branches parallel, separated medially by deep smooth slit (Figure 5E). Preanal somite slightly longer than anal somite, furnished with pair of small keel-like processes on postero-ventral surface, visible in lateral view (arrow in Figure 5C). Caudal rami subrectangular, approximately 1.3 times as long as wide, about 1.3 times as long as anal somite. Each ramus armed with five subequally long caudal setae (setae I–V) [16] plus short, slender, lightly setose caudal seta VI (Figure 5A,B).

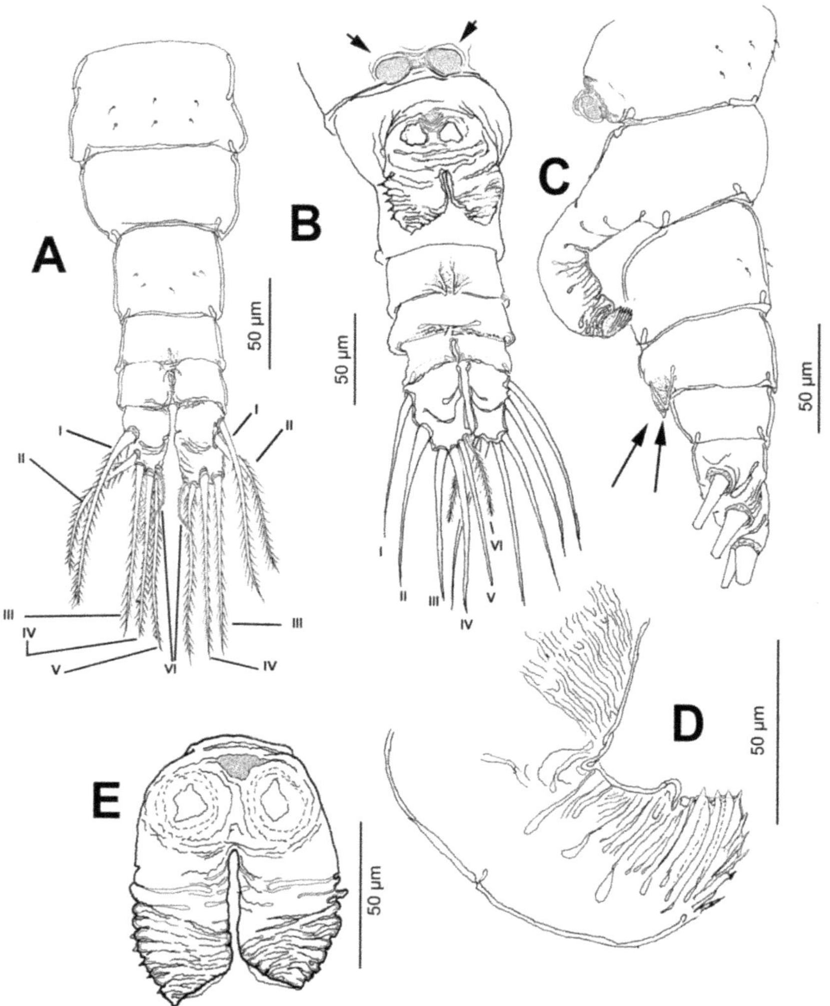

Figure 5. *Caromiobenella brasiliensis* (Dias and Suárez-Morales, 2000) comb. nov., adult male (MNRJ30136). (**A**) Urosome, dorsal view; (**B**) Urosome, ventral view showing reduced knob-like fifth legs (arrows); (**C**) Urosome, lateral view showing genital complex with ventral keel-like processes arrowed; (**D**) Detail of genital lappets, lateral view; (**E**) Genital lappets, ventral view.

3.1.5. Additional Male Specimens Measured

Five specimens: 3♂collected in 08/08/2017 (NPM-00020), 1♂10/16/2017 (NPM-00022), 1♂02/03/2018 (dissected). Body length ranged from 0.878 to 1.170 mm. Cephalothorax length ranged from 0.339 to 0.475 mm representing between 38.6% to 44.2% of the total body length (see Table 2). Antennule length between 0.330 to 0.373 mm representing 51.8% to 97.3% of the cephalothorax and between 22.9% to 37.6% of the total body length. Urosome representing 19.4–22.6% of total body length.

3.1.6. Female Specimens Measured

Based on the examination of the additional specimens: 1♀10/21/2018, 1♀09/19/, 1 ♀10/28/2018. Total body length was measured from anterior end of cephalic somite to posterior margin of anal somite. The body length of three females ranged between 2.08 and

2.35 mm. Cephalothorax length ranged between 0.863 and 1.044 mm, thus representing 41.6–44.7% of total body length. Antennule length representing 41.2–52.2% of cephalothorax length and 18.3–21.7% of total body length. Urosome representing 20.5–23.5% of total body length. Caudal rami length between 0.11 and 0.13 mm.

3.2. Molecular Analysis

Six monstrilloid copepods including four females (identified as *M. brasiliensis*) and two males (tentatively assigned to *M. brasiliensis*), underwent molecular analysis to verify that they belong to the same species. We amplified all specimens' 5′ end of the COI gene successfully, corresponding to a fragment of 681 bp (Table S1). The region showed 21 variable sites including 12 that were phylogenetically informative (shared). These copepods are conspecific, as shown by the distances among them (Table 4) with mean distances of 0.014 ± 0.003, characteristic of intraspecific individuals. The male MZ223434 clustered with MZ223430 female and the male MZ223435 clustered with the other females (Figure 6), confirming that the males belong to the same species of the studied females.

Table 4. Pair-wise genetic distances between monstrilloid copepods from Rio das Ostras (Sate of Rio de Janeiro, Brazil) based on 681 pb COI (cytochrome c-oxidase, subunit I) fragment. Distances: (below diagonal) *p*-distance; (above diagonal) Tamura-Nei distance.

	Female	**Female**	**Female**	**Female**	**Male**	**Male**
GenBank Accesion	**MZ223430**	**MZ223431**	**MZ223432**	**MZ223433**	**MZ223434**	**MZ223435**
MZ223430		0.018	0.019	0.023	0.015	0.006
MZ223431	0.018		0.010	0.010	0.009	0.018
MZ223432	0.020	0.010		0.006	0.010	0.019
MZ223433	0.022	0.010	0.006		0.013	0.019
MZ223434	0.015	0.009	0.010	0.013		0.015
MZ223435	0.006	0.018	0.019	0.019	0.015	

The branch in the ML tree (Figure 6) that contains exclusively the six Brazilian copepods (Rio das Ostras, RJ, Brazil) was 100% supported by bootstrap. This Brazilian branch appears as a sister group of *Monstrilla ilhoii* (see Table 5), followed by *Caromiobenella* branch, which includes also in its branch the species *Caromiobenella hamatapex* (*Monstrilla hamatapex* in Genbank and in Figure 6). More distantly is grouped *Caromiobenella helgolandica* (as *Monstrilla helgolandica* in Genbank and in Figure 6) and the branch of *Maemonstrilla*. The most distant group from the examined Brazilian copepods was the genus *Cymbasoma* and *Monstrillopsis longilobata* (KR049000 and KY553229).

Table 5. Mean pair-wise genetic distances over 582 pb COI fragment among different monstrilloid genera or branches as shown in Figure 6 for the comparisons with Brazilian copepods from Rio de Janeiro coast. See Table S1 for sequences obtained from GenBank. Distances: (below the diagonal) Kimura-2 Parameter; (above the diagonal) Tamura–Nei distance.

Branches/Spp	1	2	3	4	5	6	7	8
(1) *C. brasiliensis*		0.305 ± 0.026	0.308 ± 0.023	0.470 ± 0.038	0.366 ± 0.031	0.464 ± 0.030	0.505 ± 0.041	0.535 ± 0.036
(2) *M. ilhoii*	0. 299± 0.025		0.401 ± 0.028	0.459 ± 0.039	0.388 ± 0.033	0.446 ± 0.030	0.595 ± 0.054	0.576 ± 0.038
(3) Caromiobenella	0.302 ± 0.022	0.390 ± 0.027		0.443 ± 0.031	0.417 ± 0.031	0.503 ± 0.028	0.540 ± 0.041	0.545 ± 0.035
(4) *C. helgolandica*	0.452 ± 0.035	0. 435± 0.035	0.428 ± 0.029		0. 398± 0.033	0.542 ± 0.037	0.573 ± 0.054	0.620 ± 0.043
(5) *Maemonstrilla*	0.353 ± 0.028	0.374 ± 0.029	0.406 ± 0.029	**0.385 ± 0.031**		0.489 ± 0.032	0.550 ± 0.052	0.579 ± 0.041
(6) *Cymbasoma*	0.451 ± 0.028	0.434 ± 0.028	0.488 ± 0.027	0.517 ± 0.033	0.471 ± 0.030		0.596 ± 0.042	0.593 ± 0.034
(7) *M. longilobata*	0.481 ± 0.036	0.543 ± 0.041	0.515 ± 0.026	0.519 ± 0.040	0.500 ± 0.039	0. 561± 0.034		0.397 ± 0.027
(8) *External group*	0.522 ± 0.034	0.554 ± 0.034	0.532 ± 0.033	0.588 ± 0.037	0.553 ± 0.035	0.573 ± 0.032	0.388 ± 0.025	

(1) Brazilian Copepods (females: MZ223430, MZ223431, MZ223432, MZ223433 and males: MZ223434, MZ223435); (2) *Monstrilla ilhoii* (KY553214); (3) *C. hamatapex* (KR048992), *C. castorea* (KY553209, type species) and *C. polluxea* (KY553211); (4) *C. helgolandica* (KT209330 and KT209379); (5) *Maemonstrilla* sp. (KY553231, KY553232) and *Maemonstrilla simplex* (KR049003); (6) *Cymbasoma* sp. (KR048989), *Cymbasoma reticulatum* (KR048990) and *Monstrillopsis longilobata* (KY553229); (7) *Monstrillopsis longilobata* (KR049000); (8) *Tigriopus japonicus* (KR049009), *Lepeophtheirus salmonis* (KR049053) and the planktonic [22] *Calanus helgolandicus* (AY604521), *Caromiobenella hamatapex* and *C. helgolandica* appear as *Monstrilla* in GenBank.

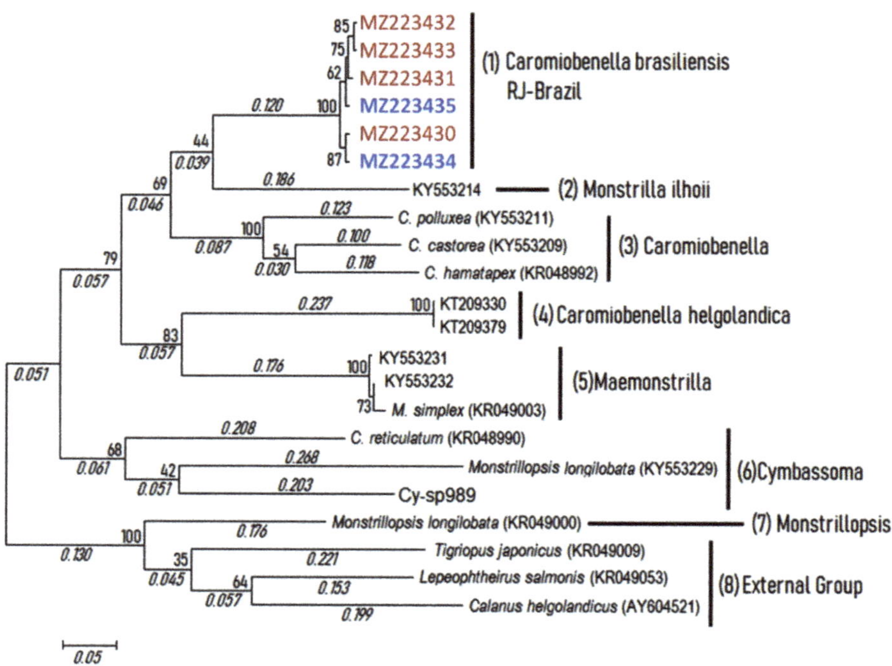

Figure 6. Maximum likelihood tree (Tamura-Nei) showing the position of Brazilian copepods from Rio de Janeiro coast in relation to other monstrilloid genera and species obtained from GenBank. Female Brazilian copepods in red and males in blue. See Table S1 for specific names and GenBank accession numbers for the partial sequences of *C. brasiliensis* used. *Caromiobenella hamatapex* and *C. helgolandica* appear as *Monstrilla* in GenBank.

4. Discussion

The males examined herein can be morphologically identified as members of *Caromiobenella* Jeon, Lee and Soh, 2018 by their possession of distinctive genus characters [3], as follows: (1) the modified distal half of the male fifth antennulary segment, with a series of brush-like processes is the main distinguishing character of *Caromiobenella* [3] and our specimens clearly have this character (Figure 3A,B); (2) the body shape and tagmosis and the urosome segmentation are also typical of *Caromiobenella* [3] (see Figure 1A–C); (3) the presence of 6 caudal setae with innermost apical caudal seta VI clearly shorter and slenderer than the others is another synapomorphic character shared with *Caromiobenella* ([3], Figure 5A,B); the male genital complex is also of the same type (type I) as that described for this genus ([3] Figure 5B–E). The results of our mtCOI analysis and comparison support the designation of our specimens as a species of *Caromiobenella.*

A reliable morphologic method to link the sexes of a monstrilloid species consists of finding particular autapomorphies shared by both sexes [2]. Thus, having a complete description of both sexes of the nominal *M. brasiliensis* ([8], present data), it was possible to use this morphologic criterion. It was previously applied to designate a male preliminarily identified as a subspecies of *M. wandelli* Suárez-Morales and Islas-Landeros, 1993 as the true male of *M. mariaeugeniae* Suárez-Morales, 1994 [23]. During our examination of the males and females from our samples, we were able to find a distinctive autapomorphy shared by both sexes of *Caromiobenella brasiliensis* comb. nov. This key character is the peculiar wrinkled protuberance on the male and female antennules; it is sexually dimorphic, present on segment 2 in the male and on segment 3 in the female ([8], Figure 4A,B); this character has not been observed in any other monstrilloid species. In addition, both sexes of *C. brasiliensis* share a short, spiniform element 1 ([8,14], Figure 3A) and a first antennulary segment partially fused with the cephalothorax ([8], Figures 3B and 5A). Another option to link both sexes is by using molecular and genetic markers, a method that was successfully tested to match males and females of Korean monstrilloid copepods [24]. Overall, *Caromiobenella brasiliensis* can be morphologically distinguished from its known congeners by the distinctive process on the second antennulary segment and by the paired keel-like processes on the ventral surface of the preanal somite.

Currently, there are nine species of *Caromiobenella* recorded from different geographic regions: *C. polluxea* Jeon, Lee and Soh, 2018, *C. castorea* Jeon, Lee and Soh, 2018, and *C. ohtsukai* Jeon, Lee and Soh, 2019 from South Korea, *C. helgolandica* (Claus, 1863) and *C. serricornis* (Sars, 1921) from Europe and Canada, *C. arctica* (Davis and Green, 1974) from northern Canada, *C. hamatapex* (Grygier and Ohtsuka, 1995) from Japan and Korea, *C. pygmaea* (Suárez-Morales, 2000) from the Mediterranean, *C. patagonica* (Suárez-Morales, Ramírez and Derisio, 2008) from the Beagle Channel, and now *C. brasiliensis* (Suárez-Morales and Dias, 2000) from off Southeast Brazil (Southwestern Atlantic).

4.1. Ecology

As far as we are aware of, monstrilloids have not been hitherto reported from tidal pools of rocky reefs. Rocky reefs are coastal shores made from solid rock and considered reef-like ecosystems. Monstrilloids parasitize of marine benthic invertebrates including benthic polychaetes, but also pyramidellid and vermetid gastropods [1,2] and mussel [4]. Interestingly, the largest known aggregations of monstrilloids have been recorded from reef-related areas [25–27].

In addition, the occurrence of monstrilloids in this coastal region, and particularly in the rocky tide pools, may be related to the abundance of host species. So far, the hosts of *C. brasiliensis* remain unknown, but a potential, locally abundant host species is the brown mussel *Perna perna* (Linnaeus, 1758), previously found, either directly [4] or indirectly [28], in association with *Monstrilla* sp. in southern Brazil.

Most (78.3%) individuals of *C. brasiliensis* from both coastal waters and rocky tidepools were collected during spring. In the rocky tidepools, the densities of *C. brasiliensis* ranged between 2.5 and 15 ind./m^3 and contributed up to 6% of the total copepod community in August (Table 1). *Caromiobenella brasiliensis* was found in water with temperatures ranging between 20.4 and 28.8 °C and salinities between 28.1 and 32.0.

4.2. Molecular Remarks

The genetic analysis based on the 681 bp COI fragment confirmed that the six copepods from Rio das Ostras (State of Rio de Janeiro, Brazil) investigated here form a single intraspecific group, and confirm that males showing the described morphology are conspecific with the females found in the surveyed area, designated here as *Caromiobenella brasiliensis* comb. nov. The phylogenetic tree (Tamura–Nei), based on 582 pb COI fragment, showed that the Brazilian species clustered in a major branch (69% bootstrap confidence) containing *Monstrilla ilhoii*, the most closely related species (D = 0.299, Kimura-2-Parameter), followed by a distance of 0.302 (D, Kimura-2-Parameter) by the branch containing three species of *Caromiobenella*, *C. polluxea*, *C. castorea* (the type species), and *C. hamatapex*. All other groups showed distances over 0.450 (D, Kimura-2-Parameter) with respect to the Brazilian copepods. The obtained phylogenetic tree (ML) suggests that the genus *Caromiobenella* is not monophyletic.

4.3. Key to Known Species of Caromiobenella (Males)

1. Antennule with wrinkled protuberance on second antennulary segment.................... *C. brasiliensis* (Dias and Suárez-Morales, 2000)

1A. Antennule lacking special processes on second segment 2

2. Element 1 (*sensu* Grygier and Ohtsuka, 1995) long, reaching well beyond midlength of succeeding second segment.................................. *C. polluxea* Jeon, Lee and Soh, 2018

2A. Element 1(*sensu* Grygier and Ohtsuka, 1995) not as long, not reaching halfway of second segment .... *C. ohtsukai* Jeon, Lee and Soh, 2019
3. Element 2d2 (*sensu* Grygier and Ohtsuka, 1995) remarkably long, reaching halfway of fourth segment... *C. castorea* Jeon, Lee and Soh, 2018
3A. Element 2d2 (*sensu* Grygier and Ohtsuka, 1995) not as long, barely reaching distal margin of third segment or shorter... 4
4. Body length (excluding caudal rami) less than 0.5 mm, 5 caudal setae.........................
.... *C. pygmaea* (Suárez-Morales, 2000)
4A. Body length (excluding caudal rami) more than 1.0 mm, 6 caudal setae................... 5
5. Elements $2v_{1\text{–}3}$ (*sensu* Grygier and Ohtsuka, 1995) relatively long, elements $2v_{2,3}$ reaching beyond halfway of succeeding third segment 6
5A. Elements 2v1–3(*sensu* Grygier and Ohtsuka, 1995) short, barely reaching halfway of third segment.. *C. serricornis* (Sars, 1921)
6. Antennulary segment 4 with proximal bulging process...
.... *C. patagonica* (Suárez-Morales, Ramírez and Derisio, 2008)
6A. Antennulary segment 4 lacking proximal bulging process..7
7. Fifth leg present.. 8
7A. Fifth leg absent... 9
8. Fifth leg 1- lobed, inner margin straight............................ *C. helgolandica* (Claus, 1863)
8A. Fifth leg 1-lobed, with two setae, inner margin produced...
.... *C. hamatapex* (Grygier and Ohtsuka, 1995)
9. Antennulary setae A–D (*sensu* Huys et al., 2007) branched.......................................
.... *C. arctica* (Davis and Green, 1974)

5. Conclusions

The integrative taxonomy approach proved to be useful in solving this particular case among the Monstrilloida by allowing us to: (1) confirm its identity as a member of *Caromiobenella* and (2) reliably match females and males of *C. brasiliensis*. This nominal species has remained as *Monstrilla brasiliensis* for the last 21 years, but the molecular analysis does not support *Caromiobenella* as a monophyletic genus. Therefore, our findings of the males and those from molecular analysis, allowed us to place the Brazilian species in *Caromiobenella,* thus representing the first discovery of this genus in the Southwestern Atlantic Ocean and the first record of monstrilloids from a restricted habitat (i.e., coastal rocky tidepools). This work represents the second successful use of integrative taxonomy (molecular + morphologic) methods to link both sexes of a monstrilloid copepod. *Caromiobenella* now includes 10 species worldwide.

Supplementary Materials: The following are available online at https://www.mdpi.com/article/10.3390/d13060241/s1, Table S1: GenBank sequences used for comparisons with Brazilian copepods sequences (present study) and for the construction of Maximum Likelihood Tree.

Author Contributions: Conceptualization, J.C.L.R., E.S.-M., C.O.D., L.I.W. and L.G.F.; methodology, J.C.L.R., E.S.-M., C.O.D., L.I.W. and L.G.F.; software, J.C.L.R. and L.I.W.; formal analysis, J.C.L.R., E.S.-M., C.O.D. and L.I.W.; investigation, J.C.L.R., E.S.-M., C.O.D., L.I.W. and L.G.F.; writing—original draft preparation, E.S.-M., J.C.L.R. and C.O.D.; writing—review and editing, E.S.-M., J.C.L.R., C.O.D., L.I.W. and L.G.F.; resources, L.G.F. and L.I.W.; supervision, L.G.F. and L.I.W.; project administration, L.G.F.; funding acquisition, L.G.F. All authors have read and agreed to the published version of the manuscript.

Funding: This research and the sample collection was carried out within the scope of the Projeto Costões Rochosos (Project Rocky Shores)/FUNBIO, and was supported by an environmental offset measure established through a Consent Decree/Conduct Adjustment Agreement between Chevron Brazil and the Brazilian Ministry for the Environment, with the Brazilian Biodiversity Fund—FUNBIO as implementer. JCLR received a Ph.D. grant from Project Rocky Shores)/FUNBIO. The Invertebrate Collection of the Institute de Biodiversity and Sustainability (NUPEM/UFRJ) of the Federal University of Rio de Janeiro is supported by the Projects Multipesca and Rocky Shores through "Programa de Apoio à Pesquisa Marinha e Pesqueira—FUNBIO", Brazil.

Institutional Review Board Statement: Not applicable.

Informed Consent Statement: Not applicable.

Data Availability Statement: The data presented in this study are available on request.

Acknowledgments: J.C.L.R. thanks the Project Rocky Shores/FUNBIO for providing a PhD fellowship during this study and to Lucas Lemos Batista, Eduardo Hugo de Morais and other project members for help in field collections. Our gratitude to Bárbara Teixeira Villarins for her valuable help in the molecular analysis.

Conflicts of Interest: The authors declare no conflict of interest. The funders had no role in the design of the study; in the collection, analyses, or interpretation of data; in the writing of the manuscript; or in the decision to publish the results.

References

1. Huys, R.; Llewellyn-Hughes, J.; Conroy-Dalton, S.; Olson, P.D.; Spinks, J.N.; Johnston, D.A. Extraordinary host switching in siphonostomatoid copepods and the demise of the Monstrilloida: Integrating molecular data, ontogeny and antennulary morphology. *Mol. Phylogenet. Evol.* **2007**, *43*, 368–378. [CrossRef]
2. Suárez-Morales, E. Diversity of the Monstrilloida (Crustacea: Copepoda). *PLoS ONE* **2011**, *6*, e22915. [CrossRef]
3. Jeon, D.; Lee, W.; Soh, H.Y. New genus and two new species of monstrilloid copepods (Copepoda: Monstrillidae): Inte-grating morphological, molecular phylogenetic, and ecological evidence. *J. Crust. Biol.* **2018**, *38*, 45–65. [CrossRef]
4. Suárez-Morales, E.; Scardua, M.P.; da Silva, P.M. Occurrence and histopathological effects of *Monstrilla* sp. (Cope-poda: Monstrilloida) and other parasites in the brown mussel Perna perna from Brazil. *J. Mar. Biol. Assoc. U. K.* **2010**, *90*, 953–958. [CrossRef]
5. Grygier, M.J. Identity of Thaumatoessa (= Thaumaleus) typica Krøyer, the first described monstrilloid copepod. *Sarsia* **1993**, *78*, 235–242. [CrossRef]
6. Grygier, M.J.; Suárez-Morales, E. Recognition and partial solution of nomenclatural issues involving copepods of the family Monstrillidae (Crustacea: Copepoda: Monstrilloida). *Zootaxa* **2018**, *4486*, 497–509. [CrossRef] [PubMed]
7. Suárez-Morales, E.; Grygier, M.J. Mediterranean and Black Sea Monstrilloid copepods (Copepoda: Monstrilloida): Rediscovering the diversity of transient zooplankters. *Water* **2021**, *13*, 1–11.
8. Suárez-Morales, E.; Dias, C. Two new species of *Monstrilla* (Copepoda: Monstrilloida) from Brazil. *J. Mar. Biol. Assoc. U. K.* **2000**, *80*, 1031–1039. [CrossRef]
9. Suárez-Morales, E.; Dias, C. A new species of Monstrilla (Crustacea: Copepoda: Monstrilloida) from Brazil with notes on *M. brevicornis* Isaac. *Proc. Biol. Soc. Wash.* **2001**, *114*, 219–228.
10. Suárez-Morales, E.; Dias, C. Taxonomic report of some monstrilloids (Copepoda: Monstrilloida) from Brazil with description of four new species. *Bull. Inst. Royal Sci. Nat. Belg. Biologie.* **2001**, *71*, 65–81.
11. Leite, N.R.; Pereira, L.C.; Abrunhosa, F.; Pires, M.A.; Da Costa, R.M. Occurrence of Cymbasoma longispinosum Bourne, 1890 (Copepoda: Monstrilloida) in the Curuçá River estuary (Amazon Littoral). *An. Acad. Bras. Ciências* **2010**, *82*, 577–583. [CrossRef] [PubMed]
12. Suárez-Morales, E.; Dias, C.O.; Bonecker, S.L. Discovery of the female of *Cymbasoma rochai* Suárez-Morales & Dias, 2001 (Copepoda, Monstrilloida, Monstrillidae), the first Brazilian member of the C. longispinosum species-group. *Crustaceana* **2020**, *93*, 1091–1101. [CrossRef]
13. Grygier, M.J.; Ohtsuka, S. A new genus of monstrilloid copepods (Crustacea) with anteriorly pointing ovigerous spines and related adaptations for subthoracic brooding. *Zool. J. Linn. Soc.* **2008**, *152*, 459–506. [CrossRef]
14. Grygier, M.J.; Ohtsuka, S. Sem Observation of the Nauplius of *Monstrilla hamatapex*, new species, from Japan and an example of Upgraded descriptive standards for monstrilloid copepods. *J. Crust. Biol.* **1995**, *15*, 703. [CrossRef]
15. Jeon, D.; Lee, W.; Soh, H.Y. New species of *Caromiobenella* Jeon, Lee & Soh, 2018 (Crustacea, Copepoda, Monstrilloida) from Chuja Island, Korea. *ZooKeys* **2019**, *814*, 33–51. [CrossRef]
16. Huys, R.; Boxshall, G.A. *Copepod Evolution*; The Ray Society: London, UK, 1991; pp. 1–468.
17. Harris, S.A.; Hoelzel, A.R. Molecular Genetic Analysis of Populations. A Practical Approach. *J. Appl. Ecol.* **1993**, *30*, 197. [CrossRef]
18. Casquet, J.; Thebaud, C.; Gillespie, R.G. Chelex without boiling, a rapid and easy technique to obtain stable amplifiable DNA from small amounts of ethanol-stored spiders. *Mol. Ecol. Resour.* **2012**, *12*, 136–141. [CrossRef] [PubMed]
19. Folmer, O.; Black, M.; Hoeh, W.; Lutz, R.; Vrijenhoek, R. DNA primers for amplification of mitochondrial cytochrome c oxidase subunit I from diverse metazoan invertebrates. *Mol. Mar. Biol. Biotechnol.* **1994**, 3, 294–299.
20. Tamura, S.; Stecher, G.; Peterson, D.; Filipski, A.; Kumar, S. MEGA6: Molecular Evolutionary Genetics Analysis version 6.0. *Mol. Biol. Evol.* **2013**, *30*, 2725–2729. [CrossRef]
21. Raupach, M.J.; Barco, A.; Steinke, D.; Beermann, J.; Laakmann, S.; Mohrbeck, I.; Neumann, H.; Kihara, T.C.; Pointner, K.; Raulovici, A.; et al. The Application of DNA Barcodes for the Identification of Marine Crustaceans from the North Sea and Adjacent Regions. *PLoS ONE* **2015**, *10*, e0139421. [CrossRef]

22. Unal, E.; Frost, B.W.; Armbrust, V.; Kideys, A.E. Phylogeography of *Calanus helgolandicus* and the Black Sea copepod *Calanus euxinus*, with notes on *Pseudocalanus elongatus* (Copepoda, Calanoida). *Deep. Sea Res. Part II Top. Stud. Oceanogr.* **2006**, *53*, 1961–1975. [CrossRef]
23. Suárez-Morales, E. On the male of *Monstrilla mariaeugeniae* Suárez-Morales & Islas-Landeros (Copepoda: Monstrilloida) from the Mexican Caribbean Sea. *Crustaceana* **1998**, *71*, 360–362.
24. Jeon, D.; Lim, D.; Lee, W.; Soh, H.Y. First use of molecular evidence to match sexes in the Monstrilloida (Crustacea: Copepoda), and taxonomic implications of the newly recognized and described, partly Maemonstrilla-like females of *Monstrillopsis longilobata* Lee, Kim & Chang, 2016. *PeerJ* **2018**, *6*, e4938. [CrossRef] [PubMed]
25. Sale, P.F.; McWilliam, P.S.; Anderson, D.T. Composition of the near-reef zooplankton at Heron Reef, Great Barrier Reef. *Mar. Biol.* **1976**, *34*, 59–66. [CrossRef]
26. Suárez-Morales, E. An aggregation of monstrilloid copepods in a western Caribbean reef area: Ecological and conceptual implications. *Crustaceana* **2001**, *74*, 689–696. [CrossRef]
27. Suárez-Morales, E. Taxonomic report on a collection of monstrilloids (Copepoda: Monstrilloida) from Banco Chinchorro, Mexico with description of a new species. *An. Inst. Biol. Univ. Nac. Autón. Mex. Ser. Zool.* **2001**, *72*, 9–28.
28. Dias, C.O.; Bonecker, S.L.C. Study of Monstrilloida distribution (Crustacea: Copepoda) in the Southwest Atlantic. *Panamjas* **2007**, *2*, 270–278.

Article

DNA Barcoding of Penaeidae (Decapoda; Crustacea): Non-Distance-Based Species Delimitation of the Most Economically Important Shrimp Family

Jorge L. Ramirez [1,*], Luisa Simbine [2], Carla G. Marques [3], Eliana Zelada-Mázmela [4], Lorenzo E. Reyes-Flores [4], Adolfo S. López [1], Jaqueline Gusmão [5], Carolina Tavares [6], Pedro M. Galetti, Jr. [3] and Patricia D. Freitas [3]

1 Facultad de Ciencias Biológicas, Universidad Nacional Mayor de San Marcos, Lima 15081, Peru; adolfo.lopez2@unmsm.edu.pe
2 Instituto Nacional de Investigação Pesqueira, Maputo 4603, Mozambique; lsimbine1978@gmail.com
3 Departamento de Genética e Evolução, Universidade Federal de São Carlos, São Carlos 13565-905, SP, Brazil; guinartc@gmail.com (C.G.M.); pmgaletti@ufscar.br (P.M.G.J.); patdf.ufscar@gmail.com (P.D.F.)
4 Laboratorio de Genética, Fisiología y Reproducción, Universidad Nacional del Santa, Nuevo Chimbote 02712, Peru; ezelada@uns.edu.pe (E.Z.-M.); eduardoreyesf@outlook.com (L.E.R.-F.)
5 Departamento de Genética, Universidade Estadual do Rio de Janeiro, Rio de Janeiro 20550-900, RJ, Brazil; gusmao.jaque@gmail.com
6 Instituto de Aplicação Fernando Rodrigues da Silveira, Universidade Estadual do Rio de Janeiro, Rio de Janeiro 20261-232, RJ, Brazil; cr_tavares@hotmail.com
* Correspondence: jramirezma@unmsm.edu.pe

Citation: Ramirez, J.L.; Simbine, L.; Marques, C.G.; Zelada-Mázmela, E.; Reyes-Flores, L.E.; López, A.S.; Gusmão, J.; Tavares, C.; Galetti, P.M., Jr.; Freitas, P.D. DNA Barcoding of Penaeidae (Decapoda; Crustacea): Non-Distance-Based Species Delimitation of the Most Economically Important Shrimp Family. *Diversity* **2021**, *13*, 460. https://doi.org/10.3390/d13100460

Academic Editors: Luc Legal and Manuel Elias-Gutierrez

Received: 31 July 2021
Accepted: 15 September 2021
Published: 23 September 2021

Abstract: The Penaeidae family includes some of the most economic and ecological important marine shrimp, comprising hundreds of species. Despite this importance and diversity, the taxonomic classification for penaeid shrimp has constantly been revised, and issues related to the species identification are common. In this study, we implemented DNA barcoding analyses in addition to single-gene species delimitation analyses in order to identify molecular operational taxonomy units (MOTUs) and to generate robust molecular information for penaeid shrimp based on the cytochrome oxidase subunit I (COI) mitochondrial gene. Our final data set includes COI sequences from 112 taxa distributed in 23 genera of penaeids. We employed the general mixed Yule coalescent (GMYC) model, the Poisson tree processes (PTP), and the Bayesian PTP model (bPTP) for MOTUs delimitation. Intraspecific and interspecific genetic distances were also calculated. Our findings evidenced a high level of hidden diversity, showing 143 MOTUs, with 27 nominal species not agreeing with the genetic delimitation obtained here. These data represent potential new species or highly structured populations, showing the importance of including a non-distance-based species delimitation approach in biodiversity studies. The results raised by this study shed light on the Penaeidae biodiversity, addressing important issues about taxonomy and mislabeling in databases and contributing to a better comprehension of the group, which can certainly help management policies for shrimp fishery activity in addition to conservation programs.

Keywords: penaeids; COI; GMYC; PTP; hidden diversity

1. Introduction

Occurring in all oceans, especially in tropical and subtropical regions, the Penaeidae family includes some of the most important marine shrimp, comprising, up until 2020, 32 genera with 224 species [1], some of which are considered the crustaceans of most major economic importance in the world [2–4]. The world production of shrimp, adding catches and shrimp farming, represents the most important fish product traded internationally in terms of value. While catches of shrimp reached new records in recent years, the world aquaculture production of crustaceans in 2018, for instance, consisted of 9.4 million tons, representing 11.4% of the world total aquaculture production of aquatic animals [3]. In several tropical developing countries, shrimp fishing represents the most valuable export product and an important employment-generating activity [5].

Due to their great commercial value, many species of the penaeid group have been economically overexploited, especially in tropical regions [3,5–11]. Such overexploitation can led to a marked decline in their natural stocks, promoting disruption of the marine environment where it occurs by affecting important ecological functions and ecosystem services, causing changes in competition and predation, loss of spawning biomass, removal of juveniles, reduction in water and habitat quality, modification of species composition and interaction, potential local extinctions, and decreasing biodiversity [12,13].

The implementation of management actions appropriate to regional realities can lead to the recovery of fishing stocks, as shown in the study that evaluated the effect of management reforms (2013–2017 period) aligned with the FAO Code of Conduct for Responsible Fisheries 2013, on Colombian shrimp stocks, employing fisheries' performance indicators. The results revealed that a regulatory reform implementation improved ecological performance by increasing stock size and decreasing bycatch, also showing positive social outcomes [14]. In India, a study conducted to evaluate the trends in penaeid shrimp landings for a period of approximately ten years (2001–2010) suggested the restriction of fishing efforts to ensure sustainability of this resource [9].

Overexploitation and depletion of key penaeid species can negatively impact higher trophic levels with a possible erosion of other fishery resources. Regional studies, such as the analysis of the population dynamics of the commercial species *Parapenaeopsis coromandelica* from the coastal waters of Teluk Penyu in Indonesia, show that the rate of exploitation per year (E) for females and males is above the sustainable level (E = 0.5) [15]. The presence of the Atlantic seabob *Xiphopenaeus kroyeri*, in high densities in coastal waters of Suriname, for example, plays an important role for trophic ecology, since this species is the main high-density epibenthic organism found up to 30 m deep, acting as a vector for energy from intertidal primary production to secondary subtidal production and serving as prey for fish species [16]. In addition, recent genetic studies show that the number of penaeid species is still underestimated, evidencing hidden cryptic diversity and taxonomic inconsistencies [1]. Delimitation analysis of penaeid shrimps from Southeast Asia indicated 94 putative species within 71 recognized species, including *Kishinouyepenaeopsis cornuta*, *K. incisa*, *Mierspenaeopsis sculptilis*, *M. hardwicki*, *Parapenaeopsis coromandelica*, and the giant tiger prawn *Penaeus monodon* [1]. These results show that even known species of great commercial importance for fishing or shrimp farming, such as *P. monodon*, can cover up the existence of several cryptic species, hindering the implementation of effective measures for the management and conservation of natural stocks and/or of genetic-based selective breeding programs [17].

Despite the ecological and economic importance of penaeid shrimp, the taxonomic classification for this group has constantly been revised, mainly because of some disagreements regarding the morphological characters and their related adopted criteria [4,18]. Morphological distinctions among some penaeid species are often quite subtle, especially involving close species that are distinguishable only due to slim differences in genitalia [2]. Consequently, penaeid species delimitation frequently requires a high level of knowledge and training on specific diagnostic characteristics. Thereby, some approaches based on molecular analyses, such as DNA barcoding, have been proposed as an efficient alternative tool to aid species identification of several crustaceans [19,20], including crabs [21,22], lobsters [23], and shrimp and prawn [24–26]. Within penaeids, DNA barcoding has been applied to identify species from a specific site [1,26], discriminating cryptic species [18,27–29], identifying juveniles and larvae individuals [30], and characterizing a given genus, as reported for *Metapenaeopsis* [25].

The implementation of the DNA barcoding can effectively contribute to a prompt identification of penaeid shrimp, decreasing or even avoiding misidentification and mislabeling products [31,32]. This approach can also disclose hidden diversity [33], revealing lineages or pointing out new species [18] that would be eventually managed inadequately if they remained unknown. Overall, in this study, we employed the DNA barcoding approach combined with three non-distance-based single-gene species delimitation analyses in order to identify consensus MOTUs (molecular operational taxonomy units) and to test the utility of this combined approach to reveal hidden diversity. Our hypothesis is that there is a degree of hidden diversity in the family Penaeidae that could be detected using species delimitation methods. Our study analyzed COI sequences for an expressive number of penaeid genera and species around the world, highlighting relevant information for this important economic and ecological resource.

2. Materials and Methods

2.1. Sampling

Sampling was performed following all legal requirements determined by the governmental laws of each country (Brazil, Mozambique, and Peru) for the ethical fishery of marine shrimp stocks. In total, we sampled 114 specimens from 18 nominal species, i.e., taxonomically valid species, that were collected in Southeast Pacific (5), Southwest Atlantic (7), and Southeast Indic (6) oceans (Table S1), comprising nine genera. Fragments of muscle tissue (about 1–2 cm^3) were fixed in 96% ethanol kept at 4 °C until DNA extraction procedures. Firstly, an initial species identification was performed by the visualization of morphological traits following taxonomic references for penaeid shrimp recognition at the species level [2,34–39]. Voucher information for the sampled species is provided in the Supplementary Material (Table S1).

Additionally, 596 DNA barcode sequences were obtained for 108 nominal species from 23 genera through data downloaded from the Barcode of Life Data System (BOLD; available at http://www.boldsystems.org/, accessed on March 2021). BOLD data were filtered by deleting data for individuals with dubious species identification, in which a single specimen was clustered with several individuals from the non-corresponding species in a preliminary phylogenetic clustering before implementing subsequent deeper analyses.

Our final data set includes a total of 710 sequences for 112 nominal species from 24 genera of penaeids (Table S1), distributed as follows: *Alcockpenaeopsis* (1), *Artemesia* (1), *Batepenaeopsis* (1), *Farfantepenaeus* (9), *Fenneropenaeus* (5), *Funchalia* (1), *Ganjampenaeopsis* (1), *Kishinouyepenaeopsis* (5), *Litopenaeus* (5), *Marsupenaeus* (2), *Megokris* (3), *Melicertus* (6), *Metapenaeopsis* (23), *Metapenaeus* (12), *Mierspenaeopsis* (2), *Parapenaeopsis* (3), *Parapenaeus* (16), *Penaeopsis* (2), *Penaeus* (3), *Rimapenaeus* (1), *Sicyonia* (1), *Trachypenaeopsis* (1), *Trachysalambria* (4), and *Xiphopenaeus* (4). Additionally, five species were included as outgroups: *Acetes chinensis* (Sergestidae), *Aristeus mabahissae* (Aristeidae), *Benthonectes filipes* (Benthesicymidae), *Robustosergia robusta* (Sergestidae), and *Solenocera crassicornis* (Solenoceridae). This dataset includes a broad sampling of Penaeidae species that account for most of the family distribution.

2.2. DNA Extraction, Amplification, and Sequencing

Total DNA was extracted using a standard phenol-chloroform method based on the protocol proposed by Sambrook et al. [40]. Fragments of the cytochrome oxidase subunit I (COI) mitochondrial gene were amplified through the polymerase chain reaction (PCR). For the species from the Southwest Atlantic and Southeast Indic oceans, we used a set of primers specifically designed for penaeid group using the Primer 3 software [41], the forward (F) and reverse (R) primer pair: COIPenF2 (3′-AGATTTACAGTCTATCGCCTA-5) and COIPenR (3′- ATACCAAATACRGCTCCYATTGA-5′). PCR was carried out on a final volume of 30 μL, using 200 μM dNTPs, 1X PCR buffer, 0.3 μM of each primer, 2.5 mM MgCl2, and 1U of Taq polymerase; using a Veriti™ Thermal Cycler (Applied Biosystems) programmed according to the following parameters: 35 cycles at 94 °C for 50 s, 51 °C for 80 s, and 72 °C for 60 s. For the species from the Southeast Pacific Ocean, we used the pair of primers LCO1490 and HCO2198 [42]. PCR was carried out on a final volume of 18 μL, using 125 μM dNTPs, 1X PCR buffer, 0.25 μM of each primer, 2 mM $MgCl_2$, and 1U of Taq polymerase, using a Veriti™ Thermal Cycler (Applied Biosystems), following the parameters: 35 cycles at 95 °C for 50 s, 49 °C for 50 s, and 72 °C for 70 s.

PCR products were purified using the PEG (polyethylene glycol 20%) protocol [43] and then COI amplicons were Sanger-sequenced for both strands using an ABI3730XL automatic sequencer (Applied Biosystems, Foster City, CA, USA). The obtained sequences were visualized and manually edited using the software Bioedit [44]. Stop codons and indels were checked, and low-quality regions were deleted. Sequence data were deposited in both public databases: GenBank (https://www.ncbi.nlm.nih.gov/genbank/) from the National Center for Biotechnology Information (NCBI) and BOLD (http://www.boldsystems.org/). GenBank accession, and BOLD record numbers for the sequences analyzed here are shown in Table S1.

2.3. DNA Barcoding and MOTU Delimitation Analyses

For the MOTUs investigation, we employed three species delimitation methods: the general mixed Yule coalescent (GMYC) model with a single threshold [45], the Poisson tree processes (PTP), and the Bayesian implementation of the PTP model (bPTP) [46]. As input for these methods, firstly an ultrametric tree was generated using Beast 2.6 [47], with a log normal relaxed clock, a birth and death model, and a GTR+I+G substitution model chosen by jModelTest 2 [48], with 200 million MCMC generations, sampled every 30,000 iterations, and a burn-in of 10%. Convergence and adequate sample size (greater than 200) were evaluated in Tracer v. 1.7 [49]. The different delimitation outputs were compared using the pipeline SPdel (https://github.com/jolobito/SPdel, accessed on July 2021) that generates a consensus delimitation (Consensus MOTUs) and provides image visualizations. Additionally, intraspecific and interspecific genetic distances, with a K2p substitution model, were calculated for nominal species and consensus MOTUs using SPdel as well.

3. Results

The alignment of COI sequences resulted in 609 base pair fragments, with 256 parsimony informative sites without gaps. The single-gene species delimitation analyses evidenced 144 MOTUs ($p < 0.0001$) for the GMYC, 143 MOTUs for the PTP, and 142 MOTUs for the bPTP analyses (Figures 1–5). SPdel summarized the previous results in 143 consensus MOTUs, of which only 85 matches with valid nominal species (Figures 1–5). The mean intra-group distances, the maximum intra-group distance, the nearest neighbor (NN), and the NN's minimum distance for both consensus MOTUS and nominal species are shown in Supplementary Material (Table S2). The overall mean of intraspecific distances was 1.3%, the maximum intraspecific distance was 19.7% (*Trachysalambria curvirostris*), and the minimum interspecific distance was 0% (*Farfantepenaeus duorarum*, *Farfantepenaeus notialis*, *Melicertus latisulcatus*, *Melicertus plebejus*, *Metapenaeopsis palmensis*, *Metapenaeopsis provocatoria*, *Metapenaeopsis toloensis*, and *Metapenaeopsis velutina*). No barcode gap was found, and the intraspecific distance for 16 nominal species was higher than the interspecific one (Table S2).

For consensus MOTUs, the overall intra-MOTU distances were 0.27%, the maximum intra-MOTU distance was 2.18% for *Trachysalambria curvirostris* from China and Egypt (Table S3), and the minimum inter-MOTU distance was 1% for both MOTUs of *Farfantepenaeus isabelae* (Table S3). Few MOTUs (two for consensus MOTUs) whose intra-MOTU distance was slightly higher than inter-MOTU distance were found and no barcode gaps were observed using any delimitation method (Table S3).

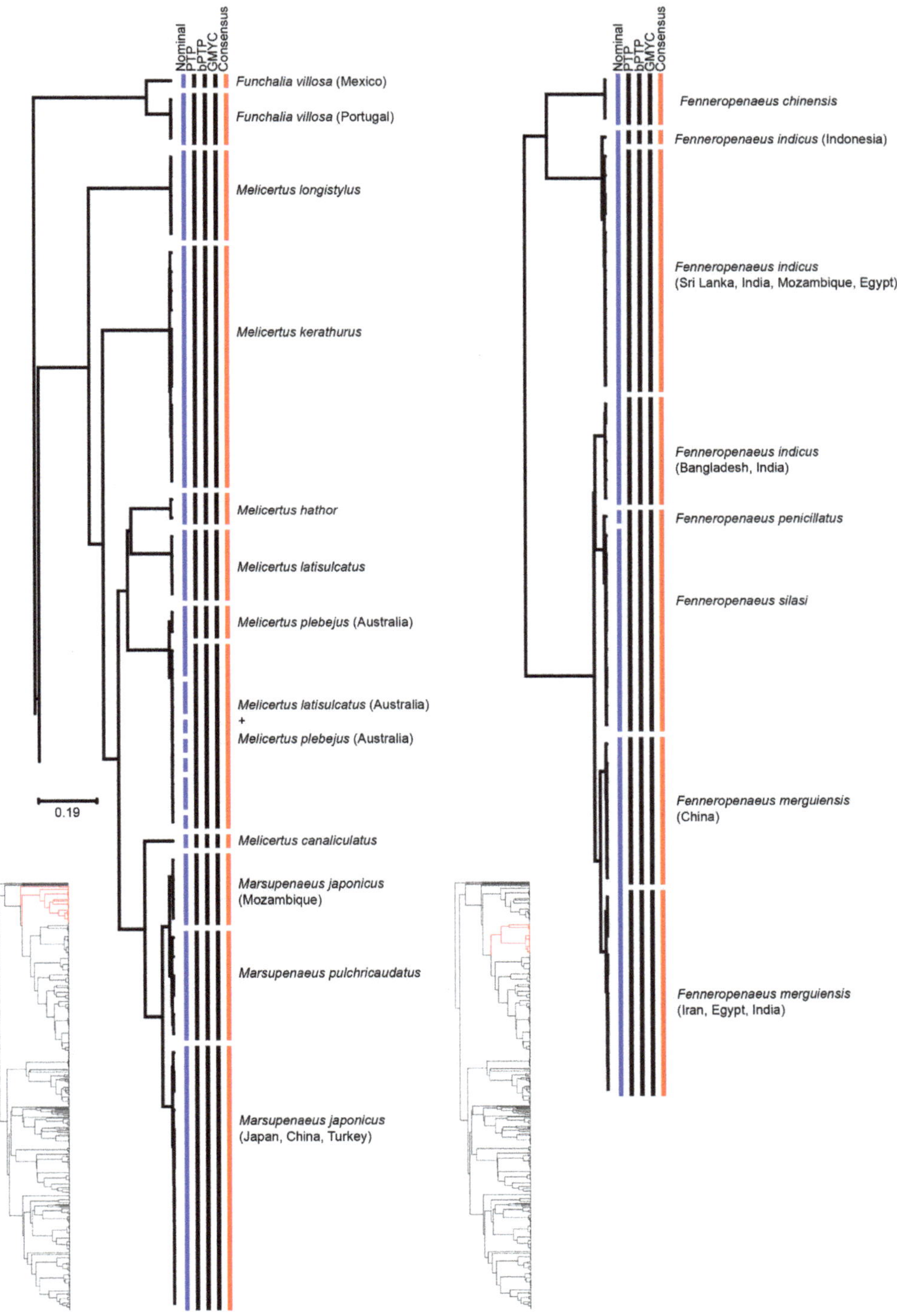

Figure 1. Bayesian tree showing the clustering of the MOTUs obtained by the species delimitation analyses for the genera *Fenneropenaeus*, *Funchalia*, *Marsupenaeus*, and *Melicertus*.

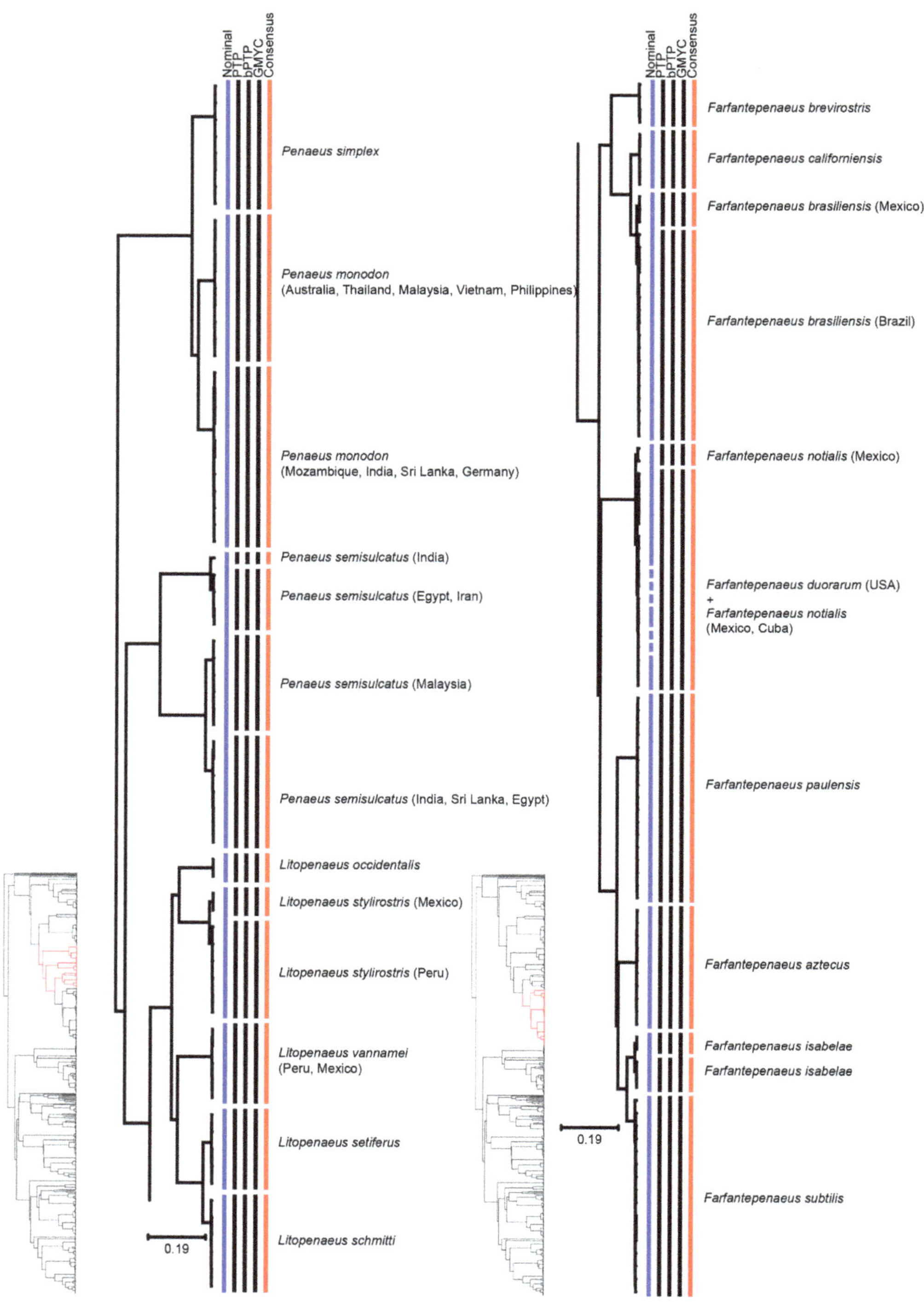

Figure 2. Bayesian tree showing the clustering of the MOTUs obtained by the species delimitation analyses for the genera *Farfantepenaeus*, *Litopenaeus*, and *Penaeus*.

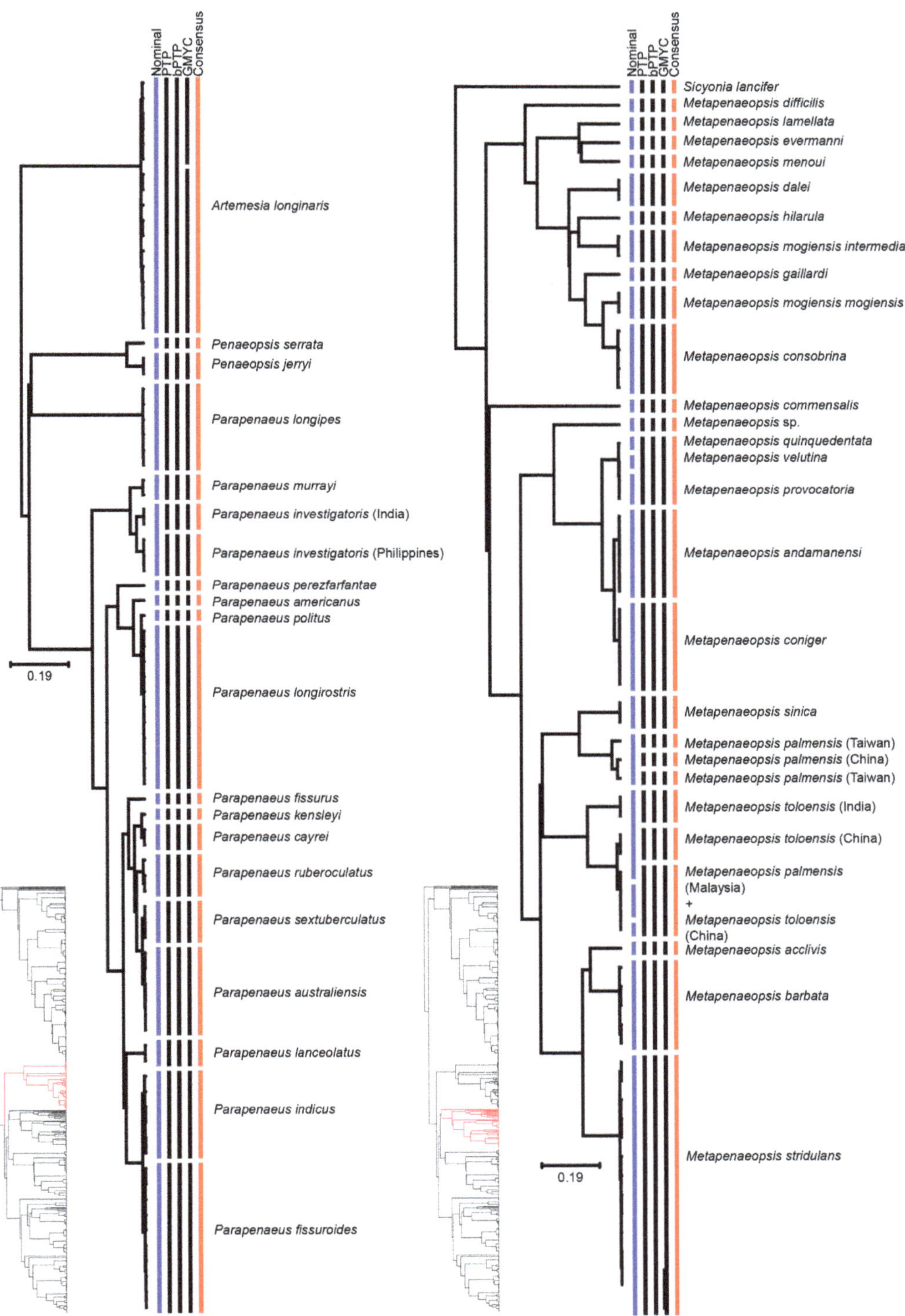

Figure 3. Bayesian tree showing the clustering of the MOTUs obtained by the species delimitation analyses for the genera *Artemesia*, *Parapenaeus*, *Penaeopsis*, *Metapenaeopsis*, and *Sicyonia*.

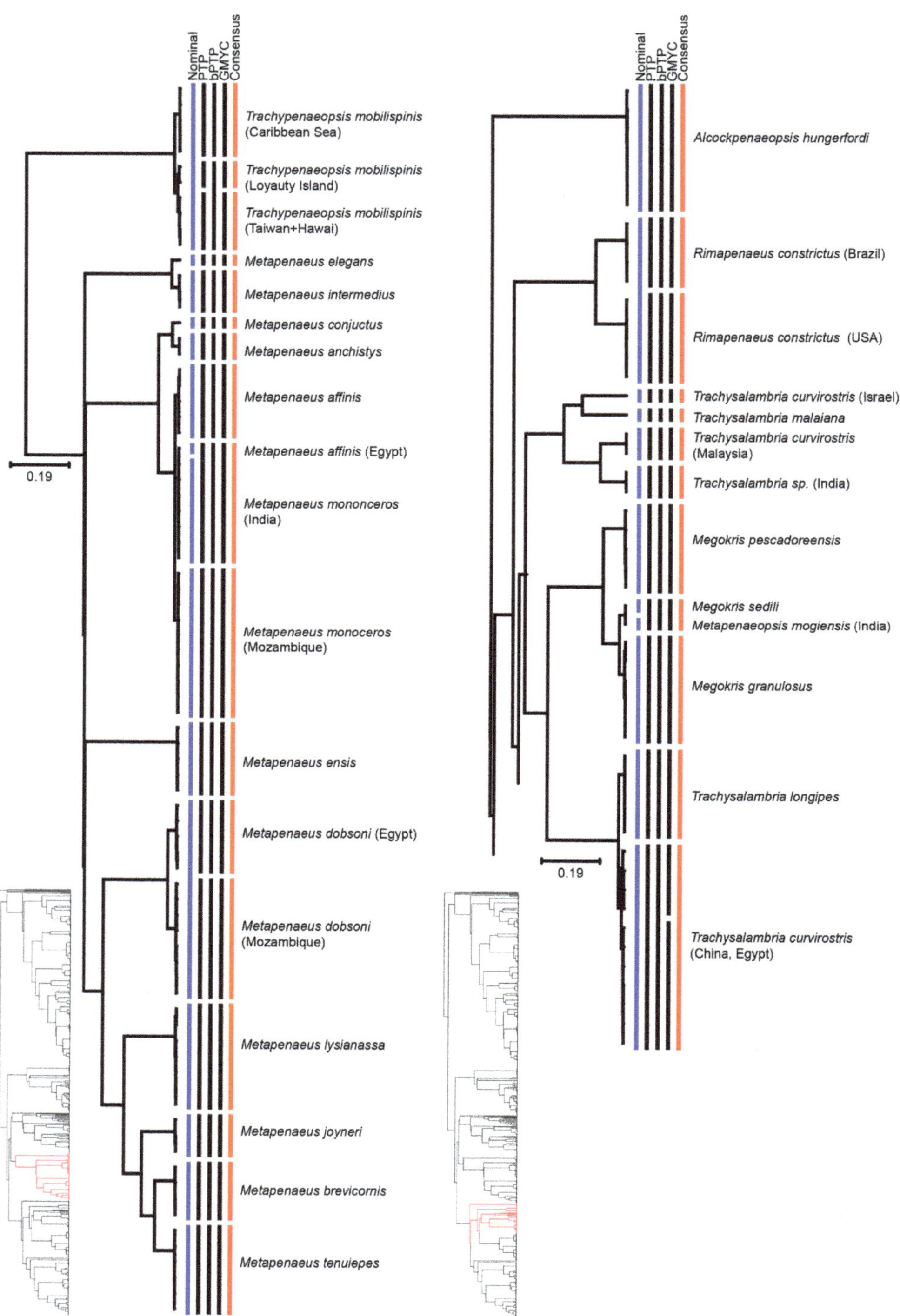

Figure 4. Bayesian tree showing the clustering of the MOTUs obtained by the species delimitation analyses for the genus *Alcockpenaeopsis*, *Megokris*, *Metapenaeus*, *Rimapenaeus*, *Trachypenaeopsis*, and *Trachysalambria*.

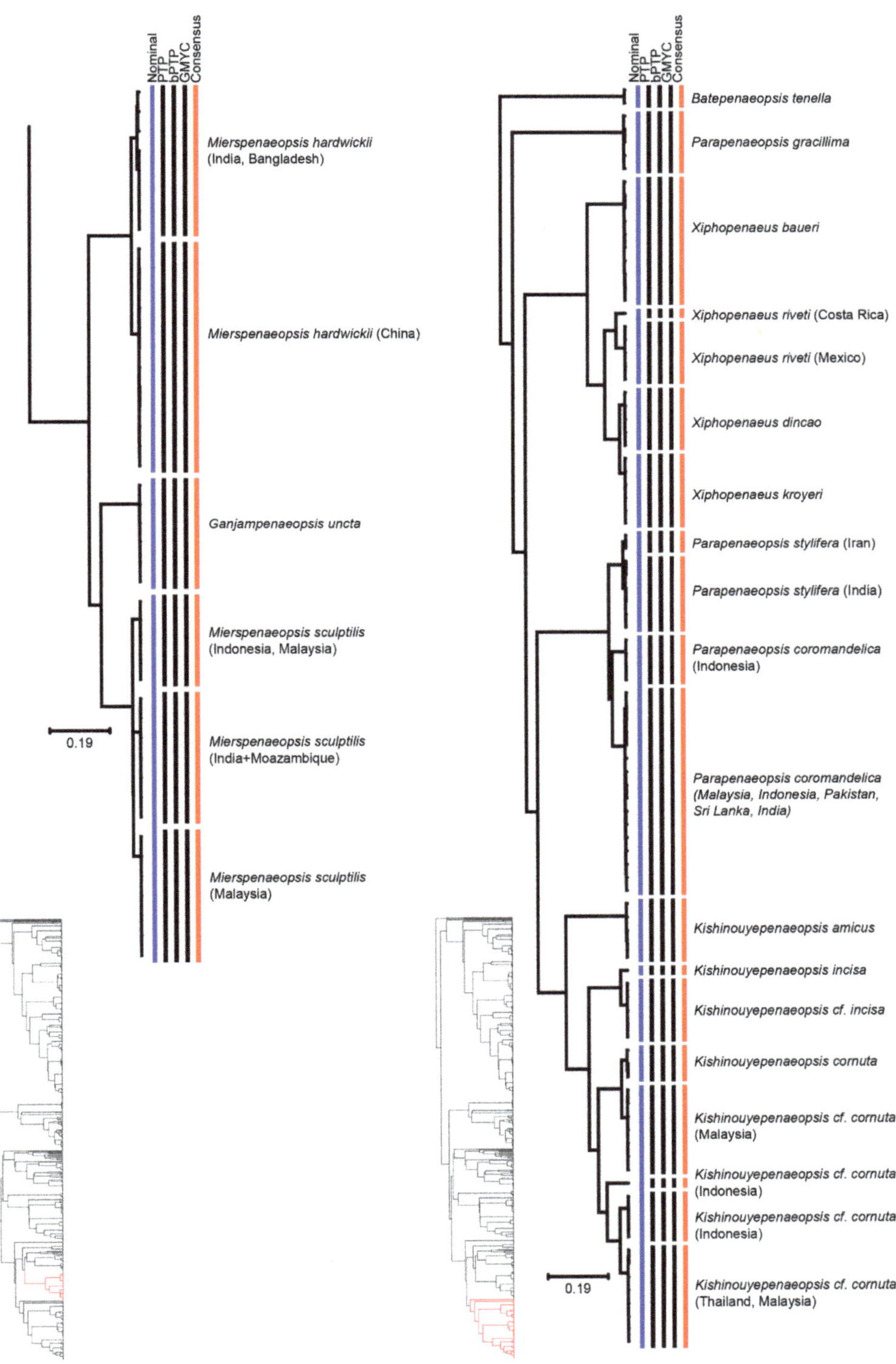

Figure 5. Bayesian tree showing the clustering of the MOTUs obtained by the species delimitation analyses for the genera *Batepenaeopsis*, *Ganjampenaeopsis*, *Mierspenaeopsis*, *Kishinouyepenaeopsis*, *Xiphopenaeus*, and *Parapenaeopsis*.

4. Discussion

Our integrated approach, combining DNA barcoding with non-distance-based single-gene species delimitation methods, was efficient in raising some issues and pointing inconsistences out for the penaeids analyzed herein. These methods are advantageous because they are independent of a distance criterion (cut-off threshold value) and do not require prior delimitation. It is known that such strategy can constitute a powerful tool for MOTU delimitation, aiding the knowledge about species diversity in different taxa [1,28,50–52]. Here, we present a meaningful COI dataset for different species from the most commercially important shrimp family, highlighting novelties on the penaeid biodiversity, including 114 new records for little-studied areas, and demonstrating the effectiveness of species delimitation (Consensus MOTUs) to accelerate the study of biodiversity.

Despite of the economic importance of the Penaeidae group, our findings evidenced a high level of hidden diversity, showing 143 MOTUs distributed in 112 nominal species, with 27 nominal species not agreeing with the genetic delimitation obtained here. These data represent potential new species or highly structured populations that have probably not been managed or protected adequately by the existing fishery policies legislation. The degree of hidden diversity found herein (143 MOTUs in 112 species, 27.6%) is similar to that found for penaeids from South-East Asian waters (94 MOTUs in 71, 32%) [1]; however, the methodology used herein is based on three different coalescent species delimitation methods and COI barcoding region, while the latter used ABGD (a species delimitation method based on distance) and bPTP in two different COI regions. Such a high level of hidden diversity likely reflects the already-mentioned difficulties in identifying and discriminating penaeid shrimp species using only morphological characters [2,18,34]. Moreover, for some penaeid genus, the species identification is commonly based on the genitalia morphology of adults, requiring a high level of expertise to correctly identify the species [2,18,34]. This fact usually implies misidentification, compromising the data reliability available in public databases (e.g., GenBank or BOLD) as observed for *Metapenaeopsis mogiensis*. Our results showed this species to be polyphyletic, with three unrelated groups, including one species more related to *Megorkis sedili* than *Metapenaeopsis* (Figure 4). These findings at least indicate that a revision of the vouchers reported in the BOLD database is needed to determine the correct identification of these specimens.

The potential presence of cryptic species also challenges the correct discrimination of species with similar morphology. For *Melicertus plebejus* and *M. latisulcatus*, for example, we observed three MOTUs (Figure 1), two of them including only specimens from one nominal species and a third one including specimens of both species. Indeed, these penaeids share a similar morphology and coloration, and they have been considered sister species [53]. In this sense, we observed this third MOTU to be more related to the MOTU including only *M. plebejus* (Figure 1), likely representing a potential cryptic species hidden by a resembling morphology.

A similar case encompasses *Farfantepenaeus duorarum* and *F. notialis* (Figure 2), which was found herein distributed in two MOTUs, one including only *F. notialis* specimens; however, a different clade clustered both *F. notialis* and *F. duorarum* species. The minimum inter-MOTU genetic distance between these groups was 1.69%. In fact, *F. notialis* and *F. duorarum* are morphologically very similar, *F. notialis* being initially described as a subspecies of *F. duorarum* [4]. In previous molecular studies, the validity of these species was questioned due to the low genetic distance and the lack of reciprocal monophyly [54]. Our results, using species delimitation methods, supports the existence of two MOTUs, suggesting the existence of different species but not supporting the current taxonomy identification of *F. notialis* and *F. duorarum* sampling. In this sense, a study including samples from the entire geographical distribution of both species and conclusive taxonomic diagnose is still necessary to correctly delimitate these taxa.

The current nominal valid species subdivided here in two or more MOTUs, correlated with the geography and with high intra-MOTU genetic distance, may include potential cryptic species, indicating therefore the need of further taxonomic studies [55]. This is the case observed in *Funchalia villosa*, *Litopenaeus stylirostris*, *Metapenaeopsis toloensis*,

Mierspenaeopsis hardwickii, *Parapenaeus investigatoris*, and *Rimapenaeus constrictus* (Table 1). Overall, an integrative taxonomic approach, also including broader sampling, is imperative to understand the meaning of the findings raised here for these species.

Table 1. Penaeid species exhibiting more of one MOTUs.

Species	Consensus MOTUs	Previous Genetic Studies	References
Farfantepenaeus brasiliensis	MOTU 1: South West Atlantic MOTU 2: North West Atlantic	South West Atlantic North West Atlantic	[54]
Farfantepenaeus isabelae	MOTU 1: Brazil MOTU 2: Brazil	One single clade	[54]
Fenneropenaeus indicus	MOTU 1: West Indian MOTU 2: East Indian Ocean	West Indian East Indian Ocean	[56,57]
Fenneropenaeus merguiensis	MOTU 1: China Sea MOTU 2: Andaman Sea, Arabian Sea, and Persian Gulf	China Sea Andaman Sea	[1,58,59]
Funchalia villosa	MOTU 1: Gulf of Mexico MOTU 2: Portugal	- -	- -
Kishinouyepenaeopsis cf. *cornuta*	MOTU1: Indian Ocean MOTU 2: Strait of Malacca MOTU 3: Andaman Sea and Strait of Malacca MOTU 4: South China Sea	Indian Ocean Strait of Malacca Andaman Sea and Strait of Malacca South China Sea	[1]
Litopenaeus stylirostris	MOTU 1: Mexico MOTU 2: Peru	- -	- -
Marsupenaeus japonicus	MOTU 1: Japan, China, Turkey MOTU 2: Mozambique	Japan Mozambique	[60]
Metapenaeopsis palmensis	MOTU 1: China MOTU 2: Taiwan MOTU 3: Taiwan	South China Sea East China Sea	[1]
Metapenaeopsis toloensis	MOTU 1: India MOTU 2: China	- -	- -
Metapenaeopsis toloensis/M. palmensis	MOTU: Malaysia, China	-	-
Metapenaeus dobsoni	MOTU 1: Mozambique MOTU 2: North West Indian	Mozambique North West Indian	[60]
Metapenaeus monoceros	MOTU 1: East Indian MOTU 2: West Indian Ocean	East Indian West Indian Ocean	[57]
Mierspenaeopsis hardwickii	MOTU 1: China MOTU 2: India and Bangladesh	- -	- -
Parapenaeus investigatoris	MOTU 1: Philippines MOTU 2: India	- -	- -
Penaeus monodon	MOTU 1: Australia MOTU 2: Mozambique, India, and Sri Lanka	South-East Africa South and South-East Asia	[57]
Penaeus semisulcatus	MOTU 1: Malaysia MOTU 2: India MOTU 3: Egypt and Iran MOTU 4: India, Sri Lanka, and Egypt	Indian Ocean Indian Ocean, South China Sea, Strait of Malacca Indian Ocean Strait of Malacca, South China Sea, Andaman Sea, Celebes Sea	[29]
Rimapenaeus constrictus	MOTU 1: Brazil MOTU 2: USA	- -	- -
Trachypenaeopsis mobilispinis	MOTU 1: Pacific MOTU 2: Loyauty Island MOTU 3: Caribbean Sea	Pacific Loyauty Island Caribbean Sea	[61]
Trachysalambria curvirostris	MOTU 1: Israel MOTU 2: China and Egypt MOTU 3: China	Yellow Sea South China Sea	[62]
Xiphopenaeus riveti	MOTU 1: Costa Rica MOTU 2: Mexico	Costa Rica Mexico	[34]

Despite such novelties, the species delimitation methods employed herein supported taxonomic uncertainties previously reported in the literature for ten species (Table 1) as follows: *Farfantepenaeus brasiliensis* [54], *Fenneropenaeus indicus* [56,57], *Fenneropenaeus merguiensis* [1], *Kishinouyepenaeopsis* cf. *cornuta* [1], *Marsupenaeus japonicus* [1], *Metapenaeus dobsoni* [60], *Metapenaeus monoceros* [57], *Penaeus semisulcatus* [1], *Trachypenaeopsis mobilispinis* [61], and *Xiphopenaeus riveti* [34]. Additionally, our species delimitation analyses supported molecular groups distinct from the molecular studies prior reported for four species (Table 1): *Farfantepenaeus isabelae* [54], *Metapenaeopsis palmensis* [1], *Penaeus monodon* [57], and *Trachysalambria curvirostris* [62]. Some of these species, such as *T. curvirostris*, are considered as morphological species complex [63].

For *Metapenaeopsis provocatoria*, *M. velutina* and *M. quinquedentate* the analyses evidenced interesting results grouping these species within a single MOTU, with a maximum intra-MOTU distance of 0.49%. As discussed by Cheng et al. [25], the three nominal species are morphologically distinguishable, but some morphological traits might vary depending on the environmental conditions. In this way, an integrative study considering more representative sampling using a larger number of molecular markers is still necessary to address the taxonomy status of these species. Additionally, the possibility of misidentification of these samples should be explored.

In sum, our findings state that the family Penaeidae still holds a large unknown diversity that was revealed here after combining the DNA barcoding approach with robust species delimitation methods, showing the importance of including this approach in biodiversity studies. The data raised by this study shed light on the penaeid biodiversity, addressing important issues about taxonomy and mislabeling in databases and contributing to a better comprehension into the group that can certainly help management policies for shrimp fishery activity, in addition to conservation programs.

Supplementary Materials: The following are available online at https://www.mdpi.com/article/10.3390/d13100460/s1, Table S1: Sampling information, BOLD process ID and GenBank accession for specimens included in the analysis. Table S2: Genetic K2P distances of Penaeidae species. The mean and the maximum of intra-group distances, the nearest neighbor (NN), and the minimum distance to the NN. Table S3: Genetic K2P distances of Penaeidae MOTUs. The mean and the maximum of intra-group distances, the nearest neighbor (NN), and the minimum distance to the NN.

Author Contributions: Conceptualization, J.L.R., P.M.G.J. and P.D.F.; methodology, J.L.R., L.S., C.G.M., E.Z.-M. and L.E.R.-F.; formal analysis, J.L.R. and A.S.L.; data curation, J.L.R. and A.S.L.; writing—original draft preparation, J.L.R.; writing—review and editing, J.L.R., J.G., C.T., P.M.G.J. and P.D.F. All authors have read and agreed to the published version of the manuscript.

Funding: This study was funded by Concytec/Fondecyt (Círculos de Investigación 023-2016) and DNA Barcoding Brazilian Network. Marine Organisms-Molecular Identification of Biodiversity (MCT/CNPq/FNDCT-50/2010). P.M.G.J. and P.D.F. thank Conselho Nacional de Desenvolvimento Científico e Tecnológico (CNPq, 303524/2019-7and 304477/2018-4, respectively) and Centro de Oceanografia Integrada INCT Mar COI (MCT/CNPq/FNDCT/FAP-71/2010, Institutos Nacionais de Ciência e Tecnologia). P.D.F. thank São Paulo State Research Support Foundation (Fapesp 2011/21384-6; 2012/17322-8). C.G.M. and L.S. received grants from Conselho Nacional de Desenvolvimento Científico e Tecnológico (CNPq), Brazil.

Institutional Review Board Statement: Not applicable.

Data Availability Statement: All sequences files are available from the BOLD database (dataset DS-PENAEID) and GenBank (Table S1).

Acknowledgments: We thank Lesly Llaja Salazar for assisting in the species identification. We also thank Allysson Pontes Pinheiro, Ana Karina de Francisco, Artur de Lima Preto, Eloize Luvesuto, Irina Álvarez Llaque, Joaquim Blanco, Luis Enrique Santos Rojas, and Thiago Buosi Silva for helping with biological sampling and sample processing.

Conflicts of Interest: The authors declare no conflict of interest.

References

1. Hurzaid, A.; Chan, T.; Nor, S.A.M.; Muchlisin, Z.A.; Chen, W. Molecular phylogeny and diversity of penaeid shrimps (Crustacea: Decapoda) from South-East Asian waters. *Zool. Scr.* **2020**, *49*, 596–613. [CrossRef]
2. Dall, W.; Hill, B.J.; Rothlisberg, P.C.; Sharples, D.J. *The Biology of the Penaeidae*; Academic Press: Cambridge, MA, USA, 1990.
3. FAO. *The State of World Fisheries and Aquaculture 2018-Meeting the Sustainable Development Goals*; FAO: Rome, Italy, 2018; pp. 1–227.
4. De Grave, S.; Fransen, C.H.J.M. Carideorum Catalogus: The Recent Species of the Dendrobranchiate, Stenopodidean, Procarididean and Caridean Shrimps (Crustacea: Decapoda). *Zoologische Mededelingen* **2011**, *9*, 85.
5. De Francisco, A.K.; Pinheiro, A.P.; Silva, T.; Galetti, P.M., Jr. Isolation and characterization of microsatellites in three overexploited penaeid shrimp species along the Brazilian coastline. *Conserv. Genet.* **2009**, *10*, 563–566. [CrossRef]
6. Gillert, R. *Global Study of Shrimp Fisheries*; FAO: Rome, Italy, 2008.
7. Borrell, Y.; Espinosa, G.; Romo, J.; Blanco, G. DNA microsatellite variability and genetic differentiation among natural populations of the Cuban white shrimp *Litopenaeus schmitti*. *Mar. Biol.* **2003**, *144*, 327–333. [CrossRef]
8. D'Incao, F.; Valentini, H.; Rodrigues, L.F. Avaliação da pesca de camarões nas regiões sudeste e sul do Brasil. *Atlantica* **2002**, *40*, 103–116.
9. Maheswarudu, G.; Sreeram, M.P.; Dhanwanthari, E.; Varma, J.B.; Sajeev, C.K.; Rao, S.S.; Rao, K.N. Trends in penaeid shrimp landings by sona boats at Visakhapatnam Fishing Harbour, Andhra Pradesh. *Indian J. Fish.* **2018**, *65*, 58–65. [CrossRef]
10. Galindo-bect, M.S.; Glenn, E.P.; Page, H.M.; Fitzsimmons, K.; Galindo-Bect, L.A.; Hernandez-Ayon, J.M.; Petty, R.L.; Garcia-Hernandez, J.; Moore, D. Penaeid shrimp landings in the upper Gulf of California in relation to Colorado River freshwater discharge. *Fish Bull* **2000**, *98*, 222–225.
11. Hartono, S.; Riani, E.; Yulianda, F.; Puspito, G. Pemanfaatan Sumber Daya Udang Penaeid di Teluk Ciletuh, Palabuhanratu Berdasarkan Analisis Kesesuaian Kawasan. *J. Ilmu Teknol. Kelaut. Trop.* **2020**, *12*, 195–209. [CrossRef]
12. Blaber, S.J.M.; Cyrus, D.P.; Albaret, J.-J.; Ching, C.V.; Day, J.W.; Elliott, M.; Fonseca, M.S.; Hoss, D.E.; Orensanz, J.; Potter, I.C.; et al. Effects of fishing on the structure and functioning of estuarine and nearshore ecosystems. *ICES J. Mar. Sci.* **2000**, *57*, 590–602. [CrossRef]
13. Pauly, D.; Christensen, V.; Guénette, S.; Pitcher, T.J.; Sumaila, U.R.; Walters, C.J.; Watson, R.A.; Zeller, D. Towards sustainability in world fisheries. *Nature* **2002**, *418*, 689–695. [CrossRef]
14. Marco, J.; Valderrama, D.; Rueda, M. Evaluating management reforms in a Colombian shrimp fishery using fisheries performance indicators. *Mar. Policy* **2021**, *125*, 104258. [CrossRef]
15. Suradi, W.S.; Solichin, A.; Taufani, W.T.; Djuwito; Sabdono, A. Population dynamics of exploited species west shrimps *Parapenaeopsis coromandelica* H. Milne. Edwards 1837 from the Teluk Penyu coastal waters, Indonesian ocean. *Egypt. J. Aquat. Res.* **2017**, *43*, 307–312. [CrossRef]
16. Willems, T.; De Backer, A.; Kerkhove, T.; Dakriet, N.N.; De Troch, M.; Vincx, M.; Hostens, K. Trophic ecology of Atlantic seabob shrimp Xiphopenaeus kroyeri: Intertidal benthic microalgae support the subtidal food web off Suriname. *Estuarine Coast. Shelf Sci.* **2016**, *182*, 146–157. [CrossRef]
17. Yudhistira, A.; Arisuryanti, D.T. Preliminary findings of cryptic diversity of the giant tiger shrimp (*Penaeus monodon* Fabricius, 1798) in Indonesia inferred from COI mitochondrial DNA. *Genettika* **2019**, *51*, 251–260. [CrossRef]
18. Tavares, C.; Gusmao, J. Description of a new Penaeidae (Decapoda: Dendrobranchiata) species, *Farfantepenaeus isabelae* sp. nov. *Zootaxa* **2016**, *4171*, 505–516. [CrossRef]
19. Costa, F.; Dewaard, J.R.; Boutillier, J.; Ratnasingham, S.; Dooh, R.T.; Hajibabaei, M.; Hebert, P. Biological identifications through DNA barcodes: The case of the Crustacea. *Can. J. Fish. Aquat. Sci.* **2007**, *64*, 272–295. [CrossRef]
20. Radulovici, A.E.; Sainte-Marie, B.; Dufresne, F. DNA barcoding of marine crustaceans from the Estuary and Gulf of St Lawrence: A regional-scale approach. *Mol. Ecol. Resour.* **2009**, *9*, 181–187. [CrossRef] [PubMed]
21. Abbas, E.M.; Abdelsalam, K.M.; Mohammed-Geba, K.; Ahmed, H.O.; Kato, M. Genetic and morphological identification of some crabs from the Gulf of Suez, Northern Red Sea, Egypt. *Egypt. J. Aquat. Res.* **2016**, *42*, 319–329. [CrossRef]
22. Apreshgi, K.P.; Dhaneesh, K.V.; Radhakrishnan, T.; Kumar, A.B. DNA barcoding of fiddler crabs Uca annulipes and U. perplexa (Arthropoda, Ocypodidae) from the southwest coast of India. *J. Mar. Biol. Assoc. India* **2016**, *58*, 101–104. [CrossRef]
23. Govender, A.; Groeneveld, J.; Singh, S.; Willows-Munro, S. The design and testing of mini-barcode markers in marine lobsters. *PLoS ONE* **2019**, *14*, e0210492. [CrossRef]
24. Jamaluddin, J.A.F.; Akib, N.A.M.; Ahmad, S.Z.; Halim, S.A.A.A.; Hamid, N.K.A.; Nor, S.A.M. DNA barcoding of shrimps from a mangrove biodiversity hotspot. *Mitochondrial DNA Part A* **2019**, *30*, 618–625. [CrossRef]
25. Cheng, J.; Sha, Z.-L.; Liu, R.-Y. DNA barcoding of genus Metapenaeopsis (Decapoda: Penaeidae) and molecular phylogeny inferred from mitochondrial and nuclear DNA sequences. *Biochem. Syst. Ecol.* **2015**, *61*, 376–384. [CrossRef]
26. Kundu, S.; Rath, S.; Tyagi, K.; Chakraborty, R.; Kumar, V. Identification of penaeid shrimp from Chilika Lake through DNA barcoding. *Mitochondrial DNA Part B* **2018**, *3*, 161–165. [CrossRef] [PubMed]
27. Gusmão, J.; Lazoski, C.; Monteiro, F.A.; Solé-Cava, A.M. Cryptic species and population structuring of the Atlantic and Pacific seabob shrimp species, *Xiphopenaeus kroyeri* and *Xiphopenaeus riveti*. *Mar. Biol.* **2006**, *149*, 491–502. [CrossRef]
28. Kerkhove, T.R.H.; Boyen, J.; De Backer, A.; Mol, J.H.; Volckaert, F.A.M.; Leliaert, F.; De Troch, M. Multilocus data reveal cryptic species in the Atlantic seabob shrimp *Xiphopenaeus kroyeri* (Crustacea: Decapoda). *Biol. J. Linn. Soc.* **2019**, *127*, 847–862. [CrossRef]

29. Chan, T.-Y.; Muchlisin, Z.A.; Hurzaid, A. Verification of a pseudocryptic species in the commercially important tiger prawn *Penaeus monodon* Fabricius, 1798 (Decapoda: Penaeidae) from Aceh Province, Indonesia. *J. Crustac. Biol.* **2021**, *41*. [CrossRef]
30. Ditty, J.G.; Bremer, J.R.A. Species Discrimination of Postlarvae and Early Juvenile Brown Shrimp (*Farfantepenaeus aztecus*) and Pink Shrimp (*F. duorarum*) (Decapoda: Penaeidae): Coupling Molecular Genetics and Comparative Morphology to Identify Early Life Stages. *J. Crustac. Biol.* **2011**, *31*, 126–137. [CrossRef]
31. Shen, Y.; Kang, J.; Chen, W.; He, S. DNA barcoding for the identification of common economic aquatic products in Central China and its application for the supervision of the market trade. *Food Control.* **2016**, *61*, 79–91. [CrossRef]
32. Nicolè, S.; Negrisolo, E.; Eccher, G.; Mantovani, R. DNA barcoding as a reliable method for the authentication of commercial seafood products. *Food Technol. Biotechnol.* **2012**, *50*, 387–398.
33. Alam, M.M.M.; De Croos, M.D.S.T.; Pálsson, S.; Pálsson, S. Mitochondrial DNA variation reveals distinct lineages in *Penaeus semisulcatus* (Decapoda, Penaeidae) from the Indo-West Pacific Ocean. *Mar. Ecol.* **2016**, *38*, e12406. [CrossRef]
34. Carvalho-Batista, A.; Terossi, M.; Zara, F.J.; Mantelatto, F.L.; Costa, R.C. A multigene and morphological analysis expands the diversity of the seabod shrimp Xiphopenaeus Smith, 1869 (Decapoda: Penaeidae), with descriptions of two new species. *Sci. Rep.* **2019**, *9*, 1–19. [CrossRef] [PubMed]
35. Pérez Farfante, I. Western Atlantic shrimps of the genus Penaeus. *Fish Bull* **1969**, *67*, 461–591.
36. Pérez Farfante, I. *Illustrated Key to the Penaeoid Shrimps of Commerce in the Americas*; Technical Report; National Marine Fisheries Service: Silver Spring, MD, USA, 1988. Available online: https://repository.library.noaa.gov/view/noaa/5793 (accessed on 12 June 2021).
37. Santamaría, J.; Carbajal-Enzian, P.; Clemente, S. Guía Ilustrada para Reconocimiento de Langostinos y Otros Crustáceos con Valor Comercial en el Perú. Available online: https://repositorio.imarpe.gob.pe/handle/20.500.12958/3311 (accessed on 12 June 2021).
38. Méndez, M. Claves de identificación y distribución de los langostinos y camarones (Crustacea: Decapoda) del mar y rios de la costa del Perú. *Boletín Inst Mar Perú* **1981**, *5*, 1–170.
39. Norma, C.F. Lista de Crustáceos del Perú (Decapoda y Stomatopoda) con Datos de su Distribución Geográfica. 1970. Available online: https://biblioimarpe.imarpe.gob.pe/handle/20.500.12958/263 (accessed on 12 June 2021).
40. Sambrook, J.; Fritish, E.F.; Maniatis, T. *Molecular Cloning: A Laboratory Manual*; Cold Spring Harbor Laboratory Press: New York, NY, USA, 1989.
41. Untergasser, A.; Cutcutache, I.; Koressaar, T.; Ye, J.; Faircloth, B.C.; Remm, M.; Rozen, S.G. Primer3—New capabilities and interfaces. *Nucleic Acids Res.* **2012**, *40*, e115. [CrossRef]
42. Folmer, O.; Black, M.; Hoeh, W.; Lutz, R.; Vrijenhoek, R. DNA primers for amplification of mitochondrial cytochrome c oxidase subunit I from diverse metazoan invertebrates. *Mol. Mar. Biol. Biotechnol.* **1994**, *3*, 294–299.
43. Lis, J.T.; Schleif, R. Size fractionation of double-stranded DNA by precipitation with polyethylene glycol. *Nucleic Acids Res.* **1975**, *2*, 383–390. [CrossRef]
44. Hall, T.A. BioEdit: A user-friendly biological sequence alignment editor and analysis program for Windows 95/98/NT. *Nucleic Acids Symp. Ser.* **1999**, *41*, 95–98.
45. Pons, J.; Barraclough, T.; Gómez-Zurita, J.; Cardoso, A.; Duran, D.P.; Hazell, S.; Kamoun, S.; Sumlin, W.D.; Vogler, A.P. Sequence-Based Species Delimitation for the DNA Taxonomy of Undescribed Insects. *Syst. Biol.* **2006**, *55*, 595–609. [CrossRef] [PubMed]
46. Zhang, J.; Kapli, P.; Pavlidis, P.; Stamatakis, A. A general species delimitation method with applications to phylogenetic placements. *Bioinformatics* **2013**, *29*, 2869–2876. [CrossRef] [PubMed]
47. Bouckaert, R.; Heled, J.; Kühnert, D.; Vaughan, T.; Wu, C.-H.; Xie, D.; Suchard, M.A.; Rambaut, A.; Drummond, A.J. BEAST 2: A Software Platform for Bayesian Evolutionary Analysis. *PLoS Comput. Biol.* **2014**, *10*, e1003537. [CrossRef] [PubMed]
48. Darriba, D.; Taboada, G.L.; Doallo, R.; Posada, D. jModelTest 2: More models, new heuristics and parallel computing. *Nat. Methods* **2012**, *9*, 772. [CrossRef] [PubMed]
49. Rambaut, A.; Drummond, A.J.; Xie, D.; Baele, G.; Suchard, M.A. Posterior Summarization in Bayesian Phylogenetics Using Tracer 1.7. *Syst. Biol.* **2018**, *67*, 901–904. [CrossRef] [PubMed]
50. Ramirez, J.L.; Birindelli, J.L.; Carvalho, D.C.; Affonso, P.R.A.M.; Venere, P.C.; Ortega, H.; Carrillo-Avila, M.; Rodríguez-Pulido, J.A.; Galetti, P.M.J. Revealing Hidden Diversity of the Underestimated Neotropical Ichthyofauna: DNA Barcoding in the Recently Described Genus Megaleporinus (Characiformes: Anostomidae). *Front. Genet.* **2017**, *8*, 149. [CrossRef]
51. Young, R.G.; Abbott, C.L.; Therriault, T.W.; Adamowicz, S.J. Barcode-based species delimitation in the marine realm: A test using Hexanauplia (Multicrustacea: Thecostraca and Copepoda). *Genome* **2017**, *60*, 169–182. [CrossRef] [PubMed]
52. Hofmann, E.P.; Nicholson, K.E.; Luque-Montes, I.R.; Köhler, G.; Cerrato-Mendoza, C.A.; Medina-Flores, M.; Wilson, L.D.; Townsend, J.H. Cryptic Diversity, but to What Extent? Discordance between Single-Locus Species Delimitation Methods within Mainland Anoles (Squamata: Dactyloidae) of Northern Central America. *Front. Genet.* **2019**, *10*, 11. [CrossRef]
53. Ma, K.Y.; Chan, T.-Y.; Chu, K.H. Refuting the six-genus classification of *Penaeus s.l.* (Dendrobranchiata, Penaeidae): A combined analysis of mitochondrial and nuclear genes. *Zool. Scr.* **2011**, *40*, 498–508. [CrossRef]
54. Timm, L.; Browder, J.A.; Simon, S.; Jackson, T.L.; Zink, I.C.; Bracken-Grissom, H.D. A tree money grows on: The first inclusive molecular phylogeny of the economically important pink shrimp (Decapoda: *Farfantepenaeus*) reveals cryptic diversity. *Invertebr. Syst.* **2019**, *33*, 488–500. [CrossRef]
55. Lukhtanov, V.A. Species Delimitation and Analysis of Cryptic Species Diversity in the XXI Century. *Entomol. Rev.* **2019**, *99*, 463–472. [CrossRef]

56. Alam, M.M.M.; Westfall, K.M.; Pálsson, S. Mitochondrial DNA variation reveals cryptic species in *Fenneropenaeus indicus*. *Bull. Mar. Sci.* **2014**, *91*, 15–31. [CrossRef]
57. Alam, M.M.M. Recent Phylogeographic Studies Revealed Distinct lineages In Penaeid Shrimps. *Oceanogr. Fish. Open Access J.* **2018**, *7*, 1–4. [CrossRef]
58. Hualkasin, W.; Sirimontaporn, P.; Chotigeat, W.; Querci, J.; Phongdara, A. Molecular phylogenetic analysis of white prawns species and the existence of two clades in *Penaeus merguiensis*. *J. Exp. Mar. Biol. Ecol.* **2003**, *296*, 1–11. [CrossRef]
59. Wanna, W.; Phongdara, A. Genetic variation and differentiation of *Fenneropenaeus merguiensis* in the Thai Peninsula. In *Phylogeography and Population Genetics in Crustacea*, 1st ed.; CRC Press: Boca Raton, FL, USA, 2011; pp. 175–189.
60. Simbine, L.; Marques, C.G.; Freitas, P.D.; Samucidine, K.E.; Gusmão, J.; Tavares, C.; Junior, P.G. Metapenaeus dobsoni (Miers, 1878), an alien Penaeidae in Mozambican coastal waters: Confirmation by mtDNA and morphology analyses. *WIO J. Mar. Sci.* **2018**, *17*, 1–12.
61. Crosnier, A.; Machordom, A.; Boisselier-Dubayle, M.-C. Les espèces du genre Trachypenaeopsis (Crustacea, Decapoda, Penaeidae). Approches morphologiques et moléculaires. *Zoosystema* **2007**, *29*, 471–489.
62. Han, Z.; Zhu, W.; Zheng, W.; Li, P.; Shui, B. Significant genetic differentiation between the Yellow Sea and East China Sea populations of cocktail shrimp *Trachypenaeus curvirostris* revealed by the mitochondrial DNA COI gene. *Biochem. Syst. Ecol.* **2015**, *59*, 78–84. [CrossRef]
63. Sakaji, H.; Hayashi, K.-I. A Review of the *Trachysalambria curvirostris* Species Group (Crustacea: Decapoda: Penaeidae) with Description of a New Species. *Species Divers.* **2003**, *8*, 141–174. [CrossRef]

Article

Integrative Taxonomy Reveals That the Marine Brachyuran Crab *Pyromaia tuberculata* (Lockington, 1877) Reached Eastern Atlantic

Jorge Lobo-Arteaga [1,2,*], Miriam Tuaty-Guerra [1,3] and Maria José Gaudêncio [1]

1 Instituto Português do Mar e da Atmosfera, I.P. (IPMA), Av. Alfredo Magalhães Ramalho, 6, 1495-165 Algés, Portugal; mguerra@ipma.pt (M.T.-G.); mgaudencio@ipma.pt (M.J.G.)
2 Centro de Ciências do Mar e do Ambiente (MARE), Universidade Nova de Lisboa, Campus de Caparica, 2829-516 Caparica, Portugal
3 Centro Interdisciplinar de Investigação Marinha e Ambiental (CIIMAR), Universidade do Porto, Terminal de Cruzeiros do Porto de Leixões, 4450-208 Matosinhos, Portugal
* Correspondence: jorge.arteaga@ipma.pt

Abstract: *Pyromaia tuberculata* is native to the north-eastern Pacific Ocean and currently established in distant regions in the Pacific Ocean and southwest Atlantic. Outside its native range, this species has become established in organically polluted enclosed waters, such as bays. The Tagus estuary, with a broad shallow bay, is one of the largest estuaries in the west coast of Europe, located in western mainland Portugal, bordering the city of Lisbon. In this study, sediment samples were collected in the estuary between 2016 and 2017. Several adult specimens of *P. tuberculata*, including one ovigerous female, were morphologically and genetically identified, resulting in accurate identification of the species. The constant presence of adults over a 16-month sampling period suggests that the species has become established in the Tagus estuary. Moreover, their short life cycle, which allows for the production of at least two generations per year, with females reaching maturity within six months after settlement, favours population establishment. Despite being referred to as invasive, there are no records of adverse effects of *P. tuberculata* to the environment and socio-economy in regions outside its native range. However, due to its expanding ability, its inclusion in European monitoring programmes would indeed be desirable.

Keywords: Decapoda; DNA barcoding; European marine waters; larval stage; macrobenthos; new record; non-indigenous species

Citation: Lobo-Arteaga, J.; Tuaty-Guerra, M.; Gaudêncio, M.J. Integrative Taxonomy Reveals That the Marine Brachyuran Crab *Pyromaia tuberculata* (Lockington, 1877) Reached Eastern Atlantic. *Diversity* **2021**, *13*, 225. https://doi.org/10.3390/d13060225

Academic Editors: Manuel Elias-Gutierrez and Eric Buffetaut

Received: 4 May 2021
Accepted: 19 May 2021
Published: 23 May 2021

1. Introduction

Brachyuran and crab-like anomuran decapods are worldwide-spread taxa and therefore easily established beyond their native regions. Brockerhoff and McLay [1] reported that 48 of the 73 (i.e., 66%) worldwide non-indigenous species of brachyuran and crab-like anomuran decapods became established. *Pyromaia tuberculata* (Lockington, 1877) is a brachyuran decapod, belonging to the family Inachoididae. This species is native to the north-eastern Pacific Ocean, from the Gulf of California to the Panama Canal [2], and lives under rocks, among sponges and seaweed, on wharf piles, and on sand and mud, from the intertidal zone to 650 m depth [1]. *P. tuberculata* is widely distributed and already established in distant regions in the Pacific Ocean, where the first record out of its native region was in Japan, before 1970 [3]. Approximately two decades after, it was found in Brazil, southwest Atlantic Ocean. However, its occurrence was considered as a natural distribution pattern [4]. More recently, Tavares and Mendonça [5] recognized it as a non-indigenous species in the southwest Atlantic Ocean, with records in Brazil and Argentina (Figure 1). Out of its native distribution range, *P. tuberculata* has been reported in organically polluted ecosystems, such as Tokyo Bay (Japan) and Guanabara Bay (Brazil) [5].

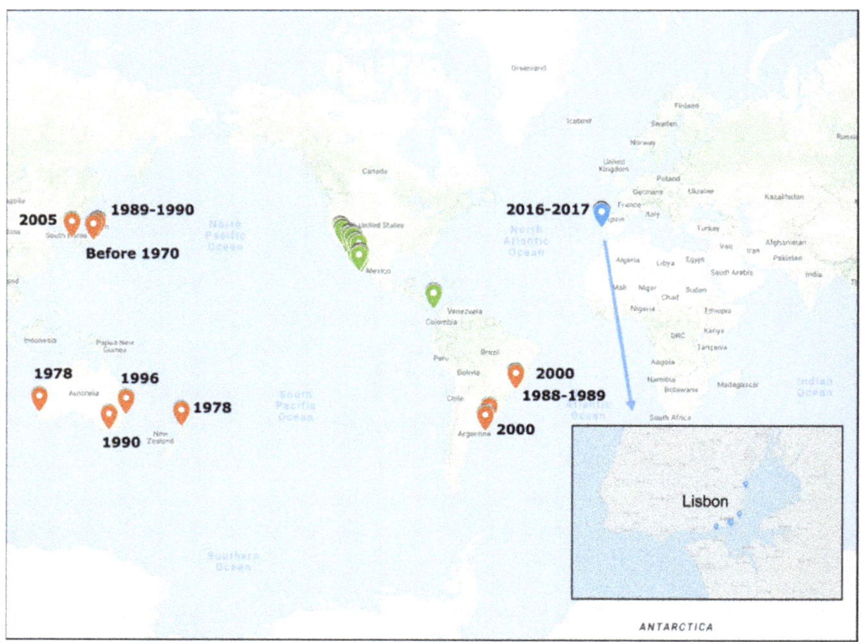

Figure 1. Location and date of collection of *P. tuberculata* specimens. Green pinpoints: native locations; orange pinpoint: non-native locations; blue pinpoints: new records. Available geographical coordinates were used for the United States, Mexico, and Panama [2], Japan and South Korea [3,6–8], Australia and New Zealand [9–12], and Argentina and Brazil [13,14]. The map was created using Google Map Maker (https://www.google.com/maps/d/) (1 March 2020) and edited with GNU Image Manipulation Program (GIMP) 2.10.8.

Due to the complexity of taxonomy, experts generally work with local or regional dichotomous keys for identifying species. Identifying non-indigenous species which are not expected to occur in a specific locality is not an easy task. In those situations, genetic tools are very useful and can help to identify species faster, easier and more accurately. The barcode region of the mtDNA gene cytochrome c oxidase I (COI-5P) is broadly used to identify and discriminate crustacean species [15–17], including decapods [18]. In this study, the species *Pyromaia tuberculata* was identified through integrative taxonomy, combining morphological and genetic approaches, which provided a robust and consistent identification of the specimens, recorded for the first time in the east Atlantic.

2. Materials and Methods

2.1. Sampling and Sample Processing

Several sediment samples were collected between 23 May 2016 and 28 September 2017, using a 0.1 m^2 Smith-McIntyre grab in the Tagus estuary, bordering the city of Lisbon in western mainland Portugal (from 38°41.773′ N, 9°10.231′ W to 38°47.245′ N, 9°5.244′ W) (Figure 1). The sampling stations were located close to the effluents of wastewater treatment plants of the city of Lisbon. Sediment subsamples were used for total organic matter determination and grain size analysis. The remaining sediment samples were sieved through a 500 μm mesh sieve, the specimens were stored in absolute ethanol and then photographed with a digital camera.

2.2. Sediment Grain Size Analysis

Grain size analysis was performed by dry sieving, following the procedure described in Gaudêncio et al. [19]. Sediment fractions, expressed as a percentage of total dry weight of each sample, were used to determine the descriptive parameters: median (P50 values) and sorting coefficient [20], calculated by the expression $\sqrt{Q3}/Q1$, where Q1 and Q3 are the 1st and 3rd quartiles, respectively. Those values, as well as the percentage of fines and gravel according to Folk's classification [21], were used to classify the sediment types according to the Udden/Wentworth scale [22].

2.3. Total Organic Matter Analysis

The total organic matter content of the sediment samples was estimated by mass Loss on Ignition (LOI) in a muffle furnace. Pre-weighed, oven-dried (100 °C) and subsequently mashed up samples were ignited at 450 °C to constant weight. After cooling in a desiccator at room temperature, they were weighed again. The total organic matter content was estimated by the difference between the weight of the oven-dried samples and the weight of the combusted ones.

2.4. Morphological Analysis

Morphological identification was carried out under a stereomicroscope Leica MZ12, using dichotomous keys [2,23,24] and descriptions of genus and species [2,13,24,25]. Specimens' measurements were obtained with a vernier calliper following Garth [24] and are expressed in millimetres (mm). Specimens' sex was determined by visual examination of the shape of the abdomen. Life stage was identified according to morphometric data and to the comparative description of adult and young specimens [2].

2.5. Molecular Analysis

A small piece of tissue (approximately 3 mm^3) was extracted from each specimen in order to perform molecular analysis. The genetic marker, Cytochrome *c* oxidase subunit I (COI-5P), was amplified with puRe Taq Ready-To-Go PCR beads (Amersham Biosciences, Little Chalfont, UK) and the primer pair, LoboF1 (5′-KBTCHACAAAYCAYAARGAYATHGG-3′) and LoboR1 (5′-TAAACYTCWGGRTGWCCRAARAAYCA-3′) [26], using a BioRad thermal cycler. The PCR thermal cycling conditions were as follows: (1) 5′ at 94 °C; (2) 5 cycles: 30″ at 94 °C, 1′30″ at 45 °C, 1′ at 72 °C; (3) 45 cycles: 30″ at 94 °C, 1′30″ at 54 °C, 1′ at 72 °C; (4) 5′ at 72 °C. Each reaction contained 1.5 μL (10 μm) of each primer, 4 μL of DNA template and was completed with sterile milliQ-grade water to a total volume of 25 μL. PCR products were stained with Greensafe (NZYTech, Lisbon, Portugal) and then separated by electrophoresis in 1.5% agarose gel in TBE buffer. Amplified products were purified using magnetic beads and subsequently sequenced bidirectionally by StabVida, using the BigDye Terminator 3 kit on an ABI 3730XL DNA analyser (Applied Biosystems™, Foster City, CA, USA). All information about the specimens was compiled in a data set created in BOLD (Data set DS-PTUBERCU): specimen details, images, taxonomy, collection data and genetic information. GenBank accession numbers for the sequences obtained are included between MZ261721 and MZ261725.

Molecular analysis was performed using MEGA version X [27]. Sequence trace files were analysed and then the sequences were manually aligned. BOLD Identification System tool (BOLD-IDS) [28] and GenBank BLASTn search [29] were used to search for matching sequences. Sequences belonging to the same species, plus other taxonomically related species, were added to the alignment for subsequent analysis. An estimation of evolutionary divergence between sequences was conducted using the Kimura 2-parameter model [30]. The evolutionary history was inferred by using the Maximum Likelihood method and Tamura 3-parameter (T92) model [31]. T92 was indicated to describe the substitution pattern the best, by having the lowest BIC score (Bayesian Information Criterion) in MEGA version X.

3. Results

3.1. Taxonomic Description

(1) Family Inachoididae Dana, 1851
(2) Subfamily Inachoidinae Dana, 1851
(3) *Pyromaia* Stimpson, 1871

(4) *Pyromaia tuberculata* (Lockington, 1877)

Diagnosis: Carapace broadly pyriform. Surface granulate and tuberculate with fine pubescence. Granulation more extensive in female. Three median tubercles: one gastric, one cardiac, and one intestinal. Tubercles more prominent in male. Gastric and cardiac tubercles enlarged; intestinal tubercle spiniform. Short spiniform tubercle on the first abdominal segment. Female abdomen broad, rounded. Male abdomen narrow, with lateral borders slightly concave and distal part subtriangular with rounded apex. Rostrum a simple spine. Supraorbital arch with a tubercle (very short in female) at summit; postorbital spine large curving around eye in male. Antero-external spine of basal antennal article incurving in male, not incurving and shorter in female. Male cheliped shorter than pereopods II to IV, and almost the same length as pereopod V. Female cheliped shorter than pereopods II to V, slenderer than in male. Chela subglobular in male, slightly inflated in female. Dactyls long, slightly incurved, unarmed.

Note: intestinal tubercle very short, almost indistinct in specimen Pt3 (male). Cheliped longer than pereopod V in specimen Pt5 (male). Cheliped much shorter than pereopod II, shorter than pereopods III and IV, almost the same length as pereopod V in specimen Pt7 (male).

3.2. Material Examined

A total of 6 specimens collected at five stations in the Tagus estuary, Portugal (Figure 1). One male and one ovigerous female are shown in Figure 2. Information about the sampling stations is presented in Table 1.

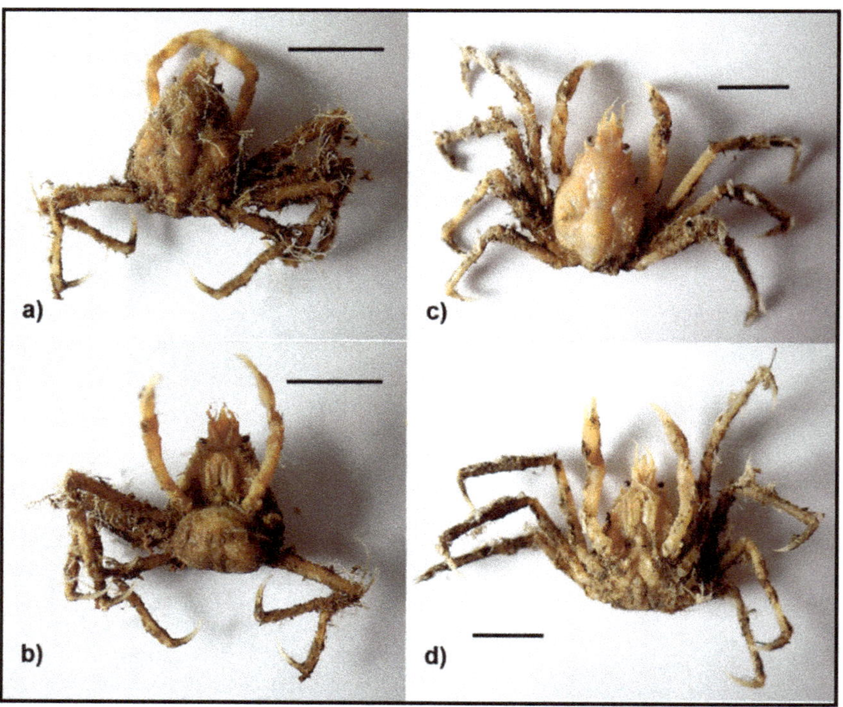

Figure 2. *Pyromaia tuberculata* collected in the Tagus estuary, Portugal. Female (Pt2) dorsal (**a**) and ventral (**b**) views; male (Pt3) dorsal (**c**) and ventral (**d**) views. Scale bar represents 10 mm.

Table 1. Characterisation of the sampling stations of *Pyromaia tuberculata* in the Tagus estuary, Lisbon.

Station	A8	T6	B11	C9	T1
WTP	Alcântara	Terreiro do Paço	Beirolas	Chelas	Terreiro do Paço
Sampling device	Smith-McIntyre grab 0.1 m^2	Smith-McIntyre grab 0.1 m^2	Smith-McIntyre grab 0.1 m^2	Smith-McIntyre grab 0.1 m^2	Beam trawl
Date	28 September 2017	27 September 2017	10 October 2016	12 October 2016	23 May 2016
Depth (m)	19.6	16.1	6.7	14.6	18–19
Latitude N	38°41.773′	38°42.277′	38°47.245′	38°43.390′	38°42.421′ to 38°42.315′
Longitude W	9°10.231′	9°7.962′	9°5.244′	9°6.251′	9°7.612′ to 9°7.887′
Surface salinity	34.69	33.79	15.30	30.40	NA
Gravel fraction (%) [a]	57.9	9.3	4.1	1.9	4.6–7.6
Sand fraction (%) [b]	34.5	66.6	95.8	37.3	76.5–69.5
Fine fraction (%) [c]	7.6	24.1	0	60.8	18.9–22.9
Median (μm)	2198	210	294	-	270–211
So	2	-	1.3	-	1.5–1.7
TOM (%)	3.6	3.6	2.7	4.7	1.6–2.9
Solids (%)	74	57	72	63	62–74

WTP: Wastewater treatment plant. [a] Particle size 8000–2000 μm. [b] Particle size 2000–63 μm. [c] Particle size < 63 μm. So: Sorting coefficient. TOM: Total Organic Matter.

3.3. Biometry

Specimens measurements, as well as gender and life stage are shown in Table 2.

Table 2. Sex, life stage and measurements (mm), of *Pyromaia tuberculata* specimens collected in the Tagus estuary, Lisbon.

Specimen	Pt1	Pt2	Pt3	Pt4	Pt5	Pt7
Sex	Male	Female	Male	Male	Male	Male
Life Stage	Adult	Adult	Adult	Adult	Adult	Adult
Length of carapace, including rostrum	15.0	16.0	17.0	19.0	20.0	13.0
Length of rostrum	4.5	4.0	4.0	6.0	4.0	4.0
Width of rostrum	3.5	3.0	4.0	4.0	4.0	4.0
Length of carapace without rostrum	10.5	12.0	13.0	13.0	16.0	11.0
Width of carapace	9.2	10.0	12.0	15.0	16.0	9.0
Length of cheliped	16.5	15.0	20.0	26.0	28.0	14.0
Length of chela	7.0 [1]	6.0	9.0	13.0	14.0	7.0
Length of dactyl	4.0 [2]	4.0	5.0	7.0	8.0	4.0
Length of first walking leg	28.0	25.0	31.0	40.0	41.0	25.0
Length of second walking leg	26.0	24.0	29.0	35.0	36.0	23.0
Length of third walking leg	24.0	23.0	27.0	31.0	33.0	21.0
Length of fourth walking leg	21.0	20.0	26.0	30.0	30.0	19.0

[1] Measure taken on the external margin (fixed finger). [2] Measure taken on the inner margin (mobile finger).

3.4. Distribution and Ecology

P. tuberculata is native to the north-eastern Pacific Ocean [2] and already established in distant regions in the Pacific Ocean and southwest Atlantic. In this study we present a new record of the species in the northeast Atlantic, Portugal, being the first one in Europe (Figure 1). In its native region, the species lives under rocks, among sponges and seaweed on wharf piles, on sand and mud, from the intertidal zone down to 650 m [1]. However, out of its native range it is found mainly in organically polluted and large shallow bays from the intertidal zone down to 80 m [5,32], close to populous urban areas and ports with dense vessel traffic, as, for example, in Tokyo (Japan) and in Rio de Janeiro (Brazil). In the Tagus estuary, it was found from 7 to 20 m depth in sediments with different physico-chemical characteristics, but always close to the effluents of wastewater treatment plants of the city of Lisbon. The total organic matter of the sediments ranged from 2 to 5%, and sediment grain size varied from sandy gravel to slightly gravely sandy mud (Table 1).

3.5. Genetic Analysis

The new COI-5P sequences, with 658 bp, obtained from all specimens found in the Tagus estuary were identical; no genetic divergence was observed. The same pattern is found when comparing our sequences to all available COI-5P sequences from Argentina, Brazil, California (USA), and Japan. A phylogenetic tree is represented in Figure 3.

Figure 3. Maximum likelihood tree, generated from COI-5P sequences of *Pyromaia tuberculata* from Argentina, Brazil, California (USA), Japan, and Portugal. 1000 bootstrap replications were applied. Data coverage of the node is 82%. Scale bar represents 0.02 substitutions per site.

4. Discussion

The first record of *P. tuberculata* out of its native region was in Japan before 1970 [3]. Afterwards it was identified in Oere Point, New Zealand and Cockburn Sound, Australia, in 1978 ([9,10], respectively), and Jukbyeon, Korea, in 1982 [33]. Later, in 1988, the species was collected in the southwest Atlantic Ocean, in Paraná, Brazil [4] and recently, in 2000, in Uruguay and in Samboronbón Bay and Mar del Plata, Argentina [13]. In the meantime, the species was also expanding its distribution over the Pacific Ocean, with records in Japan (Tokyo Bay, [6,7]), Australia (e.g., in Port Phillip Bay [11], east of Newcastle [12]) and Korea (Gijang, Busan, [8]). Not only has the species been expanding its distribution around the Pacific and southwest Atlantic, but it has successfully adapted to different environments, being quite abundant in some locations [5].

Although this is the first record of *P. tuberculata* in the northeast Atlantic, it is not likely that the species is newly arrived. It is probable that a population is already established in the Tagus estuary, since several adult specimens were collected there over 16 months (between May 2016 and September 2017), including one ovigerous female. Furota et al. [7] observed that females reach maturity within 6 months after settlement, and the life cycle of the species is short, completing at least two reproduction events per year. However, further investigation is needed in order to confirm its establishment in Portuguese waters. Since the pelagic larval stages of the species take approximately 17.5 days (two zoeal stages and one megalopa) before settlement [11], its long-distance introduction pathway is likely to be shipping, and ballast water the vector of introduction in Europe, as reported by other authors for the South Pacific and West Atlantic (e.g., [1,6]).

The COI-5P barcode sequences of *P. tuberculata* generated in this study were compared to those available in public databases and no divergences between them were found, probably because the species dispersion is relatively recent. For this reason, haplotype network analysis could not be performed in order to identify the population genetic structure to unveil the colonization process and potential population of origin, as several studies have shown [34,35]. The availability of these sequences in public databases will help to promptly detect new introductions through plankton communities, since ballast water is a primary transport vector of species with a planktonic phase [36]. The use of a high-throughput sequencing metabarcoding approach using plankton communities, environmental DNA, or sediment samples (e.g., [37,38]), has demonstrated to be reliable for application to non-indigenous species.

There are no adverse effects associated with the species so far [1]. Despite being widely established outside its native range, *P. tuberculata* continues to expand. Its short life cycle, resistance to quasi-anoxic conditions, and tolerance to a wide range of temperatures [7] give the species the ability to establish in remote locations and to rapidly colonise local habitats. Even so, it would indeed be desirable that its inclusion be considered in European monitoring programmes of non-indigenous species, as a contribution to improve the scientific knowledge required for the assessment of descriptor 2 (non-indigenous species) under the European Marine Strategy Framework Directive, namely tracking and minimising new introductions.

Author Contributions: All authors contributed to the study conception and design. M.T.-G. and M.J.G. collected the organisms and data and performed morphological analysis, as well as grain size and total organic matter analyses. J.L.-A. performed molecular analysis and analysed the data. J.L.-A. led the writing of the manuscript. All authors contributed critically to the revision of the submitted manuscript. All authors have read and agreed to the published version of the manuscript.

Funding: This study was supported by the project ProtectInvad, funded by the European Union and the Portuguese Government under the Mar2020 Programme. Other supports were provided by the Marine and Environmental Sciences Centre (MARE) which is financed by national funds from FCT/MCTES (UIDB/04292/2020) and by the Interdisciplinary Centre of Marine and Environmental Research (CIIMAR), reference UIDB/04423/2020 and UIDP/04423/2020.

Institutional Review Board Statement: Not applicable.

Informed Consent Statement: Not applicable.

Data Availability Statement: All specimens were deposited in the National Museum of Natural History and Science in Lisbon, Portugal. GenBank accession numbers for the sequences obtained in this study are included between MZ261721 and MZ261725.

Acknowledgments: The authors are grateful to António Manuel Pereira (IPMA) for helping with sediment and organisms sampling and sorting, as well as laboratory procedures for sediment physico-chemical analyses.

Conflicts of Interest: The authors declare no conflict of interest.

References

1. Brockerhoff, A.M.; McLay, C.L. Human-Mediated Spread of Alien Crabs. In *In the Wrong Place-Alien Marine Crustaceans: Distribution, Biology and Impacts*; Galil, B., Clark, P.F., Carlton, J.T., Eds.; Springer Science: Frankfurt, Germany, 2011; pp. 27–106. [CrossRef]
2. Rathbun, M.J. The spider crabs of America. *Bull. US Nat. Mus.* **1925**, *129*, 1–613. [CrossRef]
3. Sakai, T. *Crabs of Japan and the Adjacent Seas*; Kodansha: Tokyo, Japan, 1976.
4. Melo, G.A.S.; Veloso, V.G.; Oliveira, M.C. A fauna de Brachyura (Crustacea, Decapoda) do litoral do estado de Paraná. Lista preliminar. *Nerítica* **1989**, *4*, 1–31. (In Portuguese)
5. Tavares, M.; Mendonça, J.B., Jr. Introdução de crustáceos decápodes exóticos no Brasil: Uma roleta ecológica. In *Água de Lastro e Bioinvasão*; InterCiência: Rio de Janeiro, Brazil, 2004; pp. 59–76. (In Portuguese)
6. Furota, T. Life cycle studies on the introduced spider crab *Pyromaia tuberculata* (Lockington) (Brachyura: Majidae). I. Egg and larval stages. *J. Crustacean Biol.* **1996**, *16*, 71–76. [CrossRef]
7. Furota, T. Life cycle studies on the introduced spider crab *Pyromaia tuberculata* (Lockington) (Brachyura: Majidae). II. Crab stage and reproduction. *J. Crustacean Biol.* **1996**, *16*, 77–91. [CrossRef]
8. Oh, S.M.; Ko, H.S. Complete larval development of *Pyromaia tuberculata* (Crustacea: Decapoda: Majoidea: Inachoididae). *Anim. Cells Syst.* **2010**, *14*, 129–136. [CrossRef]

9. Morgan, G.J. An introduced eastern Pacific majid crab from Cockburn Sound, southwestern Australia. *Crustaceana* **1990**, *58*, 316–317. [CrossRef]
10. Webber, W.R.; Wear, R.G. Life history studies on New Zealand Brachyura, 5. Larvae of family Majidae. *N. Z. J. Mar. Freshw. Res.* **1981**, *15*, 331–383. [CrossRef]
11. Poore, G.C.B.; Storey, M. Soft sediment Crustacea of Port Phillip Bay. In *Marine Biological Invasions of Port Phillip Bay, Victoria. Centre for Research on Introduced Marine Pests Technical Report 20*; Hewitt, C., Campbell, M.L., Thresher, R.E., Martin, R.B., Eds.; CSIRO Marine Research: Hobart, Australia, 1999; pp. 150–170.
12. Ahyong, S.T. Range extension of two invasive crab species in eastern Australia: *Carcinus maenas* (Linnaeus) and *Pyromaia tuberculata* (Lockington). *Mar. Pollut. Bull.* **2005**, *50*, 460–462. [CrossRef] [PubMed]
13. Schejter, L.; Spivak, E.D.; Luppi, T. Presence of *Pyromaia tuberculata* (Lockington, 1877) adults and larvae in the Argentine continental shelf (Crustacea: Decapoda: Majoidea). *Proc. Biol. Soc. Wash.* **2002**, *115*, 605–610.
14. Fransozo, A.; Sousa, A.N.D.; Barros Rodrigues, G.F.; Telles, J.N.; Fransozo, V.; Negreiros-Fransozo, M.L. Decapod crustaceans captured along with the sea-bob shrimp fisheries on non-consolidated sublitoral from Northern coast of São Paulo, Brazil. *B. Inst. Pesca.* **2016**, *42*, 369–386. (In Portuguese with English Abstract) [CrossRef]
15. Costa, F.O.; DeWaard, J.R.; Boutillier, J.; Ratnasingham, S.; Dooh, R.T.; Hajibabaei, M.; Hebert, P.D. Biological identifications through DNA barcodes: The case of the Crustacea. *Can. J. Fish. Aquat. Sci.* **2007**, *64*, 272–295. [CrossRef]
16. Lobo, J.; Ferreira, M.S.; Antunes, I.C.; Teixeira, M.A.L.; Borges, L.M.; Sousa, R.; Gomes, P.A.; Costa, M.H.; Cunha, M.R.; Costa, F.O. Contrasting morphological and DNA barcode-suggested species boundaries among shallow-water amphipod fauna from the southern European Atlantic coast. *Genome* **2016**, *60*, 147–157. [CrossRef] [PubMed]
17. Lobo, J.; Tuaty-Guerra, M. A new deep-sea Cirripedia of the genus *Heteralepas* from the northeastern Atlantic. *Eur. J. Taxon.* **2017**, 385. [CrossRef]
18. Bartilotti, C.; Salabert, J.; Dos Santos, A. Complete larval development of *Thor amboinensis* (De Man, 1888) Decapoda: Thoridae) described from laboratory-reared material and identified by DNA barcoding. *Zootaxa* **2016**, *4066*, 399–420. [CrossRef]
19. Gaudêncio, M.J.; Guerra, M.T.; Glémarec, M. 1991 Recherches biosédimentaires sur la zone maritime de l'estuaire du Tage, Portugal: Données préliminaire. In *Estuaries and Coasts: Spatial and Temporal Intercomparisons*; Elliott, M., Ducrotoy, J.-P., Eds.; Olsen and Olsen: Fredensborg, Denmark, 1991; pp. 11–16. (In French)
20. Trask, P.D. Mechanical analyses of sediments by centrifuge. *Econ. Geol.* **1930**, *25*, 581–599. [CrossRef]
21. Folk, R.L. The distinction between grain size and mineral composition in sedimentary rock nomenclature. *J. Geol.* **1954**, *62*, 344–359. [CrossRef]
22. Wentworth, C.K. A scale of grade and class terms for clastic sediments. *J. Geol.* **1922**, *30*, 377–392. [CrossRef]
23. Lemaitre, R.; Campos, N.H.; Bermúdez, A. A new species of *Pyromaia* from the Caribbean Sea, with a redescription of *P. propinqua* Chace, 1940 (Decapoda: Brachyura: Majoidea: Inachoididae). *J. Crustac. Biol.* **2001**, *21*, 760–773. [CrossRef]
24. Garth, J.S. Brachyura of the Pacific coast of America. Oxyrhyncha. *Allan Hancock Pac. Exped.* **1958**, *21*, 1–499.
25. Hendrickx, M.E. The stomatopod and decapod crustaceans collected during the GUAYTEC 11 Cruise in the Central Gulf of California, México, with the description of a new species of *Plesionika* Bate (Caridea: Pandalidae). *Rev. Biol. Trop.* **1990**, *38*, 35–53.
26. Lobo, J.; Costa, P.M.; Teixeira, M.A.L.; Ferreira, M.S.; Costa, M.H.; Costa, F.O. Enhanced primers for amplification of DNA barcodes from a broad range of marine metazoans. *BMC Ecol.* **2013**, *13*, 34. [CrossRef]
27. Kumar, S.; Stecher, G.; Li, M.; Knyaz, C.; Tamura, K. MEGA X: Molecular Evolutionary Genetics Analysis across Computing Platforms. *Mol. Biol. Evol.* **2018**, *35*, 1547–1549. [CrossRef] [PubMed]
28. Ratnasingham, S.; Hebert, P.D.N. BOLD: The Barcode of Life Data System (http://www.barcodinglife.org). *Mol. Ecol. Resour.* **2007**, *7*, 355–364. [CrossRef]
29. Altschul, S.F.; Gish, W.; Miller, W.; Myers, E.W.; Lipman, D.J. Basic local alignment search tool. *J. Mol. Biol.* **1990**, *215*, 403–410. [CrossRef]
30. Kimura, M. A simple method for estimating evolutionary rate of base substitutions through comparative studies of nucleotide sequences. *J. Mol. Evol.* **1980**, *16*, 111–120. [CrossRef]
31. Tamura, K. Estimation of the number of nucleotide substitutions when there are strong transition-transversion and G + C-content biases. *Mol. Biol. Evol.* **1992**, *9*, 678–687. [CrossRef] [PubMed]
32. Furota, T.; Furuse, K. Distribution of the introduced spider crab, *Pyromaia tuberculata*, along the coast of Japan. *Benthos Res.* **1988**, *33*, 75–78. [CrossRef]
33. Kim, H.S. Systematic studies on crustaceans of Korea, 1. Decapods. *Proc. Coll. Nat. Sci. Seoul Natl. Univ.* **1985**, *10*, 63–94.
34. Torkkola, J.; Riginos, C.; Liggins, L. Regional patterns of mtDNA diversity in *Styela plicata*, an invasive ascidian, from Australian and New Zealand marinas. *Mar. Freshw. Res.* **2013**, *64*, 139–145. [CrossRef]
35. Pineda, M.C.; Turon, X.; Pérez-Portela, R.; López-Legentil, S. Stable populations in unstable habitats: Temporal genetic structure of the introduced ascidian *Styela plicata* in North Carolina. *Mar. Biol.* **2016**, *163*, 59. [CrossRef]
36. Carlton, J.T.; Geller, J.B. Ecological roulette: The global transport of nonindigenous marine organisms. *Science* **1993**, *261*, 78–82. [CrossRef] [PubMed]

37. Zaiko, A.; Samuiloviene, A.; Ardura, A.; Garcia-Vazquez, E. Metabarcoding approach for nonindigenous species surveillance in marine coastal waters. *Mar. Pollut. Bull.* **2015**, *100*, 53–59. [CrossRef] [PubMed]
38. Rey, A.; Basurko, O.C.; Rodriguez-Ezpeleta, N. Considerations for metabarcoding-based port biological baseline surveys aimed at marine nonindigenous species monitoring and risk assessments. *Ecol. Evol.* **2020**, *10*, 2452–2465. [CrossRef]

Article

Species Delimitation of Southeast Pacific Angel Sharks (*Squatina* spp.) Reveals Hidden Diversity through DNA Barcoding

Rosa M. Cañedo-Apolaya [1,*], Clara Ortiz-Alvarez [2], Eliana Alfaro-Cordova [2], Joanna Alfaro-Shigueto [2,3], Ximena Velez-Zuazo [4], Jeffrey C. Mangel [2], Raquel Siccha-Ramirez [5], Carmen Yamashiro [6] and Jorge L. Ramirez [1]

1 Facultad de Ciencias Biológicas, Universidad Nacional Mayor de San Marcos, 15081 Cercado de Lima, Peru; jramirezma@unmsm.edu.pe
2 ProDelphinus, 15074 Lima, Peru; clara@prodelphinus.org (C.O.-A.); eliana@prodelphinus.org (E.A.-C.); jalfaros@cientifica.edu.pe (J.A.-S.); jeff@prodelphinus.org (J.C.M.)
3 Carrera de Biología Marina, Universidad Científica del Sur, 15067 Lima, Peru
4 Center for Conservation and Sustainability, Smithsonian Conservation Biology Institute, National Zoological Park, Washington, DC 20008, USA; xvelezuazo@gmail.com
5 Laboratorio Costero de Tumbes, Instituto del Mar del Perú, 24540 Zorritos, Peru; raquelisabell@yahoo.com
6 Dirección General de Investigaciones de Recursos Demersales Y Litorales, Instituto del Mar del Perú, 07021 La Punta, Peru; cyamashirog@unmsm.edu.pe
* Correspondence: rosamlcanedo@gmail.com

Citation: Cañedo-Apolaya, R.M.; Ortiz-Alvarez, C.; Alfaro-Cordova, E.; Alfaro-Shigueto, J.; Velez-Zuazo, X.; Mangel, J.C.; Siccha-Ramirez, R.; Yamashiro, C.; Ramirez, J.L. Species Delimitation of Southeast Pacific Angel Sharks (*Squatina* spp.) Reveals Hidden Diversity through DNA Barcoding. *Diversity* **2021**, *13*, 177. https://doi.org/10.3390/d13050177

Academic Editors: Eric Buffetaut and Manuel Elias-Gutierrez

Received: 20 February 2021
Accepted: 29 March 2021
Published: 21 April 2021

Publisher's Note: MDPI stays neutral with regard to jurisdictional claims in published maps and institutional affiliations.

Abstract: Angel sharks are distributed worldwide in tropical to subtropical waters. Across the Eastern Pacific Ocean (EPO), two valid species are reported: The Pacific angelshark *Squatina californica* and the Chilean angelshark *Squatina armata*; however, there is still uncertainty about their geographic distribution, mainly along the northern Peru coast where the species have been reported to be sympatric. The aim of this study is to describe the genetic differences between the genus *Squatina* from the EPO, including samples from northern Peru, and using DNA barcoding and three species delimitation models: Poisson tree processes (PTP) model, Bayesian implementation of the PTP (bPTP) model and the general mixed Yule coalescent (GMYC) model. The three approaches summarized 19 nominal *Squatina* species in 23 consensus Molecular Taxonomic Units (MOTU). Only 16 of them were in accordance with taxonomic identifications. From the EPO, four *Squatina* MOTUs were identified, one from North America (*S. californica* USA/Mexico) and three sampled in northern Peru, *S. californica* Peru, *S. armata* and *Squatina* sp. (a potential new species). This study contributes to the management and conservation policies of angel sharks in Peru, suggesting the presence of an undescribed species inhabiting the northern Peruvian coast. The use of molecular approaches, such as DNA barcoding, has the potential to quickly flag undescribed species in poorly studied regions, including the Southeast Pacific, within groups of ecologically and economically important groups like angel sharks.

Keywords: elasmobranchii; Humboldt current; Eastern Pacific Ocean; biodiversity; mtDNA

1. Introduction

One of the most diverse groups of marine predators is the subclass Elasmobranchii (i.e., sharks, skates, and rays). Among them, there is a small but highly distinctive group of bizarrely-shaped benthic sharks, commonly called angel sharks. This group of ray-like sharks belongs to the monophyletic family Squatinidae (Bonaparte, 1838) [1–3] encompassing a unique genus, *Squatina* (Dumeril, 1805), with 22 extant species described based on morphological characters or molecular information [4–7]. Although, there are two additional species from the Gulf of Mexico described [8], *Squatina mexicana* Castro–Aguirre, Espinosa–Pérez and Huidobro–Campos, 2007, and *Squatina heteroptera* Castro–Aguirre, Espinosa–

Pérez and Huidobro–Campos, 2007, these two species are considered as not valid [9] because of uncertainty about the validity of their morphological description [6,7,10–12]. Currently, *S. mexicana* and *S. heteroptera* are considered junior synonyms of *Squatina dumeril* Lesueur, 1818 [7,10].

Angel sharks are distributed worldwide in tropical to subtropical shelf waters, although most of the species have restricted distribution in regional seas (e.g., *S. david* distributed within the Southern Caribbean Sea) [10]. Across the Eastern Pacific Ocean (EPO), two valid and sympatric species occur: the Pacific angelshark *Squatina californica* Ayres, 1859 and the Chilean angelshark *Squatina armata* (Philippi, 1887) [4]. Until recently it was considered that *S. californica* was distributed off the coast of North America, from Alaska to the Gulf of California [13]. However, some studies reported its presence in Ecuador [14,15] and Peru [15]. On the other hand, *S. armata* has been reported to inhabit waters from northern Peru to the central coast of Chile [4,16,17]. Nevertheless, the northern limit of its distribution range is not clear since some studies have suggested its presence in Ecuador [18], Colombia [19], and Costa Rica [20]. Current information such as the range of its geographic extent and abundance has been used to determine the extinction risk by the International Union for Conservation of Nature (IUCN) Red List of threatened species, classifying *S. californica* as 'Near Threatened' [21] and *S. armata* as 'Critically Endangered' [22]. Nonetheless, to establish management measures also at a national level, data concerning its biology, ecology, as well as its taxonomic status, need to be resolved.

Several studies support the validity of both species based on morphological taxonomic characters [10,16,23]; however, other studies consider these species as synonymous [1,23,24]. This conclusion that *S. californica* and *S. armata* were identical was made after the comparison of various specimens from the northern and southern hemisphere. Nonetheless, neither information about the exact number of specimens nor how this comparison was made is provided by the authors [25]. Subsequent publications [1,13,14] also do not provide more detail to explain this suggested synonymy. The first species described in the EPO was *S. californica* based on the revision of one single specimen (around 96 cm, unidentified sex) collected in San Francisco Bay, United States, in September 1857 [26]. The specimen were compared to *Squatina dumeril* and differed in qualitative characteristic and meristic traits (i.e., form of the orbits, form of teeth, size of pectorals, form of pectorals, form of the ventrals, form of the dorsal and number of teeth) [26]. *S. armata*, was described as *Rhina armata*, based also on a single male individual (103 cm) collected at Iquique, Chile (unknown sampling date) [27]. The reference species used to compare external characteristics was a specimen identified by the author as '*Rhina squatina*' collected in Rio de Janeiro, Brazil, although it might be one of the four known species of the Western Atlantic (i.e., *S. dumeril*, *S. argentina*, *S. guggenheim*, *S. occulta*, or *S. punctata*) but not the extant *S. squatina*. This latter species is excluded because its presence has been only reported at the Baltic Sea, North Sea, Black Sea, Mediterranean Sea, and Canary Islands [9]. In this case, seven taxonomic characteristics were reported to differ between *S. armata* and the specimen from Brazil (i.e., form of the pectoral fins, size of the pectoral fins, width of the head, shape and size of the two spiracles, form of the tail, and presence of enlarged 'denticles' on the pectoral fin) [27].

Around the world, delimitation and the identification of angel shark species have become important due to their biological characteristics (e.g., large size, reproductive cycle, demersal nature) and because this group is susceptible and vulnerable to fisheries and human activities [28,29]. Therefore, to achieve effective conservation and fishery management strategies, correctly delimiting sampling units (i.e., taxon 'species') is urgently needed to generate species-specific data to support these strategies [30]. Over the last decades, classical taxonomy (i.e., description and identification of species through taxonomic morphological characters) has been the main approach used to describe angel sharks [6,8,24,26,27,31–34]. Nonetheless, the use of molecular tools, such as genomic data, is increasingly being considered to support the process of species delimitation [5,30]. For instance, instead of the use of morphological characters, the molecular approach uses DNA sequences which are grouped into Molecular Taxonomic Units (MOTU). A MOTU is considered as an operational definition that groups individual DNA sequences (i.e., cluster of sequences) based on an explicit algorithm and is used to estimate diversity at the species level. A powerful tool that uses standardized gene regions (e.g., cytochrome c oxidase—COI) to delimit and identify species is DNA barcoding. This molecular tool couples a comprehensive dataset of COI DNA sequences together with validated identified voucher specimens to support taxonomic studies [35,36]. Among their benefits, DNA barcoding can assist in defining phylogenetic relationships and species geographic boundaries. Furthermore, DNA barcoding together with single locus species delimitation methods have recently shown to be an effective tool for validating elasmobranch identification and describing new species in a number of fisheries [37–43].

Stelbrink et al. [4] employed COI and 16S rRNA markers to provide a comprehensive global phylogeny of 17 *Squatina* species, including the two species found along the EPO, *S. armata* and *S. californica*. In that study, *S. armata* is the first species to branch off the clade that includes the North and South American species, indicating the existence of two different species. However, in that study, only samples from the ends of both distributions, in California and Chile, were used. Thus, the detailed distribution of both species along the EPO remains uncertain, especially in the areas of North of South America where both species have been reported [14,15,18,44]. Additionally, angel sharks are target species and are captured by small-scale fisheries [28]. However, reports of landings are rather general and likely include several species of angel sharks, thus molecular tools could serve well for the accurate identification of the species landed, their distributions, and the fisheries with which they interact [28].

In this regard, the aim of the present study is to describe the genetic differences within the genus *Squatina* from the EPO, including samples of angel sharks from an area not previously covered (i.e., northern Peru), and integrate them with mtDNA COI sequences data from Barcode of Life Data (BOLD) System to evaluate species boundaries using species delimitation methods and determine MOTUs.

2. Materials and Methods

2.1. Morphological Identification and Sample Collection

As part of the sampling campaign of the consortium for DNA barcoding of Peruvian marine species (PeMar), a total of eight specimens of angel sharks were sampled between 2017 and 2018 from fish markets and landing sites in northern Peru (Figures 1 and 2). All specimens collected were identified using external morphological characteristics following a literature review [1,10,23]. Each specimen was photographed, and one specimen was fixed and deposited in the fish collection of the San Marcos–Natural History Museum (UNMSM) for further analyses. Muscle tissue samples were extracted from all specimens and preserved in 96% ethanol at room temperature (17–20 °C). Sequences, sample records, and voucher numbers can be viewed in Table 1.

2.2. DNA Extraction, Amplification and Sequencing

Total genomic DNA was isolated using Tissue Kit (Thermo Scientific) following the manufacturer's instructions. A partial fragment (~655 base-pair) of the mitochondrial Cytochrome Oxidase subunit I (COI) gene was amplified through Polymerase Chain Reaction (PCR) using primers FishF1-FishR1 or FishF2-FishR2 [35], that amplify an overlapping region from the 5′ region of the COI gene. The PCR was performed with a final volume of 25 μL containing 16.35 μL distilled water, 2.5 μL dNTP (8 nM), 0.6 μL of each primer (5 μM), using just one pair of primers (i.e., F1/R1 or F2/R2) and 0.6 μL of Taq polymerase (5 μ/μL). PCR conditions consisted of an initial denaturation at 95 °C for 2 min, followed by 30 cycles including denaturation at 95 °C for 45 s, annealing at 52 °C for 45 s, and extension at 72 °C for 60 s, and a final extension at 72 °C for 5 min. Amplified products were checked on 1% agarose gel and both strands per amplicon were sent to Macrogen (Rockville, MD, USA) for Sanger sequencing. The sequencing was carried out using the same set of primers that was used in the PCR, however a greater number of samples were amplified and sequenced using the Fish F1 and Fish R1 primers, since these had better efficiency for our samples. Sequences were cleaned and contigs were assembled using the software CodonCode Aligner (CodonCode Corporation, Dedlham, MA, USA). Multiple alignments were done using a ClustalW algorithm [45], implemented in the software MEGA 7 [46] and were checked manually for misalignments and trimmed to the shortest common sequence length.

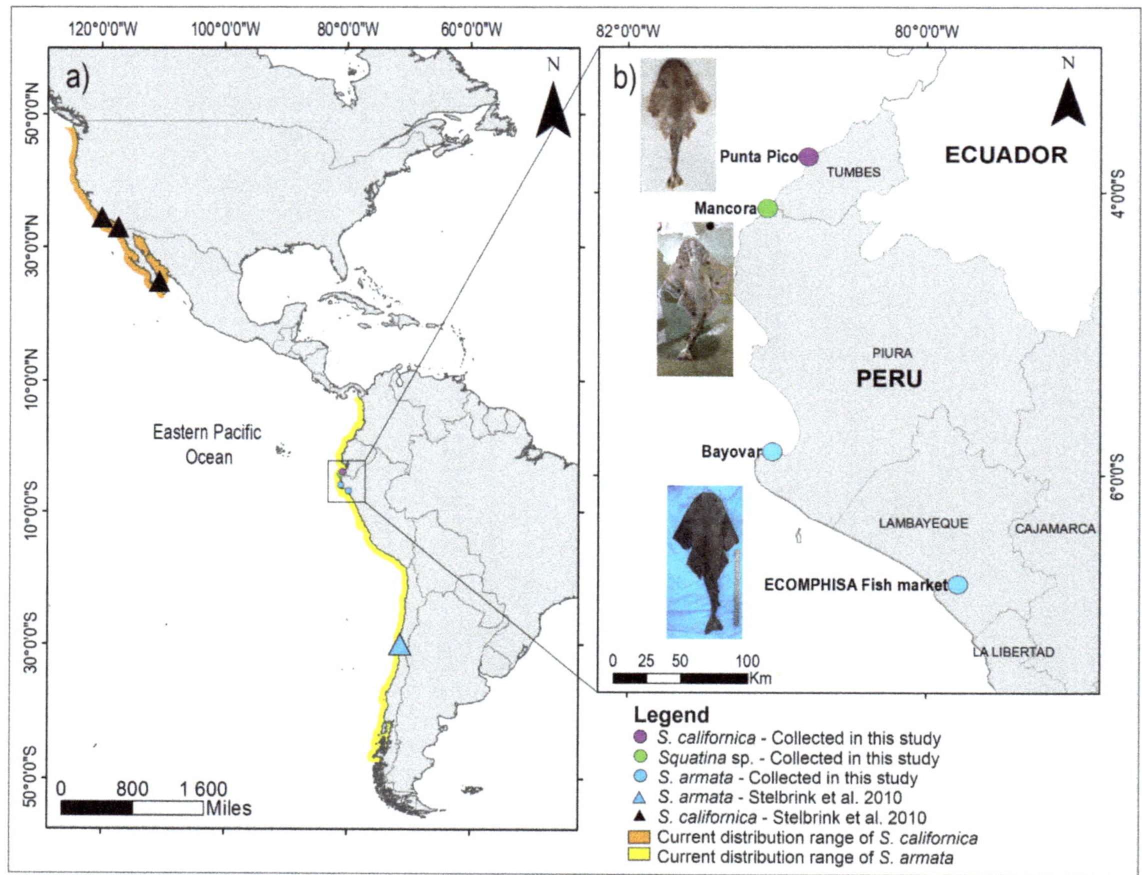

Figure 1. Geographic distribution map and sampling sites of DNA sequences of *Squatina* species along Eastern Pacific Ocean. (**a**) Current known distribution range adapted from Fricke et al., (2020) and sampling sites of *Squatina californica* (black triangle) and *Squatina armata* (light blue triangle) reported by Stelbrik et al., 2010. (**b**) Sampling sites of *Squatina californica* (purple circle), *Squatina* sp. (green circle), and *Squatina armata* (light blue circle) selected by PeMar Project in northern Peru.

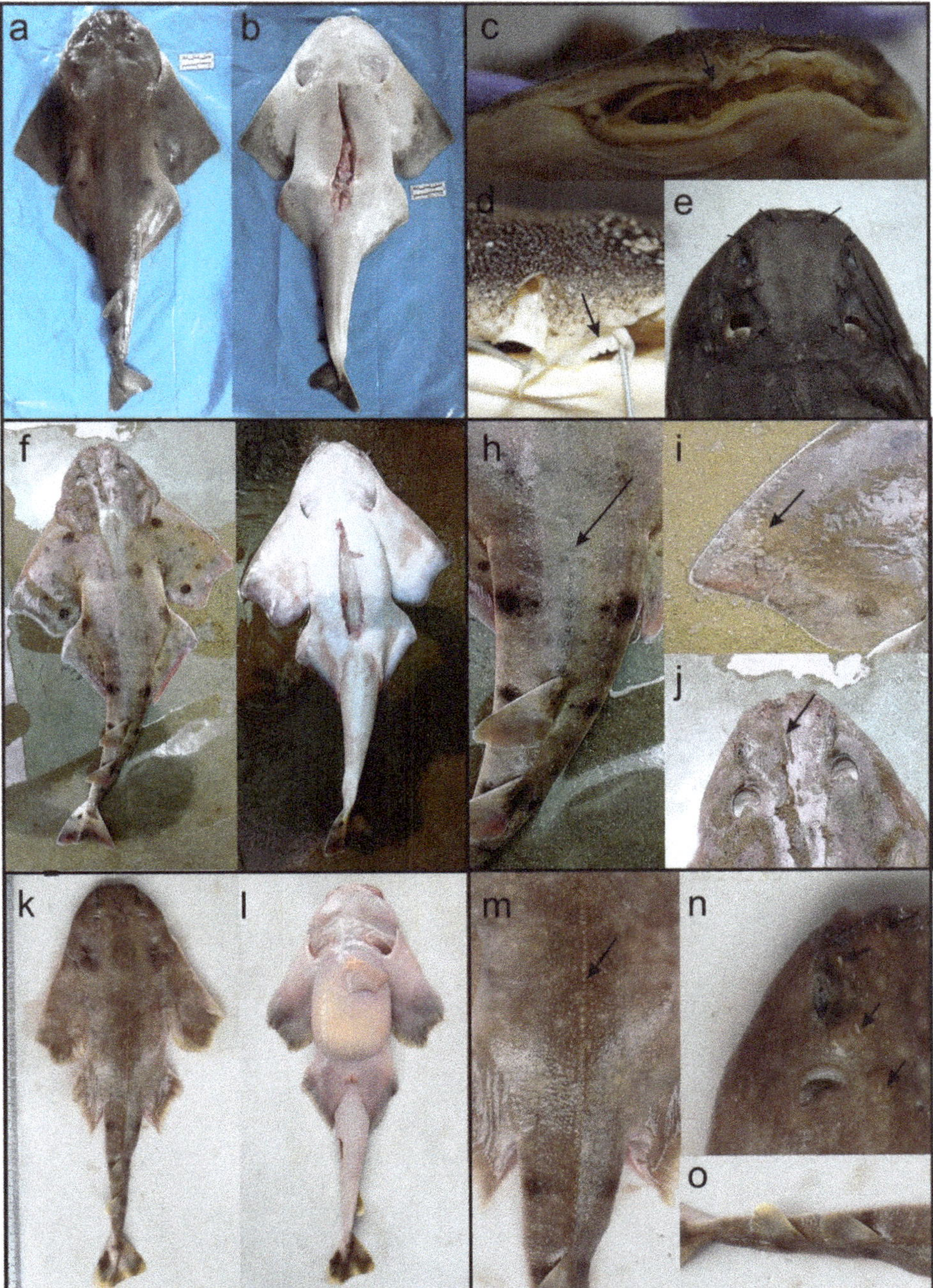

Figure 2. Specimens of angel sharks collected for this study: Dorsal (**a**) and ventral (**b**) images of one fresh specimen of *Squatina armata* (Pemar_V0173). Detailed images of barbels (**c**), anterior nasal flaps (**d**) and thorns on snout, between eyes and spiracles (**e**) observed on preserved specimen of *S. armata* (Pemar_V0174). Dorsal (**f**) and ventral (**g**) images of one fresh specimen observed on *Squatina* sp. (Pemar_V0209). Detailed images of thorns along the middle line of the back (**h**), denticles covering the edges of the pectoral fin (**i**) and concave between eyes (**j**) of specimens of *Squatina* sp. (Pemar_V0209 and Pemar_V0211). Dorsal (**k**) and ventral (**l**) images of one fresh pup of *S. californica* (LCT_2160). Detailed images of thorns (**m**,**n**) and pale dorsal fins (**o**) of *S. californica* pup (LCT_2160).

Table 1. Angel shark species sampled along the northern Peruvian coast, PeMar project code, BOLD code, voucher code, sampling site (region and exact site), sex, and total length (TL).

Species	Pemar Code	BOLD Code	Voucher Code	Sampling Site	Sex and Total Length
Squatina sp.	Pemar_V0209	PMVTB124-21	-	Piura/Mancora	Female, unknown TL
Squatina sp.	Pemar_V0210	PMVTB125-21	-	Piura/Mancora	Male, unknown TL
Squatina sp.	Pemar_V0211	PMVTB126-21	-	Piura/Mancora	Male, unknown TL
Squatina armata	Pemar_V0083	PMVTB046-20	-	Lambayeque/Terminal Pesquero ECOMPHISA	Female, 81 cm TL
Squatina armata	Pemar_V0173	PMVTB067-20	-	Piura/close to Bayovar Port	Female, 64.9 cm TL
Squatina armata	Pemar_V0174	PMVTB068-20	MUSM 65818	Piura/close to Bayovar Port	Male, 37.2 cm TL
Squatina armata	Pemar_V0086	PMVTB047-20	-	Lambayeque/Terminal Pesquero ECOMPHISA	Female, 93.4 cm TL
Squatina californica	LCT_2160	FMCT1223-19		Tumbes/805 m off Punta Pico coast	Female, 23.7 cm TL

2.3. Species Delimitation Methods

To apply several species delimitation methods to infer MOTUs, our samples were combined with COI sequences from 19 *Squatina* nominal species (Table 2, Table S1) obtained from the public repository BOLD (www.boldsystems.org, accessed on 5 February 2021), GenBank (www.ncbi.nlm.nih.gov/genbank/, accessed on 5 February 2021), and from the literature [5,35,36,43,47–56]. The set of DNA fragments were chosen following two criteria: (1) the location of sampling mentioned in the public database (e.g., country, province, region, sector, exact site, coordinates or FAO fishing zones) and (2) length of the DNA fragment (i.e., ~610 bp). To reduce computational time, only 10 COI sequences were chosen per species, when it was possible. The two sequences of *Squalus* used as an outgroup by Stelbrink et al. [4] were also included in the molecular analysis.

We performed and compared three molecular species delimitation methods using the pipeline SPdel (https://github.com/jolobito/SPdel, accessed on 8 March 2021) added to a delimitation species method based on classical taxonomy. The first method implemented was the general mixed Yule coalescent (GMYC) model [57] with a single threshold. The GMYC model identifies the threshold value at the transition of branching patterns that are characteristics of the speciation process [58] versus coalescence process [59], and identifies significant changes in the branching rates in a time-calibrated ultrametric tree. The other methods used were the Poisson tree processes (PTP), and the Bayesian implementation of the PTP model (bPTP) [60]. These two methods directly use the number of substitutions as opposed to the GMYC method that uses time. For all models, an ultrametric tree was used as an input file which was generated in BEAUti v2.1 [61], with a normal relaxed clock and a birth and death model, and a HKY+G substitution model suggested by the Bayesian Information Criterion in jModeltest 2 [62]. The analysis was implemented with 60,000,000 million MCMC generations and with a burn-in of 10%. To run the analysis, we used BEAST v2.5 [61] implemented in the Cyber Infrastructure for phylogenetic Research (CIPRES; https://www.phylo.org, accessed on 8 March 2021) [63]. Convergence and adequate sample size (greater than 130) were evaluated in Tracer v1.6.0. The pipeline SPdel used for our analysis performs a comparison between the four chosen methods and ultimately generates a consensus species delimitation considering three molecular approaches. Only MOTUs supported by at least two of the applied delimitation methods were considered consensus MOTUs. Furthermore, we quantified the degree of genetic divergences for nominal species and for each one of the species delimitation methods used, calculating the values of intragroup and intergroup Kimura 2–parameter (K2-P) genetic distances in SPdel.

Table 2. List of extant species, their current distribution (Fricke et al., 2020), number of COI sequences used in this study per species, and sampling sites reported in BOLD System per species. See Table S1 for detailed list.

Extant Species	Distribution	BOLD System	
		N° COI Sequences Used in This Study	Sampling Site (Country)
Squatina aculeata Cuvier 1829	Mediterranean Sea; eastern Atlantic: southern Portugal south to Namibia, including Selvagens Islands (Portugal)	10	Malta, Senegal, Turkey
Squatina africana Regan 1908	Western Indian Ocean: East Africa, South Africa to Madagascar	11	South Africa and Indian Ocean
Squatina albipunctata Last and White 2008 *	Australia: Queensland to Victoria	10	Australia
Squatina australis Regan 1906	Southeastern Indian Ocean: Victoria, Tasmania, South Australia and Western Australia	10	Australia
Squatina armata (Philippi 1887) **	Southeastern Pacific: Ecuador south to Chile	3	Chile
Squatina californica Ayres 1859	Eastern Pacific: Puget Sound (Washington, DC, USA.) south to Pacific coast of Baja California Sur (Mexico); Ecuador south to Chile (needs confirmation); questionable from Alaska (USA)	10	Mexico, United States
Squatina david Acero P., Tavera, Anguila and Hernández 2016 ***	Western Atlantic: Panama, Colombia, Venezuela (southern Caribbean)	3	Colombia
Squatina dumeril Lesueur 1818	Western Atlantic (including Caribbean Sea)	8	United States
Squatina formosa Shen and Ting 1972	Western North Pacific	10	Taiwan, Japan
Squatina guggenheim Marini 1936 ****	Southwestern Atlantic: Brazil south to Argentina	11	Brazil, Argentina
Squatina japonica Bleeker 1858	Northwestern Pacific	1	Japan
Squatina legnota Last and White 2008	Off southern Indonesia	6	Indonesia
Squatina nebulosa Regan 1906	Western North Pacific	2	China, South China Sea
Squatina occulta Vooren and da Silva 1991	Southwestern Atlantic: Brazil, Uruguay and Argentina	7	Brazil
Squatina oculata Bonaparte 1840	Mediterranean Sea; eastern Atlantic: Portugal south to Namibia	10	Malta, Senegal, Turkey
Squatina pseudocellata Last and White 2008	Australia: Western Australia	7	Australia
Squatina squatina (Linnaeus 1758)	Western Baltic Sea; North Sea; Mediterranean Sea; Black Sea; eastern Atlantic: Norway south to Western Sahara, including Canary Islands	10	Egypt, Ireland, Spain (Canary Islands), Turkey
Squatina tergocellata McCulloch 1914	Southern and western Australia	8	Australia
Squatina tergocellatoides Chen 1963	North Pacific: Taiwan Straits, Vietnam, Hong Kong, Malaysia	4	Malaysia, Vietnan
Squatina argentina (Marini 1930)	Southwestern Atlantic: Brazil to Uruguay and Argentina	no sequences	
Squatina caillieti Walsh, Ebert and Compagno 2011	Philippines	no sequences	
Squatina varii Vaz and Carvalho 2018	Brazil	no sequences	

* Sequences identified in BOLD as *Squatina* sp. Pi24 and *Squatina* sp. Pi26 are included within this taxon. ** It is identified in BOLD as *Squatina californica*. *** It is identified in BOLD as *Squatina* sp. JT-2016a. **** Sequences identified in BOLD as *Squatina* sp. MFSP273-09 is included within this taxon.

3. Results

3.1. Taxonomic Identification

Most of our sampled specimens were not retained because they were part of fishers' daily catch. For this reason, identification was done mainly in the field (i.e., with fresh specimens) and through photographs allowing for diagnostic taxonomic characters to be assessed. The four specimens collected at the ECOMPHISA fish market and close to the Bayovar Port were identified as *S. armata* (Table 1, Figure 1). For the identification we used an illustrated guide [10] and a taxonomic key [15], distinguishing the following combination of characteristics: reddish-brown to blackish dorsal surface (Figure 2a), white ventral surface with dark brown edged of pectoral and pelvic fins (Figure 2b), narrow and simple barbels (Figure 2c), anterior nasal flaps fringed (Figure 2d), and thorns on snout, between eyes and spiracles (Figure 2e). One angel shark pup was collected off the coast of Punta Pico (Table 1, Figure 1). It was identified as *S. californica* following the aforementioned illustrated guide, distinguishing the following characteristics: reddish-brown dorsal surface with scattered light spots (Figure 2k), white edged pectoral and pelvic fins (Figure 2l), presence of thorns (Figure 2m,n), and pale dorsal fins (Figure 2o). Finally, three specimens collected in Mancora (Table 1, Figure 1) were identified as *S. californica* in the field, however after the molecular analysis, they were allocated to the taxon *Squatina* sp. (Figure 2f–j).

3.2. MOTU Delimitation Analyses

We obtained eight sequences (610 base-pair) from two nominal species, *S. armata* (n = 4), *S. californica* (n = 1), and *Squatina* sp. (n = 3), from northern Peru with 30 parsimony informative sites. The final alignment of mtDNA COI sequences resulted in 591 bp (shortest common sequence length) comprised of a total of 19 nominal species (Result S1).

The species delimitation analyses showed that PTP and the bPTP method delimited the same 23 *Squatina* MOTUs (Figure 3), with a maximum intra-MOTU divergence of 0.99% (for the MOTU of *S. squatina*) and a minimum inter-MOTU divergence of 0.49% (between the MOTUs of *S. californica* collected in Peru and USA/Mex). On the other hand, the GMYC method delimited *Squatina* 25 MOTUs with a maximum intra-MOTU divergence of 0.99% (for the MOTU of *S. squatina*) and a minimum inter-MOTU divergence of 0.16% (between the MOTUs of *S. squatina* collected in Ireland and Turkey). Single-locus species delimitation results from PTP, bPTP, and GMYC approaches were summarized by using the pipeline SPdel, in 25 consensus MOTUs. The maximum intra-MOTU distance was 0.99% (for the MOTU of *S. squatina*) (Table 3) and the minimum inter-MOTU distance was 0.49% (between the MOTUs of *S. californica* collected in Peru and USA/Mexico) (Table 3). In contrast, if considering species delimited through traditional taxonomy, the maximum intraspecific distance was 2.51% (between specimens morphologically identified as *S. africana*) (Table 3) and the minimum interspecific distance was 0 between specimens of *S. formosa* (collected in Thailand) and *S. nebulosa* (collected in China and Southern China Sea) (Table 3).

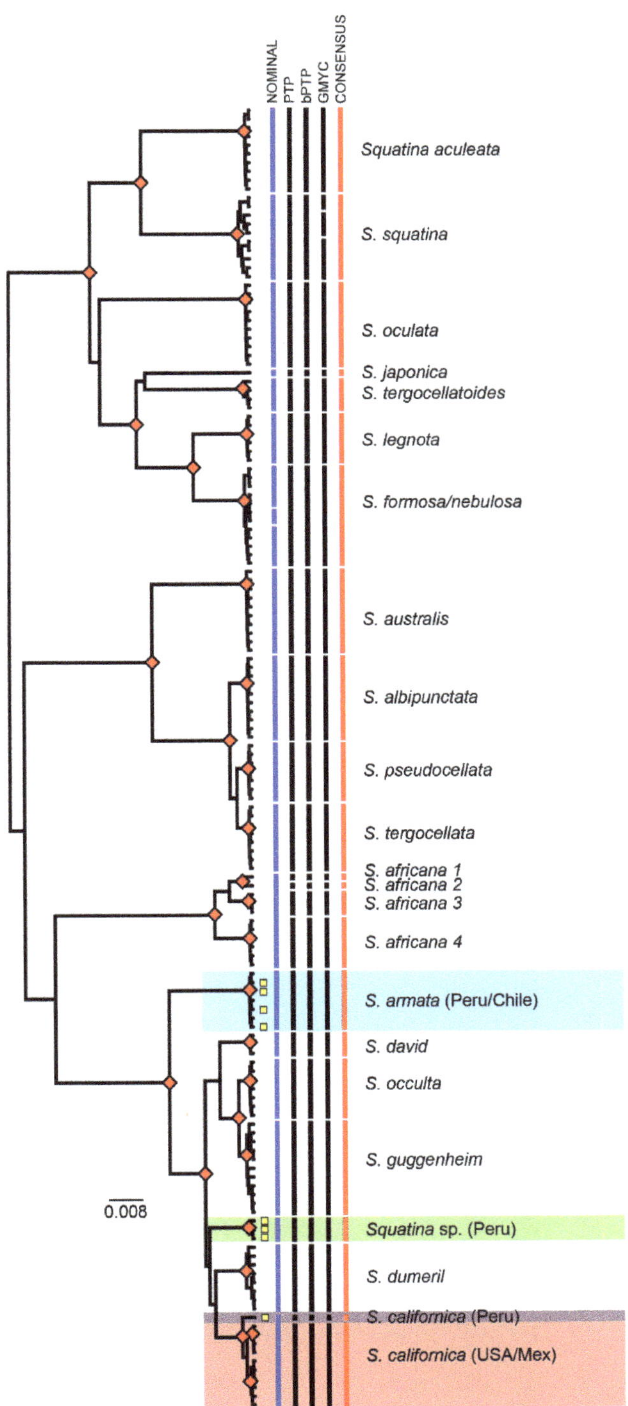

Figure 3. Bayesian tree showing the clustering of the MOTUs obtained by the species delimitation analyses (PTP, bPTP, and GYMC) and the consensus analysis. The red diamonds indicate nodes with supports higher than 0.9 Bayesian posterior probability. The scale bar indicates nucleotide substitutions per site. Samples from the Eastern Pacific Ocean are delimited by squares. Yellow squares indicate samples from northern Peru collected in this study.

Table 3. Genetic K2-P distances of MOTUs and nominal species of angel sharks: mean intra-MOTU divergence (mean intra-), maximum intra-MOTU divergence (maximum intra-), distance to the nearest neighbor (distance to NN) and the nearest neighbor (NN).

	Mean Intra-	Maximum Intra-	Distance to NN	NN
Nominal				
Squatina aculeata	0.06	0.16	6.22	*Squatina legnota*
Squatina africana	1.39	2.51	7.34	*Squatina* sp.
Squatina albipunctata	0	0	0.99	*Squatina pseudocellata*
Squatina armata	0	0	2.85	*Squatina guggenheim*
Squatina australis	0.03	0.16	5.33	*Squatina albipunctata*
Squatina californica	0.45	0.85	1.99	*Squatina david*
Squatina david	0.11	0.16	1.33	*Squatina guggenheim*
Squatina dumeril	0.43	0.99	2.52	*Squatina david*
Squatina formosa	0.13	0.33	0	*Squatina nebulosa*
Squatina guggenheim	0.31	0	6.99	*Squatina occulta*
Squatina japonica	NaN	0	6.99	*Squatina formosa*
Squatina legnota	0	0	2.51	*Squatina formosa*
Squatina nebulosa	0	0	0	*Squatina formosa*
Squatina occulta	0	0	0.50	*Squatina guggenheim*
Squatina oculata	0	0	5.87	*Squatina legnota*
Squatina pseudocellata	0	0	0.99	*Squatina albipunctata*
Squatina squatina	0.52	0.99	6.61	*Squatina aculeata*
Squatina tergocellata	0	0	0.99	*Squatina pseudocellata*
Squatina tergocellatoides	0.25	0.50	4.44	*Squatina legnota*
Squatina sp.	0.22	0.33	1.50	*Squatina guggenheim*
Squalus acanthias	NaN	0	7.70	*Squalus cubensis*
Squalus cubensis	NaN	0	7.70	*Squalus acanthias*
Consensus MOTUs				
Squatina aculeata	0.06	0.16	6.22	*Squatina legnota*
Squatina albipunctata	0	0	0.99	*Squatina pseudocellata*
Squatina armata	0	0	2.85	*Squatina guggeheim*
Squatina australis	0.03	0.17	5.33	*Squatina albipunctata*
Squatina david	0.11	0.16	1.33	*Squatina guggeheim*
Squatina dumeril	0.43	0.99	2.52	*Squatina david*
Squatina guggeheim	0.31	0.66	0.50	*Squatina occulta*
Squatina japonica	NaN	0	6.99	*Squatina formosa/nebulosa*
Squatina legnota	0	0	2.51	*Squatina formosa/nebulosa*
Squatina occulta	0	0	0.50	*Squatina guggeheim*
Squatina oculata	0	0	5.87	*Squatina legnota*
Squatina pseudocellata	0	0	0.99	*Squatina albipunctata*
Squatina tergocellata	0	0	0.99	*Squatina pseudocellata*
Squatina tergocellatoides	0.25	0.50	4.44	*Squatina legnota*
Squatina sp.	0.22	0.33	1.50	*Squatina guggeheim*
Squatina squatina	0.52	0.99	6.61	*Squatina aculeata*
Squatina californica USA/Mex	0.41	0.85	0.49	*Squatina californica* Per
Squatina californica Per	NaN	0	0.50	*Squatina californica* USA/Mex
Squatina africana 1	NaN	0	0.83	*Squatina africana* 2
Squatina africana 2	NaN	0	0.83	*Squatina africana* 1
Squatina africana 3	0.11	0.17	0.83	*Squatina africana* 2
Squatina africana 4	0	0	2.00	*Squatina africana* 2
Squatina formosa/nebulosa	0.15	0.33	2.51	*Squatina legnota*
Squalus acanthias	NaN	0	7.70	*Squalus cubensis*
Squalus cubensis	NaN	0	7.70	*Squalus acanthias*

From the 23 consensus Squatina MOTUs, 16 are in accordance with taxonomic identification: *S. aculeata*, *S. albipunctata*, *S. armata*, *S. australis*, *Squatina* sp., *S. david*, *S. dumeril*, *S. guggenheim*, *S. japonica*, *S. legnota*, *S. occulta*, *S. oculata*, *S. pseudocellata*, *S. tergocellata*, *S. ter-*

gocellatoides, and S. squatina. While seven consensus MOTUs do not match with the current taxonomy: S. californica USA/Mexico, S. californica Per, S. africana 1, S. africana 2, S. africana 3, S. africana 4, and S. formosa/nebulosa. Both nominal species, S. africana and S. californica, were split into four and two MOTUs, respectively (Table 3). The S. californica samples formed a polyphyletic group, with two MOTUs closely related, one from the Northeast Pacific (samples collected within the Gulf of California, Mexico, and in California, USA), and the other composed by samples collected in northern Peru (Figure 3). The minimum intra-MOTU divergence calculated for S. californica from Northeast Pacific was 0% and the maximum intra-MOTU value was 0.83%. Furthermore, the minimum and the maximum inter-MOTU divergences between the MOTU of S. californica from the northern hemisphere and the southern hemisphere were 0.5 and 0.83%, respectively. Samples identified initially as S. californica (Pemar_V0209, Pemar_V0210, and Pemar_V0211) from northern Peru, were grouped into a clade, separately from S. californica and S. armata and for this reason they were renamed as Squatina sp. The minimum and maximum intra-MOTU divergences for Squatina sp. were 0 and 0.33%, respectively.

4. Discussion

Only two valid species had been previously described for the EPO: Squatina californica and Squatina armata [7,10,28], and both species have been reported as sympatric for the northern Peruvian coast [15]. Nonetheless, there is controversy about the validity of these species due to the lack of taxonomic studies confirming their presence across the whole range of their geographic extents [10]. Our results show a new scenario, reporting the existence of four MOTUs along the EPO (Figure 3), revealing a hidden diversity that may include at least one new species for the genus Squatina in the Southeast Pacific.

The Bayesian phylogenetic tree obtained shows a similar topology compared with the comprehensive phylogenetic analysis carried out by Stelbrink et al. [4], and all species delimitation analyses performed (i.e., PTP, bPTP, and GMYC) yielded mostly the same result. The four MOTUs found in the EPO were grouped within the clade of North and South American species described by Stelbrink et al. [4]. However, the four MOTUs are not phylogenetically closely related with the exception of the MOTU of S. californica from North America and the MOTU of S. californica from northern Peru that are sister MOTUs (Figure 3).

All the algorithms used split the cluster of S. californica into two MOTUs. One group includes specimens collected along the Pacific coast of the United States and specimens from the Gulf of California, the other group includes sequences from Punta Pico located in the region of Tumbes, Peru (Figure 3). In some previous studies, two different population lineages originated in the Northern Pacific and in the Gulf of California were found [4,13,64], but this division was not supported by the coalescent species delimitation methods herein used, instead of reciprocal monophyly. Nonetheless, the minimum K2-P genetic distances within the MOTUs of S. californica from the northern hemisphere (0.5%) are comparable to the lowest genetic distances found between Squatina nominal species (0.5% for S. guggenheim and S. occulta). In regard to S. californica from the southern hemisphere, it was discriminated as a different MOTU from the clade of the northern hemisphere, even though the maximum K2-P genetic distances between both clades is 0.5%, similar to the values observed between the population from North America and other Squatina nominal species. This significant heterogeneity between S. californica populations may be promoted by their ecological behavior (e.g., limited ability for sustained swimming due to their morphological and anatomical characteristics) [13], their reproductive behavior (e.g., philopatric behavior) [28] but also due to geographic barriers (e.g., deep marine basins as barriers to dispersal of these populations) [13]. To elucidate if the degree of separation of these MOTUs of S. californica corresponds to a recent divergent species or a strong population structure, more studies including specimens of different ontogenetic stages collected along the Eastern Pacific, their morphological, anatomical, and ecological traits and nuclear markers to evaluate the degree of separation of these MOTUs, assessing introgression and gene flow, are necessary.

A third EPO MOTU corresponds to samples matching the original *S. armata* description (e.g., the presence of narrow and simple barbels and anterior nasal flaps fringed, and thorns on the head). This MOTU clade is the first to branch off from the American clade and includes samples from Chile and Peru. The last EPO MOTU corresponds to *Squatina* sp. from northern Peru, a new lineage that represents a potential undescribed new species. The MOTU was related to the clade formed by *S. dumeril* and the two MOTUs of *S. californica*, but with a low posterior probability support and their exact evolutionary relationships remain unknown (Figure 3). Additionally, the nearest species to *Squatina* sp. using K2-P distance was *S. guggenheim*, with a higher value (1.5%) compared to other nominal *Squatina* species. An integrative taxonomic study to evaluate if this MOTU corresponds to a new species is imperative. In addition, since there are no taxonomic studies providing a direct comparison of both EPO species, besides the two original descriptions, a revision is necessary of *S. armata* and *S. californica* to clarify characteristics and dichotomous identification keys that can help with field identification along the distribution.

The exact distributions of *S. californica* and *S. armata* or even of the four MOTUs remain uncertain. Published studies reporting the presence of angel sharks off the coasts of South America are based on literature reviews [16,17] or governmental reports that do not clearly differentiate between both species, reporting angel sharks as *Squatina* spp. [65]. Similarly, previous distributions reports could be erroneous due to the presence of the cryptic species herein described or by the difficulties during species assignment due to the lack of clear morphological taxonomic characters to distinguish these species.

The limited movement, site fidelity, and preference for coastal areas, as well as other characteristics, such as their large body size and low reproductive output, make angel sharks susceptible to overexploitation by fisheries [28]. In Peru, both *S. californica* and *S. armata* have been reported as caught or landed by small scale fisheries by the Institute of Peruvian Sea (Instituto del Mar del Peru in Spanish; IMARPE) [19,65,66]. Before 1996, landings of angel sharks were reported under their common name in Spanish "angelotes" [65]. Nonetheless, some fishing expedition reports, catalogues and identification guides mentioned only the presence of *S. armata* [19,66,67].

From 1996 to 2010, all fishery landings of angel sharks along the coast were reported by IMARPE as *S. californica* [65]. These reports showed a marked decline over the period 1996 to 2010 [65]. Currently, the fishery continues and there is still a lack of detailed landing reports of angel sharks in Peru [68]. Managing angel sharks in groups (e.g., genus level) can mask population declines and can represent an impediment to fisheries research but also it may hamper the national enforcement regulation for conservation and management [69]. Landings information along the Peruvian coast (2010–2019) is still reported as *Squatina* sp. by the Ministry of Production (Ministerio de Producción in Spanish; PRODUCE, Nro Registro 00090925-2020) indicating, at least for the northern region of Peru (i.e., Tumbes, Piura, Lambayeque, and La Libertad regions) a decline of caught specimens reported in tons. Due to the complicated taxonomy of angel sharks added to a potential new, undescribed species, landings reports might be underestimating the population decline of all the species distributed in Peruvian waters. As well as in Peru, in the United States and Mexico, a decline has also been observed [28,70–72], for instance, reported landings from US fisheries showed a large decline influenced by fisheries regulation (e.g., minimum landing size and ban on gillnets and trammel nets) [28].

In our species delimitation analysis, that included 19 from the 22 nominal *Squatina* species, some differences were observed between MOTUs and nominal data that deserve further discussion. All species delimitations split *S. africana* into four MOTUs. A recent study which included sequences from South Africa and Indian Ocean specimens, found a high number of haplotypes within *S. africana* [73] using DNA fragments of COI gene, and with a high K2-P genetic distances (up to 2.5%), when compared with values reported in this study between other pairs of nominal *Squatina* species (e.g., 0.99% between specimens of *S. albipunctata* and *S. pseudocellata*). Similar to *S. californica*, the split of *S. africana* into 4 MOTUs may be explained due to the life-traits of angel sharks and the geographical barriers. Along the South African coast several barriers to gene flow have been detected, which often coincide with patterns of ocean currents [74]. For example, the South-East and East coasts include three marine ecoregions, the Agulhas Ecoregion, Natal Ecoregion, and part of the Delagoa Ecoregion. Biogeographic forcing agents defining these ecoregions include bathymetric or coastal complexity and currents [75]. This is evident along the east side of South Africa, where a very narrow extension of the continental shelf and wide, deep areas are observed between the northern part of the Agulhas Ecoregion and the Southern part of the Natal Ecoregion. The South-East and East coasts appear to have several population genetic boundaries for several rocky shore and estuarine species (e.g., the fishes *Clinus cottoides, Clinus superciliosus,* and *Muraenoclinus dorsalis*) [74,76]. Although previous studies showed a genetic break in these areas, the effect of biogeographic barriers on demersal shark species is still largely unknown and needs further investigation [77]. Finally, the MOTU *S. africana* 1 was collected in the Indian Ocean close to the Maldives, far from the sampling area of MOTUs *S. africana* 3 and *S. africana* 4 which is characterized by different oceanographic and geomorphologic features.

Samples of *S. nebulosa* (sampling location in Japan and Thailand) and *S. formosa* (sampling location in China and Sea of South China) from BOLD System were clustered in the same MOTU sharing the same haplotype. Diagnosing both species is particularly challenging, their subtle external morphological differences may cause an overlap among many characteristics used [31]. Due to the difficulties in the correct identification of these species, the source from where we retrieved the sequences (i.e., GenBank which lacks physical voucher information) and the fact the taxonomy for these individuals were not revised, our result needs to be tested in an integrative taxonomic study including morphology and genetic analyses.

Finally, our results have far-reaching implications for the management and conservation policies of angel sharks in EPO, suggesting the presence of an undescribed species inhabiting the northern Peruvian coast, living in sympatry with the two species already reported. Many species of angel sharks are affected by artisanal and industrial fisheries and their populations have declined around the world [28]. All efforts to quantify the impact of fisheries in South American angel sharks will be unsuccessful without a precise identification of the species. The use of molecular approaches, such as DNA barcoding, has a potential to quickly identify undescribed species in poorly studied regions like the Southeast Pacific [78]; and in ecologically and economically important groups like Elasmobranchii that hold a high level of taxonomic uncertainties [78].

Supplementary Materials: The following are available online at https://www.mdpi.com/1424-2818/13/5/177/s1, Table S1: Complete list of COI sequences of angel sharks retrieved from BOLD System and used in this study, including collection location and access numbers Bold System and GenBank Result S1: Final matrix containing all the DNA sequences worked on this paper.

Author Contributions: Conceptualization, R.M.C.-A. and J.L.R.; methodology R.M.C.-A.; J.L.R.; formal analysis, R.M.C.-A. and J.L.R.; investigation, R.M.C.-A.; resources, R.S.-R. and J.C.M.; data curation, J.A.-S.; writing—original draft preparation, R.M.C.-A., C.O.-A., R.S.-R., E.A.-C., and J.L.R.; writing—review and editing, J.A.-S., J.C.M., C.O.-A., R.S.-R., E.A.-C., X.V.-Z., C.Y., and J.L.R. All authors have read and agreed to the published version of the manuscript.

Funding: This research was supported by the Consejo Nacional de Ciencia, Tecnología e Innovación Tecnológica (CONCYTEC)–Peru (Círculo de Investigación 023-2016-FONDECYT) and by the Vicerrectorado de Investigación de la Universidad Nacional Mayor de San Marcos (Número de proceso B18100094).

Institutional Review Board Statement: Not applicable.

Informed Consent Statement: Not applicable.

Data Availability Statement: Not applicable.

Acknowledgments: We are grateful to the Department of Ichthyology of at the Natural History Museum of San Marcos University for housing the specimen of angel shark collected for this study.

Conflicts of Interest: The authors declare no conflict of interest.

References

1. Compagno, L.J.V. *FAO Species Catalogue—Volume 4, Sharks of the World: An Annotated and Illustrated Catalogue of Shark Species Known to Date. Part 1—Hexanchiformes to Lamniformes*; FAO: Rome, Italy, 1984; pp. 503–512.
2. Compagno, L.J.V. *Sharks of the World: An Annotated and Illustrated Catalogue of Shark Species Known to Date—Volume 2, Bullhead, Mackerel and Carpet Sharks (Heterodontiformes, Lamniformes and Orectolobiformes)*; FAO Species Catalogue for Fishery Purposes: Rome, Italy, 2001; Volume 2, p. 278.
3. Vélez-Zuazo, X.; Agnarsson, I. Molecular Phylogenetics and Evolution Shark tales: A molecular species-level phylogeny of sharks (Selachimorpha, Chondrichthyes). *Mol. Phylogenet. Evol.* **2011**, *58*, 207–217. [CrossRef] [PubMed]
4. Stelbrink, B.; Von Rintelen, T.; Cliff, G.; Kriwet, J. Molecular systematics and global phylogeography of angel sharks (genus *Squatina*). *Mol. Phylogenet. Evol.* **2010**, *54*, 395–404. [CrossRef] [PubMed]
5. Acero, P.A.; Tavera, J.J.; Anguila, R.; Hernández, L. A New Southern Caribbean Species of Angel Shark (Chondrichthyes, Squaliformes, Squatinidae), Including Phylogeny and Tempo of Diversification of American Species. *Copeia* **2016**, *104*, 577–585. [CrossRef]
6. Vaz, D.F.B.; De Carvalho, M.R. New Species of *Squatina* (Squatiniformes: Squatinidae) from Brazil, with Comments on the Taxonomy of Angel Sharks from the Central and Northwestern Atlantic. *Copeia* **2018**, *106*, 144–160. [CrossRef]
7. Weigmann, S. Annotated checklist of the living sharks, batoids and chimaeras (Chondrichthyes) of the world, with a focus on biogeographical diversity. *J. Fish Biol.* **2016**, *88*, 837–1037. [CrossRef]
8. Castro-Aguirre, J.L.; Pérez, H.E.; Campos, L.H. Dos nuevas especies del género *Squatina* (Chondrichthyes: Squatinidae) del Golfo de México. *Rev. Biol. Trop.* **2006**, *54*, 1031–1040. [PubMed]
9. Fricke, R.; Eschmeyer, W.N.; Van Der Laan, R. (Eds.) Eschmeyer's Catalog of Fishes: Genera, Species, References. Available online: http://researcharchive.calacademy.org/research/ichthyology/catalog/fishcatmain.asp (accessed on 5 January 2021).
10. Ebert, D.A.; Fowler, S.L.; Compagno, L.J. *Sharks of the World: A Fully Illustrated Guide*; Wild Nature Press: Plymouth, UK, 2013.
11. Del Moral-Flores, L.F.; Morrone, J.J.; Durand, J.A.; Espinosa-Pérez, H.; De León Pérez-Ponce, G. Lista patrón de los tiburones, rayas y quimeras de México. *Arx. Miscellània Zoològica* **2015**, *13*, 47–163. [CrossRef]
12. Ehemann, N.R.; del González-González, L.V.; Chollet-Villalpando, J.G.; De La Cruz-Agüero, J. Updated checklist of the extant Chondrichthyes within the Exclusive Economic Zone of Mexico. *ZooKeys* **2018**, *774*, 17–39. [CrossRef] [PubMed]
13. Ramírez-Amaro, S.; Ramírez-Macías, D.; Vázquez-Juárez, R.; Flores-Ramírez, S.; Galván-Magaña, F.; Gutiérrez-Rivera, J.N. Estructura poblacional del tiburón ángel del Pacífico (*Squatina californica*) a lo largo de la costa noroccidental de México con base en la región control del ADN mitochondrial. *Ciencias Mar.* **2017**, *43*, 69–80. [CrossRef]
14. Jacquet, J.; Alava, J.J.; Pramod, G.; Henderson, S.; Zeller, D. In hot soup: Sharks captured in Ecuador's waters. *Environ. Sci.* **2008**, *5*, 269–283. [CrossRef]
15. Chirichigno, N.; Cornejo, R. *Catalogo Comentado Peces Marinos del Perú*; Instituto del Mar del Peru: Callao, Peru, 2001; p. 217.
16. Bustamante, C.; Vargas-Caro, C.; Bennett, M.B. Not all fish are equal: Functional biodiversity of cartilaginous fishes (Elasmobranchii and Holocephali) in Chile. *J. Fish Biol.* **2014**, *85*, 1617–1633. [CrossRef] [PubMed]
17. Cornejo, R.; Vélez-Zuazo, X.; González-Pestana, A.; Kouri, C.; Mucientes, G. An updated checklist of Chondrichthyes from the southeast Pacific off Peru. *Check List* **2015**, *17*. [CrossRef]
18. Calle-Morán, M.D.; Béarez, P. Updated checklist of marine cartilaginous fishes from continental and insular Ecuador (Tropical Eastern Pacific Ocean). *Cybium Rev. Int. d'Ichtyologie.* **2020**, *44*, 239–250.
19. Mejía-Gallegos, J.; Flores-Portugal, L.A.; Segura, G. *Exploración sobre Recursos Costeros y Recursos Demersales; Crucero 7104 B/I SNP 1*; Instituto del Mar del Peru: Callao, Peru, 1971; p. 16.
20. Espinoza, M.; Diaz, E.; Angulo, A.; Hernández, S.; Clarke, T.M. Chondrichthyan Diversity, Conservation Status, and Management Challenges in Costa Rica. *Front. Mar. Sci.* **2018**, *5*, 1–15. [CrossRef]
21. Cailliet, G.M.; Chabot, C.L.; Nehmens, M.C.; Carlsle, A.B. *Squatina californica* (amended version of 2016 assessment). Available online: https://www.iucnredlist.org/species/39328/177163701 (accessed on 20 February 2021).
22. Dulvy, N.K.; Acuña, E.; Bustamante, C.; Cavallos, A.; Herman, K.; Navia, A.F.; Pardo, S.A.; Velez-Zuazo, X. *Squatina armata*. Available online: https://www.iucnredlist.org/species/44571/116831653 (accessed on 20 February 2021).
23. Chirichigno, N.; Velez, J. *Clave para Identificar los Peces Marinos del Perú*; Instituto del Mar del Perú: Callao, Peru, 1998; p. 31.
24. Last, P.R.; White, W.T. Three new angel sharks (Chondrichthyes: Squatinidae) from the Indo-Australian region. *Zootaxa* **2008**, *1734*, 1–26. [CrossRef]
25. Kato, S.; Springer, S.; Wagner, M.H. *Field Guide to Eastern Pacific and Hawaiian Sharks*; United States Department of the Interior: Washington, DC, USA, 1967; Volume 271, 47p.
26. Ayres, W.O. On new fishes of the California coast. *Proc. Calif. Acad. Sci.* **1859**, *2*, 25–32.
27. Philippi, R.A. Sobre los tiburones y algunos otros peces de Chile. *An. Univ. Chile* **1887**, *71*, 535–574.
28. Ellis, J.R.; Barker, J.; McCully Phillips, S.R.; Meyers, E.K.M.; Heupel, M. Angel sharks (Squatinidae): A review of biological knowledge and exploitation. *J. Fish Biol.* **2021**, *98*, 592–621. [CrossRef]

29. Domingues, R.R.; Hilsdorf, A.W.S.; Gadig, O.B.F. The importance of considering genetic diversity in shark and ray conservation policies. *Conserv. Genet.* **2017**, *19*, 501–525. [CrossRef]
30. Hosegood, J.; Humble, E.; Ogden, R.; De Bruyn, M.; Creer, S.; Stevens, G.M.W.; Abudaya, M.; Bassos-Hull, K.; Bonfil, R.; Fernando, D.; et al. Phylogenomics and species delimitation for effective conservation of manta and devil rays. *Mol. Ecol.* **2020**, *29*, 4783–4796. [CrossRef]
31. Walsh, J.H.; Ebert, D.A. A review of the systematics of western North Pacific angel sharks, genus *Squatina*, with redescriptions of *Squatina formosa*, *S. japonica*, and *S. nebulosa* (Chondrichthyes: Squatiniformes, Squatinidae). *Zootaxa* **2007**, *1551*, 31–47. [CrossRef]
32. Walsh, J.H.; Ebert, D.A.; Compagno, L.J.V. *Squatina caillieti* sp. nov., a new species of angel shark (Chondrichthyes: Squat-iniformes: Squatinidae) from the Philippine Islands. *Zootaxa* **2011**, *59*, 49–59. [CrossRef]
33. Theiss, S.M.; Ebert, D.A. Lost and found: Recovery of the holotype of the ocellated angelshark, *Squatina tergocellatoides* Chen, 1963 (Squatinidae), with comments on western Pacific squatinids. *Zootaxa* **2013**, *3752*, 73–85. [CrossRef]
34. Vaz, D.F.B.; De Carvalho, M.R. Morphological and taxonomic revision of species of *Squatina* from the Southwestern Atlantic Ocean (Chondrichthyes: Squatiniformes: Squatinidae). *Zootaxa* **2013**, *3695*, 1–81. [CrossRef] [PubMed]
35. Ward, R.D.; Zemlak, T.S.; Innes, B.H.; Last, P.R.; Hebert, P.D. DNA barcoding Australia's fish species. *Philos. Trans. R. Soc. Lond. B Biol. Sci.* **2005**, *360*, 1847–1857. [CrossRef]
36. Ward, R.D.; Holmes, B.H.; White, W.T.; Last, P.R. DNA barcoding Australasian chondrichthyans: Results and potential uses in conservation. *Mar. Freshw. Res.* **2008**, *59*, 57–71. [CrossRef]
37. Velez-Zuazo, X.; Alfaro-Shigueto, J.; Mangel, J.; Papa, R.; Agnarsson, I. What barcode sequencing reveals about the shark fishery in Peru. *Fish. Res.* **2015**, *161*, 34–41. [CrossRef]
38. Borsa, P.; Arlyza, I.S.; Hoareau, T.B.; Shen, K.-N. Diagnostic description and geographic distribution of four new cryptic species of the blue-spotted maskray species complex (Myliobatoidei: Dasyatidae; *Neotrygon* spp.) based on DNA sequences. *J. Oceanol. Limnol.* **2017**, *36*, 827–841. [CrossRef]
39. Smart, J.J.; Chin, A.; Baje, L.; Green, M.E.; Appleyard, S.A.; Tobin, A.J.; Simpfendorfer, C.A.; White, W.T. Effects of Including Misidentified Sharks in Life History Analyses: A Case Study on the Grey Reef Shark *Carcharhinus amblyrhynchos* from Papua New Guinea. *PLoS ONE* **2016**, *11*, e0153116. [CrossRef]
40. Cariani, A.; Messinetti, S.; Ferrari, A.; Arculeo, M.; Bonello, J.J.; Bonnici, L.; Cannas, R.; Carbonara, P.; Cau, A.; Charilaou, C.; et al. Improving the Conservation of Mediterranean Chondrichthyans: The ELASMOMED DNA Barcode Reference Library. *PLoS ONE* **2017**, *12*, e0170244. [CrossRef]
41. Bernardo, C.; Corrêa de Lima Adachi, A.M.; Paes da Cruz, V.; Foresti, F.; Loose, R.H.; Bornatowski, H. The label "Cação" is a shark or a ray and can be a threatened species! Elasmobranch trade in Southern Brazil unveiled by DNA barcoding. *Mar. Policy* **2020**, *116*, 103920. [CrossRef]
42. Bunholi, I.V.; da Silva Ferrette, B.L.; De Biasi, J.B.; de Oliveira Magalhães, C.D.; Rotundo, M.M.; Oliveira, C.; Foresti, F.; Mendonça, F.F. The fishing and illegal trade of the angelshark: DNA barcoding against misleading identifications. *Fish. Res.* **2018**, *206*, 193–197. [CrossRef]
43. Vella, A.; Vella, N.; Schembri, S. A molecular approach towards taxonomic identification of elasmobranch species from Maltese fisheries landings. *Mar. Genom.* **2017**, *36*, 17–23. [CrossRef] [PubMed]
44. Navia, A.F.; Mejía-Falla, P. Checklist of marine elasmobranchs of Colombia. *Univ. Sci.* **2019**, *24*, 241–276. [CrossRef]
45. Thompson, J.D.; Higgins, D.G.; Gibson, T.J. CLUSTAL W: Improving the sensitivity of progressive multiple sequence alignment through sequence weighting, position-specific gap penalties and weight matrix choice. *Nucleic Acids Res.* **1994**, *22*, 4673–4680. [CrossRef]
46. Kumar, S.; Stecher, G.; Tamura, K. MEGA7: Molecular Evolutionary Genetics Analysis Version 7.0 for Bigger Datasets Brief Communication. *Mol. Biol. Evol.* **2016**, *33*, 1870–1874. [CrossRef]
47. Stoeckle, M.Y.; Das Mishu, M.; Charlop-Powers, Z. GoFish: A streamlined environmental DNA presence/absence assay for marine vertebrates. *bioRixv* **2018**, 331322. [CrossRef]
48. Steinke, D.; Connell, A.D.; Hebert, P.D. Linking adults and immatures of South African marine fishes. *Genome* **2016**, *59*, 959–967. [CrossRef]
49. Almerón-Souza, F.; Sperb, C.; Castilho, C.L.; Figueiredo, P.I.C.C.; Gonçalves, L.T.; Machado, R.; Oliveira, L.R.; Valiati, V.H.; Fagundes, N.J.R. Molecular Identification of Shark Meat from Local Markets in Southern Brazil Based on DNA Barcoding: Evidence for Mislabeling and Trade of Endangered Species. *Front. Genet.* **2018**, *9*, 1–12. [CrossRef]
50. Fitzpatrick, C.K.; Finnegan, K.A.; Osaer, F.; Narváez, K.; Shivji, M.S. The complete mitochondrial genome of the Critically Endangered Angelshark, *Squatina squatina*. *Mitochondrial DNA Part B Resour.* **2017**, *2*, 212–213. [CrossRef]
51. Moftah, M.; Aziz, S.H.A.; Elramah, S.; Favereaux, A. Classification of Sharks in the Egyptian Mediterranean Waters Using Morphological and DNA Barcoding Approaches. *PLoS ONE* **2011**, *6*, e27001. [CrossRef]
52. Gao, Y.; Liu, T.; Wei, T.; Geng, X.; Wang, J.; Ma, H. Complete mitochondrial genome of Clouded angelshark (*Squatina nebulosa*). *Mitochondrial DNA Part A* **2016**, *27*, 1599–1600. [CrossRef]
53. De Oliveira Ribeiro, A.; Caires, R.A.; Mariguela, T.C.; Pereira, L.H.G.; Hanner, R.; Oliveira, C. DNA barcodes identify marine fishes of São Paulo State, Brazil. *Mol. Ecol. Resour.* **2012**, *12*, 1012–1020. [CrossRef]
54. Keskin, E.; Atar, H.H. DNA barcoding commercially important fish species of Turkey. *Mol. Ecol. Resour.* **2013**, *13*, 788–797. [CrossRef]

55. Wang, Z.-D.; Guo, Y.-S.; Liu, X.-M.; Fan, Y.-B.; Liu, C.-W. DNA barcoding South China Sea fishes. *Mitochondrial DNA* **2012**, *23*, 405–410. [CrossRef]
56. Sembiring, A.; Pertiwi, N.P.D.; Mahardini, A.; Wulandari, R.; Kurniasih, E.M.; Kuncoro, A.W.; Cahyani, N.D.; Anggoro, A.W.; Ulfa, M.; Madduppa, H.; et al. DNA barcoding reveals targeted fisheries for endangered sharks in Indonesia. *Fish. Res.* **2015**, *164*, 130–134. [CrossRef]
57. Fujisawa, T.; Barraclough, T.G. Delimiting Species Using Single-Locus Data and the Generalized Mixed Yule Coalescent Approach: A Revised Method and Evaluation on Simulated Data Sets. *Syst. Biol.* **2013**, *62*, 707–724. [CrossRef] [PubMed]
58. Yule, G.U. A Mathematical Theory of Evolution, Based on the Conclusions of Dr. J.C. Willis, F.R.S. *Philos. Trans. R. Soc. London Ser. B Contain. Pap. Biol. Character* **1925**, *213*, 21–87.
59. Hudson, R.R. Gene genealogies and the coalescent process. *Oxford Surv. Evol. Biol.* **1990**, *7*, 44.
60. Zhang, J.; Kapli, P.; Pavlidis, P.; Stamatakis, A. A general species delimitation method with applications to phylogenetic placements. *Bioinformatics* **2013**, *29*, 2869–2876. [CrossRef] [PubMed]
61. Bouckaert, R.; Heled, J.; Kühnert, D.; Vaughan, T.; Wu, C.-H.; Xie, D.; Suchard, M.A.; Rambaut, A.; Drummond, A.J. BEAST 2: A Software Platform for Bayesian Evolutionary Analysis. *PLoS Comput. Biol.* **2014**, *10*, e1003537. [CrossRef] [PubMed]
62. Darriba, D.; Taboada, G.L.; Doallo, R.; Posada, D. jModelTest 2: More models, new heuristics and parallel computing. *Nat. Methods* **2012**, *9*, 772. [CrossRef] [PubMed]
63. Miller, M.A.; Pfeiffer, W.; Schwartz, T. Creating the CIPRES Science Gateway for inference of large phylogenetic trees. In Proceedings of the 2010 Gateway Computing Environments Workshop (GCE), IEEE, New Orleans, LA, USA, 14 November 2010; pp. 1–8. [CrossRef]
64. Gaida, I.H. Evolutionary Aspects of Gene Expression in the Pacific Angel Shark, *Squatina california* (Squatiniformes: Squatinidae). *Copeia* **1995**, 532–554. [CrossRef]
65. Gonzalez-Pestana, A.; Kouri, C.; Velez-Zuazo, X. Shark fisheries in the Southeast Pacific: A 61-year analysis from Peru. *F1000Research* **2014**, 3, 164. [CrossRef] [PubMed]
66. IMARPE. *Resultados Preliminares del Primer Crucero de Exploración Pesquera del SNP-1 6901*; Instituto del Mar del Perú: Callao, Peru, 1969; pp. 11–12.
67. Chirichigno, N. *Clave para identificar los peces marinos del Perú*; Instituto del Mar del Perú: Callao, Peru, 1974; p. 30.
68. Bartholomew, D.C.; Mangel, J.C.; Alfaro-Shigueto, J.; Pingo, S.; Jimenez, A.; Godley, B.J. Remote electronic monitoring as a potential alternative to on-board observers in small-scale fisheries. *Biol. Conserv.* **2018**, *219*, 35–45. [CrossRef]
69. Cashion, M.S.; Bailly, N.; Pauly, D. Official catch data underrepresent shark and ray taxa caught in Mediterranean and Black Sea fisheries. *Mar. Policy* **2019**, *105*, 1–9. [CrossRef]
70. Leet, W.S.; Dewees, C.M.; Klingbeil, R.; Larson, E.J. California's Living Marine Resources: A Status Report 2001. Available online: https://wildlife.ca.gov/Conservation/Marine/Status/2001 (accessed on 18 January 2021).
71. Pondella, D.J., II; Allen, L.G. The nearshore fish assemblage of Santa Catalina Island. In Proceedings of the Fifth California Islands Symposium, Santa Rosa Island, CA, USA, 29 March–1 April 1999; pp. 394–400.
72. Bizzarro, J.J.; Smith, W.D.; Hueter, R.E.; Villavicencio–Garayzar, C.J. Activities and Catch Composition of Artisanal Elasmobranch Fishing Sites on the Eastern Coast of Baja California Sur, Mexico. *Bull. South Calif. Acad. Sci.* **2009**, *108*, 137–151. [CrossRef]
73. Ambily, M.N.; Zacharia, P.U.; Najmudeen, T.M.; Ambily, L.; Sunil, K.T.S.; Radhakrishnan, M.; Kishor, T.G. First Record of African Angel Shark, *Squatina africana* (Chondricthyes: Squatinidae) in Indian Waters, Confirmed by DNA Barcoding. *J. Ichthyol.* **2018**, *58*, 312–317. [CrossRef]
74. Von Der Heyden, S. Why do we need to integrate population genetics into South African marine protected area planning? *Afr. J. Mar. Sci.* **2009**, *31*, 263–269. [CrossRef]
75. Spalding, M.D.; Fox, H.E.; Allen, G.R.; Davidson, N.; Ferdaña, Z.A.; Finlayson, M.; Halpern, B.S.; Jorge, M.A.; Lombana, A.; Lourie, S.A.; et al. Marine Ecoregions of the World: A Bioregionalization of Coastal and Shelf Areas. *Bioscience* **2007**, *57*, 573–583. [CrossRef]
76. Von Der Heyden, S.; Prochazka, K.; Bowie, R.C.K.; Prochazka, K. Significant population structure and asymmetric gene flow patterns amidst expanding populations of *Clinus cottoides* (Perciformes, Clinidae): Application of molecular data to marine conservation planning in South Africa. *Mol. Ecol.* **2008**, *17*, 4812–4826. [CrossRef] [PubMed]
77. Van der Merwe, A.E.B.; Gledhill, K.S. Molecular species identification and population genetics of chondrichthyans in South Africa: Current challenges, priorities and progress. *Afr. Zoöl* **2015**, *50*, 205–217. [CrossRef]
78. Ramirez, J.L.; Rosas-Puchuri, U.; Cañedo, R.M.; Alfaro-Shigueto, J.; Ayon, P.; Zelada-Mázmela, E.; Siccha-Ramirez, R.; Velez-Zuazo, X. DNA barcoding in the Southeast Pacific marine realm: Low coverage and geographic representation despite high diversity. *PLoS ONE* **2020**, *15*, e0244323. [CrossRef] [PubMed]

Article

Assessing Temporal Patterns and Species Composition of Glass Eel (*Anguilla* spp.) Cohorts in Sumatra and Java Using DNA Barcodes

Arif Wibowo [1,2], Nicolas Hubert [3,*], Hadi Dahruddin [4], Dirk Steinke [5], Rezki Antoni Suhaimi [1,2], Samuel [1,2], Dwi Atminarso [1,2,6], Dian Pamularsih Anggraeni [1,2], Ike Trismawanti [1,2], Lee J. Baumgartner [6] and Nathan Ning [6]

1 Southeast Asian Fisheries Development Center, Inland Fishery Resources Development and Management Department, Jalan H.A. Bastari No. 08, Jakabaring, Palembang 30267, Indonesia; wibarf@yahoo.com (A.W.); rezki.antoni.s@gmail.com (R.A.S.); sam_asr@yahoo.co.id (S.); dwiatminarso@gmail.com (D.A.); nebula_dpa@yahoo.com (D.P.A.); ike_trisnawati@yahoo.co.id (I.T.)
2 Research Institute for Inland Fisheries and Extensions, Agency for Marine and Fisheries Research and Human Resources, Ministry of Marine Affairs and Fisheries, Jalan H.A. Bastari No. 08, Jakabaring, Palembang 30267, Indonesia
3 UMR 5554 ISEM, IRD, Universite Montpellier, CNRS, EPHE, Université de Montpellier, Place Eugène Bataillon, CEDEX 05, 34095 Montpellier, France
4 Division of Zoology, Research Center for Biology, Indonesian Institute of Sciences (LIPI), Jalan Raya Jakarta Bogor Km 46, Cibinong 16911, Indonesia; dahruddinhadi@gmail.com
5 Department of Integrative Biology, Centre for Biodiversity Genomics, 50 Stone Rd E, Guelph, ON N1G 2W1, Canada; dsteinke@uoguelph.ca
6 Institute for Land, Water and Society, Charles Sturt University, P.O. Box 789, Albury, NSW 2640, Australia; lbaumgartner@csu.edu.au (L.J.B.); nning@csu.edu.au (N.N.)
* Correspondence: nicolas.hubert@ird.fr

Citation: Wibowo, A.; Hubert, N.; Dahruddin, H.; Steinke, D.; Suhaimi, R.A.; Samuel; Atminarso, D.; Anggraeni, D.P.; Trismawanti, I.; Baumgartner, L.J.; et al. Assessing Temporal Patterns and Species Composition of Glass Eel (*Anguilla* spp.) Cohorts in Sumatra and Java Using DNA Barcodes. *Diversity* **2021**, *13*, 193. https://doi.org/10.3390/d13050193

Academic Editor: Eric Buffetaut and Manuel Elias-Gutierrez

Received: 16 March 2021
Accepted: 23 April 2021
Published: 29 April 2021

Publisher's Note: MDPI stays neutral with regard to jurisdictional claims in published maps and institutional affiliations.

Abstract: Anguillid eels are widely acknowledged for their ecological and socio-economic value in many countries. Yet, knowledge regarding their biodiversity, distribution and abundance remains superficial—particularly in tropical countries such as Indonesia, where demand for anguillid eels is steadily increasing along with the threat imposed by river infrastructure developments. We investigated the diversity of anguillid eels on the western Indonesian islands of Sumatra and Java using automated molecular classification and genetic species delimitation methods to explore temporal patterns of glass eel cohorts entering inland waters. A total of 278 glass eels were collected from monthly samplings along the west coast of Sumatra and the south coast of Java between March 2017 and February 2018. An automated, DNA-based glass eel identification was performed using a DNA barcode reference library consisting of 64 newly generated DNA barcodes and 117 DNA barcodes retrieved from BOLD for all nine *Anguilla* species known to occur in Indonesia. Species delimitation methods converged in delineating eight Molecular Operational Taxonomic Units (MOTUs), with *A. nebolusa* and *A. bengalensis* being undistinguishable by DNA barcodes. A total of four MOTUs were detected within the glass eel samples, corresponding to *Anguilla bicolor*, *A. interioris*, *A. marmorata*, and *A. nebulosa*/*A. bengalensis*. Monthly captures indicated that glass eel recruitment peaks in June, during the onset of the dry season, and that *A. bicolor* is the most prevalent species. Comparing indices of mitochondrial genetic diversity between yellow/silver eels, originating from several sites across the species range distribution, and glass eels, collected in West Sumatra and Java, indicated a marked difference. Glass eels displayed a much lower diversity than yellow/silver eels. Implications for the management of glass eel fisheries and species conservation are discussed.

Keywords: species delimitation; DNA-based classification; genetic diversity; catadromy; conservation

1. Introduction

The freshwater eel family Anguillidae consists of 20 species and two genera [1], all well known for their catadromous life-cycles. Adults spawn in marine environments, and hatched larvae, known as leptocephalus, migrate to inland waters. Upon approaching continental shelfs, larvae become competent (glass eels) before entering inland waters, where they acquire pigmentation (elvers). Eels further grow up in freshwaters (yellow eels) until they become sexually mature (silver eels) and return to marine spawning grounds [2–6].

Anguillid eels are a significant food resource around the world, and contribute greatly to many national economies [2,7–10]. For example, approximately 150,000 tons of eels were consumed each year both in Japan (2000–2002) and China (2012–2013) [11]. In Indonesia, the annual eel trade is estimated to be worth 100 million USD [12]. Demand for anguillid species has steadily increased in recent years; however, eel farming solely relies on wild-caught juvenile eels (elvers, silver eels), as breeding in captivity is not yet commercially viable [11,13,14]. The growing demand, in combination with habitat destruction and fragmentation caused by dams and other infrastructures, has led to a precipitous decline in populations across the globe, and some species have become endangered [2,15–19]. Unfortunately, scientific knowledge regarding eel diversity, distribution and abundance is sparse [15,19]. In particular, little is known about the ecology of anguillid eels in the tropics, even though such regions typically support relatively high abundances and species richness' of anguillid eels [7,20–22]. Indeed, two-thirds of the known species of the genus, *Anguilla*, occur in the tropical Indian and Pacific Ocean, while the remaining species occur in temperate parts of both the Pacific and Atlantic Ocean [1]. For Indonesia, a total of seven to nine species has been reported, as the biological status of two species pairs (*A. bengalensis* vs. *A. nebulosa*, *A. borneensis* vs. *A. malgumora*) is still under debate [7,22–24]. This diversity is among the highest in the Indo-Pacific Ocean; however, few attempts have been made to document diversity and distribution of *Anguilla* spp. in Indonesia. Previous studies have mostly focused on either specific life stages or selected species [7,22,25,26]. Therefore, there is a clear and pressing need to assess spatial and temporal patterns of all life stages of tropical anguillid eels to gain a comprehensive understanding of their ecology.

Studies of biodiversity and distribution of tropical anguillid eels were long impeded by difficulties in identifying species using morphological characters, particularly for early life stages [27]. The development of standardized molecular approaches to automated specimen identification, such as DNA barcoding [28,29], opened new perspectives for monitoring and the exploration of species boundaries. Several studies have already successfully applied DNA barcoding to characterize the Indonesian ichthyofauna in various contexts, ranging from island inventories [30] over lineage-specific re-appraisals [31–36] to larval identification [37,38].

This study investigated the diversity of anguillid eels on the western Indonesian islands of Sumatra and Java using DNA barcoding. Anguillid glass eels are currently being intensively harvested in the western parts of Indonesia, particularly *A. bicolor* and *A. marmorata* on Java [22]. In addition, the rapid development of dams and other river infrastructures is compromising upstream glass eel migration. Therefore, this study aimed to: (1) establish a DNA barcode reference library for all known *Anguilla* species occurring in Indonesia based on both available and newly generated Cytochrome Oxidase I (COI) sequences, (2) apply this library to automated glass eel identification to examine temporal patterns of recruitment and species composition of glass eel cohorts in western Indonesia, where the fishing pressure is currently high.

2. Materials and Methods

2.1. Sampling

Sampling consisted of two distinct campaigns in Indonesia targeting freshwater life stages (yellow/silver eels) and early life stages (glass eels/elvers) entering freshwaters. Yellow/silver eels were sampled across several sites on Sumatra, Java, Bali, Lombok, Sulawesi and Ambon Islands between November 2012 and July 2019 using electrofishing and fishing rods (DS-ANGUILLA; doi:dx.doi.org/10.5883/DS-ANGUILLA). Specimens were photographed, individually labeled and voucher specimens were preserved in a 5% formalin solution. Prior to fixation, a fin clip or a muscle biopsy was taken and fixed separately in a 96% ethanol solution for further genetic analyses. Both tissues and voucher specimens were deposited in the National Collections at the Museum Zoologicum Bogoriense (MZB), Research Center for Biology (RCB), Indonesian Institute of Sciences (LIPI). Specimens were further identified using field guides [39–41]. Glass eels and elvers were sampled using a systematic and standardized procedure, developed for the purpose of this study, on Sumatra and Java between March 2017 and February 2018 (Figure 1). A trap device (dimensions 3.5 m × 2 m × 0.8 m) was used for 6-h periods after dark once every week (Figure 2). Glass eel and elver samples were collected and preserved in 70% ethanol solution for further genetic analysis.

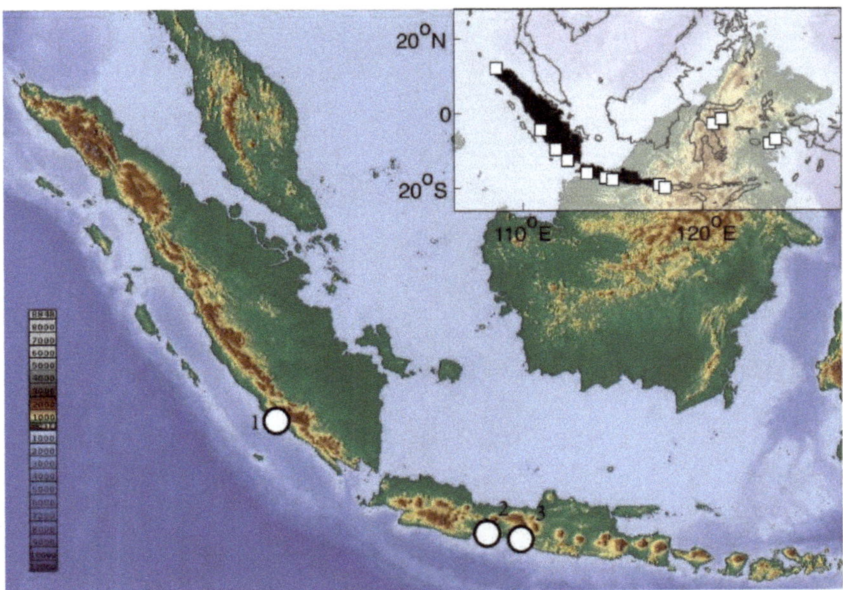

Figure 1. Collection sites for the 278 glass eels (white circle) and 64 yellow/silver eels (white squares) sampled in this study. 1: Tengah Kedurang, Kedurang river, Bengkulu Selatan, Bengkulu (−4.462, 103,067); 2: Serayu River, Purwokerto, Jawah Tengah (−7.508, 109.296); 3: Kebumen, Jawa Tengah (−7.768, 109.644).

2.2. DNA Extraction, Amplification and Sequencing

Total genomic DNA was extracted from muscle tissue using the Geneaid extraction procedure following manufacturer's specifications (www.geneaid.com, accessed on 12 March 2020). A partial fragment of 652 bp of the mitochondrial COI coding gene, corresponding to the standard DNA Barcoding fragment [42], was amplified using the universal primers, Fish-COI-F and COI-Fish-R [43]. The primer sequences were as follows: Fish-COI-F, 5′-TAA TAC GAC TCA CTA TAG GGT TCT CCA CCA ACC ACA ARG AYA TYGG-3′; COI-Fish-R, 5′-ATT AAC CCT CAC TAA AGG GCA CCT CAG GGT GTC CGA ARA AYC ARAA-3′. Amplifications were performed in a 50 µL reaction volume consisting of 16 µL of ultrapure water, 2 µL of each primer (1 µM), 25 µL of PCR ready mixture solution (KAPA), and 25 µL of genomic DNA. The polymerase chain reaction (PCR) cycling parameters included an initial DNA polymerase activation step of 15 min at 95 °C, followed by 35 cycles of 30 s at 94 °C, 90 s at 55 °C and 30 s at 72 °C and ending with a final extension of 5 min at 72 °C. The PCR products were visualized on a 1% agarose gel and purified (Thermo Scientific PCR purification kit). Yellow/silver eel amplicons were bi-directionally sequenced and larval amplicons were sequenced with the reverse primer (COI-Fish-R). Sequencing was done using the EZ-Seq service (Macrogen) for glass eels, and the Centre for Biodiversity Genomics (University of Guelph) for yellow/silver eels. Sequences were deposited in BOLD in the dataset DS_ANGUILLA (doi:dx.doi.org/10.5883/DS-ANGUILLA) and GenBank (accession numbers MN961249-MN961267; MT155391-MT155483). Additional publicly available COI sequences for the *Anguilla* species occurring in Indonesia were retrieved from BOLD [44] and aggregated in the dataset DS-ANGUILLA (doi:dx.doi.org/10.5883/DS-ANGUILLA).

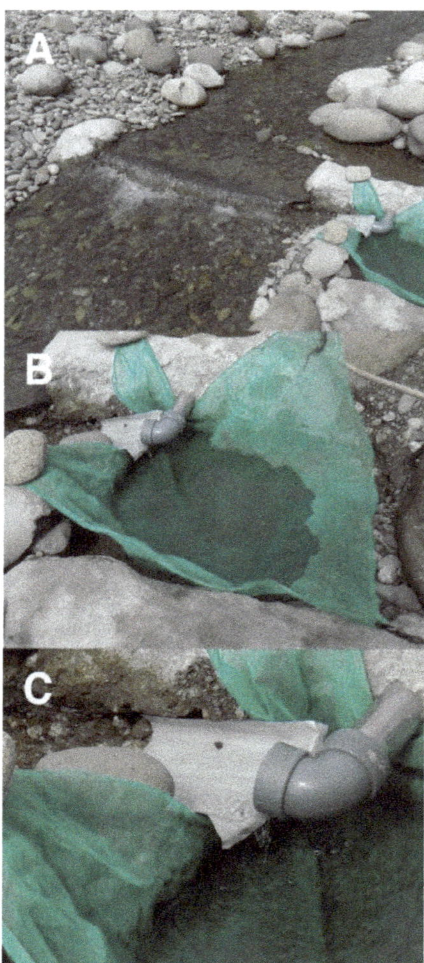

Figure 2. Trap net specifically designed to collect glass eels. (**A**) general overview of the trap; (**B**) collecting section; (**C**) close-up.

2.3. Data Analysis

As one main objective was to implement automated identification of *Anguilla* glass eels and elvers through DNA barcodes, we first established a reference library. The reference library was based both on sequences produced from yellow/silver eel specimens, identified using morphological characters, and on DNA barcode records retrieved from BOLD. To validate each reference, we examined the match of species boundaries, delineated either by morphological characters, or Molecular Operational Taxonomic Units (MOTU), representing diagnosable molecular lineages [45–47]. It is well accepted that a combination of different delimitation approaches is capable of overcoming potential pitfalls arising from uneven sampling [33,48–51]. Therefore, we used a scheme based on a 50 percent consensus between four algorithms: (1) Refined Single Linkage (RESL) as implemented in

BOLD and used to produce Barcode Index Numbers (BIN) [52], (2) Automatic Barcode Gap Discovery (ABGD) [53], (3) Poisson Tree Process (PTP) in its single (sPTP) and multiple (mPTP) rates version as implemented in the stand-alone software mptp_0.2.3 [54,55], and (4) General Mixed Yule-Coalescent (GMYC) in its single (sGMYC) and multiple (mGMYC) rates version as implemented in the R package Splits 1.0–19 [56].

As the mPTP algorithm uses a phylogenetic tree as an input file, a maximum likelihood (ML) tree was first reconstructed using RAxML [57] based on a GTR+I+Γ substitution model. An ultrametric, fully resolved tree was reconstructed using the Bayesian approach implemented in BEAST 2.6.2 [58], to be later applied with the GMYC algorithm. Duplicated sequences were pruned prior to reconstructing the ultrametric tree based on a strict-clock of 1.2% per million year [59]. Two Markov chains, each with a length of 10 million, were run independently using Yule pure birth and GTR+I+Γ substitution models. Trees were sampled every 5000 states, after an initial burnin period of 1 million. Both runs were combined using LogCombiner 2.6.2 and the maximum credibility tree was constructed using TreeAnnotator 2.6.2 [58].

A final COI gene tree was reconstructed using the SpeciesTreeUCLN algorithm of the StarBEAST2 package [60]. This approach implements a mixed-model, including a coalescent component within species and a diversification component between species, that allows accounting for variations of substitution rates within and between species [61]. StarBEAST2 jointly reconstructs gene trees and species trees, and as such requires species designations, which were determined using the consensus of our species delimitation analyses. The StarBEAST2 analysis was performed using the same parameters as the BEAST analysis described above.

Finally, unknown sequences from early life stages were classified to the species level using the R package *BarcodingR* [62]. A single sequence was randomly selected for each of the MOTUs for classification using the following algorithms: (1) BP [62], (2) fuzzy-set [63], and (3) Bayesian [64]. Similarly, the final assignment of an unknown to a known sequence was established based on a 50% consensus.

Several parameters of genetic diversity were estimated using the R package *pegas* 0.14 [65], including the number of haplotypes (*h*), haplotype diversity (*Hd*) [66], nucleotide diversity (π) [67], genetic diversity based on the number of segregating sites (θ) [68], and Tajima's *D* test of neutrality [69]. Kimura 2-parameter (K2P) [70] pairwise genetic distances were calculated using the R package *Ape* 5.4 [71]. Maximum intraspecific and nearest neighbor distances were calculated from the matrix of pairwise K2P genetic distances using the R package *Spider* 1.5 [72].

3. Results

A total of 64 and 278 COI sequences were generated for all yellow/silver and glass eels sampled, respectively. The yellow/silver eel COI sequences were combined with 117 sequences retrieved from BOLD into a DNA barcode reference library comprising 181 COI mostly full-length sequences (652 bp) for the 9 species of *Anguilla* occurring in Indonesia [1,23].

DNA-based species delimitation analyses resulted in congruent delimitation schemes with 7 MOTUS for mPTP, 8 MOTUs for RESL, ABGD and mGMYC, 10 MOTUs for sPTP and 11 MOTUs for sGMYC; with a consensus scheme consisting of 8 MOTUs (Figure 3; Table S1). All MOTUs matched the species boundaries defined by morphology-based identifications, with the exception of *Anguilla bengalensis* and *A. nebulosa*, which displayed tightly intertwined mitochondrial lineages (Figure 3). The coalescent tree of this species pair was the youngest of the *Anguilla* mitochondrial gene trees, with a Most Recent Common Ancestor (MRCA) dated around 500,000 years. By contrast the mitochondrial MRCA of the *Anguilla* analyzed in our study dated back 5.3 million years (Myr). Species showed a distinct barcoding gap, which is defined as the lack of overlap between the distributions of maximum intraspecific and minimum interspecific genetic distance. Maximum intraspecific distances were ten-fold higher on average than minimum interspecific distances (Table 1).

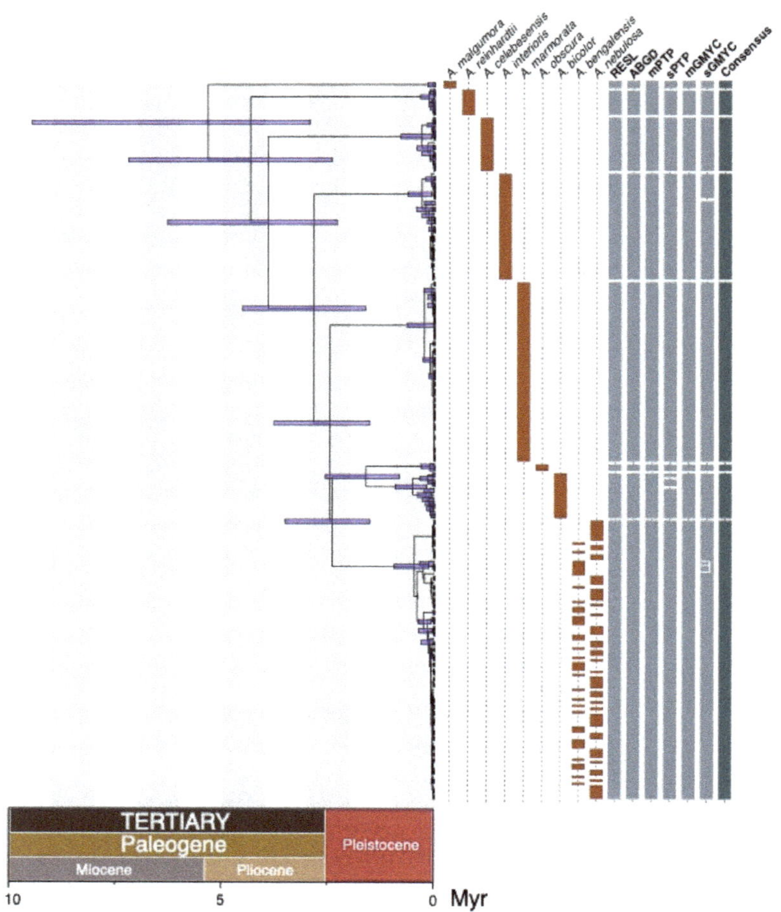

Figure 3. Mitochondrial gene tree of *Anguilla* spp. inferred with SpeciesTreeUCLN, including 95% HPD interval for node age estimates, species boundaries according to morphology-based identifications, and genetic species delimitation results for six methods and their 50% consensus.

Assignment of unknown glass eels/elvers sequences to known yellow/silver eels sequences resulted in fully congruent classifications across the three methods (Table 1 and Table S2). Fuzzy average yielded the lowest support, with assignment probabilities ranging from 0.753 to 0.989 per species on average; while Bayesian assignments were all supported by probabilities of 1 (Table 1). Among the 278 larvae sequences, 15 were assigned to *A. interioris*, 26 to *A. marmorata*, 24 to the species pair *A. bengalensis*/*A. nebulosa* and 213 to *A. bicolor* (Table S2). During the sampling period from February 2017 to March 2018, glass eels were collected only between May and November 2017 (Figure 4). A peak of abundance was observed in May and June, a period corresponding to the onset of the dry season and tightly following the first peak of annual temperatures in May (Figure 4). Multi-species glass eel assemblages were observed throughout the dry season between May and August, and later in November, with a peak of species richness in June when 4 species were collected.

Table 1. MOTUs delimitated among yellow/silver eels based on the majority rule consensus of the six delimitation methods, summary statistics of MOTU genetic distances, assignment probabilities (average, minimum, and maximum) of unknown glass eel sequences according to Fuzzy, Bayesian, and BP, and identifications of glass eels to the species level.

	K2P Genetic Distance			Assignment Probability					
BIN	Max. Intraspecific	Min. Interspecific	Species	Fuzzy Average (Min–Max)	Fuzzy Identif.	Bayesian Average (Min–Max)	Bayesian Identif.	BP Average (Min–Max)	BP Identif.
BOLDAAD2092	0.004	0.031	*A. marmorata*	0.753 (0.636–0.758)	*A. marmorata*	1	*A. marmorata*	0.949 (0.945–0.949)	*A. marmorata*
BOLDAAD9080	0.002	0.060	*A. reinhardtii*						
BOLDAAE4923	0.029	0.044	*A. bicolor*	0.735 (0.572–1)	*A. bicolor*	1	*A. bicolor*	0.885 (0.853–0.963)	*A. bicolor*
BOLDAAJ2664	0.015	0.038	*A. bengalensis*	0.989 (0.969–1)	*A. bengalensis*	1	*A. bengalensis*	0.948 (0.947–0.948)	*A. bengalensis*
			A. nebulosa		*A. nebulosa*		*A. nebulosa*		*A. nebulosa*
BOLDADC7451	0.008	0.038	*A. interioris*	0.789 (0.595–0.919)	*A. interioris*	1	*A. interioris*	0.946 (0.936–0.952)	*A. interioris*
BOLDADC7453	0	0.087	*A. malgumora*						
BOLDADC7455	0.006	0.060	*A. celebesensis*						
BOLDADC8019	0.002	0.035	*A. obscura*						

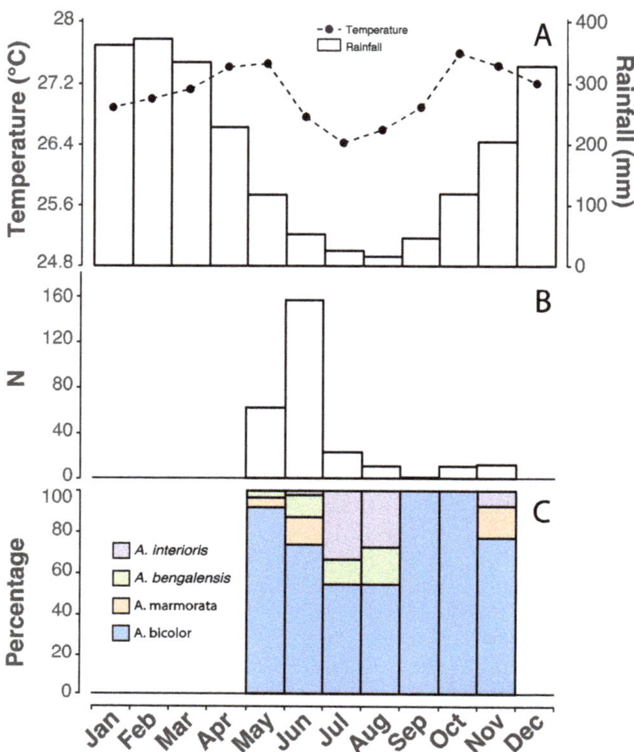

Figure 4. Abundance and species diversity of glass eels collected between March 2017 and April 2018. (**A**), monthly temperatures and rainfalls in South Sumatra and West Java between 1991 and 2016 (modified from World Bank); (**B**), number of monthly collected glass eels (N, number of individual); (**C**), monthly proportions of species among collected glass eels.

All estimates of mitochondrial genetic diversity indicated that glass eels/elvers display a much lower diversity than their species gene pools (Table 2). The number of haplotypes (h) and the haplotypic diversity (Hd) were four-fold and three-fold lower, respectively, in glass eels/elvers than in yellow/silver eels. A similar trend was observed at the nucleotide level, with nucleotide diversity (π) and Theta (θw) being two-and-a-half-fold and ten-fold lower, respectively, in glass eels/elvers than in yellow/silver eels. This result was also reflected by noticeable differences in the Tajima D test, which shifted toward a positive D value in larvae indicating a scarcity of rare alleles. However, none of the Tajima D tests performed were significant, except for the *A. bengalensis*/*A. nebulosa* species pair among yellow/silver eels with a significant negative D value.

Table 2. Summary statistics of genetic diversity per species, including yellow/silver and glass eels/elevers, for the four species detected among glass eel cohorts. *N*, number of individuals; *h*, number of haplotypes; *Hd*, haplotypic diversity; π, nucleotide diversity; θw, theta; Tajima's *D* test including *D* value and significance of the test (* significant). Regional refers to estimates at the scale of species distribution ranges. Local refers to estimates at the scale of glass eel sampling sites.

Species	Stage/Scale	*N*	*h*	*Hd*	π	θw	D	*p*-Value
A. interioris	Yellow-silver eels/Regional	35	9	0.718	0.004	2.428	−0.341	0.7330
	Glass eels/Local	15	2	0.514	0.004	1.230	2.155	0.0312
A. marmorata	Yellow-silver eels/Regional	58	6	0.539	0.003	5.381	−2.158	0.0308
	Glass eels/Local	26	2	0.077	0.001	0.262	−1.156	0.2479
A. nebulosa/A. bengalensis	Yellow-silver eels/Regional	91	16	0.667	0.002	4.132	−2.243	0.0249 *
	Glass eels/Local	24	2	0.489	0.001	0.268	1.391	0.1642
A. bicolor	Yellow-silver eels/Regional	15	11	0.952	0.013	6.766	0.050	0.9603
	Glass eels/Local	213	7	0.476	0.003	1.179	0.997	0.3189

4. Discussion

The present study further highlights the usefulness of DNA barcoding in capturing boundaries between *Anguilla* species [7,22,73–76]. Among the nine species included, seven displayed well-differentiated mitochondrial lineages, with minimum genetic distances to the nearest neighbor largely exceeding maximum intraspecific genetic distance. DNA barcodes failed to differentiate between the species-pair *Anguilla bengalensis* and *A. nebulosa*. The young age of their mitochondrial MRCA may be indicative of a recent divergence. However, the persistence of ancestral polymorphisms between diversifying lineages [77] tend to inflate genetic diversity in diverging pairs [78]. Here, genetic diversity indices show no differences between this species pair and the other species, thus questioning their distinctiveness. Although *A. nebulosa* and *A. bengalensis* are currently both considered as valid species, they have a complex taxonomic history and were synonymized in the past [23]. The lack of a declared type locality in the original description of *A. bengalensis* is further confusing its taxonomic status. From a mitochondrial perspective, this species pair behaves as a single cohesive lineage, which suggests that further studies are needed across their distribution range to establish a more comprehensive assessment of their genetic diversity.

Tropical river systems support assemblages of multiple anguillid species [26,79,80]. The co-occurrence of multiple species in the tropics, particularly Southeast Asia, is thought to be the result of shorter migration routes, easing dispersal [24], and of higher productivity leading to better growth rates and increased carrying capacities of aquatic habitats [81]. The detection of multiple species in glass eel cohorts in Sumatra and Java is consistent with the co-occurrence of multiple species of anguillid eels. Although the four species detected have been previously reported from Western Indonesia [1,23,82], little information is available on diversity patterns and the abundance of yellow/silver eels upstream. In particular, our results question how species abundance in glass eel cohorts translates into yellow/silver eel communities. Ecological dynamics during growth may translate into important shifts in species composition from larval to later stage assemblages, a trend that was observed in previous DNA barcoding studies of early life stages of marine and freshwater fishes [38,83–85]. An alternative explanation might be that anguillid eels are typical *r*-strategists, producing a large number of offspring at the cost of low survival and recruitment rates [5,21,86]. Such species are subject to important stochastic fluctuations of their demographic parameters depending on temporal heterogeneity and environmental disturbances [87]. Recent new records of amphidromous fishes (i.e., species for which the adult breeds in freshwater and the larva drifts into the sea) in Java, particularly in the family Gobiidae, based on a single or a few individuals, seem to corroborate the stochastic nature of migratory fish recruitment in Western Indonesia [30].

A. bicolor, for instance, was extremely abundant during our glass eel sampling campaign. Yet, yellow/silver eels of *A. bicolor* are known to be uncommon throughout the species' range distribution [88]. *A. bicolor* may have several spawning sites, one of which is supposed to be located off the Southwest cost of Sumatra [88]. The abundance of *A. bicolor* in glass eel samples may corroborate this hypothesis; however, it questions the supposed scarcity of yellow/silver eels. A similar trend has been observed for *A. interioris*, in which most yellow/silver eel specimen records have been reported from New Guinea, while glass eels have been observed in several western islands of Indonesia [25,81,88]. Due to this knowledge gap on distribution and biology, *A. interioris* is currently listed as data deficient in the IUCN Red List [6].

Comparison of mitochondrial genetic diversity between yellow/silver eels originating from their known species range and glass eels from local streams in Sumatra and Java indicated large differences. Several species, such as *A. marmorata*, *A. bicolor* and *A. bengalensis/A. nebulosa*, display wide range distributions throughout the Indian and Pacific oceans. These differences might be explained by the existence of several spawning sites and associated genetic population structuring [21,79]. Likewise, mechanisms inducing glass eel migration choices may account for these large differences between glass and yellow/silver eel mitochondrial genetic diversity. Natal homing, that is migration directed towards parental habitats, is currently observed for *A. anguilla* and *A. americana*, which seemingly share spawning grounds in the Atlantic Ocean and hybridize but whose yellow/silver eels do not co-occur in Europe and North America [89–91]. Alternatively, inter-annual variability in reproductive success and larval survival might be responsible for these large differences between glass and yellow/silver eel genetic diversity in *A. marmorata*, *A. bicolor* and *A. bengalensis/A. nebulosa*. This result is more intriguing for *A. interioris*, as its yellow/silver eels have been mainly observed in the northern rivers of New Guinea and it is supposed to have a single spawning site offshore of northern New Guinea. The presence of glass eels of *A. interioris* at low frequency in western Indonesia suggests random dispersal of glass eels for *A. interioris* and very low recruitment outside of New Guinea. These results suggest interspecific differences in life history traits, and phenotypic plasticity seems to be an important driver of species-specific life history traits and adaptation in *Anguilla* [92].

The general trend towards lower mitochondrial genetic diversity among glass eels in western Indonesia than among yellow/silver eels over species range distributions suggests small effective population sizes as a consequence of either fragmentation and the existence of several spawning sites or stochastic fluctuations in breeding and recruitment outcomes, as well as potential migration preferences for some species. This pattern suggests that *Anguilla* populations in South Sumatra and Java should probably be considered small populations. If commercial harvesting of glass eels were to be intensified in the future, this could bring them potentially close to an extinction vortex, i.e., a situation in which reduced population size and increased demographic variance either induce spatial fragmentation or a reduction in population adaptive potential [93,94]. Furthermore, the rapid development of infrastructure during the last two decades in this part of Indonesia has already resulted in fragmentation and reduced habitat availability, as well as water pollution associated with industrial waste and agricultural run-off [95–98].

5. Conclusions

Given the intense harvesting of glass eels in Java, the present study calls for a broader assessment of their recruitment dynamics to inform the sustainable management of these fisheries. The presence of only a small portion of the species' mitochondrial genetic diversity in Sumatra and Java glass eels sampled suggests that they have small effective population sizes and are potentially endangered in some regions. It also suggests important shifts of species abundance between glass and yellow/silver eels, pointing to major gaps in our understanding of tropical *Anguilla* life cycle ecology, and highlighting inadequacies in knowledge about glass eel recruitment per se for the sustainable management of *Anguilla* fisheries. The yellow and silver eel life stages of these species should receive high priority for conservation, given that they play an inherently important role in recruitment, particularly in the context of fragmented and disconnected aquatic landscapes.

Supplementary Materials: The following are available online at https://www.mdpi.com/article/10.3390/d13050193/s1, Table S1: Results of the genetic species delimitation analyses. Table S2: Results of the assignment analyses of unknown sequences to known species.

Author Contributions: Conceptualization, A.W., N.H., L.J.B. and N.N.; methodology, A.W., N.H., L.J.B. and N.N.; software, A.W. and N.H.; validation, A.W., N.H., D.S., L.J.B. and N.N.; formal analysis, A.W., N.H.; investigation, A.W., N.H., D.S., L.J.B., N.N., R.A.S., S., D.A., D.P.A., I.T.; resources, A.W., N.H., D.S., L.J.B. and N.H.; data curation, A.W., D.S. and N.H.; writing—original draft preparation, A.W. and N.H.; writing—review and editing, H.D., D.S. and L.J.B.; visualization, A.W., N.H., H.D., D.S., L.J.B. and N.H.; supervision, A.W., N.H., L.J.B. and N.N.; project administration, A.W., N.H., L.J.B. and N.N.; funding acquisition, A.W., N.H., D.S., L.J.B. and N.N. All authors have read and agreed to the published version of the manuscript.

Funding: This research was funded by the Research Institute for Inland Fisheries through DIPA 2017–2019, by CFREF to the University of Guelph's Food from Thought program, by the "Institut de Recherche pour le Développement" through annual funding and "Fonds d'Amorçage" (226F2ABIOS), by the French Embassy in Indonesia through the program "Science et Impact", by Campus France through a Bio-Asia grant (BIOSHOT project) and by ANR (ANR-17-ASIE-0006).

Institutional Review Board Statement: Not applicable.

Informed Consent Statement: Not applicable.

Data Availability Statement: The data presented in this study are openly available in BOLD at [dx.doi.org/10.5883/DS-ANGUILLA], reference number DS-ANGUILLA.

Acknowledgments: We would like to thank the Research Institute for Inland Fisheries and extensions for their staffing. We also thank the officers from the local fisheries districts in South Bengkulu, Cilacap, Kebumen and Pacitan for assisting with sample collection and engaging in productive discussions. We wish to thank Siti Nuramaliati Prijono, Witjaksono, Mohammad Irham, Marlina Adriyani, Ruliyana Susanti, Hari Sutrisno and the late Sri Sulandari, at Research Centre for Biology (RCB-LIPI); Joel Le Bail and Nicolas Gascoin at the French embassy in Jakarta for their continuous support. We are thankful to Daisy Wowor and Ujang Nurhaman at RCB-LIPI, Sumanta and Bambang Dwisusilo at IRD Jakarta for their help during the field sampling. Finally, we acknowledge all the reviewers who helped with improving this manuscript. A permit to collect fish was awarded to Arif Wibowo from the Research Institute for Inland Fisheries and Extensions, Ministry of Marine and Fisheries Affairs, Republic of Indonesia and Nicolas Hubert (7/TKPIPA/FRP/SM/VII/2012, 68/EXT/SIP/FRP/SM/VIII/2013, 41/EXT/SIP/FRP/SM/VIII/2014, 361/SIP/FRP/E5/Dit.KI/IX/2015, 50/EXT/SIP/FRP/E5/Dit.KI/IX/2016, 45/EXT/SIP/FRP/E5/Dit.KI/VIII/2017, 392/SIP/FRP/E5/Dit.KI/XI/2018, and 200/E5/E5.4/SIP/2019). No experimentation was conducted on live specimens during this study. This publication has ISEM number 2021-084 SUD.

Conflicts of Interest: The authors declare no conflict of interest. The funders had no role in the design of the study; in the collection, analyses, or interpretation of data; in the writing of the manuscript, or in the decision to publish the results.

References

1. Froese, R.; Pauly, D. Fishbase. Available online: http://www.fishbase.org (accessed on 15 December 2020).
2. Kuroki, M.; Miller, M.J.; Tsukamoto, K. Diversity of early life-history traits in freshwater eels and the evolution of their oceanic migrations. *Can. J. Zool.* **2014**, *92*, 749–770. [CrossRef]
3. Tesch, F.W.; Rohlf, N. Migration from continental waters to the spawning grounds. In *Eel Biology*; Springer: Berlin, Germany, 2003; pp. 223–234.
4. Elliott, J.M.; Tesch, F.-W. The Eel: Biology and Management of Anguillid Eels. *J. Anim. Ecol.* **1978**, *47*, 1033. [CrossRef]
5. Watanabe, S. Taxonomy of the Freshwater Eels, Genus *Anguilla* Schrank, 1798. In *Eel Biology*; Springer: Tokyo, Japan, 2003; pp. 3–18.
6. Jacoby, D.M.; Casselman, J.M.; Crook, V.; DeLucia, M.-B.; Ahn, H.; Kaifu, K.; Kurwie, T.; Sasal, P.; Silfvergrip, A.M.; Smith, K.G.; et al. Synergistic patterns of threat and the challenges facing global anguillid eel conservation. *Glob. Ecol. Conserv.* **2015**, *4*, 321–333. [CrossRef]
7. Sugeha, H.Y.; Suharti, S.R.; Wouthuyzen, S.; Sumadhiharga, K. Biodiversity, distribution and abundance of the tropical anguillid eels in the Indonesian waters. *Mar. Res. Indones.* **2008**, *33*, 129–138. [CrossRef]
8. Ringuet, S.; Muto, F.; Raymakers, C. Eels: Their harvest and trade in Europe and Asia. *Traffic Bull. Int.* **2002**, *19*, 80–106.
9. FAO. *Des Pêches Et De L'aquaculture*; FAO: Rome, Italy, 2018; ISBN 9789251306925.

10. Tsukamoto, K.; Kuroki, M. *Eels and Humans*; Springer: Tokyo, Japan, 2014; ISBN 4431545298.
11. Shiraishi, H.; Crook, V. *Eel Market Dynamics: An Analysis of Anguilla Production*; TRAFFIC: Tokyo, Japan, 2015.
12. BKIPM. *Eels Trade in Indonesia*; BKIPM: Jakarta, Indonesia, 2018.
13. Tanaka, H.; Kagawa, H.; Ohta, H. Production of leptocephali of Japanese eel (*Anguilla japonica*) in captivity. *Aquaculture* **2001**, *201*, 51–60. [CrossRef]
14. Tanaka, H.; Kagawa, H.; Ohta, H.; Unuma, T.; Nomura, K. The first production of glass eel in captivity: Fish reproductive physiology facilitates great progress in aquaculture. *Fish Physiol. Biochem.* **2003**, *28*, 493–497. [CrossRef]
15. Friedland, K.D.; Miller, M.J.; Knights, B. Oceanic changes in the Sargasso Sea and declines in recruitment of the European eel. *ICES J. Mar. Sci.* **2007**, *64*, 519–530. [CrossRef]
16. Wirth, T.; Bernatchez, L. Decline of North Atlantic eels: A fatal synergy? *Proc. R. Soc. B Boil. Sci.* **2003**, *270*, 681–688. [CrossRef] [PubMed]
17. Dekker, W. Management of the eel is slipping through our hands! Distribute control and orchestrate national protection. *ICES J. Mar. Sci.* **2016**, *73*, 2442–2452. [CrossRef]
18. Righton, D.; Walker, A.M. Anguillids: Conserving a global fishery. *J. Fish Biol.* **2013**, *83*, 754–765. [CrossRef] [PubMed]
19. Bonhommeau, S.; Chassot, E.; Rivot, E. Fluctuations in European eel (*Anguilla anguilla*) recruitment resulting from environmental changes in the Sargasso Sea. *Fish. Oceanogr.* **2008**, *17*, 32–44. [CrossRef]
20. Arai, T.; Kadir, S.R.A. Opportunistic spawning of tropical anguillid eels *Anguilla bicolor bicolor* and *A. bengalensis bengalensis*. *Sci. Rep.* **2017**, *7*, 1–17. [CrossRef] [PubMed]
21. Arai, T.; Limbong, D.; Otake, T.; Tsukamoto, K. Recruitment mechanisms of tropical eels *Anguilla* spp. and implications for the evolution of oceanic migration in the genus Anguilla. *Mar. Ecol. Prog. Ser.* **2001**, *216*, 253–264. [CrossRef]
22. Fahmi, M.R.; Solihin, D.D.; Soewardi, K.; Pouyaud, L.; Berrebi, P. Molecular phylogeny and genetic diversity of freshwater *Anguilla* eels in indonesian waters based on mitochondrial sequences. *Vie Milieu-Life Environ.* **2015**, *65*, 139–150.
23. Eschmeyer, W.N.; Fricke, R.; van der Laan, R. Catalog of Fishes Electronic Version. Available online: https://researcharchive.calacademy.org/research/ichthyology/catalog/fishcatmain.aspg (accessed on 15 December 2020).
24. Aoyama, J. Life History and Evolution of Migration in Catadromous Eels (Genus: *Anguilla*). *Aqua-BioScience Monogr.* **2009**, *2*, 1–42. [CrossRef]
25. Aoyama, J.; Wouthuyzen, S.; Miller, M.J.; Minegishi, Y.; Kuroki, M.; Suharti, S.R.; Kawakami, T.; Sumardiharga, K.O.; Tsukamoto, K. Distribution of leptocephali of the freshwater eels, genus *Anguilla*, in the waters off west Sumatra in the Indian Ocean. *Environ. Boil. Fishes* **2007**, *80*, 445–452. [CrossRef]
26. Ndobe, S.; Serdiati, N.; Moore, A. Species Composition Of Glass Eels Recruiting To The Palu River. *J. Agroecol.* **2015**, *1*, 12–23.
27. Shirotori, F.; Ishikawa, T.; Tanaka, C.; Aoyama, J.; Shinoda, A.; Yambot, A.V.; Yoshinaga, T. Species composition of anguillid glass eels recruited at southern Mindanao Island, the Philippines. *Fish. Sci.* **2016**, *82*, 915–922. [CrossRef]
28. Hebert, P.D.N.; Cywinska, A.; Ball, S.L.; Dewaard, J.R. Biological identifications through DNA barcodes. *Proc. R. Soc. B Boil. Sci.* **2003**, *270*, 313–321. [CrossRef] [PubMed]
29. Hebert, P.D.; Ratnasingham, S.; De Waard, J.R. Barcoding animal life: Cytochrome c oxidase subunit 1 divergences among closely related species. *Proc. R. Soc. B Boil. Sci.* **2003**, *270*, S96–S99. [CrossRef] [PubMed]
30. Dahruddin, H.; Hutama, A.; Busson, F.; Sauri, S.; Hanner, R.; Keith, P.; Hadi, D.; Hubert, N. Revisiting the ichthyodiversity of Java and Bali through DNA barcodes: Taxonomic coverage, identification accuracy, cryptic diversity and identification of exotic species. *Mol. Ecol. Resour.* **2017**, *17*, 288–299. [CrossRef] [PubMed]
31. Hutama, A.; Dahruddin, H.; Busson, F.; Sauri, S.; Keith, P.; Hadiaty, R.K.; Hanner, R.; Suryobroto, B.; Hubert, N. Identifying spatially concordant evolutionary significant units across multiple species through DNA barcodes: Application to the conservation genetics of the freshwater fishes of Java and Bali. *Glob. Ecol. Conserv.* **2017**, *12*, 170–187. [CrossRef]
32. Hubert, N.; Lumbantobing, D.; Sholihah, A.; Dahruddin, H.; Delrieu-Trottin, E.; Busson, F.; Sauri, S.; Hadiaty, R.; Keith, P. Revisiting species boundaries and distribution ranges of *Nemacheilus* spp. (Cypriniformes: Nemacheilidae) and *Rasbora* spp. (Cypriniformes: Cyprinidae) in Java, Bali and Lombok through DNA barcodes: Implications for conservation in a biodiversity hotspot. *Conserv. Genet.* **2019**, *20*, 517–529. [CrossRef]
33. Sholihah, A.; Delrieu-Trottin, E.; Sukmono, T.; Dahruddin, H.; Risdawati, R.; Elvyra, R.; Wibowo, A.; Kustiati, K.; Busson, F.; Sauri, S.; et al. Disentangling the taxonomy of the subfamily Rasborinae (Cypriniformes, Danionidae) in Sundaland using DNA barcodes. *Sci. Rep.* **2020**, *10*, 1–14. [CrossRef] [PubMed]
34. Lim, H.; Abidin, M.Z.; Pulungan, C.P.; De Bruyn, M.; Nor, S.A.M. DNA Barcoding Reveals High Cryptic Diversity of the Freshwater Halfbeak Genus *Hemirhamphodon* from Sundaland. *PLoS ONE* **2016**, *11*, e0163596. [CrossRef] [PubMed]
35. Farhana, S.N.; Muchlisin, Z.A.; Duong, T.Y.; Tanyaros, S.; Page, L.M.; Zhao, Y.; Adamson, E.A.S.; Khaironizam, Z.; De Bruyn, M.; Azizah, M.N.S. Exploring hidden diversity in Southeast Asia's *Dermogenys* spp. (Beloniformes: Zenarchopteridae) through DNA barcoding. *Sci. Rep.* **2018**, *8*, 10787. [CrossRef]
36. Hubert, N.; Hadiaty, R.K.; Paradis, E.; Pouyaud, L. Cryptic Diversity in Indo-Australian Rainbowfishes Revealed by DNA Barcoding: Implications for Conservation in a Biodiversity Hotspot Candidate. *PLoS ONE* **2012**, *7*, e40627. [CrossRef]
37. Wibowo, A.; Sloterdijk, H.; Ulrich, S.P. Identifying Sumatran Peat Swamp Fish Larvae through DNA Barcoding, Evidence of Complete Life History Pattern. *Procedia Chem.* **2015**, *14*, 76–84. [CrossRef]

38. Wibowo, A.; Wahlberg, N.; Vasemägi, A. DNA barcoding of fish larvae reveals uncharacterised biodiversity in tropical peat swamps of New Guinea, Indonesia. *Mar. Freshw. Res.* **2017**, *68*, 1079–1087. [CrossRef]
39. Kottelat, M.; Whitten, A.J.; Kartikasari, N.; Wirjoatmodjo, S. *Freshwater Fishes of Western Indonesia and Sulawesi*; PERIPLUS: Jakarta, Indonesia, 1993.
40. Keith, P.; Marquet, G.; Lord, C.; Kalfatak, D.; Vigneux, E. *Poissons et Crustacés d'Eau Douce du Vanuatu*; Société Française d'Ichtyologie: Paris, France, 2010.
41. Keith, P.; Marquet, G.; Gerbeaux, P.; Vigneux, E.; Lord, C. *Poissons et Crustacés d'Eau Douce de Polynésie*; Société Française d'Ichthyologie: Paris, France, 2013.
42. Hubert, N.; Hanner, R.; Holm, E.; Mandrak, N.E.; Taylor, E.; Burridge, M.; Watkinson, D.; Dumont, P.; Curry, A.; Bentzen, P.; et al. Identifying Canadian Freshwater Fishes through DNA Barcodes. *PLoS ONE* **2008**, *3*, e2490. [CrossRef] [PubMed]
43. Ivanova, N.V.; Zemlak, T.S.; Hanner, R.H.; Hebert, P.D.N. Universal primer cocktails for fish DNA barcoding. *Mol. Ecol. Notes* **2007**, *7*, 544–548. [CrossRef]
44. Ratnasingham, S.; Hebert, P.D.N. BOLD: The Barcode of Life Data System (www.barcodinglife.org). *Mol. Ecol. Notes* **2007**, *7*, 355–364. [CrossRef] [PubMed]
45. Avise, J.C. *Molecular Markers, Natural History and Evolution*; Hall C.& Ed.: New York, NY, USA, 1989.
46. Moritz, C. Defining "Evolutionary Significant Units" for conservation. *Trends Ecol. Evol.* **1994**, *9*, 373–375. [CrossRef]
47. Vogler, A.P.; DeSalle, R. Diagnosing Units of Conservation Management. *Conserv. Biol.* **1994**, *8*, 354–363. [CrossRef]
48. Kekkonen, M.; Mutanen, M.; Kaila, L.; Nieminen, M.; Hebert, P.D.N. Delineating Species with DNA Barcodes: A Case of Taxon Dependent Method Performance in Moths. *PLoS ONE* **2015**, *10*, e0122481. [CrossRef] [PubMed]
49. Limmon, G.; Delrieu-Trottin, E.; Patikawa, J.; Rijoly, F.; Dahruddin, H.; Busson, F.; Steinke, D.; Hubert, N. Assessing species diversity of Coral Triangle artisanal fisheries: A DNA barcode reference library for the shore fishes retailed at Ambon harbor (Indonesia). *Ecol. Evol.* **2020**, *10*, 3356–3366. [CrossRef] [PubMed]
50. Delrieu-Trottin, E.; Durand, J.; Limmon, G.; Sukmono, T.; Sugeha, H.Y.; Chen, W.; Busson, F.; Borsa, P.; Dahruddin, H.; Sauri, S. Biodiversity inventory of the grey mullets (Actinopterygii: Mugilidae) of the Indo-Australian Archipelago through the iterative use of DNA-based species delimitation and specimen assignment methods. *Evol. Appl.* **2020**, *13*, 1451–1467. [CrossRef] [PubMed]
51. Shen, Y.; Hubert, N.; Huang, Y.; Wang, X.; Gan, X.; Peng, Z.; He, S. DNA barcoding the ichthyofauna of the Yangtze River: Insights from the molecular inventory of a mega-diverse temperate fauna. *Mol. Ecol. Resour.* **2019**, *19*, 1278–1291. [CrossRef] [PubMed]
52. Ratnasingham, S.; Hebert, P.D.N. A DNA-Based Registry for All Animal Species: The Barcode Index Number (BIN) System. *PLoS ONE* **2013**, *8*, e66213. [CrossRef] [PubMed]
53. Puillandre, N.; Lambert, A.; Brouillet, S.; Achaz, G. ABGD, Automatic Barcode Gap Discovery for primary species delimitation. *Mol. Ecol.* **2011**, *21*, 1864–1877. [CrossRef] [PubMed]
54. Zhang, J.; Kapli, P.; Pavlidis, P.; Stamatakis, A. A general species delimitation method with applications to phylogenetic placements. *Bioinformatics* **2013**, *29*, 2869–2876. [CrossRef] [PubMed]
55. Kapli, P.; Lutteropp, S.; Zhang, J.; Kobert, K.; Pavlidis, P.; Stamatakis, A.; Flouri, T. Multi-rate Poisson Tree Processes for single-locus species delimitation under Maximum Likelihood and Markov Chain Monte Carlo. *Bioinformatics* **2017**, *33*, 1630–1638. [CrossRef]
56. Fujisawa, T.; Barraclough, T.G. Delimiting Species Using Single-Locus Data and the Generalized Mixed Yule Coalescent Approach: A Revised Method and Evaluation on Simulated Data Sets. *Syst. Biol.* **2013**, *62*, 707–724. [CrossRef] [PubMed]
57. Stamatakis, A. RAxML version 8: A tool for phylogenetic analysis and post-analysis of large phylogenies. *Bioinformatics* **2014**, *30*, 1312–1313. [CrossRef] [PubMed]
58. Bouckaert, R.; Heled, J.; Kühnert, D.; Vaughan, T.; Wu, C.-H.; Xie, D.; Suchard, M.A.; Rambaut, A.; Drummond, A.J. BEAST 2: A Software Platform for Bayesian Evolutionary Analysis. *PLoS Comput. Biol.* **2014**, *10*, e1003537. [CrossRef] [PubMed]
59. Bermingham, E.; McCafferty, S.S.; Martin, A.P. Fish Biogeography and Molecular Clocks: Perspectives from the Panamanian Isthmus. In *Molecular Systematics of Fishes*; Academic Press: San Diego, CA, USA, 1997; pp. 113–126.
60. Ogilvie, H.A.; Bouckaert, R.R.; Drummond, A.J. StarBEAST2 Brings Faster Species Tree Inference and Accurate Estimates of Substitution Rates. *Mol. Biol. Evol.* **2017**, *34*, 2101–2114. [CrossRef] [PubMed]
61. Ho, S.Y.W.; Larson, G. Molecular clocks: When timesare a-changin'. *Trends Genet.* **2006**, *22*, 79–83. [CrossRef] [PubMed]
62. Zhang, A.-B.; Hao, M.-D.; Yang, C.-Q.; Shi, Z.-Y. BarcodingR: An integrated r package for species identification using DNA barcodes. *Methods Ecol. Evol.* **2016**, *8*, 627–634. [CrossRef]
63. Zhang, A.-B.; Muster, C.; Liang, H.-B.; Zhu, C.-D.; Crozier, R.; Wan, P.; Feng, J.; Ward, R.D. A fuzzy-set-theory-based approach to analyse species membership in DNA barcoding. *Mol. Ecol.* **2011**, *21*, 1848–1863. [CrossRef] [PubMed]
64. Jin, Q.; Han, H.; Hu, X.; Li, X.; Zhu, C.; Ho, S.Y.W.; Ward, R.D.; Zhang, A. Quantifying species diversity with a DNA barcoding-based method: Tibetan moth species (Noctuidae) on the Qinghai-Tibetan Plateau. *PLoS ONE* **2013**, *8*, e64428. [CrossRef] [PubMed]
65. Paradis, E. pegas: An R package for population genetics with an integrated-modular approach. *Bioinformatics* **2010**, *26*, 419–420. [CrossRef] [PubMed]
66. Nei, M.; Tajima, F. DNA Polymorphism Detectable by Restriction Endonucleases. *Genetics* **1981**, *97*, 145–163. [CrossRef]
67. Nei, M. *Molecular Evolutionary Genetics*; Columbia University Press: New York, NY, USA, 1987.

68. Watterson, G. On the number of segregating sites in genetical models without recombination. *Theor. Popul. Biol.* **1975**, *7*, 256–276. [CrossRef]
69. Tajima, F. Statistical method for testing the neutral mutation hypothesis by DNA polymorphism. *Genetics* **1989**, *123*, 585–595. [CrossRef]
70. Kimura, M. A simple method for estimating evolutionary rates of base substitutions through comparative studies of nucleotide sequences. *J. Mol. Evol.* **1980**, *16*, 111–120. [CrossRef]
71. Paradis, E.; Schliep, K. ape 5.0: An environment for modern phylogenetics and evolutionary analyses in R. *Bioinformatics* **2018**, *35*, 526–528. [CrossRef] [PubMed]
72. Brown, S.D.J.; Collins, R.A.; Boyer, S.; Lefort, C.; Malumbres-Olarte, J.; Vink, C.J.; Cruickshank, R.H. Spider: An R package for the analysis of species identity and evolution, with particular reference to DNA barcoding. *Mol. Ecol. Resour.* **2012**, *12*, 562–565. [CrossRef]
73. Muchlisin, Z.A.; Batubara, A.S.; Fadli, N.; Muhammadar, A.A.; Utami, A.I.; Farhana, N.; Siti-Azizah, M.N. Assessing the species composition of tropical eels (Anguillidae) in Aceh Waters, Indonesia, with DNA barcoding gene cox1. *F1000Research* **2017**, *6*, 258. [CrossRef] [PubMed]
74. Hanzen, C.; Lucas, M.C.; O'Brien, G.; Downs, C.T.; Willows-Munro, S. African freshwater eel species (*Anguilla* spp.) identification through DNA barcoding. *Mar. Freshw. Res.* **2020**, *71*, 1543. [CrossRef]
75. Stein, F.M.; Wong, J.C.Y.; Sheng, V.; Law, C.S.W.; Schröder, B.; Baker, D.M. First genetic evidence of illegal trade in endangered European eel (*Anguilla anguilla*) from Europe to Asia. *Conserv. Genet. Resour.* **2016**, *8*, 533–537. [CrossRef]
76. Arai, T.; Wong, L.L. Validation of the occurrence of the tropical eels, *Anguilla bengalensis bengalensis* and *A. bicolor bicolor* at Langkawi Island in Peninsular Malaysia, Malaysia. *Trop. Ecol.* **2016**, *57*, 23–31.
77. Rosenberg, A.; Norborg, M. Genealogical trees, coalescent theory and the analysis of genetic polymorphisms. *Nat. Rev. Genet.* **2002**, *3*, 380–390. [CrossRef] [PubMed]
78. Hubert, N.; Hanner, R. DNA Barcoding, species delineation and taxonomy: A historical perspective. *DNA Barcodes* **2015**, *3*, 44–58. [CrossRef]
79. Arai, T.; Aoyama, J.; Limbong, D.; Tsukamoto, K. Species composition and inshore migration of the tropical eels *Anguilla* spp. recruiting to the estuary of the Poigar River, Sulawesi Island. *Mar. Ecol. Prog. Ser.* **1999**, *188*, 299–303. [CrossRef]
80. Sugeha, H.Y.; Arai, T.; Miller, M.J.; Limbong, D.; Tsukamoto, K. Inshore migration of the tropical eels *Anguilla* spp. recruiting to the Poigar River estuary on north Sulawesi Island. *Mar. Ecol. Prog. Ser.* **2001**, *221*, 233–243. [CrossRef]
81. Fahmi, M.R. Phylogeography of Tropical Eels (*Anguilla* spp.). In *Indonesian Waters*; Bogor Agricultural University: Bogor, Indonesia, 2013.
82. Hubert, N.; Kadarusman; Wibowo, A.; Busson, F.; Caruso, D.; Sulandari, S.; Nafiqoh, N.; Rüber, L.; Pouyaud, L.; Avarre, J.C.; et al. DNA barcoding Indonesian freshwater fishes: Challenges and prospects. *DNA Barcodes* **2015**, *3*, 144–169. [CrossRef]
83. Hubert, N.; Espiau, B.; Meyer, C.; Planes, S. Identifying the ichthyoplankton of a coral reef using DNA barcodes. *Mol. Ecol. Resour.* **2015**, *15*, 57–67. [CrossRef] [PubMed]
84. Collet, A.; Durand, J.-D.; Desmarais, E.; Cerqueira, F.; Cantinelli, T.; Valade, P.; Ponton, D. DNA barcoding post-larvae can improve the knowledge about fish biodiversity: An example from La Reunion, SW Indian Ocean. *Mitochondrial DNA Part A* **2018**, *29*, 905–918. [CrossRef] [PubMed]
85. Steinke, D.; Connell, A.D.; Hebert, P.D. Linking adults and immatures of South African marine fishes. *Genome* **2016**, *59*, 959–967. [CrossRef]
86. Pianka, E.R. On r and K selection. *Am. Nat.* **1970**, *104*, 592–597. [CrossRef]
87. Houde, E.D. Recruitment variability. *Fish Reprod. Biol.* **2016**, 98–187.
88. UNEP-WCMC Preliminary Overview of the Genus Anguilla. 2015. Available online: https://ec.europa.eu/environment/cites/pdf/reports/Preliminary%20overview%20of%20the%20genus%20Anguilla.pdf (accessed on 15 December 2020).
89. Nikolic, N.; Liu, S.; Jacobsen, M.W.; Jónsson, B.; Bernatchez, L.; Gagnaire, P.; Hansen, M.M. Speciation history of European (*Anguilla anguilla*) and American eel (*A. rostrata*), analysed using genomic data. *Mol. Ecol.* **2020**, *29*, 565–577. [CrossRef] [PubMed]
90. Chang, Y.-L.K.; Feunteun, E.; Miyazawa, Y.; Tsukamoto, K. New clues on the Atlantic eels spawning behavior and area: The Mid-Atlantic Ridge hypothesis. *Sci. Rep.* **2020**, *10*, 1–12. [CrossRef] [PubMed]
91. Arai, T. Ecology and evolution of migration in the freshwater eels of the genus *Anguilla* Schrank, 1798. *Heliyon* **2020**, *6*, 05176. [CrossRef] [PubMed]
92. Enbody, E.D.; Pettersson, M.E.; Sprehn, C.G.; Palm, S.; Wickström, H.; Andersson, L. Ecological adaptation in European eels is based on phenotypic plasticity. *Proc. Natl. Acad. Sci. USA* **2021**, *118*, e2022620118. [CrossRef] [PubMed]
93. Gilpin, E.; Soulé, M. Minimum viable populations: Processes of species extinction. In *Conservation Biology: The Science of Scarcity and Diversity*; Soulé, M.E., Ed.; Sinauer: Sunderland, UK, 1986; pp. 19–34.
94. Fagan, W.F.; Holmes, E.E. Quantifying the extinction vortex. *Ecol. Lett.* **2005**, *9*, 51–60. [CrossRef] [PubMed]
95. Spracklen, D.V.; Reddington, C.L.; A Gaveau, D.L. Industrial concessions, fires and air pollution in Equatorial Asia. *Environ. Res. Lett.* **2015**, *10*, 91001. [CrossRef]

96. Breckwoldt, A.; Dsikowitzky, L.; Baum, G.; Ferse, S.C.; Van Der Wulp, S.; Kusumanti, I.; Ramadhan, A.; Adrianto, L. A review of stressors, uses and management perspectives for the larger Jakarta Bay Area, Indonesia. *Mar. Pollut. Bull.* **2016**, *110*, 790–794. [CrossRef]
97. Hayati, A.; Tiantono, N.; Mirza, M.F.; Putra, I.D.S.; Abdizen, M.M.; Seta, A.R.; Solikha, B.M.; Fu'Adil, M.H.; Putranto, T.W.C.; Affandi, M.; et al. Water quality and fish diversity in the Brantas River, East Java, Indonesia. *J. Biol. Res.* **2017**, *22*, 43–49. [CrossRef]
98. Garg, T.; Hamilton, S.E.; Hochard, J.P.; Kresch, E.P.; Talbot, J. (Not so) gently down the stream: River pollution and health in Indonesia. *J. Environ. Econ. Manag.* **2018**, *92*, 35–53. [CrossRef]

Article

Mitochondrial Genetic Diversity among Farmed Stocks of *Oreochromis* spp. (Perciformes, Cichlidae) in Madagascar

Nicolas Hubert [1,*], Elodie Pepey [1,2], Jean-Michel Mortillaro [1,2,3], Dirk Steinke [4], Diana Edithe Andria-Mananjara [3] and Hugues de Verdal [1,2]

1 UMR ISEM (IRD, UM, CNRS), Université de Montpellier, Place Eugène Bataillon, CEDEX 05, 34095 Montpellier, France; elodie.pepey@cirad.fr (E.P.); jean-michel.mortillaro@cirad.fr (J.-M.M.); hugues.de_verdal@cirad.fr (H.d.V.)
2 Cirad, UMR ISEM, Université de Montpellier, 389 Avenue Agropolis, CEDEX 05, 34095 Montpellier, France
3 FOFIFA DRZVP, rue Farafaty, Antananarivo 00101, Madagascar; adianaedith@gmail.com
4 Department of Integrative Biology, Centre for Biodiversity Genomics, University of Guelph, 50 Stone Rd E, Guelph, ON N1G 2W1, Canada; dsteinke@uoguelph.ca
* Correspondence: nicolas.hubert@ird.fr

Abstract: The fast development of aquaculture over the past decades has made it the main source of fish protein and led to its integration into the global food system. Mostly originating from inland production systems, aquaculture has emerged as strategy to decrease malnutrition in low-income countries. The Nile tilapia (*Oreochromis niloticus*) was introduced to Madagascar in the 1950s, and is now produced nationally at various scales. Aquaculture mostly relies on fry harvested from wild populations and grow-out in ponds for decades. It has recently been diversified by the introduction of several fast-growing strains. Little is known how local genetic diversity compares to recently introduced strains, although high and comparable levels of genetic diversity have previously been observed for both wild populations and local stocks. Our study compares DNA barcode genetic diversity among eight farms and several strains belonging to three species sampled. DNA-based lineage delimitation methods were applied and resulted in the detection of six well differentiated and highly divergent lineages. A comparison of DNA barcode records to sequences on the Barcode of Life Data System (BOLD) helped to trace the origin of several of them. Both haplotype and nucleotide diversity indices highlight high levels of mitochondrial genetic diversity, with several local strains displaying higher diversity than recently introduced strains. This allows for multiple options to maintain high levels of genetic diversity in broodstock and provides more options for selective breeding programs.

Keywords: aquaculture; DNA barcoding; domestication; exotic species; management; tilapia

Citation: Hubert, N.; Pepey, E.; Mortillaro, J.-M.; Steinke, D.; Andria-Mananjara, D.E.; de Verdal, H. Mitochondrial Genetic Diversity among Farmed Stocks of *Oreochromis* spp. (Perciformes, Cichlidae) in Madagascar. *Diversity* **2021**, *13*, 281. https://doi.org/10.3390/d13070281

Academic Editors: Manuel Elias-Gutierrez and Michael Wink

Received: 12 May 2021
Accepted: 16 June 2021
Published: 22 June 2021

1. Introduction

In the light of a global biodiversity decline and the wide-spread depletion of wild fish stocks [1–3], aquaculture has become an increasingly important source of animal protein in tropical countries [4,5]. With a global production of more than 160 million tons of fish (in 2015) aquaculture has become an important component of the global food system [5,6], with Asia contributing nearly 90% of the freshwater production. Globally it is now producing nearly twice as much as fisheries [4], mostly through inland aquaculture. It has been frequently integrated in national strategies to prevent malnutrition, particularly in low-income, tropical countries [7].

Inland aquaculture mostly relies on three species or species groups: carp (*Cyprinus carpio*) which is the most farmed fish worldwide with 13 million tons produced in 2017, followed by tilapia (*Oreochromis* spp.) and catfishes, such as *Pangasius* spp., with 5 million tons each produced in 2017 [5]. Their success lies mostly in the fast growth rate of these species and the availability of protocols for farming and captive breeding. As a consequence, they were introduced in numerous countries outside their native distribution ranges [8].

In Madagascar, freshwater fish aquaculture has a long tradition. The first introductions of exotic species for aquaculture trace back to 1857 (giant goramy-*Osphronemus goramy*), 1861 (goldfish-*Carassius auratus*), 1912 (carp-*Cyprinus carpio*) [9], and the 1950s (nile tilapia-*Oreochromis niloticus*) [10,11]. The successful introduction of these species led to the establishment of multiple populations in the wild, further used as source of young fry for farming [12]. Nile tilapia and carp have been used in semi-intensive productions in association with rice cultivation in paddy fields since the earliest development of aquaculture in Madagascar [13]. Adults and fry are traditionally caught in the wild and introduced to paddy fields when rice plants are sown [12]. Aquaculture in ponds follows similar practices leading to similar genetic diversity found between farmed and wild tilapia populations [14]. More recently, multiple strains of *O. niloticus* and other *Oreochromis* species were introduced to upscale tilapia aquaculture by producing fry from imported strains known to have desirable properties, such as fast growth rates [15]. Some of these strains, such as the Genetically Improved Farmed Tilapia (GIFT) strain, are the result of multiple crossings between several farmed strains and wild populations, as well as selective breeding [16,17]. As a result, strains of multiple origins are currently maintained in farmed stocks of tilapia in Madagascar, but their genetic variability is not known.

The present study aims to estimate the genetic diversity of currently farmed strains of *Oreochromis* in Madagascar, which now include local and recently imported strains of *O. niloticus*, such as GIFT and JICA, as well as *O. mossambicus* and *O. macrochir* [14]. We opted for the use of a standardized mitochondrial marker, a 652 bp fragment of the cytochrome oxidase I gene known as DNA barcode [18,19], for a variety of reasons: (1) it can be easily retrieved thanks to the availability of universal primers for fish [20]; (2) a large number of DNA barcode sequences are available for *Oreochromis* (>1200 public records for most species) in the Barcode of Life Data System, BOLD [21]; (3) the mitochondrial diversity of farmed tilapia Madagascar is largely unknown.

We generated DNA barcodes for currently farmed strains of tilapia, and applied DNA-based delimitation methods to detect mitochondrial lineages among them and to estimate their diversity. By using new DNA barcode records and published data from BOLD, potential evolutionary origins of revealed lineages are discussed.

2. Materials and Methods

2.1. Sampling, Sequencing and International Repositories

In February 2016, a total of 262 specimens was collected at eight sites maintaining breeding stocks (Figure 1). Collected information, including geocoordinates, are included in the dataset DS-TILMADA (dx.doi.org/10.5883/DS-TILMADA) on BOLD. Specimens were identified to the species level based on the information of the farmers. Specimens were further photographed, individually labeled, and voucher specimens were preserved in a 5% formalin solution. A fin clip or a muscle biopsy were taken from each specimen and fixed in a 96% ethanol solution for further genetic analyses.

Genomic DNA was extracted from fin clip samples using a Qiagen DNeasy 96 tissue extraction kit following manufacturer's specifications. A 652-bp segment from the 5′ region of the cytochrome oxidase I gene (COI) was amplified using the primer cocktail C_FishF1t1/C_FishR1t1 [20]. PCR amplifications were done on a Veriti 96-well Fast thermocycler (ABI-AppliedBiosystems) with a final volume of 10.0 μL containing 5.0 μL Buffer 2X, 3.3 μL ultrapure water, 1.0 μL each primer (10 μM), 0.2 μL enzyme Phire Hot Start II DNA polymerase (5 U), and 0.5 μL of DNA template (~50 ng). The following thermocycler regime was used: initial denaturation at 98 °C for 5 min followed by 30 cycles denaturation at 98 °C for 5 s, annealing at 56 °C for 20 s, and extension at 72 °C for 30 s, followed by a final extension step at 72 °C for 5 min. PCR products were purified with ExoSap-IT (USB Corporation, Cleveland, OH, USA) and sequenced in both directions. Sequencing reactions were performed at the Centre for Biodiversity Genomics, University of Guelph, Canada, using the BigDye Terminator v3.1 Cycle Sequencing Ready Reaction kit following standard protocols. Sequencing was performed on an ABI 3730xl capillary sequencer (Applied Biosystems). Sequences and collateral information were deposited on BOLD [21], and are available as a public dataset (dx.doi.org/10.5883/DS-TILMADA, Table S1).

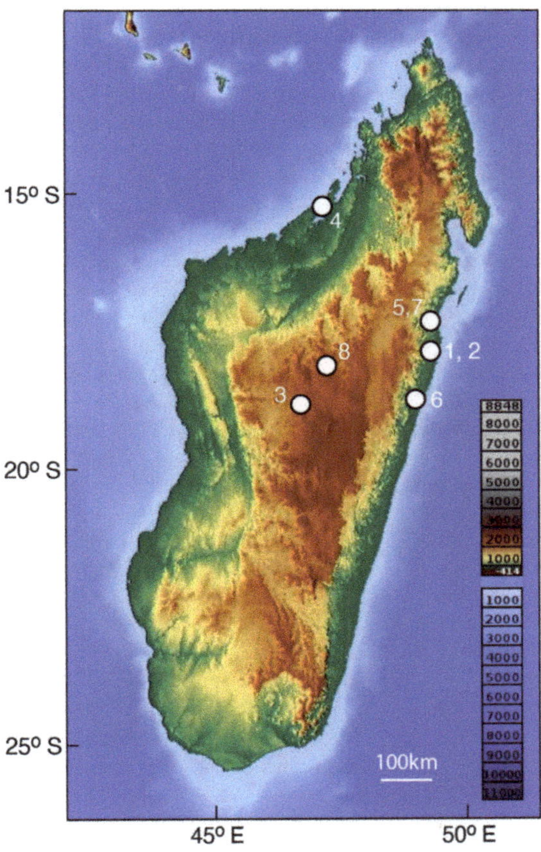

Figure 1. Sampling sites in Madagascar for the 262 DNA barcode records of *Oreochromis* spp. analyzed in this study. Site 1, *O. niloticus* GIFT strain at Matera (−18.0433, 49.3682); Site 2, *O. niloticus* GIFT strain at Tamatave (−18.449, 49.3987); Site 3, *O. niloticus* local strain at Ampefy (−19.0408, 46.7363); Site 4, *O. niloticus* JICA strain at Mahajunga (−15.6932, 46.3193); Site 5, *O. niloticus* local strain at Fenerive (−17.3666, 49.3961); Site 6, *O. niloticus* local strain at Brickaville (−18.8172, 49.0674); Site 7, *O. mossambicus* local strain at Fenerive (−17.3632, 49.3992); Site 8, *O. macrochir* local strain at Milasoa (−18.2224, 47.1461).

2.2. Mitochondrial Lineage Delimitation and Genetic Diversity

Several methods for species delineation based on DNA sequences have been proposed [22–26]. Each of these have different properties, particularly when dealing with singletons (i.e., lineages represented by a single sequence) or heterogeneous speciation rates among lineages [27]. A combination of different approaches is increasingly used to overcome potential pitfalls arising from uneven sampling [28–32]. We used six different sequence-based methods of species delimitation to identify Molecular Operational Taxonomic Units (MOTU): (1) Refined Single Linkage (RESL) as implemented in BOLD and used to generate Barcode Index Numbers (BIN) [25]; (2) Automatic Barcode Gap Discovery (ABGD) [24]; (3) Poisson Tree Process (PTP) in its single (sPTP) and multiple rates version (mPTP) as implemented in the stand-alone software mptp_0.2.3 [26,33]; (4) General Mixed

Yule-Coalescent (GMYC) in its single (sGMYC) and multiple threshold version (mGMYC) as implemented in the R package Splits 1.0–19 [34].

The mPTP algorithm and the GMYC both use phylogenetic trees as input file. We reconstructed a maximum likelihood (ML) tree for the former using RAxML [35] with a GTR + I + Γ substitution model. For the GMYC algorithm we calculated an ultrametric, fully resolved the tree using the Bayesian approach implemented in BEAST 2.6.2 [36]. Sequences were collapsed into haplotypes prior to reconstructing the ultrametric tree using RAxML, and Bayesian reconstruction was based on a strict-clock prior of 1.2% per million year [37]. Two Markov chains of 20 million each were ran independently using Yule pure birth and GTR + I + Γ substitution models, other tree priors were used as default. Stability (ESS > 200) and convergence of Markov chains was verified using Tracer 1.7.1 [36]. Trees were sampled every 5000 states, after an initial burn in period of 5 million states. Both runs were combined with trees re-sampled every 20,000 states using LogCombiner 2.6.2, and the maximum credibility tree was constructed using TreeAnnotator 2.6.2 [36].

A final COI gene tree was reconstructed using the SpeciesTreeUCLN algorithm of the StarBEAST2 package [38]. This approach implements a mixed-model including a coalescent component within species and a diversification component between species that allows accounting for variations of substitution rates within and between species [39]. SpeciesTreeUCLN jointly reconstructs gene trees and species trees, and, as such, requires the designation of species, which were determined using the consensus of our species delimitation analyses. The SpeciesTreeUCLN analysis was performed with the same parameters as mentioned above.

Several parameters of genetic diversity were estimated using the R package pegas 1.0 [40], including the number of haplotypes (h), haplotype diversity (Hd) [41], nucleotide diversity (π) [42], genetic diversity based on the number of segregating sites (θ) [43], and Tajima's D test of neutrality [44]. Kimura 2-parameter (K2P) [45] pairwise genetic distances were calculated using the R package Ape 5.4 [46]. Maximum intraspecific and nearest neighbor genetic distances were calculated from the pairwise K2P distance matrix using the R package Spider 1.5 [47].

3. Results

A total of 263 COI sequences were generated from six sites for *O. niloticus*, and one site for *O. mossambicus* and *O. macrochir*. Sequences consisted of mostly full-length sequences (652 bp) and no stop codons were detected, suggesting that the sequences collected represent functional coding regions. DNA-based species delimitation methods resulted in congruent delimitation schemes with 6 MOTUs for ABGD, RESL, and sPTP, 5 MOTUs for mPTP and sGMYC, and 8 MOTUs for mGMYC (Figure 2, Table S1). The final consensus consisted of 6 MOTUs (Figure 2). Maximum distances within MOTU ranged from 0 (BOLD:AAA8513, BOLD:ACR7163, BOLD:ADI0792) to 0.007 (BOLD:AAC9904), and minimum distances between MOTUs ranged from 0.016 (BOLD: AAA8513, BOLD: ADI0792) to 0.065 for BOLD:AAA6537 (Table 1). K2P distances were 7-fold higher on average between than within MOTUs. None of the MOTUs were restricted to a single strain, excepting BOLD:AAA8513 restricted to *O. macrochir* (Figure 2). The five remaining MOTUs were shared between strains of *O. niloticus* and *O. mossambicus* (Figure 2, Table 1).

Shared MOTUs among species of *Oreochromis* were found on BOLD for several BINs (Table 1). Records associated to: (1) BOLD:AAA6537 members mostly belong to *O. niloticus*; (2) BOLD:AAA8511 members mostly belong to *O. niloticus* and *O. mossambicus*; (3) BOLD:AAA8513 members belong to *O. macrochir*; (4) BOLD:AAC9904 members mostly belong to *O. niloticus*; (5) BOLD:ACR7163 members mostly belong to *O. urolepis*; and (6) BOLD:ADI0792 members mostly belong to *O. mossambicus*. A substantial proportion of *Oreochromis*, BOLD records have not been identified to species particularly within the BINs BOLD:AAA6537 and BOLD:AAA8513, resulting in a low frequency of sequences named. Two discrepancies in the proportion of species per MOTU were detected between BOLD records and sequences generated for the present study. BOLD:ACR7163 mostly contains records assigned to *O. urolepis* in BOLD although this species was never reported from Madagascar. BOLD:ADI0792, mostly represented by *O. niloticus*, is a new occurrence of this lineage for this species.

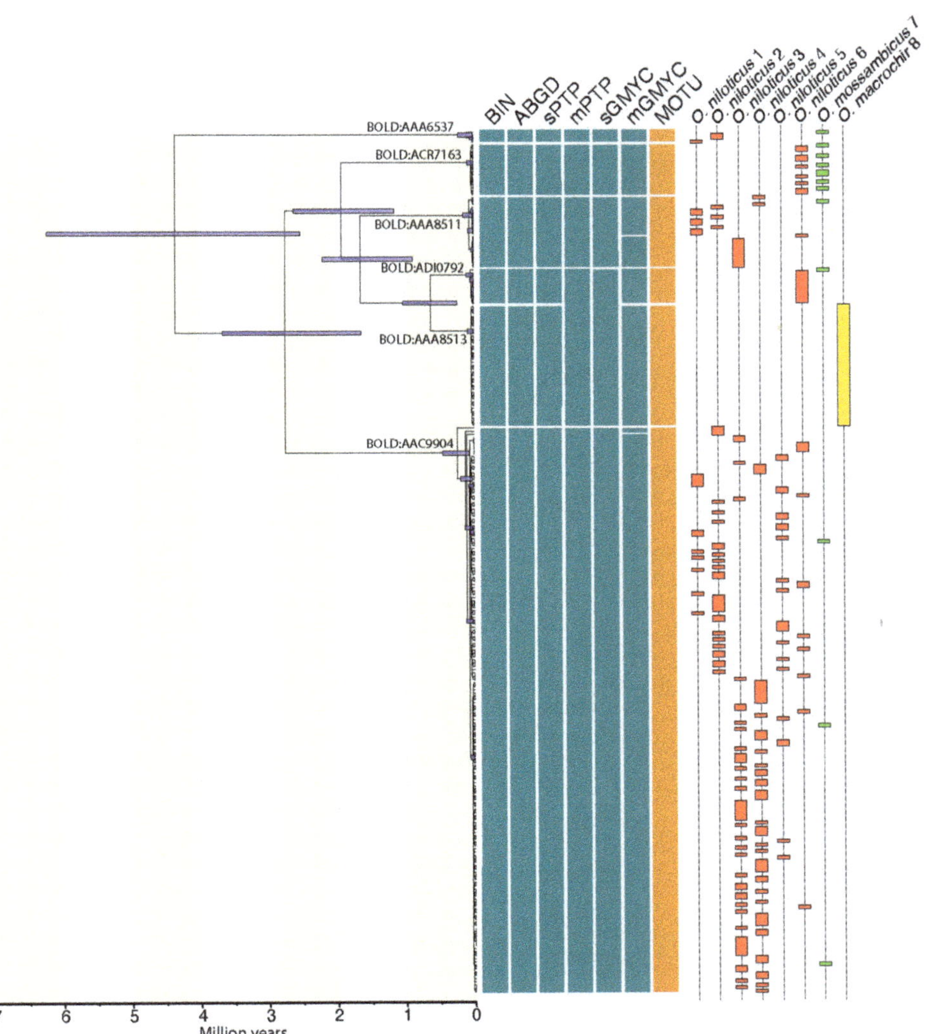

Figure 2. Mitochondrial gene tree for the 262 DNA barcodes of *Oreochromis* spp. inferred with SpeciesTreeUCLN, including 95% HPD interval for node age estimates, genetic lineage delimitation results for five methods (mGMYC discarded) and their 50% consensus, BOLD Barcode Index Numbers (BIN) for each MOTU, and distribution of farmed strain individuals.

Table 1. Summary of genetic distances and species composition of each MOTUs including their BOLD Barcode Index Number (BIN), number of individuals in present study, number of publicly available records in BOLD (not including newly generated records here), maximum intraspecific and minimum interspecific K2P genetic distances, and relative proportions of species within MOTUs according to the present study and in publicly available BOLD records.

	N		K2P Distance		*O. niloticus* (Freq.)		*O. mossambicus* (Freq.)		*O. macrochir* (Freq.)		*O. aureus* (Freq.)		*O. urolepis* (Freq.)	
BIN	**Present Study**	**BOLD**	**Within (Max)**	**Between (Min)**	**Present Study**	**BOLD**	**Present Study**	**BOLD**	**Present Study**	**BOLD**	**Present Study**	**BOLD**	**Present Study**	**BOLD**
BOLD:AAA6537	4	304	0.004	0.065	0.75	0.33	0.25	0.0001	-	-	-	0.3	-	-
BOLD:AAA8511	22	234	0.004	0.035	0.9995	0.14	0.0005	0.52	-	-	-	-	-	-
BOLD:AAA8513	37	18	0	0.016	-	-	-	-	1	0.17	-	-	-	-
BOLD:AAC9904	172	417	0.007	0.050	0.9998	0.73	0.0002	0.0001	-	-	-	-	-	-
BOLD:ACR7163	16	24	0	0.036	0.56	-	0.44	0.01	-	-	-	-	-	0.55
BOLD:ADI0792	11	28	0	0.016	0.9	-	0.1	0.29	-	-	-	-	-	-

As a result of the presence of multiple mitochondrial lineages in breeding stocks, which are estimated to have diverged between 0.6 and 2.7 million years ago (Figure 2), the nucleic diversity in most stocks is high, with π and θw above 0.015 and 8, respectively (Table 2). However, the JICA stock of *O. niloticus* (Site 4, Table 2) and *O. macrochir* at Milasoa (Site 8, Table 2) display much lower nucleic diversity. The high diversity estimates of nucleic diversity are accompanied by high haplotype diversity at sites 1, 2, 3, 6, and 7, due to the occurrence of a high number of haplotypes, peaking at 8 in site 2 (Table 2). None of the Tajima's *D* tests were significant, indicating that diversity was mostly balanced for the eight breeding stocks, despite some negative values at site 2 (*O. niloticus* GIFT at Tamatave) and site 4 (*O. niloticus* JICA at Mahajunga) indicative of a deficit of rare haplotypes.

Table 2. Summary statistics of genetic diversity per MOTU. *N*, number of individuals; *h*, number of haplotypes; *Hd*, haplotypic diversity; π, nucleotide diversity; θw, theta; Tajima's *D* test including *D* value and significance of the test (* significant at 0.025 threshold, none of the tests were significant).

Species	Strain	Locality	Site	*N*	*h*	*Hd*	π	θw	*D*	*p*-Value
O. niloticus	GIFT	Matera	1	16	3	0.575	0.031	18.082	0.434	0.664
O. niloticus	GIFT	Tamatave	2	31	8	0.539	0.018	15.519	−1.262	0.207
O. niloticus	Local	Ampefy	3	58	7	0.622	0.016	8.209	0.554	0.579
O. niloticus	JICA	Mahajunga	4	46	4	0.243	<0.001	0.455	−1.481	0.127
O. niloticus	Local	Fenerive	5	30	3	0.393	0.008	8.582	−1.719	0.086
O. niloticus	Local	Brickaville	6	31	6	0.785	0.037	14.268	2.086	0.037
O. mossambicus	Local	Fenerive	7	13	6	0.718	0.037	23.52	0.067	0.947
O. macrochir	Local	Milasoa	8	37	2	0.054	<0.001	0.240	−1.131	0.258

4. Discussion

The present study highlights high levels of mitochondrial genetic diversity, both at nucleotide and haplotype level, across most *Oreochromis* stocks assessed. Such high levels of genetic diversity originate from the co-occurrence of multiple, highly divergent MOTUs within most stocks, for both *O. niloticus* and *O. mossambicus*. However, high haplotype diversity also indicates the presence of substantial genetic diversity within each MOTU. These likely have multiple origins. In their native range distribution, *Oreochromis* species are known to easily hybridize after secondary contacts and can result in discordant evolutionary histories between genomes [48]. Multiple cases of introgression of both mitochondrial and nuclear genomes have been reported [49–53]. This likely accounts for the evolutionary success of the group. Hybrids with higher fitness than parental lineages, living outside of the ecological range of their parental species, have been detected in cichlids [54,55] and other perciformes [56]. In the context of fish farming, mitochondrial genomes resulting from ancient introgression events are likely introduced during the building of brood stock or through introductions into the wild. The occurrence of BOLD:ACR7163 and BOLD:AAA6537, mostly assigned to *O. urolepis* and *O. aureus* records on BOLD, respectively, seems to support the assumption that heterospecific mitochondrial genomes resulting from ancient introgression events were inadvertently introduced into Madagascar stocks of *O. niloticus*, as *O. urolepis* and *O. aureus* have never been reported as introduced species for Madagascar [57]. Alternatively, inadvertent and unrecorded introductions of *O. urolepis* and *O. aureus* individuals within imported batches of *O. niloticus* cannot be discarded, particularly if young, immature, and morphologically indistinguishable individuals were introduced.

The high mitochondrial diversity and multiple occurrences of MOTUs among species and strains might further reflect local breeding strategies and stock maintenance. Broodstock are known to display similar levels of genetic diversity in comparison with wild populations, which is the result of introductions in the 1950s [14]. Our study suggests tilapia production in Madagascar relies mostly on natural populations, with fry being collected in rivers and grow-out occurrence in ponds. Alternatively, fry are produced in ponds and broodstock genetic diversity is maintained by the regular addition of wild adults [14,58]. Relationships between broodstocks and wild populations depend on the scale of the fish farms. Oswald et al. [58] identified at least three type of fish farms in Madagascar: (1) small-scale, artisanal farms consisting of a single pond for fry production and grow-out with production mostly devoted to local consumption; (2) medium-scale farms where fry production and grow-out are managed in different ponds, and fry production is partly disseminated to small-scale farms for grow-out; and (3) large-scale farms devoted to fry production and distribution, constituting the main source of fry for aquaculture. Fry distribution can occur over large distances within Madagascar, and most small farms have no particular strategy to maintain strains. The similar mitochondrial composition of sampled stocks, and multiple occurrences of MOTUs in *O. niloticus* and *O. mossambicus* strains confirms that stocks used to be mixed in the past, and further contributed to increase strain mitochondrial genetic diversity. However, management of the most recently introduced strains seems to depart from this trend as the *O. macrochir* strain hosts a single and private MOTU. Observed haplotype and nucleotide diversity among imported (GIFT, JICA) and local tilapia strains are mostly similar, however, several local strains display higher mitochondrial genetic diversity. This trend suggests that the initial genetic pool of *O. niloticus* introduced in the 1950s was already diverse. Its successful introduction likely accounts for the persistence of high mitochondrial genetic diversity in farmed strains, due to their historical reliance on wild populations. Several studies have shown a relationship between individual genetic diversity and fitness in fishes [59–61]. As such, this high mitochondrial genetic diversity, also previously reported using nuclear markers [14], suggests local strains constitute good candidates for selective breeding, with genetically diverse wild populations available for maintaining high levels of genetic diversity.

5. Conclusions

The present study shows that tilapia stocks in Madagascar have genetically diverse mitochondrial genomes, likely resulting from an intricate history of ancient introgressions, introduction of genetically diverse populations in the 1950s, and local practices. The detection of higher mitochondrial genetic diversity in several local strains in comparison to introduced strains of *O. niloticus*, suggests local practices are beneficial in maintaining a high level of genetic diversity. In the context of fast development of modern aquaculture, this high genetic diversity, also previously observed for nuclear markers [14], certainly constitutes an asset. With genetically diverse wild populations, multiple options are available for maintaining high levels of genetic diversity in broodstock and several strategies are further available for selective breeding programs.

Supplementary Materials: The following are available online at https://www.mdpi.com/article/10.3390/d13070281/s1. Table S1: Results of the genetic species delimitation, including the majority-rule consensus.

Author Contributions: Conceptualization, E.P., H.d.V. and N.H.; methodology, E.P., H.d.V. and N.H.; software, E.P., H.d.V. and N.H.; validation, J.-M.M., E.P., D.S., D.E.A.-M., H.d.V. and N.H.; formal analysis, N.H.; investigation, J.-M.M., E.P., D.S., D.E.A.-M., H.d.V. and N.H.; resources, E.P., D.E.A.-M., H.d.V., D.S. and N.H.; data curation, E.P. and N.H.; writing—original draft preparation, N.H., E.P., J.-M.M. and H.d.V.; writing—review and editing, D.E.A.-M. and D.S.; visualization, J.-M.M., E.P., D.S., D.E.A.-M., H.d.V. and N.H.; supervision, E.P. and N.H.; project administration, E.P.; funding acquisition, E.P., D.S. and N.H. All authors have read and agreed to the published version of the manuscript.

Funding: This research was funded by the CFREF to the University of Guelph's Food from Thought program, by the "Institut de Recherche pour le Développement" through annual funding and "Fonds d'Amorçage" (226F2ABIOS) and through the AMPIANA project funded by the European Union (EuropeAid/135-182/DD).

Institutional Review Board Statement: Not applicable.

Informed Consent Statement: Not applicable.

Data Availability Statement: The data presented in this study are openly available in BOLD at [dx.doi.org/10.5883/DS-TILMADA], reference number DS-TILMADA.

Acknowledgments: Authors are most grateful to Lionel Dabbadie (Cirad) and Tojoharivelo Rakotomalala (FOFIFA) for the identification of the different sites studied and for organizing field sampling. The authors are also grateful to Marc Canonne (Cirad), the NGO APDRA *Pisciculture Paysanne* and FOFIFA for technical and logistic assistance before or during field work. Finally, the authors are very grateful to André Ramilison and Zarre, fish farmers, who agreed to make their fish available for this study and the staff of the Ivoloina fish farming station. This is publication ISEM 2021-134 SUD.

Conflicts of Interest: The authors declare no conflict of interest. The funders had no role in the design of the study; in the collection, analyses, or interpretation of data; in the writing of the manuscript, or in the decision to publish the results.

References

1. Worm, B.; Barbier, E.B.; Beaumont, N.; Duffy, E.; Folke, C.; Halpern, B.S.; Jackson, J.B.C.; Lotze, H.K.; Micheli, F.; Palumbi, S.R.; et al. Impacts of biodiversity loss on ocean ecosystem services. *Science* **2006**, *314*, 787–790. [CrossRef]
2. Newbold, T.; Hudson, L.N.; Arnell, A.P.; Contu, S.; De Palma, A.; Ferrier, S.; Hill, S.L.L.; Hoskins, A.J.; Lysenko, I.; Phillips, H.R.P. Has land use pushed terrestrial biodiversity beyond the planetary boundary? A global assessment. *Science* **2016**, *353*, 288–291. [CrossRef]
3. Thiault, L.; Mora, C.; Cinner, J.E.; Cheung, W.W.L.; Graham, N.A.J.; Januchowski-Hartley, F.A.; Mouillot, D.; Sumaila, U.R.; Claudet, J. Escaping the perfect storm of simultaneous climate change impacts on agriculture and marine fisheries. *Sci. Adv.* **2019**, *5*, eaaw9976. [CrossRef] [PubMed]
4. FAO. *Des Pêches Et De L'aquaculture*; FAO: Rome, Italy, 2018; ISBN 9789251306925.
5. Naylor, R.L.; Hardy, R.W.; Buschmann, A.H.; Bush, S.R.; Cao, L.; Klinger, D.H.; Little, D.C.; Lubchenco, J.; Shumway, S.E.; Troell, M. A 20-year retrospective review of global aquaculture. *Nature* **2021**, *591*, 551–563. [CrossRef] [PubMed]
6. Tacon, A.G.J. Trends in global aquaculture and aquafeed production: 2000–2017. *Rev. Fish. Sci. Aquac.* **2020**, *28*, 43–56. [CrossRef]
7. Lobell, D.B.; Burke, M.B.; Tebaldi, C.; Mastrandrea, M.D.; Falcon, W.P.; Naylor, R.L. Prioritizing Climate Change Adaptation Needs for Food Security in 2030. *Science* **2008**, *319*, 607–610. [CrossRef] [PubMed]
8. Froese, R.; Pauly, D. Fishbase. Available online: http://www.fishbase.org (accessed on 14 January 2021).
9. Kiener, A. Intérêt et perspectives de la pisciculture de la carpe à Madagascar. *Bull. Madag.* **1958**, *8*, 693–702.
10. Kiener, A. Poissons, pêche et pisciculture à Madagascar. *Centrotechique For. Trop. Fr.* **1963**, *405*, 599–606.
11. Therezien, Y. L'introduction de poissons d'eau douce à Madagascar, leur influence sur la modification du biotope. *Bull. Français Piscic.* **1960**, *199*, 45–61. [CrossRef]
12. Duchaufour, H.; Razafimbelo-Andriamifidy, T.; Rakotoarisoa, J.; Ramamonjisoa, B. ; *Rakotondravao. Recherche Interdisciplinaire pour le Développement Durable Application à Différentes Thématiques de Territoire et la Biodiversité des Espaces Ruraux Malgaches*; Cirad: Antananarivo, Madagascar, 2016.
13. Bentz, B.; Oswald, M. Respective roles of national institutions and farmers groups in the implementation of an innovation enabling smallholders to reproduce Carp inside their rice fields in betafo (Madagascar). In Proceedings of the Symposium Innovation and Sustainable Development in Agriculture and Food, ISDA 2010, Montpellier, France, 28 June 2010.
14. Ravakarivelo, M.; Pepey, E.; Benzie, J.; Raminosoa, N.; Rasamoelina, H.; Mikolasek, O.; De Verdal, H. Genetic variation in wild populations and farmed stocks of Nile tilapia (*Oreochromis niloticus*) in Madagascar. *Revue d'Elevage et de Médecine Vétérinaire des Pays Tropicaux* **2019**, *72*, 101. [CrossRef]
15. Ponzoni, R.W.; Hamzah, A.; Tan, S.; Kamaruzzaman, N. Genetic parameters and response to selection for live weight in the GIFT strain of Nile tilapia (*Oreochromis niloticus*). *Aquaculture* **2005**, *247*, 203–210. [CrossRef]
16. Ponzoni, R.W.; Nguyen, N.H.; Khaw, H.L.; Hamzah, A.; Bakar, K.R.A.; Yee, H.Y. Genetic improvement of Nile tilapia (*Oreochromis niloticus*) with special reference to the work conducted by the WorldFish Center with the GIFT strain. *Rev. Aquac.* **2011**, *3*, 27–41. [CrossRef]
17. Gupta, M.V.; Acosta, B.O. From drawing board to dining table: The success story of the GIFT project. *NAGA WorldFish Cent. Q.* **2004**, *27*, 4–14.
18. Ward, R.D.; Hanner, R.H.; Hebert, P.D.N. The campaign to DNA barcode all fishes, FISH-BOL. *J. Fish Biol.* **2009**, *74*, 329–356. [CrossRef]
19. Hubert, N.; Hanner, R.; Holm, E.; Mandrak, N.E.; Taylor, E.; Burridge, M.; Watkinson, D.; Dumont, P.; Curry, A.; Bentzen, P.; et al. Identifying Canadian freshwater fishes through DNA barcodes. *PLoS ONE* **2008**, 3. [CrossRef]
20. Ivanova, N.V.; Zemlak, T.S.; Hanner, R.H.; Hébert, P.D.N. Universal primers cocktails for fish DNA barcoding. *Mol. Ecol. Notes* **2007**, *7*, 544–548. [CrossRef]
21. Ratnasingham, S.; Hebert, P.D.N. BOLD: The Barcode of Life Data System (www.barcodinglife.org). *Mol. Ecol. Notes* **2007**, *7*, 355–364. [CrossRef]
22. Hebert, P.D.N.; deWaard, J.R.; Zakharov, E.; Prosser, S.W.J.; Sones, J.E.; McKeown, J.T.A.; Mantle, B.; La Salle, J. A DNA "barcode blitz": Rapid digitization and sequencing of a natural history collection. *PLoS ONE* **2013**, *8*, e68535. [CrossRef]

23. Pons, J.; Barraclough, T.G.; Gomez-Zurita, J.; Cardoso, A.; Duran, D.P.; Hazell, S.; Kamoun, S.; Sumlin, W.D.; Vogler, A.P. Sequence-based species delimitation for the DNA taxonomy of undescribed insects. *Syst. Biol.* **2006**, *55*, 595–606. [CrossRef]
24. Puillandre, N.; Lambert, A.; Brouillet, S.; Achaz, G. ABGD, Automatic Barcode Gap Discovery for primary species delimitation. *Mol. Ecol.* **2012**, *21*, 1864–1877. [CrossRef]
25. Ratnasingham, S.; Hebert, P.D.N. A DNA-based registry for all animal species: The barcode index number (BIN) system. *PLoS ONE* **2013**, *8*, e66213. [CrossRef] [PubMed]
26. Kapli, P.; Lutteropp, S.; Zhang, J.; Kobert, K.; Pavlidis, P.; Stamatakis, A.; Flouri, T. Multi-rate Poisson Tree processes for single-locus species delimitation under maximum likelihood and Markov chain Monte Carlo. *Bioinformatics* **2017**, *33*, 1630–1638. [CrossRef] [PubMed]
27. Luo, A.; Ling, C.; Ho, S.Y.W.; Zhu, C.-D. Comparison of methods for molecular species delimitation across a range of speciation scenarios. *Syst. Biol.* **2018**, *67*, 830–846. [CrossRef] [PubMed]
28. Kekkonen, M.; Mutanen, M.; Kaila, L.; Nieminen, M.; Hebert, P.D.N. Delineating species with DNA barcodes: A case of taxon dependent method performance in moths. *PLoS ONE* **2015**, *10*, e0122481. [CrossRef]
29. Kekkonen, M.; Hebert, P.D.N. DNA barcode-based delineation of putative species: Efficient start for taxonomic workflows. *Mol. Ecol. Resour.* **2014**, *14*, 706–715. [CrossRef]
30. Shen, Y.; Hubert, N.; Huang, Y.; Wang, X.; Gan, X.; Peng, Z.; He, S. DNA barcoding the ichthyofauna of the Yangtze River: Insights from the molecular inventory of a mega-diverse temperate fauna. *Mol. Ecol. Resour.* **2019**, *19*, 1278–1291. [CrossRef]
31. Sholihah, A.; Delrieu-Trottin, E.; Sukmono, T.; Dahruddin, H.; Risdawati, R.; Elvyra, R.; Wibowo, A.; Kustiati, K.; Busson, F.; Sauri, S.; et al. Disentangling the taxonomy of the subfamily Rasborinae (Cypriniformes, Danionidae) in Sundaland using DNA barcodes. *Sci. Rep.* **2020**, *10*, 2818. [CrossRef]
32. Limmon, G.; Delrieu-Trottin, E.; Patikawa, J.; Rijoly, F.; Dahruddin, H.; Busson, F.; Steinke, D.; Hubert, N. Assessing species diversity of Coral Triangle artisanal fisheries: A DNA barcode reference library for the shore fishes retailed at Ambon harbor (Indonesia). *Ecol. Evol.* **2020**, *10*, 3356–3366. [CrossRef]
33. Zhang, J.; Kapli, P.; Pavlidis, P.; Stamatakis, A. A general species delimitation method with applications to phylogenetic placements. *Bioinformatics* **2013**, *29*, 2869–2876. [CrossRef]
34. Fujisawa, T.; Barraclough, T.G. Delimiting species using single-locus data and the generalized mixed Yule coalescent approach: A revised method and evaluation on simulated data sets. *Syst. Biol.* **2013**, *62*, 707–724. [CrossRef]
35. Stamatakis, A. RAxML version 8: A tool for phylogenetic analysis and post-analysis of large phylogenies. *Bioinformatics* **2014**, *30*, 1312–1313. [CrossRef]
36. Bouckaert, R.; Heled, J.; Kühnert, D.; Vaughan, T.; Wu, C.H.; Xie, D.; Suchard, M.A.; Rambaut, A.; Drummond, A.J. BEAST 2: A software platform for Bayesian evolutionary analysis. *PLoS Comput. Biol.* **2014**, *10*, 1–6. [CrossRef]
37. Bermingham, E.; McCafferty, S.; Martin, A.P. Fish biogeography and molecular clocks: Perspectives from the Panamanian isthmus. In *Molecular Systematics of Fishes*; Kocher, T.D., Stepien, C.A., Eds.; CA Academic Press: San Diego, CA, USA, 1997; pp. 113–128.
38. Ogilvie, H.A.; Bouckaert, R.R.; Drummond, A.J. StarBEAST2 brings faster species tree inference and accurate estimates of substitution rates. *Mol. Biol. Evol.* **2017**, *34*, 2101–2114. [CrossRef]
39. Ho, S.Y.W.; Larson, G. Molecular clocks: When times are a-changin'. *TRENDS Genet.* **2006**, *22*, 79–83. [CrossRef]
40. Paradis, E. pegas: An {R} package for population genetics with an integrated-modular approach. *Bioinformatics* **2010**, *26*, 419–420. [CrossRef]
41. Nei, M.; Tajima, F. DNA polymorphism detectable by restriction endonucleases. *Genetics* **1981**, *97*, 145–163. [CrossRef]
42. Nei, M. *Molecular Evolutionary Genetics*; Columbia University Press: Chichester, NY, USA, 1987; ISBN 0231063210.
43. Watterson, G.A. On the number of segregating sites in genetical models without recombination. *Theor. Popul. Biol.* **1975**, *7*, 256–276. [CrossRef]
44. Tajima, F. Statistical method for testing the neutral mutation hypothesis by DNA polymorphism. *Genetics* **1989**, *123*, 585–595. [CrossRef]
45. Kimura, M. A Simple method for estimating evolutionary rates of base substitutions through comparative studies of nucleotide-sequences. *J. Mol. Evol.* **1980**, *16*, 111–120. [CrossRef]
46. Paradis, E.; Schliep, K. ape 5.0: An environment for modern phylogenetics and evolutionary analyses in R. *Bioinformatics* **2019**, *35*, 526–528. [CrossRef]
47. Brown, S.D.J.; Collins, R.A.; Boyer, S.; Lefort, C.; Malumbres-Olarte, J.; Vink, C.J.; Cruickshank, R.H. Spider: An R package for the analysis of species identity and evolution, with particular reference to DNA barcoding. *Mol. Ecol. Resour.* **2012**, *12*, 562–565. [CrossRef]
48. Ford, A.G.P.; Bullen, T.R.; Pang, L.; Genner, M.J.; Bills, R.; Flouri, T.; Ngatunga, B.P.; Rüber, L.; Schliewen, U.K.; Seehausen, O. Molecular phylogeny of Oreochromis (Cichlidae: Oreochromini) reveals mito-nuclear discordance and multiple colonisation of adverse aquatic environments. *Mol. Phylogenet. Evol.* **2019**, *136*, 215–226. [CrossRef]
49. Nyingi, D.W.; Agnèse, J. Recent introgressive hybridization revealed by exclusive mtDNA transfer from *Oreochromis leucostictus* (Trewavas, 1933) to *Oreochromis niloticus* (Linnaeus, 1758) in Lake Baringo, Kenya. *J. Fish Biol.* **2007**, *70*, 148–154. [CrossRef]
50. D'Amato, M.E.; Esterhuyse, M.M.; Van Der Waal, B.C.W.; Brink, D.; Volckaert, F.A.M. Hybridization and phylogeography of the Mozambique tilapia *Oreochromis mossambicus* in southern Africa evidenced by mitochondrial and microsatellite DNA genotyping. *Conserv. Genet.* **2007**, *8*, 475–488. [CrossRef]

51. Gregg, R.E.; Howard, J.H.; Snhonhiwa, F. Introgressive hybridization of tilapias in Zimbabwe. *J. Fish Biol.* **1998**, *52*, 1–10. [CrossRef]
52. Mojekwu, T.O.; Cunningham, M.J.; Bills, R.I.; Pretorius, P.C.; Hoareau, T.B. Utility of DNA barcoding in native *Oreochromis* species. *J. Fish Biol.* **2021**, *98*, 498–506. [CrossRef]
53. Firmat, C.; Alibert, P.; Losseau, M.; Baroiller, J.-F.; Schliewen, U.K. Successive invasion-mediated interspecific hybridizations and population structure in the endangered cichlid *Oreochromis mossambicus*. *PLoS ONE* **2013**, *8*, e63880. [CrossRef]
54. Seehausen, O. Hybridization and adaptive radiation. *Trends Ecol. Evol.* **2004**, *19*, 198–207. [CrossRef]
55. Selz, O.M.; Seehausen, O. Interspecific hybridization can generate functional novelty in cichlid fish. *Proc. R. Soc. B* **2019**, *286*, 20191621. [CrossRef]
56. Englebrecht, C.; Freyhof, J.; Nolte, A.; Rassman, K.; Schliewen, U.; Tautz, D. Phylogeography of the bullhead *Cottus gobio* (Pisces: Teleostei: Cottidae) suggests a pre-Pleistocene origin of the major central European populations. *Mol. Ecol.* **2000**, *9*, 709–722. [CrossRef]
57. Fricke, R.; Mahafina, J.; Behivoke, F.; Jaonalison, H.; Léopold, M.; Ponton, D. Annotated checklist of the fishes of Madagascar, southwestern Indian Ocean, with 158 new records. *FishTaxa* **2018**, 3, 1–432.
58. Oswald, M.R.; Ravakarivelo, M.; Mikolasek, O.; Rasamoelina, H.; de Verdal, H.; Bentz, B.; Pepey, E.; Al, E. Combining a comprehensive approach to fish-farming systems with assessment of their genetics—From planning to realization. In *Actes Projet FSP PARRUR, Recherche Interdisciplinaire pour le Développement Durable et la Biodiversité des Espaces Ruraux Malgaches. Application à Différentes Thématiques de Territoire*; Parrur, Ed.; Cirad: Antananarivo, MG, USA, 2016; pp. 219–267.
59. Borrell, Y.J.; Pineda, H.; McCarthy, I.; Vazquez, E.; Sanchez, J.A.; Lizana, G.B. Correlations between fitness and heterozygosity at allozyme and microsatellite loci in the Atlantic salmon, *Salmo salar* L. *Heredity* **2004**, *92*, 585–593. [CrossRef] [PubMed]
60. Lieutenant-Gosselin, M.; Bernatchez, L. Local heterozygosity-fitness correlations with global positive effects on fitness in threespine stickleback. *Evolution* **2006**, *60*, 1658–1668. [CrossRef] [PubMed]
61. Borrell, Y.J.; Carleos, C.E.; Sánchez, J.A.; Vázquez, E.; Gallego, V.; Asturiano, J.F.; Blanco, G. Heterozygosity–fitness correlations in the gilthead sea bream *Sparus aurata* using microsatellite loci from unknown and gene-rich genomic locations. *J. Fish Biol.* **2011**, *79*, 1111–1129. [CrossRef]

MDPI

Article

Validity of *Pampus liuorum* Liu & Li, 2013, Revealed by the DNA Barcoding of *Pampus* Fishes (Perciformes, Stromateidae)

Jiehong Wei [1,5,†], Renxie Wu [2,†], Yongshuang Xiao [1], Haoran Zhang [2], Laith A. Jawad [3], Yajun Wang [4], Jing Liu [1,*] and Mustafa A. Al-Mukhtar [6]

1 Laboratory of Marine Organism Taxonomy and Phylogeny, Institute of Oceanology, Chinese Academy of Sciences, Qingdao 266071, China; w443687230@163.com (J.W.); dahaishuang1982@163.com (Y.X.)
2 College of Fisheries, Guangdong Ocean University, Zhanjiang 524088, China; wurenxie@163.com (R.W.); zhanghaoran0619@163.com (H.Z.)
3 School of Environmental and Animal Sciences, Unitec Institute of Technology, 139 Carrington Road, Mt Albert, Auckland 1025, New Zealand; laith_jawad@hotmail.com
4 College of Marine Sciences, Ningbo University, Ningbo 315211, China; wangyajun@nbu.edu.cn
5 University of Chinese Academy of Sciences, Beijing 100049, China
6 Marine Science Centre, University of Basrah, Basrah 61004, Iraq; mam@msc-basra.com
* Correspondence: jliu@qdio.ac.cn
† These authors contributed equally to this paper.

Citation: Wei, J.; Wu, R.; Xiao, Y.; Zhang, H.; Jawad, L.A.; Wang, Y.; Liu, J.; Al-Mukhtar, M.A. Validity of *Pampus liuorum* Liu & Li, 2013, Revealed by the DNA Barcoding of *Pampus* Fishes (Perciformes, Stromateidae). *Diversity* **2021**, *13*, 618. https://doi.org/10.3390/d13120618

Academic Editor: Manuel Elias-Gutierrez

Received: 28 September 2021
Accepted: 23 November 2021
Published: 25 November 2021

Abstract: The genus *Pampus* contains seven valid species, which are commercially important fishery species in the Indo-Pacific area. Due to their highly similar external morphologies, *Pampus liuorum* has been proposed as a synonym of *Pampus cinereus*. In this study, partial sequences of COI (582 bp) and Cyt*b* (1077 bp) were presented as potential DNA barcodes of six valid *Pampus* species and the controversial species *P. liuorum*. A species delimitation of the seven *Pampus* species was performed to verify their validities. Explicit COI barcoding gaps were found in all assessed species, except for *P. liuorum* and *P. cinereus*, which resulted from their smaller interspecific K2P distance (0.0034–0.0069). A Cyt*b* barcoding gap (0.0200) of the two species was revealed, with a K2P distance ranging from 0.0237 to 0.0277. The longer Cyt*b* fragment is thus a more suitable DNA barcode for the genus *Pampus*. In the genetic tree, using concatenated Cyt*b* and COI sequences, the seven species reciprocally formed well-supported clades. Species delimitations with ABGD, GMYC, and bPTP models identified seven operational taxonomic units, which were congruent with the seven morphological species. Therefore, all of the seven analyzed species, including *P. liuorum*, should be kept as valid species.

Keywords: *Pampus liuorum* Liu & Li, 2013; DNA barcoding; species delimitation; systematics; Indo-West Pacific

1. Introduction

Pomfrets, species of genus *Pampus* Bonaparte, 1834, family Stromateidae Rafinesque, 1810, are pelagic marine fishes widely distributed along the coast of the Indo-West Pacific region. Seven valid species of genus *Pampus* have been recognized, namely, *Pampus argenteus* (Euphrasen, 1788), *P. candidus* (Cuvier, 1829), *Pampus chinensis* (Euphrasen, 1788), *Pampus cinereus* (Bloch, 1795), *Pampus minor* Liu & Li, 1998, *Pampus nozawae* (Ishikawa, 1904), and *Pampus punctatissimus* (Temminck & Schlegel, 1845) [1–9]. They contribute high commercial values to fisheries of the countries along the coast of the Indo-West Pacific region. In 2016, fishery harvests of pomfret in China reached over three million tons [10].

The taxonomy of the genus *Pampus* has long been confused by their highly similar external morphologies. *Pampus argenteus* might be the most confusing name in the genus *Pampus*. Its holotype is not available in its original description, while the vague original morphological description was found to be applicable to multiple known pomfret species [3]. Twelve available names were assigned as junior synonyms of *P. argenteus*, including *P. minor*, *P. cinereus*, *P. candidus*, and *P. punctatissimus*, which have been recognized or resurrected as valid species [1,4,5,9]. Liu et al. [11] presented a morphological comparison of *P. argenteus*, *P. cinereus*, *P. chinensis*, *P. minor*, and *P. punctatissimus*, which indicated that the five species differed from each other in numerous external and skeletal characters, e.g., skull, gill rakers, and sensory canal systems on the head and lateral lines. Liu et al. [3], based on the original description and type locality of *P. argenteus*, redescribed the species and designated its neotype, which set up a reference for verifying validities of its junior synonyms. Simultaneously, the neotype of *P. cinereus* was assigned and described by Liu et al. [6] as a substitution of its lost holotype. Liu and Li [2] described a novel species, *Pampus liuorum* Liu & Li, 2013, based on its distinct morphology compared with six known pomfret species. However, the phylogenetic tree by Yin et al. [7], inferred from numerous nuclear gene loci, indicated that the specimens of *P. cinereus* and *P. liuorum* formed a mixing clade, refusing monophyly of the two species. *Pampus liuorum* is thus suspected to be a junior synonym of *P. cinereus* [7], and its monophyly and exclusiveness await further verification. Li et al. [12] proposed the resurrection of *P. echinogaster* from *P. argenteus* because of their distinct cytochrome *c* oxidase subunit I (COI) gene sequences. However, a morphological comparison indicated that *P. echinogaster sensu* Li et al. [12] is similar to the neotype of *P. argenteus* designated in Liu et al. [3], and thus could be a misidentification. *Pampus nozawae* used to be considered as a junior synonym of *P. cinereus* [6]. Its validity was recently proposed based on its distinct axial skeletal morphology comparing to its congeners [8], although a redescription and neotype designation of this species are currently unavailable. Therefore, the validities of *P. nozawae* and *P. echinogaster* are still uncertain. Radhakrishnan et al. [9] resurrected *P. candidus* based on its distinct morphological and genetic characteristics compared to *P. argenteus*, *P. cinereus*, and *P. liuorum*.

DNA barcoding, the idea of using short segments of genes to enable the precise identification of species, was proposed as an alternative way to clarify the species and genetic diversity of the genus *Pampus* [3,13,14]. Guo et al. [13] carried out preliminarily explorations on the genetic diversity of the genus *Pampus* using partial sequences of 16S ribosomal RNA (16S rRNA) and COI genes, and confirmed that *P. minor* was genetically distinct from its congeners. Cui et al. [14], using mitogenomic data, identified five species among specimens collected from the coast of China, i.e., *P. minor*, *P. punctatissimus*, *P. chinensis*, *P. cinereus*, and *Pampus* sp. (possibly *P. argenteus* or *P. echinogaster*). Li et al. [15] reported a new species, *Pampus* sp. nov., claiming its mitogenome to be different from its congeners. Radhakrishnan et al. [16] reported two new species, *Pampus* sp1. and *Pampus* sp2., from the Indian Ocean. Li et al. [17] presented an integrative comparison of morphological and genetic differences in seven *Pampus* species from the Indo-Pacific region. Neighbor-joining trees inferred from COI sequences suggested that *Pampus* sp1. and *Pampus* sp2. *sensu* (Radhakrishnan et al. [16]) are identical to *P. argenteus* and *Pampus* sp. nov. *sensu* (Li et al. [15]), respectively [17]. Despite the huge efforts, the misidentifications and mislabelings of the pomfret species frequently occur, especially on NCBI (National Center for Biotechnology Information) GenBank, which could hinder the application of DNA barcoding for *Pampus* species identification [7,17].

To establish reliable references for pomfret species identification, partial COI and cytochrome *b* gene (Cyt*b*) sequences of seven pomfret species are presented in this study as potential DNA barcodes. To verify the validity of the pomfret species, we performed phylogenetic inference and species delimitation with well-identified *Pampus* specimens collected from the Indo-Pacific region, including type specimens of *P. argenteus* and *P. liuorum* deposited in the Museum of Marine Biology, Institute of Oceanology, Chinese Academy of Sciences (IOCAS).

2. Materials and Methods

2.1. Sampling and Species Identification

In this study, seven pomfret species (74 specimens) were assessed (Figure 1): *Pampus argenteus*, *P. candidus*, *P. chinensis*, *P. cinereus*, *P. minor*, *P. liuorum*, and *P. punctatissimus*.

Due to a lack of specimens, *Pampus nozawae* was not included in this study. Six of the assessed *Pampus* species, including a total of seventy specimens, were collected from nine localities along the coast of China from August 2009 to January 2014 using commercial fishing trawl boats or gillnet fishing. Two paratypes of *P. liuorum* (i.e., IOCAS20120541 and 0542) were derived from Liu and Li [2], where the species was first described. Three specimens of *P. argenteus* (i.e., IOCAS120413, 0423, 0435) were derived from Liu et al. [6], where *P. argenteus* was redescribed. All specimens were carefully identified based on the type of specimen and our previous work on *Pampus* taxonomy [2–6]. Four specimens of *P. candidus* were collected from coastal Iraq in the northern Indian Ocean and identified based on morphological descriptions and the Cyt*b* sequences of Radhakrishnan et al. [9]. Muscle tissues of the specimens were taken and preserved in 95% ethanol for further experiments. All voucher specimens of the barcodes were deposited at the Museum of Marine Biology, IOCAS, Qingdao, China. Sequences of *Peprilus medius* (COI, AB205449; Cyt*b*, AB205471) from Doiuhi and Nakabo [18] were obtained from NCBI GenBank and selected as an outgroup for molecular analyses.

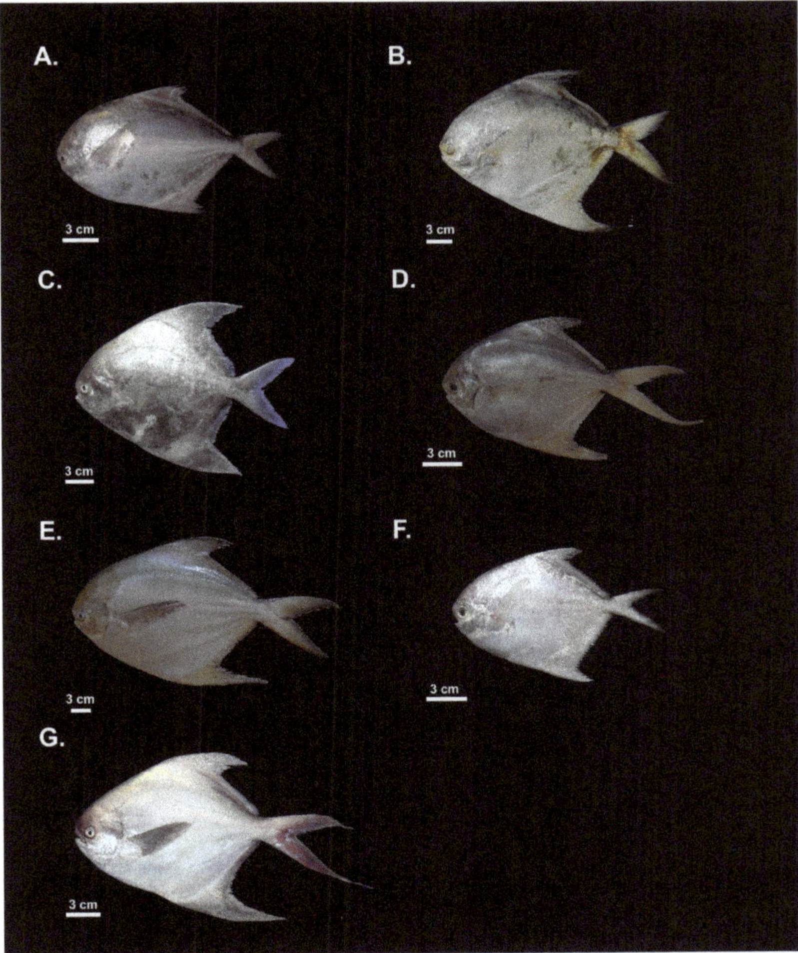

Figure 1. Photographs of the seven studied *Pampus* species of this study. (**A**) *P. argenteus* (PA-IOCAS120435); (**B**) *P. candidus* (PCA-2015004); (**C**) *P. chinensis* (PCH-201006003); (**D**) *P. cinereus* (PCI-20120520); (**E**) *P. liuorum* (PL-IOCAS120542); (**F**) *P. minor* (PM-2012504); (**G**) *P. punctatissimus* (PP-2013129).

2.2. DNA Extraction, Amplification and Sequencing

Total genomic DNA was extracted from muscle tissues, following the protocol of Sambrook et al. [19]. The COI barcode sequence was amplified by two pairs of fish-specific primers (FishF1 and FishR1; FishF2 and FishR2) [20]. Based on mitochondrial genome sequences of *Pampus* in Cui et al. [14], a new primer (Thr20-Pam) was designed, and three primers reported by Doiuhi and Nakabo [18] were modified to form three pairs of primers for Cyt*b* sequence amplification of the genus *Pampus*. The primer names and sequences are as follows: one forward primer: L14724-Pam (5′-GACTTGAAAAACCATCGTTG-3′); three reverse primers: Thr20-Pam (5′-GTTTACAAGACCGGCGCTCT-3′), H15915-Pam (5′-TTCCGACGTCCGGTTTACAAGAC-3′), and H15973-Strdei (5′-TTGGGAGYYRGTGGTAG-GAGTT-3′). Polymerase chain reactions (PCR) were performed in a 50 μL volume with 50 ng template DNA, 5 μL of 10 × reaction buffer, 1.5 mM $MgCl_2$, 200 μM dNTP mixture, 0.2 μM of each primer, and 2.5 U *Taq* DNA polymerase (Transgen Biotech Co., Ltd., Beijing, China). PCR cycles were conducted on a Veriti™ 96-Well Thermal Cycler (Applied Biosystems, USA) under the following protocol: initial denaturation for 4 min at 94 °C, followed by 35 cycles of 45 s at 94 °C, 45 s at 50–52 °C, 45 s at 72 °C, and a final 10 min extension at 72 °C. PCR amplification without the addition of the template DNA was used as a negative control reaction to ensure no cross-contamination during the experiments. PCR products were separated on 1.2% agarose gel, and then sent to Sangon Biotech Co., Ltd. (Shanghai, China) for bidirectional DNA sequencing with the corresponding forward and reverse primers in PCR reactions, using the ABI Prism 3730 automatic sequencer (Applied Biosystems, Foster City, CA, USA).

2.3. Phylogenetic Inference and Barcoding Gaps

The raw sequences were first assembled in EditSeq V7.1.0 (Lasergene, DNASTAR, Madison, WI, USA), and only high-quality bases with clear signals were retained for analyses. Sequence alignment was carried out in MegAlign V7.1.0 (Lasergene, DNASTAR) using the ClustalW algorithm with default settings. The sequences were trimmed to obtain uniform lengths for subsequent analyses. The COI and Cyt*b* sequences were deposited in NCBI GenBank. Sampling information, specimen photos (whenever available), and corresponding COI and Cyt*b* sequences of the specimens were also archived on the Barcoding of Life Database (BOLD) under a public project coded by IOCAS (https://www.boldsystems.org/index.php/Public_SearchTerms?query=IOCAS, access on 1 November 2021). Sampling information, voucher specimen numbers (Museum ID of BOLD), and NCBI GenBank accession numbers of COI and Cyt*b* of the specimens are summarized in Table 1. Sequence variation indices of COI and Cyt*b* sequences among *Pampus* species, including base composition and number of polymorphic sites, parsimony informative sites, and indels, were calculated using DnaSP v6 [21]. COI and Cyt*b* sequences of the specimens were concatenated to form another dataset for tree inferences and species delimitations.

Table 1. Sampling data, BOLD sample IDs, and GenBank accession numbers of the *Pampus* species used in this study. The"✓" sign indicates that the specimen photo is available on BOLD.

Species	Sampling Date	Sampling Location (Number of Specimens)	BOLD Specimen Voucher		GenBank Accession Number	
			Museum ID	Photo Reference	COI	Cyt*b*
Pampus argenteus	April 2012	Zhuhai, Guangdong, China (3)	PA-IOCAS120413	✓	MK300954	MK301024
			PA-IOCAS120423	✓	MK300957	MK301027
			PA-IOCAS120435	✓	MK300958	MK301028

Table 1. *Cont.*

Species	Sampling Date	Sampling Location (Number of Specimens)	BOLD Specimen Voucher		GenBank Accession Number	
			Museum ID	Photo Reference	COI	Cyt*b*
	April–May 2012	Shenzhen, Guangdong, China (6)	PA-20120418	✓	MK300955	MK301025
			PA-20120419	✓	MK300956	MK301026
			PA-20120443	✓	MK300959	MK301029
			PA-20120444	✓	MK300960	MK301030
			PA-20120445	✓	MK300961	MK301031
			PA-20120447	✓	MK300962	MK301032
	May 2012	Zhanjiang, Guangdong, China (3)	PA-20120531	✓	MK300963	MK301033
			PA-20120532	✓	MK300964	MK301034
			PA-20120533	✓	MK300965	MK301035
	January 2014	Weihai, Shandong, China (4)	PA-201401001		MK300988	MK301058
			PA-201401002		MK300989	MK301059
			PA-201401003		MK300990	MK301060
			PA-201401004		MK300991	MK301061
	April 2012	Qingdao, Shandong, China (7)	PA-20120401	✓	MK300981	MK301051
			PA-20120402	✓	MK300982	MK301052
			PA-20120403	✓	MK300983	MK301053
			PA-20120404	✓	MK300984	MK301054
			PA-20120405		MK300985	MK301055
			PA-20120406		MK300986	MK301056
			PA-20120409		MK300987	MK301057
	May 2012	Zhoushan, Zhejiang, China (3)	PA-EZ2012003	✓	MK300992	MK301062
			PA-EZ2012004	✓	MK300993	MK301063
			PA-EZ2012005	✓	MK300994	MK301064
Pampus candidus	January 2015	Iraq (4)	PCA-2015004	✓	MZ604279	MZ604560
			PCA-2015005		MZ604280	MZ604561
			PCA-2015006		MZ604281	MZ604562
			PCA-2015007		MZ604282	MZ604563
Pampus chinensis	August 2009	Xiamen, Fujian, China (1)	PCH-200908009	✓	MK300966	MK301036
	May 2010	Zhuhai, Guangdong, China (5)	PCH-2010050025	✓	MK301037	MK300967
			PCH-2010050027	✓	MK301038	MK300968
			PCH-201006001	✓	MK301039	MK300969
			PCH-201006002	✓	MK301040	MK300970
			PCH-201006003	✓	MK301041	MK300971
Pampus cinereus	April–May 2012	Shenzhen, Guangdong, China (3)	PCI-20120457	✓	MK300972	MK301042
			PCI-20120459	✓	MK300973	MK301043
			PCI-20120460	✓	MK300974	MK301044
	April–May 2012	Zhuhai, Guangdong, China (3)	PCI-20120464	✓	MK300975	MK301045
			PCI-20120465	✓	MK300976	MK301046
			PCI-20120481	✓	MK300977	MK301047
	May 2012	Zhanjiang, Guangdong, China (3)	PCI-20120520	✓	MK300978	MK301048
			PCI-20120521	✓	MK300979	MK301049
			PCI-20120522	✓	MK300980	MK301050
Pampus liuorum	May 2012	Zhuhai, Guangdong, China (2)	PL-IOCAS120541	✓	MK300995	MK301065
			PL-IOCAS120542	✓	MK300996	MK301066

Table 1. *Cont.*

Species	Sampling Date	Sampling Location (Number of Specimens)	BOLD Specimen Voucher		GenBank Accession Number	
			Museum ID	Photo Reference	COI	Cyt*b*
	July–August 2013	Dongshan, Fujian, China (9)	PL-20130726061 PL-20130726062 PL-20130726063 PL-20130726064 PL-20130726065 PL-20130810031 PL-20130726066 PL-20130810029 PL-20130810030		MK300997 MK300998 MK300999 MK301000 MK301001 MK301002 MK301003 MK301004 MK301005	MK301067 MK301068 MK301069 MK301070 MK301071 MK301072 MK301073 MK301074 MK301075
Pampus minor	October 2013	Zhoushan, Zhejiang, China (1)	PM-2013159		MK301013	MK301083
	April 2012	Shenzhen, Guangdong, China (1)	PM-20120430	✓	MK301006	MK301076
	May 2010	Zhuhai, Guangdong, China (2)	PM-S20-098 PM-S20-102	 ✓	MK301014 MK301015	MK301084 MK301085
	May 2012	Zhanjiang, Guangdong, China (3)	PM-20120503 PM-20120504 PM-20120513	✓ ✓ ✓	MK301007 MK301008 MK301009	MK301077 MK301078 MK301079
	April 2013	Beihai, Guangxi, China (3)	PM-2013065 PM-2013066 PM-2013067		MK301010 MK301011 MK301012	MK301080 MK301081 MK301082
Pampus punctatissimus	June 2013	Zhoushan, Zhejiang, China (2)	PP-20130618 PP-20130619	✓ ✓	MK301017 MK301018	MK301087 MK301088
	October 2013	Xiamen, Fujian, China (5)	PP-2013129 PP-2013138 PP-2013139 PP-2013146 PP-2013154	✓	MK301019 MK301020 MK301021 MK301022 MK301023	MK301089 MK301090 MK301091 MK301092 MK301093
	April 2012	Zhuhai, Guangdong, China (1)	PP-20120427	✓	MK301016	MK301086

Due to more genetic distance references for Kimura's two-parameter model (K2P) [22], we calculated pairwise K2P distances to estimate barcoding gaps of each species. K2P distances among and within the identified *Pampus* species, namely, interspecific and intraspecific K2P distances, were calculated in MEGA7 using the COI and Cyt*b* datasets [23]. Interspecific and intraspecific K2P distances of each species were visualized using boxplots in OriginPro 2020 (OriginLab ©, Northampton, MA, USA). The barcoding gap for each species was then calculated as the difference between the minimum interspecific distance and the maximum intraspecific distance [24,25].

Three datasets were used for phylogenetic inference, i.e., the COI dataset, the Cyt*b* dataset, and concatenated datasets of the two genes. Specially, COI and Cyt*b* sequences were treated as two partitions in the concatenated dataset. Best-fit models available in IQtrees v 1.6.12 [26] and MrBayes v 3.2 [27] were selected in jModelTest 2 [28] using the Akaike information criterion [29]. The best fit models for COI and Cyt*b* were HKY + G + I [30] and GTR + G [31]. Maximum likelihood trees were inferred in IQtrees v1.6.12 [26], with 1000 bootstrap replicates to estimate the bootstrap values (BSs) of nodes. For BI trees, two independent Markov chain Monte Carlo (MCMC) runs were performed in MrBayes v3.2, with four chains for 500,000 generations, sampling every 100 generations and discarding the first 25% of samples as burn-ins [27]. Sufficient convergence of the runs was evaluated with summary statistics in MrBayes v3.2 (effective sampling size > 200, potential scale reduction factors≈1). All phylogenetic trees were rooted by the outgroup *Peprilus medius*.

2.4. Species Delimitation

Species delimitation was performed with the concatenated dataset of COI and Cyt*b* using a distance-based method, i.e., automatic barcode gap discovery (ABGD) [32], and two tree-based methods, i.e., the single threshold Bayesian Poisson tree processes (bPTP) model and the generalized mixed Yule-coalescent (GMYC) model [33–35]. The ABGD attempts to identify the barcoding gap as the first significant gap in pairwise distances among a given sequence dataset and uses the detected gap to partition the data [32]. The ABGD was performed on an online ABGD interface of Muséum National d'Histoire Naturelle, France (https://bioinfo.mnhn.fr/abi/public/abgd/abgdweb.html, access on 1 November 2021), scanning a range of prior intraspecific divergence values from 0.1% to 10% with 50 search steps and default settings, although applying K2P distances [22] instead of Jukes-Cantor distances [36].

The single-threshold GMYC identifies speciation events by detecting apparent branching rate increases at the transition of interspecific diversification to population-level coalescence. The GMYC model requires inputs of ultrametric trees; therefore, the ultrametric tree of the concatenated dataset was generated using BEAST2 v 2.5.1 [37], applying prior best-fit models of the two genes, the lognormal relax clock model, and constant population size coalescent tree. Specially, the root node height was constrained to an arbitrary age of 1. Two parallel MCMC runs were performed for 50,000,000 generations, with sampling trees and parameters every 1000 generations. Logfiles were combined in LogCombiner v. 2.5.1 and subsequently analyzed with Tracer v. 1.7 of the BEAST2 package. Sufficient convergence of the two runs was checked by the convergence of parameter values, and ESS should be greater than 200. Trees were summarized with TreeAnnotator v. 2.5.1 and visualized in FigTree v 1.4.4 (http://tree.bio.ed.ac.uk/software/figtree/, access on 1 November 2021). The Newick ultrametric tree was uploaded to the Exelixis Lab web interface for GMYC modeling (https://species.h-its.org/gmyc/, access on 1 September 2021).

bPTP modeling was also performed on the Exelixis Lab web interface (https://species.h-its.org/, access on 1 September 2021). The bPTP model is an updated version of the original maximum likelihood PTP model, with both the implementation of maximum likelihood searches and Bayesian analyses. Similar to GMYC modeling, the bPTP model delimitates speciation events based on a shift in the number of substitutions between internal nodes instead of time [38,39]. It requires a distance-based phylogram instead of a time-based ultrametric tree [40], and thus might eliminate an error-prone step of divergence time inference that potentially affects the previous method. The Newick tree file for bPTP modeling was generated in MrBayes v 3.2 using the concatenated dataset of COI and Cyt*b*. The settings for MrBayes v 3.2 were the same as those described in Section 2.3.

3. Results

3.1. Sequence Variation Indices and Barcoding Gaps of COI and Cytb

For COI and Cyt*b*, 582 and 1077 bp sequences were retrieved from each specimen collected in this study, respectively; no indel was found in either dataset. The two datasets were concatenated and formed a 1659 bp dataset. Average base compositions (A:G:C:T) of the COI and Cyt*b* datasets were 0.248:0.175:0.246:0.330 and 0.266:0.131:0.292:0.311. Among the seven assessed *Pampus* species, the 582 bp COI dataset contained 167 polymorphic sites, including 132 parsimony informative sites. The 1077 bp Cyt*b* dataset contained 361 polymorphic sites, including 284 parsimony informative sites. Pairwise COI K2P distances among the seven *Pampus* species (i.e., interspecific distances) ranged from 0.0034 to 0.1823, and pairwise COI K2P distances within each species (i.e., intraspecific distances) ranged from 0.0000 to 0.0052 (Table 2). COI barcoding gaps have been well identified in five species, i.e., *P. argenteus*, *P. candidus*, *P. chinensis*, *P. minor*, and *P. punctatissimus*

(Figure 2), with their values ranging from 0.0104 to 0.1221 (Table 2). In contrast, the COI barcoding gaps of *P. cinereus* and *P. liuorum* were found to be very small (0.0017 and 0.0000, respectively; Figure 2 and Table 2), which resulted from smaller pairwise K2P distances comparing sequences of *P. cinereus* and *P. liuorum* (0.0034–0.0069). For the Cyt*b* dataset, interspecific K2P distances among the seven species ranged from 0.0237 to 0.1850, whereas intraspecific K2P distances ranged from 0.0000 to 0.0065. The Cyt*b* barcoding gaps have been well identified in all seven species, with the values being 0.0200–0.1452. The smallest Cyt*b* barcoding gap (0.0200) has been observed in *P. cinereus* and *P. liuorum*.

Table 2. Interspecific and intraspecific K2P distances of the seven analyzed *Pampus* species.

Species	COI			Cyt*b*		
	Interspecific	**Intraspecific**	**Barcoding Gap**	**Interspecific**	**Intraspecific**	**Barcoding Gap**
Pampus argenteus	0.1273–0.1572	0.0000–0.0052	0.1221	0.1508–0.1809	0.0000–0.0056	0.1452
Pampus candidus	0.0139–0.1556	0.0000–0.0034	0.0105	0.0355–0.1849	0.0009–0.0065	0.0290
Pampus chinensis	0.0580–0.1823	0.0000–0.0034	0.0545	0.0555–0.1790	0.0000–0.0028	0.0527
Pampus cinereus	0.0034–0.1799	0.0000–0.0017	0.0017	0.0237–0.1850	0.0000–0.0037	0.0200
Pampus liuorum	0.0034–0.1572	0.0000–0.0034	0.0000	0.0237–0.1811	0.0000–0.0037	0.0200
Pampus minor	0.1318–0.1572	0.0000–0.0034	0.1283	0.1698–0.1850	0.0000–0.0047	0.1651
Pampus punctatissimus	0.0580–0.1427	0.0000–0.0034	0.0545	0.0555–0.1777	0.0000–0.0056	0.0499
Overall	0.0034–0.1823	0.0000–0.0052	−0.0018	0.0237–0.1850	0.0000–0.0065	0.0172

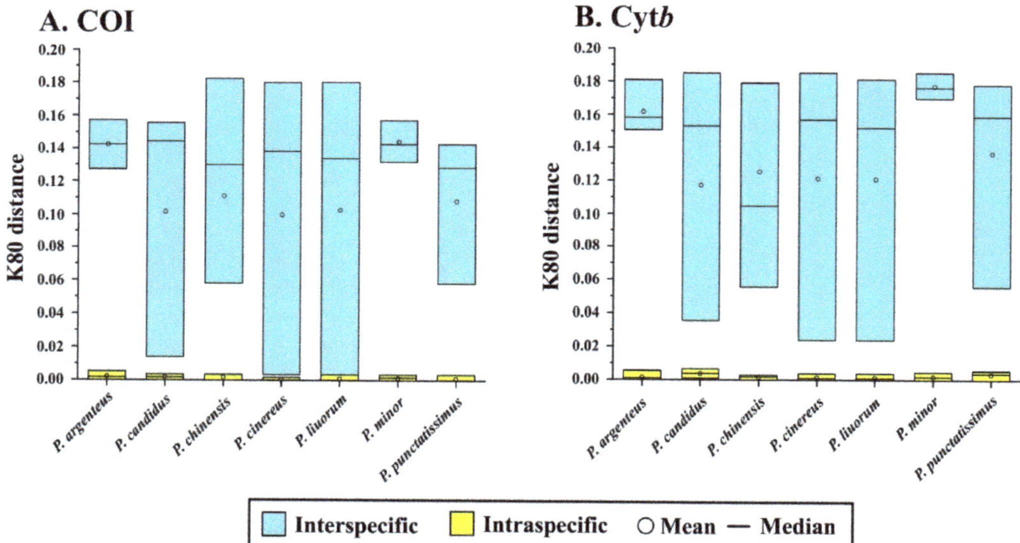

Figure 2. Boxplot showing the COI (**A**) and Cyt*b* (**B**) barcoding gaps of the studied species in genus *Pampus*. Pairwise interspecific and intraspecific K2P distances of each species are annotated with blue and yellow. Mean and median values of the interspecific and intraspecific distances are indicated with circles and lines, respectively.

3.2. Phylogenetic Inference

Maximum likelihood and BI trees retrieved from COI and Cyt*b* datasets singly recovered well-supported clades, corresponding to the morphologically identified species. In the COI tree (Figure 3A), five well-supported clades (BS = 81–100; posterior probabilities, PP = 1) can be identified, which are, based on their morphological identification, *P. argenteus*, *P. minor*, *P. chinensis*, *P. punctatissimus*, and a mix clade of *P. liuorum*, *P. cinereus*, and *P. candidus*. The COI sequences of *P. liuorum*, *P. cinereus*, and *P. candidus* do not form monophyla reciprocally. Instead, sequences of the three species form a single well-supported (BS = 81; PP = 1) clade, with the COI sequences of *P. candidus* and *P. cinereus* being two monophyla nested within it (Figure 3A). For Cyt*b* trees (Figure 3B), the sequences of the seven morphological species, i.e., *P. argenteus*, *P. minor*, *P. chinensis*, *P. punctatissimus*, *P. liuorum*, *P. cinereus*, and *P. candidus*, form monophyla reciprocally, which are well supported by BS values of 93–100, and a PP value of 1 (Figure 3B). *Pampus liuorum* has been resolved as a sister species of *P. cinereus* (BS = 74; PP = 0.84), whereas *P. candidus* is closely linked to the two species (BS = 100; PP = 1, Figure 3B). A sister relationship between *P. argenteus* and *P. minor* is indicated in the Cyt*b* tree, although it is supported by a relatively low PP value (PP = 0.87, Figure 3B).

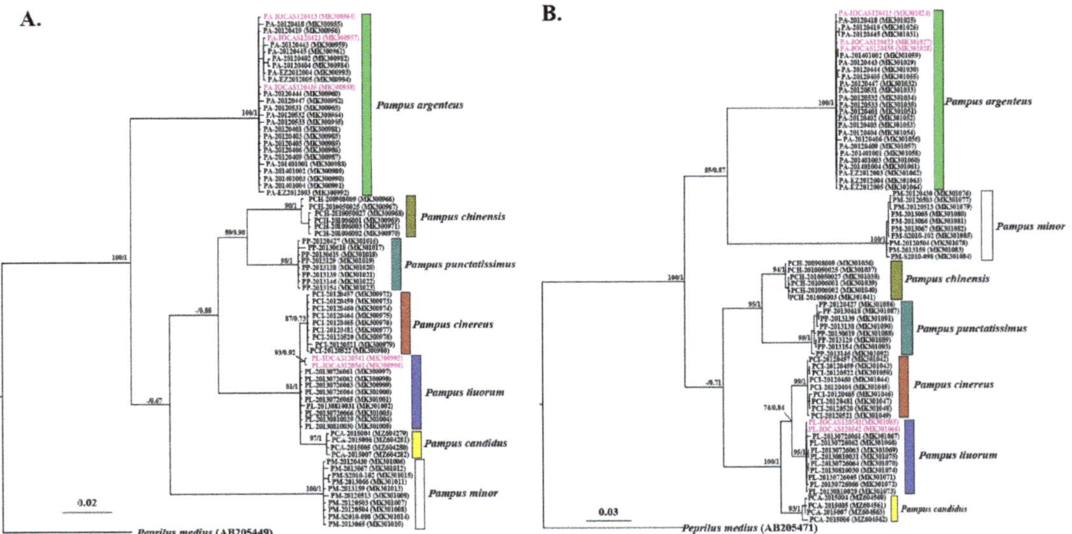

Figure 3. Maximum likelihood tree of the *Pampus* species inferred from COI (**A**) and Cyt*b* (**B**) datasets. Bootstrap values (before slash) and posterior probabilities (after slash) are shown on each node; "-" on the node indicates that the node was not included in the maximum likelihood or Bayesian analyses. Specimens in purple are specimens derived from Liu and Li [2] and Liu et al. [3].

Similar to the Cyt*b* trees, phylogenetic trees retrieved from concatenated datasets of COI and Cyt*b* well support the monophyly of all seven morphological species (BS = 98–100; PP = 1, Figure 4). The topology of the ML and BI trees is almost identical, except for the different relationships of *P. candidus*, *P. liuorum*, and *P. cinereus*. In the ML tree, *Pampus liuorum* is a sister to *P. cinereus* (BS = 60), whereas *P. candidus* is closely linked to the two species (Figure 4). In the BI tree, *Pampus candidus* is resolved as a sister species of *P. cinereus* (PP = 0.51). In both the ML and BI trees, *Pampus argenteus* is resolved as a sister of *P. minor* (BS = 78, PP = 0.64, Figure 4).

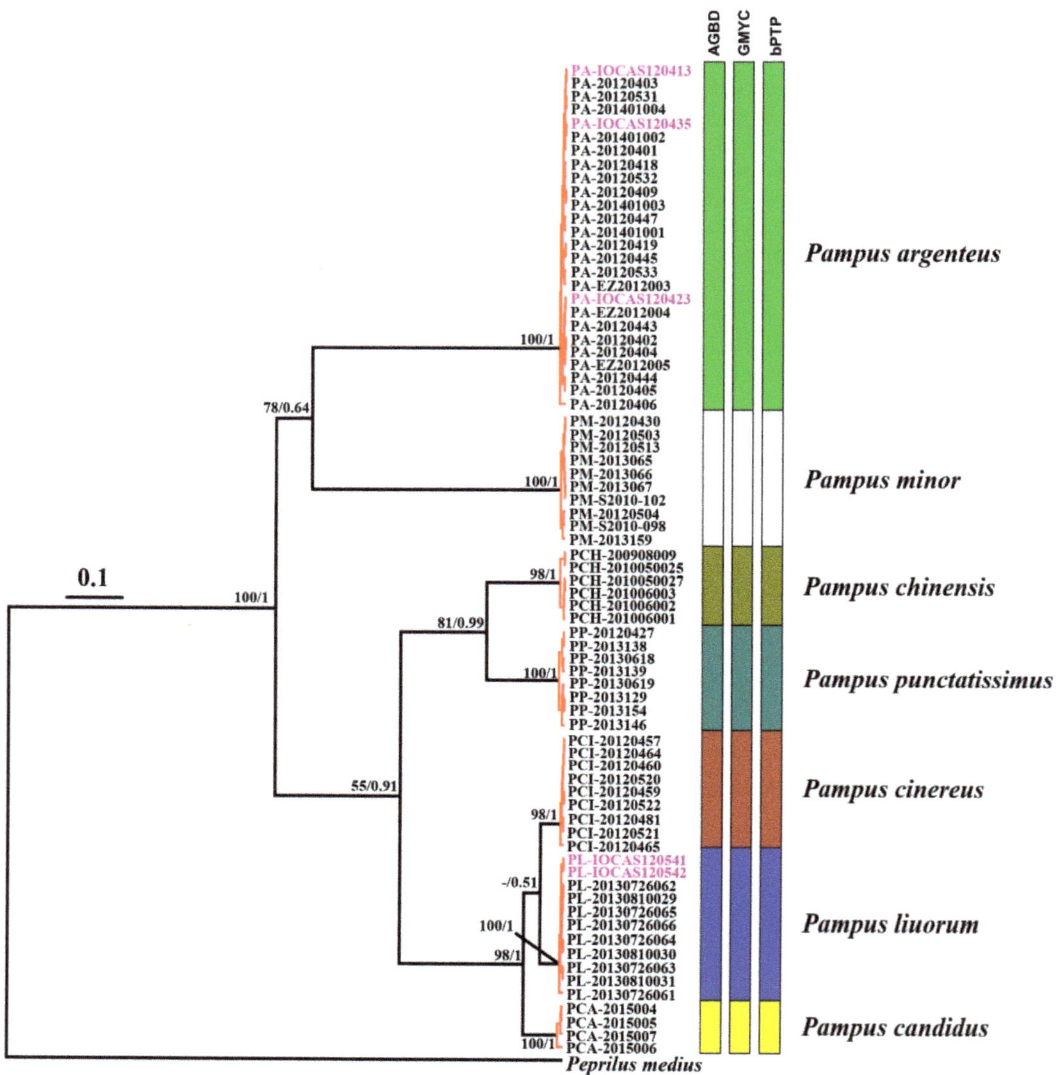

Figure 4. Maximum likelihood tree of the *Pampus* species inferred from concatenated dataset of COI and Cyt*b* sequences. Bootstrap values (before slash) and posterior probabilities (after slash) are shown on each node; "-" on the node indicates that the node was not included in the maximum likelihood or Bayesian analyses. Results of the two species delimitation methods, i.e., the GYMC and bPTP model, are shown on the left. Specimens in purple are specimens derived from Liu and Li [2] and Liu et al. [3].

3.3. Species Delimitation

Species delimitation with the ABGD, GMYC, and bPTP methods using the concatenated dataset consensually concluded seven operational taxonomic units (OTUs) among the analyzed *Pampus* specimens, which are congruent with the seven morphological species (Figure 4). The ABGD method indicated that the first detected significant barcoding gap was 0.0166. The number of OTUs was reduced from seven to five when applying a larger prior maximum intraspecific K2P distance, e.g., the next maximum intraspecific K2P distance value scanned by the ABGD, 0.0184, which suggested that seven putative species were delimitated with the first barcoding gap detected. The GMYC model delimited seven

OTUs as the maximum likelihood solution, which was also the only solution in the confidence interval. The likelihood ratio test of the GMYC model showed highly significant differences ($p < 0.001$) between the maximum likelihood (-Log $L_{GMYC\text{-}max} = 671.454$) of the GMYC model and likelihood (-Log $L_{Null} = 649.49$) of the null model (i.e., assuming only one species among all analyzed specimens). The likelihood ratio test therefore refuted the null model and supported the alternative hypothesis, i.e., the seven species delimitation. The bPTP modeling detected the seven most supported partitions among all analyzed specimens. The delimitation support values for each morphological species are as follows: *P. argenteus*, 0.95; *P. minor*, 0.91; *P. chinensis*, 0.97; *P. punctatissimus*, 0.88; *P. candidus*, 0.82; *P. cinereus*, 0.96; and *P. liuorum*, 0.92.

4. Discussion

4.1. *Pampus cinereus, Pampus liuorum, and Pampus candidus as Distinct Valid Species*

Both phylogenetic inferences of COI and Cyt*b* implied a relatively closer evolutionary relationship of *P. candidus*, *P. liuorum*, and *P. cinereus*. The K2P distances between each of these three species (COI, 0.0034–0.0210; Cyt*b*, 0.0237–0.0277) were relatively small compared with those of other species pairs (COI, 0.0580–0.1572; Cyt*b*, 0.0555–0.1850), which might imply a close phylogenetic relationship and more recent origin of these three species. Phylogenetic trees retrieved from COI, Cyt*b*, and the concatenated dataset of the two genes congruently resolved the three species as a monophyletic group, well supported by BS values of 81–100 and PP values of 1 (Figures 3 and 4). A close relationship of *P. cinereus* and *P. candidus* was also supported by the phylogenetic tree in Radhakrishnan et al. [9]. However, our phylogenetic inference is based on only two mitochondrial gene fragments, which might account for the low support values in the trees and the inconsistency between the BI and ML trees (Figure 4). The phylogeny of the genus *Pampus* needs to be clarified with larger genetic datasets in the future.

Despite their close genetic relationships, the three species are clearly delineated as different species in the ABGD, GMYC, and bPTP models (Figure 4). Liu and Li [2] illustrated that *P. liuorum* could be distinguished from *P. cinereus* by the following characteristics: shorter pectoral fins [31.5–41.7% standard length (SL) vs. 42.0–47.2% SL]; more vertebrae (38 vs. 36); when alive, with golden bronze or yellowish blue color on its back (vs. *P. cinereus*, whole body silvery grey, anal fin and ventral side sometimes yellow). Although the total vertebral counts of *P. cinereus* and *P. liuorum* were claimed to be identical (37 vertebrae) in Jawad and Liu [8], the actual numbers of total vertebrae counted from their radiographs were 36 (*P. cinereus*, Figure 3C in Jawad and Liu [8]) and 38 (*P. liuorum*, Figure 1A in Jawad and Liu [8]), which agrees with the descriptions in Liu and Li [2]. The recently resurrected *P. candidus* possesses an intermediate number of total vertebrae (37 vertebrae) between *P. cinereus* (36 vertebrae) and *P. liuorum* (38 vertebrae) [9]. It could also be discriminated from *P. liuorum* by having fewer vertebrae (14 vs. 15) between the first pterygiophore of dorsal and anal fins [9]. Therefore, the total vertebral count is an exclusive and conservative characteristic in identifying the three species. Yin et al. [7] proposed a synonymy of *P. cinereus* and *P. liuorum*, because their phylogenetic analysis using numerous nuclear genes indicated a mixing clade of *P. cinereus* with *P. liuorum*. In fact, the mixing clade of *P. liuorum* and *P. cinereus* contains three well-supported clades (BS = 100), i.e., a clade of *P. cinereus*, a clade of *P. liuorum*, and a mixed clade formed of two "*P. cinereus*" and "*P. liuorum*" specimens. The genetic distances among the three clades (approximately 0.0056–0.0100) were similar to those between *P. chinensis* and *P. punctatissimus* (approximately 0.0073–0.0144, Figure 1 in Yin et al. [7]), implying that the three clades might contain three species. The total vertebral counts of *P. cinereus* (36–37) and *P. liuorum* (36–38) varied between the estimated specimens in Yin et al. [7], which was incongruent with those recorded in Liu and Li [2]. Yin et al.'s [7] conclusion on the synonymy of *P. cinereus* and *P. liuorum* might be based on misidentified specimens, and might therefore be incorrect. Our analyses indicate that the well-identified specimens of *P. liuorum*, including the paratypes of the species (i.e., IOCAS120541, 0542), are delineated as a single species, which is clearly distinct from *P. cinereus* and *P. candidus* (Figure 4). It supports that *P. liuorum* described in Liu and Li [2] is a valid species. On this basis, the genus *Pampus* now contains eight recognized valid species: *Pampus argenteus* (Euphrasen, 1788), *P. candidus* (Cuvier, 1829), *Pampus chinensis* (Euphrasen, 1788), *Pampus cinereus* (Bloch, 1795), *Pampus liuorum* Liu & Li, 2013, *Pampus minor* Liu & Li, 1998, *Pampus nozawae* (Ishikawa, 1904), and *Pampus punctatissimus* (Temminck & Schlegel, 1845); however, *P. nozawae* needs further taxonomic revision in order to clarify its validity.

4.2. Species Delimitation and Validity of Pampus argenteus

Our phylogenetic trees and species delimitation analyses (ABGD, GMYC, and bPTP model) also support the validity of *P. argenteus* (Figure 4). The identity of *P. argenteus* used to be disputed because of its lack of holotype and the vague original morphological description upon its first publication, which could be applied to multiple known pomfret species [6,10,14]. To solve this taxonomic problem, Liu et al. [6] redescribed the species and designated its neotype—the neotype is assigned as the new name-bearing type for *P. argenteus*. Concurrently, Liu et al. [6] listed a series of non-type specimens identified as *P. argenteus*, which are alternative morphological references of *P. argenteus*. In our genetic analyses (Figures 3 and 4), three of these non-type specimens (i.e., IOCAS120413, IOCAS120423, and IOCAS120435) formed a well-supported clade with the other *P. argenteus* specimens, which were delineated as a single species in ABGD, GMYC and bPTP modeling (Figure 4). *Pampus argenteus* could be distinguished from its congeners by having a combination of the following characters: mouth subterminal (vs. mouth terminal, *P. chinensis* and *P. punctatissimus*); eyes small, with an eye diameter 24.6–27.1% of head length (vs. 27.3–36.4% of head length, *P. minor*); more vertebrae, a total vertebral count of 40 (vs. 32–38, other *Pampus* species); and dorsal and anal fins with short falcate lobes (vs. fins with long falcate lobes, *P. cinereus*, *P. liuorum*, *P. candidus*, and *P. punctatissimus*) [2,6,9,11]. *Pampus argenteus* redescribed in Liu et al. [6] is thus a valid species with exclusive morphological and genetic characteristics.

4.3. Verification of COI and Cytb as Potential DNA Barcodes for Pomfret Identification

In this study, both COI and Cyt*b* exhibited certain abilities to identify species of the genus *Pampus*, although the shorter fragment of COI failed to distinguish the closely related species *P. candidus*, *P. liuorum*, and *P. cinereus*. The anterior region of COI (~600 bp, amplified from universal primer pairs for fish, e.g., FishF1 and Fish R1; Fish F2 and Fish R2 [20]; VF1 and VR1 [41]) is a common DNA barcode for fish identification [42,43]. It has widely been applied in various areas, including fishery management [42,44,45] and the forensic investigation of smuggled fish products [46]. Barcoding gaps between intraspecific and interspecific genetic distance have frequently been reported in mitochondrial barcodes among a vast number of fish taxa, with the intraspecific genetic difference rarely exceeding 2% [47–49]. The 2% genetic difference in mitochondrial genes could thus be empirically accepted as a general boundary and standard for distinguishing interspecific and intraspecific divergence [42,50,51]. In our study, 582 bp of the common COI barcodes were obtained for *Pampus* species using the two primer pairs from Ward et al. [20]. Explicit barcoding gaps (Figure 2) were found in five of the analyzed species, i.e., *P. argenteus*, *P. candidus*, *P. chinensis*, *P. minor*, and *P. punctatissimus*. Nevertheless, our result showed no obvious COI barcoding gap for *P. cinereus* and *P. liuorum*, although there were only 2–4 bp differences among their COI sequences. The shorter traditional COI barcode (586 bp) could contain insufficient variant information to distinguish the two species. The fragments of Cyt*b* have recently been applied as alternative barcodes for fish identification [52,53]. Comparative analyses of COI, Cyt*b*, 16S rRNA, and 18S rRNA suggested that Cyt*b* possesses a higher level of sequence variation among fish species [53]. In this study, analyses on 1077 bp partial Cyt*b* sequences clearly verified barcoding gaps for all seven pomfret species, with the maximum intraspecific K2P distance and minimum interspecific K2P distance being 0.0065 and 0.0237, respectively (Figure 2 and Table 2). Phylogenetic inference using Cyt*b* sequences supported the monophyly of each analyzed species (Figure 3B). This suggests that the longer fragments of Cyt*b* could provide more variant information than the traditional barcoding region of COI in identifying *Pampus* species. Therefore, adopting longer fragments of Cyt*b* as the DNA barcode could be a recommended strategy to ascertain the accurate identification of pomfret species.

5. Conclusions

In this study, we have evaluated partial sequences of the COI (582 bp) and Cyt*b* (1077 bp) of seven *Pampus* species as their potential DNA barcodes. Cyt*b* barcoding gaps have been identified in all assessed species, whereas COI barcoding gaps were not identified in *P. cinereus* and *P. liuorum*, which suggests that the longer fragment of Cyt*b* would be a more suitable barcode for the genus *Pampus*. Species delimitations have been performed with GMYC and bPTP models to assess the validities of the seven collected species. Both delimitation methods identified seven OTUs, which were congruent with the seven morphological species. Therefore, we proposed the seven analyzed species, including the controversial species *Pampus liuorum* Liu & Li, 2013, as valid species.

Author Contributions: Conceptualization, J.L. and R.W.; software, J.W. and R.W.; formal analysis, J.W., R.W. and H.Z.; data curation, J.W., R.W., J.L. and M.A.A.-M.; investigation, R.W., Y.X., L.A.J., J.L. and M.A.A.-M.; writing—original draft preparation, J.W. and R.W.; writing—review and editing, J.W., R.W., Y.X., L.A.J., Y.W., J.L. and M.A.A.-M.; visualization, J.W.; supervision, J.L. and Y.W.; funding acquisition, R.W., Y.W. and J.L. All authors have read and agreed to the published version of the manuscript.

Funding: This study is funded by National Natural Science Foundation of China (Nos. 31872195 and 31772869), Strategic Priority Research Program of the Chinese Academy of Sciences (No. XDB42000000), and the Science and Technology Planning Project of Guangdong Province, China (No. 2017A030303077).

Institutional Review Board Statement: The experimental animal protocols in the present study have been reviewed and approved by the Animal Experimental Ethics Committee of Institute of Oceanology, Chinese Academy of China (approval number: 0928-2021). Experiment procedures were performed in accordance with the Provisions and Regulations for the National Experimental Animal Management Regulations (China, July 2013).

Informed Consent Statement: Not applicable.

Data Availability Statement: All of the COI and Cyt*b* data used in this study have been deposited in NCBI GenBank (https://www.ncbi.nlm.nih.gov, access on 1 November 2021) with the accession numbers MK300954–MK301093, MZ604279–MZ604282, and MZ6042560–MZ6042563. The COI and Cyt*b* sequences have also been deposited in BOLD under project code IOCAS and the title of this study (https://www.boldsystems.org/index.php/Public_SearchTerms?query=IOCAS, access on 1 November 2021).

Acknowledgments: The authors are thankful to Chun-sheng Li of the Institute of Oceanology, Chinese Academy of Sciences, China, for sharing part of his *Pampus* collections and useful information on species identification with us.

Conflicts of Interest: The authors declare no conflict of interest.

References

1. Fricke, R.; Eschmeyer, W.N.; Van der Laan, R. Eschmeyer's Catalog of Fishes: Genera, Species, References. Available online: http://researcharchive.calacademy.org/research/ichthyology/catalog/fishcatmain.asp (accessed on 17 July 2021).
2. Liu, J.; Li, C. A new species of the genus *Pampus* (Perciformes, Stromateidae) from China. *Acta Zootaxonomica Sin.* **2013**, *38*, 885–890.
3. Liu, J.; Li, C.S.; Niu, P. Identity of silver pomfret *Pampus argenteus* (Euphrasen, 1788) based on specimens from its type locality, with a neotype designation (Teleostei, Stromateidae). *Acta Zootaxonomica Sin.* **2013**, *38*, 171–177.
4. Liu, J.; Li, C. A new pomfret species, *Pampus minor* sp. nov. (Stromateidae) from Chinese waters. *Chin. J. Oceanol. Limnol.* **1998**, *16*, 280–285.
5. Liu, J.; Li, C. Redescription of a stromateid fish *Pampus punctatissimus* and comparison with *Pampus argenteus* from Chinese coastal waters. *Chin. J. Oceanol. Limnol.* **1998**, *16*, 161–166.

6. Liu, J.; Li, C.; Ning, P. A redescription of grey pomfret *Pampus cinereus* (Bloch, 1795) with the designation of a neotype (Teleostei: Stromateidae). *Chin. J. Oceanol. Limnol.* **2013**, *31*, 140–145. [CrossRef]
7. Yin, G.X.; Pan, Y.L.; Sarker, A.; Baki, M.A.; Kim, J.-K.; Wu, H.L.; Li, C.H. Molecular systematics of *Pampus* (Perciformes: Stromateidae) based on thousands of nuclear loci using target-gene enrichment. *Mol. Phylogenet. Evol.* **2019**, *140*, 106595. [CrossRef]
8. Jawad, L.A.; Liu, J. Comparative osteology of the axial skeleton of the genus *Pampus* (Family: Stromateidae, Perciformes). *J. Mar. Biol. Assoc. U. K.* **2016**, *97*, 277–287. [CrossRef]
9. Radhakrishnan, D.P.; Kumar, R.G.; Mohitha, C.; Rajool Shanis, C.; Kinattukara, B.K.; Saidumohammad, B.V.; Gopalakrishnan, A. Resurrection and Re-description of *Pampus candidus* (Cuvier), Silver Pomfret from the Northern Indian Ocean. *Zool. Stud.* **2019**, *58*, e7.
10. Guo, Y.; Zhao, W.; Gao, H.; Wang, S.; Yu, P.; Yu, H.; Wang, D.; Wang, Q.; Wang, J.; Wang, Z. China Fishery Statistical Yearbook. In *Bureau of Fisheries and Fishery Management*; The Ministry of Agriculture of the PRC, Ed.; China Agriculture Press: Beijing, China, 2017; p. 143.
11. Liu, J.; Li, C.; Li, X. Studies on Chinese pomfret fishes of the genus *Pampus* (Pisces: Stromateidae). *Stud. Mar. Sin.* **2002**, *44*, 240–252.
12. Li, Y.; Zhang, Y.; Gao, T.; Han, Z.; Lin, L.; Zhang, X. Morphological characteristics and DNA barcoding of *Pampus echinogaster* (Basilewsky, 1855). *Acta Oceanol. Sin.* **2017**, *36*, 18–23. [CrossRef]
13. Guo, E.; Liu, Y.; Liu, J.; Cui, Z. DNA barcoding discriminates *Pampus minor* (Liu et al., 1998) from *Pampus* species. *Chin. J. Oceanol. Limnol.* **2010**, *28*, 1266–1274. [CrossRef]
14. Cui, Z.X.; Liu, Y.; Li, C.P.; Chu, K.H. Species delineation in *Pampus* (Perciformes) and the phylogenetic status of the Stromateoidei based on mitogenomics. *Mol. Biol. Rep.* **2011**, *38*, 1103–1114. [CrossRef] [PubMed]
15. Li, Y.; Zhang, Z.; Song, N.; Gao, T. Characteristics of complete mitogenome of *Pampus* sp. nov. (Perciformes: Stromateidae). *Mitochondrial DNA Part A* **2016**, *27*, 1640–1641.
16. Radhakrishnan, P.D.; Mohitha, C.; Kumar Rahul, G.; Rajool Shanis, C.P.; Basheer, V.S.; Gopalakrishnan, A. Molecular based phylogenetic species recognition in the genus *Pampus* (Perciformes: Stromateidae) reveals hidden diversity in the Indian Ocean. *Mol. Phylogenet. Evol.* **2017**, *109*, 240–245.
17. Li, Y.; Zhou, Y.D.; Li, P.F.; Gao, T.X.; Lin, L.S. Species identification and cryptic diversity in *Pampus* species as inferred from morphological and molecular characteristics. *Mar. Biodivers.* **2019**, *49*, 2521–2534. [CrossRef]
18. Doiuchi, R.; Nakabo, T. Molecular phylogeny of the stromateoid fishes (Teleostei: Perciformes) inferred from mitochondrial DNA sequences and compared with morphology-based hypotheses. *Mol. Phylogenet. Evol.* **2006**, *39*, 111–123. [CrossRef]
19. Sambrook, J.; Russell, D.W. *Molecular Cloning: A Laboratory Manual*, 3rd ed.; Cold Spring Harbor Laboratory Press: New York, NY, USA, 2001; ISBN 978-0-87969-576-7.
20. Ward, R.D.; Zemlak, T.S.; Innes, B.H.; Last, P.R.; Hebert, P.D. DNA barcoding Australia's fish species. *Philos. Trans. R. Soc. B* **2005**, *360*, 1847–1857. [CrossRef]
21. Rozas, J.; Ferrer-Mata, A.; Sanchez-DelBarrio, J.C.; Guirao-Rico, S.; Librado, P.; Ramos-Onsins, S.E.; Sanchez-Gracia, A. DnaSP 6: DNA Sequence Polymorphism Analysis of Large Data Sets. *Mol. Biol. Evol.* **2017**, *34*, 3299–3302. [CrossRef]
22. Kimura, M. A simple method for estimating evolutionary rates of base substitutions through comparative studies of nucleotide sequences. *J. Mol. Evol.* **1980**, *16*, 111–120. [CrossRef]
23. Kumar, S.; Stecher, G.; Tamura, K. MEGA7: Molecular Evolutionary Genetics Analysis Version 7.0 for Bigger Datasets. *Mol. Biol. Evol.* **2016**, *33*, 1870–1874. [CrossRef]
24. Meier, R.; Zhang, G.Y.; Ali, F. The use of mean instead of smallest interspecific distances exaggerates the size of the "barcoding gap" and leads to misidentification. *Syst. Biol.* **2008**, *57*, 809–813. [CrossRef]
25. Cardoni, S.; Tenchini, R.; Ficulle, I.; Piredda, R.; Simeone, M.C.; Belfiore, C. DNA barcode assessment of Mediterranean mayflies (Ephemeroptera), benchmark data for a regional reference library for rapid biomonitoring of freshwaters. *Biochem. Syst. Ecol.* **2015**, *62*, 36–50. [CrossRef]
26. Nguyen, L.T.; Schmidt, H.A.; von Haeseler, A.; Minh, B.Q. IQ-TREE: A fast and effective stochastic algorithm for estimating maximum-likelihood phylogenies. *Mol. Biol. Evol.* **2015**, *32*, 268–274. [CrossRef] [PubMed]
27. Ronquist, F.; Teslenko, M.; van der Mark, P.; Ayres, D.L.; Darling, A.; Hohna, S.; Larget, B.; Liu, L.; Suchard, M.A.; Huelsenbeck, J.P. MrBayes 3.2: Efficient Bayesian phylogenetic inference and model choice across a large model space. *Syst. Biol.* **2012**, *61*, 539–542. [CrossRef] [PubMed]
28. Darriba, D.; Taboada, G.L.; Doallo, R.; Posada, D. jModelTest 2: More models, new heuristics and parallel computing. *Nat. Methods* **2012**, *9*, 772. [CrossRef]
29. Akaike, H. A new look at the statistical model identification. *IEEE Trans. Autom. Control.* **1974**, *19*, 716–723. [CrossRef]
30. Hasegawa, M.; Kishino, H.; Yano, T.A. Dating of the human-ape splitting by a molecular clock of mitochondrial DNA. *J. Mol. Evol.* **1985**, *22*, 160–174. [CrossRef]
31. Tavare, S. Some probabilistic and statistical problems in the analysis of DNA sequences. *Lect. Math. Life Sci.* **1986**, *17*, 57–86.
32. Puillandre, N.; Lambert, A.; Brouillet, S.; Achaz, G. ABGD, Automatic Barcode Gap Discovery for primary species delimitation. *Mol. Ecol.* **2012**, *21*, 1864–1877. [CrossRef]

33. Pons, J.; Barraclough, T.G.; Gomez-Zurita, J.; Cardoso, A.; Duran, D.P.; Hazell, S.; Kamoun, S.; Sumlin, W.D.; Vogler, A.P. Sequence-based species delimitation for the DNA taxonomy of undescribed insects. *Syst. Biol.* **2006**, *55*, 595–609. [CrossRef]
34. Fontaneto, D.; Herniou, E.A.; Boschetti, C.; Caprioli, M.; Melone, G.; Ricci, C.; Barraclough, T.G. Independently evolving species in asexual bdelloid rotifers. *PLoS Biol.* **2007**, *5*, e87. [CrossRef]
35. Fujisawa, T.; Barraclough, T.G. Delimiting species using single-locus data and the Generalized Mixed Yule Coalescent approach: A revised method and evaluation on simulated data sets. *Syst. Biol.* **2013**, *62*, 707–724. [CrossRef] [PubMed]
36. Jukes, T.H.; Cantor, C.R. Evolution of protein molecules. *Mamm. Protein Metab.* **1969**, *3*, 21–132.
37. Bouckaert, R.; Heled, J.; Kuhnert, D.; Vaughan, T.; Wu, C.H.; Xie, D.; Suchard, M.A.; Rambaut, A.; Drummond, A.J. BEAST 2: A software platform for Bayesian evolutionary analysis. *PLoS Comput. Biol.* **2014**, *10*, e1003537. [CrossRef]
38. Zhang, J.J.; Kapli, P.; Pavlidis, P.; Stamatakis, A. A general species delimitation method with applications to phylogenetic placements. *Bioinformatics* **2013**, *29*, 2869–2876. [CrossRef] [PubMed]
39. Ahrens, D.; Fujisawa, T.; Krammer, H.J.; Eberle, J.; Fabrizi, S.; Vogler, A.P. Rarity and Incomplete Sampling in DNA-Based Species Delimitation. *Syst. Biol.* **2016**, *65*, 478–494. [CrossRef]
40. Le Ru, B.P.; Capdevielle-Dulac, C.; Toussaint, E.F.A.; Conlong, D.; Van den Berg, J.; Pallangyo, B.; Ong'amo, G.; Chipabika, G.; Molo, R.; Overholt, W.A.; et al. Integrative taxonomy of *Acrapex* stem borers (Lepidoptera: Noctuidae: Apameini): Combining morphology and Poisson Tree Process analyses. *Invertebr. Syst.* **2014**, *28*, 451–475. [CrossRef]
41. Ivanova, N.V.; Zemlak, T.S.; Hanner, R.H.; Hebert, P.D. Universal primer cocktails for fish DNA barcoding. *Mol. Ecol. Notes* **2007**, *7*, 544–548. [CrossRef]
42. Ward, R.D. FISH-BOL, a case study for DNA barcodes. In *DNA Barcodes*; Kress, W.J., Erickson, D.L., Eds.; Humana Press: New York, NY, USA, 2012; pp. 423–439.
43. Wu, R.; Zhang, H.; Liu, J.; Niu, S.; Xiao, Y.; Chen, Y. DNA barcoding of the family Sparidae along the coast of China and revelation of potential cryptic diversity in the Indo-West Pacific oceans based on COI and 16S rRNA genes. *J. Oceanol. Limnol.* **2018**, *36*, 1753–1770. [CrossRef]
44. Becker, S.; Hanner, R.; Steinke, D. Five years of FISH-BOL: Brief status report. *Mitochondrial DNA Part A* **2011**, *22*, 3–9. [CrossRef]
45. Rathipriya, A.; Karal Marx, K.; Jeyashakila, R. Molecular identification and phylogenetic relationship of flying fishes of Tamil Nadu coast for fishery management purposes. *Mitochondrial DNA Part A* **2019**, *30*, 500–510. [CrossRef] [PubMed]
46. Carvalho, D.C.; Palhares, R.M.; Drummond, M.G.; Frigo, T.B. DNA Barcoding identification of commercialized seafood in South Brazil: A governmental regulatory forensic program. *Food Control* **2015**, *50*, 784–788. [CrossRef]
47. Zhang, J.B. Species identification of marine fishes in china with DNA barcoding. *Evid. Based Complement. Altern. Med.* **2011**, *2011*, 1–10. [CrossRef] [PubMed]
48. Zhang, Y.-H.; Qin, G.; Zhang, H.-X.; Wang, X.; Lin, Q. DNA barcoding reflects the diversity and variety of brooding traits of fish species in the family Syngnathidae along China's coast. *Fish. Res.* **2017**, *185*, 137–144. [CrossRef]
49. Xu, L.; Wang, X.; Van Damme, K.; Huang, D.; Li, Y.; Wang, L.; Ning, J.; Du, F. Assessment of fish diversity in the South China Sea using DNA taxonomy. *Fish. Res.* **2021**, *233*, 105771. [CrossRef]
50. Shen, Y.; Guan, L.; Wang, D.; Gan, X. DNA barcoding and evaluation of genetic diversity in Cyprinidae fish in the midstream of the Yangtze River. *Ecol. Evol.* **2016**, *6*, 2702–2713. [CrossRef]
51. Liu, K.Y.; Zhao, S.; Yu, Z.C.; Zhou, Y.J.; Yang, J.Y.; Zhao, R.; Yang, C.X.; Ma, W.W.; Wang, X.; Feng, M.X.; et al. Application of DNA barcoding in fish identification of supermarkets in Henan province, China: More and longer COI gene sequences were obtained by designing new primers. *Food Res. Int.* **2020**, *136*, 109516. [CrossRef] [PubMed]
52. Lemer, S.; Aurelle, D.; Vigliola, L.; Durand, J.D.; Borsa, P. Cytochrome *b* barcoding, molecular systematics and geographic differentiation in rabbitfishes (Siganidae). *Comptes Rendus Biol.* **2007**, *330*, 86–94. [CrossRef]
53. Zhang, J.; Hanner, R. Molecular approach to the identification of fish in the South China Sea. *PLoS ONE* **2012**, *7*, e30621. [CrossRef]

Article

The Mitogenome Structure of Righteye Flounders (Pleuronectidae): Molecular Phylogeny and Systematics of the Family in East Asia

Alexander D. Redin and Yuri Ph. Kartavtsev *

A.V. Zhirmunsky National Scientific Center of Marine Biology (NSCMB), Far Eastern Branch, Russian Academy of Sciences, 690041 Vladivostok, Russia

* Correspondence: yuri.kartavtsev48@hotmail.com

Abstract: This paper reports the first complete sequence of the mitochondrial genome (mitogenome) of the yellow-striped flounder *Pseudopleuronectes herzensteini* (Pleuronectoidei: Pleuronectidae). Mitogenome evolution, and molecular phylogenetic reconstruction based on four to six techniques, including coalescent analysis, were performed for flatfish. The genome size of the specimen sampled was 16,845 bp, including 13 protein-coding genes, 22 tRNA genes, *12S*, and *16S* rRNA genes, and the control region, CR. The composition and arrangement of the genes are similar to those in other teleost fish, including the second mitogenome reported in this paper. The frequency of A, C, G, and T nucleotides in the *P. herzensteini* mitogenome is 27%, 29.2%, 17.6%, and 26.2%, respectively. The ratio of complementary nucleotides in the mitogenome of this and other species of the family was A+T:G+C (53.2: 46.8%) and do not deviate significantly from the expected equilibrium proportion. The submission to the global database (GenBank) of two new mitogenomes along with 106 analyzed GenBank sequences will contribute to phylogenetic studies of flounders at the family and suborder levels. Based on 26 and 108 nucleotide sequences of protein-coding genes (PCGs), we investigated the molecular phylogeny of flounders and performed analysis for two sets of sequences, including those of members of the family Pleuronectidae and the suborder Pleuronectoidei and estimated their importance in establishing the taxonomy at these two levels. Data obtained by up to six techniques of multigene phylogenetic reconstructions support monophyly within the family Pleuronectidae with high statistical confidence; however, conclusions regarding the phylogenetics at the suborder level require further investigation. Our results also revealed paraphyletic and weakly supported branches that are especially numerous at the suborder level; thus, there is a clear need for taxonomic revisions at the suborder, and possibly family levels. Genetic distance analysis reveals the suitability for DNA barcoding of species specimens at single genes as well as at whole mitogenome data.

Keywords: mitogenome evolution; flounder; molecular diversity; phylogenomics; systematics; DNA barcoding; multigene phylogenetic reconstructions; divergence time; protein-coding genes (PCGs); genetic distance; coalescent analysis; Bayesian skyline

Citation: Redin, A.D.; Kartavtsev, Y.P. The Mitogenome Structure of Righteye Flounders (Pleuronectidae): Molecular Phylogeny and Systematics of the Family in East Asia. *Diversity* **2022**, *14*, 805. https://doi.org/10.3390/d14100805

Academic Editor: Manuel Elías-Gutierrez

Received: 26 August 2022
Accepted: 21 September 2022
Published: 27 September 2022

1. Introduction

The righteye flounder, family Pleuronectidae (Osteichthyes, Carangiformes, Pleuronectoidei), which is the main focus of this study, comprises one of the largest families within the suborder Pleuronectoidei (formerly order Pleuronectiformes), including 59 nominal species that are distributed in marine waters of the Northern Hemisphere [1,2]. Based on ten synapomorphies in morphological characters, Cooper and Chapleau [2] treated the family Pleuronectidae as a monophyletic taxon. Although previous research has attempted to classify the flounders by various approaches the morphological, anatomic, cytological, chromosome, and molecular-and-genetic, a consensus on the taxonomy of these fish is still lacking. In this paper, the authors would like to shed light on the systematics of some questionable flatfish taxa.

The yellow-stripe flounder *Pseudopleuronectes herzensteini* (Jordan and Snyder, 1901), for which one of the two mitogenomes reported in this paper is describing in more detail below, belongs to the well-established genus of the family Pleuronectidae. It is a bottom-dwelling marine fish found in temperate waters of the northwestern Pacific, from the Sea of Japan to the Kuril Islands, Sakhalin, Korea, the Yellow Sea, Bohai Bay, and the East China Sea [2]. Due to the fishery importance of this and other flounder species and the need to manage these valuable bioresources, both the accurate classification of specimens of species within genera and the upper taxa relationships for Pleuronectidae and other families of the suborder are vital.

Several classifications of flounders of the family Pleuronectidae were proposed by different authors [2–5]. There is also some controversy regarding the phylogenetic relationships of flounders inferred from morphological and molecular genetic data [2,6–13]. Complications regarding flatfish specimen identification, speciation, and evolutionary diversification resulting in the support for monophyly of Pleuronectidae and Pleuronectoidei/Pleuronectiformes indicate that these views are not universal nor have received clear support in phylogenetic studies [12,13]. Evidence for flatfish paraphyly was considered quite long ago [3,14,15] and later developed into a phyletic generalization that supports the monophyly of this taxon [6]. Chapleau's [6] conclusion of pleuronectiform monophyly was accepted by many researchers and received certain molecular support [8,11,13,16]. Other molecular-based studies offered also opposite evidence, indicating flatfish paraphyly [9,10,17–23]. The complications surrounding DNA sequence analysis and judgments about the monophyly of flatfish are continuing and papers to validate these points have been written [12,13,16,24], including this paper.

In the current paper, we report the results of a thorough examination of the phylogenetic signal in the mitochondrial genome (mitogenome) to infer pleuronectiform relationships, mostly for Russian Far Eastern Pleuronectidae but with particular insight into the suborder level. Because of the limited space of the paper, we focus on these two issues and do not discuss higher taxa such as the Carangimorpha or the clade L *sensu* [17]. Mitogenomes offer several advantages for phylogenetic inference. They are highly conserved in organization and have uniparental/haploid inheritance and a large number of characters (variable nucleotides or amino acids, if translated) that are inherited as a single unit due to a circular DNA (mtDNA), with no, or a very low, recombination. Because mtDNA sequences show faster rates of substitution and a smaller effective population size if compared to nuclear DNA (nDNA) [25–27], they are often more suitable for recovering a phylogenetic signal for diversification events in lineages up to the order level (when the accumulated number of reverse mutations is not high). Previous studies showed that sequences of protein-coding genes (PCGs) in mitogenomes give very reliable information for recovering the diversity of flatfish lineages because tree topologies do not differ significantly from those based on complete mitogenome sequence [12,16]. For these reasons, this study exclusively uses PCGs for inferring a phylogenetic signal.

To our knowledge, this is the first study to present the composition of the complete mitochondrial genome of *P. herzensteini*. Also, the mitogenome of a new specimen of flounder *Platichthys stellatus* was sequenced and analyzed. A molecular phylogenetic study was performed based on the original nucleotide sequences of mtDNA of these two species, as well as on sets of GenBank sequences [28], a total of 26 and 108 sequences for the family Pleuronectidae and the suborder Pleuronectoidei, respectively. From these data, several types of gene trees were reconstructed and the divergence of taxa among recent members of the family Pleuronectidae and the suborder Pleuronectoidei were estimated. An approximation of these data into time scale by Bayesian skyline was performed as well.

2. Materials and Methods

2.1. Materials and General Analysis of Approaches

A total of 108 sequences belonging to the suborder Pleuronectoidei of the order Carangiformes were analyzed, including two presented in this paper (see below). Latin names are given in accordance with the classification [2].

Two specimens of *Pseudopleuronectes herzensteini* and *Platichthys stellatus* (pieces of muscle tissue fixed in 95% ethanol) were derived from the collection of the Laboratory of Molecular Systematics. The voucher specimens, 7K *Pseudopleuronectes herzensteini* stripe-yellow flounder fished by gill net in Vostok Bay Peter the Great Bay, Sea of Japan, and *Platichthys stellatus* labeled Ps2-011 obtained from bottom trawling in the Okhotsk Sea; both are stored at the Museum of the A.V. Zhirmunsky National Scientific Center of Marine Biology of the Far East Center of Russian Academy of Sciences (NSCMB FEB RAS). DNA was isolated using commercial kits (DNA Extran-2, Sintol, Moscow, Russia). Then, 350 ng of total DNA was collected for both samples and sent to Novogene (China) for sequencing. Sequencing was carried out on the Illumina platform (Novaseq 6000 sequencer, Peking, China).

According to the sequencing technique, the length of nucleotide fragment reads along the mitogenome was 150 bp. The fragments were assembled into a complete mitogenome sequence using the NOVOPlasty4.2.1 software (https://github.com/ndierckx/NOVOPlasty) on the Ubuntu 20.04 LTS subsystem [29]. Protein-coding genes, rRNAs, and tRNAs were annotated and mapped using the MitoAnnotator WEB bench [30].

Analysis of variability and divergence was carried out starting with relatively simple software packages, DNAsp-5 [31] and MEGA-X [32]. Molecular phylogenetic analysis was performed mainly on the basis of nucleotide sequences (below referred to as sequences) of PCGs using the software MrBayes 3.2.1 or 3.2.7 [33,34], MEGA-X [32], and BEAST-2 [35–39] (including the latest updates at: http://www.beast2.org/; accessed on 24 July 2021). Protein-coding genes were extracted from complete mitochondrial genomes based on the FishMitoPipe script (https://github.com/Sturcoal/FishMitoPipe#fishmitopipe-the-pipeline-for-fish-mitochondrial-genome-manipulation-before-phylogenetic-analysis; Vladivostok, Russia, accessed on 1 January 2020), then combined into a super-matrix of sequences using SequenceMatrix [40]. Sequences were aligned using the ClustalW program in MEGA-X (http://www.megasoftware.net; Tokyo, Japan) [32]. The gap opening and gap extension penalties were set at 15.0 and 5.0, respectively (for other settings of the alignment program, the default parameters were used). After the first alignment step, large gaps were manually removed; the final alignment in the second step was performed with reduced penalty levels (5.0 and 0.5 for the two options, respectively). All gaps were then manually removed again.

For a comparative analysis of mitogenomes, PhyloSuite software was additionally used [41]. To work in the PhyloSuite software, complete mitogenome sequences were previously downloaded from GenBank in the ID.gb format (where ID is, the sequence access number on the site with the extension code for the GenBank file, .gb). Then, all information about the sequences of PCGs, rRNAs, tRNAs, control region (CR), and other information was extracted from mitogenomes. After that, the resulting sequences were aligned in a program block (utility), MAFFT. Alignment was carried out in two stages. In the first stage, PCGs were aligned, and in the second stage, rRNAs, tRNAs, and CR were aligned. Next, the resulting fasta files (.fas, .fasta) for protein-coding, rRNA, and tRNA sequences were moved to one folder and then stitched into a single file using another program block, Concatenate Sequence.

The obtained concatenated sequences were analyzed using the software utility, PartitionFinder, to select the most appropriate mitogenome partition schemes and to define optimal models for the molecular substitution along sequences. For subsequent phylogenetic analysis, within this block, the model fitting options for the MrBayes software were selected (the desired option is selected in the menu window instead of the default option "all") with an economical ("greedy") search method. After working in PartitionFinder, the results were sent to the MrBayes software package integrated with PhyloSuite software.

When running MrBayes in the PhyloSuite software, additional options, such as the choice of an outgroup, the number of generations, and others that determine the probabilistic parameters of the tree reconstruction, are determined by software and manually. So, for the last case in the menu window, when starting this block, we set the number of generations (n) equal to $n = 2 \times 10^6$, and the number of Markov chains in digital Monte Carlo simulation (mcmc) equal to 4. However, for the former case, the tree special models for each gene were selected by PartitionFinder utility and automatically recorded in the command block of BA analysis.

2.2. Molecular Phylogenetic Analysis

The molecular phylogenetic analysis is aimed basically at building gene trees. Phylograms based on PCG sequences were generated using several approaches. Initially, the optimal substitution model for nucleotides in the lineages (their evolution) was estimated based on the sequence's matrices that formed for the analysis. The best-suited model, as determined using MEGA-X software, was the GTR+G+I model (General Time Reversible, with G, Gamma mode variation across sites, and I, Invariable fraction of nucleotides). This model was defined as best for both 26 sequences that were chosen for the analysis of the family Pleuronectidae, as well as for 108 sequences of suborder Pleuronectoidei. Phylogenetic trees were constructed using four methods: Bayesian analysis (BA), maximum likelihood (ML), neighbor-joining (NJ), and maximum parsimony (MP). These techniques were performed by using an original software package (SP) MrBayes-3.2.7 (http://nbisweden.github.io/MrBayes/download.html; accessed on 2 June 2021) for BA [33,34], or by SP MEGA-X [32] for ML-, NJ- and MP-techniques; for the set of 13 PCGs and 26 Pleuronectidae sequences the additional gene tree reconstructions were performed using SP PhyloSuite and BEAST-2.

SP MrBayes-3.2.7 was used to do the BA analysis, as stated above. Before tunning BA, the SP SequenceMatrix-8.1 [42] was used and the super-matrix for the BA analysis was obtained as one of its output files (Fl-26seq-pt4.nex). Next, the numerical simulation for tree reconstruction by SP MrBayes-3.2.7 was run. Program parameters for such runs included: applying one million generations ($n = 10^6$), four parallel Markov chains using the program utility 'mcmc', the definitions of partitions for 13 PCGs, descriptors for coding of nucleotide positions within codons, that defined by SequenceMatrix, and several other options used in SP MrBayes-3.2.7; the output have the mode of BA consensus tree. Three other tree reconstructions ML, NJ, and MP run with k = 1000 bootstrap replications (providing bootstrap support for the branch nodes). As an outgroup for tree rooting one taxon for the family Pleuronectidae, *Paralichthys olivaceus*, and the two taxa *Tetraodon mbu* and *Acrossocheilus monticola* for suborder were used, which are known from literary sources as the most recent common ancestors (mrca), to *Pleuronectidae* I and *Paralichthodidae*, correspondingly [13], Table 1 in it; see more details in Results and Discussion sections). Dating of divergence time on paleontological records for the mrca pairs comprise reference points for calibration of molecular divergence. Calibration points for molecular divergence are 27.83 and 46.19 million years (Mya), correspondingly to Pleuronectidae and Pleuronectoidei from the two above taxa [13], Table 1 in it. Below in the second following paragraph, more details are given on this point.

As previously stated, molecular phylogenetic reconstructions for the Pleuronectidae family were undertaken using the base SPs the MrBayes and MEGA-X involving PCGs and four tree-building techniques: BA, ML, NJ, and Mp. Topology and time divergence using coalescent analysis (CA) were reconstructed by SP BEAST-2 in addition to those four for all 13 PCGs and 26 sequences in the family, including the outgroup. CA parameters from four fundamental models were integrated for this: (I) Yule CA (Yule, 1924), (II) Bayesian Skyline CA, (III) CA for a population of constant size, and (IV) Extended Bayesian Skyline CA [38,39]. For each of the four CA models several files that designed in BEAUti2.6.6. utility, were run as explained below. Also, in one of the PS BEAST-2 simulation models the running file contained partitions and nucleotide positions that were created by the PhyloSuite software and its utility PartitionFinder.

Table 1. Species list used in the study with the GenBank accession numbers.

Species	GenBank Number
Acrossocheilus monticola	KT367805
Achirus lineatus	JQ639067
Trinectes maculatus	JQ639070
Neoachiropsetta milfordi	AP014593
Arnoglossus polyspilus	AP014586
Arnoglossus tenuis	KP134337
Asterorhombus intermedius	MK256952
Bothus myriaster	KJ433563
Bothus pantherinus	AP014587
Chascanopsetta lugubris	AP017455
Chascanopsetta lugubris	KJ433561
Crossorhombus azureus	JQ639068
Crossorhombus kobensis	AP014589
Crossorhombus valderostratus	KJ433566
Grammatobothus polyophthalmus	MK770643
Laeops lanceolata	AP014591
Lophonectes gallus	KJ433567
Psettina iijimae	KP134336
Citharoides macrolepidotus	AP014588
Lepidoblepharon ophthalmolepis	AP014592
Cynoglossus abbreviatus	GQ380410
Cynoglossus abbreviatus	JQ349004
Cynoglossus bilineatus	JQ349000
Cynoglossus gracilis	KT809367
Cynoglossus interruptus	LC482306
Cynoglossus itinus	JQ639062
Cynoglossus joyneri	KU497492
Cynoglossus joyneri	KU754054
Cynoglossus joyneri	KY008569
Cynoglossus nanhaiensis	MT117229
Cynoglossus puncticeps	JQ349003
Cynoglossus robustus	LC482305
Cynoglossus roulei	MK574671
Cynoglossus roulei	MN966658
Cynoglossus semilaevis	EU366230
Cynoglossus semilaevis	GQ380409
Cynoglossus senegalensis	MH709122
Cynoglossus trulla	JQ348998

Table 1. *Cont.*

Species	GenBank Number
Cynoglossus trigrammus	KP057581
Cynoglossus zanzibarensis	KJ433559
Paraplagusia bilineata	JQ349001
Paraplagusia bleekeri	JQ349002
Paraplagusia japonica	JQ639066
Symphurus orientalis	KP992899
Symphurus plagiusa	JQ639061
Cyclopsetta fimbriata	AP014590
Paralichthys adspersus	MW288827
Paralichthys dentatus	KU053334
Paralichthys lethostigma	KT896534
Paralichthys olivaceus	AB028664
Pseudorhombus cinnamoneus	JQ639069
Pseudorhombus dupliciocellatus	KJ433562
Cleisthenes pinetorum	KT223828
Clidoderma asperrimum	MK210570
Colistium nudipinnis	JQ639063
Hippoglossoides platessoides	MN122825
Hippoglossus hippoglossus	AM749122
Hippoglossus hippoglossus	AM749123
Hippoglossus hippoglossus	AM749124
Hippoglossus stenolepis	AM749126
Hippoglossus stenolepis	AM749127
Hippoglossus stenolepis	AM749128
Hippoglossus stenolepis	AM749129
Limanda aspera	KP013094
Limanda limanda	MN122886
Pelotretis flavilatus	KC554065
Peltorhamphus novaezeelandiae	JQ639065
Platichthys stellatus	EF424428
Platichthys stellatus	**MZ365029**
Pleuronichthys cornutus	JQ639071
Pleuronichthys cornutus	KY038655
Pseudopleuronectes herzensteini	**MW713061**
Pseudopleuronectes yokohamae	KT224485
Pseudopleuronectes yokohamae	KT878309
Reinhardtius hippoglossoides	AM749130
Reinhardtius hippoglossoides	AM749131
Reinhardtius hippoglossoides	AM749132
Reinhardtius hippoglossoides	AM749133

Table 1. *Cont.*

Species	GenBank Number
Verasper moseri	EF025506
Verasper moseri	LC583747
Verasper variegatus	DQ403797
Verasper variegatus	MK210571
Psettodes erumei	FJ606835
Samaris cristatus	JQ700101
Samariscus latus	KF494223
Scophthalmus maximus	EU419747
Zeugopterus punctatus	MT410862
Aesopia cornuta	KF000065
Aseraggodes kobensis	KJ601760
Brachirus orientalis	KJ433558
Brachirus orientalis	KJ513134
Heteromycteris japonicus	JQ639060
Liachirus melanospilos	KF573188
Pardachirus pavoninus	AP006044
Pardachirus pavoninus	KJ433565
Pardachirus pavoninus	KJ461620
Zebrias japonicus	KJ433482
Zebrias japonicus	KJ433568
Solea ovata	KF142459
Solea ovata	KJ496338
Solea senegalensis	AB270760
Zebrias crossolepis	KJ433564
Zebrias crossolepis	KT367804
Zebrias quagga	JQ348999
Zebrias zebra	JQ700100
Zebrias zebrinus	KC491209
Zebrias zebrinus	KC519737
Tetraodon mbu	AP011923

Note. The original sequences that were submitted by our team are in a bold font.

SP BEAST-2, v2.6.5 [38,39] and its newest update v2.6.6 were applied to the 26 sequences matrix of 13 PCGs for the estimation of node ages in simulated trees. An independent GTR+G+I model of nucleotide substitution with gamma-distributed rate variation across sites (defined previously in MEGA as described above) with n = 5–15 categories and an uncorrelated relaxed exponential clock and lognormal relaxed clock [38,39,43] were selected in different runs. The random option for initial phylogenetic trees was used to generate the final set. Priors that followed a Yule CA branching model, Bayesian Skyline, Extended Bayesian Skyline, and CA for a constant size population were employed. Two points for fossil calibration were used in this analysis. The first point taken from the nearly oldest flatfish stem fossil, *Eobothus minimus* (Agassiz, 1833) from the Upper Eocene (50 Mya) of Monta Bolca (Italy) dates the time for the most recent common ancestor, TMRCA of Bottidae, Pleuronectidae, and Paralichthyidae [13]. In this paper, we used as the first reference

TRMCA the date 46.19 Mya, close to the above dating back to the Paralichthodidae, the other superfamily Soleoidea representative, as given in Table 1 [13]. The second setting points to a more recent age constraint for the clade (Pleuronectidae I, Paralichthyidae I), equal to 27.83 Mya [13] Table 1 in it. These calibration points were modeled with a normal distribution with a mean of 46.2 Mya and a standard deviation of 1.0 Mya. Simulations were run by setting the option monophyly for the whole tree and the option outgroup definition to *Paralichthys olivaceus*. At least, six, fifteen, seven, and eleven independent runs for four tested basic model sets (I–IV) were performed using 50–70 million generations and sampling every 1000th tree with the specific sets of settings. All runs were checked for sufficient mixing, stable convergence on a unimodal posterior and tree priors, and with effective sample sizes (ESS) exceeded the score of 100–200 for all meaningful parameters using TRACER v1.5 [35,39] and its update TRACER v1.7. After 50% of the resulting trees were removed as burn-in, the remaining trees were summarized in a Maximum Clade Credibility consensus tree with TreeAnnotator v2.6.5 [39] and the update v2.6.6. Along with the SP BEAST-2.6.5-2.6.6, the BEAUTY-2.6.5-2.6.6 as its main utility was involved in the building of the main framework file for calculations in BEAST (BEAUTY-file performed in .xml format). Also, the BEAGLE database (Beagle 5.2; washington.edu) was used in most runs as recommended by SP BEAST-2 developers (Drummond, Bouckaert, 2015; Bouckaert et al., 2019).

Phylogenetic trees were visualized and edited, when necessary, using SP FigTree 1.4.0 [44] and MEGA-X. Additionally, beyond five basic gene tree reconstruction techniques (BA-, ML-, NJ-, MP-, and CA-trees), the IQ-TREE version 2.1.2 software (http://www.iqtree.org; Wien, Austria) [45] was used for ML-tree reconstructions that run with the default parameters and auto-detection the sequence type as well as with the best-fitting substitution model definition. IQ-TREE performed the Ultrafast Bootstrap [46] and the SH-aLRT branch test [47] to estimate the scores for nodes' support; in this case, runs made with n = 2000–5000 replicates.

Sequences of complete mitogenome obtained by our team and presented here for two flounder species, *Pseudopleuronectes herzensteini* and *Platichthys stellatus* have been submitted to GenBank [28] and are listed in Table 1 along with sampled GenBank sequences. For the sake of brevity, the structure of the mitogenome is visually represented only for the species *P. herzensteini*. However, sequences of both species were used for molecular phylogenetic analysis as well as for comparison of mitogenome structure for other representatives of the family Pleuronectidae. The map of circular mitogenome of yellow-stripe flounder *P. herzensteini* was obtained with the usage of MitoFish WEB bench [48], CLOROBOX WEB resource, and the utility of the late, GeSeq; MPI-MP CHLOROBOX-GeSeq (mpg.de).

For the analysis of variability and divergence of sequences, several SP or their special utilities are used. The list included six main SP: MEGA-X, DNAsp, MrBayes, PhyloSuite, BEAST-2, and IQ-TREE. Ending the current section, it is suitable to exemplify the analytical resources developed for them. The amounts of calculations could be represented partly by the information capacity in the folders and files with their sizes in mega-bites, MB. For simplicity, let us take only the family Pleuronectidae. MEGA-X: The folder Pleuronectidae (created 8 June 2021) has the size 15 MB. This folder is comprised of four subfolders, including 88 files. DNAsp: The folder Pleuronectidae (created 7 September 2021), has the size 38 MB. In the calculations, 31 files were involved. MrBayes: The folder Flound2021-Pleuronectidae (created 2 June 2021), has the size 235 MB. The folder comprised of 18 subfolders, including 367 files. PhyloSuite: The folder PhyloSuite (created 14 September 2021), has the size 18.7 GB. The folder is comprised of 662 subfolders, including 2,187,119 files (here big fraction of files are comprised of the SP itself but not the calculation files). BEAST-2: The folder Pleuronectidae (created 24 July 2021), has the size 37.8 GB. The folder is comprised of 75 subfolders, including 1075 files. Remarkably, the most interesting results were obtained by CA simulations for a population of constant size (CA analysis, model III), but computing resources used were greatest for the CA model IV.

3. Results

3.1. Structure and Variability of the Mitochondrial Genome of the Yellow-Stripe Flounder Pseudopleuronectes Herzensteini and Other Members of the Family Pleuronectidae

The complete mitogenome of *P. herzensteini* is 16,845 bp long (GenBank accession No: MW713061). It is including 13 protein-coding genes, 22 tRNAs, 12S rRNA and 16S rRNA genes, and a control region, CR (Figure 1). Most of the genes are located in the "+" strand, except ND6 and eight tRNA genes, which are located in the "−" strand (Figure 1). For greater clarity, data on the structure of the mitogenome are given in a separate table for three members of the Pleuronectidae, including the two species we describe herein (Table 2).

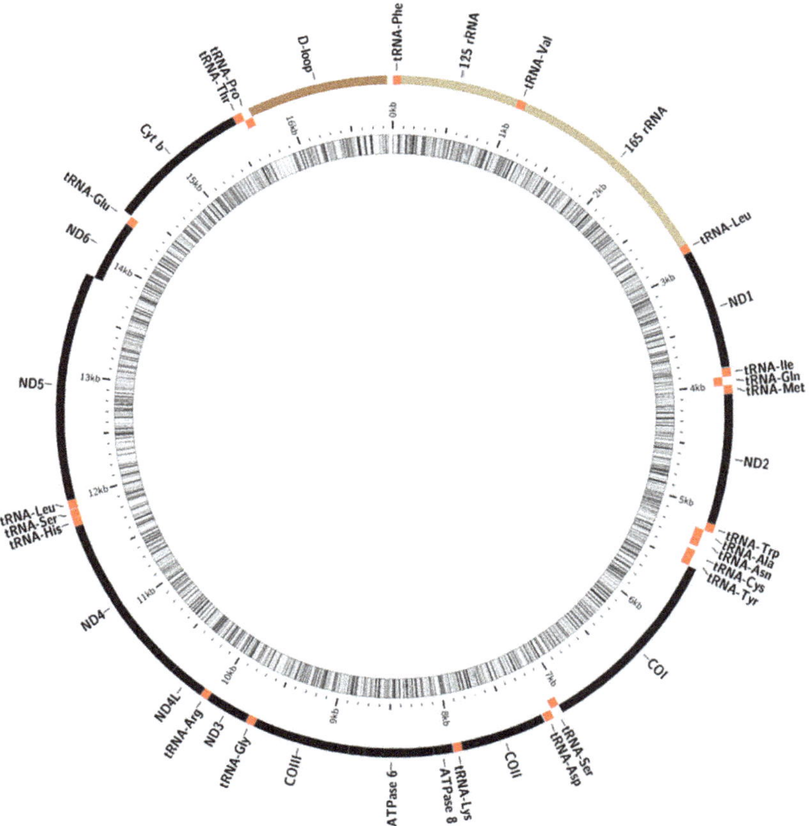

Figure 1. Map of the circular mitochondrial genome of the yellow-stripe flounder *Pseudopleuronectes herzensteini*. The external ring displays the abbreviations and composition for the main components of the mitogenome. It includes: 13 protein-coding genes (ATPase6, ATPase8, COI, COII, COIII, Cyt-b, ND1, ND2, ND3, ND4, ND4L, ND5, and ND6), 2 rRNA genes (*12S* rRNA and *16S* rRNA), and 22 tRNA genes (tRNA-Val, -Leu, -Ile, -Met, -Trp, -Ala, -Asn, -Cys, -Tyr, -Ser, -Asp, -Lys, -Gly, -Arg, -His, -Ser, -Leu, -Glu, and -Pro). Shifted inside line display the components of the genome located in the "−"-chain. Most genes are located in the "+"-chain. Inside ring mapping, the whole mitogenome length (kb), spanning orientation, and its longevity.

Table 2. Mitochondrial genome information on two flatfish sequences presented in the current paper (*P. herzensteini* and *P. stellatus*) and the third (*P. yokohamae*), retrieved from GenBank.

Genome Content/Sequences	***Pseudopleuronectes herzensteini*** **MW713061**	***Platichthys stellatus*** **MZ365029**	***Pseudopleuronectes yokohamae*** **KT224485**
Size (bp)	16,845	16,992	17,383
Gene number, PCGs	13	13	13
Gene number, rRNA	2	2	2
Gene number tRNA	22	22	22
tRNA-Phe	1.68 (+)	1.68 (+)	1.68 (+)
12S rRNA	69.1017 (+)	69.1017 (+)	69.1017 (+)
tRNA-Val	1018.1090 (+)	1018.1090 (+)	1018.1090 (+)
16S rRNA	1091.2806 (+)	1091.2805 (+)	1091.2806 (+)
tRNA-Leu	2807.2880 (+)	2806.2879 (+)	2807.2880 (+)
ND1	2881.3855 (+)	2880.3854 (+)	2881.3855 (+)
tRNA-Ile	3861.3931 (+)	3860.3930 (+)	3861.3931 (+)
tRNA-Gln	3931.4001 (−)	3930.4000 (−)	3931.4001 (−)
tRNA-Met	4001.4069 (+)	4000.4068 (+)	4001.4069 (+)
ND2	4070.5114 (+)	4069.5113 (+)	4070.5114 (+)
tRNA-Trp	5115.5186 (+)	5114.5185 (+)	5115.5186 (+)
tRNA-Ala	5188.5256 (−)	5187.5255 (−)	5188.5256 (−)
tRNA-Asn	5258.5330 (−)	5257.5329 (−)	5258.5330 (−)
tRNA-Cys	5368.5432 (−)	5368.5432 (−)	5369.5433 (−)
tRNA-Tyr	5433.5500 (−)	5433.5500 (−)	5434.5501 (−)
COI	5502.7061 (+)	5502.7061 (+)	5503.7062 (+)
tRNA-Ser	7062.7132 (−)	7062.7132 (−)	7063.7133 (−)
tRNA-Asp	7147.7217 (+)	7147.7217 (+)	7148.7218 (+)
COII	7224.7914 (+)	7224.7914 (+)	7225.7915 (+)
tRNA-Lys	7915.7987 (+)	7915.7987 (+)	7916.7988 (+)
ATPase 8	7989.8156 (+)	7989.8156 (+)	7990.8157 (+)
ATPase 6	8147.8829 (+)	8147.8829 (+)	8148.8830 (+)
COIII	8830.9614 (+)	8830.9614 (+)	8831.9615 (+)
tRNA-Gly	9615.9686 (+)	9615.9686 (+)	9616.9687 (+)
ND3	9687.10035 (+)	9687.10035 (+)	9688.10036 (+)
tRNA-Arg	10,036.10104 (+)	10,036.10104 (+)	10,037.10105 (+)
ND4L	10105.10401 (+)	10,105.10401 (+)	10,106.10402 (+)
ND4	10,395.11775 (+)	10,395.11775 (+)	10,396.11776 (+)
tRNA-His	11,776.11845 (+)	11,776.11845 (+)	11,777.11846 (+)
tRNA-Ser	11,846.11912 (+)	11,846.11912 (+)	11,847.11913 (+)

Table 2. *Cont.*

Genome Content/Sequences	*Pseudopleuronectes herzensteini* **MW713061**	*Platichthys stellatus* **MZ365029**	*Pseudopleuronectes yokohamae* **KT224485**
tRNA-Leu	11,917.11989 (+)	11,917.11989 (+)	11,918.11990 (+)
ND5	11,990.13828 (+)	11,990.13828 (+)	11,991.13829 (+)
ND6	13,825.14346 (−)	13,825.14346 (−)	13,826.14347 (−)
tRNA-Glu	14,347.14415 (−)	14,347.14415 (−)	14,348.14416 (−)
Cyt b	14,420.15560 (+)	14,420.15560 (+)	14,421.15561 (+)
tRNA-Thr	15,561.15633 (+)	15,561.15633 (+)	15,562.15634 (+)
tRNA-Pro	15,633.15703 (−)	15,634.15704 (−)	15,634.15704 (−)
control region	15,704.16845 (+)	15,705.16992 (+)	15,705.17383 (+)

Note. Abbreviations are as follows: PCGs, protein-coding genes; NCR, noncoding region; +/−, location of genes at the "+/−" strand; tRNA genes are designated by three-letter amino acid codes.

The 22 tRNA genes studied are located between the rRNA genes and the PCGs. Their length varies from 66 bp (tRNA-Cys) to 74 bp (tRNA-Leu, Lys, Thr) (Figure 1, Table 2). All tRNAs chains are capable of forming a typical clover-leaf structure, with the exception of tRNA-Cys, which forms a different secondary structure. The secondary structure of the studied tRNAs was clarified using the tRNAscan-SE software [49,50].

Most protein-coding genes (12) use the ATG start codon. The exception is the COI gene, which uses GTG. A complete three-nucleotide stop codon, TAA, is used in four protein-coding genes, ND5, COI, ND1, and ATP6. The ND4, Cyt-b, ND2, and COII genes have an incomplete stop codon using only T. The ND4L gene terminates with A; ATP8, with G; COIII, with TA; and ND3, with a TC combination. The ND6 gene has the TAG stop codon. Thirteen protein-coding genes of the *P. herzensteini* mitogenome encode 3708 amino acids. The most commonly used amino acid is leucine (17.53%), and cysteine is the least used (0.62%). The control region (CR, D-loop) 1142 bp long is located between tRNA-Pro and tRNA-Phe (Figure 1), as was the case in the study [51].

The arrangement of genes in the studied taxa of the Pleuronectidae is conserved, and the changes within the family are due only to sporadic rearrangements and duplications of tRNA genes (Figure 2). The analysis of the properties of the sequences presented showed very high variability and informative capacity of the 13 PCGs of the studied members of the flounder family Pleuronectidae. The overall heterogeneity of nucleotide frequencies of different types with a prevalence of purines (T+C) over pyrimidines (A+G) is well known for PCGs due to its hydrophobic impact on polypeptides, but herein it was provided with necessary statistical evaluation (Table S4). Nucleotide diversity along sequences of the 13 PCGs varied widely (Figure S3); however, it was fundamentally similar across genes (Table S5). The analysis showed that nucleotide diversity did not differ significantly between the 13 PCG sequences, averaging about 12%: Pi = 0.12 ± 0.03. In general, the structure of the mitogenome of 26 studied members of the Pleuronectidae with a representative of the outgroup is very conserved, which is illustrated in more detail with numerical data for three pleuronectids (Table 2). The visual representation for all 26 sequences clearly demonstrates differences for only one of two specimens of the genus *Verasper*, *V. moseri* (Figure 2). In this specimen, three amino acid sites are lost, which may result from an error by the authors during this mitogenome annotation or the SP ITOL [52], the online service itself; because when checking the sequence by the MitoAnnotator of the MitoFish online services and by GenBank itself, this sequence has the typical content of amino acid sites.

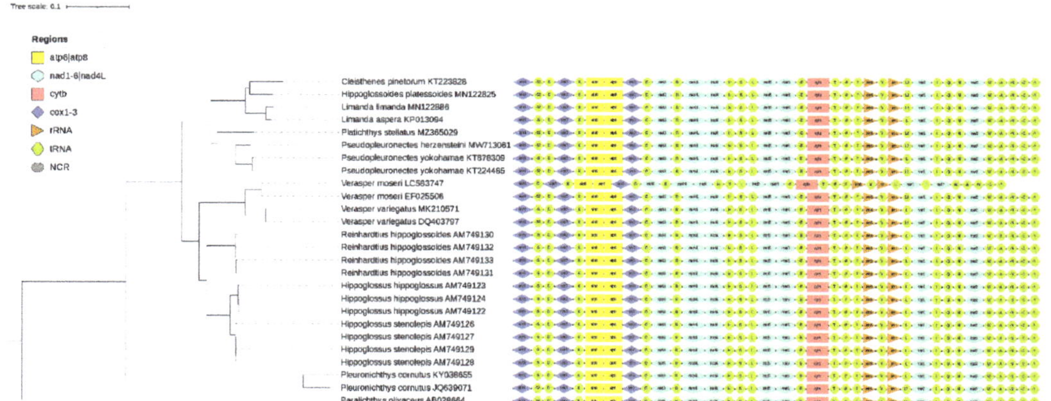

Figure 2. A map of the mitogenomes in linear mode for the 26 species of flounder of the family Pleuronectidae along with BA-tree. For simplicity, CRs are excluded from the comparison because of their variable numbers in flounders' mitogenomes. Details of the phylogenetic reconstruction and tree topology will be presented in the next sections. Probabilities for all nodes in the gene tree that are depicted on the left are equal to 1.0 for all interspecies branches. The number of generations simulated in this case is equal to $n = 2 \times 10^6$.

3.2. Analysis of Properties of Sequences

Shortly, the output information on the sequences analyzed by the DNAsp-5.10.02 software is listed as follows. Selected region: 1–11,401 bp, Number of sites: 11,401, Total number of sites (excluding sites with gaps/missing data): 11,400, Sites with alignment gaps or missing data: 1, Invariable (monomorphic) sites: 7200, Variable (polymorphic) sites: 4200 (Total number of mutations: 5729), Singleton variable sites: 457, Parsimony informative sites: 3743.

The ratio of pyrimidines (T, C) and purines (A, G) in aligned sequences deviated from the 50:50 ratio (Table S4) toward pyrimidines, thus indicating the heterogeneity of the composition of nucleotides with the predominance of C- and T-nucleotides (Table S4). The overall heterogeneity of nucleotide frequencies in each of the two sets (unaligned sequences and aligned) is significant: Wilk's Lambda = 0.0054, F = 801, d.f. = 6; 380, $p < 0.0001$ (Table S4). The average values of nucleotide frequencies between the two sets of sequences do not differ significantly: Wilk's Lambda = 0.9981, F = 0, d.f. = 6; 380, $p < 0.9992$. The proportion of G+C nucleotides varies in the range of 0.4024–0.5001 and totals G+Ctot = 0.46 ± 0.04, i.e., close to an expected value of 50% (Table S5; here and below, after the "±" sign, the standard errors of the mean values are given, SE). In this case, the proportions for the 13 coding sequences (G+Cc) and totals (G+Ctot) coincide, since the G+C proportion was not estimated for non-coding sequences (Table S5).

A general characterization of sequence variability for each of the 13 PCGs, including the analysis of 15 variables such as the number of variable sites (S), nucleotide diversity (π, for simplicity denoted as, Pi), etc., as well as the total values for these variables for PCGs, is presented in the Supplement table (Table S5). The data obtained indicate that, in general, the sequence variability of the 13 PCGs is quite high: the haplotype or gene diversity, Hd, varies between 13 PCG mitogenome sections in the range of 0.957–0.997, with a total value of Hd = 1; the number of variable sites, S, is rather large for the studied set of PCGs, S = 4200. The nucleotide diversity per site, Pi, which is the most representative measure of gene variability (Nei, 1987, equation 10.5), totaled to Pi = 0.12 (this is calculated value by DNAsp-5; Table S5). Our calculation of the average for this index based on data in Table S5 showed that Pi does not differ significantly between 13 PCGs: mean Pi = 0.12 ± 0.04. Tajima's D values are negative for all 13 PCGs, with a total D value of −0.205, suggesting either cut-off or eliminative selection against non-synonymous substitutions (mutations).

The possibility of such an interpretation of the data is evidenced by 2–3 times higher proportions of synonymous substitutions in codons, Pi(s), compared to non-synonymous ones, Pi(a): Pi(s) = 0.3450, while Pi(a) = 0.0536; thus, Pi(a)/Pi(s) ratio is 0.120. Recalculation of pairwise scores between all set of sequences in terms of distances or more precisely number of nucleotide substitutions (or segregating sites, K) Ks and Ka [53,54] (p. 219) yields similar to the above estimates of the range of variation but permit to evaluate approximately the degree of difference between these values: Ks = 0.4999 ± 0.0282 (n = 300, where n is the sample size) and Ka = 0.0536 ± 0.0097 (n = 300).

The genetic distances between intrageneric and intrafamily groups differ significantly (see discussion below in Section 3.3). Notably, the interspecies distance in the genus *Pleuronichthys*, which is represented by two specimens of *Pleuronichthys cornutus* and *Pleuronichthys japonicus*, the latter being considered a synonym of *P. cornutus* [55] is too large for intraspecific values.

3.3. Reconstruction of Gene Trees and Analysis of Molecular Phylogenetic Relationships

A generalized characterization of molecular phylogenetic relationships based on protein-coding gene (PCGs) sequences between the studied species of the Pleuronectidae and the chronology of divergence is presented, as noted earlier, for the 26 PCG sequences (Figures 2–4). The topology of gene trees and the molecular systematics of the suborder Pleuronectoidei are considered separately also based on PGGs (Figure 5).

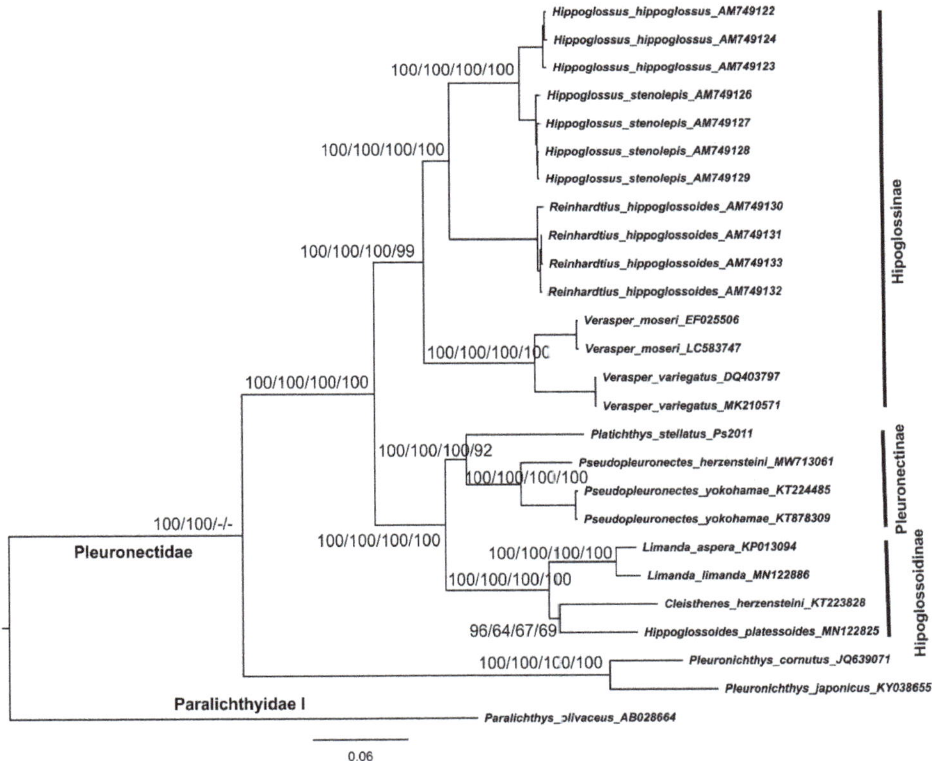

Figure 3. Molecular phylogenetic relationships of flounders of the family Pleuronectidae reconstructed using four approaches: BA, ML, NJ, and Mp. Support values (%) at the tree nodes are shown in the direction: BA/ML/NJ/Mp. For BA reconstructions, posterior probabilities for model generations, n = 10^6 as well as for the other three techniques, bootstrap replicas, k = 1000 are given. Supports for intraspecific nodes are omitted. The tree is rooted in the outgroup *Paralichthys olivaceus*.

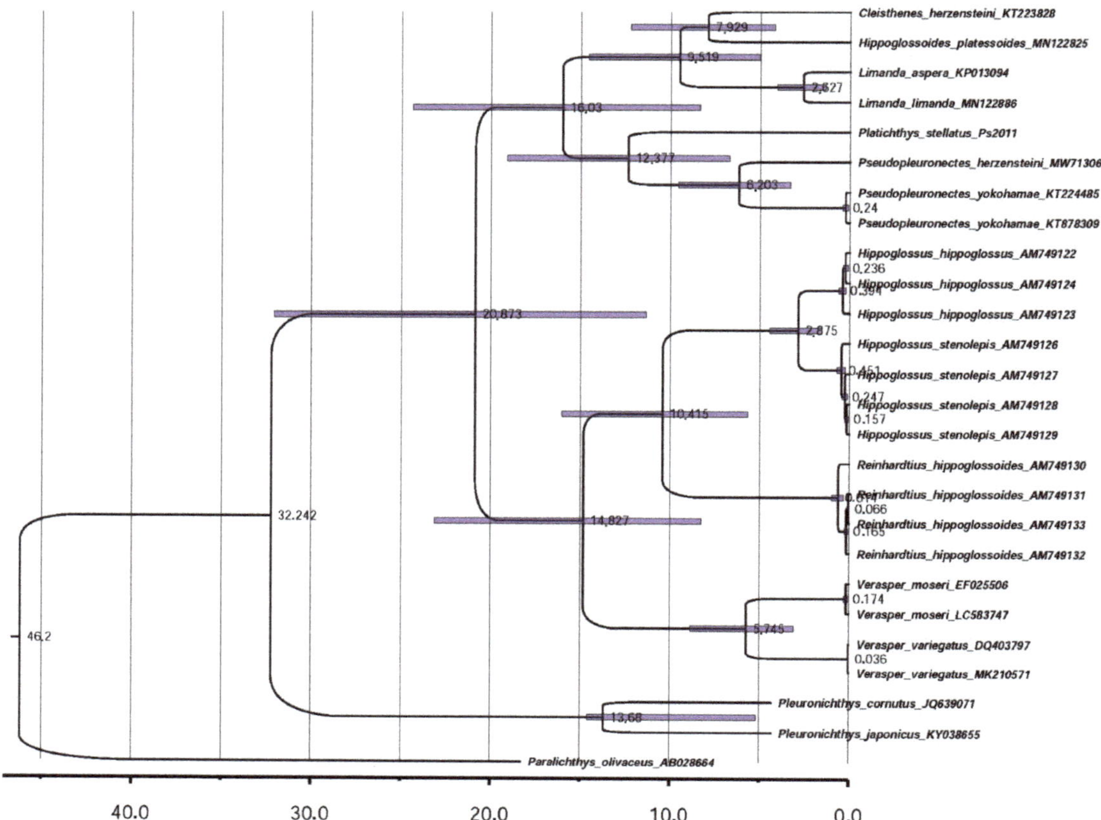

Figure 4. CA-based time-tree reconstruction via BEAST-2 and FigTree utilities that based on 13 PCG sequences of 26 analyzed representatives of flounder family Pleuronectidae and the outgroup. Details for the simulation of the tree in the current image are given in the text. Besides the nodes, their ages are given after rooting the tree with the outgroup taxon *Paralichthys olivaceus* and converting the scale in node ages to root age, which equated to 46.2 Mya. Bars are representing CA 95%HDP for the node ages.

3.3.1. Molecular Phylogenetics and Dating of Divergence of Flounders of the Family Pleuronectidae

Reconstruction of molecular phylogenetic relationships based on the 26 PCGs of pleuronectids was performed using five different approaches: BA, ML, NJ, MP, and CA, as described in the Materials and Methods section. For brevity, reconstructions of gene trees visualized on the basis of the BA-tree and concordance of BA-inferred topology with other topologies represented by bootstrap support scores for nodes (Figure 3; BA-reconstruction using MrBayes 3.2.1–3.2.7 and ML-, NJ-, and MP-reconstructions using the MEGA-X software).

One other reconstruction of the topology of tree branches (nodes) based on CA of the 26 pleuronectids using BEAST-2 yields information that is completely congruent to the previous four depicted in Figure 3 (Figure S1, Supplement). A CA-analysis with divergence dating at the nodes of the gene tree is presented separately (Figure 4). The reconstructions based on the 26 PCG sequences show that the family Pleuronectidae has one highly supported node (100% for two variants of topology reconstruction, BA and ML) or monophyly, with the nearest close relative *Paralichthys olivaceus* from the family Paralichthyidae (more precisely, with representatives of its branch I; Figures 2–4 and data of the next subsection). The internal topology includes three subfamilies and is well supported by all four methods in this case of tree reconstruction: Pleuronectinae, 92–100%, Hippoglossoidinae, 100%, and Hippoglossinae, 100% (Figure 3). It is important to note a very well-supported (100%) common, rather compact branch of the first two subfamilies (Figure 3). In addition to the tree topology data, the monophyly of the family Pleuronectidae is supported by the common structure of the mitogenome and the direction of genes' location in the mitogenomes for all studied representatives, except one of the two specimens of *V. moseri*, which, as noted earlier, is rather due to a technical error in the description of one tRNA in this sequence (Figure 2).

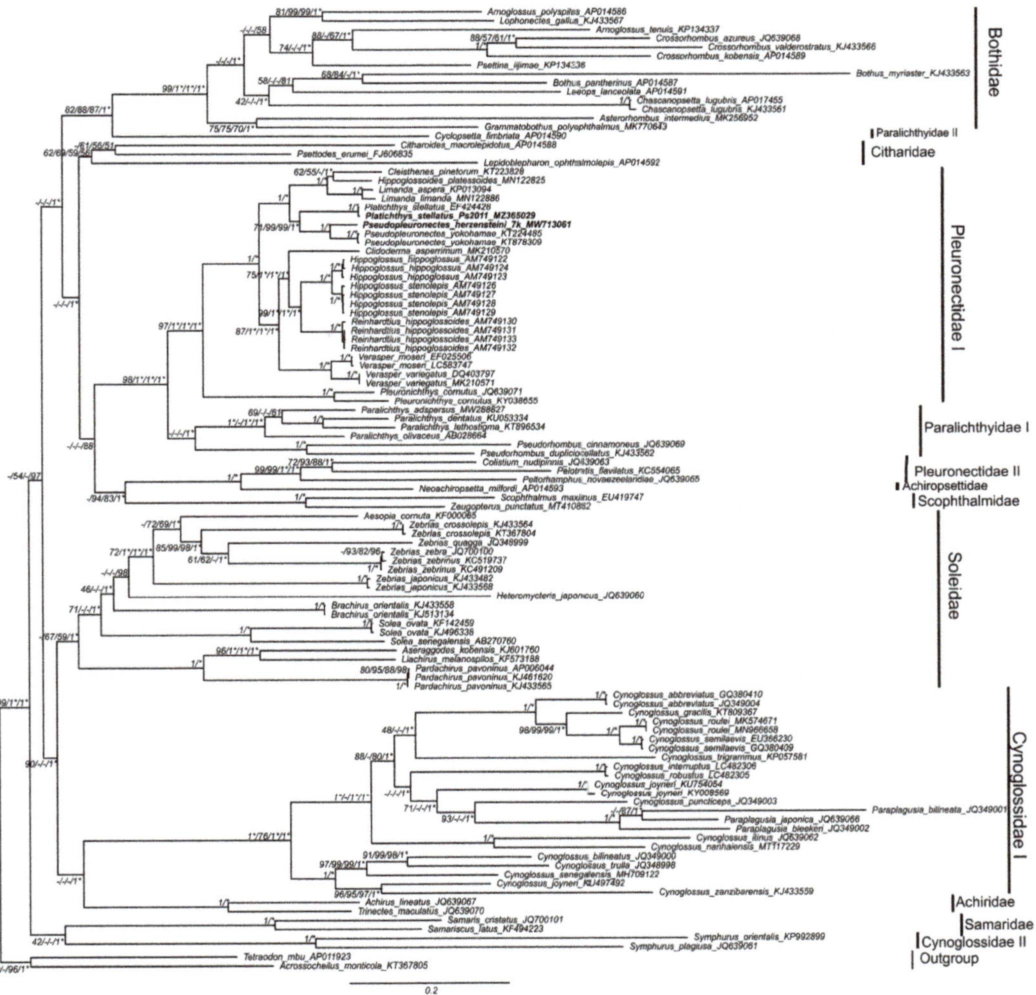

Figure 5. Rooted gene tree of the studied representatives of the suborder Pleuronectoidei constructed from PCGs of 108 mitogenomes. The topology of the gene tree is reconstructed using the BA approach. The numbers at the nodes are support values for the four tree reconstruction techniques that are placed in the order: MP/ML/NJ/BA. The posterior probabilities (%, n = 10^6 generations) are shown for the BA tree, the bootstrap support values (k = 1000 replicas) are given for the other three reconstructions. Dash means absence of support for the current node in the reconstruction by a certain technique. Support values that equal to 100% shown for convenience by the asterisks.

Reconstruction of the time of divergence of phyletic lineages based on the sequences of 13 PCGs for the 26 representatives of taxa reveals an exact match of the gene tree topology with the previous four reconstructions given in Figures 2 and 3 and yields a date of 32.293 Mya for the divergence of the family Pleuronectidae from the mrca, *Paralichthys olivaceus* (Figure 4). The common root of these two taxa is dated at 46.2 Mya and is calibrated to the time of divergence of the common ancestor of the Pleuronectidae-Paralichthyidae, that is, Paralichthodidae, as noted in the Materials and Methods section.

For building the tree depicted in Figure 4, special files with the sequence matrix (Fl26seq-pt8-11401-123ps4.nex; Table S1, Supplement) and the whole set of parameters that were used for tree simulation by CA of the constant population are used (Table S2, Supplement; File: Fl26seq-pt8-123ps4-tip-r24b1-n = 5E7-fix-pop-hm.xml). Table S2 in the supplementary file was built by BEAUti v2.6.6 utility of BEAST-2 software starting with exporting by BEAUti the file Fl26seq-pt8-11401-123ps4.nex. Basic model parameters could be read from this file using BEAUti v2.6.6. After running the main SP BEAST v2.6.6 utility implementing $n = 5 \times 10^7$ generations and other parameters necessary for appropriate simulation, sets of trees and other estimators were obtained; a brief description of the simulation procedure and parameters are given below for six items. The total number of trees was 50,002; 25,001 of them were used. So, 25,001 trees were processed after ignoring the first 50% = 25,000 trees. The final tree contained a total of 25 unique clades. A maximum credibility tree was constructed using TreeAnnotator v2.6.4 based on the file Fl26seq-pt8-11401-123ps4.trees (Table S3, Supplement) that are suitable for further processing in FigTree software, as recommended by the SP BEAST-2 creators. Properties of the quality of the model parameters for simulation were retrieved from several runs of Tracer v1.7.2 utility of SP BEAST-2. Principal files from Tracer for the simulation are placed in the Supplement in a special folder that includes .pdf- vs. .txt-files, and xml-file: Tracer_out_for_Fig4. The sequence partitions of the simulation run that was used for building Figure 4 and the main properties of the simulation schedule are as follows: (i) Sequence partitions of 26 specimens for all 13 PCGs are comprised main data set; details are implemented in the file Fl26seq-pt8-123ps4-tip-r24b1-n = 5E7-fix-pop-hm.xml (Table S2; Sequence partitions, see the menu folder) and can be viewed for inspection via BEAUti; (ii) Priors for the model of the Coalescent Constant Population are defined in the same file (Table S2; Priors); (iii) Tip dates are set numerically as scores of years for two calibration dates, 32.293 Mya vs. 46.19 Mya for mrca Pleuronectidae-Paralichthyidae as given above and three sequences were used: Paralichthys_olivaceus_AB028664 (age 4.619E7), Pleuronichthys_japonicus_KY038655 (age 2.783E), Pleuronichthys_cornutus_JQ639071 (age 2.783E7) (Table S2; Tip dates); (iv) Gamma Site Model is implemented for calculations (Table S2; Site Model: substitution rate = 2.0, G category count = 4, I = 0.477, shape = 1); (v) Clock Model is implemented by following options (Table S2; Relaxed Clock Exponential: Clock.rate = 2.0); (vi) mcmc setting is performed (Table S2; MCMC: Chain Length = 50000000). An independent analysis supports prior in the item (ii) indicating the appropriate choosing the model of the Coalescent Constant Population (Figure S4). Empirical data agreed with the expectation curve on constant population growth (changes) as depicted at miss-match distribution for 25 mitogenome sequences set of the Pleuronectidae flounders (Figure S4).

As noted above, the five tree building methods (BA, ML, NJ, MP, and CA) provide virtually the same topologies for the 26 pleuronectids when rotating branches within and between subfamilies in the images (Figures 2–4). Data on the node ages in Figure 4 are fully concordant with data on the node probabilities and bootstrap supports given in Figure 3. Node ages for the sequences belonging to the same species do not differ judging on large sampling or standard errors (SEs), while ages for inter-genera (8–15 Mya), inter-subfamilies (21 Mya), and family (32 Mya) levels are more realistic (Figure 4). Other details on the tree lineage divergence estimated by the ultra-fast ML technique as implemented in SP IQTREE are given in Figure S1 and are provided in the Discussion section with the representation of confidence intervals for nodes/branches. In concluding the current section, we should emphasize the fine concordance of the five molecular genetic reconstructions with simulated lineage diversification in time.

3.3.2. Phylogenetic Relationships and Molecular Systematics of the Studied Representatives of the Suborder Pleuronectoidei

The main results of the molecular genetic reconstruction of the relationships between members of the suborder are shown in Figure 5

Family Pleuronectidae. According to the data of Section 2.1, the branches of three subfamilies Pleuronectinae, Hippoglossoidinae and Hippoglossinae are very well supported (100%) within the main representatives of the family (denoted as Pleuronectidae I) for all variants of tree reconstruction, with a separate external position of two members of the genus *Pleuronichthys* (Figure 5). The branch of species in the genus *Limanda* forms a common node with *Cleisthenes pinetorum* and *Hippoglossoides platessoides*, being placed in the subfamily Hippoglossoidinae (Figure 5). The species *Colistium nudipinnis, Pelotretis flavilatus*, and *Peltorhamphus novaezeelandiae*, formally belonging to the family Pleuronectidae (Pleuronectidae II), form a common branch with *Neoachiropsetta milfordi* from the family Achiropsettidae (Figure 5). These four species, in turn, form a single branch with members of the family Scophthalmidae (Figure 5).

Pleuronichhyinae branch. The divergence between *Pleuronicthys cornutus* (JQ639071) and *Pleuronicthys japonicus* (KY038655) are thought to have diverged around 6.5–13 Mya (Figure 4). As noted above, according to a recent revision [55] the current status of *P. japonicus* is defined as being a synonym of *P. cornutus* (Official status of *Pleuronicthys japonicus*: Synonym of *Pleuronicthys cornutus* (Temminck & Schlegel 1846). Basic taxa are Pleuronectidae: Pleuronichhyinae. Distribution: Sea of Japan and Pacific coast of Japan, to the southern East China Sea and the Seto Inland Sea [if valid]; CAS–Eschmeye's Catalog of Fishes: Species; calacademy.org)). However, the above divergence dates and genetic distances for this pair are greater than some of the interspecies values (Tables S6 and S7). This is clearly evident for the genus *Verasper* data and for other taxa of the family Pleuronectidae (Figure 4). However, the confidence intervals for the divergence times overlap significantly (Supplement, Figure S1), hindering reliable interpretation.

Family Paralichthyidae. This group forms by two separate branches, Paralichthyidae I and Paralichthyidae II, i.e., is basically polyphyletic (Figure 5). Paralichthyidae I, as noted above, comprises the external branch to the Pleuronectidae with high levels of support (100%) for three of the four building techniques for the common node (Figure 5). The inner node for *P. adspersus* is not well-supported (Figure 5).

Family Cynoglossidae. This group is basically polyphyletic, as it is made up of two different branches, Cynoglossidae (I) and Cynoglossidae (II) (Figure 5). The primary branch, Cynoglossidae (I), is strongly supported by four tree-building techniques in this case with a single root, i.e., being monophyletic, but it is divided into two subdivisions, one of which contains partially African roots (Figure 5). The family Achiridae branch is attached as an external branch to the main branch of the family Cynoglossidae (I) (Figure 5). The branch of two representatives of the genus *Symphurus*, which is included in a separate paraphyletic branch of the family Cynoglossidae (II), forms a separate node with the family Samaridae. This complex of taxa constitutes the outer branch for the entire suborder, and is located immediately before members of the outgroup (Figure 5).

Family Citharidae. In all our reconstructions this group does not form a well-supported branch external to the family Pleuronectidae (Figure 5). According to the data presented, *Psettodes erumei* from the family Psettodidae forms a mixed cluster with the family Citharidae (Figure 5). However, the support levels for this node of topology are not high (60–70%) that require new investigation on this point in the future.

Family Bothidae. In all four reconstructions, this group forms well-supported branch with a variety of genera, with an external and also sharp branch comprised of *Cyclopsetta fimbriata* (Figure 5). The genus *Arnoglossus* is paraphyletic; one of its members, *A. polyspilus*, constitutes a common branch with a member of the genus *Lophonectes*, while another species, such as *A. tenuis*, forms a separate branch combined with the genus *Crossorhombus*. *Cyclopsetta fimbriata* which currently represents the family Paralichthyidae II, is an external and also sharp branch of the family Bothidae (Figure 5).

4. Discussion

4.1. The Structure and Variability of Mitogenome Yellow-Stripe Flounder Pseudopleuronectes Herzensteini and Other Studied Representatives of the Family Pleuronectidae

The structure of the mitogenome described herein (Figure 1) is the same as in other Teleosts; the mitogenome has a CR with a replication origin, 13 PCGs, two rRNA genes, and 22 tRNA genes [56–58]. These data, in combination with the signal on the topology of threes including ND6 gene usage and without it, which did not find topology differences [16], allow us to use all 13 PCGs in phylogenetic reconstructions in this paper. The total estimates of synonymous and non-synonymous substitutions in codons were: Pi(s) = 0.3450 and Pi(a) = 0.0536. The assessment of the degree of this difference could be calculated somewhat differently, for Ks and Ka or pairwise values between all sequence variants [54]. Such estimation, for pairwise estimates of the degree of difference between all sequence variants, showed that the variation rows of these values do not overlap: Ks = 0.4999 $\pm$ 0.0282 (n = 300), Ka = 0.0536 $\pm$ 0.0097 (n = 300) and that the Ka/Ks ratio is 0.122. A Student's t-test revealed the statistical significance of the difference between the mean values of Ks and Ka: $t_{Ks/Ka} = (0.4999 - 0.0536)/\sqrt{(0.02822 + 0.00972)} = 14.88$, d.f. = 598, $p < 0.001$. However, selective neutrality testing of variability of 13 PCG using SP DNAsp did not reveal significant deviations: Tajima's D = -0.20499, $p > 0.10$ (Statistical significance: Not significant, NS); test statistic Fu and Li's D* = 0.62300, $p > 0.10$ (NS); test statistic Fu and Li's F* = 0.41880, $p > 0.10$ (NS). Testing for the neutrality of PCGs of intraspecific clusters of three available different species gives a similar result: $p > 0.10$ (NS). That is, in accordance with the widely accepted [54] and logical hypothesis of natural cutoff selection, which acts against nucleotide substitutions in codons (deleterious mutations) leading to less active (ineffective) macromolecules. In other words, data on the significance $t_{Ks/Ka}$ might be the evidence for a normalizing selection acting against mutations with a phenotypic effect in mtDNA sequences. This effect was derived from the relatively homogeneous material of PCG sequences of flounders of a single family. Unfortunately, one test of our data supported the hypothesis, while another did not. The proof appears to be insufficient.

4.2. Gene tree Topology Analysis of the Molecular Phylogenetic Relationships in the Family Pleuronectidae and in the Suborder Pleuronectoidei, and Levels of Genetic Divergence in the Hierarchy of Evolutionary Units (Populations of Species and Ranked Taxa)

Family Pleuronectidae

Topological and chronological reconstructions for the family Pleuronectidae are well supported by various methods, as demonstrated in the Results (see Figures 2–5, Figure S1, Supplement), and are consistent with relatively recent publications on molecular phylogenetics of flatfish [11–13,16,59]. Divergence dates obtained from simulation and CA-analysis in the BEAST-2 software, both visualized in the traditional format (Figure 4, Figure S1, Supplement) and via a DensiTree2.6.4 utility of BEAST-2 software in a more modern representation of phylogenetic relationships (Figure 6), indicate the origin of the main branch of the family Pleuronectidae I from a common ancestor with Paralichthyidae I (represented in this case by *P. olivaceus*) at about 32 million years ago. The previously reported data from a joint analysis of molecular divergence, combined with morphological and paleontological data [11,13], convincingly prove the reliability of this conclusion. There is a very close dating of 42.7–49.4 Mya for similar taxa [60], which, taking into account the topological and time estimation errors (see Figures 4 and 6), coincides with the value presented in our work. The diversification of flounders of Pleuronectidae I occurred from ancestors from the Indo-West Pacific basin, and it was followed by two stages of migration and geographic radiation of the modern Pleuronectidae I species in the northeastern Atlantic and northern Pacific basins [13]. At the end of the Results section, significant differences in the time of divergence of taxa of the species rank are reported. Unfortunately, large sampling errors (Figures 4 and 6) prohibit broad conclusions on the chronology of diversification within the Pleuronectidae I. This will be a task for future research based on a more representative sample of genes and taxa. However, intraspecific differences in the level of divergence, taking into account 95% HPDs of some taxa, differ significantly (see Figure 4). Moreover, for a pair of members of the genus *Pleuronichthys*, the differences from others in intraspecific divergence are so great that no doubt is left concerning their at least species rank, in contrast to the introduced synonymizing to single species (CAS—Eschmeyer's Catalog of Fishes: Species; calacademy.org).

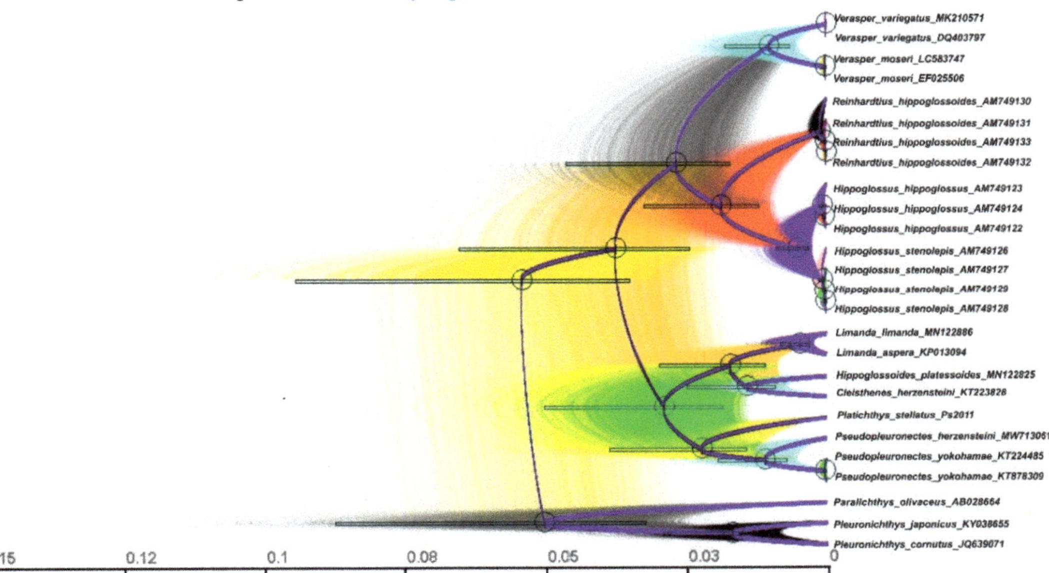

Figure 6. Phylogenetic lineages reconstructed via BEAST-2 and visualized with DensyTree software based on 26 sequences of 13 PCGs of the flounder family Pleuronectidae. Simulated lineages are naturally rooted in three presumed ancestral taxa including the predefined outgroup taxon *Paralichthys olivaceus*. Wider violet lines depict the consensus trees constructed by computer simulations of coalescent process of molecular evolution in constant populations during 5×10^7 generations by BEAST-2 as explained in detail in the Results and in the above paragraph in the main text. Thin lines show all possible trees that occurred during the time span as depicted in the scale given in node ages. DensyTree reconstructed time-tree interrelationships based on the same BEAST run and the output tree file as that used for building Figure 4. Source tree file available for use from Table S3, Supplement (File: Fl26seq-pt8-11401-123ps4.trees). Bars represent CA 95%HDP for the node ages. Circles with a dot inside show the support area and averages for clades. The main branches are stained with different colors.

Our independent analysis of the genetic distances (*TrN*-distances) of 13 PCGs of 25 sequences of mitogenomes for (1) intraspecific comparisons, (2) interspecific comparisons within genera, and (3) intergeneric comparisons within the family Pleuronectidae revealed statistically significant differences for all three groups (Figure 7A). However, the data presented in Figure 7A also demonstrate a strong overlap of the average distances for groups 1 and 2, supporting the above doubt about the validity of combining two taxa, *P. cornutus–p. japonicus*, into one species, *Pleuronicthys cornutus*. These data convincingly show a slightly lower (although non-significant, $p > 0.05$) interspecific divergence of mitogenomes in the flounder genera of the Pleuronectidae, compared with other animals. This is based on two estimates from which the distances were estimated in the three comparison groups. Thus, for the Pleuronectidae flounders the *TrN*-distances are: $0.76 \pm 0.87\%$, (2) $3.34 \pm 0.85\%$, (3) $14.24 \pm 0.23\%$ (Figure 7A; F = 176.26, d.f. = 2; 296, $p < 0.0001$); for representatives of eight different groups of animals, the *p*-distance values were for three corresponding comparison groups: (1) $0.79 \pm 0.04\%$, (2) $8.23 \pm 0.22\%$, (3) $16.47 \pm 0.29\%$ [61] (Arthropods, Chordates, Echinoderms, Flatworms, Mollusks, Nematodes, Segmented worms, and Sponges included). Divergence values similar to those of flounders were reported in another review of whole mitogenome coding genes for two comparison groups in five different taxa of animals: (1) 0.92 ± 0.94%, (2) 4.64 ± 1.90% [62] (our approximate numerical estimate of *K2P*-distances from Figure 3 of the authors). Values obtained in the current study for PCGs of four comparison groups of all studied members of the suborder Pleuronectoidei are as follow: (1) 0.54 ± 0.78%, (2) 14.99 ± 0.48%, (3) 16.51 ± 0.22%, (4) 33.57 ± 0.07% (Figure 7B; comparison groups are representing four different hierarchies of the suborder taxa; F = 2719.4, d.f. = 3; 5040, $p < 0.0001$). Evidently, interspecies estimates of distances within genera (group 2) in above cases including the *p*-, *K2P*- and *TrN*-distance measures vary from 4–8% to 15%. As we revealed, minimal differences for flatfish between comparison groups 1 vs. 2 (Figure 7A) and 2 vs. 3 (Figure 7B), could create difficulties in determining molecular genetic delimitation of species, and obscure the systematics of this fish taxon. We will come once more to the latter matter in the ongoing paragraphs below.

The genetic divergence in the comparison groups (within species and in the hierarchy of taxa) for individual genes [59,61,63–66] corresponds well to divergence estimates based on mitogenomes [61,62,67–69]. As estimated elsewhere, distance estimates by different models including simple *p*-distance below 15–17% correspond with other measures and are consistent with simulated expectations based on random drift with time [63,70,71]. Thus, the sequences of individual mtDNA genes, such as COI, Cyt-b, 16S rRNA, quite well represent the divergence inferred from the analysis of complete mitogenomes or their PCGs. Furthermore, the near linear relationship of genetic divergence and the hierarchy of comparison groups (taxa) that was found for both mtDNA (Figure 7A,B and the above-cited works) and nDNA genes [68,69,72] supports at the molecular level, the current evolutionary paradigm: the Synthetic Theory of Evolution (STE) or Neo-Darwinism. This is well compatible with the predominance of the geographic model of speciation in nature [61,66,69].

This conclusion is critically important for understanding the fundamental mechanisms of speciation within the realm of evolutionary biology and evolutionary genetics [62,64–66]. Furthermore, our conclusions aid in practical needs such as identifying specimens in systematics, within the activity in international programs for biodiversity deciphering, e.g., iBOL [73], as well as in the fields of biomedicine and trade, where current erroneous identification (accidental or intentional) can lead to significant economic loss, both public and private [62,68,74,75]. This is far from being a complete list of applications of the approach we used in this study [61,68,69,73,76].

Below, we discuss the taxonomy of flounders from the standpoint of tree topologies. As noted above in Section 2.1 and evident in the tree topologies in Figure 3, members of the genus *Limanda* are included in the branch of the subfamily Hippoglossoidinae. A comparative anatomical study by Cooper and Chapleau (1998) did not confirm the monophyly of this genus within the family Pleuronectidae. In our study, as in other molecular phylogenetic investigations [8,11,16,59,77], some representatives of the genus *Limanda* were definitely placed in the subfamily Hippoglossoidinae, and some of them were included in the subfamily Pleuronectinae. Therefore, it is appropriate, following the opinion by Cooper and Chapleau [2] on *L. sakhalinensis* and our observations, to recommend a revision of the family and three of its genera, establishing a new taxon of a tribe rank Hippoglossoidini, and including in it the representatives of Far Eastern *Limanda*, as well as the genera *Cleisthenes* and *Hippoglossoides*, leaving the latter in the subfamily Hippoglossoidinae. Such a transformation is consistent with molecular genetic data on several mitochondrial and nuclear genes [11,16,59].

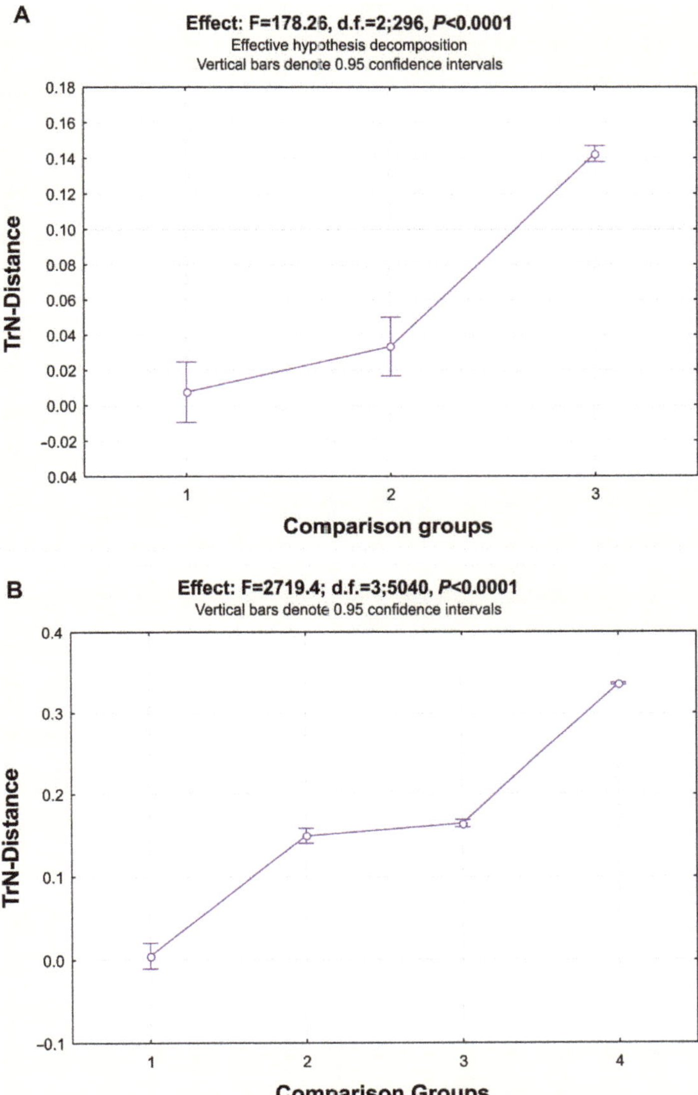

Figure 7. Univariate ANOVA showing the variability of the mean values of genetic distances (*Y* axis) for the comparison groups of the sequences in sampled taxa for 13 PCGs of 26 flatfish mitogenomes in Pleuronectidae (**A**, top) and 106 flatfish mitogenomes in Pleuronectoidei (**B**, bottom). *Y* axis, Tamura-Nei (TrN) variation in the mean values of distances (in frequencies) among three comparison groups for flatfish: (1) *TrN* distances within the species, between individuals of the same species; (2) *TrN* distances within genera, between individuals of different species of the same genus; (3) *TrN* distances within the family, between species of different genera of the same family, (4) *TrN* distances within suborder, between individuals of different families of the same suborder. Data on these two analyses were obtained on the sequences of the complete mitogenome of flatfish from GenBank in 2021.

The interpretation of the topology and system of the family Pleuronectidae is generally similar to interpretations presented in previous studies [10,11,13,23,24,59,69,78]. According to the data presented above (see Figures 2–6), as well as other reconstructions based on fast ML algorithms and complex models that take into account most demands to tree building, like gene partitions and nucleotide substitutions models using PhyloSuite software and its ultra-fast IQ-TREE utility (implementing the ultrafast bootstrap approximation results) (Figure S3, Supplement), the monophyly of the main branch of the family (Pleuronectidae I) is well-supported and agrees with other data [11,13,16,24,59,60]. Representatives of Pleuronectidae II (see Figure 5) are combined into a well-supported branch together with species of the family Scophthalmidae and a representative of the Achropsettidae, which certainly requires further analysis for disproving or support, and then giving taxonomic revision. The latter idea of placing Psettus-like flounders into separate suborder Psettoidei has already been put forward on the basis of valid data for a different set of taxa and markers, including nDNA sequences [13,78].

The topology of the gene tree of the entire suborder (Figure 5) is similar to the topology in some other studies [12,13,16,24] etc. However, there are clear differences since investigators used different markers, particularly in our case, only protein-coding regions of mtDNA. Moreover, in addition to the properties of genes and the informative capacity of sequences, also of importance are methods of analysis, how representative the species sampled within their taxa are, the degree of heterochrony of phyletic lineages for selected genes, and environmental and other factors [12,13,79], etc. The results presented in the paper are convincing evidence that the heterogeneity of the studied sequences, representing the phyletic lineages of flounders, were not responsible for any substantial errors in the molecular phylogenetic reconstructions in this work. The high congruence of tree topologies (similar support for most nodes) obtained using four methods of reconstruction for the suborder and five methods for the family Pleuronectidae (see Figures 2–6, Figures S1, S2, and S5) points to the representativeness of the molecular phylogenetic reconstructions for the taxa discussed. Additional information about the sufficient compactness of the studied mitogenomes was obtained from estimates of the compositional distance (bias) in the sequences (Figure S5; numbers below the branches of the ML tree).

In addition to this conclusion, the values of the compositional distance for the two nominal taxa in the genus *Pleuronichthys,* which stand out in the within species range, confirm their reassignment to species rank. All the data concerning the structural conservatism of most of the analyzed mitogenomes (see Figure 2, Table 2, and the corresponding paragraphs in the context) indicate the validity of the evolutionary signal presented in this study and its significance for the taxonomy of the family Pleuronectidae and partly for the entire suborder. An important point for such reasoning is the saturation effect among the mitogenome sequences in our investigation for flounders that was calculated. Nucleotide composition saturation was firstly evaluated by comparing the Iss and Iss.c indices for the 26 mitogenomes of the family Pleuronectidae (Iss = 0.8021, Iss.c = 0.8463, t = 18.2051, d.f. = 5977, $p < 0.0001$; two-sided *t*-test, SP DAMBI [80,81]; significant differences between Iss and Iss.c define the absence of composition saturation and its impact on topology signal). For the suborder, Pleuronectoidei similar results were obtained on the 13 PCGs of 108 representatives (Table S8). For other flatfish mitogenomes, there is the study of the saturation of the nucleotide composition that involved a different, wider set of representatives of the order [12]. Thus, neither our findings nor literature sources [12] indicated any significant influence of nucleotide composition saturation on the topology at the family and even suborder/order level.

The data support a close-to-linear relationship between genetic distances and taxon rank for mitogenomes (Figure 7) [61,62,67–69] and individual genes of not only mtDNA but also nuclear DNA for a vast set of taxa [62,64–66,69,72]. To clarify the system of the suborder and individual families, a great deal of research is still required. This obviously follows from the paraphyletic nature of a number of branches denoted by duplication of the family names: Pleuronectidae I and II, Paralichthyidae I and II, etc. (see Figure 5), as was also noted earlier for these and other taxa [13], Figure 1. In our study, members of *Pleuronichthys* occupy a separate position relative to other members of the family, being an external branch and uniting with *Paralichthys olivaceus*, which was used as an external taxon for the family Pleuronectidae (see Figures 2–6, Figures S1, S2, and S5). These data support the recently advanced (CAS—Eschmeyer's Catalog of Fishes: Genera (calacademy.org)) idea of creating an independent pleuronichthoid subfamily, the Pleuronichthyinae, in contrast to the traditional view [82]. Nevertheless, the monophyly of the family Pleuronectidae and most of Pleuronectoidei leaves no serious doubt, despite the weak support for the monophyly of the suborder Pleuronectoidei, 22–46% in 18 out of 23 different assessments of mitogenome signal for the pattern marker's combinations given in Figure 1 and Table 1 [13] and in other references [8,11,12,59,77].

The data shown in the Results section and discussed in the previous paragraph demonstrate the reliability of the gene tree topology and molecular phylogenetic reconstructions in our study. This view is based on the consistency of several tree reconstruction methodologies, as well as their analytic algorithms and numerical simulations. We did not include any topology restrictions or the influence of markers' mitogenome partitions on the phylogenetic signal in the analysis, with the exception of outgroup (however, gene partitions and accounting for nucleotide positions in codons were used in BA- and ML-techniques in some SP, as explained in the Material and Methods section), as was conducted, for example, in [14] Tables 1 and 2. Our findings on tree topology for the family Pleuronectidae and the suborder Pleuronectoidei, on the other hand, were consistent with the above-mentioned article. Consistent with Campbell and colleagues [12] Table 1, no serious discrepancies between topologies were observed during phylogenetic reconstructions that were completed with and without gene partitions and positions of nucleotides in codons (these data are not presented in the paper). This paper, as well as other recent works [78,83,84] suggest that, the compactness and composition within the Pleuronectoidei/Pleuronectiformes remain unresolved. E.g., in our observation, despite clustering with the family Citharidae (Figure 5), *Psettodes ereumei* is currently attributed to the family Psettodidae (CAS—Eschmeyer's Catalog of Fishes: Genera (calacademy.org)) and placed in the suborder Pleuronectoidei according to some other molecular genetic evaluations [12,84], that were consistent with our own low-level support for this topology node as noted before.

Supplementary Materials: The following supporting information can be downloaded at: https://www.mdpi.com/article/10.3390/d14100805/s1, Figure S1: BEAST-2 and FigTree topology reconstruction based on 13 PCG-sequences of 26 analyzed flounder representatives of the family Pleuronectidae with posterior probabilities implemented. Figure S2: The phylogram built by PhyloSiut software and its IQ-TREE utility for gene tree reconstruction based on 13 PCG-sequences of 26 analyzed flounder representatives of the family Pleuronectidae with the out-group taxon *Paralichthys olivaceus*. Figure S3: Plot of the distribution on nucleotide diversity per site (Pi) along the whole length at 13 PCGs of flatfish mitogenome in Pleuronectidae. On the Y-axis the diversity scores are given, Pi. The red line presents Pi variation at nucleotide positions along DNA chain, X-axis. Figure S4: Plot of pairwise differences for 13 PCGs of flatfish mitogenome in the family Pleuronectidae. On Y-axis the frequency of pairwise differences for 12 sliding windows of divergence estimates are given. On X-axis with the red lines the observed scores of k presented. With green line shows the expected distribution for k. The average number of pairwise differences comprised, k = 1440.113. Figure S5: The phylogram built by MEGA-X software and its utility for ML-gene tree reconstruction based on 13 PCG-sequences of 26 analyzed flounder representatives of the family Pleuronectidae with the out-group taxon *Paralichthys olivaceus*; Table S1: Supplement File: Fl26seq-pt8-11401-123ps4.nex; Table S2: Supplement File: Fl26seq-pt8-123ps4-tip-r24b1-n=5E7-fix-pop-hm.xml; Table S3: Supplement File: Fl26seq-pt8-11401-123ps4.trees; Table S4: Nucleotide content of 25 mitogenome sequences of PCGs among Pleuronectidae; Table S5: Perdomain diversity and DNA varion data for 13 PCGs of 25 selected mitogenome sequences among representatives of Pleuronectidae family; Table S6: TrN-distance-mtx-13PCRs-Pleuronectidae-taxa-ranked; Table S7: TrN-distance-mtx-13PCGs-suborder-Pleuronectoidei-taxa-ranked; Table S8: Test of substitution saturation for Pleuronectoidei PCGs mitogenome sequences.

Author Contributions: Conceptualization, the authors are solely responsible for the content and the writing of this paper. Y.P.K. made an impact in all sections of the research: planning, funding, specimen's collection, data analysis, MS writing, etc. A.D.R. took part in specimen's collection, sequencing and work with GenBank, data analysis, MS writing. Methodology, Y.P.K. made an impact in it. Software, Y.P.K. and A.D.R. made a nearly equal impact in it. Validation, Y.P.K. and A.D.R. made a nearly equal impact in it. Formal analysis, Y.P.K. and A.D.R. made a nearly equal impact in it. Investigation, A.D.R. took part in specimen's collection, sequencing and work with GenBank, data analysis, mostly for Pleuronectoidei, while Y.P.K. have impact mostly in this par to Pleuronectidae. Resources, Y.P.K. made most impact in it. Data curation, Y.P.K. made most impact in it. Writing—original draft preparation, Y.P.K. and A.D.R. made a nearly equal impact in it. Writing—review and editing, Y.P.K. made most impact in it. Supervision, Y.P.K. made most impact in it. Project administration and funding acquisition, Y.P.K. made most impact in it. All authors have read and agreed to the published version of the manuscript.

Funding: This research was supported by the Ministry of Science and Higher Education of the Russian Federation in the framework of the Federal Project #13.1902.21.0012 "Basic problems in the research and conservation of deep-water ecosystems ... " (agreement no. 075-15-2020-796).

Institutional Review Board Statement: Not applicable.

Data Availability Statement: Data supporting reported results can be found in the Supplement and resources that given there, including links to publicly archived datasets analyzed or generated during the study.

Acknowledgments: Our thanks to J. Eimes, K. Saitoh for MS proofreading and advises. We like send our sincere gratitude to S.V. Turanov for sequencing support.

Conflicts of Interest: The authors have no conflicts of interest to declare.

References

1. Keast, A.; Chapleau, F. A phylogenetic reassessment of the monophyletic status of the family Soleidae, with comments on the suborder Soleoidei (Pisces; Pleuronectiformes). *Can. J. Zool.* **1988**, *66*, 2797–2810. [CrossRef]
2. Cooper, J.A.; Chapleau, F. Monophyly and intrarelationships of the family Pleuronectidae (Pleuronectiformes), with a revised classification. *Fish. Bull.* **1998**, *96*, 686–726.
3. Norman, I.R. *A Systematic Monograph of the Flatfishes (Heterosomata). Volume I. Psettodidae, Bothidae, Pleuronectidae*; British Museum: London, UK, 1934; 459p.
4. Sakamoto, K. Interrelationships of the family Pleuronectidae (Pisces: Pleuronectiformes). In *Memoirs of the Faculty of Fisheries, Hokkaido University*; Hokkaido University: Sapporo, Japan, 1984; Volume 31, pp. 95–215.
5. Lindberg, G.U.; Fedorov, V.V. *Fishes of Japan Sea and Nearby Parts of Okhotsk and Yellow Seas. Part 6. Teleostomi. Osteichthyes. Actinopterigii. XXXI. Pleuronectiformes*; Sankt-Petersburg University Press: Sankt-Petersburg, Russia, 1993; p. 272.
6. Chapleau, F. Pleuronectiform relationships: A cladistic reassessment. *Bull. Mar. Sci.* **1993**, *52*, 516–540.
7. Vernau, O.; Moreau, C.; Catzeflis, F.M.; Renaud, F. Phylogeny of flatfishes (Pleuronectiformes): Comparisons and contradictions of molecular and morpho-anatomical data. *J. Fish Biol.* **1994**, *45*, 685–696. [CrossRef]
8. Kartavtsev, Y.P.; Park, T.-J.; Vinnikov, K.A.; Ivankov, V.N.; Sharina, S.N.; Lee, J.-S. Cytochrome b (Cyt-b) gene sequence analysis in six flatfish species (Teleostei, Pleuronectidae), with phylogenetic and taxonomic insights. *Mar. Biol.* **2007**, *152*, 757–773. [CrossRef]
9. Betancur-R, R.; Broughton, R.E.; Wiley, E.O.; Carpenter, K.; López, J.A.; Li, C.; Holcroft, N.I.; Arcila, D.; Sanciangco, M.; Cureton, J.C., II; et al. The tree of life and a new classification of bony fishes. *PLoS Curr.* **2013**, *5*. [CrossRef]
10. Betancur-R, R.; Li, C.; Munroe, T.A.; Ballesteros, J.A.; Ortí, G. Addressing gene tree discordance and non-stationarity to resolve a multi-locus phylogeny of the flatfishes (Teleostei: Pleuronectiformes). *Syst. Biol.* **2013**, *62*, 763–785. [CrossRef]
11. Vinnikov, K.A.; Thomson, R.C.; Munroe, T.A. Revised classification of the righteye flounders (Teleostei: Pleuronectidae) based on multilocus phylogeny with complete taxon sampling. *Mol. Phylogenet. Evol.* **2018**, *125*, 147–162. [CrossRef]
12. Campbell, M.A.; López, J.A.; Satoh, T.P.; Chen, W.J.; Miya, M. Mitochondrial genomic investigation of flatfish monophyly. *Gene* **2014**, *551*, 176–182. [CrossRef]
13. Campbell, M.A.; Chanet, B.; Chen, J.-N.; Lee, M.-Y. Origins and relationships of the Pleuronectoidei: Molecular and morphological analysis of living and fossil taxa. *Zool. Scr.* **2019**, *48*, 640–656. [CrossRef]
14. Amaoka, K. Studies on the sinistral flounder found in the waters around Japan. Taxonomy, anatomy, and phylogeny. *Shimonoseki Univ. Fish.* **1969**, *18*, 65–340.
15. Chabanaud, p. Le problème de la phylogénèse des Heterosomata. *Bull. De L'institut Océanographique De Monaco* **1949**, *950*, 1–24.
16. Kartavtsev, Y.P.; Sharina, S.N.; Saitoh, K.; Imoto, J.M.; Hanzawa, N.; Redin, A.D. Phylogenetic relationships of Russian Far Eastern Flatfish (Pleuronectiformes, Pleuronectidae) based on two mitochondrial gene sequences, Co-1 and Cyt-b, with inferences in order phylogeny using complete mitogenome data. *Mitochondrial DNA* **2014**, *27*, 667–678. [CrossRef] [PubMed]
17. Chen, W.-J.; Bonillo, C.; Lecointre, G. Repeatability of clades as a criterion of reliability: A case study for molecular phylogeny of Acanthomorpha (Teleostei) with larger number of taxa. *Mol. Phylogenet. Evol.* **2003**, *26*, 262–288. [CrossRef]

18. Dettai, A.; Lecointre, G. Further support for the clades obtained by multiple molecular phylogenies in the acanthomorph bush. *Comptes Rendus Biol.* **2005**, *328*, 674–689. [CrossRef] [PubMed]
19. Smith, W.L.; Wheeler, W.C. Venom evolution widespread in fishes: A phylogenetic road map for the bioprospecting of piscine venoms. *J. Hered.* **2006**, *97*, 206–217. [CrossRef]
20. Li, B.; Dettaï, A.; Cruaud, C.; Couloux, A.; Desoutter-Meniger, M.; Lecointre, G. RNF213, a new nuclear marker for acanthomorph phylogeny. *Mol. Phylogenet. Evol.* **2009**, *50*, 345–363. [CrossRef]
21. Near, T.J.; Eytan, R.I.; Dornburg, A.; Kuhn, K.L.; Moore, J.A.; Davis, M.P.; Wainwright, P.C.; Friedman, M.; Smith, W.L. Resolution of ray-finned fish phylogeny and timing of diversification. *Proc. Natl. Acad. Sci. USA* **2012**, *109*, 13698–13703. [CrossRef]
22. Near, T.J.; Dornburg, A.; Eytan, R.I.; Keck, B.P.; Smith, W.L.; Kuhn, K.L.; Moore, J.A.; Price, S.A.; Burbrink, F.T.; Friedman, M.; et al. Phylogeny and tempo of diversification in the superradiation of spiny-rayed fishes. *Proc. Natl. Acad. Sci. USA* **2013**, *110*, 12738–12743. [CrossRef]
23. Campbell, M.A.; Chen, W.-J.; López, J.A. Are flatfishes (Pleuronectiformes) monophyletic? *Mol. Phylogenet. Evol.* **2013**, *69*, 664–673. [CrossRef]
24. Betancur-R, R.; Ortí, G. Molecular evidence for the monophyly of flatfishes (Carangimorpharia: Pleuronectiformes). *Mol. Phylogenet. Evol.* **2014**, *73*, 18–22. [CrossRef] [PubMed]
25. Nei, M. *Molecular Evolutionary Genetics*; Columbia University Press: New York, NY, USA, 1987; 512p.
26. Felsenstein, J. *Inferring Phylogenies*; Sinauer Associates, Inc.: Sunderland, MA, USA, 2004.
27. Charlesworth, B. Effective population size and patterns of molecular evolution and variation. *Nat. Rev. Genet.* **2009**, *10*, 195–205. [CrossRef] [PubMed]
28. GenBank NCBI. National Center for Biotechnology Information. Available online: https://www.ncbi.nlm.nih.gov/ (accessed on 1 January 2021).
29. Dierckxsens, N.; Mardulyn, P.; Smits, G. NOVOPlasty: De novo assembly of organelle genomes from whole genome data. *Nucleic Acids Res.* **2016**, *45*, e18. [CrossRef]
30. Iwasaki, W.; Fukunaga, T.; Isagozawa, R.; Yamada, K.; Maeda, Y.; Satoh, T.P.; Sado, T.; Mabuchi, K.; Takeshima, H.; Miya, M.; et al. MitoFish and MitoAnnotator: A Mitochondrial Genome Database of Fish with an Accurate and Automatic Annotation Pipeline. *Mol. Biol. Evol.* **2013**, *30*, 2531–2540. [CrossRef]
31. Rozas, J.; Sánchez-DelBarrio, J.C.; Messegyer, X.; Rozas, R. DnaSP, DNA polymorphism analyses by the coalescent and other methods. *Bioinformatics* **2003**, *19*, 2496–2497. [CrossRef]
32. Kumar, S.; Stecher, G.; Li, M.; Knyaz, C.; Tamura, K. MEGA X: Molecular Evolutionary Genetics Analysis across computing platforms. *Mol. Biol. Evol.* **2018**, *35*, 1547–1549. [CrossRef] [PubMed]
33. Huelsenbeck, J.P.; Ronquist, F. MRBAYES: Bayesian inference of phylogenetic trees. *Bioinformatics* **2001**, *17*, 754–755. [CrossRef]
34. Ronquist, F.; Huelsenbeck, J.p. MrBayes 3: Bayesian phylogenetic inference under mixed models. *Bioinformatics* **2003**, *19*, 1572–1574. [CrossRef]
35. Drummond, A.J.; Rambaut, A. BEAST: Bayesian evolutionary analysis by sampling trees. *BMC Evol. Biol.* **2007**, *7*, 214. [CrossRef]
36. Drummond, A.J.; Suchard, M.A.; Xie, D.; Rambaut, A. Bayesian phylogenetics with BEAUti and the BEAST 1.7. *Mol. Biol. Evol.* **2012**, *29*, 1969–1973. [CrossRef]
37. Bouckaert, R.; Heled, J.; Kühnert, D.; Vaughan, T.; Wu, C.-H.; Xie, D.; Suchard, M.A.; Rambaut, A.; Drummond, A.J. BEAST 2: A software platform for Bayesian evolutionary analysis. *PLoS Comput. Biol.* **2014**, *10*, e1003537. [CrossRef] [PubMed]
38. Drummond, A.J.; Bouckaert, R.R. *Bayesian Evolutionary Analysis with BEAST*; Cambridge University Press: Cambridge, UK, 2015.
39. Bouckaert, R.; Vaughan, T.G.; Barido-Sottani, J.; Duchêne, S.; Fourment, M.; Gavryushkina, A.; Heled, J.; Jones, G.; Kühnert, D.; De Maio, N.; et al. BEAST 2.5: An advanced software platform for Bayesian evolutionary analysis. *PLoS Comput. Biol.* **2019**, *15*, e1006650. [CrossRef] [PubMed]
40. Meier, R.; Kwong, S.; Vaidya, G.; Peter, K.; Ng, L. DNA Barcoding and Taxonomy in Diptera: A Tale of High Intraspecific Variability and Low Identification Success. *Syst. Biol.* **2006**, *55*, 715–723. [CrossRef] [PubMed]
41. Zhang, D.; Gao, F.; Jakovlić, I.; Zou, H.; Zhang, J.; Li, W.X.; Wang, G.T. PhyloSuite: An integrated and scalable desktop platform for streamlined molecular sequence data management and evolutionary phylogenetics studies. *Mol. Ecol. Resour.* **2020**, *20*, 348–355. [CrossRef] [PubMed]
42. Vadya, G.; Lohman, D.J.; Meier, R. SequenceMatrix: Concatenation software for the fast assembly of multi-gene datasets with character set and codon information. *Cladistics* **2011**, *27*, 171–180. [CrossRef]
43. Drummond, A.J.; Ho, S.Y.W.; Phillips, M.J.; Rambaut, A. Relaxed Phylogenetics and Dating with Confidence. *PLoS Biol.* **2006**, *4*, e88. [CrossRef] [PubMed]
44. Rambaut, A. FigTree v1.4.4. 2016. Available online: http://tree.bio.ed.ac.uk/software/figtree/ (accessed on 26 August 2022).
45. Trifinopoulos, J.; Nguyen, L.T.; von Haeseler, A.; Minh, B.Q. W-IQ-TREE: A fast online phylogenetic tool for maximum likelihood analysis. *Nucleic Acids Res.* **2016**, *44*, W232–W235. [CrossRef]
46. Hoang, D.T.; Chernomor, O.; Von Haeseler, A.; Minh, B.Q.; Vinh, L.S. UFBoot2: Improving the ultrafast bootstrap approximation. *Mol. Biol. Evol.* **2017**, *35*, 518–522. [CrossRef]
47. Guindon, S.; Dufayard, J.F.; Lefort, V.; Anisimova, M.; Hordijk, W.; Gascuel, O. New algorithms and methods to estimate maximum-likelihood phylogenies: Assessing the performance of PhyML 3.0. *Syst. Biol.* **2010**, *59*, 307–321. [CrossRef]
48. MitoFish WEB Bench. Available online: http://mitofish.aori.u-tokyo.ac.jp/annotation/input.html (accessed on 1 January 2022).

49. Lowe, T.M.; Chan, P.p. tRNAscan-SE On-line: Search and Contextual Analysis of Transfer RNA Genes. *Nucleic Acids Res.* **2016**, *44*, W54–W57. [CrossRef]
50. Lowe, T.M.; Eddy, S.R. tRNAscan-SE: A Program for Improved Detection of Transfer RNA Genes in Genomic Sequence. *Nucleic Acids Res.* **1997**, *25*, 955–964. [CrossRef] [PubMed]
51. Saitoh, K.; Hayashizaki, K.; Yokoyama, Y.; Asahida, T.; Toyohara, H.; Yamashita, Y. Complete nucleotide sequence of Japanese flounder (Paralichthys olivaceus) mitochondrial genome: Structural properties and cue for resolving teleostean relationships. *J. Hered.* **2000**, *91*, 271–278. [CrossRef] [PubMed]
52. Letunic, I.; Bork, p. Interactive Tree Of Life (iTOL) v5: An online tool for phylogenetic tree display and annotation. *Nucleic Acids Res.* **2021**, *49*, W293–W296. [CrossRef]
53. Nei, M.; Gojobori, T. Simple methods for estimating the numbers of synonymous and nonsynonymous nucleotide substitutions. *Mol. Biol. Evol.* **1986**, *3*, 418–426. [PubMed]
54. Li, W.H. *Molecular Evolution*; Sinauer Ass.: Sunderland, MA, USA, 1997.
55. Fricke, R.; Eschmeyer, W.N.; Van der Laan, R. (Eds.) Eshmayer's Catalog of Fishes: Genera, Species, References. California Academy of Sciences. 2022. Available online: https://researcharchive.calacademy.org/research/ichthyology/catalog/fishcatmain.asp (accessed on 15 February 2022).
56. Clayton, D.A. Replication and transcription of vertebrate mitochondrial DNA. *Annu. Rev. Cell Biol.* **1991**, *7*, 453–478. [CrossRef]
57. Boore, J.L. Animal mitochondrial genomes. *Nucleic Acids Res.* **1999**, *27*, 1767–1780. [CrossRef]
58. Shi, W.; Kong, X.Y.; Wang, Z.M.; Jiang, J.X. Utility of tRNA genes from the complete mitochondrial genome of Psetta maxima for implying a possible sister-group relationship to the Pleuronectiformes. *Zool Stud.* **2011**, *50*, 665–681.
59. Redin, A.D.; Kartavtsev, Y.p. Phylogenetic relationships of flounders from the family Pleuronectidae (Ostichties: Pleuronectiformes) based on S16 rRNA gene. *Russ. J. Genet.* **2021**, *57*, 348–360. [CrossRef]
60. Shi, W.; Chen, S.; Kong, X.; Si, L.; Gong, L.; Zhang, Y.; Yu, H. Flatfish monophyly refereed by the relationship of Psettodes in Carangimorphariae. *BMC Genom.* **2018**, *19*, 400. [CrossRef]
61. Kartavtsev, Y.P.; Redin, A.D. Estimates of genetic introgression, gene tree reticulation, taxon divergence, and sustainability of DNA barcoding based on genetic molecular markers. *Biol. Bull. Rev.* **2019**, *9*, 275–294. [CrossRef]
62. Stoeckle, M.Y.; Thaler, D.S. Why should mitochondria define species? *Hum. Evol.* **2018**, *33*, 1–30. [CrossRef]
63. Kartavtsev, Y.p. Chapter 1: Analysis of sequence diversity at mitochondrial genes on different taxonomic levels. Applicability of DNA Based Distance Data in Genetics of Speciation and Phylogenetics. In *Genetic Diversity*; Mahoney, C.L., Springer, D.A., Eds.; Nova Science Publishers, Inc.: New York, NY, USA, 2009; pp. 1–50.
64. Kartavtsev, Y.p. Sequence divergence at mitochondrial genes in animals: Applicability of DNA data in genetics of speciation and molecular phylogenetics. *Mar. Genom.* **2011**, *49*, 71–81. [CrossRef] [PubMed]
65. Kartavtsev, Y.p. Sequence divergence at Co-1 and Cyt-b mtDNA on different taxonomic levels and genetics of speciation in animals. *Mitochondrial DNA* **2011**, *2*, 55–65. [CrossRef] [PubMed]
66. Kartavtsev, Y.p. Sequence Diversity at Cyt-b and Co-1 mtDNA Genes in Animal Taxa Proved Neo-Darwinism. *Phylogenet. Evol. Biol.* **2013**, *1*, 4. [CrossRef]
67. Zolotova, A.O.; Kartavtsev, Y.p. Analysis of sequence divergence in redfin (Cypriniformes, Cyprinidae, Tribolodon) based on mtDNA and nDNA markers with inferences in systematics and genetics of speciation. *Mitochondrial DNA Part A* **2018**, *29*, 975–992. [CrossRef]
68. Kartavtsev, Y.p. Sequence divergence provide a fit between molecular evolution, Neo-Darwinism and DNA barcoding. In *HydromediT2018, Proceedings of the 3rd International Congress on Applied Ichthyology & Aquatic Environment, Volos, Greece, 8–11 November 2018*; Berillis, P., Karapanagiotidis, I., Eds.; Department of Ichthyology and Aquatic Environment, School of Agricultural Sciences, University of Thessaly: Volos, Greece, 2018; pp. 463–479. ISBN 978-618-80242-5-0. Available online: www.hydromedit.gr (accessed on 10 March 2021).
69. Kartavtsev, Y.p. Some examples of the use of molecular markers for needs of basic biology and modern society. *Animals* **2021**, *11*, 1473. [CrossRef] [PubMed]
70. Nei, M.; Kumar, S. *Molecular Evolution and Phylogenetics*; Oxford University Press: Oxford, UK, 2000; 333p.
71. Kartavtsev, Y.P.; Lee, J.-S. Analysis of nucleotide diversity at genes *Cyt-b* and *Co-1* on population, species, and genera levels. *Russ. J. Genet.* **2006**, *42*, 341–362, (In Russian, Translated in English). [CrossRef]
72. Hedges, S.B.; Marin, J.; Suleski, M.; Madeline, P.; Kumar, S. Tree of life reveals clock-like speciation and diversification. *Mol. Biol. Evol.* **2015**, *32*, 835–845. [CrossRef]
73. Ratnasingham, S.; Hebert, P.D.N. BOLD: The Barcode of Life Data System. *Mol. Ecol. Notes* **2007**, *7*, 355–364. [CrossRef]
74. Naaum, A.M.; Hanner, R. Community engagement in seafood identification using DNA barcoding reveals market substitution in Canadian seafood. *DNA Barcodes* **2015**, *3*, 74–79. [CrossRef]
75. Nedunoori, A.; Turanov, S.V.; Kartavtsev, Y.p. Fish product mislabeling identified in the Russian far east using DNA barcoding. *Gene Rep.* **2017**, *8*, 144–149. [CrossRef]
76. Shneyer, V.S.; Rodionov, A. Plant DNA Barcodes. *Biol. Bull. Rev.* **2019**, *9*, 295–300. [CrossRef]
77. Kartavtsev, Y.P.; Sharina, S.N.; Goto, T.; Chichvarkhin, A.Y.; Balanov, A.A.; Vinnikov, K.A.; Ivankov, V.N.; Hanzawa, N. Cytochrome oxidase 1 gene sequence analysis in six flatfish species (Teleostei, Pleuronectidae) of Far East Russia with inferences in phylogeny and taxonomy. *Mitochondrial DNA* **2008**, *19*, 479–489. [CrossRef] [PubMed]

78. Atta, C.J.; Yuan, H.; Li, C.; Arcila, D.; Betancur-R, R.; Hughes, L.C.; Ortí, G.; Tornabene, L. Exon-capture data and locus screening provide new insights into the phylogeny of flatfishes (Pleuronectoidei). *Mol. Phylogenet. Evol.* **2022**, *166*, 107315. [CrossRef] [PubMed]
79. Hillis, D.M.; Mable, B.K.; Moritz, C. Application of molecular systematics: The state of the field and a look to the future. In *Molecular Systematics*; Hillis, D.M., Moritz, C., Mable, B., Eds.; Sinauer Associates, Inc.: Sunderland, MA, USA, 1996; pp. 515–543.
80. Xia, X.; Xie, Z.; Salemi, M.; Chen, L.; Wang, Y. An index of substitution saturation and its application. *Mol. Phylogenet. Evol.* **2003**, *26*, 1–7. [CrossRef]
81. Xia, X.; Lemey, p. Assessing substitution saturation with DAMBE. In *The Phylogenetic Handbook*; Lemey, P., Salemi, M., Vandamme, A.-M., Eds.; Cambridge University Press: Cambridge, UK, 2009; pp. 615–630.
82. Nelson, C.S.; Beck, J.N.; Wilson, K.A.; Pilcher, E.R.; Kapahi, P.; Brem, R.B. Cross-phenotype association tests uncover genes mediating nutrient response in Drosophila. *BMC Genom.* **2016**, *17*, 867. [CrossRef] [PubMed]
83. Luo, H.; Kong, X.; Chen, S.; Shi, W. Mechanisms of gene rearrangement in 13 bothids based on comparison with a newly completed mitogenome of the threespot flounder, Grammatobothus polyophthalmus (Pleuronectiformes: Bothidae. *BMC Genom.* **2019**, *20*, 792. [CrossRef]
84. Lü, Z.; Gong, L.; Ren, Y.; Chen, Y.; Wang, Z.; Liu, L.; Li, H.; Chen, X.; Li, Z.; Luo, H.; et al. Large-scale sequencing of flatfish genomes provides insights into the polyphyletic origin of their specialized body plan. *Nat. Genet.* **2021**, *53*, 742–751. [CrossRef]

Article

Revisiting the Diversity of *Barbonymus* (Cypriniformes, Cyprinidae) in Sundaland Using DNA-Based Species Delimitation Methods

Hadi Dahruddin [1], Arni Sholihah [2], Tedjo Sukmono [3], Sopian Sauri [1], Ujang Nurhaman [1], Daisy Wowor [1], Dirk Steinke [4] and Nicolas Hubert [5,*]

[1] Museum Zoologicum Bogoriense, Research Center for Biology, Indonesian Institute of Sciences (LIPI), Gd. Widyasatwaloka, Jalan Raya Jakarta Bogor Km 46, Cibinong 16911, Indonesia; dahruddinhadi@gmail.com (H.D.); sopiansr@gmail.com (S.S.); unurhaman001@gmail.com (U.N.); dwowor@gmail.com (D.W.)

[2] Instut Teknologi Bandung, School of Life Sciences and Technology, Bandung 40132, Indonesia; rasborinae@gmail.com

[3] Department of Biology, Universitas Jambi, Jalan Lintas Jambi—Muara Bulian Km 15, Jambi 36122, Indonesia; sukmonotedjo@gmail.com

[4] Department of Integrative Biology, Centre for Biodiversity Genomics, 50 Stone Rd E, Guelph, ON N1G 2W1, Canada; dsteinke@uoguelph.ca

[5] Institut des Sciences de l'Evolution Montpellier, IRD, Univ Montpellier, CNRS, EPHE, Place Eugène Bataillon, CEDEX 05, 34095 Montpellier, France

* Correspondence: nicolas.hubert@ird.fr

Citation: Dahruddin, H.; Sholihah, A.; Sukmono, T.; Sauri, S.; Nurhaman, U.; Wowor, D.; Steinke, D.; Hubert, N. Revisiting the Diversity of *Barbonymus* (Cypriniformes, Cyprinidae) in Sundaland Using DNA-Based Species Delimitation Methods. *Diversity* **2021**, *13*, 283. https://doi.org/10.3390/d13070283

Academic Editors: Manuel Elías-Gutiérrez and Michael Wink

Received: 11 May 2021
Accepted: 17 June 2021
Published: 23 June 2021

Publisher's Note: MDPI stays neutral with regard to jurisdictional claims in published maps and institutional affiliations.

Abstract: Biodiversity hotspots often suffer from a lack of taxonomic knowledge, particularly those in tropical regions. However, accurate taxonomic knowledge is needed to support sustainable management of biodiversity, especially when it is harvested for human sustenance. Sundaland, the biodiversity hotspot encompassing the islands of Java, Sumatra, Borneo, and Peninsular Malaysia, is one of those. With more than 900 species, its freshwater ichthyofauna includes a large number of medium- to large-size species, which are targeted by inland fisheries. Stock assessment requires accurate taxonomy; however, several species groups targeted by inland fisheries are still poorly known. One of those cases is the cyprinid genus *Barbonymus*. For this study, we assembled a consolidated DNA barcode reference library for *Barbonymus* spp. of Sundaland, consisting of mined sequences from BOLD, as well as newly generated sequences for hitherto under-sampled islands such as Borneo. A total of 173 sequences were analyzed using several DNA-based species delimitation methods. We unambiguously detected a total of 6 Molecular Operational Taxonomic Units (MOTUs) and were able to resolve several conflicting assignments to the species level. Furthermore, we clarified the identity of MOTUs occurring in Java.

Keywords: Southeast Asia; inland fisheries; type locality; genetic diversity; phylogeography

1. Introduction

Sundaland, comprising the islands of Java, Bali, Sumatra, Borneo, and peninsular Malaysia, constitutes one of the world's largest biodiversity hotspots [1,2]. With circa 900 freshwater fish species, half of which are endemic, the ichthyofauna of this biogeographical region is particularly rich, with a density of 0.8 species per km^2, a value twice as large as that observed in Brazil and the Democratic Republic of Congo [3]. This large diversity is critically threatened, mostly due to the alarming rate of deforestation over the past few decades [4–6], in conjunction with pollution [7] and watershed fragmentation through the development of dams for irrigation and hydroelectric power [8]. Furthermore, the diversity of freshwater fishes in Sundaland is still not sufficiently understood [9], hampering conservation efforts.

Freshwater fishes constitute a major source of animal protein in Southeast Asia, where inland fisheries rank among the world's most productive, with Indonesia repeatedly at the top [10–13]. Sundaland hosts a substantial amount of medium to large-size species, all above 30 cm in maximum standard length [3], which are targeted by inland fisheries [11]. Common targets include genera such as snakeheads (*Channa* spp.), catfishes (*Ompok* spp., *Hemibagrus* spp.), and various cyprinid genera including *Barbonymus* spp. and *Leptobarbus* spp. [11,14]. Although common, the taxonomy of these genera is poorly understood. Species boundaries are unclear, and the diversity for a number of genera has likely been underestimated [15–17].

A good example for this is the cyprinid genus *Barbonymus*, for which species numbers range from 5 in Fishbase [18] to 10 in Eschmeyer's Catalog of Fishes [19], the latter following the checklist of Southeast Asian freshwater fishes by Kottelat [17] (Table 1). All *Barbonymus* species occur in Sundaland, except *B. altus* (Günther, 1868), which was described from the Chao Phraya River in Thailand (Text S1 in Supplementary Material). Of the nine species occurring in Sundaland, three, *B. sunieri* (Weber and de Beaufort, 1916), *B. strigatus* (Boulenger, 1894), and *B. platysoma* (Bleeker, 1855), have been described based on a single specimen, and none of them have ever been observed since their original description [20]. Among the six remaining species, *B. balleroides* (Valenciennes, in Cuvier and Valenciennes, 1842), *B. gonionotus* (Bleeker, 1849) and *B. schwanefeldii* (Bleeker, 1864) are widespread in watersheds flowing to the Java Sea, where they are frequently targeted by inland fisheries. *Barbonymus collingwoodii* (Günther, 1868) and *B. mahakkamensis* (Ahl, 1922) are endemic to North and East Borneo, respectively, and *B. belinka* (Bleeker, 1860) is an endemic species of the west coast of Sumatra.

Table 1. List of available nominal species of *Barbonymus* including species names in original descriptions, authors, species names in Eschmeyer's Catalog of Fishes and Fishbase, and current status following [17].

Original Description	Authors	Eschmeyer Catalog of Fished	Fishbase	Status
Barbus altus	Günther 1868	*Barbonymus altus*	*Barbonymus altus*	Valid as *Barbonymus altus* (Günther 1868)
Barbus amblycephalus	Bleeker 1855	*Barbonymus balleroides*	*Barbonymus balleroides*	Valid as *Barbonymus balleroides* (Valenciennes 1842)
Barbus balleroides	Valenciennes 1842	*Barbonymus balleroides*	*Barbonymus balleroides*	Valid as *Barbonymus balleroides* (Valenciennes 1842)
Barbus boulengerii	Popta 1905	*Barbonymus collingwoodii*	*Barbonymus collingwoodii*	Valid as *Barbonymus collingwoodii* (Günther 1868)
Barbus bramoides	Valenciennes 1842	*Barbonymus balleroides*	*Barbonymus balleroides*	Valid as *Barbonymus balleroides* (Valenciennes 1842)
Barbus erythropterus	Bleeker 1849	*Barbonymus balleroides*	*Barbonymus balleroides*	Valid as *Barbonymus balleroides* (Valenciennes 1842)
Barbus foxi	Fowler 1937	*Barbonymus altus*	*Barbonymus altus*	Valid as *Barbonymus altus* (Günther 1868)
Barbus gonionotus	Bleeker 1849	*Barbonymus gonionotus*	*Barbonymus gonionotus*	Valid as *Barbonymus gonionotus* (Bleeker 1849)
Barbus hypsylonotus	Valenciennes 1842	*Barbonymus balleroides*	*Barbonymus balleroides*	Valid as *Barbonymus balleroides* (Valenciennes 1842)
Barbus javanicus	Bleeker 1855	*Barbonymus gonionotus*	*Barbonymus gonionotus*	Valid as *Barbonymus gonionotus* (Bleeker 1849)
Barbus koilometopon	Bleeker 1857	*Barbonymus gonionotus*	*Barbonymus gonionotus*	Valid as *Barbonymus gonionotus* (Bleeker 1849)
Barbus macrophthalmus	Bleeker 1855	*Barbonymus balleroides*	*Barbonymus balleroides*	Valid as *Barbonymus balleroides* (Valenciennes 1842)
Barbus mahakkamensis	Ahl, 1922	*Barbonymus mahakkamensis*	*Barbodes mahakkamensis*	Valid as *Barbus mahakkamensis* (Ahl 1922)
Barbus platysoma	Bleeker 1855	*Barbonymus platysoma*	*Barbodes platysoma*	Valid as *Barbodes platysoma* (Bleeker 1855)
Barbus schwanefeldi rubra	Vaillant 1902	*Barbonymus schwanefeldii*	*Barbonymus schwanenfeldii*	Valid as *Barbonymus schwanefeldii* (Bleeker 1854)
Barbus schwanenfeldii	Bleeker 1854	*Barbonymus schwanefeldii*	*Barbonymus schwanenfeldii*	Valid as *Barbonymus schwanefeldii* (Bleeker 1854)
Barbus strigatus	Boulenger 1894	*Barbonymus strigatus*	*Barbodes strigatus*	Valid as *Barbonymus strigatus* (Boulenger 1894)
Barbus wadon	Bleeker 1849	*Barbonymus balleroides*	*Barbonymus balleroides*	Valid as *Barbonymus balleroides* (Valenciennes 1842)
Puntius jolamarki	Smith 1934	*Barbonymus gonionotus*	*Barbonymus gonionotus*	Valid as *Barbonymus gonionotus* (Bleeker 1849)
Puntius viehoeveri	Fowler 1943	*Barbonymus gonionotus*	*Barbonymus gonionotus*	Valid as *Barbonymus gonionotus* (Bleeker 1849)
Systomus belinka	Bleeker 1860	*Barbonymus belinka*	*Puntius belinka*	Valid as *Barbonymus belinka* (Bleeker 1860)

DNA barcoding, the use of cytochrome oxidase I (COI) as a species tag for automated identification, opened new perspectives for the characterization of Sundaland's ichthyofauna by helping to clarify taxonomic confusion within several groups [15,21,22], by identifying discrepancies in historical species records [9] and by detecting a substantial amount of morphologically similar, yet highly divergent lineages (i.e., cryptic diversity) within numerous species [9,15,21–28]. Several molecular studies that aimed at characterizing patterns of genetic diversity in *Barbonymus* led to conflicting species identities associated with sequences submitted to international repositories [9,16,29–36].

As part of an ongoing project that seeks to build a DNA barcode reference library for the ichthyofauna of Sundaland [9,22], we generated new barcode records for *Barbonymus* species, which, together with previously published sequences, cover the diversity of the genus in the region. The objective of the present study is to re-examine *Barbonymus* species boundaries and their distribution ranges using DNA-based species delimitation methods. By including DNA barcode records of specimens collected near type localities, we are also reappraising published *Barbonymus* sequences.

2. Materials and Methods

2.1. Sampling and Collection Management

Specimens were captured using various methods including electrofishing, seine nets, cast nets and gill nets, as well as by visiting fish markets in Sundaland (Figure 1; DS-BARBONYM, dx.doi.org/10.5883/DS-BARBONYM, accessed on 15 January 2021). Specimens were photographed and individually labeled, and voucher specimens were preserved in a 5% formalin solution. A fin clip or a muscle biopsy was taken for each specimen and fixed in a 96% ethanol solution for further genetic analyses. Both tissue and voucher specimens were deposited in the national collections at the Museum Zoologicum Bogoriense (MZB) in the Research Centre for Biology (RCB) of the Indonesian Institute of Sciences (LIPI).

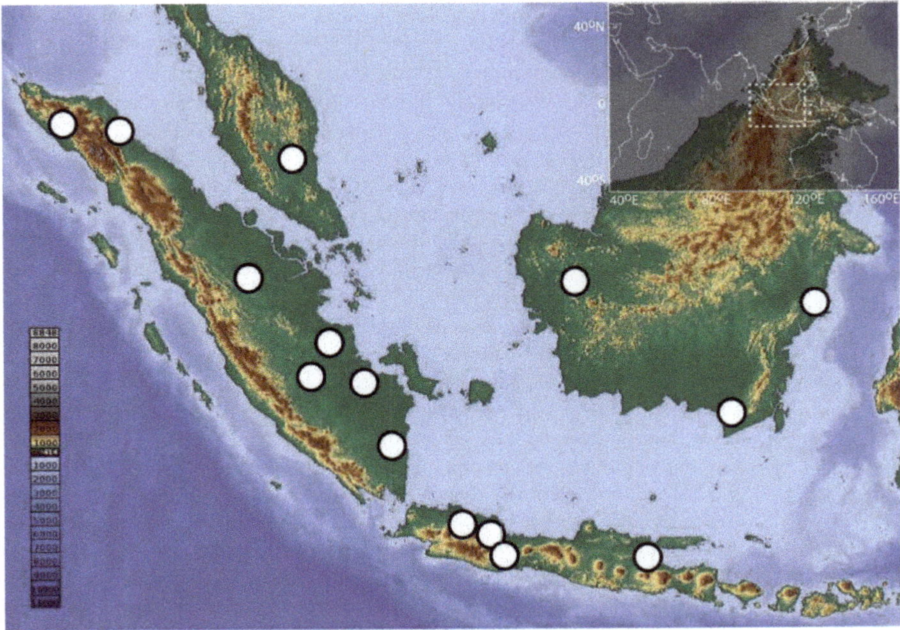

Figure 1. Collecting sites in Sundaland for the 173 DNA barcode records of *Barbonymus* analyzed in this study.

2.2. Sequencing and International Repositories

Genomic DNA was extracted from the muscle tissue samples using a Qiagen DNeasy 96 tissue extraction kit following manufacturer's specifications. A 651 bp segment from the 5′ region of the cytochrome oxidase I gene (COI) was amplified using the primer cocktail C_FishF1t1/C_FishR1t1 [37]. Polymerase Chain Reaction (PCR) amplifications were done on a Veriti 96-well Fast thermocycler (ABI—Applied Biosystems) with a final volume of 10.0 µL containing 5.0 µL buffer 2X, 3.3 µL ultrapure water, 1.0 µL each primer (10 µM), 0.2 µL enzyme Phire Hot Start II DNA polymerase (5U), and 0.5 µL of DNA template (~50 ng). The following thermocycler regime was used: initial denaturation at 98 °C for 5 min followed by 30 cycles denaturation at 98 °C for 5 s, annealing at 56 °C for 20 s, and extension at 72 °C for 30 s, followed by a final extension step at 72 °C for 5 min. PCR products were purified with ExoSap-IT (USB Corporation, Cleveland, OH, USA) and sequenced in both directions. Sequencing reactions were performed at the Centre for Biodiversity Genomics, University of Guelph, Canada, using the BigDye Terminator v3.1 Cycle Sequencing Ready Reaction kit following standard protocols described in [38]. Sequencing was performed on an ABI 3730xl capillary sequencer (Applied Biosystems). Sequences and collateral information were deposited on BOLD [39] and are available as a public data set (dx.doi.org/10.5883/DS-BARBONYM, accessed on 15 January 2021, Table S1).

2.3. Genetic Species Delimitation and Phylogenetic Inferences

Several methods for species delineation based on DNA sequences have been proposed [40–43]. Each of these have different properties, particularly when dealing with singletons (i.e., lineages represented by a single sequence) or heterogeneous speciation rates among lineages [44]. A combination of different approaches is increasingly used to overcome potential pitfalls arising from uneven sampling [22,45–48]. We used six different sequence-based methods of species delimitation to identify the Molecular Operational Taxonomic Unit (MOTU): (1) Refined Single Linkage (RESL) as implemented in BOLD and used to generate Barcode Index Numbers (BIN) [42], (2) Automatic Barcode Gap Discovery (ABGD) [41], (3) Poisson Tree Process (PTP) in its single (sPTP) and multiple rates version (mPTP) as implemented in the stand-alone software mptp_0.2.3 [43,49], and (4) General Mixed Yule-Coalescent (GMYC) in its single (sGMYC) and multiple threshold version (mGMYC) as implemented in the R package Splits 1.0–19 [50].

Both the mPTP algorithm and the GYMC use phylogenetic trees as input files. We reconstructed a maximum likelihood (ML) tree for the former using RAxML [51] based on a GTR + I + Γ substitution model. For the GYMC algorithm, we calculated an ultrametric, fully resolved tree using the Bayesian approach implemented in BEAST 2.6.2 [52]. Sequences were collapsed into haplotypes prior to reconstructing the ultrametric tree using RAxML, and Bayesian reconstruction was based on a strict-clock prior of 1.2% per million year [53]. Two Markov chains of 20 million each were run independently using Yule pure birth and GTR + I + Γ substitution models. Trees were sampled every 5000 states, after an initial burn-in period of 5 million. Both runs were combined with trees resampled every 20,000 states using LogCombiner 2.6.2, and the maximum credibility tree was constructed using TreeAnnotator 2.6.2 [52].

A final COI gene tree was reconstructed using the SpeciesTreeUCLN algorithm of the StarBEAST2 package [54]. This approach implements a mixed-model including a coalescent component within species and a diversification component between species that allows accounting for variations of substitution rates within and between species [55]. SpeciesTreeUCLN jointly reconstructs gene trees and species trees and therefore requires the designation of species, which were determined using the consensus of our species delimitation analyses. The SpeciesTreeUCLN analysis was performed with the same parameters as mentioned above.

Kimura 2-parameter (K2P) [56] pairwise genetic distances were calculated using the R package Ape 5.4 [57]. Maximum intraspecific and nearest neighbor genetic distances were calculated from the pairwise K2P distance matrix using the R package Spider 1.5 [58]. Haplotype extraction and haplotype network reconstruction were performed for the most widespread species using the R package pegas 1.0 [59].

3. Results

The total of 173 DNA barcodes used for this study comprised 154 sequences mined from BOLD and 19 sequences generated for *Barbonymus* specimens originating from Sumatra and Borneo. The newly generated sequences represent the first DNA barcode records of *Barbonymus* for Borneo. All the sequences were above 500 bp in length and no stop codons were detected, suggesting that the sequences collected represent functional coding regions. DNA-based species delimitation methods resulted in congruent delimitation schemes with 6 MOTUs for mPTP, sPTP, ABGD, RESL, and sGMYC (Figure 2; Table S1). However, mGMYC resulted in the delimitation of 31 highly incongruent MOTUs. Therefore, the mGMYC partitioning scheme was discarded. The final consensus scheme consisted of six MOTUs (Figure 2; Table 2) showing a distinct barcoding gap, which is defined as the lack of overlap between maximum intraspecific and minimum interspecific genetic distance. Maximum intraspecific distances ranged from 0 (BOLD:ADN2907, BOLD:AED2516) to 0.018 (BOLD:AAD1940) (Table 1). Minimum interspecific distances ranged from 0.026 for the two *Barbonymus* MOTUs (BOLD:AAE2136 and BOLD:AEB4313) to 0.08 for BOLD:AAD1940 (Table 1).

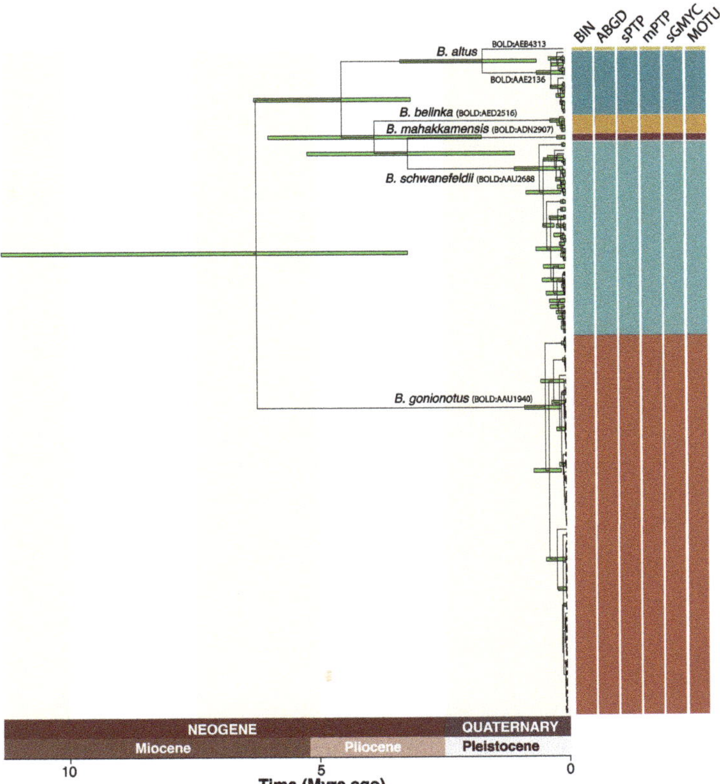

Figure 2. Mitochondrial gene tree for the 173 DNA barcodes of *Barbonymus* inferred with SpeciesTreeUCLN, including 95% HPD interval for node age estimates, genetic species delimitation results for five methods (mGMYC discarded) and their 50% consensus, BOLD Barcode Index Numbers (BIN) for each MOTU, and revised species names.

Table 2. Summary of genetic distances and MOTUs including species names, number of individuals analyzed, BOLD Barcode Index Number (BIN), maximum intraspecific and minimum interspecific K2P genetic distances.

			K2P Genetic Distance	
Species	N	BIN	Max. Intraspecific	Min. Interspecific
Barbonymus altus	17	BOLD:AAE2136	0.010	0.026
	1	BOLD:AEB4313	-	0.026
Barbonymus belinka	5	BOLD:AED2516	0.000	0.069
Barbonymus gonionotus	98	BOLD:AAD1940	0.018	0.080
Barbonymus mahakkamensis	2	BOLD:ADN2907	0.000	0.050
Barbonymus schwanefeldii	50	BOLD:AAU0688	0.013	0.050

Conflicting species-level assignments were detected, particularly for previously published records from Java [9], where BIN BOLD:ADD1940 and BOLD:AAU0688, initially assigned to *B. balleroides* and *B. gonionotus*, match *B. gonionotus* and *B. schwanefeldii*, respectively (Table S1). These results extend the occurrence of *B. schwanefeldii* to Java Island and question the occurrence of *B. balleroides* in Java (Figure 3). Along the same line, the BIN BOLD:AED2516, initially assigned to *B. gonionotus* and highlighted as a potentially new taxon [16], likely corresponds to *B. belinka*, because *B. collingwoodii* is endemic to North Borneo, *B. mahakkamensis* corresponds to a distinct lineage (BOLD:ADN2907, Figure 2) restricted to East Borneo (Figure 3), and the occurrence of *B. balleroides* in Sumatra is uncertain (Text S1).

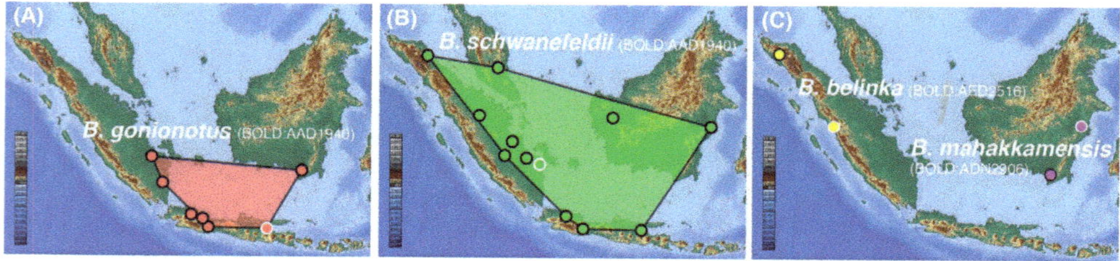

Figure 3. Revised range distribution and type localities (white circle) for (**A**) *B. gonionotus* (only records in Sundaland with geocoordinates are shown, its range distribution expands northward to Thailand and India), (**B**) *B. schwanefeldii* (only records in Sundaland are shown), (**C**) *B. belinka*, and *B. mahakkamensis*. *B. altus* occurs in mainland Asia.

The Bayesian gene tree based on delimited MOTUs suggests close phylogenetic affinities between the Sundaland MOTUs corresponding to *B. belinka*, *B. mahakkamensis*, and *B. schwanefeldii*, with a Most Recent Common Ancestor (MRCA) dated at 3.8 Ma (Figure 2). This group is more closely related to MOTUs of *B. altus* from mainland Asia, with a MRCA dated at 4.5 Ma, than the MOTU assigned to *B. gonionotus*, which diverged from other *Barbonymus* MOTUs about 6 Ma.

Intraspecific phylogeographic patterns were further explored for *B. gonionotus* and *B. schwanefeldii* using sequences with revised species assignment (Figure 4). A total of 40 haplotypes was observed for both *B. gonionotus* and *B. schwanefeldii*. Haplotype networks were markedly different for both species, with a star-like structure for *B. schwanefeldii* (Figure 4B) and a more scattered network for *B. gonionotus* (Figure 4A). Most islands of Sundaland host haplotypes scattered across networks for both species; however, mainland Asia is much more represented in the haplotype network of *B. schwanefeldii* (Figure 4B) than in *B. gonionotus* (Figure 4A).

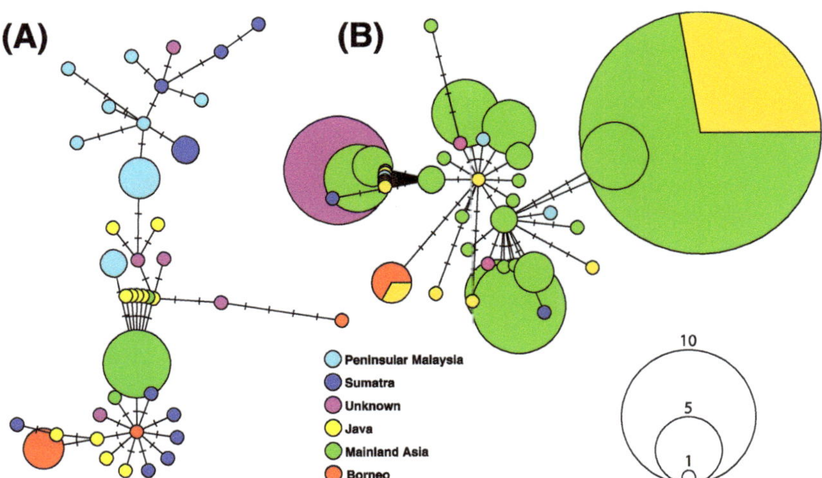

Figure 4. Haplotype networks reconstructed based on 98 and 50 DNA barcodes available for *B. gonionotus* (**A**) and *B. schwanefeldii* (**B**), respectively. Numbers of changes are indicated by small segments on links. Circles represent haplotypes; the size of a circle is proportional to the haplotype frequency.

4. Discussion

The present study provides an update of the *Barbonymus* diversity in Sundaland through the aggregation of newly generated and recently published DNA barcode records, resulting in a reference library consisting of 173 sequences largely distributed across mainland Asia and Sundaland (the haplotype MK978151 was observed in five individuals in [16], resulting in a dataset in BOLD consisting of 169 sequences). DNA-based species delimitation methods largely agreed on the delineation of six MOTUs, except mGMYC with a much higher estimate, a fact that was already reported in other studies [22,27,46,48,60]. Aside from this noticeable exception, methods were concordant in their delimitations, and a barcoding gap was observed between all six MOTUs, with maximum K2P distances mostly <1% within MOTUs, thereby falling into previously observed ranges for Cypriniformes in Sundaland [9,16,21,22]. Along the same line, inferred ages of divergence among MOTUs are consistent with previous age estimates among Sundaland freshwater fishes, with most species originating during the last 5 Ma [21,22,26,28].

Discrepancies between BIN and species designation were observed within three MOTUs, corresponding to (1) BOLD:ADD1940 with sequences assigned to both *B. gonionotus* and *B. balleroides*, (2) BOLD:AAU0688 with sequences assigned to both *B. gonionotus* and *B. schwanefeldii*, and (3) BOLD:AED2516 with sequences assigned initially to *B. gonionotus*. Most discrepancies could be related to *B. gonionotus*, which appears scattered across these three MOTUs.

Most sequences of BOLD:ADD1940 from mainland Asia were attributed to *B. gonionotus* [29,32,33,36], while sequences from Sundaland were called *B. balleroides* [9,16]. Sequences from BOLD:AAU0688 from Java were assigned to *B. gonionotus*, while previously published and newly generated sequences from Sumatra and Borneo were assigned to *B. schwanefeldii*. These results suggest multiple misidentifications or large-scale introgressive hybridization between *B. gonionotus* and *B. schwanefeldii*, as the former is not reported to occur in Java [3,18–20]. While hybridization and introgression have been previously reported for Cypriniformes [61–63], such large-scale mitochondrial introgressions have never been reported for Sundaland fishes [22,23,25]. Along the same line, shared polymorphism through recent common ancestry might be responsible for these discrepancies; however, as inferred by this study, *B. gonionotus* is the most divergent species in Sundaland, and ancestral polymorphisms are usually detected across whole species distribution ranges [64], which is not the case here. The most likely hypothesis is that *B. schwanefeldii* occurs in Java, extending its range of distribution over all islands of Sundaland. Cases of translocation of *B. schwanefeldii* outside its distribution range have been reported, e.g., to Papua New Guinea [65], as a strategy to improve local fisheries by introducing species with fast growth rates. Numerous introductions have been reported in Java of both Sundaland species, non-native to Java, and exotic species [9]. This makes an introduction of *B. schwanefeldii* through translocations from either Sumatra or Borneo populations likely. However, the haplotype network of BOLD:AAU0688 indicates that Java specimens consist of mostly private haplotypes (Figure 4). If the *B. schwanefeldii* occurrence in Java resulted from recent introductions, most haplotypes would be shared with other populations in Sumatra and Borneo, which is not the case here. Furthermore, cases of MOTUs widely distributed in Sundaland were previously detected [15,28] and most BOLD:AAU0688 specimens from Java initially identified as *B. gonionotus* are juveniles and sub-adults, which makes misidentifications likely. These results suggest that Java populations previously assigned to mboxemphB. gonionotus actually correspond to *B. schwanefeldii* (Figure 3).

The present study further questions the occurrence of *B. balleroides* in Java, which was comprehensively sampled recently [9,21], resulting in the discovery of two distinct *Barbonymus* MOTUs. The MOTU BOLD:AAD1940 was sampled at the type locality of *B. gonionotus* in Surabaya (Text S1, Figure 3), and all mainland Asia samples of this MOTU were previously assigned to *B. gonionotus*. This result suggests that *Barbonymus* populations previously assigned to *B. balleroides* [9,16,20] actually belong to *B. gonionotus* (Figure 4).

The MOTU BOLD:AED2516 likely corresponds to *B. belinka*. *Barbonymus sunieri*, *B. strigatus*, and *B. platysoma* have been described based on single specimens from either Java or North Borneo, none of which have been observed for decades. In addition, *B. collingwoodii* is endemic to North Borneo and *B. mahakkamensis* belongs to a distinct lineage (BOLD:ADN2907, Figure 2) restricted to East Borneo (Figure 3). The type locality of *B. balleroides* is unknown, but the holotype refers to the Indo-Australian region (Text S1). However, synonymies suggest ample distribution of *B. balleroides* in Java and Borneo, though its occurrence in Sumatra remains to be confirmed. Thus, *B. belinka* is the only name available for this MOTU in Sumatra, considering known distribution ranges of recently observed *Barbonymus* species.

A single case of unrecognized diversity is detected within *B. altus*, with two MOTUs detected including BOLD:AAE2136 and BOLD:AEB4313. The MOTU BOLD:AEB4313 corresponds to a singleton mined from GenBank and originating from a specimen sampled in the Mekong River (Vietnam), which is consistent with the identification as *B. altus* considering its type locality is the neighboring Chao Phraya River in Thailand. Raw electropherograms are not available for this sequence, which makes its quality assessment impossible; however, its placement within *B. altus* seems to confirm that this singleton does not result from poor sequence quality.

5. Conclusions

The present study confirms the utility of DNA barcoding for clarifying species identity and distribution ranges in cases of conflicting records. This is particularly evident in Java, where large conflicts between historical records and recent reappraisals based on DNA sequences were recently detected [9,21,66], suggesting large knowledge gaps. The lack of historical records for *B. schwanefeldii* in Java seems to indicate perpetuated misidentifications of *Barbonymus* populations. Individual translocations of *B. schwanefeldii* from Sumatra and Borneo to Java alone fail to account for its new occurrence. Our study suggests that both MOTUs reported from Java correspond to *B. gonionotus* and *B. schwanefeldii* and highlights the degree to which *Barbonymus* species are morphologically similar and difficult to distinguish based on meristic counts alone. However, DNA barcodes are clustered into MOTUs, which are unambiguously captured by most DNA-based species delimitation methods. Our revised DNA barcode reference library opens new perspectives for the management of the inland fisheries of Sundaland by enabling fast and reliable species level identification of *Barbonymus* spp. Considering the importance of *Barbonymus* spp. in local artisanal fisheries, and the difficulty of performing stock assessments at species level due to overlapping meristic counts among species, this library can be readily used as an alternative.

Supplementary Materials: The following are available online at https://www.mdpi.com/article/10.3390/d13070283/s1, Table S1: Results of the genetic species delimitation analyses. Text S1: Nomenclature of the ten nominal species of *Barbonymus* following Eschmeyer et al., 2018.

Author Contributions: Conceptualization, H.D., A.S., T.S., D.W., D.S. and N.H.; methodology, A.S., H.D., D.S. and N.H.; software, A.S. and N.H.; validation, H.D., A.S., T.S., D.W., D.S. and N.H.; formal analysis, H.D., A.S. and N.H.; investigation, H.D., A.S., T.S., D.W., D.S. and N.H.; resources, H.D., A.S., T.S., S.S., U.N., D.S. and N.H.; data curation, H.D., D.S. and N.H.; writing—original draft preparation, H.D., A.S., D.S. and N.H.; writing—review and editing, T.S., D.W. and S.S.; visualization, H.D., A.S., T.S., D.W., D.S. and N.H.; supervision, D.W., D.S. and N.H.; project administration, N.H.; funding acquisition, D.S., D.W. and N.H. All authors have read and agreed to the published version of the manuscript.

Funding: This research was funded by the CFREF to the University of Guelph's Food from Thought program, by the "Institut de Recherche pour le Développement" through annual funding and "Fonds d'Amorçage" (226F2ABIOS), by the French Embassy in Indonesia through the program "Science et Impact", by Campus France through a Bio-Asia grant (BIOSHOT project) and by ANR (ANR-17-ASIE-0006).

Institutional Review Board Statement: Not applicable.

Informed Consent Statement: Not applicable.

Data Availability Statement: The data presented in this study are openly available in BOLD at [dx.doi.org/10.5883/DS-BARBONYM], accessed on 15 January 2021, reference number DS-BARBONYM.

Acknowledgments: We wish to thank Siti Nuramaliati Prijono, Witjaksono, Mohammad Irham, Marlina Adriyani, Ruliyana Susanti, Hari Sutrisno, Yuli Fitriana, Muhamad Syamsul Arifin Zein, late Sri Sulandari, and late Renny K. Hadiaty, at Research Centre for Biology (RCB-LIPI), as well as Joel Le Bail and Nicolas Gascoin at the French embassy in Jakarta for their continuous support. We are thankful to Daisy Wowor and Ujang Nurhaman at RCB-LIPI, and Sumanta and Bambang Dwisusilo at IRD Jakarta for their help during the field sampling. Finally, we acknowledge the reviewers who helped to improve this manuscript. A permit to collect fish was awarded to Nicolas Hubert (7/TKPIPA/FRP/SM/VII/2012, 68/EXT/SIP/FRP/SM/VIII/2013, 41/EXT/SIP/FRP/SM/VIII/2014, 361/SIP/FRP/E5/Dit.KI/IX/2015, 50/EXT/SIP/FRP/E5/Dit.KI/IX/2016, 45/EXT/SIP/FRP/E5/Dit.KI/VIII/2017, 392/SIP/FRP/E5/Dit.KI/XI/2018, and 200/E5/E5.4/SIP/2019). No experimentation was conducted on live specimens during this study. This publication has ISEM number 2021-135 SUD.

Conflicts of Interest: The authors declare no conflict of interest. The funders had no role in the design of the study; in the collection, analyses, or interpretation of data; in the writing of the manuscript, or in the decision to publish the results.

References

1. Myers, N.; Mittermeier, R.A.; Mittermeier, C.G.; da Fonseca, G.A.B.; Kent, F. Biodiversity hotspots for conservation priorities. *Nature* **2000**, *403*, 853–858. [CrossRef]
2. Mittermeier, R.A.; Turner, W.R.; Larsen, F.W.; Brooks, T.M.; Gascon, C. Global biodiversity conservation: The critical role of hotspots. In *Biodiversity Hotspots*; Springer: Berlin/Heidelberg, Germany, 2011; pp. 3–22.
3. Hubert, N.; Wibowo, A.; Busson, F.; Caruso, D.; Sulandari, S.; Nafiqoh, N.; Pouyaud, L.; Rüber, L.; Avarre, J.C.; Herder, F.; et al. DNA barcoding Indonesian freshwater fishes: Challenges and prospects. *DNA Barcodes* **2015**, *3*, 144–169. [CrossRef]
4. Imai, N.; Furukawa, T.; Tsujino, R.; Kitamura, S.; Yumoto, T. Factors affecting forest area change in Southeast Asia during 1980–2010. *PLoS ONE* **2018**, *13*, e0197391. [CrossRef]
5. Laumonier, Y.; Uryu, Y.; Stüwe, M.; Budiman, A.; Setiabudi, B.; Hadian, O. Eco-floristic sectors and deforestation threats in Sumatra: Identifying new conservation area network priorities for ecosystem-based land use planning. *Biodivers. Conserv.* **2010**, *19*, 1153–1174. [CrossRef]
6. Gaveau, D.L.A.; Sloan, S.; Molidena, E.; Yaen, H.; Sheil, D.; Abram, N.K.; Ancrenaz, M.; Nasi, R.; Quinones, M.; Wielaard, N. Four decades of forest persistence, clearance and logging on Borneo. *PLoS ONE* **2014**, *9*, e101654. [CrossRef] [PubMed]

7. Garg, T.; Hamilton, S.E.; Hochard, J.P.; Kresch, E.P.; Talbot, J. (Not so) gently down the stream: River pollution and health in Indonesia. *J. Environ. Econ. Manag.* **2018**, *92*, 35–53. [CrossRef]
8. Mulligan, M.; van Soesbergen, A.; Sáenz, L. GOODD, a global dataset of more than 38,000 georeferenced dams. *Sci. Data* **2020**, *7*, 1–8. [CrossRef] [PubMed]
9. Dahruddin, H.; Hutama, A.; Busson, F.; Sauri, S.; Hanner, R.; Keith, P.; Hadiaty, R.; Hubert, N. Revisiting the ichthyodiversity of Java and Bali through DNA barcodes: Taxonomic coverage, identification accuracy, cryptic diversity and identification of exotic species. *Mol. Ecol. Resour.* **2017**, *17*, 288–299. [CrossRef]
10. FAO. *Des Pêches et de L'aquaculture*; FAO: Rome, Italy, 2018; ISBN 9789251306925.
11. Koeshendrajana, S.; Cacho, O.J. *Management Options for the Inland Fisheries Resource in South Sumatra, Indonesia: I Bioeconomic Model*; Working Papers 12932; University of New England: Armidale, Australia, 2001.
12. Coates, D. Inland capture fishery statistics of Southeast Asia: Current status and information needs. *RAP Publ.* **2002**, *11*, 114.
13. Welcomme, R.L.; Baird, I.G.; Dudgeon, D.; Halls, A.; Lamberts, D.; Mustafa, M.G. *Fisheries of the Rivers of Southeast Asia*; Jon Wiley & Sons: Hoboken, NJ, USA, 2016; pp. 363–376.
14. Muthmainnah, D.; Makmur, H.R.; Sawestri, S.; Supriyadi, F.; Fatah, K. *The Features of Inland Fisheries in Southeast Asia: Inland Capture Fisheries and Its Status*; Inland Fishery Resources Development and Management Department, Southeast: Samut Prakan, Thailand, 2019; ISBN 6024408161.
15. Conte-Grand, C.; Britz, R.; Dahanukar, N.; Raghavan, R.; Pethiyagoda, R.; Tan, H.H.; Hadiaty, R.K.; Yaakob, N.S.; Rüber, L. Barcoding snakeheads (Teleostei, Channidae) revisited: Discovering greater species diversity and resolving perpetuated taxonomic confusions. *PLoS ONE* **2017**, *12*, e0184017. [CrossRef]
16. Batubara, A.S.; Muchlisin, Z.A.; Efizon, D.; Elvyra, R.; Fadli, N.; RIZAL, S.; Siti-Azizah, M.N.; Wilkes, M. DNA barcoding (COI genetic marker) revealed hidden diversity of Cyprinid fish (*Barbonymus* spp.) from Aceh Waters, Indonesia. *Biharean Biol.* **2013**, *15*, 1.
17. Kottelat, M. The fishes of the inland waters of Southeast Asia: A catalog and core bibliography of the fishes known to occur in freshwaters, mangroves and estuaries. *Raffles Bull. Zool.* **2013**, (Suppl. 27), 1–663.
18. Froese, R.; Pauly, D. Fishbase. Available online: http://www.fishbase.org (accessed on 15 January 2021).
19. Eschmeyer, W.N.; Fricke, R.; van der Laan, R. Catalog of Fishes Electronic Version. Available online: http://www.calacademy.org/research/ichthyology/catalog/fishcatsearch.html (accessed on 15 January 2021).
20. Kottelat, M.; Whitten, A.J.; Kartikasari, N.; Wirjoatmodjo, S. *Freshwater Fishes of Western Indonesia and Sulawesi*; Periplus Publishing Group: Hong Kong, China, 1993; ISBN 0945971605.
21. Hubert, N.; Lumbantobing, D.; Sholihah, A.; Dahruddin, H.; Delrieu-Trottin, E.; Busson, F.; Sauri, S.; Hadiaty, R.; Keith, P. Revisiting species boundaries and distribution ranges of *Nemacheilus* spp. (Cypriniformes: Nemacheilidae) and *Rasbora* spp. (Cypriniformes: Cyprinidae) in Java, Bali and Lombok through DNA barcodes: Implications for conservation in a biodiversity hotspot. *Conserv. Genet.* **2019**, *20*, 517–529. [CrossRef]
22. Sholihah, A.; Delrieu-Trottin, E.; Sukmono, T.; Dahruddin, H.; Risdawati, R.; Elvyra, R.; Wibowo, A.; Kustiati, K.; Busson, F.; Sauri, S.; et al. Disentangling the taxonomy of the subfamily Rasborinae (Cypriniformes, Danionidae) in Sundaland using DNA barcodes. *Sci. Rep.* **2020**, *10*, 2818. [CrossRef] [PubMed]
23. Lim, H.-C.; Abidin, M.Z.; Pulungan, C.P.; De Bruyn, M.; Mohd Nor, S.A. DNA barcoding reveals high cryptic diversity of freshwater halfbeak genus Hemirhamphodon from Sundaland. *PLoS ONE* **2016**, *11*, e0163596. [CrossRef] [PubMed]
24. Beck, S.V.; Carvalho, G.R.; Barlow, A.; Ruber, L.; Hui Tan, H.; Nugroho, E.; Wowor, D.; Mohd Nor, S.A.; Herder, F.; Muchlisin, Z.A.; et al. Plio-Pleistocene phylogeography of the Southeast Asian Blue Panchax killifish, Aplocheilus panchax. *PLoS ONE* **2017**, *12*, e0179557. [CrossRef] [PubMed]
25. Nurul Farhana, S.; Muchlisin, Z.A.; Duong, T.Y.; Tanyaros, S.; Page, L.M.; Zhao, Y.; Adamson, E.A.S.; Khaironizam, M.Z.; de Bruyn, M.; Siti Azizah, M.N. Exploring hidden diversity in Southeast Asia's *Dermogenys* spp. (Beloniformes: Zenarchopteridae) through DNA barcoding. *Sci. Rep.* **2018**, *8*, 10787. [CrossRef]
26. Hutama, A.; Dahruddin, H.; Busson, F.; Sauri, S.; Keith, P.; Hadiaty, R.K.; Hanner, R.; Suryobroto, B.; Hubert, N. Identifying spatially concordant evolutionary significant units across multiple species through DNA barcodes: Application to the conservation genetics of the freshwater fishes of Java and Bali. *Glob. Ecol. Conserv.* **2017**, *12*, 170–187. [CrossRef]
27. Delrieu-Trottin, E.; Durand, J.; Limmon, G.; Sukmono, T.; Sugeha, H.Y.; Chen, W.; Busson, F.; Borsa, P.; Dahruddin, H.; Sauri, S. Biodiversity inventory of the grey mullets (Actinopterygii: Mugilidae) of the Indo-Australian Archipelago through the iterative use of DNA-based species delimitation and specimen assignment methods. *Evol. Appl.* **2020**, *13*, 1451–1467. [CrossRef]
28. Sholihah, A.; Delrieu-Trottin, E.; Condamine, F.L.; Wowor, D.; Rüber, L.; Pouyaud, L.; Agnèse, J.-F.; Hubert, N. Impact of Pleistocene Eustatic Fluctuations on Evolutionary Dynamics in Southeast Asian Biodiversity Hotspots. *Syst. Biol.* **2021**. [CrossRef]
29. Rahman, M.M.; Norén, M.; Mollah, A.R.; Kullander, S.O. Building a DNA barcode library for the freshwater fishes of Bangladesh. *Sci. Rep.* **2019**, *9*, 9382. [CrossRef] [PubMed]
30. Aquino, L.M.G.; Tango, J.M.; Canoy, R.J.C.; Fontanilla, I.K.C.; Basiao, Z.U.; Ong, P.S.; Quilang, J.P. DNA barcoding of fishes of Laguna de Bay, Philippines. *Mitochondrial DNA* **2011**, *22*, 143–153. [CrossRef] [PubMed]
31. Meganathan, P.; Austin, C.M.; Tam, S.M.; Chew, P.C.; Siow, R.; Rashid, Z.A.; Song, B.K. *An Application of DNA Barcoding to the Malaysian Freshwater Fish Fauna: mtDNA COI Sequences Reveal Novel Haplotypes, Cryptic Species and Field-Based Misidentification*; Monash University Malaysia: Selangor, Malaysia, 2015.

32. Panprommin, D.; Soontornprasit, K.; Tuncharoen, S.; Iamchuen, N. Efficacy of DNA barcoding for the identification of larval fish species in the Upper and Middle Ing River, Thailand. *Gene Rep.* **2021**, *23*, 101057. [CrossRef]
33. Barman, A.S.; Singh, M.; Singh, S.K.; Saha, H.; Singh, Y.J.; Laishram, M.; Pandey, P.K. DNA barcoding of freshwater fishes of Indo-Myanmar biodiversity hotspot. *Sci. Rep.* **2018**, *8*, 8579. [CrossRef] [PubMed]
34. Barman, A.S.; Singh, M.; Pandey, P.K. DNA barcoding and genetic diversity analyses of fishes of Kaladan River of Indo-Myanmar biodiversity hotspot. *Mitochondrial DNA Part A* **2018**, *29*, 367–378. [CrossRef]
35. Esa, Y.B.; Siraj, S.S.; Daud, S.K.; Ryan, J.J.R.; Rahim, K.A.A.; Tan, S.G. Molecular systematics of mahseers (Cyprinidae) in Malaysia inferred from sequencing of a mitochondrial Cytochrome C Oxidase I (COI) gene. *Pertanika J. Trop. Agric. Sci.* **2008**, *31*, 263–269.
36. Lakra, W.S.; Singh, M.; Goswami, M.; Gopalakrishnan, A.; Lal, K.K.; Mohindra, V.; Sarkar, U.K.; Punia, P.P.; Singh, K.V.; Bhatt, J.P. DNA barcoding Indian freshwater fishes. *Mitochondrial DNA Part A* **2016**, *27*, 4510–4517. [CrossRef]
37. Ivanova, N.V.; Zemlak, T.S.; Hanner, R.H.; Hébert, P.D.N. Universal primers cocktails for fish DNA barcoding. *Mol. Ecol. Notes* **2007**, *7*, 544–548. [CrossRef]
38. Hebert, P.D.N.; deWaard, J.R.; Zakharov, E.; Prosser, S.W.J.; Sones, J.E.; McKeown, J.T.A.; Mantle, B.; La Salle, J. A DNA "barcode blitz": Rapid digitization and sequencing of a natural history collection. *PLoS ONE* **2013**, *8*, e68535. [CrossRef]
39. Ratnasingham, S.; Hebert, P.D.N. BOLD: The Barcode of Life Data System. *Mol. Ecol. Notes* **2007**, *7*, 355–364. [CrossRef]
40. Pons, J.; Barraclough, T.G.; Gomez-Zurita, J.; Cardoso, A.; Duran, D.P.; Hazell, S.; Kamoun, S.; Sumlin, W.D.; Vogler, A.P. Sequence-based species delimitation for the DNA taxonomy of undescribed insects. *Syst. Biol.* **2006**, *55*, 595–606. [CrossRef]
41. Puillandre, N.; Lambert, A.; Brouillet, S.; Achaz, G. ABGD, Automatic Barcode Gap Discovery for primary species delimitation. *Mol. Ecol.* **2012**, *21*, 1864–1877. [CrossRef]
42. Ratnasingham, S.; Hebert, P.D.N. A DNA-based registry for all animal species: The barcode index number (BIN) system. *PLoS ONE* **2013**, *8*, e66213. [CrossRef] [PubMed]
43. Kapli, P.; Zhang, J.; Kobert, K.; Pavlidis, P.; Stamatakis, A.; Flouri, T. Multi-rate Poisson Tree Processes for single-locus species delimitation under Maximum Likelihood and Markov Chain Monte Carlo. *Bioinformatics* **2017**, *33*, 1630–1638. [CrossRef] [PubMed]
44. Luo, A.; Ling, C.; Ho, S.Y.W.; Zhu, C.-D. Comparison of methods for molecular species delimitation across a range of speciation scenarios. *Syst. Biol.* **2018**, *67*, 830–846. [CrossRef]
45. Kekkonen, M.; Mutanen, M.; Kaila, L.; Nieminen, M.; Hebert, P.D.N. Delineating Species with DNA Barcodes: A Case of Taxon Dependent Method Performance in Moths. *PLoS ONE* **2015**, *10*, e0122481. [CrossRef] [PubMed]
46. Kekkonen, M.; Hebert, P.D.N. DNA barcode-based delineation of putative species: Efficient start for taxonomic workflows. *Mol. Ecol. Resour.* **2014**, *14*, 706–715. [CrossRef] [PubMed]
47. Shen, Y.; Hubert, N.; Huang, Y.; Wang, X.; Gan, X.; Peng, Z.; He, S. DNA barcoding the ichthyofauna of the Yangtze River: Insights from the molecular inventory of a mega-diverse temperate fauna. *Mol. Ecol. Resour.* **2019**, *19*, 1278–1291. [CrossRef]
48. Limmon, G.; Delrieu-Trottin, E.; Patikawa, J.; Rijoly, F.; Dahruddin, H.; Busson, F.; Steinke, D.; Hubert, N. Assessing species diversity of Coral Triangle artisanal fisheries: A DNA barcode reference library for the shore fishes retailed at Ambon harbor (Indonesia). *Ecol. Evol.* **2020**, *10*, 3356–3366. [CrossRef]
49. Zhang, J.; Kapli, P.; Pavlidis, P.; Stamatakis, A. A general species delimitation method with applications to phylogenetic placements. *Bioinformatics* **2013**, *29*, 2869–2876. [CrossRef]
50. Fujisawa, T.; Barraclough, T.G. Delimiting species using single-locus data and the generalized mixed Yule coalescent approach: A revised method and evaluation on simulated data sets. *Syst. Biol.* **2013**, *62*, 707–724. [CrossRef]
51. Stamatakis, A. RAxML version 8: A tool for phylogenetic analysis and post-analysis of large phylogenies. *Bioinformatics* **2014**, *30*, 1312–1313. [CrossRef] [PubMed]
52. Bouckaert, R.; Heled, J.; Kühnert, D.; Vaughan, T.; Wu, C.H.; Xie, D.; Suchard, M.A.; Rambaut, A.; Drummond, A.J. BEAST 2: A Software Platform for Bayesian Evolutionary Analysis. *PLoS Comput. Biol.* **2014**, *10*, e1003537. [CrossRef] [PubMed]
53. Bermingham, E.; McCafferty, S.; Martin, A.P. Fish biogeography and molecular clocks: Perspectives from the Panamanian isthmus. In *Molecular Systematics of Fishes*; Kocher, T.D., Stepien, C.A., Eds.; Academic Press: San Diego, CA, USA, 1997; pp. 113–128.
54. Ogilvie, H.A.; Bouckaert, R.R.; Drummond, A.J. StarBEAST2 brings faster species tree inference and accurate estimates of substitution rates. *Mol. Biol. Evol.* **2017**, *34*, 2101–2114. [CrossRef]
55. Ho, S.Y.W.; Larson, G. Molecular clocks: When timesare a-changin'. *Trends Genet.* **2006**, *22*, 79–83. [CrossRef]
56. Kimura, M. A Simple Method for Estimating Evolutionary Rates of Base Substitutions through Comparative Studies of Nucleotide-Sequences. *J. Mol. Evol.* **1980**, *16*, 111–120. [CrossRef]
57. Paradis, E.; Schliep, K. ape 5.0: An environment for modern phylogenetics and evolutionary analyses in R. *Bioinformatics* **2019**, *35*, 526–528. [CrossRef]
58. Brown, S.D.J.; Collins, R.A.; Boyer, S.; Lefort, C.; Malumbres-Olarte, J.; Vink, C.J.; Cruickshank, R.H. Spider: An R package for the analysis of species identity and evolution, with particular reference to DNA barcoding. *Mol. Ecol. Resour.* **2012**, *12*, 562–565. [CrossRef]
59. Paradis, E. pegas: An {R} package for population genetics with an integrated–modular approach. *Bioinformatics* **2010**, *26*, 419–420. [CrossRef] [PubMed]
60. Blair, C.; Bryson, J.R.W. Cryptic diversity and discordance in single-locus species delimitation methods within horned lizards (Phrynosomatidae: *Phrynosoma*). *Mol. Ecol. Resour.* **2017**, *17*, 1168–1182. [CrossRef] [PubMed]

61. Tang, Q.; Liu, S.; Yu, D.; Liu, H.; Danley, P.D. Mitochondrial capture and incomplete lineage sorting in the diversification of balitorine loaches (Cypriniformes, Balitoridae) revealed by mitochondrial and nuclear genes. *Zool. Scr.* **2012**, *41*, 233–247. [CrossRef]
62. Hopkins, R.L.; Eisenhour, D.J. Hybridization of Lythrurus fasciolaris and *Lythrurus umbratilis* (Cypriniformes: Cyprinidae) in the Ohio River basin. *Copeia* **2008**, *2008*, 162–171. [CrossRef]
63. Atsumi, K.; Nomoto, K.; Machida, Y.; Ichimura, M.; Koizumi, I. No reduction of hatching rates among F1 hybrids of naturally hybridizing three Far Eastern daces, genus *Tribolodon* (Cypriniformes, Cyprinidae). *Ichthyol. Res.* **2018**, *65*, 165–167. [CrossRef]
64. Hubert, N.; Hanner, R. DNA barcoding, species delineation and taxonomy: A historical perspective. *DNA Barcodes* **2015**, *3*, 44–58. [CrossRef]
65. Wibowo, A.; Atminarso, D.; Baumgartner, L.; Vasemagi, A. High prevalence of non-native fish species in a remote region of the Mamberamo River, Indonesia. *Pac. Conserv. Biol.* **2020**, *26*, 293–300. [CrossRef]
66. Keith, P.; Lord, C.; Darhuddin, H.; Limmon, G.; Sukmono, T.; Hadiaty, R.; Hubert, N. Schismatogobius (Gobiidae) from Indonesia, with description of four new species. *Cybium* **2017**, *41*, 195–211.

MDPI

Article

Identifying Early Stages of Freshwater Fish with DNA Barcodes in Several Sinkholes and Lagoons from the East of Yucatan Peninsula, Mexico

Adrián Emmanuel Uh-Navarrete [1], Carmen Amelia Villegas-Sánchez [2], José Angel Cohuo-Colli [1], Ángel Omar Ortíz-Moreno [3] and Martha Valdez-Moreno [1,*]

1 El Colegio de la Frontera Sur, Department of Aquatic Ecology and Systematics, Av. Centenario Km. 5.5, Chetumal 77014, Quintana Roo, Mexico; adrian.uh@posgrado.ecosur.mx (A.E.U.-N.); jose.cohuo@ecosur.mx (J.A.C.-C.)

2 Tecnológico Nacional de México/I. T. Chetumal, Av. Insurgentes No. 330, Chetumal 77013, Quintana Roo, Mexico; carmen.vs@chetumal.tecnm.mx

3 Comisión Nacional de Áreas Naturales Protegidas, Reserva de la Biosfera Sian Ka'an, Calle 61 # 767, Colonia Centro, Felipe Carrillo Puerto 77200, Mexico; omortiz@conanp.gob.mx

* Correspondence: mvaldez@ecosur.mx

Abstract: Our work shows the efficacy of DNA barcoding for recognizing the early stages of freshwater fish. We collected 3195 larvae and juveniles. Of them, we identified 43 different morphotypes. After DNA barcodes of 350 specimens, we ascertained 7 orders, 12 families, 19 genera, 20 species, and 20 Barcode Index Numbers, corresponding to putative species. For the first time, we reported the presence of the brackish species, *Gobiosoma yucatanum* in Lake Bacalar. Specimens of the genus *Atherinella* sp. and *Anchoa* sp. are possibly new species. Using both methods, morphology, and DNA barcodes, we identified 95% of the total larvae collected (2953 to species, and 78 to genus), and all of them were native. From them, the order Gobiiformes represented 87%. The most abundant species were *Lophogobius cyprinoides* and *Dormitator maculatus*, followed by *Gobiosoma yucatanum* and *Ctenobius fasciatus*. The Muyil and Chuyanché lagoons have the highest number of species. We present for the first time a short description of *Cyprinodon artifrons* and *Floridichthys polyommus*. This information conforms an indispensable baseline for ecological monitoring, to evaluate impacts, and developing management and conservation plans of biodiversity, principally in areas under human pressure such as Sian Ka'an, and Lake Bacalar, where tourism is high and growing in disorder.

Keywords: biodiversity; Quintana Roo; Bacalar; Sian Ka'an; fish larvae

Citation: Uh-Navarrete, A.E.; Villegas-Sánchez, C.A.; Cohuo-Colli, J.A.; Ortíz-Moreno, Á.O.; Valdez-Moreno, M. Identifying Early Stages of Freshwater Fish with DNA Barcodes in Several Sinkholes and Lagoons from the East of Yucatan Peninsula, Mexico. *Diversity* **2021**, *13*, 513. https://doi.org/10.3390/d13110513

Academic Editor: Manuel Elias-Gutierrez

Received: 17 September 2021
Accepted: 20 October 2021
Published: 22 October 2021

1. Introduction

Quintana Roo state, in the east of the Yucatan Peninsula, Mexico, is among the places with higher levels of biodiversity. For this reason, on average this region receives 5 million visitors annually [1], and in the last decade contributed with the highest tourism income for the whole country [2]. Among the most visited natural attractions of the state are freshwater bodies such as lagoons and cenotes (sinkholes that connect to one of the most complex underground water systems in the world) [3]. However, the richness of the fishes that inhabit such ecosystems is still poorly known. Specific studies about their larvae are almost absent; the few fish larvae studies developed here have focused on marine organisms and estuarine environments [4–8].

The knowledge of the early stages of fish is essential as it provides information about the recruitment rates of juveniles and the size of the adult population [9,10]. Besides, these stages are helpful in the characterization of taxonomic diversity, times, and locations of spawning, and the assessment of connectivity between ecosystems [11].

The early stages of fish are considered as ichthyoplankton. They are characterized by a high rate of development and a great diversity of forms that usually differ from adults [9].

Furthermore, the morphology of the same species can change rapidly and significantly during its growth from pre-flexion to post-flexion larvae and later to the juvenile stage [12]. Hence, the identification using morphology tends to be difficult.

Given the difficulties in identifying fish larvae by traditional methods, applying multiple and complementary perspectives is advisable to identify them more precisely [13,14]. Integrative taxonomy proposes using morphological characteristics and molecular identification [15,16]. This type of work has been used successfully before, as evidenced by the study of Valdez-Moreno et al. [7], Hubert et al. [17], Baldwin, et al. [18], and Ko et al. [12], among others.

Considering the lack of information and the relevance of the studies that help understand biodiversity, this research has the goal to analyze the identities of the fish larvae that inhabit freshwater ecosystems from Quintana Roo, based on integrative taxonomy. We consider this work as a starting point towards the conservation and sustainable use of these systems.

2. Materials and Methods

2.1. Study Site and Field Sampling

We collected in 18 places; 3 of them in Sian Ka'an, a Biosphere Reserve, 9 nearby the reserve, 4 in Lake Lake Bacalar, and 2 lagoons Xul-Ha, and Huay Pix associated with Bacalar (Figure 1 and Table 1).

We collected the samples using light traps according to the methodology proposed by Elías-Gutiérrez et al. [11]. We placed only one light trap in most sites, except in Muyil, Chunyanxche, Del Padre, Siijil Noh Ha, and Chancah Veracruz, where due to their depth and area, two traps were used (one littoral and one limnetic). The sampling sites and collection dates are summarized in Table 1. Figure 1 shows the geographic locations of all sampled sites.

Once we got the samples, we immediately filtered them on a 50 µm sieve, then we fixed the material with ethanol 96% and placed it on ice. In the lab, all samples were stored for at least one week at −18 °C before processing [11,19].

2.2. Morphological Analysis

In the laboratory, the larvae were separated by morphotypes and stored in 5 mL vials with 4 mL of 96% ethanol.

For the morphological identification of the larvae, we used different identification keys as Richards [9], Fahay [20], and studies of freshwater fishes where their larval stages are described [21]. For information about what species could be found in the study area, we used Miller et al. [22], Schmitter-Soto [23], and Valdez-Moreno [24,25].

All larvae also were measured and separated according to the stage of development (preflexion, flexion, postflexion, transition, and juvenile) following the previous criteria proposed [9].

2.3. DNA Barcode Analysis

We selected 353 specimens for this study (from 1 to 5 organisms of each morphotype). All of them were photographed under a Nikon SM2 745T stereomicroscope with an Eos Rebel T7i camera.

We used a small piece of muscle (1–3 mm^3) or the right eye to extract the DNA for molecular analysis. We sterilized the forceps and the material using chlorine diluted in water in a proportion 1:5 and subsequently neutralized it with 96% ethanol between each tissue or eye extraction.

For each sample's tissue digestion, a lysis buffer was used with proteinase K, and they were allowed to digest overnight at 56 °C. The extraction was carried out through 1.0 mm PALL glass fiber plates [26]. A Cytochrome Oxidase I (COI) gene segment with an approximate length of 650 Bp [27] was amplified using the FishF1 and FishF2 primers [28,29]. Amplification was carried out with a final volume of 12.5 µL, prepared as follows: 6.5 µL of

10% trehalose, 2 µL of ultrapure water, 1.25 µL PCR buffer X10, 0.625 µL MgCl2 (50 mM), 0.125 µL of each Primer (0.01 mM), 0.06525 µL dNTP mix (10 mM), 0.625 µL Taq polymerase, and 2 µL of template of DNA. The reactions we cycled at 94 °C for a 1 min, followed by five cycles at 94 °C for 30 s, 45–50 °C for 40 s and 72 °C for 1 min, followed by 35 cycles at 94 °C for 30 s, 51–54 °C for 40 s and 72 °C for 1 min, and finally by one last cycle of 72 °C for 10 min. We visualized the PCR products in agarose gel Invitrogen TM with 4 µL of sample and 16 µL of water. We sequenced the PCR products of Cenote Cocalitos samples at the Canadian Center for DNA Barcoding (Guelph, ON, Canada), and the rest of PCR products were sent to sequence at Eurofins Scientific (Louisville, KY, USA). Finally, we edited the sequences with Codon code v.3.0.1 and uploaded them to BOLD (www.boldsystems.org) in the dataset DS-FLYP Fish larvae from Yucatan(dx.doi.org/10.5883/DS-FLYP).

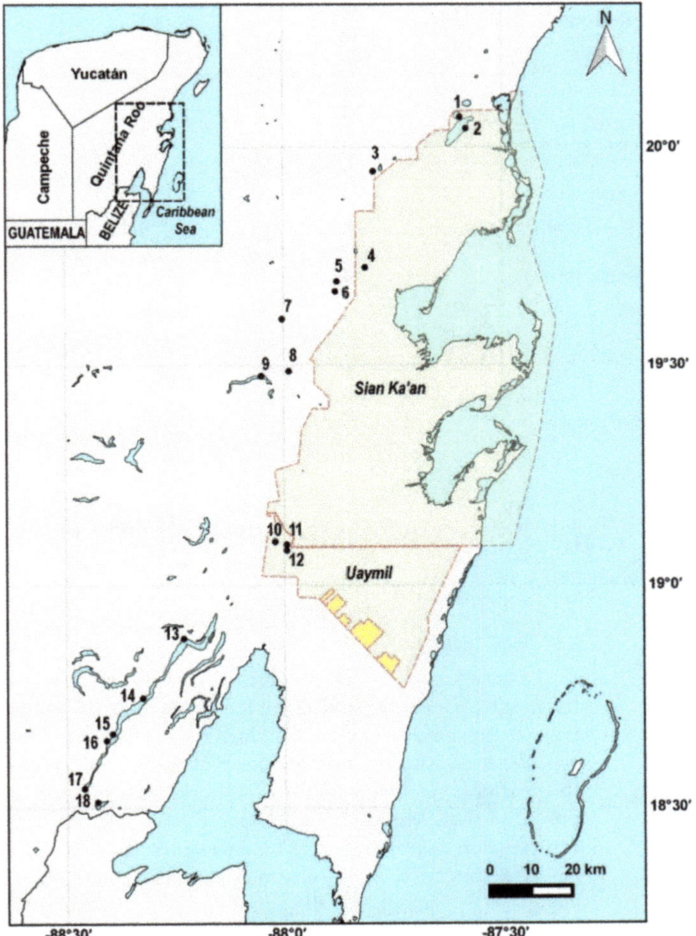

Figure 1. Study area, with the 18 sampling points: (1) Muyil, (2) Chunyaxche, (3) Km 48, (4) Santa Teresa, (5) Tres Reyes 2, (6) Tres Reyes 1, (7) Del Padre, (8) Chancah Veracruz, (9) Siijil Noh Ha, (10) El Toro, (11) Pucté 2, (12) Pucté-Cafetal, (13) Buena Vista, (14) Cayuco Maya, (15) Brujas, (16) Cocalitos, (17) Xul-ha, and (18) Huay Pix. The shaded area represents the Sian Ka'an reserve and the Uaymil protection area.

Table 1. Collection sites * inside Sian Ka'an reserve, ** nearby of the reserve, *** inside Lake Bacalar.

Number	Site	Lat N	Long W	Number of Light Traps Per Day	Date
1	Muyil Lagoon *	20.069	−87.594	2	24 August 2019
2	Chunyaxche lagoon *	20.042	−87.581	2	24 August 2019
3	Km 48 sinkhole **	19.943	−87.794	1	24 August 2019
4	Santa Teresa sinkhole *	19.723	−87.813	1	23 August 2019
5	Tres Reyes 2 sinkhole **	19.692	−87.877	1	23 August 2019
6	Tres Reyes 1 sinkhole **	19.668	−87.881	1	23 August 2019
7	Del Padre sinkhole **	19.604	−88.003	2	23 August 2019
8	Chancah Veracruz sinkhole **	19.486	−87.988	2	22 August 2019
9	Siijil Noh Ha sinkhole **	19.475	−88.052	2	22 August 2019
10	El Toro sinkhole **	19.098	−88.021	1	22 August 2019
11	Pucté 2 sinkhole **	19.091	−87.994	1	25 August 2019
12	Pucté-Cafetal sinkhole **	19.079	−87.994	1	25 August 2019
13	Buena Vista ***	18.88	−88.231	1	17 August 2015
14	Cayuco Maya ***	18.746	−88.325	1	18 August 2015
15	Brujas sinkhole ***	18.666	−88.395	1	28 June 2015
					1 August 2019
16	Cocalitos sinkhole ***	18.651	−88.409	1	2 August 2015
					18 April 2015
					28 June 2015
					19 July 2015
					20 August 2015
					1 December 2015
17	Xul-Ha lagoon	18.543	−88.46	1	15 August 2015
18	Huay Pix lagoon	18.512	−88.43	1	14 August 2015

2.4. Data Analysis

The sequences obtained were compared with sequences previously published using the specimen identification tool in the Barcode of Life Data System (BOLD) [30]. In addition, these sequences reached the standard to get a Barcode Index Number (BIN) [31].

We used Kimura 2-parameter model (K2P) to get the genetic divergences between the species [32] and a maximum likelihood (ML) tree, using 500 bootstrap replicates [33], provided with MEGA 7.0 software. Finally, the ML tree was simplified by using the compression feature provided by the same software [34].

The criteria to assign taxonomic level identification using BOLD was a similarity value ≥99%. The resulting BIN number allowed to assign the specimens to species level [35]. Similarities with values ≥ 94% to ≤98.4% were identified to genus, and similarities <94% were assigned to family [36].

3. Results and Discussion

3.1. Species Identification

We collected a total of 3195 larvae and juveniles. Using only morphology, we recognized 43 different morphotypes and only one species, *Bathygobius soporator*. We assigned the remaining 42 morphotypes to genus (*Gobiosoma*, *Ctenogobius*, and *Astyanax*) or family (Cyprinodontidae, Engraulidae, Gobiidae Atherinopsidae, Cichlidae, Poeciliidae, and Characidae).

These results showed how difficult the identification of the early stages of fish using only morphological characters is. There is no information available for most freshwater fish larvae [37,38]. In addition, the fragile specimens can be damaged during the manipulation and fixation process, which makes the identification more difficult [36,39]. In some cases, the larvae can lose some body parts by predation when all zooplankton are together in the light trap (obs. pers.). The result is either a high probability of erroneous taxonomic identification or simply making it impossible.

For DNA barcode analyses, we processed 350 specimens from the 43 morphotypes. We obtained 347 sequences. Their length varied between 600 and 658 base pairs (bp), after trimming the sequences to remove un-formative nucleotide sites at the 3′ and 5′ ends, except for five, which had from 171 to 585 bp. Only three specimens could not be sequenced. We did not observe insertions, deletions, or stop codons in any of the sequences.

We considered a mini-barcode sequence with 171 bp [40], belonging to *Gobiosoma yucatanum* with 100% similarity. This technique has been used previously, demonstrating its efficacy in identifying degraded DNA [41–44].

Our sequencing success rate was 99%. This value is similar to that reported by Frantine-Silva et al. [39], who registered a 99.81% success rate on fish eggs and larvae from the Paranapanema River in Brazil. In contrast, Almeida et al. [36] had a lower value, reporting 79.6% success in the same river with similar material. This difference could be explained because the latter authors used a low alcohol percentage to fix the samples (70%) and did not keep them in cold storage. Several authors suggest optimizing DNA fixation [11,35]. The length and quality of the sequences were similar to other studies that worked with this gene.

When comparing the 347 sequences with Bold identification system (http://www.boldsystems.org/index.php/IDS_OpenIdEngine. Accessed on 15 September 2021), 324 (93.4%) matched in BOLD reference library with >99% similarity, allowing the species-level identification [35,36] (Table 2). For the remaining 23 (6.6%), we identified them to genus level, 22 *Atherinella* sp., and one *Anchoa* sp.

All sequences represented 7 orders, 12 families, 19 genera, 20 species, and 20 BINs (Table 2 and Supplementary Materials. Gobiiformes were the order with more species (five species), followed by Cyprinodontiformes, Clupeiformes (four), and Cichliformes (three).

All species identified were based on similarity values that confirmed their placement under different numbers of BINs [31]. These were congruent with the K2P distance tree, and the patterns are seen in the genetic distance.

The ID tree does not show overlapping between species clusters (Figure 2). These results allowed us to make reliable species assignments [45,46].

Each BIN number was associated with one species, except *Astyanax aeneus* and *Cyprinodon artifrons*. In these two cases the BINs cannot distinguish them, due to both having congeners closely related that have recently evolved. Consequently, the genetic distance between them is small [28,35,47,48].

Most of the fish species found in this study have been reported in different aquatic systems in Yucatán and the Quintana Roo state [22,24]. Valdez Moreno et al. [25], using metabarcoding in the same places as us (except Minicenote), corroborated the presence of all the species found by us in these sites [24,25]. However, we found some interesting cases, explained in the following paragraphs.

Table 2. Species and genus list detected with Barcodes with their BINs numbers.

Order	Family	Species	# Specimens	% Similarity	BIN
Atheriniformes	Atherinopsidae	*Atherinella* sp.	22	100	BOLD:AAI4788
Beloniformes	Belonidae	*Strongylura notata*	2	100	BOLD:AAC4691
	Hemiramphidae	*Chriodorus atherinoides*	1	100	BOLD:AAD0222
Characiformes	Characidae	*Astyanax aeneus*	3	100	BOLD:AAA6360
Cichliformes	Cichlidae	*Mayaheros urophthalmus*	3	100	BOLD:AAB5118
		Thorichthys meeki	8	100	BOLD:AAA4760
		Vieja melanura	3	100	BOLD:AAB9907
Clupeiformes	Clupeidae	*Dorosoma petenense*	1	100	BOLD:AAC3463
	Engraulidae	*Anchoa lyolepis*	1	99.85	BOLD:AAR3806
		Anchoa sp.	1	100	BOLD:AAE1085
		Anchovia clupeoides	26	100	BOLD:ACV0719
Cyprinodontiformes	Cyprinodontidae	*Cyprinodon artifrons*	56	100	BOLD:AAA8182
		Floridichthys polyommus	48	100	BOLD:AAA6554
		Garmanella pulchra	11	100	BOLD:AAD5728
	Poeciliidae	*Gambusia yucatana*	5	100	BOLD:AAA4520
Gobiiformes	Eleotridae	*Dormitator maculatus*	20	99.36–100	BOLD:AAC2209
	Gobiidae	*Bathygobius soporator*	7	100	BOLD:AAA7195
		Gobiosoma yucatanum	50	100	BOLD:ACV0831
		Lophogobius cyprinoides	45	99.83–100	BOLD:AAB6671
	Oxudercidae	*Ctenogobius fasciatus*	34	100	BOLD:AAE7730

We collected *Anchoa lyolepis* in the Cocalitos sinkhole in Lake Bacalar. It is considered a marine species and has been reported in the northern Gulf of Mexico and from Yucatán to Brazil [49]. In the BOLD database, there are eight specimens, one is from the same sinkhole, but the specimen is incomplete. It has no head, so morphological determination is difficult. The other samples are dried fishes collected from a Mexico City market, whose morphology appears to be this species. So, it is probable that our larvae seem to be well identified. This observation will have to be confirmed with more specimens.

The *Gobiosoma yucatanum* holotype was collected from the south side of the pier in Chetumal city, Quintana Roo [50]. Its distribution range includes rivers, estuaries, and inland lagoons from Mexico to Belize and Honduras [51,52]. Elías-Gutiérrez et al. [11] reported the presence of *Gobiosoma* sp. in the larval stage in Lake Bacalar. In the BOLD database, there are adults from Chetumal Bay, all matched with our larvae. It is the first report of the presence of *G. yucatanum* in Lake Bacalar. This result confirms the connectivity between the Chetumal bay and Lake Bacalar. Something similar was reported for *Cyprinodon artifrons*, whose adults are located in the in the reef lagoon in Xcalak and Chetumal Bay, and the larvae which were in Lake Bacalar [11].

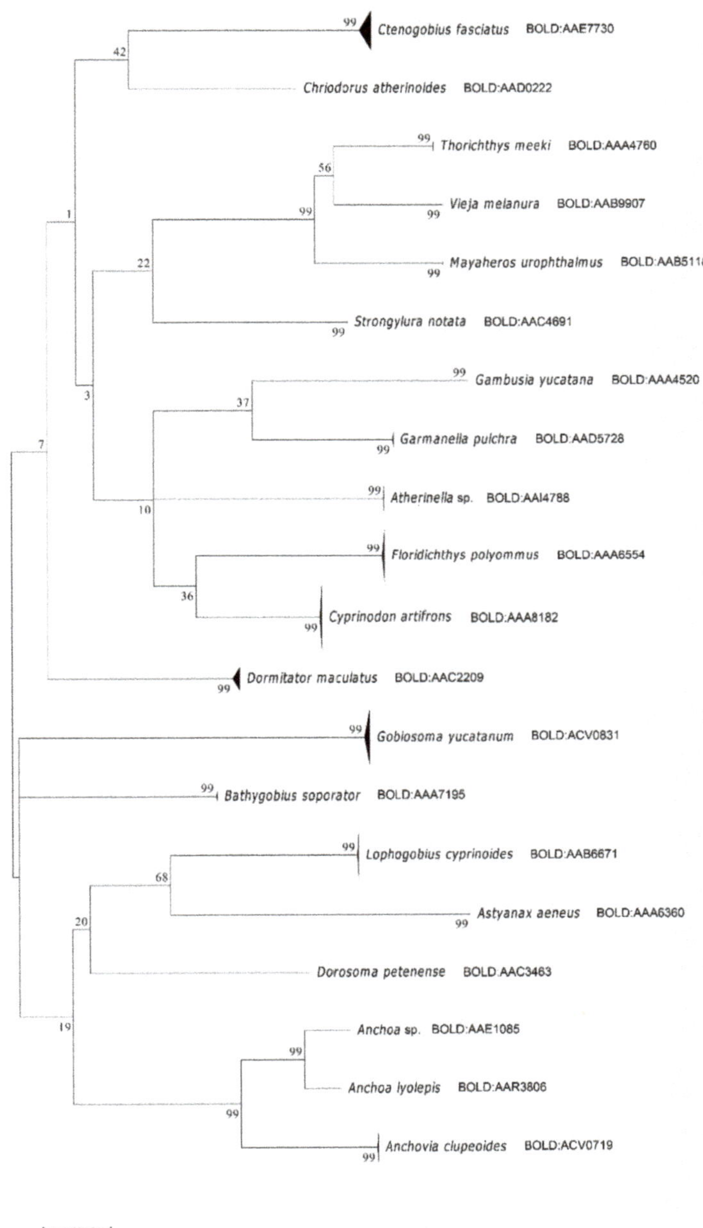

Figure 2. ML tree showing the clustering of the 20 identified species. The numbers on the branches are the bootstrap support after 500 replicates. Numbers after the names are the BINs.

Some specimens of the genus *Atherinella* matched with two species, *A. alvarezi* and *Atherinella* sp., with a 100% similarity value (Table 2). Both species have also been reported in inland waters of the Yucatán peninsula [22–24,53]. However, the morphological characters overlap between them. The taxonomy of freshwater atherinopsids of this region needs further studies because they possibly belong to an undescribed species [23].

One specimen collected in Huay Pix near Lake Bacalar matched with *Anchoa* sp. Four species of this genus were found in freshwater. However, *Anchoa parva* is the only one reported in the Yucatan Peninsula and Bacalar, Quintana Roo [22,23]. We do not have any barcodes for it yet. It is necessary to collect and sequence this species and compare it with our larvae to confirm its identity.

With the barcode results, we were able to recognize the different morphotypes. Then, it allowed us to identify 2681 larvae that did not enter the molecular analysis. These results are excellent because we identified 95% (3031) from all the collected larvae and juveniles. In addition, it allowed us to know the stage of development in each of them (Table 3).

Table 3. Species, size range, number of specimens, stage of development and collection sites: (1) Muyil, (2) Chunyaxche, (3) Km 48, (4) Santa Teresa, (5) Tres Reyes 2, (6) Tres Reyes 1, (7) Del Padre, (8) Chancah Veracruz, (9) Siijil Noh Ha, (10) El Toro, (11) Pucté 2, (12) Pucté-Cafetal, (13) Buena Vista•, (14) Cayuco Maya•, (15) Brujas•, (16) Cocalitos•, (17) Xul-ha, (18) Huay Pix. • inside Lake Bacalar. * LT = total length. ** LS = standard length.

Specie	Size Range (mm) * LT ** LS	Number of Specimens	Stage	Collecting Sites														
				1	2	3	7	8	9	10	11	12	13	14	15	16	17	18
Atherinella sp.	3.2–7.2 *	58	Preflexion larvae						x		x	x					x	
	11.5	1	Postflexion larvae								x						x	
	21–22 *	3	Transition stage								x						x	
	27.5–42.5 *	15	Juvenil								x							
Strongylura	12 **	1	Postflexion larvae													x		
notata	95 **	1	Juvenil													x		
Chriodorus atherinoides	15 **	1	Postflexion larvae										x					
Astyanax aeneus	33.7–40.2 *	3	Juvenil							x								
Mayaheros	6.5–6.9 *	2	Postflexion larvae	x														
urophthalmus	14.4–17.50 *	2	Transition stage	x														
Thorichthys	4.5–5	3	Flexion larvae									x						
meeki	6.5–7.4 *	7	Postflexion larvae				x											
Vieja melanura	9.8 *	1	Postflexion larvae		x													
	14.4–15 *	2	Transition stage	x														
Dorosama petenense	4 **	4	Preflexion larvae													x		
Anchoa lyolepis	30 **	1	Transition stage													x		
Anchoa sp.	21 **	1	Transition stage															x
Anchovia	3 **	1	Preflexion larvae													x		
clupeoides	7 **	6	Flexion larvae													x		
	11–23 **	52	Postflexion larvae											x	x	x		
	37–47	5	Juvenile												x			
Cyprinodon	4.2–6.9 *	49	Postflexion larvae	x	x											x		
artifrons	9–11 *	11	Transition stage	x	x													
Floridichthys	4.5–8.5 *	84	Postflexion larvae	x	x													
polyommus	9–10 *	3	Transition stage		x													
	24.2 *	1	Juvenil		x													
Garmanella	4.2–4.5 *	3	Postflexion larvae	x	x													
pulcara	11–12 *	2	Transition stage		x													
	14.2–20 *	6	Juvenil	x	x													
Gambusia yucatana	7.9–8.2 *	6	Postflexion larvae					x										
Dormitator	5.2–13.6 *	529	Postflexion larvae	x	x	x						x						
maculatus	14.2–16.8 *	26	Transition stage	x	x	x												
Bathygobius	2–2.2 **	38	Preflexion larvae													x		
soporator	7 **	2	Postflexion larvae													x		
Gobiosoma	2.2–3.5 **	72	Preflexion larvae													x		x

Table 3. *Cont.*

Specie	Size Range (mm) * LT ** LS	Number of Specimens	Stage	Collecting Sites														
				1	2	3	7	8	9	10	11	12	13	14	15	16	17	18
yucatanum	3.8–4.6 **	216	Flexion larvae													x		
	4.8–11 **	270	Posflexion larvae											x		x		
	13 **	1	Transition stage													x		
Lophogobius	2.8–3.2 **	157	Preflexion larvae												x	x		
cyprinoides	3.4–4 **	420	Flexion larvae										x		x	x		
	4.3–8 **	415	Postflexion larvae										x		x	x		
Ctenogobius	3.3–3.4 **	29	Preflexion larvae													x		
fasciatus	3.5–3.8 **	94	Flexion larvae													x		
	4.5–11 **	427	Postflexion larvae													x	x	
unidentified organisms		164		x												x		
Total		3195																

Most of the larvae identified belonged to the order Gobiiformes, representing 87% of the total. The most abundant species were *Lophogobius cyprinoides* (992 specimens), *Dormitator maculatus* (655 specimens), *G. yucatanum* (556 specimens), and *Ctenobius fasciatus* (550 specimens). The rest of the species had less than 88 to 1 individual. The most widely distributed species were *D. maculatus* and *Atherinella* sp. located in four different sites (Table 3).

These results are different from those reported by other studies, based in adults. In three previous research about the fish community structure in the Sian Ka'an reserve, it was found that Cichlidae and Poeciliidae are the families with more species in this area [54–56]. Their presence and other families have been found in samples from these aquatic environments (obs. pers.) and using environmental DNA [24,25,57]. These differences are most likely due to different collection methods and objectives (minnow traps, dip, cast nets, hanging nets, DNA in water, and none of them collected larvae).

In our results, we reported larvae from the poeciliid *Gambusia yucatana*. However, their larvae were not collected with the use of the light traps. We collected some adults with a hand net in Del Padre sinkhole and put them in an aquarium. The females gave birth to some young, which were barcoded and included in this study (Table 3).

These results show that light traps have limitations due to their selectivity [58]. Some species seem to be less attracted to light, such as the poeciliids. Most likely, the parental care of cichlids such as *Mayaheros uruphtalmus*, *Thorichthys meeki*, and *Vieja melanura* [59] prevents their larvae from dispersing and reaching the light traps (obs. pers.). It is necessary to perform more studies about fish behavior to confirm these ideas.

The results regarding the number of species per site were variable. Cocalitos Sinkhole had nine species, Muyil lagoon had six, and Chuyanché lagoon had five. The rest of the localities presented from three to one species, while no larvae were collected in Santa Teresa, Tres Reyes 1, and Tres Reyes 2.

Although Cocalitos has the highest number of species, it is necessary to consider that this place was sampled six times compared to the others, which were visited once or twice (Table 1). Based on our results and previous studies [25], Muyil and Chuyanché lagoons have the highest number of species. The differences among the richness reported can be associated with several factors such as vegetation, shore area, and collecting method [54,56].

The analysis of the different early development stages showed that the larval stage was predominant in all species, except *A. aeneus*, represented by the juvenile stage. *A. lyolepis* and *Anchoa* sp. were found in the transition stage.

The most common larval stage was postflexion in 12 species, followed by the flexion stage in 6 species and preflexion in 4 species.

Atherinella sp., *Anchovia clupeoides*, and *G. yucatanum* were the species that showed four different stages during the same sampling day. *Chriodorus atherinoides*, *A. aeneus*, *Dorosoma petenense*, *A. lyolepis*, *Anchoa* sp., and *G. yucatana* were found in a single stage. The other 11 species had 2 or 3 stages (Table 3).

The high percentage of larvae in postflexion indicates that there has been a recent reproduction period. Schmitter-Soto [23] mentions that the reproduction period of *C. artifrons* is during spring and autumn. For *T. meeki*, *M. urophthalmus*, *V. melanura* and *A. aeneus*, it is from March to June, while *D. maculatus* is from September to October. In thecase of *Floridichthys polyommus*, its reproduction is in spring and summer [60]. Miller et al. [22] reported that the genus *Atherinella* and *G. yucatana* have long reproductive periods. Therefore, most of the species found here have reproduction periods close to or during the summer.

Some authors report that the caudal and pectoral fins are the first to develop because they are the main ones involved in the locomotion of the larvae for feeding and movement [9,61–64]. These ideas agree with the results presented here. Most larvae were found in postflexion, which developed these fins, allowing them to enter the light trap. It also explains the small number of larvae in the preflexion stage since they have more limited movement [9].

Identifying larvae of *C. artifrons*, *F. polyommus*, and *Garmanella pulchra* (Cyprinodontidae) is challenging because they are morphologically similar. The DNA barcodes allowed us to distinguish them.

In the following paragraphs, we present a short description of the larvae from the first two species. The minimum number of specimens reviewed for each of the stages was five.

3.2. *Cyprinodon Artifrons (Hubbs, 1936)*

Postflexion (TL 4.2–4.6 mm): presence of finfold; up to 9 pectoral rays. Robust head and dorsally pigmented without a distinct pattern. Eyes with a round pupil. Incipient lateral pigmentation of the body with 5–7 patches of melanophores, the patches beginning above or slightly forward the anus and ending at the base of the caudal fin; some larvae with prominent abdomen have marked ventral pigmentation (linear or slightly branched), larvae without prominent abdomen show less ventral pigmentation and more isolated melanophores (Figures 3A and 5A).

Postflexion (TL 5.0–5.5 mm): presence of finfold; 10–13 pectoral rays. The pupil with a small depression (barely noticeable), with 1–3 melanophores below the eye (Figure 3C1). Slight or absence of ventral pigmentation (Figures 3B and 5A).

Postflexion (TL 6.0–6.9 mm): presence or absent vestigial finfold; 10–14 pectoral rays; 2–6 dorsal rays; 3–4 anal rays. Lateral pigmentation of the body with five patches of melanophores; melanophores on the dorsal part of the head and body; a patch of pigments between the fourth and fifth dorsal fin rays (Figure 3C).

Transition stage (TL 9–11 mm): no finfold; 13–15 pectoral rays; 6–10 dorsal rays; 6–8 anal rays; 4–7 pelvic rays. Head pigmented dorsally and laterally. A slightly oval pupil with further pronounced upper depression; a patch of melanophores below the eye (usually three to four). Incipient scales; pigments and melanophores on the dorsal part of the head and body; five well defined lateral bands, the second band is at the level of the anus, at this same level, is located a patch of pigments in the dorsal fin from the fourth to the seventh radius (Figure 3D).

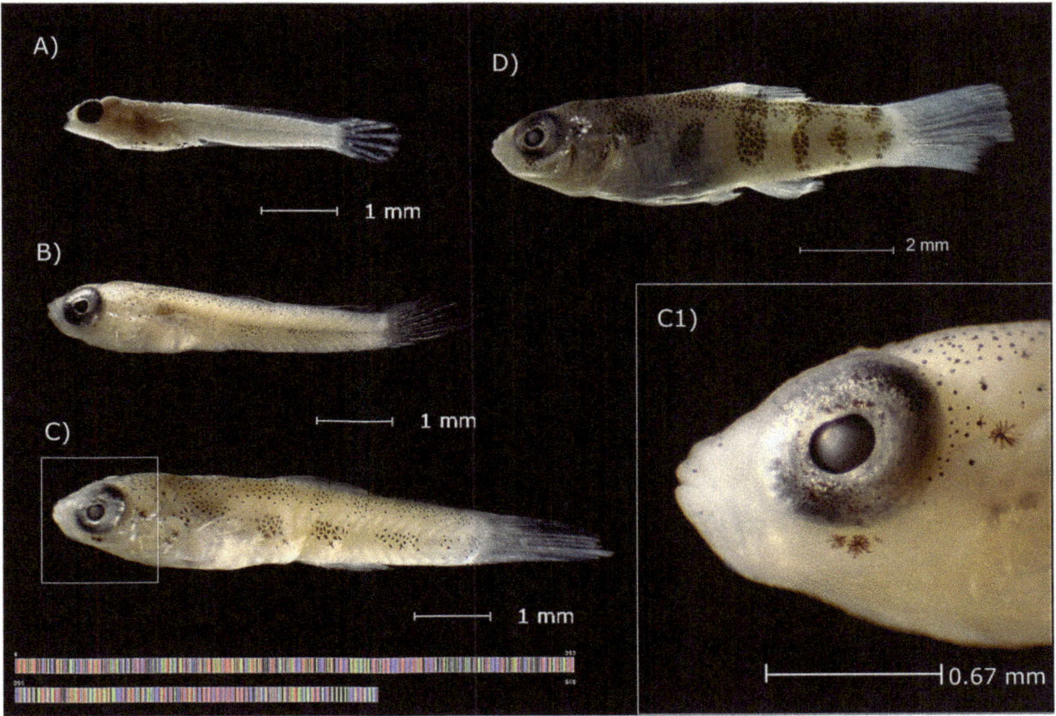

Figure 3. Early stages of development of *Cyprinodon artifrons* with their DNA barcode. (**A**) LT = 4.6 mm; (**B**) LT = 5.5mm; (**C**) LT = 6.9 mm; (**C1**) Pigments below the eye and pupil shape. (**D**) LT = 11 mm.

3.3. Floridichthys Polyommus (Hubbs, 1936)

Postflexion (TL 4.5–5.5 mm): presence of finfold; up to 10 pectoral rays. Robust head and dorsally pigmented; 4.5 mm larvae without pigment below the eye, from 5.4 mm pigmentation appears below the eye. Lateral pigmentation without pattern or with four to six patches of melanophores from anus to the base of the caudal fin; prominent abdomen with large branching melanophores (Figures 4A and 5B).

Postflexion (TL 6.0–6.5 mm): presence of finfold; 11–16 pectoral rays; up to 6 dorsal rays; up to 5 anal rays. A slightly oval pupil with an invagination of approximately 1/5 of the pupil diameter (Figure 4C1); pigmentation below the eye with 3–12 melanophores. Lateral pigmentation without pattern or with four to six patches of melanophores; the size of the abdomen and the ventral pigmentation begins to reduce from 6.5 mm (Figure 4B).

Postflexion (TL 7.0–8.5 mm): presence or absent vestigial finfold; 15 to 18 pectoral rays; 5–10 dorsal rays; 3–7 anal rays; up to 4 pelvic rays. The main characteristics of the eye are the oval shape of the pupil with invagination and the pigmentation below the eye. Lateral pigmentation without pattern or with four to five patches of melanophores; more dispersed pigmentation than *C. artifrons* in similar sizes already show pigmentation in the form of lateral bands. The abdomen with slight or totally absent pigmentation; pigment at the base of the first dorsal ray (dorsal view) (Figure 4C).

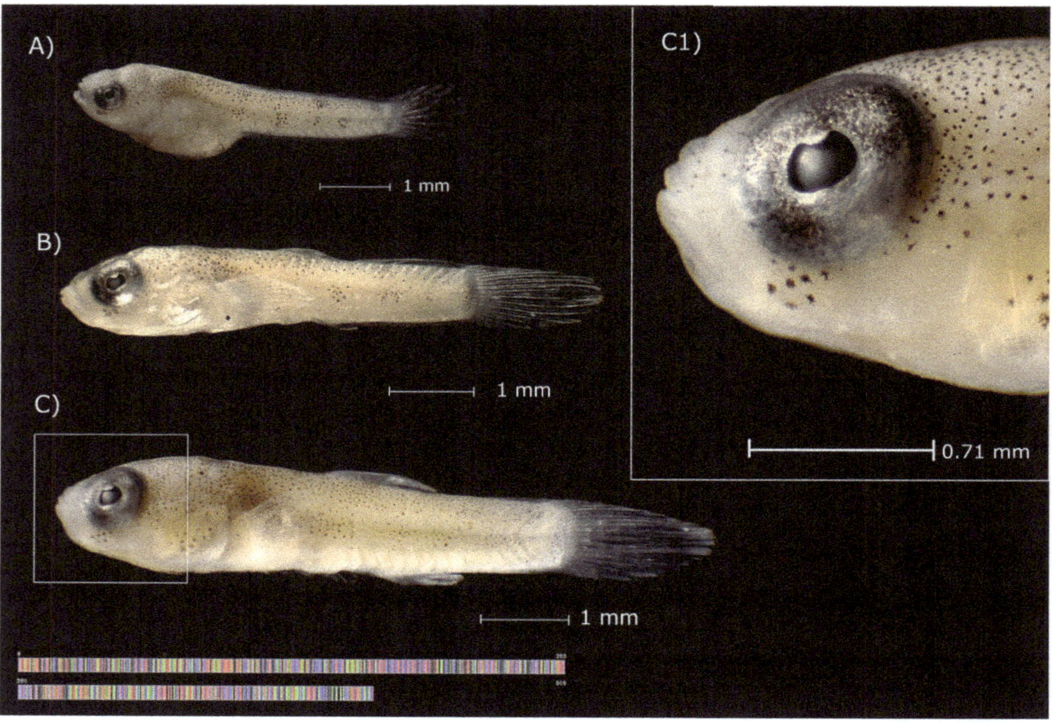

Figure 4. Early stages of development of *Floridichthys polyommus* with their DNA barcode. (**A**) LT = 5.4 mm; (**B**) LT = 6.4 mm; (**C**) LT = 7.5 mm. (**C1**) Pigments below the eye, invagination, and pupil shape.

Compared to *Cyprinodon artifrons*, *Floridichthys polyommus* has a prominent abdomen and large branching melanophores (Figure 5).

The characters used to separate and describe the larvae of *C. artifrons* and *F. polyommus* principally were pigmentation below the eye, lateral pigmentation, the pupil shape and ventral pigmentation. However, we consider that the last two are the most notable and relevant.

We observed that *F. polyommus* presented an invagination in the pupil in sizes greater than 5.5 mm, unlike *C. artifrons* had minor depression and was distinguishable only up to 9.0 mm.

The pigmentation below the eyes was different for both species. In the case of *C. artifrons*, most of the specimens had no more than three melanophores (Figure 3C1), while in *F. polyommus* up to 11 melanophores were observed (Figure 4C1).

In the smallest sizes, ventral pigmentation and the belly prominent was found for both species. The pigmentation in *C. artifrons* showed a linear pattern (Figure 5A), while *F. polyommus* showing this pigmentation more intense and with a branched pattern (Figure 5B). This character has been used previously to delimit some species of the Cyprinidae family [65].

The melanophores patches in the lateral side of the body, at the size of 6.5 mm in *C. artifrons* are well defined; from 9.0 mm, the melanophores are grouped into bands (Figure 3). In the case of *F. polyommus*, the patches of lateral pigments were less distinguishable from 6.5 mm (Figure 4).

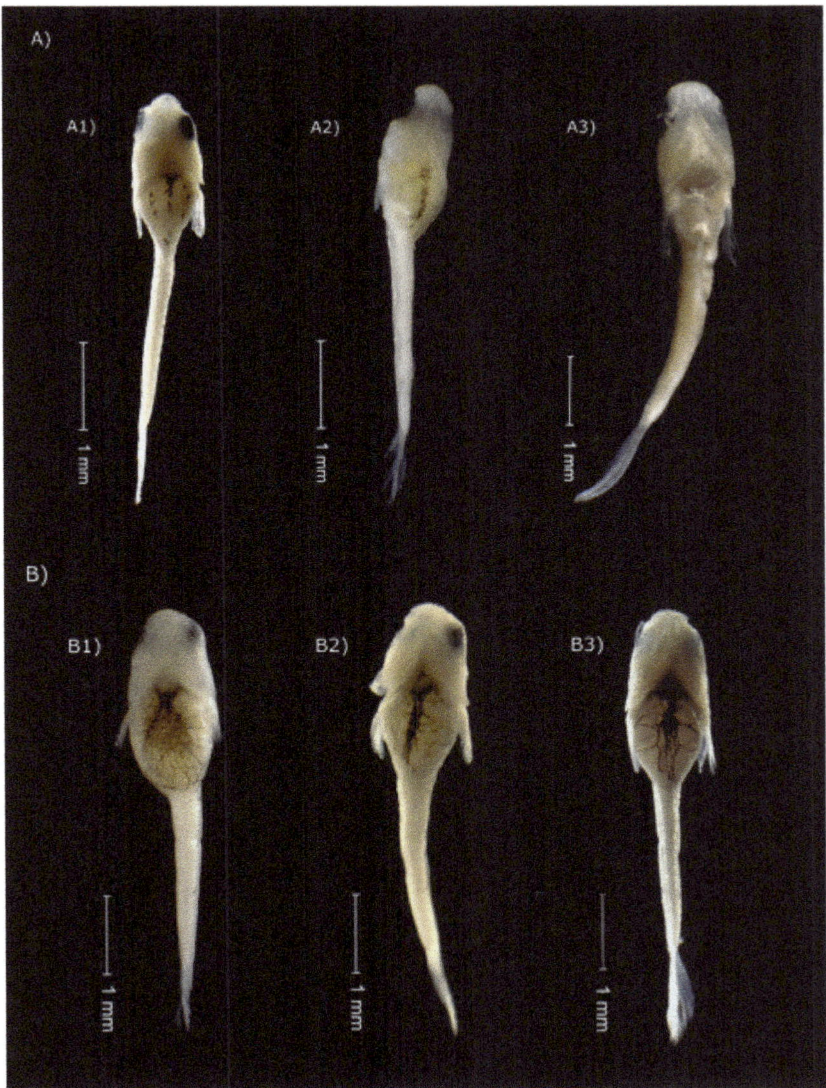

Figure 5. Ventral pigmentation patterns of (**A**) *Cyprinodon artifrons,* (**A1**) LT = 4.6 mm; (**A2**) LT = 4.5 mm; (**A3**) LT= 6.0 mm; and (**B**) *Floridichthys polyommus,* (**B1**) LT = 5.4 mm; (**B2**) LT = 5.5 mm; (**B3**) LT = 6.0 mm.

The smallest sizes were the most difficult to identify because most of the characteristics used to delimit these species appear after 5.0 mm.

It is essential to consider identifying the larvae of these species correctly. It is necessary to take all these characteristics into account as a whole and not individually.

4. Conclusions

In the present study, 95% of the specimens were identified employing DNA barcodes, providing the first report on the taxonomic composition of freshwater ichthyoplankton and their distribution in epicontinental systems in Quintana Roo state.

We report, for first time, the presence of *G. yucatanum* in the Lake Bacalar and the first larval description of *C. artifrons* (postflexion of and transition) and *F. polyommus* (postflexion).

Although the use of light traps has limitations due to selectivity, they have the advantage of catching species that often escape other sampling gears. However, the best approach is using various sampling gears to estimate species richness, measure biodiversity, and plan conservation strategies.

These results are the first report about some aspects of reproductive biology, and development of several species. This information is indispensable for ecological monitoring, to evaluate impacts, and developing management and conservation plans of biodiversity.

Our study confirmed the utility of the DNA barcodes for identifying larvae and juveniles by comparing their sequences with adults uploaded in BOLD.

This work confirms the importance of building DNA barcode reference libraries for all Mexican ichthyofauna.

This research is essential in areas under human pressure from different activities such as the Sian Ka'an Biosphere Reserve and the Lake Bacalar, where tourism is growing mostly in disordered ways.

Supplementary Materials: The following are available online at https://www.mdpi.com/article/10.3390/d13100513/s1, Supplementary Figure S1: ID tree of all specimens found in the dataset used for this paper. Include species name, sample ID, locality collection, BIN number.

Author Contributions: Conceptualization and methodology, M.V.-M. and A.E.U.-N.; software, M.V.-M. and A.E.U.-N.; validation, M.V.-M., A.E.U.-N. and C.A.V.-S.; formal analysis, A.E.U.-N., M.V.-M., C.A.V.-S. and J.A.C.-C.; investigation, A.E.U.-N. and M.V.-M.; resources, M.V.-M., C.A.V.-S. and Á.O.O.-M.; data curation, M.V.-M., A.E.U.-N. and J.A.C.-C.; writing—original draft preparation, M.V.-M., A.E.U.-N., C.A.V.-S. and J.A.C.-C.; writing—review and editing, M.V.-M., A.E.U.-N., C.A.V.-S. and J.A.C.-C.; visualization, A.E.U.-N., M.V.-M., C.A.V.-S. and Á.O.O.-M.; supervision, A.E.U.-N., M.V.-M. and C.A.V.-S.; project administration, M.V.-M. All authors have read and agreed to the published version of the manuscript.

Funding: This study was financed by the Global Environment Fund through the United Nations Development Programme (UNDP, Mexico), Comisión Nacional para el Conocimiento y Uso de la Biodiversidad (CONABIO) and Comisión Nacional de Áreas Naturales Protegidas (CONANP) as part of the investigation called: *Programa de detección temprana piloto de especies acuáticas invasoras a través de los métodos de código de barras de la vida y análisis de ADN ambiental en la Reserva de la Biosfera Sian Ka'an* within Project 00089333 "Aumentar las capacidades de México para manejar especies exóticas invasoras a través de la implementación de la Estrategia Nacional de Especies Invasoras". Part of genetic analyzes were funded by a Tecnológico Nacional de México project (serial: 10091.21-P).

Institutional Review Board Statement: Ethical review and approval were waived for this study due to it involving material collected under the permit issued by Dirección General de Ordenamiento Pesquero y Acuícola No P2F/DGOPA-063/20.

Data Availability Statement: All data from this study are available in dataset DS-FLYP Fish larvae from Yucatan Peninsula (dx.doi.org/10.5883/DS-FLYP). In addition, all sequences are available on GenBank (www.ncbi.nlm.nih.gov).

Acknowledgments: Jorge Vidal Quijada, Heliseo Torres Otero, Gilberto Paredes Pat, Isaías Tun Santiaog, and Alejandro Tuyub Tuyub from Limones town guided us within and in the surroundings of the Sian Ka'an Reserve. Edilberto Sosa Martín supported our sampling. All ejidal commissars assisted us in their communities. Alma Estrella García Morales from the Mexican Barcode of Life (MEXBOL) node Chetumal assisted with DNA extraction, PCR reactions, and sequence edition of all material presented here. Janneth Padilla assisted us with the map.

Conflicts of Interest: The authors declare no conflict of interest. The funders had no role in the study design, analyses, interpretation of data, the writing of the manuscript, or the decision to publish the results.

References

1. SECTUR. Results of Tourism Activity Mexico. Available online: http://www.acs-aec.org/sites/default/files/20150515_rat_a_marzo_15_en.pdf (accessed on 31 August 2021).
2. Sánchez-Flores, E. La Contribución Del Turismo Al Crecimiento Económico: Análisis Regional En México. *Transitare* **2016**, *2*, 183–204.
3. Martos-López, L.A. Underwater Archaeological Exploration of the Mayan Cenotes. *Mus. Int.* **2008**, *60*, 100–110. [CrossRef]
4. Campbell, M.; Withers, K.; Tolan, J. Occurrence of Larval and Juvenile Fish in Mangrove Habitats in the Sian Ka'an Biosphere Reserve, Quintana Roo, Mexico. *Gulf Caribb. Res.* **2008**, *20*, 81–86. [CrossRef]
5. Quintal-Lizama, C.; Vásquez-Yeomans, L. Asociaciones de Larvas de Peces En Una Bahía Del Caribe Mexicano. *Rev. Biol. Trop.* **2001**, *49*, 11.
6. Sanvicente-Añorve, L.; Hernández-Gallardo, A.; Gómez-Aguirre, S.; Flores-Coto, C. *Fish Larvae from a Caribbean Estuarine System in Book The Big Fish Bang*, 1st ed.; Browman, H.I., Skiftesvik, A.B., Eds.; The Instituto Marine Research: Bergen, Norway, 2003; pp. 365–379.
7. Valdez-Moreno, M.; Vásquez-Yeomans, L.; Elías-Gutiérrez, M.; Ivanova, N.; Hebert, P.D.N. Using DNA Barcodes to Connect Adults and Early Life Stages of Marine Fishes from the Yucatan. *Mar. Freshw. Res.* **2010**, *61*, 665–671. [CrossRef]
8. Victor, B.; Vásquez-Yeomans, L.; Valdez-Moreno, M.; Wilk, L.; Jones, D.; Shivji, M.; Lara, M.; Caldow, C. The Larval, Juvenile, and Adult Stages of the Caribbean Goby, *Coryphopterus Kuna* (Teleostei: Gobiidae): A Reef Fish with a Pelagic Larval Duration Longer than the Post-Settlement Lifespan. *Zootaxa* **2010**, *2346*, 53–61. [CrossRef]
9. Richards, W.J. *Early Stages of Atlantic Fishes*, 1st ed.; CRC Press: Boca Raton, FL, USA, 2005.
10. Rodríguez, J.M.; Alemany, F.; García, A. *A Guide to the Eggs and Larvae of 100 Common Western Mediterranean Sea Bony Fish Species*; The Food and Agriculture Organization: Rome, Italy, 2017.
11. Elías-Gutiérrez, M.; Valdez-Moreno, M.; Topan, J.; Young, M.; Cohuo-Colli, J. Improved Protocols to Accelerate the Assembly of DNA Barcode Reference Libraries for Freshwater Zooplankton. *Ecol. Evol.* **2018**, *8*, 3002–3018. [CrossRef]
12. Ko, H.L.; Wang, Y.T.; Chiu, T.S.; Lee, M.A.; Leu, M.Y.; Chang, K.Z.; Chen, W.Y.; Shao, K.T. Evaluating the Accuracy of Morphological Identification of Larval Fishes by Applying DNA Barcoding. *PLoS ONE* **2013**, *8*, 3–9. [CrossRef]
13. Dayrat, B. Towards Integrative Taxonomy. *Biol. J. Linn. Soc. Lond.* **2005**, *85*, 407–415. [CrossRef]
14. Steele, P.R.; Pires, J.C. Biodiversity Assessment: State-of-the-Art Techniques in Phylogenomics and Species Identification. *Am. J. Bot.* **2011**, *98*, 415–425. [CrossRef]
15. Baldwin, C.C.; Johnson, G.D. Connectivity across the Caribbean Sea: DNA Barcoding and Morphology Unite an Enigmatic Fish Larva from the Florida Straits with a New Species of Sea Bass from Deep Reefs off Curacao. *PLoS ONE* **2014**, *9*, e97661. [CrossRef]
16. Krishnamurthy, K.P.; Francis, R.A. A Critical Review on the Utility of DNA Barcoding in Biodiversity Conservation. *Biodivers. Conserv.* **2012**, *21*, 1901–1919. [CrossRef]
17. Hubert, N.; Delrieu-Trottin, E.; Irisson, J.O.; Meyer, C.; Planes, S. Identifying Coral Reef Fish Larvae through DNA Barcoding: A Test Case with the Families Acanthuridae and Holocentridae. *Mol. Phylogenet. Evol.* **2010**, *55*, 1195–1203. [CrossRef]
18. Baldwin, C.; Brito, B.; Smith, D.; Weigt, L.; Escobar-Briones, E. Identification of Early Life-History Stages of Caribbean Apogon (Perciformes: Apogonidae) through DNA Barcoding. *Zootaxa* **2011**, *3133*, 1–36. [CrossRef]
19. Prosser, S.; Martínez-Arce, A.; Elías-Gutiérrez, M. A New Set of Primers for COI Amplification from Freshwater Microcrustaceans. *Mol. Ecol. Resour.* **2013**, *13*, 1151–1155. [CrossRef]
20. Fahay, M.P. *Early Stages of Fishes in the Western North Atlantic Ocean: Davis Strait, Southern Greenland and Flemish Cap to Cape Hatteras*; Northwest Atlantic Fisheries Organization: Dartmouth, NS, Canada, 2007.
21. Beeching, S.C.; Pike, R.E. Ontogenetic Color Change in the Firemouth Cichlid, *Thorichthys Meeki*. *Copeia* **2010**, *2010*, 189–195. [CrossRef]
22. Miller, R.R.; Minckley, W.L.; Norris, S.M. *Peces Dulceacuícolas de México*, 1st ed.; CONABIO: Tlalpan, DF, Mexico, 2009.
23. Schmitter-Soto, J.J. *Catalogo de Los Peces Continentales de Quintana Roo*; El Colegio de la Frontera Sur: San Cistóbal de las Casas, Chiapas, Mexico, 1998.
24. Valdez-Moreno, M.; Ivanova, N.V.; Elías-Gutiérrez, M.; Pedersen, S.L.; Bessonov, K.; Hebert, P.D.N. Using eDNA to Biomonitor the Fish Community in a Tropical Oligotrophic Lake. *PLoS ONE* **2019**, *14*, e0215505. [CrossRef]
25. Valdez-Moreno, M.; Elías-Gutiérrez, M.; Mendoza-Carranza, M.; Rendón-Hernández, E.; Alarcón-Chavira, E. DNA Barcodes Applied to a Rapid Baseline Construction in Biodiversity Monitoring for the Conservation of Aquatic Ecosystems in the Sian Ka'an Reserve (Mexico) and Adjacent Areas. *Diversity* **2021**, *13*, 292. [CrossRef]
26. Ivanova, N.V.; Dewaard, J.R.; Hebert, P.D.N. An Inexpensive, Automation-Friendly Protocol for Recovering High-Quality DNA. *Mol. Ecol. Notes* **2006**, *6*, 998–1002. [CrossRef]
27. Hebert, P.D.N.; Cywinska, A.; Ball, S.L.; Dewaard, J.R. Biological Identifications through DNA Barcodes. *Proc. R. Soc. B* **2003**, *270*, 313–321. [CrossRef]
28. Ward, R.D.; Zemlak, T.S.; Innes, B.H.; Last, P.R.; Hebert, P.D.N. DNA Barcoding Australia's Fish Species. *Philos. Trans. R. Soc. London B Biol. Sci.* **2005**, *360*, 1847–1857. [CrossRef]
29. Ivanova, N.V.; Zemlak, T.S.; Hanner, R.H.; Hebert, P.D.N. Universal Primer Cocktails for Fish DNA Barcoding. *Mol. Ecol. Notes* **2007**, *7*, 544–548. [CrossRef]

30. Ratnasingham, S.; Hebert, P.D.N. Bold: The Barcode of Life Data System (Www.Barcodinglife.Org). *Mol. Ecol. Notes* **2007**, *7*, 355–364. [CrossRef]
31. Ratnasingham, S.; Hebert, P.D. A DNA-Based Registry for All Animal Species: The Barcode Index Number (BIN) System. *PLoS ONE* **2013**, *8*, e66213. [CrossRef] [PubMed]
32. Kimura, M. A Simple Method for Estimating Evolutionary Rates of Base Substitutions through Comparative Studies of Nucleotide Sequences. *J. Mol. Evol.* **1980**, *16*, 111–120. [CrossRef]
33. Saitou, N.; Nei, M. The Neighbor-Joining Method: A New Method for Reconstructing Phylogenetic Trees. *Mol. Biol. Evol.* **1987**, *4*, 406–425. [CrossRef]
34. Kumar, S.; Stecher, G.; Tamura, K. MEGA7: Molecular Evolutionary Genetics Analysis Version 7.0 for Bigger Datasets. *Mol. Biol. Evol.* **2016**, *33*, 1870–1874. [CrossRef]
35. Sarmiento-Camacho, S.; Valdez-Moreno, M. DNA Barcode Identification of Commercial Fish Sold in Mexican Markets. *Genome* **2018**, *61*, 457–466. [CrossRef]
36. Almeida, F.S.; Frantine-Silva, W.; Lima, S.C.; Garcia, D.A.Z.; Orsi, M.L. DNA Barcoding as a Useful Tool for Identifying Non-Native Species of Freshwater Ichthyoplankton in the Neotropics. *Hydrobiologia* **2018**, *817*, 111–119. [CrossRef]
37. Pegg, G.; Sinclair, B.; Briskey, L.; Aspden, W. MtDNA Barcode Identification of Fish Larvae in the Southern Great Barrier Reef, Australia. *Sci. Mar.* **2006**, *70*, 7–12. [CrossRef]
38. Loh, W.K.W.; Bond, P.; Ashton, K.J.; Roberts, D.T.; Tibbetts, I.R. DNA Barcoding of Freshwater Fishes and the Development of a Quantitative QPCR Assay for the Species-Specific Detection and Quantification of Fish Larvae from Plankton Samples. *J. Fish Biol.* **2014**, *85*, 307–328. [CrossRef] [PubMed]
39. Frantine-Silva, W.; Sofia, S.H.; Orsi, M.L.; Almeida, F.S. DNA Barcoding of Freshwater Ichthyoplankton in the Neotropics as a Tool for Ecological Monitoring. *Mol. Ecol. Resour.* **2015**, *15*, 1226–1237. [CrossRef] [PubMed]
40. Meusnier, I.; Singer, G.A.C.; Landry, J.-F.; Hickey, D.A.; Hebert, P.D.N.; Hajibabaei, M. A Universal DNA Mini-Barcode for Biodiversity Analysis. *BMC Genom.* **2008**, *9*, 214. [CrossRef] [PubMed]
41. Bhattacharjee, M.J.; Ghosh, S.K. Design of Mini-Barcode for Catfishes for Assessment of Archival Biodiversity. *Mol. Ecol. Resour.* **2014**, *14*, 469–477. [CrossRef]
42. Fields, A.T.; Abercrombie, D.L.; Eng, R.; Feldheim, K.; Chapman, D.D. A Novel Mini-DNA Barcoding Assay to Identify Processed Fins from Internationally Protected Shark Species. *PLoS ONE* **2015**, *10*, e0114844. [CrossRef]
43. Shokralla, S.; Hellberg, R.; Handy, S.; King, I.; Hajibabaei, M. A DNA Mini-Barcoding System for Authentication of Processed Fish Products OPEN. *Sci. Rep.* **2015**, *5*, 15894. [CrossRef]
44. Dhar, B.; Ghosh, S.K. Mini-DNA Barcode in Identification of the Ornamental Fish: A Case Study from Northeast India. *Gene* **2017**, *627*, 248–254. [CrossRef]
45. Bénard-Capelle, J.; Guillonneau, V.; Nouvian, C.; Fournier, N.; Le Loët, K.; Dettai, A. Fish Mislabelling in France: Substitution Rates and Retail Types. *PeerJ* **2015**, *2*, e714. [CrossRef]
46. Hanner, R.; Becker, S.; Ivanova, N.V.; Steinke, D. FISH-BOL and Seafood Identification: Geographically Dispersed Case Studies Reveal Systemic Market Substitution across Canada. *Mitochondrial DNA* **2011**, *22*, 106–122. [CrossRef]
47. Valdez-Moreno, M.; Pool-Canul, J.; Contreras-Balderas, S. A Checklist of the Freshwater Ichthyofauna from El Petén and Alta Verapaz, Guatemala, with Notes for Its Conservation and Management. *Zootaxa* **2005**, *1072*, 43–60. [CrossRef]
48. Anbarasi, G.; Vignesh, R.; Arulmoorthy, M.P.; Rathiesh, A.C.; Srinivasan, M. Barcoding Profiling and Intra Species Variation Within the Barcode Region of Two Estuarine Fishes Oreochromis Mossambicus and Oreochromis Niloticus. *GJBB* **2015**, *4*, 59–68.
49. Whitehead, P.J.P.; Nelson, G.J.; Wongratana, T. *FAO Species Catalogue, Clupeoid Fishes of the World (Engraulidae)*; FAO: Rome, Italy, 1988; pp. 305–579.
50. Dawson, C.E. *Gobiosoma (Garmannia) Yucatanum*, a New Seven-Spined Atlantic Goby from Mexico. *Copeia* **1971**, *1971*, 432–439. [CrossRef]
51. Greendfield, D.W.; Thomerson, J.E. *Fishes of the Continental Waters of Belize*, 1st ed.; University Press of Florida: Gainesville, FL, USA, 1997.
52. FishBase. Available online: http://www.fishbase.org (accessed on 29 July 2021).
53. López-Vila, J.M.; Valdez-Moreno, M.; Schmitter-Soto, J.J.; Mendoza-Carranza, M.; Herrera-Pavón, R.L. Composición y Estructura de La Ictiofauna Del Río Hondo, México-Belice, Con Base En El Uso Del Arpón. *Rev. Mex. Biodivers.* **2014**, *85*, 866–874. [CrossRef]
54. Zambrano, L.; Vázquez-Domínguez, E.; Garcia-Bedoya, D.; Loftus, W.; Trexler, J. Fish Community Structure in Freshwater Karstic Water Bodies of the Sian Ka'an Reserve in the Yucatan Peninsula, Mexico. *Ichthyol. Explor. Freshw.* **2006**, *17*, 193–206.
55. Escalera-Vázquez, L.H.; Zambrano, L. The Effect of Seasonal Variation in Abiotic Factors on Fish Community Structure in Temporary and Permanent Pools in a Tropical Wetland. *Freshw. Biol.* **2010**, *55*, 2557–2569. [CrossRef]
56. Camargo-Guerra, T.; Escalera-Vázquez, L.H.; Zambrano, L. Fish Community Structure Dynamics in Cenotes of the Biosphere Reserve of Sian Ka'an, Yucatán Peninsula, Mexico. *Rev. Mex. Biodivers.* **2013**, *84*, 901–911. [CrossRef]
57. Montes-Ortiz, L.; Elías-Gutiérrez, M. Faunistic Survey of the Zooplankton Community in an Oligotrophic Sinkhole, Cenote Azul (Quintana Roo, Mexico), Using Different Sampling Methods, and Documented with DNA Barcodes. *J. Limnol.* **2018**, *77*, 428–440. [CrossRef]

58. Vásquez-Yeomans, L.; Vega-Cendejas, M.E.; Montero, J.L.; Sosa-Cordero, E. High Species Richness of Early Stages of Fish in a Locality of the Mesoamerican Barrier Reef System: A Small-Scale Survey Using Different Sampling Gears. *Biodivers. Conserv.* **2011**, *20*, 2379–2392. [CrossRef]
59. Magalhães, A.L.B.; Orsi, M.L.; Pelicice, F.M.; Azevedo-Santos, V.M.; Vitule, J.R.S.; Lima-Junior, D.P.; Brito, M.F.G. Small Size Today, Aquarium Dumping Tomorrow: Sales of Juvenile Non-Native Large Fish as an Important Threat in Brazil. *Neotrop. Ichthyol.* **2017**, *15*, e170033. [CrossRef]
60. Trujillo-Jiménez, P.; Sedeño-Díaz, J.E.; López-López, E. Reproductive Traits of the "Ocellated Killifish" *Floridichthys Polyommus* Hubbs, 1936 (Pisces: Cyprinodontidae) Inhabiting Estuary of the Champotón River (Campeche, Mexico). *J. Appl. Ichthyol.* **2018**, *34*, 806–814. [CrossRef]
61. Koumoundouros, G.; Gagliardi, F.; Divanach, P.; Boglione, C.; Cataudella, S.; Kentouri, M. Normal and Abnormal Osteological Development of Caudal Fin in *Sparus Aurata* L. Fry. *Aquaculture* **1997**, *149*, 215–226. [CrossRef]
62. Faustino, M.; Power, D.M. Development of the Pectoral, Pelvic, Dorsal and Anal Fins in Cultured Sea Bream. *J. Fish Biol.* **1999**, *54*, 1094–1110. [CrossRef]
63. López, N.; Ruíz, C.; Landines, M. Descripcion Macroscopica Del Desarrollo Larval Del Coporo (*Prochilodus Mariae*). *Rev. Fac. Med. Vet. Zootec.* **2005**, *52*, 110–119. [CrossRef]
64. Thorsen, D.H.; Hale, M.E. Development of Zebrafish (*Danio Rerio*) Pectoral Fin Musculature. *J. Morphol.* **2005**, *266*, 241–255. [CrossRef]
65. Fuiman, L.A.; Conner, J.V.; Lathrop, B.F.; Buynak, G.L.; Snyder, D.E.; Loos, J.J. State of the Art of Identification for Cyprinid Fish Larvae from Eastern North America. *Trans. Am. Fish. Soc.* **1983**, *112*, 319–332. [CrossRef]

Article

Molecular Characterization of the Common Snook, *Centropomus undecimalis* (Bloch, 1792) in the Usumacinta Basin

Jazmín Terán-Martínez [1], Rocío Rodiles-Hernández [2], Marco A. A. Garduño-Sánchez [3] and Claudia Patricia Ornelas-García [4,*]

1 Doctorado en Ecología y Desarrollo Sustentable, Departamento de Conservación de la Biodiversidad, El Colegio de la Frontera Sur, Carretera Panamericana y Periférico sur s/n, Barrio Ma. Auxiliadora, San Cristóbal de Las Casas C.P. 29290, Mexico; jateran@ecosur.edu.mx
2 Departamento de Conservación de la Biodiversidad, El Colegio de la Frontera Sur, Carretera Panamericana y Periférico sur s/n, Barrio Ma. Auxiliadora, San Cristóbal de Las Casas C.P. 29290, Mexico; rrodiles@ecosur.mx
3 Posgrado en Ciencias Biológicas, Departamento de Zoología, Instituto de Biología, Universidad Nacional Autónoma de Mexico, Tercer Circuito Exterior S/N, Ciudad de Mexico C.P. 045110, Mexico; marco.garduno@st.ib.unam
4 Departamento de Zoología, Instituto de Biología, Universidad Nacional Autónoma de Mexico, Tercer Circuito Exterior S/N, Ciudad de Mexico C.P. 045110, Mexico
* Correspondence: patricia.ornelas.g@ib.unam.mx; Tel.: +52-55-4940-9215

Citation: Terán-Martínez, J.; Rodiles-Hernández, R.; Garduño-Sánchez, M.A.A.; Ornelas-García, C.P. Molecular Characterization of the Common Snook, *Centropomus undecimalis* (Bloch, 1792) in the Usumacinta Basin. *Diversity* **2021**, *13*, 347. https://doi.org/10.3390/d13080347

Academic Editor: Bert W. Hoeksema

Received: 28 April 2021
Accepted: 26 May 2021
Published: 29 July 2021

Publisher's Note: MDPI stays neutral with regard to jurisdictional claims in published maps and institutional affiliations.

Abstract: The common snook is one of the most abundant and economically important species in the Usumacinta basin in the Gulf of Mexico, which has led to overfishing, threatening their populations. The main goal of the present study was to assess the genetic diversity and structure of the common snook along the Usumacinta River in order to understand the population dynamics and conservation status of the species. We characterized two mitochondrial markers (mtCox1 and mtCytb) and 11 microsatellites in the Usumacinta basin, which was divided into three zones: rainforest, floodplain and river delta. The mitochondrial data showed very low diversity, showing some haplotypic diversity differences between the rainforest and delta zones. In contrast, we consistently recovered two genetic clusters in the Usumacinta River basin with the nuclear data in both the DAPC and STRUCTURE analyses. These results were consistent with the AMOVA analyses, which showed significant differences among the genetic clusters previously recovered by DAPC and STRUCTURE. In terms of diversity distribution, the floodplain zone corresponded to the most diverse zone according to the mitochondrial and nuclear data, suggesting that this is a transition zone in the basin. Our results support the relevance of the molecular characterization and monitoring of the fishery resources at the Usumacinta River to better understand their connectivity, which could help in their conservation and management.

Keywords: gene flow; hydrological connectivity; Usumacinta Basin; Gulf of Mexico; tropical rainforest; *Centropomus undecimalis*

1. Introduction

Ecosystem integrity and aquatic biodiversity are largely determined by hydrologic connectivity [1,2]. For freshwater ecosystems, connectivity involves the exchange of matter, energy and organisms along the river; thus, species can move among feeding, spawning and refuge habitats [3]. Connectivity comprises four dimensions: longitudinal, lateral, vertical and temporal [4]. Longitudinal connectivity is considered one of the most important dimensions of freshwater fish species' connectivity [5] because it allows upstream and downstream fish migration cycles to occur [6]. For migratory species, the maintenance of longitudinal connectivity is very important; therefore, being able to evaluate their presence and extension is of great importance to better understand what the threats to their conservation could be [5,7–10].

Genetic data represent a valuable tool for assessing connectivity [11] by providing relevant information about gene exchange within and across populations [12]. Due to next-generation sequencing, DNA barcodes have been used to monitor and explore biological diversity with molecular markers like never before. In this sense, the Cytochrome c oxidase subunit 1 (mtCox1) barcoding region has been widely used as a valuable marker in vertebrates for phylogeography and conservation biology [13]. In this regard, previous studies using DNA sequences have served not only to characterize cryptic diversity but also to diagnose population variants within species [14,15], which, combined with nuclear markers, could be useful for species management and conservation [16].

The common snook, *Centropomus undecimalis*, is an euryhaline fish, which means that the species breeds at river mouths or in estuarine environments and then migrates to river environments to feed [17]. The species *C. undecimalis* is widely distributed along the Atlantic slope, from the coast of North Carolina, USA, to Brazil. In the Gulf of Mexico, it is found at the mouths of the main basins [18,19], including the Usumacinta River basin, where the species coexists with three other species of the genus: *C. poeyi*, *C. parallelus* and *C. mexicanus* [20]. However, *C. undecimalis* is the most abundant, largest and most economically relevant of these species due to its high commercial value [19,21]; therefore, the species is a key resource for the local fisheries, which has led to overfishing, threatening its populations in some regions [19,22–24]. Thus, we consider that molecular characterization of the common snook populations in the Usumacinta River basin could shed light on the conservation and management of the species.

Previous genetic studies of the common snook in the Usumacinta River basin recovered a single genetic pool, including samples from the San Pedro River, at the basin's floodplain, to the coastal area at Tabasco, Mexico [25]. However, other studies using freshwater species diversity and molecular data have suggested that the upper and lower part of the Usumacinta basin are different biogeographic units [10,26–29]. In this sense, Ornelas-García et al. [10] reported that despite the connectivity within the basin, the genetic diversity could be heterogeneously distributed, at least in *Astyanax aeneus* species, where the upper and lower basin present different levels of haplotype diversity, while the middle part of the basin presents the highest diversity, tentatively associated with a transition zone. Similarly, Elías et al. [29] suggested that the upper and lower Usumacinta River basin does not correspond to a single biogeographic unit, based on endemic species diversity as well as the phylogeographic patterns obtained with some representative fish groups in the basin (i.e., cichlids and poeciliids).

In the present study, we assessed *C. undecimalis* genetic diversity and structure in the Usumacinta River basin by means of the genetic characterization of two mitochondrial markers (mtCox1 and mtCytb) and 11 nuclear microsatellite loci. For this purpose, we conducted a sampling in the Usumacinta River basin, within a region of more than 600 km along its course, from the upper part of the basin in Mexico, at the tropical rainforest (in the Lacandon forest, Chiapas Mexico), throughout the Usumacinta River's course until reaching the river mouth in the Biosphere Reserve Pantanos de Centla, Tabasco, and the coastal lagoon of Terminos, Campeche, in Mexico.

2. Materials and Methods

2.1. Sample Collection and DNA Extraction

In total, 81 individuals of *C. undecimalis* were collected from 15 sampling localities along the Usumacinta River basin in Mexico, during the rainy and dry seasons between February 2015 and March 2019 (Figure 1). We used the hydrological subdivision proposed by Soria-Barreto et al. [20] with some modifications; thus, 3 geographical zones were defined, considering river sub-basins as well as previously described fish diversity. The first was the rainforest zone (RZ), which is the upper zone of the basin in the Mexican portion, with most of the sampling points being included within the Montes Azules Biosphere Reserve, except for the Benemerito location; thus, we collected samples from the Lacantun River towards the Benemerito location and its confluence with the Usumacinta River. The floodplain zone (FZ) included six sampling points, from the Emiliano Zapata location to the Jonuta location, following the course of the Usumacinta River, including the floodplain lagoons of Canitzan and Catazajá. Additionally, two tributaries in this zone were sampled: the San Pedro River (at the border between Mexico and Guatemala) and the Chacamax River. Finally, the Usumacinta River delta (RD) was divided into three branches, and we sampled two of them at five sampling points: Salsipuedes at the confluence between the Usumacinta River and the Grijalva River, at the Pantanos de Centla Biosphere Reserve, the Palizada River and Pom Lagoon (both in a coastal lagoon of Terminos (a RAMSAR site)), and at the sea, in front of the Campeche coastline (see Figure 1, Supplementary Material S1, Table S1: Sampling localities from the Usumacinta River basin). The specimens were collected using gill nets and harpoons. All specimens were identified using the keys of Castro-Aguirre [22] and Miller et al. [24].

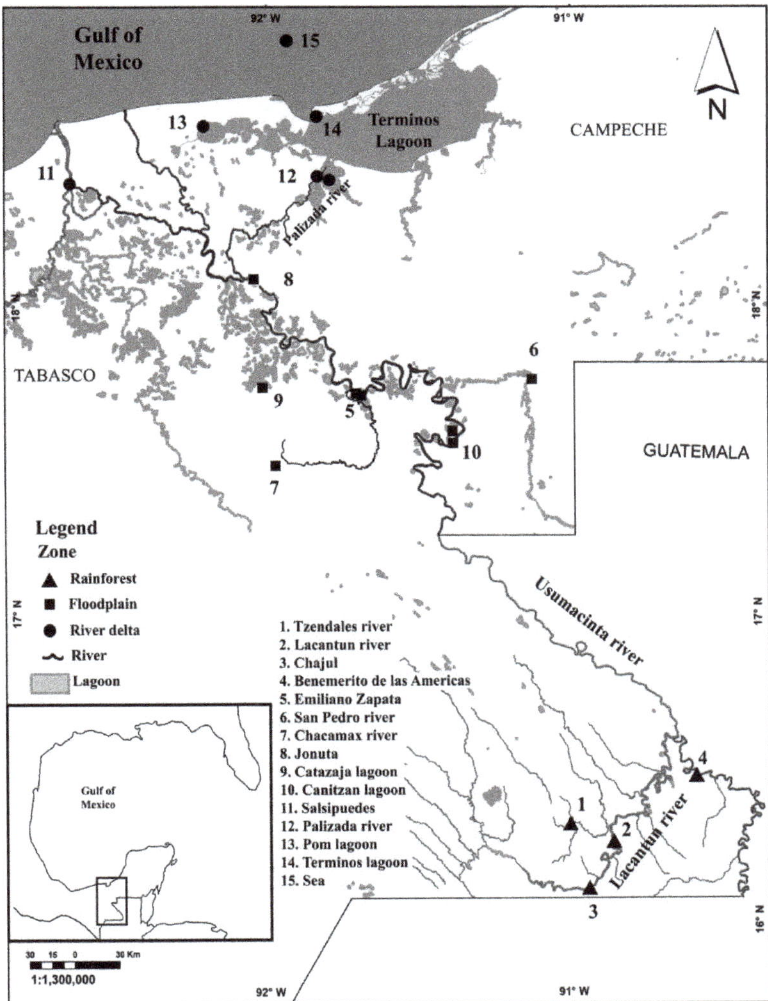

Figure 1. Map of the sampled localities for *Centropomus undecimalis*. Triangles correspond to the rainforest (RZ), squares correspond to the floodplain (FZ) and the circles correspond to the river delta (RD). Lagoons are represented by the shaded polygons.

Fin clip samples were taken from all individuals and preserved in 90% ethanol and stored at −20 °C. Some individuals were preserved in formalin (10%) as voucher specimens for future morphological analyses. The voucher samples were deposited at the Fish Collection (ECOSC) at ECOSUR in San Cristobal de las Casas, Chiapas, Mexico. From a fin clip, the DNA was extracted using a standard protocol involving proteinase-K in SDS/EDTA digestion and NaCl (4.5 M) and chloroform, as described by Sonnenberg et al. [30]. Both DNA quality and concentrations were measured using a Nanodrop 1000 device (Thermo Scientific, Mexico city, Mexico).

2.2. Mitochondrial and Nuclear Amplification

Out of the 81 collected samples (see Supplementary Material S1, Table S1), a subset of 72 samples was successfully amplified for a fragment of the cytochrome oxidase mitochondrial gene (mtCox1) with the Fish F (5′-TTC TCA ACT AAC CAY AAA GAY ATY GG-3) and Fish R (5′-TAG ACT TCT GGG TGG CCR AAR AAY CA-3) primers [31]. However, only 34 samples were successfully amplified for the mitochondrial cytochrome b gene (mtCytb) fragment; the primers used were GLU DG (5′-TGACCTGAAR-AACCAYCGTTG-3′) and H1690 (5′-CGAYCTTCGGATTACAAGACCG-3′) [32]. For both fragments, amplification was performed in a 10 µL reaction containing 2 µL of template DNA, 2 µL of a buffer, 3–4 mM MgCl2, 0.6 mM dNTPs, 0.1 µL of Taq DNA polymerase and 0.2 µL of each primer. The cycling conditions included an initial denaturation at 94 °C for 4 min, followed by 30 cycles of 94 °C for 45 s, a primer annealing temperature of 51 °C for mtCox1 and a primer annealing temperature was 48 °C for mtCytb for 30 s, and 72 °C for 30 s, with a final extension at 72 °C for 10 min. For verification of DNA quality, electrophoresis was performed in 1% agarose gels. The amplified fragments were analyzed using the Applied Biosystems 3730xl with 96 capillaries (at the National Biodiversity Laboratory (LANABIO) IBUNAM, Mexico City, Mexico).

Microsatellite loci were selected from a previous study in *C. undecimalis* [33]. Among the loci described therein, 11 loci were chosen (*Cun*01, *Cun*02, *Cun*04A, *Cun*06, *Cun*09, *Cun*10A, *Cun*14, *Cun*20, *Cun*21A, *Cun*21B and *Cun*22) and amplified for a total of 81 samples (see Supplementary Material S1, Table S2: Eleven microsatellite loci used in determining genetic variation among the *Centropomus undecimalis* samples from 13 sampling localities along the Usumacinta River basin). The forward primers of these 11 primer pairs were fluorescently labeled with the 6-FAM and HEX dyes (Macrogen, Inc., Seoul, South Korea). The loci were amplified in 2 multiplex reactions using a Multiplex PCR kit (QIAGEN) in a 5 µL final reaction volume following the kit instructions. The PCR amplification procedure consisted of 1 cycle of denaturation at 95 °C for 5 min, then 35 cycles of 94 °C for 30 s, annealing for 30 s at 52–60 °C and extension at 72 °C for 45 s, followed by a final 7-min extension at 72 °C. To verify which microsatellites were amplified successfully, the PCR products were visualized in a 2% agarose gel in 1× TAE buffer at 120 V for 30 min and stained with the GelRed Nucleic Acid Gel Stain (Biotium, San Francisco, CA, USA). Allele sizes were determined by comparing the fragments with the LIZ 500 Size Standard (Applied Biosystems, Waltham, MA, USA). Allele scoring was performed using Geneious v 10.6, USA (www.geneious.com, accessed on 28 April 2021).

2.3. mtDNA Diversity and Genetic Differentiation

We performed independent analyses for each mtDNA fragment, due to the great difference in the number of amplified samples between the 2 mitochondrial markers (72 for mtCox1 and 34 for mtCytb). A *de novo* alignment was performed using the BIOEDIT Sequence Alignment Editor [34]. We checked each chromatogram to verify each position by eye and corrected sequencing errors if necessary. We calculated the number of haplotypes (*h*), haplotypic diversity (*Hd*), nucleotide diversity (*pi*) and the number of polymorphic sites with DnaSP v.6.12.03 [35]. The haplotype networks were constructed for each mitochondrial fragment independently (for mtCOX1 and mtCytb) with PopArt software V. 1.7 [36] using the ML topology. We constructed an ML phylogenetic tree with RAxML v8.2.X software [37], implemented via the Cipres web portal [38]. We used JMODELTEST v2.1.1 [39] to identify the most appropriate model of sequence evolution for both mtDNA fragments (HKY + G). We performed a hierarchical analysis of molecular variance (AMOVA) [40] to partition the mtDNA genetic variation into a geographical context, including the variation among zones (RZ, FZ and RD). The significance of the variance components associated with the different levels of genetic structure was tested using nonparametric permutation procedures as implemented in Arlequin V3.5.2.2 [40].

2.4. Microsatellite Diversity and Genetic Differentiation

We genotyped 81 individuals of *C. undecimalis*, with 11 microsatellite loci (see Supplementary Material S1, Table S3: Genotypes of nine loci of *Centropomus undecimalis* populations). The fragment length was standardized with an internal size marker, GeneScan-500 Liz (Applied Biosystems), in Genious R.10.6, USA (www.geneious.com, accessed on 28 April 2021). Allele frequency tables for individuals were created using the Bin utility in Genious R.10.6, USA. Genotypes were checked with Micro-Checker v2.23 [41] for null alleles, large allele dropout and stutter bands. Departure from Hardy–Weinberg equilibrium and linkage disequilibrium between loci were tested with GENEPOP v4.7 (online version) [42] using a Markov chain algorithm with 10,000 iterations for dememorization, 100 batches and 5000 iterations per batch. The following basic genetic statistics were calculated using GenAlEX [43]: number of alleles per locus (*Na*), effective number of alleles per locus (*Ne*), observed (*Ho*) and expected (*He*) heterozygosity for each locus, fixation index (F_{IS}) values per locus and the number of private alleles. We used Arlequin V3.5. [40] to calculate *F*-statistics to measure the genetic differentiation among populations from different sites [44].

To identify the genetic structure of *C. undecimalis* in the Usumacinta River basin, we carried out a Bayesian clustering analysis in STRUCTURE v2.3.3 [45]. This method allowed us to determine the optimal number of groups or clusters (K), assigning each individual to 1 or more groups. We applied the admixture model and the uncorrelated allele frequency model. To determine the optimal number of clusters, without prior information, the program was run 10 times for different K values (K from 1 to 13 + 1); for each run, the MCMC algorithm was run with 1 M replicates and a burn-in of 200,000 replicates. To determine the most likely K value based on the DK method, also known as the Evanno method [46], the STRUCTURE results were analyzed with STRUCTURE Harvester [47].

In addition, a discriminant analysis of principal components (DACP) [48,49] was performed in RStudio [50]. DAPC is a multivariate analysis designed to identify and describe clusters of genetically related individuals. DAPC relies on data transformation using PCA as a prior step to discriminant analysis (DA), ensuring that the variables submitted to DA are perfectly uncorrelated. The DA method defines a model in which genetic variation is partitioned into a between-group and a within-group component, and yields synthetic variables which maximize the first while minimizing the second [48].

DAPC was performed with the microsatellite data, using the individual clustering assignment found in STRUCTURE. In this regard, we used the individual cluster assignments based on the 2 best Structure K values (K = 2 and K = 3). We also ran the K-means clustering algorithm (which relies on the same model as DA) with different numbers of clusters, each of which gave rise to a statistical model and an associated likelihood. With the *find.clusters* function of the Adegenet package v. 2.1.3 [48], we evaluated K = 1 to K = 13 possible clusters in 10 different iterations (DAPC) [49]. In both cases, the selection of the number of principal components was carried out with a cross-validation analysis. The validation set was selected via stratified random sampling, which guaranteed that at least 1 member of each conglomerate or cluster was represented in both the training set and the validation set [51]. The clusters or conglomerates resulting from the DAPC were visualized in a scatter diagram, using the first 2 discriminant functions, representing individuals as points, whereas genetic groups were enclosed by inertia ellipses, with a positive coefficient for the inertia ellipse size of 1.5. The clusters were grouped by their proximity in the discriminant space through a minimum spanning tree (MST). The proportions of intermixes, obtained from the membership probability based on the retained discriminant functions, were plotted for each individual. The DAPC admixture index was plotted using Structurly [52]. Additionally, we tested the possible presence of substructures within the estimated clusters with the *find.clusters* function in Rstudio [50].

Using the microsatellite data and the genetic groups estimated from the DAPC analysis, we calculated a genetic distance matrix, based on the number of allelic differences between individuals (Hamming distances), to construct minimum expansion networks with the *poppr* package, version 2.8.5 [53]. In the estimated minimum expansion network, each node represents the multilocus genotypes of the different samples, and the edges represent the genetic distances connecting the multilocus genotypes [53].

We estimated gene flow based on recent migration rates (m) among the genetic clusters obtained with STRUCTURE software (K = 2 and K = 3) and DAPC (K = 3) using an assignment test in BayesASS 3.0.3 [54]. First, we ran BayesASS 3.0.3 for 10,000,000 iterations with a burn-in of 1,000,000 to adjust the mixing parameters, in order to have acceptance rates for the proposed changes in the parameters between 20% and 60%, according to the BayesASS user manual [54]. The delta values used for allele frequency (a), migration rate (m) and the inbreeding (f) coefficient were 0.4, 0.2 and 0.55, respectively. With those parameters, we then ran BayesASS iteratively 10 times, with different starting seeds, and the total log likelihood was plotted on tracer to assess convergence within runs. The number of times each outcome was achieved over the 10 runs was recorded, and the mean migration rates were calculated for each of these outcomes. Migration rates with lower 95% confidence intervals below $m = 0.02$ were not considered significant and were also omitted. The effective size (Ne) of the populations was estimated in the web version of NeEstimator v2.1 via the linkage disequilibrium (LD) method and with 95% CI [55].

Finally, 2 different groupings were tested in the hierarchical AMOVA using data from the 11 microsatellite loci: (1) geographical criteria, including variation among basin zones (RZ, FZ and RD), among localities within the zones and within localities; and (2) according to the clusters obtained with the DACP analysis (K = 3) and STRUCTURE software (K = 2 and K = 3). A sequential Bonferroni test was performed to adjust the critical value of significance [56].

2.5. Isolation by Distance

To detect the effect of isolation by geographical distance (IBD), we compared the correlation of genetic distance (F_{ST}), RE = $F_{ST}/(1 - F_{ST})$ [57] with geographical distance [58]. The distances were obtained by following the channels of the rivers sampled with the measure tool of the ArcGIS program. IBD was estimated using the correlation coefficient (R^2) for all pairs of populations for the 2 mitochondrial fragments and the 11 microsatellite loci with the Mantel test in the *vegan* package in RStudio [50].

3. Results

A fragment of 574 bases of mtCox1 ($n = 72$) and a fragment of 454 bases of mtCytb ($n = 34$) were obtained from three zones within the Usumacinta basin. In the dataset for the mtCox1 fragment, we recovered eight haplotypes with a total of seven variable sites, with a low haplotypic diversity of $Hd = 0.281$ and a low nucleotide diversity with a π value of 0.0006 (Table 1). The haplotype network for mtCox1 showed a star-like shape (Figure 2A), with Hap3 showing the highest frequency and being present at the 80% the sampled zones (Figure 2B). Despite this, the results indicated the presence of exclusive haplotypes for each zone: rainforest, Hap 1; floodplain, Hap 4, 5 and 6; river delta, Hap 7 and 8.

Table 1. Genetic diversity estimations of mtDNA for each zone of the common snook (*Centropomus undecimalis*): n = number of sequences; h = number of haplotypes; H_d = haplotypic diversity; S = number of variable sites; k = average pairwise nucleotide differences; π = nucleotide diversity. Summary statistics for nine polymorphic microsatellite loci: n = number of individuals with amplification; N_a = number of alleles per locus; N_e = effective number of alleles per locus; H_O = observed heterozygosity; H_e = expected heterozygosity; F_{IS} = fixation index given for each locus. Values with the asterisk represent significant deviations from Hardy–Weinberg equilibrium.

		mtDNA						***Locus***						
Group		**mtCox**	**Cyt-b**		***Cun20***	***Cun 10A***	***Cun 02***	***Cun 21A***	***Cun 04A***	***Cun 01***	***Cun 021B***	***Cun 06***	***Cun 14***	
Rainforest zone n = 23	n	17	10	n	20	20	21	22	18	23	22	22	22	
	h	3	2	N_a	14.000	15.000	8.000	5.000	5.000	8.000	5.000	7.000	10.000	
	H_d	0.323	0.466	N_e	7.619	11.429	6.300	1.770	3.927	3.348	3.796	4.990	5.378	
	S	2	1	H_o	0.950	0.900	0.571	0.500	0.889	0.739	0.864	0.636	0.864	
	k	0.338	0.001	H_e	0.869	0.913	0.841	0.435	0.745	0.701	0.737	0.800	0.814	
	π	0.0005	0.466	F_{IS}	−0.094 *	0.014	0.321	−0.150	−0.193	−0.054	−0.173	0.204 *	−0.061	All locus F_{IS} −0.011
Floodplain zone n = 29	n	29	10	n	24	27	29	29	23	27	29	26	29	
	h	5	4	N_a	15.000	19.000	8.000	3.000	6.000	11.000	6.000	6.000	10.000	
	H_d	0.369	0.822	N_e	6.776	12.678	6.570	1.597	3.792	3.455	3.103	4.711	6.029	
	S	3	2	H_o	0.708	0.815	0.448	0.483	0.696	0.556	0.724	0.577	0.862	
	k	0.458	1.08	H_e	0.852	0.921	0.848	0.374	0.736	0.711	0.678	0.788	0.834	
	π	0.0008	0.002	F_{IS}	0.169	0.115	0.471 *	−0.291	0.055	0.218 *	−0.068	0.268 *	−0.033 *	All locus F_{IS} 0.085
River delta n = 29	n	26	14	n	29	29	29	29	29	29	29	28	29	
	h	3	2	N_a	6.000	17.000	3.000	3.000	5.000	11.000	6.000	4.000	10.000	
	H_d	0.150	0.538	N_e	4.890	7.410	2.683	1.279	3.266	2.448	4.102	1.744	6.207	
	S	3	1	H_o	0.690	0.862	0.552	0.241	0.621	0.586	0.793	0.571	0.759	
	k	0.230	0.538	H_e	0.795	0.865	0.627	0.218	0.694	0.592	0.756	0.427	0.839	
	π	0.0004	0.001	F_{IS}	0.133	0.003	0.120	−0.106	0.105	0.009 *	−0.049	−0.339	0.096	All locus F_{IS} 0.011
All	n	72	34	n	73	76	79	80	70	79	80	76	80	
	h	8	4	N_a	18	21	8	6	6	14	6	9	14	
	H_d	0.281	0.615	N_e	4.216	5.933	3.250	1.562	3.154	2.937	3.111	2.370	4.390	
	S	7	2	H_o	0.784	0.868	0.495	0.410	0.732	0.652	0.772	0.568	0.822	
	k	0.354	0.722	H_e	0.743	0.815	0.647	0.276	0.674	0.635	0.661	0.527	0.756	
	π	0.0006	0.001	F_{IS}	0.071	0.049	0.331	−0.203	−0.015	0.067	−0.097	0.113	0.001	

For the mtCytb fragment, four haplotypes were identified with two variable sites, with a relatively larger haplotype diversity (*Hd* = 0.615) but with a low π value of 0.001 (Table 1). In the mtCytb haplotype network, we observed that the four haplotypes were distributed more homogeneously among the sampled populations, with two haplotypes present in all the sampled localities (Figure 2B), while the other two haplotypes were only present in the FZ (that is, Jonuta, Emiliano Zapata and Catazaja). For both mitochondrial markers, the highest number of private haplotypes was found in the FZ (Table 1), which also showed the highest genetic diversity; thus, for mtCytb, the genetic diversity was *Hd* = 0.822 and for mtCox1, *Hd* = 0.36, while the lowest diversity was observed for the RZ: *Hd* = 0.46 and 0.32 for mtCytb and mtCox1, respectively.

In the hierarchical AMOVA for mtCox1 and mtCytb, most of the variation was recovered within populations: 86.61% and 92.99%, respectively (Table 2), with very low but not significant ΦCT values among groups (i.e., mtCox1 ΦCT = −0.032; mtCytb ΦCT = 0.061). Similarly, the differences between populations within groups were low but not significant (i.e., mtCox1 ΦSC = −0.080; mtCytb ΦSC = 0.009; mtCox1 ΦST = −0.044; mtCytb ΦST = 0.07 (Table 2)).

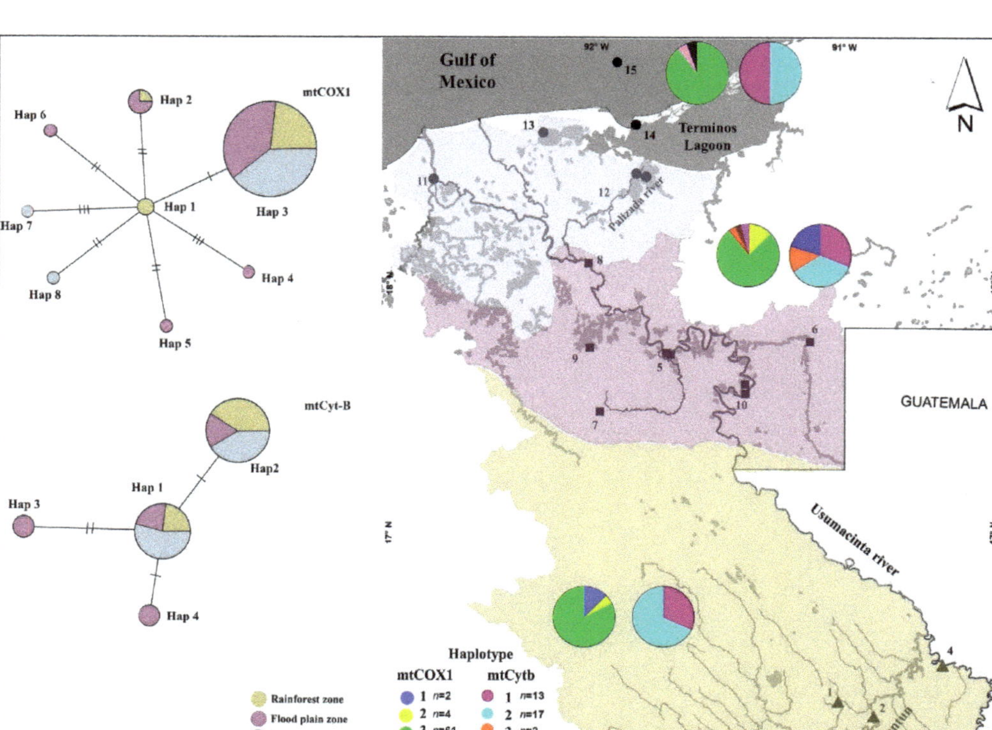

Figure 2. (**A**) ML haplotype networks for mtCox1 and mtCytb. The area of the circles is proportional to the number of sequences found for each haplotype. Partitions and colors inside the circles represent the proportion of each zone within each haplotype. (**B**) Frequencies and geographic distribution of mtCox1 and mtCytb haplotypes determined by combining allelic variants. The polygon indicates the distribution within the sampled area (rainforest, floodplain and river delta); n represents the number of sequences by haplotype.

Table 2. Analysis of molecular variance (AMOVA) for two mitochondrial and nine microsatellite loci, among zones (rainforest, floodplain and river delta) and groups obtained with DAPC and STRUCTURE. Significant values are shown with the asterisk ($p < 0.05$).

Marker	Variance Among Groups (%)	Variance among Populations within Groups (%)	Variance within Populations (%)	Among Groups F_{CT}/Φ_{CT}	Among Populations F_{SC}/Φ_{SC}	Within Populations F_{ST}/Φ_{ST}
mtCox1	3.29	16.00	86.61	0.03295	−0.08010	−0.04451
mtCytb	6.13	0.88	92.99	0.06134	0.00936	0.07013
Microsatellite K = 3 DACP	11.11	−0.69	89.57	0.111 *	−0.007	0.104 *
Microsatellite K = 2 STRUCTURE	5.79010	3.17138	91.03	0.0579	0.0336	0.089
K = 3 STRUCTURE	1.69	4.27	94.04	0.016 *	0.043 *	0.059 *

3.1. Microsatellite Structure

The *Cun*-22 and *Cun*-09 loci were excluded from the analysis because the *Cun*-22 locus showed more than 51% missing data, while the *Cun*-09 and *Cun*-10 loci were linked; therefore, we kept *Cun*-10, since it was the marker with least missing data (see Supplementary Material S1, Table S4: Linkage disequilibrium results of nine microsatellites from *Centropomus undecimalis*). The summary statistics for all microsatellite loci are presented in Table 1. The number of alleles per locus ranged from 6 to 21. Five loci exhibited departure from Hardy–Weinberg equilibrium (HWE), due, in some cases, to a statistically significant deficit of heterozygotes.

We found evidence that null alleles may be present at five loci (*Cun*-20, *Cun*-10-A, *Cun*-02, *Cun*-01 and *Cun*-06) but we found no evidence for allele dropout or stuttering during PCR amplification. We found no evidence of scoring error, no large allele dropout and no null alleles at the *Cun*-21-A, *Cun*-04-A, *Cun*-21-B and *Cun*-14 loci. In general, the number of private alleles was low across the sampling sites (Table 3), with four private alleles being found at Chajul, Lacantun and Canitzan.

Table 3. Descriptive statistics for the microsatellite data in the Usumacinta River basin. Na = number of different alleles; Na (Freq $\geq$ 5%) = number of different alleles with a frequency $\geq$ 5%; Ne = number of effective alleles; Np = number of private alleles or the number of alleles unique to a single population.

Population	Na	Na Freq. $\geq$ 5%	Ne	Np
Tzendales River	3.556	3.556	3.115	1
Lacantun River	5.889	5.889	4.236	4
Chajul	5.111	5.111	4.012	4
Benemerito	3.667	3.667	2.889	1
Emiliano Zapata	4.333	4.333	3.426	2
San Pedro River	5.333	5.333	4.082	1
Chacamax River	3.778	3.778	3.165	2
Jonuta	4.000	4.000	3.421	0
Canitzan Lagoon	5.222	5.222	4.016	4
Pom Lagoon	5.111	4.000	3.415	1
Palizada River	4.556	4.556	3.168	2
Terminos Lagoon	3.000	3.000	2.495	1
Sea	4.556	4.556	3.225	0

A very small amount of genetic differentiation was detected among the *Centropomus undecimalis* populations studied, as revealed by significant pairwise F_{ST} values for 12 pairwise comparisons out of 77 (Table 4). A geographical pattern of the distribution of these differences was not recognized, but it could be identified that most of the significant differences were between the RD and the FZ.

Bayesian clustering via Evanno's method (i.e., STRUCTURE) [46] indicated that the most likely number of clusters was K = 2 (Supplementary Material S2: Plots generated in STRUCTURE Harvester and DAPC). The first cluster included Tzendales, Canitzan, most of the individuals from Chacamax River and some individuals from the Lacantun River and San Pedro River (Figure 3A, individuals in green), suggesting a genetic differentiation of the populations from the RZ and FZ, while the second cluster included a mixture of individuals from the RD and some individuals from the RZ and FZ (Figure 3A, individuals in red).

Table 4. Paired F_{ST} values among the 13 sampling localities of *Centropomus undecimalis* in the Usumacinta basin. Values with the asterisk represent significant values after Bonferroni correction.

	Tzendales	Lacantun River	Chajul	Benemerito	Emiliano Zapata	San Pedro River	Chacamax	Jonuta	Canitzan	Pom Lagoon	Palizada River	Terminos Lagoon	Sea
Tzendales	0												
Lacantun River	0.015	0											
Chajul	0.117	0.054	0										
Benemerito	0.003	0.034	0.090	0									
Emiliano Zapata	0.089	0.054	−0.013	0.067	0								
San Pedro River	0.020	0.025	0.088	0.085 *	0.081 *	0							
Chacamax	0.093	0.065	0.063	0.104 *	0.064	0.043	0						
Jonuta	0.043	0.014	0.090	0.132	0.107	−0.009	0.04	0					
Canitzan	0.046	0.027	0.069	0.078	0.082 *	−0.007	0.017	−0.004	0				
Pom Lagoon	0.113	0.082	0.019	0.127	−0.066	0.111	0.112	0.121	0.097	0			
Palizada River	0.041	0.0002	0.015	0.091	−0.004	−0.002	0.037	−0.003	−0.001	0.011	0		
Terminos Lagoon	0.004	0.025	0.101 *	−0.034	0.064	0.088 *	0.115	0.084	0.094 *	0.104 *	0.068 *	0	
Sea	0.051	0.038	0.106	−0.02	0.102	0.125	0.103	0.141	0.114 *	0.148 *	0.118 *	−0.038	0

Figure 3B shows the grouping obtained by STRUCTURE through Evanno's test considering three genetic clusters (K = 3). One of the clusters included all individuals from the RD zone, some individuals from the RZ (some individuals from Lacantun River) and from the FZ (some individuals from Emiliano Zapata, San Pedro River and Chacamax) (individuals in red). A second cluster joined some individuals from the RZ (Tzendales River and Lacantun River) and from the FZ (Chacamax River, San Pedro River and Canitzan Lagoon) (individuals in green). The remaining cluster was a mixture of populations from the RZ and FZ.

After we removed the missing data, the microsatellite data matrix used for the analysis of the population structure via the DAPC analysis had 64 individuals. The first analysis in the DAPC considered two genetic clusters (K = 2), corresponding to the groups obtained with STRUCTURE through Evanno's test (see Supplementary Material S3A: Discriminant analysis of principal components (DAPC): scatterplots of the discriminant analysis of principal components of the microsatellite data for three zones) and, after a cross-validation test, retained 10 principal components (PCs) with an accumulated variance of 60.1% for the total data. As can be observed in the graph, only a small number of individuals were grouped in the second cluster, which corresponded to a mixture of populations from the RZ and FZ.

In the DAPC, we considered the three genetic clusters (K = 3), obtained by STRUCTURE through Evanno's test (see Supplementary Material S3B). Thus, in the discriminant analysis of principal components (DAPC), the scatterplots retained five principal components (PCs) with an accumulative variance of 39.5% for the total data, after a cross-validation test. Thus, one of the three clusters included all individuals from the RD zone, some individuals from the RZ (Lacantun River) and from the FZ (Canitzan Lagoon and San Pedro River). A second cluster joined individuals from the RZ (Tzendales River) and from the FZ (Chacamax River and Canitzan Lagoon). The third cluster grouped a mixture of populations from the RZ and FZ.

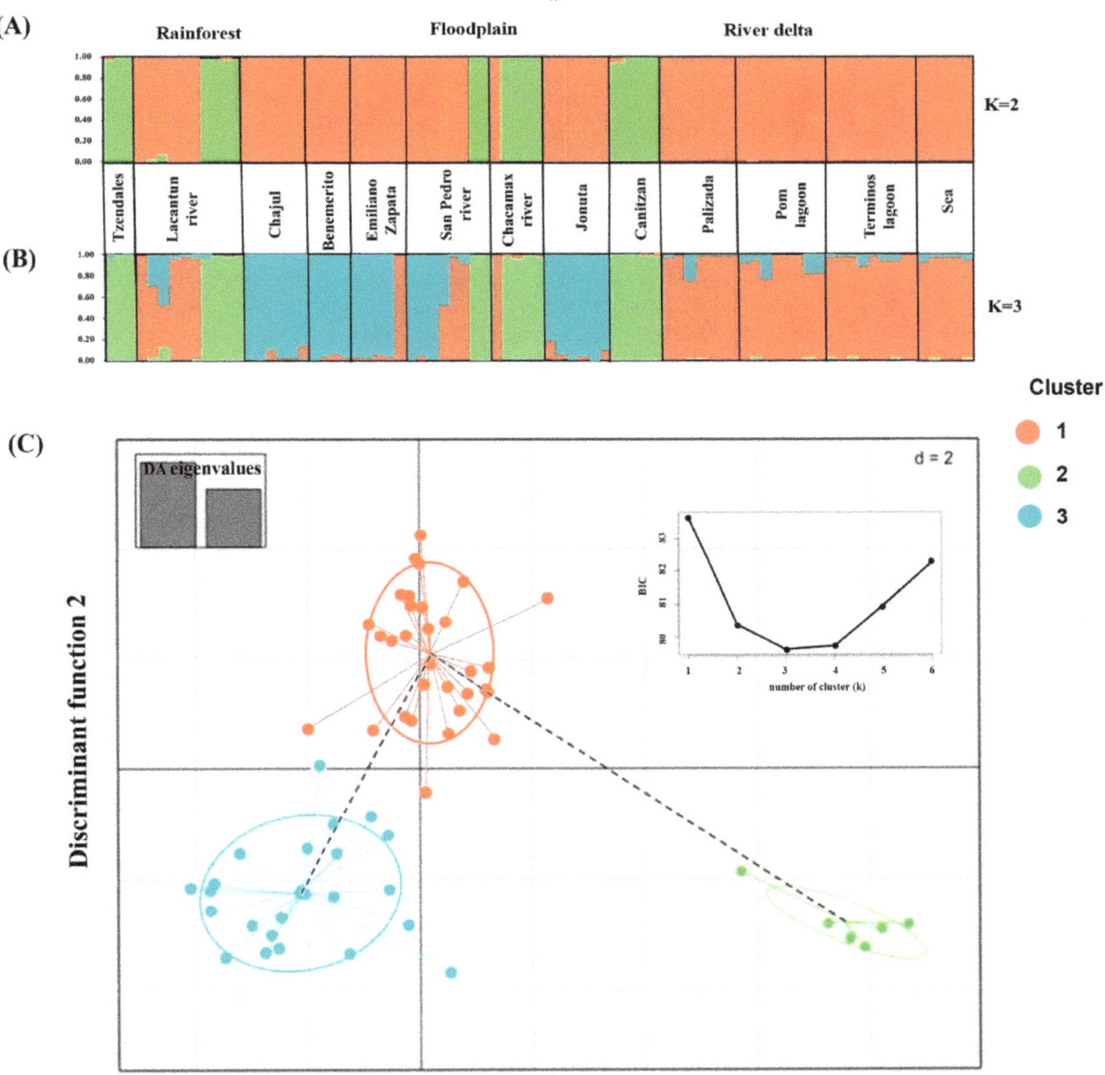

Figure 3. Assessment of population genetic structure by Bayesian cluster analysis and DAPC based on the microsatellite data. Bayesian Analysis was inferred at $K = 2$ based on locus data. Each single vertical line represents an individual and its proportional membership probability among the K clusters. (**A**) K = 2; (**B**) K = 3. (**C**) Scatterplots of the discriminant analysis of principal components of the microsatellite data, using *find.clusters*, for three zones. The axes represent the first two linear discriminants (LD). Each circle represents a cluster and each dot represents an individual.

Finally, the *find.clusters* algorithm retrieved three different genetic clusters of the 13 populations of *C. undecimalis* analyzed, for all the runs performed (i.e., 10), showing the lowest BIC value (i.e., 79). For K = 3, the cross-validation test resulted in the retention of five principal components (PCs) with an accumulative variance of 39.5% for the total data. In this case, the three clusters were mostly the same as the ones previously described for the STRUCTURE software, with the only difference that in the third cluster, an individual from the RD zone was included. The scatterplot of individuals on the two main components of

DAPC showed that they formed three groups, and no overlapping between the a priori defined groups (Figure 3C). We did not find evidence of substructuring among the clusters analyzed. In the hierarchical AMOVA, low but significant values of differentiation were recovered (Table 2). Most of the genetic variance was observed within the populations (94.04% by zones and 89.57% by STRUCTURE with K = 3). The differentiation among groups was low (F_{ST} = 0.059 by zones), as was that among populations within groups (F_{SC} = 0.043); both differences were significant. We estimated the migration rate (*m*) using BayesASS as an indicator of gene flow among the genetic groups. The BayesASS average results using the groupings obtained with DAPC analysis (K = 3) and STRUCTURE software (K = 2 and K = 3) are shown in Supplementary Material S1, Table S5 (BayesASS results showing the average migration rate (mprom) by cluster obtained); for more information, see Supplementary Material S1, Table S6 (Results of 10 runs of the BayesASS algorithm, with the average and total number of times the results were achieved). In general, the *m*-values were low among the genetic clusters. The *m*-value from k = 1 to k = 2 was the lowest (DAPC = 0.009 and STRUCTURE = 0.008). The values of the product of *m* and Ne, which represents the number of migrating individuals, are shown in Table 5.

Table 5. The estimated numbers of migrating individuals, calculated as the product of the average migration rate (*m*) and effective population size (*Ne*).

	From k = 1	From k = 2	From k = 3	Ne
DAPC				
To K = 1	(*m* = 0.9745) 86.9254	(*m* = 0.042) 3.7464	(*m* = 0.0244) 2.17648	89.2
To K = 2	(*m* = 0.009) 0.0531	(*m* = 0.9159) 5.40381	(*m* = 0.0254) 0.14986	5.9
To K = 3	(*m* = 0.0164) 0.19516	(*m* = 0.042) 0.4998	(*m* = 0.9501) 11.30619	11.9
STRUCTURE K = 3				
To K = 1	(*m* = 0.95581) 94.434028	(*m* = 0.07456) 7.366528	(*m* = 0.04342) 4.289896	98.8
To K = 2	(*m* = 0.00872) 0.142136	(*m* = 0.88829) 14.479127	(*m* = 0.02753) 0.448739	16.3
To K = 3	(*m* = 0.03545) 0.57429	(*m* = 0.03719) 0.602478	(*m* = 0.92904) 15.050448	16.2
STRUCTURE K = 2				
To K = 1	(*m* = 0.98839) 23.128326	(*m* = 0.0834) 1.95156	N/A	23.4
To K = 2	(*m* = 0.01161) 0.189243	(*m* = 0.9166) 14.94058	N/A	16.3

3.2. Isolation by Distance

Although we observed differences in the level of genetic diversity across the designated zones (RZ, FZ and RD), we did not find a correlation between the geographic and genetic distances for either mitochondrial marker (R^2 = 0.005; p = 0.4265 and R^2 = 0.0478; p = 0.3916 for mtCox1 and mtCytb, respectively) nor the nuclear loci (Figure 4, R^2 = 0.1143, p = 0.1668).

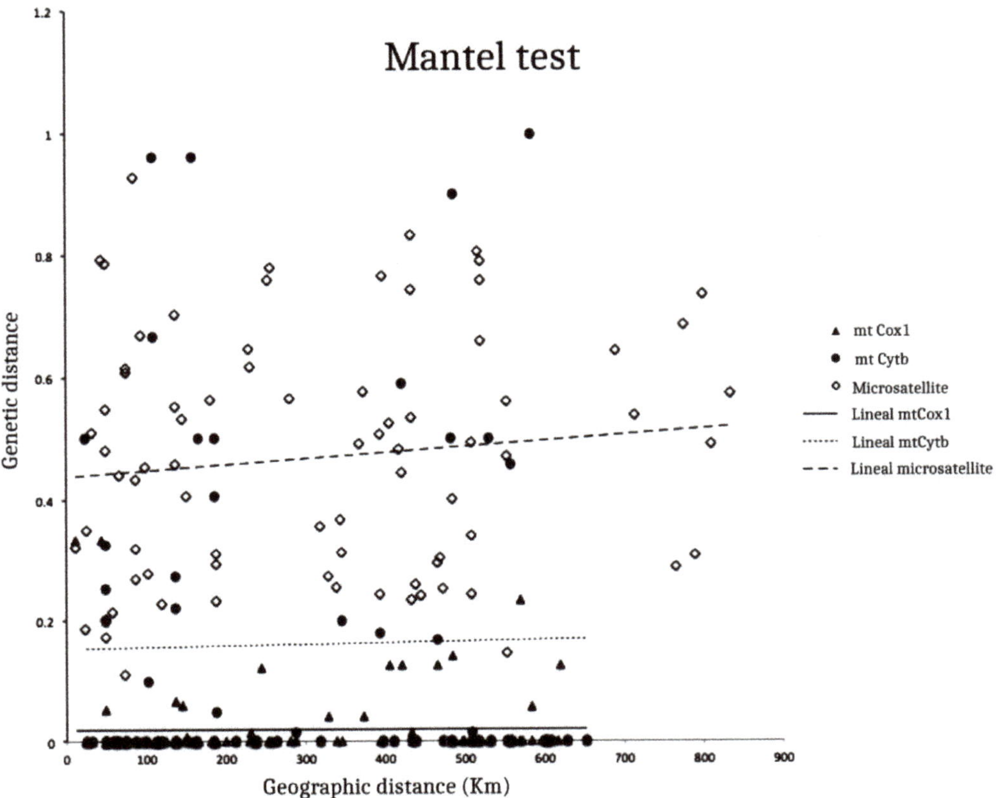

Figure 4. Mantel test. Correlation between genetic distance via Slatkin's linearized method and the geographic distance (km), including the three markers (mtCox1, mtCytb and microsatellites). $R^2 = 0.005$ ($p = 0.4265$) for mtCox1; $R^2 = 0.0478$ ($p = 0.3916$) for mtCytb and $R^2 = 0.1143$ ($p = 0.1668$) for microsatellites. The dotted line corresponds to the adjusted mtCytb data, the solid line corresponds to the mtCox1 data and dashed line corresponds to the adjusted microsatellite data.

4. Discussion

The Usumacinta River basin is one of the largest basins in the Gulf of Mexico, whose biological diversity is outstanding for the Mesoamerican region [29,59], with more than 170 species, including 50 fish families, making it one of the most diverse river basins in Mexico [20]. Its hydrological connectivity allows the common snook to complete its life cycle. The species is one of the most abundant, largest and most economically important, with a high commercial value, which has led to overfishing of the species, threatening their populations [60–62]. Testing the genetic structure of an euryhaline species such as the common snook provides a unique opportunity to demonstrate the importance of the hydrological connectivity of the Usumacinta River basin, providing a better understanding of its conservation status.

4.1. mtDNA Genetic Structure

Our mtDNA results in the common snook support a degree of connectivity among the three zones in the Usumacinta River basin, a pattern expected due to the migratory nature of the species, as well as previous observations of some of its populations in a smaller geographic area [25]. The star-like haplotype network is consistent with a lack of geographical structure, showing haplotypes with low levels of sequence divergence and a high frequency of singletons. A similar pattern can be related to rapid population expansion or selection, which caused the rapid spread of a mitochondrial lineage [63].

Despite the lack of a clear geographic differentiation, haplotype frequencies and exclusive haplotypes differ among the zones analyzed. Particularly, with the mtCox1 marker, we identified differences in haplotype occurrence and their frequencies between the rainforest and delta zones (the upper and lower parts of the basin, respectively). These results are consistent with previous comparative phylogeographic analyses of the basin, which considered endemic species data and phylogeographic analyses for a selected group of organisms (mainly cichlids and poecilids), and report differences between the upper and lower parts of the Usumacinta River basin [29]. These results provide additional evidence that could indicate the existence of geographic structure for some of the freshwater fauna in the basin (e.g., *Centropomus undecimalis*) and support the hypothesis that the Usumacinta does not correspond to a single biogeographic unit [29].

The floodplain was recovered as the most diverse zone (see Table 1), showing a higher number of exclusive haplotypes (i.e., Hap 4, 5 and 6 for Cox1 and Hap3 and 4 for Cytb, Figure 2A,B). Thus, the mtDNA variations suggested that the floodplain acts as a confluence zone between the rainforest and river delta zones for the common snook. This pattern of genetic transition was previously reported in the *Astyanax aeneus* species complex on the basis of mitochondrial markers [10], where the floodplain zone was also found to be a transition zone between the rainforest and delta zones of the Usumacinta River drainage.

Despite the aforementioned differences, we did not recover significant differences among zones according to the hierarchical AMOVA, suggesting that the levels of differentiation were fairly low among the zones. However, the molecular phylogeographic patterns gave additional evidence about the relevance of using barcodes as valuable tools to characterize and reveal cryptic diversity in widely distributed fish species. In accordance, the geographic structure found inside the basin shows the utility of genetic information for characterizing the diversity patterns in a region previously considered to be a single unit [13].

4.2. nucDNA Genetic Structure

Microsatellite loci showed low but significant levels of differentiation among the zones tested in this study. The heterozygote deficiency represents a deviation from the Hardy–Weinberg Equilibrium (HWE) when the observed heterozygosity (*Ho*) is less than the expected heterozygosity (*He*) [64]. These deviations from the HWE proportions can be generated by the presence of null alleles and by an irregular system of inbreeding or by population structure [64,65]. In our study, the heterozygote deficiency observed could be explained by the presence of null alleles in some loci, but they were not shared across populations. When we considered K = 2, one of the clusters was distributed in the rainforest and floodplain zones, which included individuals from the populations from Tzendales, Lacantun, Chacamax, San Pedro and Canitzan. The second group included a cluster that shows a wider distribution across the basin, from the delta to the rainforest zones (Figure 3). These results support the confluence of two different genotypic clusters of the common snook in the Usumacinta River basin, one of which was widely distributed in the lower part of the basin (the delta zone) and reached the uplands (the rainforest zone), while the second cluster, which showed a more restricted distribution, occurred in the rainforest and floodplain zones, possibly extending to the upper part of the Usumacinta basin in Guatemala. A similar pattern was previously reported in the white turtle (*Dermatemys mawii*) [26–28], in which two genetic clusters were recovered along the Usumacinta River basin.

The DAPC results suggest a high probability for K = 3. One of these genetic clusters is widely distributed across the Usumacinta basin and is dominant in the delta zone. Additionally, the rainforest and floodplain zones contain two different genotypic clusters; the first one coincides with the cluster detected for the K = 2 analysis and the second one is distributed among the remaining individuals from both zones, hierarchically dividing the upper (rainforest and floodplain) from the lower part of the basin (delta zone).

The ocurrence of different genetic clusters occurring in the upper Usumacinta (rainforest and floodplain zones) but not detected in the lower Usumacinta cluster (delta zone) is in agreement with a previous study which supported the existence of two biogeographic units within the Usumacinta River basin [29]. This biogeographic pattern could be the result of both recent and historical events influencing fish diversification in the basin [29]. Further studies including adjacent areas on the distribution of the common snook would allow a better understanding of the extent of the genetic clusters found here.

Additionally, the presence of two different genetic clusters in the upper part of the Usumacinta basin could be related to the occurrence of different migratory contingencies in the common snook populations. Previous studies on catadromous fish have shown that they can present alternative migratory tactics within a conditional strategy [66]; in this case, the individuals could make migratory decisions depending on the following factors: individual status (e.g., body condition, growth rate), interactions with other organisms and environmental conditions (e.g., habitat availability or river flow). Previous studies have shown a high environmental similarity between the delta and the floodplain zones, the latter being considered as a reservoir for diversity where fish species can reproduce [20,67,68]. Additionally, the common snook individuals collected from the floodplain zone presented advanced stages of gonadal development [17]. The fact that the two genetic clusters from the rainforest and floodplain were not present in the delta zone could reflect an alternative migratory contingence, in which the floodplain zone could provide the environmental conditions necessary for the common snook to complete their life cycle, as happens with other migratory species [68]. Further studies exploring the existence of migratory contingents in the life cycle of *C. undecimalis* based on otolith $^{87}Sr/^{86}Sr$ analysis could shed light about the variation in life cycles among individuals.

4.3. Gene Flow and Isolation by Distance

In this work, it was found that the migration rates for the three structuring models showed a low genetic flow between the estimated clusters (Figure 3), with a low number of possible migrants among them (Table 5). These low migration rates can promote differential segregation between genetic clusters, supporting the idea that these groups are well-discriminated units. However, in accordance with our previous results, neither type of examined markers (mtDNA and nucDNA) showed an isolation by distance (IBD) pattern, suggesting a more complex genetic structure in the *C. undecimalis* populations.

In this regard, the IBD, mitochondrial haplotypes and nuclear results are in agreement with a previous study by Hernández-Vidal et al. [25], in which the lack of an IBD pattern was described. However, in that study, the authors only compared two marine populations (referred to as the sea population here) vs. individuals from the San Pedro River near the Guatemala border, for a total of 79 individuals. Their results showed a 2% variance between the sea and the San Pedro River basin (in the floodplain zone), very similar to our results for those groups (nucDNA = 2.1%, mtCox1 = 3.29% and mtCytb = 6.13%; Table 2).

Based on the BayesASS results, we found a low gene flow among the genetic clusters recovered. However, since one of the genetic clusters recovered was found throughout the Usumacinta River basin, the structuring pattern related to geographic distance was not observed. In this regard, a previous study involving allozymes in *C. undecimalis* recovered strong differentiation between the Gulf of Mexico waters and the Caribbean populations, suggesting that these populations could correspond to different management units [18]. Additionally, recent studies of *C. undecimalis*, from the Gulf of Mexico to Brazil, showed very high genetic differentiation associated with the geographical distance between the populations [69,70]. The results for genetic structuring and the estimation of migration rates suggest that two to three different populations of the common snook converge in the Usumacinta River basin.

Hydrological connectivity, together with historical processes, could have played a major role in the genetic structure of the common snook population in the Usumacinta basin over different temporal and geographical scales [26–28]. Further studies including wider geographical sampling in the Gulf of Mexico and the Caribbean could help us to test these differential patterns of connectivity and genetic structure within the common snook across different regions.

4.4. Implications for Species Conservation

Our results could have important implications for species conservation and management. First, it is clear that the common snook migrates and disperses throughout the basin, from the rainforest zone in Mexico (Tzendales River and Lacantun River) and its border with Guatemala to the river delta zone. Thus, river connectivity is essential to allow the species to maintain its life cycle. On the other hand, the genetic differences found in the common snook, for which the floodplain zone was identified as the most diverse zone according to both mitochondrial and nuclear data, suggesting that this zone corresponds to a confluence between the rainforest and river delta zones in the Usumacinta River basin; thus, the floodplain zone corresponds to a relevant unit for the conservation and management of the species. In particular, the San Pedro River represents a unique region, due to the environmental conditions, which could provide particular biological dynamics that allow the species to reproduce. Thus, future studies could shed more light in this regard [71]. Alterations in river connectivity will impact the life history of the common snook, including its migratory and dispersal behavior and population size contractions, affecting the fisheries in the region and ultimately species conservation. The identification of the high-diversity unit zones (i.e., the FZ) through the use of barcodes could favor the implementation of a responsible management program in these zones by decision-makers for preserving not only the species but also its genetic diversity.

Regarding the genetic cluster recovered with the nuclear markers, we suggest that these could represent alternative reproductive stocks of the species within the Usumacinta basin; thus, even though we did not recover a geographic structure, we consider that the basin represents a very important system for the conservation of the species' genetic diversity, where alternative reproductive strategies could have been taking place.

Finally, our study also recovered the diversity information of mitochondrial and nuclear data that in contrast with previous studies that also shed light on the current status of the species. With the two mitochondrial markers, the genetic diversity recovered was lower (i.e., mtCox1 Hd = 0.28 and mtCytb Hd = 0.62) than previously reported for the species [18,72], and also in comparison with other euryhaline species [73,74]; for mtCox1, Hap3 was present in 80% of the samples analyzed. Similarly, in the nuclear data, the heterozygosity values obtained in our study were lower than those previously reported for 5 of our 11 microsatellite loci [25]. This information could be explained by our sample size; however, the two types of data provide evidence that is consistent with biogeographic patterns. Moreover, our results could also be related to an overexploitation of the species by the local fisheries, urging the local authorities to implement conservation and management programs to preserve the species' evolutionary history.

Supplementary Materials: The following are available online at https://www.mdpi.com/article/10.3390/d13080347/s1. Supplementary Material S1: Table S1: Sampling localities from the Usumacinta River basin. Table S2: Eleven microsatellite loci used in determining genetic variation among the *Centropomus undecimalis* samples from 13 sampling localities along the Usumacinta River basin. Table S3: Genotyping of nine loci of *Centropomus undecimalis*. Table S4: Linkage disequilibrium results of nine microsatellites from *Centropomus undecimalis*. Table S5: BayesASS results showing the average migration rate (mprom) by cluster, obtained with DAPC (k = 3) and STRUCTURE (k = 2, k = 3). SD, standard deviation; CI, 95% confidence interval; N, the number of times this outcome was reached over 10 runs with varying starting seeds. Table S6: Results of 10 runs of the BayesASS algorithm, with the average and total number of times the results were achieved. Supplementary Material S2. Plots generated in STRUCTURE Harvester and DAPC. (A) The mean log likelihood of the data [L(K)]. (B) Estimation of population clustering levels from seven microsatellite genotypes following Evanno's test [46]. (C) BIC value changes (*find.clusters* function) of DAPC. Supplementary

Material S3. Discriminant analysis of principal components (DAPC): scatterplots of the discriminant analysis of principal components of the microsatellite data for three zones.

Author Contributions: Conceptualization: J.T.-M., C.P.O.-G. and R.R.-H.; Methodology: J.T.-M., C.P.O.-G. and M.A.A.G.-S.; Formal analysis: J.T.-M., C.P.O.-G. and M.A.A.G.-S.; Writing—original draft preparation: J.T.-M. and C.P.O.-G.; Writing—review and editing: C.P.O.-G., R.R.-H. and M.A.A.G.-S.; Visualization: C.P.O.-G. and R.R.-H.; Resources: C.P.O.-G. and R.R.-H.; Funding acquisition: C.P.O.-G. and R.R.-H. All authors have read and agreed to the published version of the manuscript.

Funding: This research was support from PAPIIT, UNAM, Project number IN212419 and the project "Conectividad y Diversidad Funcional de la Cuenca del Río Usumacinta" (FID-ECOSUR 784-1004).

Institutional Review Board Statement: Field collection permit and protocols related to the animal care ethics was approved by the Secretaria del Medio Ambiente y Recursos Naturales (SEMARNAT: PPF/DGOPA-249/14).

Informed Consent Statement: Not applicable.

Data Availability Statement: The data of the sequences used will be available at https://www.ncbi.nlm.nih.gov/genbank/.

Acknowledgments: We thank Posgrado de El Colegio de la Frontera Sur for the support given to Jazmín Terán-Martínez during her PhD study. We thank Miriam Soria-Barreto and Alfonso Gonzalez-Díaz for their comments and suggestions on the study. We also thank Alberto Macossay-Cortéz, Abraham Aragón-Flores and Mayra Contreras-Flores for help with collecting fish during field trips; Cesar Maya-Bernal and Cinthya Mendoza for their support in the laboratory; and Pablo Sandoval-Rivera for his contribution to the elaboration of the cartographic content. Jazmín Terán-Martínez was funded by Consejo Nacional de Ciencia y Tecnología, Mexico (CONACyT), scholarship number 562922. SEMARNAT Fishing permit No. PPF/DGOPA-249/14.

Conflicts of Interest: The authors declare no conflict of interest. The funders had no role in the design of the study; in the collection, analyses, or interpretation of data; in the writing of the manuscript; or in the decision to publish the results.

References

1. Pringle, C.M.; Freeman, M.C.; Freeman, B.J. Regional Effects of Hydrologic Alterations on Riverine Macrobiota in the New World: Tropical–Temperate Comparisons. *BioScience* **2000**, *50*, 807–823. [CrossRef]
2. Pringle, C. *What Is Hydrologic Connectivity and Why Is It Ecologically Important?* Hydrological Process: Hoboken, NJ, USA, 2003; Volume 17, pp. 2685–2689.
3. Calles, O.; Greenberg, L. Connectivity is a two-way street-the need for a holistic approach to fish passage problems in regulated rivers. *River Res. Appl.* **2009**, *25*, 1268–1286. [CrossRef]
4. Ward, J.V. The Four-Dimensional Nature of Lotic Ecosystems. *J. N. Am. Benthol. Soc.* **1989**, *8*, 2–8. [CrossRef]
5. Branco, P.; Segurado, P.; Santos, J.; Pinheiro, P.; Ferreira, M. Does longitudinal connectivity loss affect the distribution of freshwater fish? *Ecol. Eng.* **2012**, *48*, 70–78. [CrossRef]
6. Lucas, M.C.; Baras, E. *Migration of Freshwater Fishes*; Lucas, M., Baras, E., Thom, T., Duncan, A., Slavík, O., Eds.; Blackwell Science: Maden, MA, USA, 2001.
7. Yi, Y.; Yang, Z.; Zhang, S. Ecological influence of dam construction and river-lake connectivity on migration fish habitat in the Yangtze River basin, China. *Procedia Environ. Sci.* **2010**, *2*, 1942–1954. [CrossRef]
8. Collins, S.M.; Bickford, N.; McIntyre, P.B.; Coulon, A.; Ulseth, A.J.; Taphorn, D.C.; Flecker, A.S. Population Structure of a Neotropical Migratory Fish: Contrasting Perspectives from Genetics and Otolith Microchemistry. *Trans. Am. Fish. Soc.* **2013**, *142*, 1192–1201. [CrossRef]
9. Hand, B.K.; Lowe, W.H.; Kovach, R.P.; Muhlfeld, C.C.; Luikart, G. Landscape community genomics: Understanding eco-evolutionary processes in complex environments. *Trends Ecol. Evol.* **2015**, *30*, 161–168. [CrossRef]
10. Garcia, C.P.O.; De Biología, U.I.; Bernal, C.F.M.; Hernández, R.R.; Biomédicas, U.I.D.I. Ecosur Evaluación de la Diversidad de Linajes en Sistemas Dulceacuícolas tropicales (D-LSD): El Sistema Usumacinta como caso de estudio. In *Antropización: Primer Análisis Integral*; Universidad Nacional Autónoma de México, Centro de Investigaciones en Geografía Ambiental: Mexico City, Mexico, 2019; pp. 125–148.
11. Oosthuizen, C.J.; Cowley, P.D.; Kyle, S.R.; Bloomer, P. High genetic connectivity among estuarine populations of the riverbream Acanthopagrus vagus along the southern African coast. *Estuar. Coast. Shelf Sci.* **2016**, *183*, 82–94. [CrossRef]
12. Sork, V.L.; Smouse, P.E. Genetic analysis of landscape connectivity in tree populations. *Landsc. Ecol.* **2006**, *21*, 821–836. [CrossRef]
13. DeSalle, R.; Goldstein, P. Review and Interpretation of Trends in DNA Barcoding. *Front. Ecol. Evol.* **2019**, *7*, 302. [CrossRef]

14. Ornelas-García, C.P.; Domínguez-Domínguez, O.; Doadrio, I. Evolutionary history of the fish genus Astyanax Baird & Girard (1854) (Actinopterygii, Characidae) in Mesoamerica reveals multiple morphological homoplasies. *BMC Evol. Biol.* **2008**, *8*, 340. [CrossRef]
15. Valdez-Moreno, M.; Ivanova, N.V.; Elías-Gutiérrez, M.; Pedersen, S.L.; Bessonov, K.; Hebert, P.D.N. Using eDNA to biomonitor the fish community in a tropical oligotrophic lake. *PLoS ONE* **2019**, *14*, e0215505. [CrossRef]
16. Frankham, R. Stress and adaptation in conservation genetics. *J. Evol. Biol.* **2005**, *18*, 750–755. [CrossRef] [PubMed]
17. Hernández-Vidal, U.; Chiappa-carrara, X.; Contreras-Sánchez, W.M. Reproductive variability of the common snook, *Centropomus undecimalis*, in environments of contrasting salinities interconnected by the Grijalva–Usumacinta fluvial system. *Cienc. Mar.* **2014**, *40*, 173–185. [CrossRef]
18. Tringali, M.; Bert, T. The genetic stock structure of common snook (*Centropomus undecimalis*). *Can. J. Fish. Aquat. Sci.* **1996**, *53*, 974–984. [CrossRef]
19. Mendonça, J.T.; Chao, L.; Albieri, R.J.; Giarrizzo, T.; da Silva, F.M.S.; Castro, M.G.; Brick-Peres, M.; Villwock de Miranda, L.; Vieira, J.P. *Centropomus Undecimalis*. The IUCN Red List of Threatened Species 2019: E.T191835A82665184. 2019. Available online: https://dx.doi.org/10.2305/IUCN.UK.2019-2.RLTS.T191835A82665184.en (accessed on 30 May 2021).
20. Soria-Barreto, M.; González-Díaz, A.A.; Castillo-Domínguez, A.; Álvarez-Pliego, N.; Rodiles-Hernández, R. Diversity of fish fauna in the Usumacinta Basin, Mexico. *Rev. Mex. Biodivers.* **2018**, *89*, 100–117.
21. Garcia, M.A.P.; Mendoza-Carranza, M.; Contreras-Sánchez, W.; Ferrara, A.; Huerta-Ortiz, M.; Gómez, R.E.H. Comparative age and growth of common snook *Centropomus undecimalis* (Pisces: Centropomidae) from coastal and riverine areas in Southern Mexico. *Rev. Biol. Trop.* **2013**, *61*, 807–819. [CrossRef]
22. Castro-Aguirre, J.L.; Espinosa-Pérez, H.; Schmitter-Soto, J.J. *Ictiofauna Estuarino-Lagunar y Vicaria de México*; Editorial Limusa: La Paz, Baja California Sur, Mexico, 1999; ISBN 9681857747.
23. Adams, A.J.; Wolfe, R.K.; Layman, C.A. Preliminary Examination of How Human-driven Freshwater Flow Alteration Affects Trophic Ecology of Juvenile Snook (*Centropomus undecimalis*) in Estuarine Creeks. *Chesap. Sci.* **2009**, *32*, 819–828. [CrossRef]
24. Miller, R.R. *Peces dulceacuícolas de México*; CONABIO, SIMAC, ECOSUR, Desert Fishes of Council: Mexico City, Mexico, 2009.
25. Hernández-Vidal, U.; Lesher-Gordillo, J.; Contreras-Sánchez, W.M.; Chiappa-Carrara, X. Variabilidad genética del robalo común *Centropomus undecimalis* (Perciformes: Centropomidae) en ambiente marino y ribereño interconectados. *Rev. Biol. Trop.* **2014**, *62*, 627–636. [CrossRef]
26. González-Porter, G.P.; Hailer, F.; Flores-Villela, O.; García-Anleu, R.; Maldonado, J.E. Patterns of genetic diversity in the critically endangered Central American river turtle: Human influence since the Mayan age? *Conserv. Genet.* **2011**, *12*, 1229–1242. [CrossRef]
27. Gonzalez-Porter, G.P.; Maldonado, J.E.; Flores-Villela, O.; Vogt, R.C.; Janke, A.; Fleischer, R.C.; Hailer, F. Cryptic Population Structuring and the Role of the Isthmus of Tehuantepec as a Gene Flow Barrier in the Critically Endangered Central American River Turtle. *PLoS ONE* **2013**, *8*, e71668. [CrossRef]
28. Martínez-Gómez, J. *Sistemática Molecular e Historia Evolutiva de la Familia Dermatemydidae*; Universidad de Ciencias y Artes de Chiapas: Tuxtla Gutiérrez, Chiapas, Mexico, 2017.
29. Elías, D.J.; Mcmahan, C.D.; Matamoros, W.A.; Gómez-González, A.E.; Piller, K.R.; Chakrabarty, P. Scale(s) matter: Deconstructing an area of endemism for Middle American freshwater fishes. *J. Biogeogr.* **2020**, *47*, 2483–2501. [CrossRef]
30. Sonnenberg, R.; Nolte, A.W.; Tautz, D. An evaluation of LSU rDNA D1-D2 sequences for their use in species identification. *Front. Zoöl.* **2007**, *4*, 6. [CrossRef]
31. Ward, R.D.; Zemlak, T.S.; Innes, B.H.; Last, P.R.; Hebert, P.D. DNA barcoding Australia's fish species. *Philos. Trans. R. Soc. B Biol. Sci.* **2005**, *360*, 1847–1857. [CrossRef] [PubMed]
32. Zardoya, R.; Doadrio, I. Phylogenetic relationships of Iberian cyprinids: Systematic and biogeographical implications. *Proc. R. Soc. B Boil. Sci.* **1998**, *265*, 1365–1372. [CrossRef] [PubMed]
33. Seyoum, S.; Tringali, M.D.; Sullivan, J.G. Isolation and characterization of 27 polymorphic microsatellite loci for the common snook, *Centropomus undecimalis*. *Mol. Ecol. Notes* **2005**, *5*, 924–927. [CrossRef]
34. Hall, T.A. BioEdit: A User-Friendly Biological Sequence Alignment Editor and Analysis Program for Windows 95/98/NT. Nucleic Acids Symp. *Series* **1999**, *41*, 95–98.
35. Rozas, J.; Ferrer-Mata, A.; Sánchez-DelBarrio, J.C.; Guirao-Rico, S.; Librado, P.; Ramos-Onsins, S.E.; Sánchez-Gracia, A. DnaSP 6: DNA Sequence Polymorphism Analysis of Large Data Sets. *Mol. Biol. Evol.* **2017**, *34*, 3299–3302. [CrossRef] [PubMed]
36. Leigh, J.W.; Bryant, D. Popart: Full-feature software for haplotype network construction. *Methods Ecol. Evol.* **2015**, *6*, 1110–1116. [CrossRef]
37. Stamatakis, A. RAxML version 8: A tool for phylogenetic analysis and post-analysis of large phylogenies. *Bioinformatics* **2014**, *30*, 1312–1313. [CrossRef]
38. Miller, M.A.; Pfeiffer, W.; Schwartz, T. Creating the CIPRES Science Gateway for inference of large phylogenetic trees. *Gatew. Comput. Environ. Workshop (GCE)* **2010**, 1–8. [CrossRef]
39. Darriba, D.; Taboada, G.L.; Doallo, R.; Posada, D. jModelTest 2: More models, new heuristics and parallel computing. *Nat. Methods* **2012**, *9*, 772. [CrossRef]
40. Excoffier, L.; Lischer, H.E.L. Arlequin suite ver 3.5: A new series of programs to perform population genetics analyses under Linux and Windows. *Mol. Ecol. Resour.* **2010**, *10*, 564–567. [CrossRef]

41. Van Oosterhout, C.; Hutchinson, W.F.; Wills, D.P.M.; Shipley, P. micro-checker: Software for identifying and correcting genotyping errors in microsatellite data. *Mol. Ecol. Notes* **2004**, *4*, 535–538. [CrossRef]
42. Raymond, M.; Rousset, F. GENEPOP (Version 1.2): Population Genetics Software for Exact Tests and Ecumenicism. *J. Hered.* **1995**, *86*, 248–249. [CrossRef]
43. Peakall, R.; Smouse, P.E. Genalex 6: Genetic analysis in Excel. Population genetic software for teaching and research. *Mol. Ecol. Notes* **2006**, *6*, 288–295. [CrossRef]
44. Weir, B.S.; Cockerham, C.C. Estimating F-Statistics for the Analysis of Population Structure. *Evolution* **1984**, *38*, 1358. [CrossRef] [PubMed]
45. Pritchard, J.K.; Stephens, M.; Donnelly, P. Inference of population structure using multilocus genotype data. *Genetics* **2000**, *155*, 945–959. [CrossRef] [PubMed]
46. Evanno, G.; Regnaut, S.; Goudet, J. Detecting the number of clusters of individuals using the software structure: A simulation study. *Mol. Ecol.* **2005**, *14*, 2611–2620. [CrossRef]
47. Earl, D.A.; Vonholdt, B.M. STRUCTURE HARVESTER: A website and program for visualizing STRUCTURE output and implementing the Evanno method. *Conserv. Genet. Resour.* **2011**, *4*, 359–361. [CrossRef]
48. Jombart, T.; Devillard, S.; Balloux, F. Discriminant analysis of principal components: A new method for the analysis of genetically structured populations. *BMC Genet.* **2010**, *11*, 94. [CrossRef]
49. Jombart, T.; Ahmed, I. adegenet 1.3-1: New tools for the analysis of genome-wide SNP data. *Bioinformatics* **2011**, *27*, 3070–3071. [CrossRef]
50. Team, R. *A Language and Environment for Statistical Computing*; R Core Team: Vienna, Austria, 2006.
51. Jombart, T.; Collins, C. *A Tutorial for Discriminant Analysis of Principal Components (DAPC) Using Adegenet 2.0.0*; Imperial College London, MRC Centre for Outbreak Analysis and Modelling: London, UK, 2015.
52. Criscuolo, N.G.; Angelini, C. StructuRly: A novel shiny app to produce comprehensive, detailed and interactive plots for population genetic analysis. *PLoS ONE* **2020**, *15*, e0229330. [CrossRef] [PubMed]
53. Kamvar, Z.N.; Tabima, J.F.; Grünwald, N.J. Poppr: An R package for genetic analysis of populations with clonal, partially clonal, and/or sexual reproduction. *PeerJ* **2014**, *2*, e281. [CrossRef]
54. Wilson, G.A.; Rannala, B. Bayesian Inference of Recent Migration Rates Using Multilocus Genotypes. *Genetics* **2003**, *163*, 1177–1191. [CrossRef] [PubMed]
55. Do, C.; Waples, R.S.; Peel, D.; Macbeth, G.M.; Tillett, B.J.; Ovenden, J.R. NeEstimatorv2: Re-implementation of software for the estimation of contemporary effective population size (Ne) from genetic data. *Mol. Ecol. Resour.* **2013**, *14*, 209–214. [CrossRef] [PubMed]
56. Rohlf, F.J. Comparative methods for the analysis of continuous variables: Geometric interpretations. *Evolution* **2001**, *55*, 2143–2160. [CrossRef] [PubMed]
57. Slatkin, M. A measure of population subdivision based on microsatellite allele frequencies. *Genetics* **1995**, *139*, 457–462. [CrossRef]
58. Wright, S. ISOLATION BY DISTANCE. *Genetics* **1943**, *28*, 114–138. [CrossRef]
59. Yáñez-Arancibia, A.; Day, J.W.; Currie-Alder, B. Functioning of the Grijalva-Usumacinta River Delta, Mexico: Challenges for Coastal Management. *Ocean Yearb. Online* **2009**, *23*, 473–501. [CrossRef]
60. Mendoza-Carranza, M.; Hoeinghaus, D.J.; Garcia, A.M.; Romero-Rodriguez, Á. Aquatic food webs in mangrove and seagrass habitats of Centla Wetland, a Biosphere Reserve in Southeastern Mexico. *Neotrop. Ichthyol.* **2010**, *8*, 171–178. [CrossRef]
61. Barrientos, C.; Quintana, Y.; Elías, D.J.; Rodiles-Hernández, R. Peces nativos y pesca artesanal en la cuenca Usumacinta, Guatemala. *Rev. Mex. Biodivers.* **2018**, *89*, 118–130. [CrossRef]
62. Barba-Macías, E.; Juárez-Flores, J.; Trinidad-Ocaña, C.; Sánchez-Martínez, A.D.J.; Mendoza-Carranza, M. Socio-ecological Approach of Two Fishery Resources in the Centla Wetland Biosphere Reserve. In *Socio-Ecological Studies in Natural Protected Areas*; Springer Science and Business Media LLC: Berlin, Germany, 2020; pp. 627–656.
63. Mirol, P.M.; Routtu, J.; Hoikkala, A.; Butlin, R.K. Signals of demographic expansion in Drosophila virilis. *BMC Evol. Biol.* **2008**, *8*, 59. [CrossRef] [PubMed]
64. Rousset, F.; Raymond, M. Testing heterozygote excess and deficiency. *Genetics* **1995**, *140*, 1413–1419. [CrossRef]
65. De Meeûs, T. Revisiting FIS, FST, Wahlund effects, and null alleles. *J. Hered.* **2018**, *109*, 446–456. [CrossRef]
66. Crook, D.A.; Buckle, D.J.; Allsop, Q.; Baldwin, W.; Saunders, T.M.; Kyne, P.M.; Woodhead, J.D.; Maas, R.; Roberts, B.; Douglas, M.M. Use of otolith chemistry and acoustic telemetry to elucidate migratory contingents in barramundi Lates calcarifer. *Mar. Freshw. Res.* **2017**, *68*, 1554. [CrossRef]
67. Ochoa-Gaona, S.; Ramos-Ventura, L.J.; Moreno-Sandoval, F.; Jiménez-Pérez, N. del C.; Haas-Ek, M.A.; Muñiz-Delgado, L.E. Diversidad de flora acuática y ribereña en la cuenca del río Usumacinta, México. *Rev. Mex. Biodivers.* **2018**, *89*, 3–44.
68. Vaca, R.A.; Golicher, D.J.; Rodiles-Hernández, R.; Castillo-Santiago, M. Ángel; Bejarano, M.; Navarrete-Gutiérrez, D.A. Drivers of deforestation in the basin of the Usumacinta River: Inference on process from pattern analysis using generalised additive models. *PLoS ONE* **2019**, *14*, e0222908. [CrossRef] [PubMed]
69. Carvalho-Filho, A.; De Oliveira, J.; Soares, C.; Araripe, J. A new species of snook, Centropomus (Teleostei: Centropomidae), from northern South America, with notes on the geographic distribution of other species of the genus. *Zootaxa* **2019**, *4671*, 81–92. [CrossRef]

70. De Oliveira, J.N.; Gomes, G.; Rêgo, P.S.D.; Moreira, S.; Sampaio, I.; Schneider, H.; Araripe, J. Molecular data indicate the presence of a novel species of Centropomus (Centropomidae–Perciformes) in the Western Atlantic. *Mol. Phylogenet. Evol.* **2014**, *77*, 275–280. [CrossRef]
71. Castillo-Domínguez, A.; Barba Macías, E.; Navarrete, A. de J.; Rodiles-Hernández, R.; Jiménez Badillo, M. de L. Ictiofauna de los humedales del río San Pedro, Balancán, Tabasco, México. *Rev. Biol. Trop.* **2011**, *59*, 693–708. [PubMed]
72. Anderson, J.; Williford, D.; González-Barnes, A.; Chapa, C.; Martinez-Andrade, F.; Overath, R.D. Demographic, Taxonomic, and Genetic Characterization of the Snook Species Complex (Centropomus spp.) along the Leading Edge of Its Range in the Northwestern Gulf of Mexico. *N. Am. J. Fish. Manag.* **2020**, *40*, 190–208. [CrossRef]
73. Rocha-Olivares, A.; Garber, N.M.; Stuck, K.C. High genetic diversity, large inter-oceanic divergence and historical demography of the striped mullet. *J. Fish Biol.* **2000**, *57*, 1134–1149. [CrossRef]
74. Song, C.Y.; Sun, Z.C.; Gao, T.X.; Song, N. Structure Analysis of Mitochondrial DNA Control Region Sequences and its Applications for the Study of Population Genetic Diversity of Acanthogobius ommaturus. *Russ. J. Mar. Biol.* **2020**, *46*, 292–301. [CrossRef]

MDPI
St. Alban-Anlage 66
4052 Basel
Switzerland
www.mdpi.com

Diversity Editorial Office
E-mail: diversity@mdpi.com
www.mdpi.com/journal/diversity

www.ingramcontent.com/pod-product-compliance
Lightning Source LLC
LaVergne TN
LVHW071615170726
843515LV00009B/2401

* 9 7 8 3 0 3 6 5 8 5 3 0 7 *